ISBN 978-1-330-41537-5
PIBN 10057517

FB &c Ltd, Dalton House, 60 Windsor Avenue, London, SW19 2RR.
Company number 08720141. Registered in England and Wales.

THE

GEOGRAPHICAL

AND

HISTORICAL

DICTIONARY

OF

AMERICA AND THE WEST INDIES.

CONTAINING

AN ENTIRE TRANSLATION OF THE SPANISH WORK

OF

COLONEL DON ANTONIO DE ALCEDO,

CAPTAIN OF THE ROYAL SPANISH GUARDS, AND MEMBER OF THE ROYAL ACADEMY OF HISTORY:

WITH

Large Additions and Compilations

FROM MODERN VOYAGES AND TRAVELS,

AND FROM

ORIGINAL AND AUTHENTIC INFORMATION.

BY

G. A. THOMPSON, ESQ.

IN FIVE VOLUMES.

VOL. III.

——— *Magna modis multis miranda videtur*
Gentibus humanis regio, visendaque fertur,
Rebus opima bonis. LUCRETIUS, *lib. I. line* 727.

London:

PRINTED FOR JAMES CARPENTER, OLD BOND-STREET; LONGMAN, HURST, REES, ORME, AND BROWN, PATERNOSTER-ROW; WHITE, COCHRANE, AND CO. FLEET-STREET, AND MURRAY, ALBEMARLE-STREET, LONDON; PARKER, OXFORD; AND DEIGHTON, CAMBRIDGE.

1812.

THE

GEOGRAPHICAL AND HISTORICAL DICTIONARY OF AMERICA AND THE WEST INDIES.

MASSACHUSETTS.

[CHAP. VI.

From the arrival of Governor Shute in 1716, to the arrival of Governor Belcher in 1730.

(*Anno* 1716.)—COLONEL Shute arrived at Boston, October the 4th, 1716, in a merchant ship, and was received with the usual parade. He made the opposers of the bank his first acquaintance, the old governor's family in particular, and took his lodgings at Mr. Paul Dudley's. He had received very unfavourable impressions of the other party from Mr. Belcher and Mr. Dummer, in England, and was considered, from his first arrival, as an enemy to the scheme, and the heads of the party were the heads of an opposition during the whole of his administration. In his first speech to the general court, November 7, he puts them in mind of the bad state of the trade of the province, an important article of any people's happiness, owing, as he supposed, to the great scarcity of money, and recommends the consideration of some effectual measures to supply this want, and thereby to restore trade to a flourishing condition. This was pointing out to them a further emission of government's bills, and the representatives, pleased with so easy a method of obtaining money, soon determined upon a second loan of 100,000*l.* for 10 years, to be put into the hands of commissioners appointed for each county in proportion to their taxes. This provision being made by the government, there was the less pretence for private persons or companies issuing their bills; but it gave no relief to the trade, the whole currency soon depreciating to that degree, as, with this addition, to answer the purposes of money very little more than if it had not been made. The governor became sensible of it, and recommended to them to provide against it, which they were not able to do, and many of them would not have been willing if they had been able, being in debt, and by means of the depreciation discharging their debts by a nominal sum, perhaps of not more than one half of the real value of the debts. He soon found the effects of it upon his own salary, which they refused to advance as the bills sunk, and having recommended this measure in a public speech it became more difficult afterwards to refuse repeating it.

The province had been at war with the *e.* Indians, except some short intervals, for about 40 years. The prospect of a long peace between Great Britain and France encouraged it to hope for the like with the Indians, who had always been under French influence, but their father, Rallé, a jesuit, was constantly instigating them to insult and annoy the new settlers, who, he pretended, encroached upon the lands of the Indians, and by supplying them with strong drink debauched their morals and prevented the progress of the good work he had began among them. A] [treaty or conference was thought expedient to confirm them in their friendship with the English, and, if possible, to draw them from the Roman Catholic to the Protestant religion.

(*Anno* 1717.)—The governor, therefore, the first summer after his arrival, in August, attended by several of the council both of Massachusetts and New Hampshire, and other gentlemen, met the Indians at Arowsick island. At the beginning of the conference he delivered them an English and an Indian bible, which he told them contained the religion of the English, and at the same time recommended to them Mr. Baxter, a minister who went down as a missonary, and told them he would explain the bible, and instruct them in the principles of religion. They were at no loss for an answer. "All people, they said, loved their own ministers; and as for the bible, they desired to be excused from keeping it: God had given them teaching, and if they should go from that they should displease God." They were fixed in their religion, and it would have been a loss of time to attempt to move them. The rest of the conference was upon the right of the English to settle in that part of the country. Upon complaint made by the Indians of encroachments upon their lands, the governor produced one of the original deeds which had been given by their sachems. They acknowledged the lands to the *w.* of Kennebeck belonged to the English, but they were sure no sale had ever been made of any lands to the *e.* The governor told them the English would not part with an inch of land which belonged to them. The Indians were so offended that they rose immediately, and, without any ceremony, took to their canoes and went to another island where they had their head-quarters, leaving behind an English flag which the governor had given them. In the evening several of them returned to Arowsick with a letter from Rallé to the governor, acquainting him, that the French king did not allow that in any treaty he had given away the land of the Indians to the English, and would protect the Indians against the English encroachments. The governor let them know, that he highly resented the insolence of the Jesuit, and the next morning ordered the signal for sailing. Rallé, in his letters, often laments the unsteadiness of the Indians. They were afraid at this time of a new war. The old men were loth to quit their villages at Norridgewock and Penobscot, where they lived at ease, and encamp in the woods, or, which was much worse, depend upon the French, who, they would often say, treated them like dogs when there was no immediate occasion for their service: This consideration induced them to send two of their number with a message to the governor, acknowledging that yesterday they had been rude and unmannerly, and earnestly desiring to see him again. He let them know he would see them upon no terms, unless they quitted their pretensions to the lands which belonged to the English. This the messengers promised should be done, and desired that the English colours which they had slighted might be returned them. In the evening they came again to the conference, and appointed a new speaker as a mark of resentment against the former, who, they said, had behaved ill the day before, and, without entering into any dispute about particular limits or bounds, declared they were willing the English should settle where their predecessors had settled, desired to live in peace and to be supplied with necessaries, in a way of trade, confessed that some of their inconsiderate young men had offered injuries to the English, and violated the treaty of Portsmouth in 1713. After renewing that treaty, the conference ended.

The beginning of an administration in the colonies is generally calm and without ruffle. Several months passed, after Colonel Shute's arrival, without open opposition to any measures. The town of Boston, at the first election of their representatives, left out such as had been bank men, and chose such as were of the other party, but Mr. Cooke, who was at the head of the first party, had interest enough to obtain a place in council. It was, soon after, insinuated that the governor was a weak man, easily led away, and that he was in the hands of the Dudleys, men of high principles in government, and it behoved the people to be very careful of their liberties. Mr. Cooke, who had the character of a fair and open enemy, was free in expressing his sentiments, and the governor was informed of some contemptuous language in private company, with which he was so much offended as to procure Mr. Cooke's removal from the place of clerk to the superior court. A dispute happening about the same time between Mr. Bridges, surveyor of the woods, and the inhabitants of the province of Maine, concerning the property of the white pine trees within that province; Mr. Cooke immediately inserted himself in the controversy, publicly patronized the inhabitants, and in a memorial to the house of representatives charged the surveyor with mal-conduct in threatening to prosecute all who without licence from him shall cut any pine trees in their own ground, which Mr. Cooke alleged they had good right to do, and] [he further charged the surveyor with permitting such persons as would pay him for it, to cut down the trees which were said to belong to the king.

The surveyor thereupon preferred his memorial to the governor and council, justifying himself in the discharge of his trust, and complaining of Mr. Cooke, one of the members of the council, for officiously concerning himself with the affairs of the surveyor's office, and obstructing his measures for the service of the crown. Mr. Cooke had many friends in the house ready to support him, and this dispute was the beginning of the public controversy which continued until the end of Colonel Shute's administration; parties were formed, new subjects for contention from time to time were furnished, until at length the governor was forced to leave the province.

We do not find any vote of council upon this memorial, but the governor espoused the cause of the surveyor, and, to shew his resentment against Mr. Cooke, when the list of counsellors was presented at the next election, directed his speech to him in particular, and let him know he would excuse him from attending at the board for the ensuing year.

(*Anno* 1718.)—Mr. Cooke, soon after, presented his memorial to the council, in which he justified his own conduct, and charged Mr. Bridges with "using his utmost efforts to invade the rights and properties of the people in the province of Maine by his exorbitant actions, as well as basely betraying the trust the crown had invested him with, by daily selling and bartering the very logs and timber, which he gave out was the king's, his master, whose bread he then eat." The council suffered the memorial to lay upon their table, but acted nothing upon it. Afterwards, upon the appointment of a committee by the house, they joined a committee of council to consider in general of Mr. Bridges's conduct. This committee, in their report, justified Mr. Cooke, and condemned the proceedings of the surveyor. The council put off the consideration of this report also, but the house voted their acceptance of it. The governor, of course, transmitted to the board of trade an account of all these proceedings, and very soon received an answer, censuring the house of representatives for countenancing and encouraging Mr. Cooke. This being laid before the house, they by a vote declared that a censure of the board of trade was occasioned "by sending home the papers on one side only, whereby their lordships were informed *ex parte*." The house had avoided any direct attack upon the governor until this vote, many of the principal members this year being well affected to him, but the party, without doors, especially in Boston, had been increasing against him, and, at the next election for that town, they sent all new members, and a change was made in many other towns, unfavourable to the governor's interest.

The famous projector Captain Coram, in the year 1718, was busy in a scheme for settling Nova Scotia and the lands between Nova Scotia and the province of Maine, and a petition was preferred by Sir Alexander Cairnes, James Douglass, and Joshua Gee, in behalf of themselves and others, praying for a grant upon the sea-coast five leagues *s. w.* and five leagues *n. e.* of Chibuctow harbour, where they proposed to build a town, and to improve the country round it in raising hemp, in making pitch, tar, and turpentine, and they undertook to settle a certain number of families to consist of 200 persons in three years, the rest of his Majesty's subjects not to be prohibited fishing on the coasts under regulations. To this petition, Mr. Dummer, the Massachusetts agent, objected, because of the last clause, which laid a restraint upon the fishery. The lords of trade, however, reported in favour of it, but it stopped in council.

Another petition was preferred by Willam Armstrong and others who had been officers and soldiers in the army, "praying for a grant of the lands between Nova Scotia and the province of Maine, the said tract of land having been conquered by the French in 1696 and possessed by them until 1710, when it was recovered by the English, and by the treaty of Utrecht was with Nova Scotia given up by France to the British crown." The conquest in 1696 was the taking Pemaquid fort, and holding possession of the harbour two or three days. This, however, was made a serious affair, and the agent, Mr. Dummer, was several times heard before the lords of trade. The general court being restrained from conveying these lands without consent of the crown, it was proposed, that if they would consent to resign the jurisdiction between Kennebeck and Penobscot, the crown should confirm the property of the soil, but upon the proposal's being communicated to the court, they instructed their agent to make no concessions.

One Sarah Watts, setting forth that she was heir at law to Thomas Goffe, deputy governor and one of the 26 patentees of the old colony, claimed a 26th part of the colony, and the issues and profits for 80 or 90 years. She filed a bill of complaint in chancery against the province, and there was a commission of sequestration for several New England ships in the river, which cost the owners several guineas each to the sharpers who] [had urged the woman to the suit. The agent was required to answer the bill, which he did by declaring that if the complainant could make it appear that Thomas Goffe was once seized of a 26th part of the colony, and that she was heir at law to him, which he did not believe she was able to do, yet he verily believed that when the patentees, with others, were incorporated into a body politic, their respective rights ceased and passed to the corporation, who had granted the lands away. The poor woman was at last arrested for debt and sent to Newgate, where she perished.

(*Anno* 1719.)—The governor, in the beginning of the year 1718, had consented to an impost bill which laid a duty not only upon West India goods, wines, &c. but also upon English manufactures, and a duty of tonnage upon English ships. Before the session in May, the next year, he had received an instruction from the king to give all encouragement to the manufactures of Great Britain. The house, however, passed a bill of the same tenor with that of last year, and sent it to the council for their concurrence. An amendment was proposed, viz. to leave out the duty upon English vessels and goods, but the house adhered to their bill. A conference ensued, for the house were not then so exact as they have been since, in refusing to confer upon money bills. This produced nothing more than a proposal from the house to alter the word English to European, which, being trivial, was refused. It seems, the governor, a little out of time, had taken the opinion of the council upon this question, whether, consistent with his instruction, he could give his consent to the bill?—which they determined they could not, if it should be offered to him. The house then tried the council with the following resolve, "the house insist on their vote, forasmuch as the royal charter of this province gives power to the government to impose and levy proportionable and reasonable assessments, rates, and taxes upon the estates and persons of all and every the proprietors and inhabitants of the same, which this government has been in the free and uninterrupted exercise of ever since the enjoyment of the said charter. Sent to the upper house for their concurrence." The upper house was a new name for the council, and designed as a fleer and to intimate that they might consider themselves in another capacity than as a privy council. Perhaps if Cromwell's epithet for his house of lords had come into their minds, it would have been *the other house*. Taunts and language which tends to irritate, can upon no occasion be justifiable from one branch of the legislature to the other. Upon an agreement and harmony the interest of the people depends. Upon different apprehensions of this interest, if it be the real object, the several branches, by the persuasive voice of reason, will strive to convince each other, and be willing to be convinced as truth shall appear.

The council thought themselves unkindly treated, and, by a message, desired the house to alter their vote, but they refused to do it, and gave their reasons for the new form. "The house have received new and unusual treatment from the board. 1st, It is new and unusual for the council to give his excellency their advice upon a bill, till they have acted in concert with the house in concurring or non-concurring. 2d, It is likewise new and unusual for the council to desire a free conference, upon a subject matter, and then, at the management, to inform the house that by a previous vote they had so far engaged themselves that they could not recede from it. 3d, It is likewise a new and unusual method for the honourable board, after a message to the house desiring several amendments to a bill of rates and duties which were in a great measure agreed to by the house, immediately to non-concur the bill. 4th, It is likewise new and unusual for the honourable board to intermeddle so much with the grants and funds, which this house take to be their peculiar province."

The house having in this manner expressed their resentment returned to their old style, and then the council, by message, let them know that they would not give their concurrence to any bill laying a duty upon European goods, denied the charge made against them by the house, of innovations, and intimated that any further messages would only tend to increase the misunderstanding and retard the affairs of the government, and desired the house rather to join with them in a diligent endeavour to bring the session to such a conclusion, as should promote his Majesty's honour and the interest of the province.

Several weeks having been spent in these altercations, the governor thought it time to interpose, and, sending for the house to the council chamber, he made the following mild and healing speech to them.

"Gentlemen,

"My design in sending for you up at this time, is to let you know how concerned I am at the unhappy misunderstandings that have been for many years between the council and your house relating to the impost bill, and to assure you that no person here present can be more desirous of preserving the privileges of this people than myself, so far as is consistent with the late instruc-]

[tions I have received from my royal master, which have by his special direction been laid before this court. I am fully persuaded, that to act any way contrary thereto, after the many debates and votes which have been upon that head, would rather destroy than preserve those privileges we justly prize. Gentlemen, I desire your serious consideration of what I have hinted, that so the important affairs of the province yet lying before you may have a speedy and happy conclusion."

This speech which, a year or two after, when the prejudices against the governor were at the height, would have been excepted to as irregular and anticipating matters, which it would have been time enough for the governor to have declared his sense of when they came to be laid before him, had now a good effect, and the house, the same day, resolved that a new impost bill should be brought in, and that the controverted clause in the former bill should be left out, but in the preamble to their resolve they make a heavy charge against the council for not concurring in their former bill.

"Whereas this house have voted and passed a bill granting to his Majesty several rates and duties of impost and tonnage of shipping, in which was included one per cent. on European merchandize, for which article or clause the honourable council have several times non-concurred the said bill, notwithstanding all proper endeavours have been used by this house to attain the same, which have hitherto proved fruitless, whereby a considerable part of the revenue, which would have accrued to this province, is for this present session foregone, which also tends to the depriving this government of their just rights, powers, and privileges granted by the royal charter, Resolved," &c.

The council were fond of peace, and as soon as this resolve came to their knowledge, they sent a message to the house, desiring they would not print the resolve in their votes, as it would have an ill effect and would oblige the council, in their own vindication, to reply, although they wished that all controversy between the two houses might cease. The house printed it notwithstanding, and the next day the council sent the following answer.

"The board are very much concerned to find, among the votes of the honourable house, a declaration as if the council in non-concurring the bill of impost as it was first framed, had done that whereby a considerable part of the revenue, which would have accrued to this province, is for this present session foregone, which also tends to the depriving this government of their just rights, powers, and privileges granted by the royal charter.

"This declaration contains or implies such a charge as the council can, by no means, suffer themselves to lie under, without asserting and solemnly declaring their integrity, and they are more surprised at the imputation of doing a thing which tends to deprive this government of their just rights, powers, and privileges granted by the royal charter, because on the 23d current the board sent down a message to the honourable house, "that they were always ready and desirous to concur with the honourable house of representatives in such proposals relating to an impost, as may not tend to alter or expose our present happy constitution under the royal charter;" so that it was from a sincere and just regard to the rights, powers, and privileges of this government granted by the royal charter, that the council chose rather to omit the duty of one per cent. on English goods for this session.

"That the council apprehended the duty of one per cent. on English goods affected the trade of Great Britain, and so came within the meaning of his Majesty's late additional instruction is certain: and being of that opinion, it would have been inconsistent for the board to concur the bill of impost as it was sent up, however, they can boldly and truly say, they have acted from a principle of duty to his Majesty, love and fidelity to their country, and have nothing more at heart than the just, wise, and careful preservation of those invaluable rights, powers, and privileges granted by the royal charter, which God long continue."

This controversy being over, the court was prorogued.

Before the next sessions in November, the governor received a reprimand from the lords justices, the king being absent, for consenting to the duty on English goods, &c. by the impost act in 1718. This he laid before the court. The same house which had so long contended with the council the session before, for this clause in the bill, now "readily acknowledge the exceptions taken to it are just and reasonable." An instruction to the governor to support the surveyor of the woods in the execution of his office, which was communicated to the house at the same time, was not so favourably received, and in an answer or remonstrance occasioned by the governor's speech they charge the surveyor with instances of very gross mal-conduct. What evidence they had of it does not now fully appear. The governor, by a message, desired they would not print their] [remonstrance. They sent a committee to acquaint him, they must insist upon the right they had to make it public. He made a very great mistake, and told the committee, that his Majesty had given him the power of the press, and he would not suffer it to be printed. This doctrine would have done well enough in the reigns of the Stuarts. In the present age it is justly exceptionable; although by the liberty of the press we are not to understand a liberty of printing every thing, however criminal, with impunity. The house had no opportunity to take notice of this declaration. Upon another occasion they let him know they had not forgot it. The governor was so displeased with the proceedings of the house that he put an end to the session, and they never met again.

(*Anno* 1720.)—We are now arrived to the memorable year 1720. The contests and dissensions in the government rose to a greater height than they had done since the religious feuds in the year 1636 and 1637.

The public affairs, in general, were in a very indifferent state. The Indians upon the *e.* frontiers were continually insulting and menacing the English inhabitants, so that but little progress had been made in settling the country since the peace, and this year, most of the settlements which had been begun were deserted and a new war was every day expected.

The trade of the province declined. There was a general cry for want of money, and yet the bills of credit, which were the only money, were daily depreciating. The depreciation was grievous to all creditors, but particularly distressing to the clergy and other salary men, to widows and orphans whose estates consisted of money at interest, perhaps just enough to support them, and being reduced to one half the former value, they found themselves on a sudden in a state of poverty and want. Executors and administrators, and all who were possessed of the effects of others in trust, had a strong temptation to retain them. The influence a bad currency has upon the morals of the people is greater than is generally imagined. Numbers of schemes for private and public emissions of bills were proposed as remedies: the only perhaps effectual one, the utter abolition of the bills, was omitted.

By these calamities the minds of the people were prepared for impressions from pamphlets, courants, and other newspapers, which were frequently published, in order to convince them, that their civil liberties and privileges were struck at, and that a general union was necessary. These did not pass without answers, attributing all the distress in public affairs to the wrath and resentment, the arts and sinister views of a few particular persons, but the voice of the people in general was against the governor. In the mother country, when disputes arise between the branches of the legislature upon their respective rights, parties are formed, and the body of the people are divided; for in a well constituted government it is of importance to the people that the share even of the popular part of the constitution should not be unduly raised to the suppression of the monarchical or aristocratical parts. From a regard to the common interest, therefore, in a dispute concerning prerogative and privilege, the people, ordinarily, are divided in sentiment. The reason is obvious why it is less frequently so in a colony. There the people, in general, consider the prerogative as an interest, without them, separate and distinct from the interior interest of the colony. This takes their attention from the just proportion of weight due to each branch in the constitution, and causes a bias in favour of the popular art. For the same reason, men fond of popular applause are more sure of success, with less degree of part, in a colony, than in a state not so connected, and consequently men who with unbiassed judgments discern and have virtue enough to pursue the real interest of their country, are more likely to be reproached and vilified.

The first act of the house of representatives was the choice of Mr. Cooke for their speaker. A committee was sent to the governor, at his house, to acquaint him with the choice. They reported at their return, that his excellence said, "it was very well." In the afternoon, the governor being in council, sent the secretary to acquaint the house, that he was now in the chair and ready to receive their message respecting the choice of a speaker. They sent back an answer that his excellency, upon being informed of the choice in the morning, had said, "it was very well," and they had recorded his answer in the books of the house. The governor replied, that he would receive no message from the house but when he was in the chair. The house then proposed, by message to the council, to join with them in the business of the day, the choice of counsellors; but upon the governor's telling their committee, who carried up the message, that no election should be made until he was acquainted who was chosen speaker, the house sent a new committee to acquaint him with the choice they had made. The governor replied to this committee, that Mr. Cooke had treated him] [ill as the king's governor, and therefore, according to the power given him by the royal charter, he negatived the choice, and desired they would proceed to choose another person. They sent back their answer, that they had chosen a speaker, according to their known and legal privileges, and therefore insisted upon the choice, and at the same time they renewed their motion to the council to join with them in the election. The governor told the committee, that he had received a message from the house, acquainting him with the choice they had made of a speaker, which choice had been negatived and he was no speaker. Upon this the house sent their committee to the board to acquaint them, that two messages having been sent to propose to the board to join in the choice of counsellors and no answer having been given, they now desired to know whether the board would join in the election or not.

If there had been any further delay on the part of the board, it is very probable the house would have proceeded without them, which must have increased the perplexity. The governor therefore left the board, having first charged the secretary with the following message to the house, "His excellency orders me to acquaint you, he is informed that Governor Dudley did, in the time of his government, disallow of a speaker chosen by the house, and that his proceedings therein were approved by the commissioners of trade and plantations, and that he was thereupon directed by the said commissioners to acquaint the council, that it would not be thought fit that her Majesty's right of having a negative upon the choice of a speaker be given up, which was reserved to her Majesty, as well by the charter, as by the constitution of England."

Notwithstanding the warm disputes in the preceding year, between the two houses, only one new counsellor was chosen, John Burrill, Esq. of Lynn, who had been many years speaker of the house, but this year was sent to the board, in the room of Mr. Higginson. The house had been as fond of this Mr. Burrill as of their eyes. His temperate spirit until now had engaged the whole house in his favour, and from year to year, procured him a general vote, but this year the house were willing to part with him for a gentleman obnoxious to the governor, which measure, it was easy to forsee, must give a further occasion of controversy.

Two of the newly elected counsellors were negatived, Nathaniel Byfield, who had been soliciting in England for the government when Colonel Shute was appointed, and John Clark, who was a person of many valuable qualities, and obnoxious only, for being strongly attached to Mr. Cooke, and having been a great supporter of the cause.

After the election, the governor made a further attempt to bring the house to a compliance by the following speech. "Gentlemen, at the opening of this session you thought fit to make choice of Elisha Cooke, Esq. for your speaker, and upon your reporting of it to me, I did declare my disacceptance of that election, and am firm in my opinion that I had good right so to do, by virtue of his Majesty's commission and the powers reserved by the royal charter, and am also confirmed in it by what I find transacted by the late Governor Dudley, during his administration, and also by the opinion of the right honourable the lords of trade and plantations in that matter. I must further observe to you, that the person you have chosen had invaded the king my master's rights in the woods of the province of Maine, though confirmed to his Majesty by an act of the British parliament, and I have received the thanks of the right honourable the lords of trade and plantations for removing him out of the council. He has ill treated me, who am the king's governor, and has been censured by the council for it, which stands upon record in the council books. How acceptable this matter will be at home, considering the warning we have lately had from the court of Great Britain upon the account of passing the impost bill, will be worthy of your serious reflection. These things I thought necessary to acquaint you with, and advise you to return to your house and choose some other person speaker, with a reservation of your own rights, until you shall send to the court of Great Britain for the explanation of that part of your charter relating to the affair of a speaker."

The house, immediately upon their return to their chamber, entered into a debate upon this speech, and the question being put, whether, for the reasons assigned by his excellency, the house will proceed to the choice of a new speaker?—it passed in the negative, *nemine contradicente*.

The governor gave them no opportunity to proceed on any other business, for the next day he sent for them up again, and after another short speech dissolved the court. "Gentlemen, out of a tender regard I have for the welfare of this province, I shall give you the following advice before we part; that when it shall please God we meet again in a general assembly, which shall be as soon as possible, you will not let this province suffer by the perverse temper of a particular person, but that you will choose one for a speaker that has no other view but that of the public good, one] that fears God and honours the king. It is irksome and disagreeable to me to dissolve an assembly, but as matters now stand, I am forced to do it, or must give up the king my master's prerogative, which nothing shall ever oblige me to do, who am the king's governor. Gentlemen, I do not think it for the honour of his Majesty's government that this assembly should sit any longer, and therefore I shall dissolve you."

Writs were issued for a new assembly to meet the 13th of July. The governor had no great reason to hope for a more favourable house. The people in general thought their privileges were attacked. The charter indeed was silent upon this point. In a dispute between the crown and the house of commons in the reign of king Charles II. an expedient was found which seemed to avoid the acknowledgment of the right of the crown to refuse a speaker, but a provincial law was principally relied upon, which declares, "that the representatives assembled in any great and general court shall be the sole judges of the elections and qualifications of their own members, and may from time to time settle, order, and purge their own house, and make such necessary orders for the due regulation thereof as they shall see occasion." Whether the legislators had in contemplation the right of the house to choose a speaker, exempt from the governor's negative, might well be questioned, but it was urged that the due regulation of the house might very well include this right.

The towns in general sent the former members. Boston discovered how they stood affected by leaving out Mr. Tay, who was one of those persons who serve upon a pinch, when a favourite cannot be carried by a party, to stop the gap and prevent an opposite candidate, and he came in several times upon such occasions. In his room the town now chose Mr. Clark, the negatived counsellor.

The house was willing to sit and do business, which the choice of the former speaker would have prevented. They therefore pitched upon a person less attached to party, Timothy Lindall, one of the representatives of Salem, to whom no exception was taken. The governor in his speech recommended a peaceable session, but the house could not forget the late dissolution. They began with a warm message or remonstrance to the governor, in which they tell him, "the last assembly took no great pleasure in being dissolved, before they had gone through the usual necessary business; their asserting and maintaining their just right and ancient privilege of choosing their speaker, and not owning his excellency's power to negative him, was nothing but what they were strictly obliged to, and the new house are humbly of opinion, that whoever was of advice to his excellency, in the matter, did not consult his Majesty's interest, nor the public weal and quiet of the government, but officiously endeavoured to beget unhappy misunderstandings between his excellency and the house, and break off that desirable harmony which every one ought to keep up; we earnestly hope and desire the province may never have an assembly that will willingly forego such a valuable privilege as King William and Queen Mary, of ever blessed memory, graciously favoured the province with, when they gave their royal assent to a law directing and governing that affair."

All the subsequent proceedings of this short session shew how much the house was out of temper. An Indian war used to be universally dreaded. To prevent it, the governor and council had been treating with three of the Penobscot tribe, who were sent for or came to Boston, and the house were desired to make a grant for a present to them, but by a vote they refused to do it. Some time after they ordered a small sum, ten pounds only. To the controversy with the governor and the oppositiou made to the proposals which came from him, the war, which soon after broke out, was by the governor's friends attributed.

There had been no public notaries in the province, except such as derived their authority from the archbishop of Canterbury. The house now first observed, that a notary public was a civil officer, which by the charter was to be chosen by the general court, and sent a message desiring the council to join with the house in the choice of such an officer in each port of the province. To all instruments which were sent abroad, not only the attestation of the notary himself would be necessary, but a certificate under the province seal, to shew the authority to attest; the council therefore took time to consider of the expediency of appointing such an officer, and referred the matter to the next session, but the house immediately proceeded and chose the officers by their own votes. The arguments to prove that an officer to be chosen by the whole court could derive an authority from the majority of the members of the house of representatives have not been preserved.

Being offended with the council, the house sent a message desiring, "that considering the low circumstances of the province, no draught be made upon the treasury for expences at times of public rejoicing for the future."

It had been usual to make a grant to the governor, for the salary of half the year, at the beginning of the session. The house deferred it until the [close, and then reduced it from 600*l.* to 500*l.* although the currency was depreciated. To the lieutenant-governor, they used to make a present once a year, never less than 50*l.* they now reduced it to 35*l.* Mr. Dummer had so much spirit, that he inclosed the vote in a letter to the speaker, acquainting him that "having the honour to bear the king's commission for lieutenant-governor of the province, and having been annually more than 50*l.* out of pocket in that service, he did not think it for his honour to accept of their grant."

The governor took no public notice of the proceedings of the house. On the 23d of July he put an end to the session.

During the recess of the court (August 7th) a part of the *e.* Indians fell upon Canso, within the province of Nova Scotia, but peopled every summer from the Massachusetts. The Indians surprised the English in their beds and stripped them of every thing, telling them they came to carry away what they could find upon their own land. Three or four of the English were killed. Some of the French of Cape Breton were in confederacy, and came with their vessels the next night, and carried off the plunder, together with about 2000 quintals of fish. The English vessels in the harbour were not attempted. A sloop happening to arrive the next day, the master offered his service to go out and make reprisals, and being furnished with a number of men and two or three small vessels for his consorts, for want of more ample authority, he took a commission from one Thomas Richards, a Canso justice, and went after the French and soon brought in six or seven small fishing vessels, having all of them more or less of the English property aboard.

Mr. Henshaw of Boston, a principal merchant at Canso, went to Louisbourg with a complaint to the French governor, who excused himself from intermeddling, the Indians not being French subjects, nor under his controul. The French prisouers were sent to Annapolis Royal. The loss sustained by the English was estimated at 20,000*l.* currency.

The fears of the people, in the *e.* parts of the Massachusetts, were increased by this stroke upon Canso. In a short time after, the cattle were destroyed and the lives of the owners threatened. The governor was still desirous of preserving peace, and by the advice of council, sent orders to Colonel Walton, the commanding officer of such forces as upon the alarm had been sent there, to inform the Indians, that commissioners should be sent to treat with them. The Indians liked the proposal and promised to attend the treaty.

Before the time appointed, the general court met and the house passed a resolve, "that 150 effective men, under suitable officers, be forthwith ordered to march up to Norridgewock and compel the Indians that shall be found there, or in other those parts, to make full satisfaction for the damage they have done the English, by killing their swine and sheep or carrying them away, or stealing provisions, clothing, or any other way wronging them; and that a warrant be directed to Captain John Leighton, high sheriff of the county of York, who is to accompany the forces, for the apprehending and safe bringing Mr. Rallé to Boston, who is at present resident at or near Norridgewock in Kennebeck river in this province, and, if he be not to be found, that then the sheriff direct and command the Indians there, or in the parts adjacent, to bring in and surrender up the Jesuit to him the sheriff; and, upon their refusal to comply with either of the said demands, that the commanding officer is to take the best and most effectual way to apprehend and secure the Indians so refusing, and safe conduct them to Boston."

The governor looked upon this resolve to be, in effect, a declaration of war and an invasion of the prerogative; it necessarily prevented a treaty he had agreed to hold with the Indians, and a new war must be the consequence of such a measure. The council were fond of peace, and, when the resolve was sent to them for concurrence, they rejected it. The house were less averse to war. The charge of carrying it on, it was said, would be no burden to the province, the French now durst not join the Indians, and this would be the most favourable opportunity which could be expected to subdue or utterly extirpate them. That the charge should be no burden, seems to be a paradox, but a wild opinion had filled the minds of great part of the people of the province, that if bills of credit could be issued, the advantage to trade would be so great, that the taxes by which, at distant periods, they were to be drawn in again, would not be felt. Many schemes of public expence were projected, and, among the rest, a bridge over Charles river, broader and much deeper than the Thames at London or Westminster.

We shall take no pleasure in relating the proceedings of the general court in this and the two next years. The best excuse we can make for the house is, that the attempt made to deprive them of the exclusive right of choosing their own speaker, was deemed by them a grievance, that the royal governments insist upon this right, and there was nothing in Massachusetts charter which took it from them, that this attempt raised in their minds] [a jealousy of a design against their privileges in general, and, in this state of mind, they were more easily prevailed upon by their principal directors, whose principal views were to distress the governor, to agree to such measures as under other circumstances they must have disapproved. The rule, perhaps, holds stronger with political bodies than with individuals, that when just bounds are once exceeded, the second step is as easy as the first, and so on, until at length they are drawn by degrees to such excesses as, *per saltum*, they would have been incapable of.

The public records of the general court are always open to the inspection of any of the members, but, that the house might have them under their more immediate view and charge, they passed a vote, that the secretary should make duplicates of all public records, and that one set should be lodged in such place as the house should appoint. The council, willing to have duplicates for greater security, concurred with an amendment, viz. in such place as the general assembly should direct, but this amendment the house rejected.

The house, finding the council a bar to their attempts, resolved, in one instance, to act by themselves. There was a complaint or suggestion, that false musters were made by some of the officers in the pay of province. The house taking the affair into consideration, resolved, "that one or more meet persons be appointed by this house clerk of the check, who shall, from time to time, have an inspection into the forts, garrisons, and forces, and take care that every one have their complement of men; and the better to enable them to execute the trust reposed in them, that when and so often as they shall see reason, the commanders of the forts, garrisons, and captains of any of the companies, in the pay of this government, shall call forth their men before them, and, if any do not appear, the commanding officer to give the reason of such absent men, and that no muster roll shall be accepted, and paid by the treasurer, unless approved of by the clerk of the check." The governor did not intend to admit this officer, appointed by the house, into the forts, garrisons, &c. which by the charter the crown had reserved to the governor, but he kept silent.

To another act of the house the council took exception. A message was sent by the house to the council to let them know they had appointed a committee to prepare a bill for levying soldiers, "taking it to be their peculiar care." Lest it should be understood that this was to exclude the council from concurring or non-concuring to such bill, or from advising to the levying soldiers upon an emergency in the recess of the court, the council desired the house to withdraw those words, "taking it to be their peculiar care," which they agreed to.

At this session, the house again withheld 100*l*. from the governor's usual half-year's salary. He had passed it over without notice before, but now he thought it proper to lay before them a royal instruction to recommend to the assembly to establish a sufficient allowance for him by a fixed salary. They sent him a reply, "that they humbly conceived what was granted him was an honourable allowance, and the affair of settling salaries being a matter of great weight and wholly new to the house, and many of the members absent, they did not think it proper to enter upon the consideration of it, but desired the court might rise." The governor complied with their request.

(*Anno* 1721.)—At the opening the next session, March 15th, the governor in his speech recommended measures to prevent the depreciation of the currency, to suppress a trade carried on with the French at Cape Breton, and to punish the authors of factious and seditious papers, to provide a present for the Five Nations, and to enlarge his salary.

They refused, directly or virtually, every proposal. To the first the house tell him, in their answer, "they had passed a bill for issuing 100,000*l*. more in bills of credit. This alone had a direct tendency to increase the mischief, but they add, that "to prevent their depreciation they had prohibited the buying, selling, and bartering silver, at any higher rates than set by act of parliament." This certainly could have no tendency to lessen it. Such an act can no more be executed than an act to stop the ebbing and flowing of the sea. It would probably carry away and keep out all silver and gold. The depreciation of their currency, would, notwithstanding, have been as visible by the rise of exchange with foreign countries, and have been as sensibly felt by every creditor among themselves. To his other proposals they say, "they know of no trade carried on by any people of the province with Cape Breton, and do not think any law to prevent a trade there is necessary; and for seditious and scandalous papers, the best way to suppress or prevent them is, for the executive part of the government to bring the authors to condign punishment, and if proper measures had been taken to discover and punish the authors of a libel called News from Robinson Crusoe's Island, wherein the members of the house are grossly reflected upon, few or none would have dared, afterwards, to publish any thing of that nature or tendency; but to suffer no books to be printed without] [licence from the governor will be attended with innumerable inconveniences and danger. As to the Five Nations, the house do not know enough of their number, nor what the other governments intend to give, and therefore cannot judge what is proper for them to do; and for the allowance to the governor, they think it as much as the honour and service of the government calls for, and believe the inhabitants of the several towns through the province are of the same mind."

There never had been an instance of any governor's refusing or neglecting, at the beginning of the year, to appoint a fast, in conformity to the practice of the country, but the house now endeavoured to anticipate the governor, and appointed a committee to join with a committee of council to prepare a proclamation for a public fast. The council refused to join, and acquainted the house they could find no precedent; but the house replied, that if such days had not the sanction of the whole court, people would not be liable to punishment for working or playing. The governor, willing to conform to the house so far as would consist with maintaining his right of issuing proclamations, mentioned in the proclamation which he soon after published, that the appointment was by advice of council, and upon a motion from the house of representatives; but the house refused to meet him, and declared they had never made any such motion, and ordered that no members of the house should carry any proclamations to their towns, for the present. The day was, however, observed as usual, except that one of the representatives of Boston would not attend the public worship, but opened his warehouse as upon other days.

Certain persons had cut pine trees upon that part of the province of Maine which had not been granted by the general court as private property. A deputy to the surveyor of the woods gave licence to cut the trees, as belonging to the king. The house appointed a committee to join with a committee of council, which joint committee were to seize and secure, for the province, the same logs which had been cut by licence. The council concurred, with a "saving to his Majesty all such rights as are reserved by the royal charter and acts of parliament to trees for the royal navy."

The house desired this saving might be withdrawn, not that they apprehended the reservation made in the charter or the provision by act of parliament were of no force, but they alleged that the trees they designed to seize were cut by one deputed by the deputy of the surveyor of the woods, and cut, not for the royal navy, but for other uses, and therefore they did not come within reason of the reservation or provision.

Finally, upon the council's refusing to join, the house appointed a committee of their own to seize the logs, and directed the attorney-general to prosecute those who had trespassed and made spoil upon the province lands. After they were seized, the house again desired the council to concur in a vote or order for securing and converting the logs to the benefit of the province. This, without any judicial determination, was still more irregular, and the council declined meddling with them.

As the time approached for issuing writs for a new assembly, the governor made the following speech to them before their dissolution.

"Gentlemen of the house of representatives,—In my speech at the beginning of the session, I gave you the reasons of my meeting you at this time. I have since received your answer, which I shall take care to transmit, by the first conveyance, that his Majesty may see, not only how his governor of this province is treated and supported, but what sort of regard is paid to his own royal instructions. I shall also lay before the right honourable the lords commissioners of trade and plantations, the bill for prohibiting a trade to Cape Breton, which I recommended to you several sessions, and which had twice the concurrence of his Majesty's council, but was as often thrown out in your house, notwithstanding the message that accompanied that bill.

"I am very much surprised you should refuse two other bills, which came down from the council, the one to prevent riots, the other to prohibit the making and publishing libels and scandalous pamphlets, the passing of which would, in my opinion, have tended both to the honour of the government and the public peace.

"But what gives me the greatest concern is, that the proceedings of your house, with respect to the woods in the province of Maine, are directly contrary to the reservation of his Majesty's right in the royal charter and an act of parliament, which were both set forth in my proclamation, dated the 1st of November 1720, for preventing the destruction and spoil of his Majesty's woods.

"I could heartily wish, that instead of obliging me to make such representations to the lords of trade, as I fear will not be to your advantage, you had acted with that calmness and moderation, which becomes the subjects of a prince, who possesses those qualities in an eminent degree, and which becomes the representatives of a province, that, without any encroachment on the royal pre-]

[rogative, enjoys as many and as high privileges, as the greatest advocates for liberty can desire or expect.

"I must therefore recommend to you a loyal and peaceable behaviour, and to lay aside those misunderstandings and animosities that of late prevail so much among you, which you will find to be your truest and best interest."

Doctor Noyes, one of the representatives of Boston, died while the court was sitting, March 16th, after a short illness. He was very strongly attached to the popular party, and highly esteemed by them, was of a very humane obliging disposition, and, in private life, no man was more free from obloquy. Mr. William Hutchinson who succeeded him was also a gentleman of a very fair character, sensible, virtuous, discreet, and of an independent fortune. He began his political life at a time when persons, thus qualified, were wanted for the service of their country, to moderate the passions of those who were less temperate and who had the lead in the house. In general, he adhered to the popular party also, but lived but a little while. Longer experience might probably have convinced him, that he would have shewn his gratitude to his constituents, more by endeavouring to convince them that they were running to an extreme, than by encouraging the same extremities himself.

The session of the general court, in May, this year, began as unfavourable as any former session. The house chose for their speaker John Clarke, Esq. who the year before had been negatived by the governor as a counsellor. To prevent a negative, as a speaker, they projected a new form of message, directed to the governor and council jointly, to acquaint them "that John Clarke, Esq. is chosen speaker of the house, and is now sitting in the chair." This was undoubtedly a very extraordinary contempt of the governor. Mr. John White, a gentleman of unspotted character, had been clerk of the house for many years. He was no zealous party man, but his most intimate friends, who esteemed him and sought his company for the sake of his valuable accomplishments, were strongly attached to the governor. This alone was enough to dismiss him, and Mr. William Payne, brother by marriage to Mr. Cooke, and who had formerly been of the bank party, was appointed clerk in his stead.

The governor was more wroth than upon any occasion before. He came to council in the afternoon, and sent immediately for the house, no doubt, with an intent to dissolve the court. He had several faithful advisers about him, and, whilst the house were preparing to come up, he sent a message to stop them and to let them know he accepted their choice of a speaker. This was giving a construction to their message which they did not intend, and it was giving his consent before it was asked, but it was to be preferred to a dissolution; for a dissolution of the court, before the election of counsellors, according to the construction the house have sometimes put upon the charter, would have been a dissolution of the government, for one year at least, because the time mentioned for the first election was the last Wednesday in May. The counsellors named in the charter were to continue until others were chosen and appointed in their stead. We do not know of any words in the charter which would make the choice upon another day invalid, although that be the day more particularly designed for that business. The house showed their resentment against the lieutenant-governor and Mr. Belcher, who were both left out of the council. The rest were continued.

The next step was the appointing a committee to carry a list of the new-elected counsellors to the governor; but the committee was not to desire his approbation, though this form had never been omitted in any one instance. The governor sent the list back, and took notice of the omission. The house thereupon resolved, "that considering the small-pox was in Boston, and they were very desirous the court should be removed to Cambridge, they would send the list in the usual terms, saving their right to assert their privileges at a more convenient time." What privileges they had in their minds it is difficult now to discover. Surely they could not imagine the election would have been valid without his consent. The governor negatived Colonel Byfield, the rest he consented to.

The court was adjourned to Cambridge. The governor, in his speech, took no notice of past differences. All was fair and smooth, and all was fair in the house also, the first fortnight, but on the 19th of June, the governor's speech at the dissolution of the last assembly was ordered to be read, and a committee was appointed "to vindicate the proceedings of the house from the insinuations made by the governor of their want of duty and loyalty to his Majesty." This committee made a report, not in the form of an address or message to the governor, but of a narrative and justification of the proceedings of the last assembly, and the house accepted it and ordered it to be printed.]

[To vindicate the past proceedings about the pine trees, a full consideration was now had of the several acts of parliament and the reservation to the crown in the province charter. The house did not deny a right in the crown to the trees, whilst they were standing and fit for masts, but supposed that, as soon as they were felled and cut into lengths fit for boards or timber only, the right of the crown ceased, and the owners of the soil recovered or acquired a new property in them. This, it was said, would render the provision made for the preservation of the trees, which at best was insufficient, to be of no effect, nothing being more easy than for the owners of the soil to procure the trees to be felled and cut into short logs, without possibility of discovery. However, they came to the following resolution, viz. "That inasmuch as a great number of pine trees have been cut in the province of Maine, which when standing were fit for masts for the royal navy, but are now cut into logs of about 20 feet in length, and 'although the cutting them should be allowed to be an infringement of his Majesty's rights reserved in the charter,' yet in the condition they are now in, being no longer capable of being used for masts, it is lawful for and behoves this government to cause such logs to be seized and converted to their own use, and to bring the persons who cut down the trees to punishment." In consequence and for the purposes of this resolve a committee was appointed.

The reservation in the charter is in these words, "For the better providing and furnishing of masts for our royal navy, we do hereby reserve to us, our heirs and successors, all trees of the diameter of 24 inches, and upwards of 12 inches from the ground, growing upon any soil or tract of land not heretofore granted to private persons. And we restrain and forbid all persons whatsoever from felling, cutting, or destroying any such trees without the royal licence of us, our heirs and successors, first had and obtained, upon penalty of forfeiting 100*l.* sterling unto us, our heirs and successors, for every such tree."

It was said further upon this occasion, that although the crown reserved the trees and restrained all persons from cutting them, which the necessity of the trees for national use and service might be sufficient to justify, yet it was not equitable to take them without a valuable consideration. The crown had made an absolute grant of the province of Maine to Gorges, from whom the Massachusetts purchased. The Massachusetts charter indeed was declared forfeited. Where the right was, after that, might be disputed, but this was a hard judgment, and it was the plain intent of the charter in general to restore rights, except that of the form of jurisdiction or administration of government, to the former state.

Be this as it may, it has however been thought by some judicious persons, that considering the extreme difficulty then existing of convicting trespassers of cutting the king's trees, and by such conviction putting a stop to the spoil and havoc continually making, it would have been good policy to allow the owners or proprietors of these lands a sum for every tree felled for a mast, equal to what it would be worth when cut into logs for boards or timber. This would scarce have been a tenth part of the value of the tree for a mast, and yet would have taken away the temptation to cut it for logs, and would have encouraged the preserving and cultivating the young trees, which were then of lesser dimensions. Trees that are incapable of ever serving for masts, either from decay or other defects, although of sufficient diameter, might have been allowed to be cut for logs, and it would have been no prejudice to the crown. The trespassers used to make no distinction, and trees were continually cut worth 20*l.* sterling for a mast, which when cut into logs were scarce worth 20 shillings. Very few trees were cut for masts by trespassers. The notoriety of halling, felling, and shipping masts has rendered it very difficult, when the burden of proof where the masts were cut lay upon the person who claimed the property, to escape discovery and conviction, where among the multitude of logs which were continually let loose to float down rivers to saw mills, the greatest part of the trespasses used to escape notice.

There are great numbers of white pines growing in parts of the country so remote from the sea or any river by which they can be floated to the sea, that the expence of bringing them thither would be 20 times the value of a mast in England. It seems unnecessary to have deprived the inhabitants of such places from making use of the trees for timber and boards, when they must infallibly have decayed and died in the ground, if they were to serve for no other purpose but masts.

But to return.

The house neglected making any provision for the support of the governor, or the other officers of the government who depend upon the court for their salaries. They waited to see how far the governor would consent to their several acts and votes. On the other hand, the two houses having chosen the treasurer, impost officer, and other civil officers, the governor laid by the list and neither approved nor disapproved. When the] [house sent a message to the council, to inquire whether the governor had passed upon the list, he directed the committee to tell the house that he should take his own time for it. This occasioned a reply from the house, and divers messages and answers passed upon the subject. At length the house, by a vote, determined they would not go into the consideration of grants and allowances before his excellency had passed upon the acts, resolves, and election of that session. This was in plain terms avowing what the governor at first charged them with tacitly intending. To have occurred to this instance would have laid the house under disadvantage in the dispute, some years after, about a fixed salary. To compel the governor to any particular measure, by making his support, in whole or in part, depend upon it, is said to be inconsistent with that freedom of judgment, in each branch of the legislature, which is the glory of the English constitution. This was not all. The house withheld the support of all the other salary men, because the governor would not comply with the measures of the house.

Resentment was shewn against some of the governor's friends. The agent in England, Mr. Dummer, in some of his letters, had informed the court of the sentiments of the ministry upon the proceedings of the house of representatives, and of the general approbation in England of the governor's conduct. A faithful agent would rather tell them the truth, than recommend himself to them by flattery and false representations. He lost the favour of the house, who upon the receipt of these letters voted, that it was not for the interest of the province Mr. Dummer should be continued agent any longer, and therefore it was ordered that he should be dismissed. This vote they sent to the council for concurrence, who desired the house to inform them of the grounds and reasons of this dismission. The house voted the message to be unprecedented, and insisted that the council had nothing else to do but to concur or non-concur, and then they non-concurred the vote; but this was much the same with a dismission, at least for a time, for an agent having no fixed salary depends upon temporary grants, which the house refused to make after this vote.

Paul Dudley, Esq. another of the governor's friends, had the misfortune also of falling under the displeasure of the house. He had been chosen, by a small majority, counsellor for Sagadahoc. By the charter, it was necessary for him to have been an inhabitant or proprietor of that part of the province for which he was chosen. He dwelt in the old colony of Massachusetts. It was suggested in the house that he had no lands at Sagadahoc, and they appointed a committee to inquire into this fact. Upon their applying to Mr. Dudley for evidence of his title, he told them it was too late, they should have inquired before the election. Perhaps he was in an error. He went on and told the committee he had a deed which he would not expose to the house, but he would shew it to two or three of the members. Upon this they sent another committee to inform him it was expected he should produce his deed, the next morning, to be laid upon the speaker's table. He replied that he would not produce his deed before the house, for they might possibly vote it insufficient. In this part of the province there are scarce any lands which have not more than one claimer, and it is not improbable some of the members of the house claimed the lands in Mr. Dudley's deed. The vote of the house would not have determined his title, but might have undue influence upon a jury in a judicial proceeding.

Mr. Dudley's answer was unsatisfactory, and the house voted that it was an affront, that his declining to produce his deeds gave sufficient grounds to believe that he was no proprietor, and it was therefore resolved that his election be declared null and void. This vote being sent to the council was by them unanimously non-concurred.

No grants had been made and no officers for the ensuing year had been constituted; the house, notwithstanding, sent a message to the governor to desire the court might rise. He refused to gratify them. Thursday the 13th of July had been appointed for a public fast. The members desired to be at home with their families, and on Wednesday, by a vote, they adjourned themselves to Tuesday in the next week. The house of commons adjourn for as long time, without any immediate act of royal authority, but we presume, never contrary to a signification of the mind of the king; and the adjournments over holidays are as much established, by ancient usage, as the ordinary adjournments from day to day, and being conformed to by both houses of parliament, no inconvenience can arise. But the charter was urged by the governor to be the rule in this assembly, not the analogy between a Massachusetts house of representatives and the commons of Great Britain. The governor, by charter, has the sole power of adjourning, proroguing, and dissolving the general court. Taken strictly, it would be extremely inconvenient, for the act of the governor would be necessary every day Upon a rea-]

[sonable construction, therefore, the house had always adjourned from day to day, but never for so great a number of days. The council, who were obliged to spend near a week without business, unanimously voted, upon hearing the house had adjourned, that such adjournment without his excellency's knowledge and consent, was irregular and not agreeable to the charter.

The governor afterwards made this adjournment one of the principal articles of complaint against the house.

Upon Tuesday, like the first day of a session, there was scarcely a house for business. The next morning some votes passed, which were offered to the governor, and which he would not suffer to be laid before him until he had sent for the house and told them they had made a breach upon his Majesty's prerogative, which he was under oath to take care of, and he insisted upon an acknowledgment of their error before they proceeded to business.

The house, by a vote or resolve, declared they had no design to make any breach upon the prerogative, but acknowledged they had made a mistake in not acquainting his excellency and the board with the adjournment.

The governor observed to them, that they had industriously avoided acknowledging that the sole power of adjourning, as well as proroguing and dissolving the general assembly, is vested in his Majesty's governor, by the royal charter. They thereupon agreed to the following message: "The house of representatives do truly acknowledge, that by the royal charter your excellency and the governor for the time being have the sole power and authority to adjourn, prorogue, and dissolve the general court, and the house further acknowledge, that your excellency ought to have been acquainted with the design and intention of the house in their adjournment from Wednesday the 12th to Tuesday the 18th instant, before they did adjourn, and that it was so designed and casually omitted."

The house carefully distinguished between the power of adjourning the general court and adjourning the house of representatives, one branch only, and seem to suppose, that their only mistake was their not acquainting the governor and the board with their intention, which was by no means satisfactory to the governor, and he immediately ordered the house to attend him in the council chamber. The speaker ordered all the members of the house to be called in, and expecting a dissolution, they resolved, "that all the votes of the house in the present session, more especially relating to any misunderstanding or difference that hath arisen between his excellency and the house, shall be prepared to be sent home, and that the speaker transmit them to William Tailer, Esq. now resident in London, or in his absence, to such meet persons as he shall think fit, desiring them to lay the same before his Majesty in council, or any where else, if need require, to obviate any complaint that may be made by his excellency the governor against the proceedings of this house, for their just and necessary vindication." So much time was taken up in this vote or resolve, that the governor was highly offended and sent a second time, requiring them to attend him forthwith. It has always been the practice of the house, before and since, upon a message from the governor, to stop all business and go up without delay. The speaker, at this time, was among the forwardest in the opposition. There was no need of four or five members to hold him, as the speaker of the house of commons was once held, in the chair until a number of strong resolutions had passed the house.

The governor directed his speech to the house only. "Gentlemen of the house of representatives,—I am very much concerned to find in the printed journal of the house, first, an order to appoint a committee to draw a memorial upon, or representation of my speech, made before the dissolution of the assembly in March last, and afterwards, the memorial itself, signed by Mr. Cooke in the name of the committee.

"This treatment is very surprising, from a house of representatives that profess so much loyalty and respect to his Majesty's government. It appears to me to be very irregular that the present house of representatives, whereof John Clarke, Esq. is speaker, and which consists of a majority of new members, should take upon them to answer my speech made to a former house of representatives, whereof Timothy Lindall, Esq. was speaker. These proceedings are not only improper, but without precedent from any former assembly.

"I must also observe to you, that you have not shewn that respect which is due to me as governor of this province, by suffering this order or memorial to go into the press before it was communicated to me, which if you had done, I could have convinced you, that it would have been very much for the service of your constituents, that neither the order nor the memorial should have appeared in print.

"It is my opinion, that you will quickly be convinced how much you have been wanting in] [your duty and interest, by disowning the authority of the right honourable board which his Majesty has constituted to superintend the affairs of the province and all the other plantations.

"For these reasons I should have dissolved the general court when the memorial first appeared, but I was in hopes the house might have been brought to correct or expunge it. Instead of making this use of my tenderness, you have gone on in the most undutiful manner to withdraw from his Majesty's and your country's service, by adjourning yourselves for near a week, without my knowledge and consent, contrary to the royal charter, which absolutely vests in the governors of this province the power of adjourning, proroguing, and dissolving; and that at a time when I thought it for the interest of the colony to adjourn you for two days only, having an affair of the greatest consequence to communicate to the house, which was to persuade you to take some effectual measures to prevent the plague coming among us, there being nothing so likely to bring it in as the French silk and stuffs which are constantly brought into this province.

"These your unwarrantable proceedings oblige me to dissolve this assembly."

This speech, and the dissolution which followed, further alienated the minds of the people from the governor. Some of his friends wished he had carried his resentment no further than putting an end to the session and giving time to deliberate. There was no room to expect a change for the better, upon a new election.

There was yet no open war with the Indians, but they continued their insults. The French instigated them and furnished them with ammunition and provisions. Governor Shute published a proclamation requiring the inhabitants to remain upon their estates and keep possession of the country. No wonder the proclamation was not obeyed. We know no authority he had to require them to remain. If the preservation of their own property was not sufficient to keep them there, it could not be expected they would remain merely as a barrier for the rest of the province.

In the month of August, 200 Indians with two French Jesuits came to George town upon Arowsick island, armed and under French colours, and, after some parley with the inhabitants, left a letter to be delivered to the governor, in which they make a heavy charge against the English for unjustly invading the property of the Indians and taking from them the country which God had given them.

Rallé, their spiritual father, was their patron also in their temporal concerns. Either from a consciousness of their having conveyed the country to the English, or from a deire of peace and quiet, they were averse to engaging in war. When they were at their villages, the priests were continually exciting them to act vigorously and drive all the English to the w. of Kennebeck, and such was their influence over them that they would often set out from home, with great resolution to persist in their demands, and in their parleys with the commanders of forts, as well as at more public treaties, would appear, at first, to be very sturdy, but were soon softened down to a better temper, and to agree that the English should hold the lands without molestation. When they returned home, they gave their father an account of great firmness they had shewn in refusing to make any concessions, and to this we are to impute the erroneous relation of these treaties by Charlevoix and others.

But about this time Toxus, the Norridgewock chief, died. When they came to choose another Toxus, the old men, who were averse to war, contrary to Rallé's mind, pitched upon Ouikouiroumenit, who had always been of the pacific party. They took another very disagreeable step, and submitted to send four hostages to Boston, sureties for their good behaviour and for the payment of the damages the English had sustained. Vaudreuil, the governor of Canada, was alarmed, and thought it necessary to exert himself upon this occasion. He writes to Father Rallé, of the 15th of June, "I was at Montreal, my reverend father, when your letters of the 16th and 18th of May came to my hands, informing me of the bad step taken by the Norridgewocks, in choosing Ouikouiroumenit successor to the deceased Toxus, of the great loss which the whole Abenakis nation hath sustained by his death, and the divisions prevailing among the Norridgewocks, many of whom, and especially their chiefs, have betrayed the interest of their tribe in openly favouring the pretensions of the English to the country of Norridgewock. The faint hearts of your Indians in giving hostages to the English, to secure payment of the damage they have sustained, and the audacious language which they have used to the Indians, in order to keep possession of their country and to drive you out of it, fully convinced me that every advantage would be taken, of the present state of affairs, to subject them to the English, if the utmost care should not be immediately taken to prevent so great a misfortune. Without a moment's delay, I set out, in order to apply myself to the business at Montreal, and from thence to St. François and Becancour, where I prevailed] with the Indians of those villages vigorously to support their brethren of Norridgewock, and to send two deputies for that purpose, to be present at the treaty and to let the English know, that they will not have to do with the Norridgewocks alone if they continue their injuries to them. The intendant and I have joined in a letter, to desire Father le Chase to take a journey to Norridgewock, in order to keep those Indians in their present disposition and to encourage them to behave with firmness and resolution. He will also go to Penobscot, to engage them to send some of their chiefs also, to be present on this occasion and to strengthen their brethren."

Begoir, the intendant, writes at the same time to Rallé, "I wrote, my reverend father, to Monsieur de Vaudreuil, who is at Montreal, the sentiments of Father de la Chase and my own, viz. what we think convenient to be done, until we hear from the council of the marine whether the French shall join the Indians to support them openly against the English, or shall content themselves with supplying ammunition, as the council has advised that M. Vaudreuil might do, in case the English should enterprise any thing against them. He thought it more proper to send the reverend Father la Chase, than Monsieur de Croisil, lieutenant, &c. because the English can have no room to except to one missionary's visiting another, the treaty of peace not forbidding it, whereas, if a French officer was sent, they might complain that we sent Frenchmen into a country, which they pretend belongs to them, to excite the Indians to make war upon them.

"It is to be wished that you and your Indians may be suffered to live in quiet until we know the king's intentions whether we shall openly join the Indians if they are attacked wrongfully; in the mean time we shall assist them with ammunition, which they may be assured they shall not want.

"P. S. Since I wrote the foregoing the Indians of St. François and Becancour have desired M. Vaudreuil that M. de Croisil may go with them to be a witness of their good disposition, and he has consented to join him with Father de la Chase."

The Massachusetts people made heavy complaints of the French governor, for supporting and stirring up enemies against them in time of peace between the two crowns, but he justified himself to his own master. Rallé was ranked by the English among the most infamous villains, and his scalp would have been worth an hundred scalps of the Indians. His intrepid courage and fervent zeal to promote the religion he professed, and to secure his *neophytes* or converts to the interest of his sovereign, were the principal causes of these prejudices. The French, for the same reasons, rank him with saints and heroes. He had been, near 40 years, a missionary among the Indians, and their manner of life had become quite easy and agreeable to him. They loved and idolized him, and were always ready to hazard their own lives to preserve his. His letters, upon various subjects, discovered him to have been a man of superior natural powers, which had been improved by an education in a college of Jesuits. The learned languages he was master of. His Latin is pure, classical, and elegant. He had taught many of his converts, male and female, to write, and corresponded with them in their own language, and made some attempts in Indian poetry. When he was young he learned to speak Dutch, and so came more easily to a smattering of English, enough to be understood by traders and tradesmen who had been employed in building a church and other work at Norridgewock. He corresponded, in Latin, with one or more of the ministers of Boston, and had a great fondness for shewing his talent at controversy. Pride was his foible, and he took great delight in raillery. The English idiom and the flat and bald Latin, in some of his correspondent letters, afforded him subject. Some of his contemporaries, as well as Cotton, Norton, Mitchel, and others of the first ministers of the country, would have been a match for him. He contemned and often provoked the English, and when threatened with destruction by them, if they should ever take Norridgewock, he replied—if—. We shall see, by and by, that he met with the same fate with others long before him, who by the like Laconic and insulting answers had unnecessarily provoked their enemies.

The English charged the Indians with perfidy and breach of the most solemn engagements. The Jesuit denies it and justifies their conduct, from their being under *duresse*, at such times, and compelled to agree to whatever terms are proposed to them; particularly, when they met Governor Shute, at Arowsick, in 1717, he says, the body of the Norridgewocks had fully determined, that the English should settle no farther upon Kennebeck river than a certain mill; for all the pretence they had to go beyond that, was a bargain of this sort, made by some Englishman with any Indian he happened to meet with, "I will give you a bottle of rum if you will give me leave to settle here, or if you will give me such a place; give me the bottle, says the Indian, and take as much land as you have a mind to: The Englishman asks his name, which he writes down and the bargain is [finished. Such sort of bargains being urged against the Indians, at the treaty, they rose in a body and went away in great wrath, and, although they met again the next day and submitted to the governor's terms, yet when they came home all they had done was disallowed by the body of the nation and rejected." Whilst the English kept within the mill the Jesuit forbad the Indians molesting them, but if any settled beyond those bounds he allowed and encouraged the Indians to kill their cattle and to make other spoil.

The consideration made by the purchasers of Indian lands was not always so inconsiderable as the Jesuit mentions, and the purchases were from chiefs or reputed chiefs or sachems, and possession had been taken and improvements made scores of miles beyond the limits he would restrain the English to, more than sixty years before.

The French governor, Vaudreuil, in his manuscript letters, and the French historian, Charlevoix, in print, suppose the English settlers to be mere intruders, and charge the English nation with great injustice in dispossessing the Abanakis of their country. The European nations which have their colonies in America, may not reproach one another upon this head. They all took possession contrary to the minds of the natives, who would gladly have been rid of their new guests. The best plea, viz. that a small number of families laid claim to a greater part of the globe than they were capable of improving, and to a greater proportion than the general proprietor designed for so few people, who therefore had acquired no such right to it as to exclude the rest of mankind, will hold as well for the English as any other nation. The first settlers of the Massachusetts and Plymouth were not content with this, but made conscience of paying the natives to their satisfaction for all parts of the territory which were not depopulated or deserted and left without a claimer. Gorges, the original patentee of the province of Maine, made grants or conveyance of great part of the sea-coast and rivers of that province without purchase from the natives, other parts had been purchased from them by particular persons, and the remaining part, as well as the country e. of it, the government claimed by conquest; but it must be confessed that in the several treaties of peace this right had not been acknowledged by the Indians nor insisted upon by the English, this controversy being about those parts of the country which the English claimed by purchase, and no mention made of a right to the whole by conquest.

The governor, immediately after the dissolution of the general court, issued writs for a new house of representatives, and the court met, the 23d of August, at the George tavern, the extreme part of Boston, beyond the isthmus or neck, the small-pox then prevailing in the town. The house chose Mr. Clarke, their former speaker, and informed the governor of it by message, and he sent his approbation, in writing, to the house. They passed a resolve, that they intended no more by their message than to inform the governor and council of the choice they had made, and that they had no need of the governor's approbation.

The first act of the house gave new occasion for controversy. They were so near the town as to be in danger, and, instead of desiring the governor to adjourn or prorogue the court to some other place, they passed a vote for removing the court to Cambridge and sent it to the council for concurrence. The council non-concurred in the vote. The governor let the house know, that he should be very ready to gratify them if he was applied to in such manner as should consist with the sole right in him of adjourning, proroguing, and dissolving the court. They replied, that they were very willing to acknowledge his right, so far as respected time; but as to place, by the law of the province the court was to be held in Boston, and therefore an act or order of the three branches was necessary to remove it to any other place. They let the governor know further, that although they had convened in consequence of his summons, yet, as many of the members apprehended their lives in danger, they would leave the court and go home. There was a quorum, however, who chose to risk their lives rather than concede that the governor had power, by his own act, to remove the court from Boston to any other town in the province, or risk the consequence of refusing to remain a sufficient number to make a house.

The governor had received from England the opinion of the attorney-general, that he had good right to negative the speaker, and the lords commissioners of trade and plantations had wrote to him and signified their approbation of his proceedings. These papers he caused to be laid before the house. The house drew up a remonstrance, in which they justify their own conduct and that of former assemblies, in their controversies with the governor, and with a great deal of decency declare, that, with all deference to the opinion of the attorney-general, they must still claim the right of solely electing and constituting their speaker, and they humbly presumed that their so doing could not be construed a slight of or disrespect to his Majesty's instructions, or bearing upon the] [royal prerogative. The governor gave them a short and very moderate answer; that he had made his Majesty's instructions and the royal charter the rule of his administration, that he did not desire to be his own judge, the former house had voted to send an account of their proceedings to England, and it would be very acceptable to him, if the present house would state the case and send it home to persons learned in the law, and give them directions to appear for the house, that his Majesty might judge between his governor and them, but in the mean time it was his duty to follow his instructions until they were countermanded.

Here seems to have been a calm interval. The flame was abated but the fire not extinguished. Fresh fuel soon caused a fresh flame. The grant to the governor afforded proper matter. It was said the house were bad economists. To save 100*l.* in the governor's salary they put their constituents to the expence of 500*l.* for their own wages. If the governor's demand was unreasonable, the house might be justified, although the wages of the members for the time spent in the debate amounted to much more than the sum in dispute. The currency also continued to depreciate, but this is a consideration which never had its just weight. Twenty shillings one year must be as good as 20*s.* another. They received and paid their private dues and debts in bills of credit according to their denominations, why should not the government's debts be paid in the same manner? A majority of the house were prevailed upon to vote no more than 500*l.* for half a year's salary, equal to about 180*l.* sterling.

The governor was irritated, instead of obtaining an established salary of 1000*l.* per annum, which he had been instructed to insist upon, his whole perquisites from the government would not afford him a decent support, and they were growing less every day by the sinking of the currency in its value.

The house, from an expectation that the governor would, from time to time, make complaints to the ministry, voted 500*l.* sterling, to be paid into the hands of such persons as should be chosen to defend their rights in England, but the council refused to concur the vote, because it was not expressed by whom the persons should be chosen.

At the close of the session, the house and council came into a vote, and the governor was prevailed with to consent to it, "that 300 men should be sent to the head-quarters of the Indians, and that proclamation should be made commanding them, on pain of being prosecuted with the utmost severity, to deliver up the Jesuits and the other heads and fomenters of their rebellion, and to make satisfaction for the damage they had done, and, if they refused to comply, that as many of their principal men as the commanding officer should judge meet should be seized, together with Rallé, or any other Jesuit, and sent to Boston, and, if any opposition should be made, force should be repelled by force." Judge Sewall, one of the council, scrupled the lawfulness of this proceeding against the Indians and entered his dissent. After the general court was prorogued, the governor, notwithstanding he had consented to the vote, suspended the prosecution until the Indian hostages escaped from the castle, but a war being then deemed inevitable, orders were given for raising the men. The hostages were taken and sent back to their confinement, and the orders were recalled.

A promise had been made, by the governor, to the Indians, that trading houses should be built, armourers or smiths sent down, at the charge of the province, and that they should be supplied with provisions, clothing, &c. for their furs and skins. The compliance with this promise was expected from the general court, and, at any other time, it would have been thought a well judged measure, but the unhappy controversy with the governor would not suffer any thing, from him, to be approved of, and the private traders provoked the Indians by their frauds and other injuries, and it seems the governor, as well as good Mr. Sewall, scrupled whether a declaration of war against them was just or prudent. This house and council chose to call the proceedings against them a prosecution for rebellion; but, if a view be taken of all the transactions between the English and them from the beginning, it will be difficult to say what sort of subjects they were, and it is not certain that they understood that they had promised any subjection at all.

The house, dissatisfied with the governor for not carrying into execution a vote of the whole court, resolved at the beginning of the next session, "that the government has still sufficient reason for prosecuting the e. Indians for their many breaches of covenant." The vote being sent up for concurrence, the council desired the house to explain what they intended by prosecution, but they refused to do it, and desired the council either to concur or non-concur. The house refusing to explain their meaning, the board undertook to explain it, and concurred the vote with a declaration that they understood it to be such a prosecution as had been determined the former session. This no doubt was irregular in the council, and left room] [to question whether it was a vote of the court, the house not having agreed to it as the council qualified it. However, in consequence of it, a party of men were ordered up to Norridgewock, and returned with no other success than bringing off some of Rallé's books and papers, his faithful disciples having taken care to secure his person and to fly with him into the woods. This insult upon their chief town and the spoil made upon their priest will not long remain unrevenged.

The session began at Boston the 3d of November. The governor prorogued the court to meet at Cambridge, the 7th; and before they proceeded to business, to avoid any dispute about the place of meeting, which would have obstructed the important affairs of the province, he gave his consent to a vote of the two houses, that by this instance of the governor's adjourning the court, no advantage should be taken in favour of his sole power of removing the court from place to place. In his speech he had taken no notice of party disputes, and only recommended to them to raise money for the service of the government, and particularly of their exposed frontiers.

The house, in their vote for supply of the treasury, brought in a clause which had not been in former votes, and which the council supposed would lay such restraint upon the money in the treasury, that it would not be in the governor's power, with their advice and consent, so much as to pay an express without a vote of the whole court; they therefore non-concurred in the vote, and the house refused any provision without that clause. In the midst of the dispute, Mr. Hutchinson, one of the members for Boston, was seized with the small-pox and died in a few days. The speaker, Mr. Clarke, was one of the most noted physicians in Boston, and, notwithstanding all his care to cleanse himself from infection after visiting his patients, it was supposed brought the distemper to his brother member, which so terrified the court, that after the report of his being seized it was not possible to keep them together, and the governor found it necessary to prorogue them. At the next session in March, the house insisting upon the form of supply which they had voted in the last session, the council concurred.

An affair happened during this session which shewed the uncertainty of the relation the Indians stood in to the English. Castine, son by an Indian woman to the Baron de St. Castine, who lived many years in the last century at Penobscot, had appeared among the Indians, who were in arms at Arowsick. By an order of court he had been afterwards seized in the e. country and brought to Boston, and put under close confinement.

The house ordered that he should be brought upon trial in the county of Suffolk, before the superior court, and that the witnesses who saw him in arms should be summoned to attend. This, no doubt, would have been trying in one country a fact committed in another. The council non-concurred and voted to send for witnesses, that the court might judge in what manner to proceed against him, but this was not agreed to by the house. Some time after a committee was appointed to examine him. Castine was a very subtle fellow and made all fair with the committee. He professed the highest friendship for the English, and affirmed that he came to Penobscot to prevent the Indians from doing mischief, and promised to endeavour to influence all that tribe to keep peace. The committee, therefore, reported, and the two houses accepted the report, that he should be set at large. The governor approved of this proceeding. He had yet hopes of preserving peace. To have punished him as a traitor would have destroyed all hopes of an accommodation. It might also be very well questioned whether it would have been justifiable. The tribe or nation with which he was mixed has repeatedly, in words of which they had no adequate ideas, acknowledged themselves subjects; but, in fact, in concomitant as well as precedent and subsequent transactions with them, had always been considered as free and independent, and, although they lived within the limits of the charter, the government never made any attempt to exercise any civil authority or jurisdiction over them, except when any of them came within the English settlements, and disputes had arisen between them and the English subjects.

The house, who, the last session, were for prosecuting the Indians, and could not reasonably have supposed that they would bury, as they express themselves, the late march of the English to Norridgewock, seem, notwithstanding, to be suddenly changed from vigorous measures for bringing them to terms, to schemes for appeasing and softening them; and a present was ordered to be sent to Bomaseen, the Norridgewock captain, to engage him in favour of the English.

The small-pox this year made great havoc in Boston and some of the adjacent towns. Having been prevented spreading for near 20 years, all born within that time, besides many who had escaped it before, were liable to the distemper. Of 5889 which took it in Boston, 844 died. Inoculation was introduced upon this occasion, con-]

[trary to the minds of the inhabitants in general, and not without hazard to the lives of those who promoted it, from the rage of the people. Dr. C. Mather, one of the principal ministers of Boston, had observed, in the Philosophical Transactions, a letter of Timonius, from Constantinople, and a treatise of Pylarinus, Venetian consul at Smyrna, giving a very favourable account of the operation, and he recommended a trial to the physicians of the town, when the small-pox first began to spread, but they all declined it except Dr. Boylston, who made himself very obnoxious. To shew the confidence he had of success he began with his own children and servants. Many sober pious people were struck with horror, and were of opinion that if any of his patients should die he ought to be treated as a murderer. The vulgar were enraged to that degree that his family was hardly safe in his house, and he often met with affronts and insults in the streets.

The faculty, in general, disapproved his conduct, but Dr. Douglass made the most zealous opposition. He had been regularly bred in Scotland, was assuming even to arrogance, and in several fugitive pieces, which he published, treated all who differed from him with contempt. He was credulous, and easily received idle reports of persons who had received the small-pox by inoculation taking it a second time in the natural way, of others who perished in a most deplorable manner from the corrupt matter, which had so infected the mass of blood as to render the patient incurable. At other times, he pronounced the eruption from inoculation to be only a pustulary fever, like the chicken or swine pox, nothing analogous to the small-pox, and that the patient, therefore, had not the least security against the small-pox afterwards by ordinary infection.

Another practiser, Lawrence Dalhonde, who had been a surgeon in the French army, made oath, that at Cremona, about the year 1696, the operation was made upon 13 soldiers, four of whom died, three did not take the distemper, the other six hardly escaped, and were left with tumors, inflammations, gangrenes, &c. and that about the time of the battle of Almanza, the small-pox being in the army, two Muscovians were inoculated, one without any immediate effect, but six weeks after was seized with a frenzy, swelled all over his body, and was supposed to be poisoned, and, being opened after his death, his lungs were found ulcerated, which it was determined was caused by inoculation.

The justices of the peace and select men of the town called together the physicians, who, after mature deliberation, came to the following conclusions. "That it appears by numerous instances, that inoculation has proved the death of many persons soon after the operation, and brought distempers upon many others, which, in the end, have proved deadly to them. That the natural tendency of infusing such malignant filth in the mass of blood is to corrupt and putrefy it, and if there be not a sufficient discharge of that malignity, by the place of incision or elsewhere, it lays a foundation for many dangerous diseases. That the continuing the operation among us is likely to prove of the most dangerous consequence." The practice was generally condemned.

The common people imbibed the strongest prejudices, and such as died by inoculation were no more lamented than self-murderers. Dr. Mather, the first mover, after having been reproached and vilified in pamphlets and newspapers, was at length attacked in a more violent way. His nephew, Mr. Walter, one of the ministers of Roxbury, having been privately inoculated in the doctor's house in Boston, a villain, about three o'clock in the morning, set fire to the fuse of a granado shell, filled with combustible stuff, and threw it into the chamber where the sick man was lodged. The fuse was fortunately beat off by the passing of the shell through the window, and the wild-fire spent itself upon the floor. It was generally supposed that the bursting of the shell by that means was prevented. A scurrilous menacing writing was fastened to the shell or fuse.

The moderate opposers urged that the practice was to be condemned as trusting more to the machination of men than to the all-wise providence of God in the ordinary course of nature, and as tending to propagate distempers to the destruction of mankind, which proved it to be criminal in its nature and a species of murder. The magistrates, we mean those in Boston, supposed it had a tendency to increase the malignity and prolong the continuance of the infection, and that therefore it behoved them to discountenance it.

At length the house of representatives laid hold of it, and a bill was brought in and passed to prohibit all persons from inoculation for the small-pox, but the council were in doubt and the bill stopped.

Such is the force of prejudice. All orders of men, in that day, in greater or lesser proportion, condemned a practice which is now generally approved, and to which many thousands owe the preservation of their lives.

Boylston continued the practice in spite of all the opposition. About 300 were inoculated in]

[Boston and the adjacent towns. It is impossible to determine the number which died by it. Douglass would have it there was one in 14, whilst the favourers of the practice would not allow more than one in 70 or 80. It was evident from the speedy eruption, that many had taken the distemper before they were inoculated. Indeed, where persons have continued in an infected air for months together, no true judgment can ever be made of the experiment.

(*Anno* 1722.)—The new house of representatives, in May, chose the former speaker, and the governor declared his approbation in the same manner he had done before. He negatived two of the counsellors elect, Colonel Byfield and Mr. William Clark. Mr. Clark, being a member of the house for Boston, had ever adhered closely to Mr. Cooke. The governor shewed his resentment by refusing to admit him to the council, but did not serve his own interest, Mr. Clark's opposition being of greater consequence in the house.

The Indians were meditating mischief from the time the English were at Norridgewock, but committed no hostilities until June following. They came then with about 60 men, in 20 canoes, into Merrymeeting bay, and took prisoners nine families, but gave no marks of their usual rage and barbarity. Some of their prisoners they released immediately, and others in a short time after. Enough were retained to be a security for the return of their hostages from Boston. Another small party of Indians made an attempt upon a fishing vessel belonging to Ipswich, as she lay in one of the *e.* harbours, but the fishermen being armed they killed two or three of the Indians and the rest retreated. The collector of the customs at Annapolis Royal, Mr. Newton, with John Adams, son of one of the council for Nova Scotia, were coming from thence with Captain Blin to Boston, and putting into one of the Passimaquadies, went ashore with other passengers, and were all seized and made prisoners by about 12 Indians and as many French; the people left on board the sloop cut their cables and fled to Boston.

Another party of the Indians burned a sloop at St George's river, took several prisoners, and attempted to surprise the fort.

Intelligence of these several hostile acts came to Boston whilst the general court was sitting, but there seemed to be no disposition to engage in war. Instead of the former vigorous resolves, upon lesser provocations, the house proposed that a message should be sent to the Norrigewock Indians to demand the reasons of this behaviour, restitution of the captives, and satisfaction for damages, and acquaint them that if they refused, effectual methods would be taken to compel them. The hostages given by the Indians were sent down to the *e.* and upon the restoring the English captives they were to be set at liberty.

The friends of the English captives were importunate with the government to take measures for their redemption, and a view to effect this seems to have been the chief reason which delayed a declaration of war. But soon after the prorogation of the court news came that the Indians had burnt Brunswick, a village between Caseo bay and Kennebeck, and that Captain Harman, with part of the forces posted upon the frontiers, had pursued the enemy, killed several, and taken 15 of their guns. Immediately after this news, July 25, the governor, by advice of council, caused a declaration of war to be published.

Foreign wars often delivered Greece and Rome from their intestine broils and animosities, but this war furnished a new subject for contention. The governor often charged the party in the house with assuming the direction of the war, and taking into their hands that power which the charter gives to the governor. He gave them a hint in his speech, August 8th, at the opening the next session. "One thing I would particularly remark to you, which is, that if my hands and the council's be not left at a much greater liberty than of late they have been, I fear our affairs will be carried on with little or no spirit. Surely every person who wishes well to his country will think it high time to lay aside all animosities, private peeks, and self-interest, that so we may unanimously join in the vigorous prosecution of the weighty affairs which are now upon the carpet."

The house, in an address to the governor, signified their sentiments of the necessity of this declaration of war, and promised "all necessary and chearful assistance." A committee of the two houses settled the rates of wages and provisions for the forces, to which no exception was taken, but they went further and determined the service in which they were to be employed, 300 men to be sent upon an expedition to Penobscot, and the rest to be posted at different places on the frontiers, and qualified their report by desiring the governor to give orders accordingly. He let them know that the king his master and the royal charter had given him the sole command and direction of the militia and all the forces which might be raised on any emergency, and that he would not suffer them to be under any direction but his own and those of-]

[ficers he should think fit to appoint. The house made him no answer. The destination of the military forces in this manner, and making the establishment of their wages depend upon a compliance with it, had not been the practice in former wars and administrations, but the governor found he must submit to it, or the frontiers would be without defence. He gave up his own opinion with respect to the Penobscots, and had laid the same plan which the committee had reported, and he intended to prosecute it, which made his compliance more easy. The house being dissatisfied with Major Moody, who had the command of the forces, passed the vote desiring the governor to dismiss him. The council non-concurred in this vote, "because he was condemned unheard," and substituted another vote to desire the governor to send for him that he might attend the court, but this the house would not agree to, and sent a separate message to the governor to desire him to suspend the major from his post. The governor told them he was surprised they should desire so high a piece of injustice as the punishing a man without hearing what he had to say for himself, and let them know he would inquire into the grounds of their complaint. Several other votes passed relative to the forces, which the governor did not approve.

At the next session, November 15, he recommended a law to prevent mutiny and desertion, for want of which the men were daily running away. The house thought it necessary to be first satisfied whether the desertion in the army was not owing to the unfaithfulness of the officers, and appointed two committees, one to repair to the head-quarters on the *e.* and other on the *w.* frontiers, with powers to require the officers to muster their companies, when an exact list was to be taken of the men that appeared, an account of all deserters, and of all such as were absent upon furlough, or had been dismissed, or had been exchanged, together with divers other powers. They then applied to the governor to give orders to all in command to pay a proper deference to the vote and order of the house respecting repeated abuses and mismanagements among the forces, &c.

This the governor thought he had good right to except to, and he made the vote itself, as well as the manner in which it was to be executed, an article of complaint against the house to the king, but he was prevailed upon to consent to it, and either made or intended to make this condition, that the committees should make report to him. The house urged this consent against him, but in England it was not thought a sufficient justification.

The conceding in one point naturally led to a demand of the like concessions in others.

It was thought a salutary measure to send for delegates from the Iroquois, who were in friendship with us, and to desire them to use the influence they had over the *e.* Indians, in order to their making satisfaction for the injuries done, and to their good behaviour for the time to come. When the delegates came to Boston, the house voted that the speech to be made to them by the governor should be prepared by a committee of the two houses. The governor had prepared his speech, and he directed the secretary to read it to the house of representatives, but this was not satisfactory, and they sent a message to desire that what the secretary had read might be laid before the house. The governor refused at first, but upon further consideration consented, desiring they would speedily return it. They sent it back to him, and let him know they would not agree to it, unless he would speak in the name of the general court, and the house of representatives might be present when the speech was delivered. This was disagreeable to him and a novelty to the Indians, who had always considered in their treaties the governor of Pennsylvania as well as the governor of New York, to be treating with them in their own names, or the name of the king, and not of their respective assemblies, but he submitted.

In consequence of the vote of the house in the last session, the governor had directed an expedition to Penobscot, although it was not altogether agreeable to his own judgment. It seems he had hopes of an accommodation with that tribe at least. Colonel Walton, who had the command on the *e.* frontiers, selected forces proper for the purpose, and they had actually begun their march when intelligence arrived to the colonel that Arowsick was attacked by a great number of Indians. He immediately sent an express with orders to the forces to return, and acquainted the governor with his proceedings. The council advised to keep the whole forces for the defence of our own inhabitants, and to suspend acting upon the offensive until winter, which they judged a more proper season for the expedition; and the men, in consequence of this new advice, were employed in marches upon the back of the frontiers. But the house were dissatisfied, and sent a message to the governor "to desire him to order, by express, Colonel Walton to appear forthwith before the house, to render his reasons why the orders relat-]

[ing to the expedition had not been executed." This was not only to take Walton from the command, as long as the house should think fit to detain him, but the orders "relating to the expedition" might be understood to mean the orders which had been given by the house, and not what he had received from the captain-general. The governor told the committee that he would take no notice of the message from the house unless it was otherwise expressed; besides, he and the council were well satisfied, and he thought every body else was. He added, that he intended the officers should give an account "to him" of their conduct. The next day, Nov. 20th, they sent another message to him to desire him to inform the house whether he would send for Walton as they bad desired. He then told the committee he would send his answer to the house when they thought proper. Upon this they seem to have appointed a messenger to go to the *e.* upon what occasion does not appear, and the next day passed the following extraordinary vote, "Whereas this house did on Thursday last appoint a committee to wait on his excellency the governor, praying his orders for Colonel Walton's appearance before the house, and renewed their request to him yesterday, and his excellency has not yet seen cause to comply with that vote, and the denial of Colonel Walton's being sent for has extremely discouraged the house in projecting any further schemes for carrying on the war, under any views of success. And this house being zealously inclined to do what in them lies to bring this people out of the calamities and perplexities of the present war, and to spare no cost and charge to effect so great a good, were some things at present remedied; We do, therefore, once more, with the greatest sincerity and concern for our country's good, apply to your excellency for your speedy issuing your orders concerning Colonel Walton, to be dispatched by the messenger of this house going into those parts." The governor did not like to be so closely pressed, and when the committee came to his house he told them, he would not receive the vote, and, as it is inserted in the report and journal of the house, "he went his way." They then appointed their speaker and eight principal members, a committee to wait upon the governor and desire him to return to the chair, "on some important affairs which lay before the house," but he refused to see the committee, and directed his servant to tell them he would not then be spoke to by any body.

Walton was a New Hampshire man, at the head of the forces, a small part only of which were raised in that government. This might prejudice many, but there was a private grudge against him in some of the leading men of the house, and they never left pursuing him until they effected his removal.

The house finding the governor would not comply, all their messages to him being exceptionable, as founded upon a supposed right in the house to call the officers out of the service to account before them whensoever they thought proper, and also to order the particular services in which the forces should be employed, without leaving it in the governor's power to vary, they made some alterations in the form of their request, and (Dec. 4th) passed the following vote, "Whereas this house have been informed of divers miscarriages in the management of the war in the *e.* country, voted that his excellency the governor be desired to express to Colonel Walton, that he forthwith repair to Boston, and when he hath attended upon his excellency, that he would please to direct him to wait on this house, that they may examine him, concerning his late conduct in prosecuting the war, more especially referring to the late intended expedition to the fort of Penobscot." This being more general, and not confined to the laying aside the expedition, which was known to be in consequence of orders, the governor was willing it should be construed favourably, and sent for Walton.

The council having steadily adhered to the governor, he took this opportunity to recommend to the house to act jointly with the council in messages to him of general concern, and at the same time, in a verbal message by the secretary, endeavoured to soften the temper of the house.—"Mr. speaker, his excellency commands me to acquaint this honourable house that he has taken into consideration the several messages relating to Colonel Walton, and thinks it most agreeable to the constitution, and what would tend to keep up a good agreement between the council and house of representatives, for all their messages, of a public nature and wherein the whole government is concerned, to be sent up to the council for their concurrence and not immediately to himself; however, that he will give order for Colonel Walton's coming up to town, and, when he has received an account of his proceedings, the whole court shall have the hearing of him if they desire it." In this way the governor intended to guard against any undue proceeding, there being no danger of the council's condemning a measure to which a little while before they had given their advice and consent, but the house improved the hint to a very] [different purpose, and on the 5th December voted, "that a committee, to consist of 11 members of the two houses, seven of the house of representatives and four of the council, shall meet in the recess of the court, once in 14 days, and ottener if occasion should require, to concert what steps and methods shall be put in practice, relative to the war, and having agreed upon any projections or designs, to lay them before his excellency for his approbation, who is desired to take effectual care to carry them into speedy execution." In affairs of government, of what nature soever, this was an innovation in the constitution, but in matters relative to the war it was taking the powers from the governor, which belonged to him by the constitution, and vesting them in a committee of the two houses. The council unanimously non-concurred in the vote, and altercation ensued between them and the house, but the council persevered. In the mean time the governor was engaged with the house in fresh disputes.

The committee of the house which had been sent to the e. frontiers returned, and, instead of making their report to the governor, which was the condition of his consent to their authority and of his orders to the officers to submit to them, they made their report to the house. This was disingenuous. It would not do to urge that he had no right to make conditions to their votes, for he had given no consent, unless it was conditional, and without his consent they could have no authority. As soon as he heard of the report, he sent to the house for his original order, which he had delivered to the committee. They answered that they were not possessed of it, but the chairman of the committee had left an attested copy on their files, which he might have if he pleased, but he refused the copy and insisted upon the original. He then sent for John Wainwright, the chairman of the committee, to attend him in council, and there demanded the return of the original order. Wainwright, in general, was what was called a prerogative man, but the house had enjoined him not to return the order. He acknowledged he had the order in his possession, but desired to be excused from delivering it, the house having directed him to deliver no original papers. The original vote of the house, and the governor's order in consequence of it, are as follows:

"In the house of representatives, Nov. 11, 1712.

"Whereas this house have been informed of repeated abuses and mismanagements among the officers now in pay, tending greatly to the dishonour and damage of the government, and are desirous to use all proper and suitable methods for the full discovery thereof,—and, to effect the same, have sent a committee from the house to inquire into these rumours and report how they find things,—we the representatives do most earnestly desire your excellency's orders, by the same committee, to the commanding officer and all others in command there, to pay all proper deference to the vote and order of this house respecting that matter.

JOHN CLARKE, speaker."

"Boston, Nov. 17th, 1722. } To the officer commanding in chief at the eastward.

"I do hereby give orders to the commanding officers and all other inferior officers to pay deference to the committee, and do expect that the committee lay first before me their report as captain-general, and afterwards, upon the desire of the house of representatives, it shall be laid before them.

SAMUEL SHUTE."

The house expected the governor would complain of them for usurping a military power, and might refuse to part with the original votes or orders by which he had signified his consent to it, the condition not preceding the exercise of such power.

Soon after (Dec. 18th) Colonel Walton came to town, and the house sent their committee to desire the governor to direct him to attend the house the next morning, but the governor refused to give such orders, and told the committee, that if his officers were to answer for their conduct, it should be before the whole court. They then sent their door-keeper and messenger to Walton, to let him know the house expected his attendance. He went immediately, but refused to give any account of his proceedings, without leave from the governor. The next day, Walton was ordered to appear before the whole court, and the governor sent a message by the secretary, to acquaint the house, that they might then ask any questions they thought proper, relative to his conduct, but they resolved, that their intent in sending for him was, that he should appear before them. The next day, he sent another message to acquaint the house that Walton was then before the governor and council, with his journal, and if the house inclined to it, he desired them to come up, and ask any questions they thought proper. They returned for answer, that they did not think it expedient, for they looked upon it not only their privilege, but duty, to demand, of any officer in the pay and service of the government, an account of his management while employed by the public.

This perhaps, in general, was not the cause of]

[dispute, but the question was, whether he was culpable for observing the orders which the governor had given contrary to the declared mind and order of the house? They then passed an order for Walton forthwith to lay his journal before the house. This was their last vote relative to this affair whilst the governor was in the province. He had, without making it public, obtained his Majesty's permission to leave the province and go to England. The prejudice, in the minds of the common people, increased every day. It was known to his friends, that as he sat in one of the chambers of his house, the window and door of a closet being open, a bullet entered, through the window and door passages, and passed very near him. If some thought this a mere accident, yet as he knew he had many virulent enemies, he could not be without suspicion of a wicked design; but his principal intention in going home, was to represent the conduct of the house, to call them to answer before his Majesty in council, and to obtain a decision of the points in controversy, and thereby to remove all occasion or pretence for further disputes. His departure was very sudden. The Seahorse man of war, Captain Durrell, lying in Nantasket, bound to Barbadoes to convoy the Saltortugas fleet, the governor went on board her December 27th, intending to go from Barbadoes the first opportunity for London. Not one member of the court was in the secret, nor indeed any person in the province except two or three of his domestics. The wind proved contrary for three or four days, during which the owners of the ship Ann, Captain Finch, which was then loading for London, by employing a great number of hands, had her fitted for sea and sent her to Nantasket, and offered the governor his passage in her, and he went on board and sailed the first of January.

Upon a review of this controversy with Governor Shute, we are apprehensive some of our readers will be apt to doubt the impartiality of the relation. Such steps and so frequent by one party, without scarce any attempts by the other, are not usual, but we have made the most diligent search into the conduct of the governor, as well as the house, and we are not sensible of having omitted any material fact, nor have we designedly given a varnish to the actions of one party, or high colouring to those of the other. Colonel Shute had the character of being humane, friendly, and benevolent, but somewhat warm and sudden upon provocations received, was a lover of ease and diversions, and for the sake of indulging his inclinations in those respects, would willingly have avoided controversy with particular persons or orders of men in the government; but it was his misfortune to arrive when parties ran high and the opposition had been violent. With great skill in the art of government, it might not have been impossible for him to have kept both parties in suspense, without interesting himself on either side, until he had broke their respective connections or the animosity had subsided; but, void of art, with great integrity, he attached himself to that party which appeared to him to be right, and made the other his irreconcilable enemies. His negativing Mr. Cooke, when chosen to the council, was no more than what he had an undoubted right to do by charter; but the refusal to accept him as speaker, perhaps, was impolitic, the country in general supposing it to be an invasion of the rights of the house, and it would have been less exceptionable to have dissolved them immediately, which he had a right to do, than to dissolve them after an unsuccessful attempt to enforce his negative when his right was doubtful in the province, although not so with the attorney and solicitor general, who supposed the house of representatives claimed a privilege, which the house of commons did not. The leading men in the house of representatives did not think so. That point had not been in question in England since the reign of King Charles II. when it was rather avoided than determined, and it was not certain that the house of commons in the reign of King George I. would more readily have given up the point than their predecessors in the time of King Charles. The house, in the other parts of the controversy, had less to say for themselves, and with respect to the attempts upon his military authority, were glad to be excused by an acknowledgment of their having been in the wrong. The clipping his salary, which at the highest would no more than decently support him, was highly resented by him, and his friends were heard to say, that he would have remained in the government and waited the decision of the other points, if the 200*l*. equal to about 50*l*. sterling, the deduction made, had been restored.

Under an absolute monarch the people are without spirit, wear their chains despairing of freedom. A change of masters is the sum of their hopes, and alter insurrections and convulsions, they still continue slaves. In a government founded upon the principle of liberty, as far as government and liberty can consist, such are the sweets of liberty, that we often see attempts for a greater degree of it than will consist with the established constitution, although anarchy, the greatest and worst of tyrannies may prove the consequence, until the eyes of the people are opened and they see the ne-]

[cessity of returning to their former happy state of government and order.

The lieutenant-governor took the chair, under the disadvantage of being obliged to maintain the same cause which had forced his predecessor out of it. Personal prejudice against the governor was the cause of assuming rights reserved by charter to the crown. The cause now ceased, but power once assumed is not willingly parted with. Mr. Dummer had demeaned himself very discreetly. His attachment to the cause of the governor lost him some friends and proved a prejudice to him and to his successors, for it had been usual to make an annual grant or allowance to the lieutenant-governor, in consideration of his being at hand, or, as they expressed it, ready to serve the province, in case of the governor's absence, but after the two or three first years from his arrival, they withheld it. Without any mention of the unhappy state of affairs, in a short speech to the two houses, he let them know that he would concur with them in every measure for his Majesty's service and the good of the province. An aged senator, Mr. Sewall, the only person alive who had been an assistant under the old charter, addressed himself to the lieutenant-governor with great gravity and simplicity, in primitive style, which, however obsolete, may be worth preserving. "If your honour and the honourable board please to give me leave, I would speak a word or two upon this solemn occasion. Although the unerring providence of God has brought your honour to the chair of government in a cloudy and tempestuous season, yet you have this for your encouragement, that the people you have to do with are a part of the Israel of God, and you may expect to have of the prudence and patience of Moses communicated to you for your conduct. It is evident, that our almighty Saviour counselled the first planters to remove hither and settle here, and they dutifully followed his advice, and therefore he will never leave nor forsake them nor theirs; so that your honour must needs be happy in sincerely seeking their happiness and welfare, which your birth and education will incline you to do. *Difficilia quæ pulchra.* I promise myself, that they who sit at this board will yield their faithful advice to your honour, according to the duty of their place."

The house thought it necessary to take immediate measures for their defence and vindication in England. The governor had mentioned nothing more to their lieutenant-governor, than that he was embarked and intended to return to his government early in the fall. This the lieutenant-governor communicated to the council, and the council to the house. They sent a committee immediately to the lieutenant-governor, to pray him to inform them what he knew of the governor's intended voyage; but he could tell them no more. They then appointed another committee "to prepare and lay before the house what they think proper to be done in this critical juncture, in their just and necessary vindication at the court at home," and a ship, Captain Clark, then ready to sail for London, was detained until the dispatches were ready. Anthony Sanderson, a merchant of London, had been recommended by Mr. Popple, of the plantation office, in a letter to the speaker, as a proper person for the province agent. To him the house sent their papers, to be improved as they should order.

(*Anno* 1723.)—The house was loth suddenly to recede, and the day after the governor sailed, they appointed a committee, to join with a committee of council, to consider of proper ways for carrying into execution the report of a committee of war. This was the province of the captain-general, and the council refused a concurrence. The house then passed another vote, protesting against carrying on an offensive war unless Walton, the colonel, and Moody, the major, should be removed and other suitable persons appointed. Before the council passed upon this vote, the two obnoxious persons were prevailed upon to write to the lieutenant-governor and desire a dismission, provided they might be paid their wages, and the letters being communicated to the council, they passed another vote, desiring the lieutenant-governor to dismiss the officers, agreeable to the letters received from them. In this vote the house non-concurred, and insisted upon their own vote, in which the council then non-concurred. The house then passed a resolve, that, unless Walton and Moody were dismissed, they should be necessiated to draw off part of the forces, and sent their "resolve to be laid upon the council table.' The lieutenant-governor, by a message, let the house know, that the king had appointed him general of the forces, and that he only had the power to draw them off, and added, that he expected all messages from the house should be properly addressed to him, otherwise he should pay no regard to them. The house were sensible they had gone too far, and appointed a committee to wait upon the lieutenant-governor, to desire they might have leave to withdraw their resolve, and declared that, however expressed, they intended only that they would not vote any further pay and subsistence. They persisted, however, in their refusal to provide for the pay of the two officers, whose dismission they required, nor] would they make provision for further carrying on the war until other officers were appointed.

Among the other instances of additional power to the house, they had by degrees acquired from the governor and council the keys of the treasury, and no moneys could be issued without the vote of the house for that purpose. This is no more than some colonies, without charters, claim and enjoy, but by the charter, all moneys are to be paid out of the treasury " by warrant" from the governor with advice and consent of the council. The right of the house to originate all acts and orders for raising moneys from the people, and to appropriate such moneys to such services as they thought proper, was not disputed, but they went further and would not admit that payment should be made for such services until they had judged whether they were well performed, and had passed a special order for such payment. Thus they kept every officer dependent, and Walton, because he had not observed their orders to go to Penobscot, but had conformed to the governor's orders, from whom he derived all the authority he had to march any where, was denied his pay. Other matters were alleged against Walton in the course of the dispute, but this seems to have been the principal.

The exposed state which the frontiers must have been in, if the forces had been drawn off, and they could not be kept there without pay, induced the lieutenant-governor to dismiss Walton and to appoint Thomas Westbrooke colonel and commander in chief, whereupon an establishment was settled by the house, premiums were granted for Indian scalps and prisoners, and an end was put to the session.

The Indians, we have observed, were instigated by the French to begin the war. The old men were averse to it. Rallé with difficulty prevailed upon the Norridgewocks. The Penobscots were still more disinclined, and after hostilities began, expressed their desires of an accommodation. The St. François Indians, who lived upon the borders of Canada, and the St. John's, as also the Cape Sable Indians, were so remote as not to fear the destruction of their villages by the English. They mixed with the Norridgewocks and Penobscots, and made the war general. In the latter part of July the enemies surprised Canso and other harbours near to it, and took 16 or 17 sail of fishing vessels, all belonging to the Massachusetts. Governor Phillips happened to be at Canso, and caused two sloops to be manned, partly with volunteer sailors from merchant vessels which were lading with fish, and sent them, under the command of John Eliot of Boston, and John Robinson of Cape Ann, in quest of the enemy. Eliot, as he was ranging the coast, espied seven vessels in a harbour called Winnepaug, and concealed all his men, except four or five, until he came near to one of the vessels, which had about 40 Indians aboard, who were in expectation of another prize falling into their hands. As soon as he was within hearing, they hoisted their pennants and called out, " strike English dogs and come aboard, for you are all prisoners." Eliot answered, that he would make all the haste he could. Finding he made no attempt to escape, they began to fear a tarter and cut their cable, with intent to run ashore, but he was too quick for them and immediately clapped them aboard. For about half an hour they made a brave resistance, but at length, some of them jumping into the hold, Eliot threw his hand granadoes after them, which made such havoc, that all which remained alive took to the water, where they were a fair mark for the English shot. From this or a like action, probably took rise a common expression among English soldiers and sometimes English hunters, who, when they have killed an Indian, make their boast of having killed a black duck. Five only reached the shore.

Eliot received three bad wounds, and several of the men were wounded and one killed. Seven vessels, with several hundred quintals of fish, and 15 of the captives, were recovered from the enemy. They had sent many of the prisoners away, and nine they had killed in cold blood. The Nova Scotia Indians had the character of being more savage and cruel than the other nations.

Robinson retook two vessels, and killed several of the enemy. Five other vessels the Indians had carried so far up the bay, above the harbour of Malagash, that they were out of his reach, and he had not men sufficient to land, the enemy being very numerous.

The loss of so many men enraged them, and they had determined to revenge themselves upon the poor fishermen, above 20 of whom yet remained prisoners at Malagash harbour, and they were all destined to be sacrificed to the manes of the slain Indians. The powowing and other ceremonies were performing when Captain Blin, in a sloop, appeared off the harbour, and made the signal or sent in a token which had been agreed upon between him and the Indians, when he was their prisoner, should be his protection. Three of the Indians went aboard his vessel, and agreed for the ransom both of vessels and captives, which were delivered to him and the ransom paid. In his way to Boston he made prisoners of three or four In-

[dians near cape Sable, and about the same time, Captain Southack took two canoes with three Indians in each, one of which was killed and the other five brought to Boston.

. This Nova Scotia affair proved very unfortunate for the Indians. The Massachusetts frontiers afforded them less plunder, but they were in less danger. On the 16th of September, between 400 and 500 Indians were discovered upon Arowsick island, by a party of soldiers employed as a guard to the inhabitants while at their labour. They immediately made an alarm, by firing some of their guns, and the inhabitants of the island, by this means, had sufficient notice to shelter themselves in the fort or garrison-house, and also to secure part of their goods, before the enemy came upon them.

They fired some time upon the fort and killed one man, after which they fell to destroying the cattle, about 50 head, and plundering the houses, and set fire to 26 houses, the flames of which the owners beheld from the fort, lamenting the insufficiency of their numbers to sally out and prevent the mischief.

These were the Indians which put a stop to the march to Penobscot. There were in the fort about 40 soldiers, under Captain Robert Temple and Captain Penhallow. Captain Temple was a gentleman, who came over from Ireland with an intent to settle the country with a great number of families from the *n.* of Ireland, but this rupture with the Indians broke his measures, and having been an officer in the army, Colonel Shute gave him a command here. Walton and Harman, upon the first alarm, made all the dispatch they could, and before night, came to the island in two whaleboats with 30 men more. With their joint force the English made an attempt to repel the enemy, but the disproportion in numbers was such, that in a bush-fight or behind trees, there was no chance, and the English retreated to the fort. The enemy drew off the same night, and passing up Kennebeck river, met the province sloop, and firing upon her, killed the master, Bartholomew Stretton, and then made an attempt upon Richmond fort, and from thence went to the village of Norridgewock, their head quarters.

A man was killed at Berwick, which was the last mischief done by the enemy this first year of the war.

When the general court met in May, next year, no advice had been received of any measures taken by the governor in England. The house chose their speaker and placed him in the chair without presenting him to the lieutenant-governor, which he took no notice of. They continued their claim to a share in the direction of the war, and insisted that if any proposals of peace should be made by the Indians, they should be communicated to the house and approved by them. They repeated also a vote for a committee of the two houses to meet in the recess of the court, and to settle plans for managing the affairs of the war, which the lieutenant-governor was to carry into execution, but in this the council again non-concurred. The lieutenant-governor's seal being affixed to a belt given to the delegates from the Iroquois, who came to Boston to a conference, the house passed a resolve, "that the seal be defaced and that the seal of the province be affixed to the belt, as the committee of the two houses have agreed," and sent the resolve to the council for their concurrence. The council, instead of concurring, voted, as well they might, that the resolve contained just matter of offence, and therefore they desired the house to withdraw it. This produced another resolve from the house still higher, "that the affixing a private seal, contrary to the agreement of a committee, was a high affront and indignity to them, and therefore they very justly expected the advisers and promoters thereof to be made known to the house." There was a double error in this transaction of the house, the lieutenant-governor having the unquestionable right of ordering the form of proceeding in treaties or conferences of this kind, and the house having no authority to direct the king's seal to be applied to any purpose, the governor being the keeper of the seal, and although in common parlance called the province seal, which we suppose led to the mistake, yet was it properly speaking the king's seal for the use of the province.

The lieutenant-governor took no public exception to any votes of the house this session, which we must presume to be owing to his apprehensions that in a short time, a full consideration would be had in England of matters of the same nature during Colonel Shute's administration. Before the next session of the general court (Oct. 23d) the agent Mr. Sanderson transmitted to the speaker, copy of the heads of complaint exhibited against the house for encroaching upon his Majesty's prerogative in seven instances.

"1st, In their behaviour with respect to the trees reserved for masts for the royal navy.

"2d, For refusing to admit the governor's negative upon their choice of a speaker.

"3d, Assuming power in the appointment of days for fasting and thanksgiving.

"4th, Adjourning themselves to a distant day by their own act.]

[" 5th, Dismantling forts and directing the artillery and warlike stores to other than the custody of the captain-general or his order.

" 6th, Suspending military officers and refusing their pay.

" 7th, Appointing committees of their own to direct and muster his Majesty's forces."

The house voted the complaint groundless, and ordered 100*l.* sterling to be remitted to Sanderson, to enable him to employ counsel to justify the proceedings of the house. The vote being sent to the council was unanimously non-concurred in.

The house then prepared an answer to the several articles of complaint and an address to the king, to which they likewise desired the concurrence or approbation of the council, but they were disapproved and sent back with a vote or message, that " in faithfulness to the province and from a tender regard to the house of representatives, the board cannot but declare and give as their opinion, that the answer is not likely to recommend this government and people to the grace and favour of his Majesty, but, on the contrary, has a tendency to render us obnoxious to the royal displeasure."

The house, however, ordered the answer and address to be signed by the speaker and forwarded to Mr. Sanderson, to be improved as they should order.

The council thereupon prepared a separate address to his Majesty and transmitted it to the governor. The non-concurrence of council with these measures of the house was resented, and the house desired to know what part of their answer had a tendency to render the government and people obnoxious. Here the council very prudently avoided engaging in controversy with the house. " It was not their design to enter into a detail, but only to intimate their opinion, that considering the present circumstances of affairs, some better method might be taken than an absolute justification." They had shewn their dissatisfaction with the conduct of the house, in every article which furnished matter for the complaint, except that of the speaker, and did all in their power to prevent them, but now this conduct was impeached, the arguments used by the council in a dispute with the house might be sufficient to justify the council, and set their conduct in an advantageous light, but they would strengthen and increase the prejudice against the country in general. This was an instance of public spirit worthy of imitation.

The house then resolved, " that being apprehensive that the liberties and privileges of the people are struck at by Governor Shute's memorial to his Majesty, it is therefore their duty as well as interest to send some suitable person or persons from hence, to use the best method that may be to defend the constitution and charter privileges." They had no power over the treasury without the council, and therefore sent this vote for concurrence, but it was refused, and the following vote passed in council instead of it: " The liberties and privileges of his Majesty's good subjects of this province being in danger, at this present critical conjuncture of our public affairs at the court of Great Britain, and it being our duty as well as interest to use the best methods that may be in defence of the same; and whereas Jeremiah Dummer, Esq. the agent of this court, is a person of great knowledge and long experience in the affairs of the province, and has greatly merited of this people, by his printed defence of the charter, and may reasonably be supposed more capable of serving us in this exigence than any person that may be sent from hence, voted, that the said Mr. agent Dummer be directed to appear in behalf of the province for the defence of the charter, according to such instructions as he shall receive from this court." This vote plainly intimated, that by the late conduct of the house the charter of the province was in danger, but the house seem to have overlooked it and concurred with an amendment, " that Mr. Sanderson and a person sent from hence be joined with Mr. Dummer." The council agreed that a person should be sent home, but refused to join Sanderson. Before the house passed upon this amendment, they made a further trial to obtain an independency of the council, and voted, that there should be paid out of the treasury, to the speaker of the house, 300*l.* sterling, to be applied as the house should order. Near three weeks were spent in altercations upon this subject, between the council and the house, at length it was agreed that 100*l.* should be at the disposal of the house, and 200*l.* to be paid to such agents as should be chosen by the whole court. The house were in arrears to Sanderson, which they wanted this money to discharge, and then were content to drop him.

The manner of choosing civil officers had been by a joint vote or ballot of council and house. This gives a great advantage to the house, who are four times the number of the board. But to be more sure of the person the majority of the house were fond of, they chose Mr. Cooke for agent, and sent the vote to the board for concurrence. The council non-concurred, and insisted on proceeding in the usual way, which the house were obliged to comply with. The choice, however, fell upon] [the same person, and he sailed for London the 18th of January.

Colonel Westbrook with 230 men set out from Kennebeck the 11th February this year, with small vessels and whale-boat, and ranged the coast as far e. as mount Desert. Upon his return he went up Penobscot river, where, about 32 miles from the anchoring place of the transports, he discovered the Indian castle or fortress, walled with stockadoes, about 70 feet in length and 50 in breadth, which inclosed 23 well finished wigwams. Without, was a church 60 feet long and 30 broad, very decently finished within and without, also a very commodious house in which the priest dwelt. All was deserted, and all the success attending this expedition was the burning the village. The forces returned to St. George's the 20th of March.

Captain Harman was intended, with about 120 men, for Norridgewock at the same time, and set out the 6th February, but the rivers were so open and the ground so full of water, that they could neither pass by water nor land, and having with great difficulty reached to the upper falls of Amascoggin, they divided into scouting parties and returned without seeing any of the enemy.

An attempt was made to engage the Six Nations and the Scatacook Indians in the war, and commissioners were sent to Albany empowered to promise a bounty for every scalp if they would go out against the enemy, but they had no further success than a proposal to send a large number of delegates to Boston.

The commissioners for Indian affairs in Albany had the command of the Six Nations, and would not have suffered them to engage in war if they had inclined to it. The Massachusetts commissioners were amused, and a large sum was drawn from the government in valuable presents to no purpose. No less than 63 Indians came to Boston, August 21st, the general court then sitting. A very formal conference was held with them, in the presence of the whole court, but the delegates would not involve their principals in war; if any of their young men inclined to go out, with any parties of the English, they were at liberty and might do as they pleased. Two young fellows offered their service, and were sent down to fort Richmond on Kennebeck river. Captain Heath the commander ordered his ensign, Coleby, and three of the garrison, to go up the river with them. After they had travelled a league from the fort, they judged by the smell of fire, that a party of the enemy must be near. The Mohawks would go no further until they were strengthened by more men, and sent to the fort for a whale-boat, with as many men as she could carry. Thirteen men were sent, and soon after they had joined the first party, about 30 of the enemy appeared, and, after a smart skirmish, fled to their canoes, carrying off two of their company dead, or so badly wounded as to be unable to walk, and leaving their packs behind. Coleby, who commanded the party, was killed and two others wounded. The Mohawks had enough of the service, and could not be prevailed on to tarry any longer, and were sent back to Boston.

Small parties of the enemy kept the frontiers in constant terror, and now and then met with success.

In April, they killed and took eight persons at Scarborough and Falmouth. Among the dead was the serjeant of the fort, Chubb, whom the Indians took to be Captain Harman, and no less than 15 of them aimed at him at the same time, and lodged 11 bullets in his body. This was lucky for the rest, many more escaping to the fort than would otherwise have done. In May, they killed two at or near Berwick, one at Wells, and two travelling between York and Wells. In June, they came to Roger Dering's garrison at Scarboborough, killed his wife and took three of his children, as they were picking berries, and killed two other persons. In July, Dominicus Jordan, a principal inhabitant and proprietor of Saco, was attacked in his field by five Indians, but keeping his gun constantly presented without firing, they did not care to close in with him, and after receiving three wounds he recovered the garrison. In August, the enemy appeared w. and the 13th killed two men at Northfield, and the next day a father and four of his sons, making hay in a meadow at Rutland, were surprised by about a dozen Indians. The father escaped in the bushes, but the four sons fell a prey to the enemy. Mr. Willard, the minister of Rutland, being abroad, armed, fell into their hands also, having killed one and wounded another before he was slain himself. The last of the month, they killed a man at Cocheco, and killed or carried away another at Arundel. The 11th of October, about 70 of the enemy attacked the block-house above Northfield, and killed and wounded four or five of the English. Colonel Stoddard marched immediately with 50 men from Northampton to reinforce Northfield, 50 men belonging to Connecticut having been drawn off the day before. Justice should be done to the government of Connecticut. Their frontiers were covered by the Massachusetts, and if they had not contributed to the charge of the war, it] [was not probable that the Massachusetts people would have drawn in and left Connecticut frontiers exposed. Nevertheless, they generally, at the request of the Massachusetts, sent forces every year during the summer in this and former wars, and paid their wages, the provisions being furnished by this government.

In October, the enemy surprised one Cogswell and a boat's crew which were with him at mount Desert. December 25th, about 60 Indians laid siege to the fort at Muscongus or St. George's. They surprised and took two of the garrison, who informed them the fort was in a miserable condition, but the chief officer there, —— Kennedy, being a bold resolute man, the garrison held out until Colonel Westbrook arrived with force sufficient to scatter the besiegers and put them to flight.

This summer also, July 14th, the Indians surprised one Captain Watkins, who was on a fishing voyage at Canso, and killed him and three or four of his family upon Durell's island.

Douglass and other writers applaud the administration for conducting this war with great skill. The French could not join the Indians, as in former wars. Parties of the English kept upon the march, backwards and forwards, but saw no Indians. Captain Moulton went up to Norridgewock and brought away some books and papers of the Jesuit Rallé, which discovered that the French were the instigators of the Indians to the war, but he saw none of the enemy. He came off without destroying their houses and church. Moulton was a discreet as well as brave man, and probably imagined this instance of his moderation would provoke, in the Indians, the like spirit towards the English.

(*Anno* 1724.)—The next year was unfavourable to the English in the former part of it, and the losses, upon the whole, exceeded those of the enemy; but a successful stroke or two against them in the course of the year made them weary of war, and were the means of an accommodation. The 23d of March, they killed one Smith, serjeant of the fort at cape Porpoise. In April, one Mitchell was killed at Black point and two of his sons taken, and about the same time John Felt, William Wormwell, and Ebenezer Lewis were killed at a saw mill on Kennebeck river, and one Thomson at Berwick met with the same fate in May, and one of his children was carried into captivity, another child was scalped and left on the ground for dead, but soon after was taken up and carried home alive. In the same month they killed elder Knock, at Lamprey river, George Chapley and a young woman, at Oyster river, as they were going home from public worship, and took prisoners a man and three boys at Kingston. The beginning of June, a scout of 30 men from Oyster river, were attacked before they left the houses, and two men were shot down: the rest ran upon the Indians and put them to flight, leaving their packs and one of their company who was killed in the skirmish. One Englishman was killed and two taken prisoners at Hatfield, another with a friend Indian and their horses were killed between Northfield and Deerfield.

This month news was brought to Boston of the loss of Captain Josiah Winslow and 13 of his company, belonging to the fort at St. George's river. There went out 17 men in two whale-boats, April 30. The Indians, it seems, watched their motions and waited the most convenient time and place to attack them. The next day, as they were upon their return, they found themselves on a sudden surrounded with 30 canoes, whose complement must be 100 Indians. They attempted to land but were intercepted, and nothing remained but to sell their lives as dear as they could. They made a gallant defence, and the bravery of their captain was in an especial manner applauded. Every Englishman was killed. Three Indians, we suppose of those called the Cape Ann Indians, who were of the company, made their escape and carried to the fort the melancholy news.

Encouraged by this success, the enemy made a still greater attempt by water, seized two shallops at the isles of Shoals, and afterwards other fishing vessels in other harbours, and among the rest, a large schooner with two swivel guns, which they manned, and cruised about the coast. A small force was thought sufficient to conquer these raw sailors, and the lieutenant-governor commissioned Dr. Jackson, of the province of Maine, in a small schooner with 20 men, and Silvanus Lakeman, of Ipswich, in a shallop with 16 men, to go in quest of them. They soon came up with them, and not long after returned with their rigging much damaged by the swivel guns, and Jackson and several of his men wounded, and could give no other account of the enemy than that they had gone into Penobscot.

The Seahorse man of war, Captain Durell, being then upon the Boston station, the lieutenant, master, and master's mate, each of them took the command of a small vessel with 30 men each, and went after the Indians, but it is probable they were soon tired of this new business, for they were not to be found, nor do we meet with any further intelligence about them. They took 11 vessels] with 45 men, 22 of whom they killed, and carried 23 into captivity.

At Groton they killed one man and left dead one of their own number. August 3d, they killed three, wounded one, and made another prisoner at Rutland. The 6th, four of them came upon a small house in Oxford, which was built under a hill; they made a breach in the roof, and as one of them was attempting to enter, he received a shot in his belly from a courageous woman, the only person in the house, but who had two muskets and two pistols charged, and was prepared for all four, but they thought fit to retreat, carrying off the dead or wounded man. The 16th, a man was killed at Berwick, another wounded, and a third carried away. The 26th, one was killed and another wounded at Northampton, and the 27th, the enemy came to the house of John Hanson, one of the people called Quakers, at Dover, and killed or carried away his wife, maid, and six children, the man himself being at the Friends meeting.

Discouraged with the ineffectual attempts to intercept the enemy, by parties of the forces marching upon the back of the frontiers, another expedition was resolved upon, in order to surprise them in their principal village at Norridgewock.

Four companies, consisting in the whole of 208 men, were ordered up the river Kennebeck, under Captain Harman, Captain Moulton, Captain Bourn, and Lieutenant Bean. Three Indians of the Six Nations, were prevailed with to accompany our forces. The different accounts given by the French and English of this expedition may afford some entertainment. Charlevoix, who we suppose was about that time in Canada, and might receive there or from thence the account given by the Indians themselves, relates it in this manner: "The 23d of August 1724, 1100 men, part English, part Indians, came up to Norridgewock. The thickets, with which the Indian village was surrounded, and the little care taken by the inhabitants to prevent a surprise, caused that the enemy were not discovered, until the very instant when they made a general discharge of their guns and their shot had penetrated all the Indian wigwams. There were not above 50 fighting men in the village. These took to their arms and ran out in confusion, not with any expectation of defending the place against an enemy who were already in possession, but to favour the escape of their wives, their old men and children, and to give them time to recover the other side of the river, of which the English had not then possessed themselves.

"The noise and tumult gave Father Rallé notice of the danger his converts were in. Not intimidated, he went to meet the enemy, in hopes to draw all their attention to himself and secure his flock at the peril of his own life. He was not disappointed. As soon as he appeared, the English set up a great shout, which was followed by a shower of shot, and he fell down dead near to a cross which he had erected in the midst of the village, seven Indians, who accompanied him to shelter him with their own bodies, falling dead round about him. Thus died this kind shepherd, giving his life for his sheep, after a painful mission of 37 years. The Indians, who were all in the greatest consternation at his death, immediately took to flight and crossed the river, some swimming and others fording. The enemy pursued them until they had entered far into the woods, where they again gathered together to the number of 150. Although more than 2000 shot had been fired upon them, yet there were no more than 30 killed and 14 wounded. The English, finding they had nobody left to resist them, fell first to pillaging and then burning the wigwams. They spared the church, so long as was necessary for their shamefully profaning the sacred vessels and the adorable body of Jesus Christ, and then set fire to it. At length they withdrew, with so great precipitation that it was rather a flight, and they seemed to be struck with a perfect panic. The Indians immediately returned to their village, where they made it their first care to weep over the body of their holy missionary, whilst their women were looking out for herbs and plants for healing the wounded. They found him shot in a thousand places, scalped, his skull broke to pieces with the blows of hatchets, his mouth and eyes full of mud, the bones of his legs fractured, and all his members mangled an hundred different ways. Thus was a priest treated in his mission, at the foot of a cross, by those very men who have so strongly exaggerated the pretended inhumanity of our Indians, who have never made such carnage upon the dead bodies of their enemies. After his converts had raised up and oftentimes kissed the precious remains, so tenderly and so justly beloved by them, they buried him in the same place where, the evening before, he had celebrated the sacred mysteries, namely, where the altar stood before the church was burnt."

Besides the great error in the number of the English forces, there are many embellishments in this relation in favour of the Indians and injurious to the English. Not satisfied with the journal alone which was given in by Captain Harman, we took from Captain Moulton as minute and cir-

[cumstantial an account as he could give of this affair.

The forces left Richmond fort on Kennebeck river, the 8th of August, O. S. The 9th, they arrived at Taconick, where they left their whale-boats, with a lieutenant and 40 of the 208 men to guard them. With the remaining forces, the 10th, they began their march by land for Norridgewock. The same evening, they discovered and fired upon two Indian women; one of them, the daughter of the well known Bomazeen, they killed, the other, his wife, they took prisoner. From her they received a full account of the state of Norridgewock. The 12th, a little after noon, they came near to a village. It was supposed that part of the Indians might be at their corn-fields, which were at some distance, and therefore it was thought proper to divide this small army. Harman, with about 80 men, chose to go by the way of the fields, and Moulton, with as many more, were left to march straight to the village, which about three o'clock suddenly opened upon them. There was not an Indian to be seen, being all in their wigwams. The men were ordered to advance softly and to keep a profound silence. At length an Indian came out of one of the wigwams, and as he was making water, looked round him and discovered the English close upon him. He immediately gave the war whoop and ran in for his gun. The whole village, consisting of about 60 warriors, besides old men, women, and children, took the alarm, and the warriors ran to meet the English, the rest fled to save their lives. Moulton, instead of suffering his men to fire at random through the wigwams, charged every man not to fire, upon pain of death, until the Indians had discharged their guns. It happened as he expected; in their surprise they overshot the English and not a man was hurt. The English then discharged in their turn and made great slaughter, but every man still kept his rank. The Indians fired a second volley and immediately fled towards the river. Some jumped into their canoes, but had left their paddles in their houses, others took to swimming, and some of the tallest could ford the river, which was about 60 feet over, and the waters being low, it was no where more than six feet deep. The English pursued, some furnished themselves with paddles and took the Indian canoes which were left, others waded into the river. They soon drove the Indians from their canoes into the river, and shot them in the water, and they conjectured that not more than 50 of the whole village landed on the other side, and that some of them were killed before they reached the woods.

The English then returned to the town, where they found the Jesuit in one of the wigwams, firing upon a few of our men who had not pursued after the enemy. He had an English boy in the wigwam with him, about 14 years of age, who had been taken about six months before. This boy he shot through the thigh, and afterwards stabbed in the body, but by the care of the surgeons he recovered. We find this act of cruelty in the account given by Harman upon oath. Moulton had given orders not to kill the Jesuit, but by his firing from the wigwam, one of our men being wounded, a lieutenant Jaques stove open the door and shot him through the head. Jaques excused himself to his commanding officer, alleging that Rallé was loading his gun when he entered the wigwam, and declared that he would neither give nor take quarter. Moulton allowed that some answer was made by Rallé which provoked Jaques, but doubted whether it was the same as reported, and always expressed his disapprobation of the action. Mog, a famous old chief among the Indians, was shut up in another wigwam, and firing from it killed one of the three Mohawks. His brother was so enraged that he broke down the door and shot Mog dead. The English, in their rage, followed and killed the poor squaw and two helpless children. Having cleared the village of the enemy, they then fell to plundering and destroying the wigwams. The plunder of an Indian town consisted of but a little corn, it being not far from harvest, a few blankets, kettles, guns, and about three barrels of powder, all which was brought away. New England Puritans thought it no sacrilege to take the plate from the altars of the Roman Catholic church, and this we believe was all the profaneness offered to the sacred vessels. There were some expressions of zeal against idolatry, in breaking the crucifixes and other imagery which were found there. The church itself, a few years before, had been built by carpenters from New England. Beaver and other Indian furs and skins set up the church, and a zeal against a supposed false religion destroyed the ornaments of it.

Harman and the men who went to the corn-fields did not come up till near night, when the action was over. They all of both parties lodged in the wigwams, keeping a guard of 40 men, the next morning they found 26 dead bodies, besides that of the Jesuit, and had one woman and three children prisoners. Among the dead were Bomazeen, Mog, Job, Carabesett, Wissememet, and Bomazeen's son-in-law, all noted warriors. They marched early for Taconick, being in some pain for their men and whale-boats, but found all safe.]

[Christian, one of the Mohawks, was sent, or of his own accord returned, after they had began their march, and set fire to the wigwams and to the church, and then joined the company again. The 16th, they all arrived at Richmond fort. Harman went to Boston with the scalps, and being chief in command, was made a lieutenant-colonel for an exploit in which Moulton was the principal actor, who had no distinguishing reward, except the applause of the country in general. This has often been the case in much more important service. The Norridgewock tribe never made any figure since this blow.

Encouraged by this success Colonel Westbrook was ordered to march with 300 men across from Kennebeck to Penobscot, which he performed with no other advantage than exploring the country, which before was little known. Other parties were ordered up Amasecónti Amariscoggin, and a second attempt was made upon Norridgewock, but no Indians were to be found.

The frontiers, however, continued to be infested. September the 6th, an English party of 14 went from Dunstable in search of two men who were missing. About 30 Indians lay in wait and shot down six and took three prisoners. A second party went out and lost two of their number. The *w.* frontier seems to have been better guarded, for although often alarmed, they were less annoyed.

(*Anno* 1725.)—The government increased the premium for Indian scalps and captives to 100*l.* This encouraged John Lovewell to raise a company of volunteers, to go out upon an Indian hunting. January 5th, he brought to Boston a captive and a scalp, both which he met with above 40 miles beyond Winnepesiaukee lake. Going out a second time, he discovered ten Indians round a fire all asleep: he ordered part of his company to fire, who killed three; the other seven, as they were rising up, were sent to rest again by the other part of the company reserved for that purpose. The ten scalps were brought to Boston 3d of March. Emboldened by repeated success, he made a third attempt and went out with 33 men. Upon the 8th of May, they discovered an Indian upon a point of land which joined to a great pond or lake. They had some suspicion that he was set there to draw them into a snare, and that there must be many Indians near, and therefore laid down their packs that they might be ready for action, and then marched near two miles round the pond to come at the Indian they had seen. The fellow remained, although it was certain death to him, and when the English came within gun-shot, discharged his piece, which was loaded with beaver shot, and wounded Lovewell and one of his men, and then immediately fell himself and was scalped. His name ought to have been transmitted as well as that of M. Curtius, who jumped into the gulf or chasm, upon less rational grounds, to save his country.

The Indians who lay concealed seized all the English packs, and then waited their return at a place convenient for their own purpose. One of the Indians being discovered, the rest, being about 80, rose, yelled and fired, and then ran on with their hatchets with great fury. The English retreated to the pond to secure their rear, and although so unequal in numbers, continued five or six hours till night came on. Captain Lovewell, his lieutenant Farewell, and ensign Robins were soon mortally wounded, and with five more were left dead on the spot. Sixteen escaped and returned unhurt, but were obliged to leave eight of their wounded companions in the woods without provisions; their chaplain, Mr. Fry of Andover, was one, who had behaved with great bravery and scalped one Indian in the heat of the action, but perished himself for want of relief.

One of the eight afterwards came into Berwick, and another to Saco. This misfortune discouraged scalping parties. But Indians as well as English wished to be at peace. After Rallé's death, they were at liberty to follow their inclinations. The Penobscot tribe, however, being best disposed, were first sounded. An Indian hostage and a captive were permitted, upon their parole, to go home in the winter of 1724, and they came back to the fort at St. George's the 9th of February, accompanied with two of the tribe, one a principal sachem or chief. They brought an account that, at a meeting of the Penobscots, it was agreed to make proposals of peace. The sachem or chief was sent back, with the other Indian, and promised to return in 23 days, and bring a deputation, to consist of several other chiefs, with him, but Captain Heath, having gone out upon a march from Kennebeck, across the country, to Penobscot, fell upon a deserted village of about 50 Indian houses, which he burned, but saw none of the inhabitants. The Indians who went from St. George's knew nothing of this action until they came home, and it seems to have discouraged them from returning according to their promise, and the treaty, by this means, was retarded. But upon new intimations, in June following, John Stoddard and John Wainwright, Esqs. were commissioned by the lieutenant-governor and sent] [down to St. George's, to treat with such Indians as should come in there, and settle preliminaries of peace.

A cessation of arms was agreed upon, and four delegates came up soon after to Boston, and signed a treaty of peace, and the next year, the lieutenant-governor in person, attended by gentlemen of the court and others, and the lieutenant-governor of New Hampshire, with gentlemen from that province, ratified the same at Falmouth in Casco bay. This treaty has been applauded as the most judicious which has ever been made with the Indians. A long peace succeeded it.

The pacific temper of the Indians, for many years after, cannot be attributed to any peculiar excellency in this treaty, there being no articles in it of any importance, differing from former treaties. It was owing to the subsequent acts of government in conformity to the treaty. The Indians had long been extremely desirous of trading houses to supply them with necessaries and to take off their furs, skins, &c. This was promised by Governor Shute, at a conference, but the general court, at that time, would make no provision for the performance. Mr. Dummer promised the same thing. The court then made provision for trading houses at St. George's, Kennebeek and Saco rivers, and the Indians soon found that they were supplied with goods upon better terms than they could have them from the French, or even from private English traders. Acts or laws were made, at the same time, for restraining private trade with the Indians, but the supplies, made by the province at a cheaper rate than private traders could afford, would have broke up their trade without any other provision, and laws would have signified little without that. Mr. Dummer engaged that the Indians should be supplied with goods at as cheap rates as they were sold at Boston. This was afterwards construed favourably for the government. The goods, being bought by wholesale, were sold to the Indians at the retail price in Boston, and a seeming profit, by the commissary's account, accrued to the government; but, when the charge of trading houses, truckmasters, garrisons, and a vessel employed in transporting goods, was deducted, the province was still a tributary to the Indians every year. However, it was allowed to be a well judged measure, tended to preserve peace, and was more reputable than if a certain pension had been every year paid for that purpose.

Delegates from all the tribes of Indians, particularly the Norridgewocks, not having been present at this first treaty, another was thought necessary the next year, when the former was renewed and ratified. It was most acceptable to the Indians to hold their treaties near their own settlements, and, in a proper season of the year, it was an agreeable tour to the governors or commanders in chief and the gentlemen accompanying them.

To bring this war to a close, we have passed over the other affairs of the government for a year or two past. Soon after Mr. Cooke's arrival in London, Governor Shute exhibited a second memorial against the house of representatives, for matters transacted after he left the province. The principal articles of complaint were the several orders relative to the forts and forces, which, he says, the house had taken out of the hands of the lieutenant-governor, and the affront offered to the lieutenant-governor in ordering his seal to be effaced upon the belt of wampum. Several other things seem to be brought in to increase the resentment against them, as their choosing Mr. Cooke, who had been at the head of all the measures complained of in the first memorial, for their agent; their refusing to confer with the council upon a money bill; their endeavouring by their votes to lessen the members of the council in the esteem of the people; their withholding his salary in his absence; and their assuming more and more the authority of government into their hands. The council, in this memorial, are also complained of, they having put their negative to the vote for choosing Mr. Cooke, and yet afterwards joined in election with the house, when they had reason to suppose, by the great superiority of the house in number, that he would be the person.

Mr. agent Dummer, who was to act jointly with Mr. Cooke, made an attempt to reconcile the governor to him, but he refused to see him, and the attempt offended Mr. Cooke also, and occasioned warm discourse between him and Dummer, which caused the latter to refuse to act in concert, especially as Mr. Cooke had shewn him a private instruction from the house, by which their defence against the charge of invading the royal prerogative was committed to Mr. Cooke and Mr. Sanderson, to the exclusion of Mr. Dummer.

After divers hearings upon the subject matter of the complaints, the reports of the attorney and solicitor general, of the lords committee, and finally the determination of his Majesty in council, were all unfavourable to the house of representatives.

The several acts or votes of the house relative to the king's woods, and to the forts and forces;]

[seem to have been generally deemed indefensible, the agents were advised to acknowledge them to be so, and it was so far relied upon, that they would be so acknowledged in the province, as that no special provision was thought necessary for the regulation of their future conduct, the charter being express and clear. But the governor's power to negative the speaker, and the time for which the house might adjourn, were points not so certain. What was called an explanatory charter was therefore thought necessary, and such a charter accordingly passed the seals. By this charter, the power of the governor to negative a speaker is expressly declared, and the power of the house to adjourn themselves is limited to two days. With respect to the latter, perhaps, this new charter may properly enough be called explanatory, the governor having the power, by the principal charter, of adjourning the assembly; and yet, from the nature of the thing, it was necessary that the house, a part of that assembly, should have the power of adjourning themselves, for a longer or shorter time; but the power of negativing a speaker seems to be a new article, wherein the charter is silent; so that whatever right it might be apprehended the king had to explain his own patents, where there was ambiguity, yet when an alteration is to be made in the charter, or a new rule established in any point wherein the charter is silent, the acceptance of the people, perhaps, is necessary. This seems to have been the reason of leaving it to the option of the general court, either to accept or refuse the explanatory charter. It was intimated at the same time that, if the charter should be refused, the whole controversy between the governor and the house of representatives would be carried before the parliament. Had the two points mentioned in the explanatory charter, or the conduct of the house relative to them, been all that was to be carried into parliament, the general court, probably, would not have accepted this charter. They would have urged that it was not certain that a house of commons would have determined that the king, by his governor, had a right to negative the speaker of a house of representatives in the colonies, especially as the attorney-general had inferred this right from the right of negativing the speaker of the house of commons; but it was their misfortune that in the other articles of complaint the house was generally condemned in England, the ministry were highly incensed, and it was feared the consequence of a parliamentary inquiry would be an act to vacate the charter of the province. The temper of the house was much changed, and although there were several members, who had been active in all the measures which brought this difficulty upon the country, still resolute to risk all, rather than by their own act give up any one privilege, yet a major vote was carried in the house for accepting the charter, and in such terms as would induce one to imagine it rather the grant of a favour than the deprivation of a right. It has been said that the English are islanders, and therefore inconstant. Transplanted to the continent, they are nevertheless Englishmen. When we reflect upon the many instances of frequent sudden changes, and from one extreme to the other, in ancient times, in the parliament of England, we may well enough expect, now and then, to meet with the like instances in the assemblies of the English colonies. This was the issue of the unfortunate controversy with Governor Shute, unless we allow that it was the occasion also of the controversy with his successor, which is not improbable.

The governor was offended with Mr. Dummer, for receiving grants from the court made to him for his service as commander in chief, it being expected that when the governor is absent with leave, his salary should be continued, one half of which, by a royal instruction, is to be allowed to the lieutenant-governor; but the house took a more frugal method, and made grants, of little more than one half of the governor's usual salary, to the lieutenant-governor immediately, any part of which he could very ill afford to spare from his own support. His pacific measures and accommodation or suspension of some of the controverted points might be another cause of coldness, at least, between the governor and him.

Another affair occasioned a mark of royal displeasure upon the lieutenant-governor. Synods had been frequent under the first charter, either for suppressing errors in principles, or immoralities in practice, or for establishing or reforming church government and order, but under a new charter no synod had ever been convened. A convention of ministers had been annually held at the time for election of the council. This might have been in many respects useful, but it was thought could not have that weight for promoting any of the forementioned purposes which a synod convened, and perhaps their result ratified by the government, would have. There were divers ancient members in both houses who had not then lost their affection for the platform, and an application made by the ministers for calling a synod was granted in council, but the house did not concur. Afterwards, by a vote of both houses, it] [was referred to the next session, to which the lieutenant-governor gave his consent. Opposition was made by the Episcopal ministers, but a doubt of success, in the province, caused them to apply in England, we suppose to the bishop of London. The king being abroad, an instruction came from the lords justices to cease all proceedings, and the lieutenant-governor received a reprimand for "giving his consent to a vote of reference, and neglecting to transmit an account of so remarkable a transaction." A stop was put to any further proceeding in the affair, nor has any attempt for a synod been made since.

(*Anno* 1726.)—The remainder of Mr. Dummer's short administration was easy to him. The war being over, the principal ground of dispute, the ordering the forces, ceased. Other affairs relative to the treasury, the passing upon accounts and the form of supplies, he suffered to go on according to the claim of the house. Mr. Cooke, the first election after his return from England, May 1726, was chosen of the council. This was a mark of the house's approbation of his conduct in the agency, although it had not been attended with success. The lieutenant-governor did not think it convenient to offend the house by a negative. The small allowance made him as a salary, about 250*l.* sterling per annum, he also acquiesced in for the sake of peace. The governor was expected by almost every ship for a year or two together, but by some means or other was delayed until the summer of 1727, when he was upon the point of embarking, but the sudden death of the king prevented. The principal cause of delay seems to have been the insufficiency of the salary which had been granted for his support, and the uncertainty whether the assembly would make an addition to it.

Upon the accession of King George II. a gentleman who, it is said, was in particular esteem with the king himself, was appointed governor of New York and the Jerseys, in the room of Mr. Burnet, whose administration had, in general, been very acceptable to those colonies and approved in England. The bishop, his father, had likewise been a most steady friend to the house of Hanover. Governor Burnet's fortune being reduced in the general calamity of the year 1720, he parted with a place in the revenue of 1200*l.* per annum, and received commissions for these governments, with a view to his retrieving his fortune in a course of years. He thought it hard, in so short a time, to be superseded, for although the Massachusetts and New Hampshire were given to him, yet he was to part with very profitable posts for such as, at best, would afford him no more than a decent support, an easy administration for one which he foresaw would be extremely troublesome. He complained of his hard fate, and it had a visible effect upon his spirits. Colonel Shute was provided for, more to his satisfaction than if he had returned to his government, a pension of 400*l.* sterling per annum being settled upon him, to be paid out of the 4½ per cent. duty raised in the W. India islands. The W. Indians, who would perhaps have been content if it had been applied to one of their own governors who had been superseded, had taken exception to the payment of it to a governor of the *n.* colonies.

(*Anno* 1727.)—The earthquake on the 29th of October 1727, although not confined to the Massachusetts, was so remarkable an event in Providence that we may be excused if we give a circumstantial account of it. About 40 minutes after 10 at night, when there was a serene sky and calm but sharp air, a most amazing noise was heard, like to the roaring of a chimney when on fire, as some said, only beyond comparison greater; others compared it to the noise of coaches upon pavements, and thought that 10,000 coaches together would not have exceeded it. The noise was judged by some to continue about half a minute before the shock began, which increased gradually, and was thought to have continued the space of a minute before it was at the height, and, in about half a minute more, to have been at an end by a gradual decrease. When the terror is so great, no dependence can be placed upon the admeasurement of time in any person's mind, and we always find very different apprehensions of it. The noise and shock of this and all earthquakes which preceded it in New England were observed to come from the *w.* or *n. w.* and go off to the *e.* or *s. e.* At Newbury and other towns upon Merrimack river the shock was greater than in any other part of Massachusetts, but no buildings were thrown down, part of the walls of several cellars fell in, and the tops of many chimneys were shook off. At New York it seems to have been equal to what it was in the Massachusetts, but at Philadelphia it was very sensibly weaker, and in the colonies *s.* it grew less and less until it had spent itself or become insensible. The seamen upon the coast supposed their vessels to have struck upon a shoal of loose ballast. More gentle shocks were frequently felt in most parts of New England for several months after. There have seldom passed above 15 or 20 years without an earthquake, but there had been none, very violent, in the memory of any then living. There was a general apprehension of danger of destruc-]

[tion and death, and many, who had very little sense of religion before, appeared to be very serious and devout penitents, but, too generally, as the fears of another earthquake went off, the religious impressions went with them, and they, who had been the greetest penitents, returned to their former course of life.

The trade of the province being in a bad state, and there being a general complaint of scarcity of money, the old spirit revived for increasing the currency by a further emission of bills of credit. It would be just as rational, when the blood in the human body is in a putrid corrupt state, to increase the quantity by luxurious living, in order to restore health. Some of the leading men among the representatives were debtors, and a depreciating currency was convenient for them. A bill was projected for fortifying the sea-ports. The town of Boston was to expend 10,000*l.* in forts and stores, and to enable them to do it, 30,000*l.* was to be issued in bills, and lent to the town for 13 years: Salem, Plymouth, Marblehead, Charlestown, Glocester, and even Truro, on the cape, were all to be supplied with bills of credit for the like purposes. After repeated non-concurrence and long altercation, the council were prevailed upon to agree to the bill. When it came to the lieutenant-governor, he laid the king's instruction before the council, and required their opinion, upon their oaths, whether consistent with the instruction he could sign the bill, and they answered he could not. Not only the lieutenant-governor, but several of the council, were dependent upon the house for the grant of their salaries, and this dependence was improved, as in divers instances it had been formerly. The house referred the consideration of allowance to the next session, and desired the court might rise. The lieutenant-governor let them know, by a message, that he apprehended his small support was withheld from him because he would not sign a bill contrary to his instructions. They replied, that he had recommended to them the making provision for fortifying the province, and now they had passed a bill for that purpose he refused to sign it, and they were obliged, in prudence and faithfulness to their principles, to come into a vote referring allowances and other matters to another session, when a way might be found to enable the inhabitants to pay into the treasury again such sums as should be drawn out for gratuities and allowances. After a recess of about a fortnight an expedient was found. Instead of a bill for fortifying, another was prepared with a specious title, "An act for raising and settling a public revenue for and towards defraying the necessary charges of the government by an emission of 60,000*l.* in bills of credit." This was done to bring it within the words of the instruction, which restrained the governor from consenting to the issuing bills of credit, except for charges of government. The interest of four per cent. or 2400*l.* was to be applied annually to the public charges, and gave colour for issuing the principal sum of 60,000*l.* The lieutenant-governor was prevailed upon to sign it, and the same day the house made the grant of his salary and the usual allowance to the judges, most of whom were members of the council, and to the other officers of the government. This was afterwards alleged to be a compulsion of the lieutenant-governor and such of the members of council as were salary men, to comply with the house of representatives, by withholding from them their subsistence. The eagerness of the body of the people for paper bills, more easily acquired in this way than the righteous way of industry and frugality, no doubt facilitated a compliance.

The council upon this occasion declined answering upon their oath as counsellors, when the lieutenant-governor asked their advice. They swear, that to the best of their judgment they will at all times freely give their advice to the governor for the good management of the public affairs of the government. The lieutenant-governor proposed the following question to them in writing, "Gentlemen, I find it necessary, in order to my signing the bill entitled, 'An act for raising and settling a revenue,' &c. which has passed both houses, to have your advice whether I can sign the said bill without the breach of the instruction of the lords justices of Great Britain, dated the 27th of September 1720, and the order of the lords commissioners of trade and plantations, dated the 8th of February 1726-27. W. DUMMER, February 17, 1727." Upon which the council came to the following vote, "In council, February 19, 1727, Read, and as the council have already, as they are one part of the general court, passed a concurrence with the honourable house of representatives upon the said bill, they cannot think it proper for them to give your honour any further advice thereupon, nor do they apprehend the oath of a counsellor obliges them thereto. At the same time they cannot but think it will be for the good and welfare of the province, and the necessary support of the government thereof, if the bill be consented to by your honour.

J. WILLARD, Secretary."

They had given their advice or opinion, the same session, upon the bill for fortifying, after] [they had passed it, that it was contrary to the instruction, and instances of the like kind have been frequent before and since this time.

The lieutenant-governor had a further opportunity, before Mr. Burnet's arrival, of meeting the assembly in May for election of counsellors.

The house discovered, in one instance, this session, a desire to amplify their jurisdiction. The council and house had made it a practice, ever since the charter, to unite in the choice of the treasurer, impost officer, and other civil officers, the appointment whereof is reserved to the general assembly. The council, being in number less than a third part of the house, had by this means no weight in such elections except when there were two or more candidates for an office, set up by the house, and then the balance of power, if they were united themselves, might be with them. This seems to have been an old charter practice, and handed down. The two houses, when parties to any petition or cause desire to be heard, often meet in one house, which no doubt also came from the old charter, but after they are separated, they vote separately upon the subject matter of the hearing. In this session, after a hearing of this sort, the house passed a vote, "that when a hearing shall be had on any private cause before both houses together, the subject matter shall be determined by both houses conjunctly." They might as well have voted, that after a conference between the two houses, the subject matter should be determined conjunctly. The council were sensible this was taking from the little weight they had, and unanimously non-concurred the vote.

The manner of choosing civil officers is a defect in the constitution which does not seem to have been considered at the framing the charter, and as, by charter, officers must annually be elected, it is a defect which must be submitted to. If either house should elect by themselves and send to the other for concurrence, the right of nomination would be such an advantage as neither would be willing to concede to the other. In the early days of the charter, it had been made a question, whether in any acts of government the council had a negative voice, and were not rather to vote in conjunction with the house of representatives?—and Constantine Phips gave his opinion that they had no negative. He seems not to have considered that the charter and the commissions to governors of other colonies evidently intended a legislature after the pattern of the legislature of England, as far as the state and circumstances of the colonies would admit.

The government, under the old charter and the new, had been very prudent in the distribution of the territory. Lands were granted for the sake of settling them. Grants for any other purpose had been very rare, and, ordinarily, a new settlement was contiguous to an old one. The settlers themselves, as well as the government, were inclined to this for the sake of a social neighbourhood, as well as mutual defence against an enemy. The first settlers on Connecticut river, indeed, left a great tract of wilderness between them and the rest of the colony, but they went off in a body, and a new colony, Connecticut, was settling near them at the same time. Rivers were also an inducement to settle, but very few had ventured above Dunstable upon the fine river Merrimack, and the rivers in the province of Maine had no towns at any distance from the sea into which they empty. But all on a sudden plans were laid for grants of vast tracts of unimproved land, and the last session of Mr. Dummer's administration, a vote passed the two houses, appointing a committee to lay out three lines of towns, each town of the contents of six miles square, one line to extend from Connecticut river above Northfield to Merrimack river above Dunstable, another line on each side Merrimack as far as Penicook, and another from Nichewanock river to Falmouth in Casco bay.

Pretences were encouraged, and even sought after, to entitle persons to be grantees. The posterity of all the officers and soldiers who served in the famous Naraganset expedition in 1675, were the first pitched upon, those who were in the unfortunate attempt upon Canada in 1690, were to come next. The government of New Hampshire supposed these grants were made in order to secure the possession of a tract of country challenged by them as within their bounds. This might have weight with some leading men who were acquainted with the controversy, but there was a fondness for granting land in any part of the province. A condition of settling a certain number of families in a few years, ordinarily, was annexed to the grants, but the court, by multiplying their grants, rendered the performance of the condition impracticable, there not being people enough within the province willing to leave the old settled towns, and the grantees not being able to procure settlers from abroad.

The settlement of the province was retarded by it; a trade of land-jobbing made many idle persons, imaginary wealth was created, which was attended with some of the mischievous effects of the paper currency, viz. idleness and bad economy, a real expence was occasioned to many persons, besides the purchase of the grantees title,] [for every township by law was made a proprietary, and their frequent meetings, schemes for settlement, and other preparatory business, occasioned many charges. In some few towns houses were built and some part of the lands cleared. In a short time, a new line being determined for the *n.* boundary of the Massachusetts colony, many of these townships were found to be without it. The government of New Hampshire, for the crown, laid claim to some of them; and certain persons, calling themselves proprietors under Mason, to others, and the Massachusetts people, after a further expence in contesting their title, either wholly lost the lands or made such compositiou as the new claimers thought fit to agree to.

(*Anno* 1728.)—Mr. Burnet was received with unusual pomp. Besides a committee of the general court, many private gentlemen went as far as Bristol to wait upon him; and, besides the continual addition that was making in the journey, there went out of Boston to meet him at a small distance, such a multitude of horses and carriages that he entered the town with a greater cavalcade than had ever been seen before or since. Like one of the predecessors, Lord Bellamont, he urged this grand appearance, in his first speech to the assembly, as a proof of their ability very honourably to support his Majesty's government, and at the same time acquainted them with the king's instruction to him to insist upon an established salary, and his intention firmly to adhere to it. He had asked the opinion of a New England gentleman, who was then the minister of the presbyterian church at New York, whether the assembly would comply with his instruction, and received a discouraging answer, which caused him to reply, that he would not engage in a quarrel, or to that effect; but he either received different advice upon his arrival, or for some other reason altered his mind. The assembly seemed from the beginning determined to withstand him. To do it with better grace and a more reasonable prospect of success, the quantum of the salary, it was agreed, was not worth disputing. It bore no proportion to the privilege and right of granting it for such time as they thought proper. The same persons, therefore, who six or seven years before refused to make Governor Shute, and, perhaps, the government easy, by granting not more than 500*l.* sterling a year, now readily voted for 1000*l.* or a sum which was intended to be equal to it. As soon as addresses from the council and house, the usual compliments upon the first arrival of a governor, had passed, the house made a grant of 1700*l.* towards his support and to defray the charge of his journey. In a day or two the governor let them know he was utterly unable to give his consent to it, being inconsistent with his instruction. After a week's deliberation, a grant was made of 300*l.* for the charge of his journey, which he accepted, and another of 1400*l.* towards his support, which was accompanied with a joint message from the council and house, prepared by a committee, wherein they assert their undoubted right as Englishmen, and their privilege by the charter, to raise and apply moneys for the support of government, and their readiness to give the governor an ample and honourable support, but they apprehended it would be most for his Majesty's service, &c. to do it without establishing a fixed salary. The governor was always very quick in his replies, and once, when a committee came to him with a message, having privately obtained a copy of it, gave the same committee an answer, in writing, to carry back. The same day this message was delivered he observed to them, in answer, "that the right of Englishmen could never entitle them to do wrong, that their privilege of raising money by charter was expressed to be 'by wholesome and reasonable laws and directions,' consequently not such as were hurtful to the constitution and the ends of government; that their way of giving a support to the governor could not be honourable, for it deprived him of the right of an Englishman, viz. to act according to his judgment, or obliged him to remain without support, and he appealed to their own consciences, whether they had not formerly kept back their governor's allowance until other bills were passed, and whether they had not sometimes made the salary depend upon the consent to such bills; that if they really intended, from time to time, to grant an honourable support, they could have no just objection to making their purposes effectual by fixing his salary, for he would never accept a grant of the kind they had then made." We shall be convinced that Mr. Burnet was not a person who could be easily moved from a resolution he had once taken up.

Upon the receipt of this message and the peremptory declaration of the governor, the house found this was like to be a serious affair, and that they should not so easily get rid of it as they had done of the like demands made by Dudley and Shute, and again appointed a committee to join with a committee of council to consider of this message. The exclusive right of the house in originating grants, they have often so far given up as to join with the council, by committees, to consider and report the expediency of them, the re-]

[ports generally being sent to the house, there to be first acted upon. The report of this committee was accepted in council and sent to the house, but there rejected, and not being able to unite in an answer, the house tried the council with a resolve, and sent to them for concurrence, the purport of which was, that fixing a salary on the governor or commander in chief for the time being would be dangerons to the inhabitants, and contrary to the design of the charter in giving power to make wholesome and reasonable orders and laws for the welfare of the province. This vote, in so general terms, the council did not think proper to concur, and declared that, although they were of opinion it might prove of ill consequence to settle a salary upon the governor for the time being, yet they apprehended a salary might be granted, for a certain time, to the present governor, without danger to the province, or being contrary to the design of the charter, &c.

This occasioned a conference, without effect, both houses adhering to their own votes, and from this time the house were left to manage the controversy themselves. They sent a message to the governor to desire the court might rise. He told them, that if he should comply with their desire he should put it out of their power to pay an immediate regard to the king's instruction, and he would not grant them a recess until they had finished the business for which the court was then sitting. They then, in a message to him, declared that, in faithfulness to the people of the province, they could not come into an act for establishing a salary on the governor or commander in chief for the time being, and therefore they renewed their request that the court might rise.

Both the governor and the house seem to have had some reserve in their declarations. Perhaps a salary during his administration would have satisfied him, although he demanded it for the commander in chief for the time being; and the house were scrupulous of saying that they would not settle a salary for a limited time. Each desired that the other would make some concessions. Both declined, and both by long altercation were irritated, and at length, which is often the ease, instead of closing, as seemed probable at first, widened the breach until they fixed at the opposite extremes. The major part of the council, and about a sixth part of the house, were willing to settle a salary upon Mr. Burnet for a term not exceeding three years; possibly even some who were finally the most zealous in the opposition would have submitted to this if they could have been sure of its being accepted, and had been at liberty to act according to their judgment. Mr. Cooke had experienced the ill success of the controversy with Governor Shute, and seemed desirons of being upon terms with his successor, who, upon his first arrival and until the province house could be repaired, lodged at Mr. Cooke's house, but a friendship could not long continue between two persons of so different opinions upon civil government. The language of the governor's messages was thought too dictatorial by the people, and particularly by the inhabitants of Boston, and he had been somewhat free in his jokes upon some of the shopkeepers and principal tradesmen, who were then the directors of the councils of the town, and very much influenced those of the house. An intimation in the governor's next message, that if they did not comply with the instruction, the legislature of Great Britain would take into consideration the support of the government, and, perhaps, something besides, meaning the charter, increased the prejudices against him. The house now thought themselves obliged to be more particular than they had yet been, fully to assert their rights. This was what the governor desired, and without any delay he sent them an answer. As these two messages seemed to be much in earnest, the argument on each side of the question afforded a serious topic of conversation. Not long after, the house, instead of any advances towards a compliance, which the governor wished to obtain, came to resolutions upon two questions, which shewed still more fully their sense of the point in controversy. The first question was, Whether the house will take under consideration the settling a temporary salary upon the governor or commander in chief for the time being?—This passed in the negative. Then this question was put, Whether the house can with safety to the people come into any other method for supporting the governor or commander in chief for the time being, than what has been heretofore practised?—This also passed in the negative, and was the first instance of the house's declaring they would make no advances; for in their message last preceding, they only say they do not think it advisable to pass an act for fixing a salary as prescribed. These votes caused the governor to put them in mind of a letter from their agent in the year 1722, wherein he mentions that Lord Carteret, in conversation, desired him to write to the assembly not to provoke the government in England to bring their charter before the parliament; for if they did, it was his opinion, it would be dissolved without opposition, and the governor advised them to take care their proceedings did not bring their charter into danger at that time. This caution did not prevent the house].

[from preparing a state of the controversy between the governor and them, concerning his salary, to transmit to their several towns, in the conclusion of which they say, that they dare neither come into a fixed salary on the governor for ever nor for a limited time, for the following reasons:

"1st, Because it is an untrodden path, which neither they nor their predecessors have gone in, and they cannot certainly foresee the many dangers that may be in it, nor can they depart from that way which has been found safe and comfortable.

"2dly, Because it is the undoubted right of all Englishmen, by Magna Charta, to raise and dispose of money for the public service, of their own free accord, without compulsion.

"3dly, Because it must necessarily lessen the dignity and freedom of the house of representatives in making acts and raising and applying taxes, &c. and, consequently, cannot be thought a proper method to preserve that balance in the three branches of the legislature which seems necessary to form, maintain, and uphold the constitution.

"4thly, Because the charter fully empowers the general assembly to make such laws and orders as they shall judge for the good and welfare of the inhabitants, and if they or any part of them judge this not to be for their good, they neither ought nor could come into it; for, as to act beyond or without the powers granted in the charter might justly incur the king's displeasure, so not to act up and agreeable to those powers might justly be deemed a betraying the rights and privileges therein granted, and if they should give up this right, they would open a door to many other inconveniences."

This representation was prepared to be carried home by the several members, upon the rising of the court, in order to their towns giving their instructions, but the house being kept sitting, it was printed and sent through the province. The governor sent a message to the house, a few days after, in which he takes their representation to pieces, and, in the close of his message, appeals to them whether he had not answered all their objections, "except the unknown inconveniences to which a door would be opened," which could not be answered until they could tell what they were; and charges them with calling for help from what they had not mentioned, from a sense of the imperfection of what they had, and with sending to their several towns for advice, and declaring at the same time they did not dare follow it.

It would be tedious to recite at length the several messages which passed, during the remainder of the controversy, from the chair to the house, and from the house to the chair, which followed quick one upon the back of another. The sum of the argument, upon the part of the governor, was as follows: That it was highly reasonable he should enjoy the free exercise of his judgment in the administration of government, but the grants, made for a short time only by the house, were thus limited for no other reason than to keep the governor in a state of dependence, and with design to withhold from him the necessary means of subsistence, unless he would comply with their acts and resolves, however unreasonable they might appear to him; that, in fact, they had treated Governor Shute in this manner, and no longer since than the last year the house had refused to make the usual grants and allowances, not only to the lieutenant-governor but to other officers, until they had compelled him to give his consent to a loan of 60,000*l.* in bills of credit; that a constitution which, in name and appearance, consisted of three branches, was in fact reduced to one; that it was a professed principle, in the constitution of Great Britain, to preserve a freedom in each of the three branches of the legislature, and it was a great favour shewn the province, when King William and Queen Mary established, by the royal charter, a form of government so analogous to the government of Great Britain; a principle of gratitude and loyalty, therefore, ought to induce them to establish a salary for the governor of this province, in order to his supporting his dignity and freedom, in like manner as the parliament always granted to the king what was called the civil list, not once in six months or from year to year, but for life; that this was no more than other provinces which had no charters had done for their governors; that there was nothing in the province charter to exempt them from the same obligation which other his Majesty's colonies were under to support the government; to be sure, they had no pretence to greater privileges by charter than the people of England enjoyed from Magna Charta, and yet no clause of that was ever urged as an objection against granting to the king a revenue for life; and a power by charter to grant moneys could not be a reason against granting them either for a limited or unlimited time.

On the part of the house, the substance of their defence against the governor's demand and his reasons in support of it was, that an obligation upon an assembly in the plantations could not be inferred from the practice of the house of commons in Great Britain; the king was the common father of] [all his subjects, and their interests were inseparably united, whereas a plantation governor was affected neither by the adversity nor prosperity of a colony when he had once left it; no wonder then a colony could not place the same confidence in the governor which the nation placed in the king; however, the grants to the governor always looked forward and were made not for service done, but to be done. It must be admitted the governor is in some measure dependent upon the assembly for his salary, but he is dependent in this instance only, whereas he has a check and controul upon every grant to any person in the government and upon all laws and acts of government whatsoever; nor can an exact parallel be drawn between the constitution of Britain and that of the province, for the council are dependent upon the governor for their very being, once every year, whereas the house of lords cannot be displaced unless they have criminally forfeited the rights of peers. The house were not to be governed by the practice of assemblies in some of the other colonies, nor were they to be dictated to and required to raise a certain sum for a certain time and certain purposes; this would destroy the freedom which the house apprehended they had a right to in all their acts and resolves, and would deprive them of the powers given to them by charter, to raise money and apply it when and how they thought proper. Different judgments will undoubtedly be formed upon the weight of these reasons ou the one side and the other.

The messages of the house at first were short, supposed to have been drawn by Mr. Cooke, who never used many words in his speeches in the house, which generally discovered something manly and open, though sometimes severe and bitter, and often inaccurate. In the latter part of the controversy they were generally drawn by Mr. Welles, another member from Boston, the second year of his coming to the house. These were generally more prolix, and necessarily so from the length of the messages to which they were an answer. The house had justice done them by their committees who managed this controversy, and they were then willing to allow that the governor maintained a bad cause with as plausible reasons as could be.

The contending parties, for a little while, endeavoured to be moderate and to preserve decorum, but it was impossible to continue this temper.

On the 4th of September the house repeated to the governor the request they had formerly made to rise, but he refused to grant it, and told them that unless his Majesty's pleasure had its due weight with them, their desires should have very little weight with him.

The council, who had been for some time out of the question, now interposed, and passed a vote, "that it is expedient for the court to ascertain a sum as a salary for his excellency's support, as also the term of time for its continuance." This was sent to the house for concurrence. The council seem to have gone a little out of their line, but the house took no other notice of the vote than to nonconcur it. The house being kept sitting against their will, employed part of their time in drawing up the state of the controversy, which we have mentioned.

This was not occasioned by any doubt they had themselves, but to convince the governor that the people throughout the province were generally of the same mind with the house, and for this purpose they thought it necessary to obtain from their towns an express approbation of their conduct. It was well known that not a town in the province would then have instructed their representatives to fix a salary upon the governor for the time being.

One of the king's governments (Barbadoes) was at this time warmly contending with its governor against fixing a salary. The assembly of that island, some years before, had settled a very large salary upon a governor, against whom they afterwards made heavy complaints, charging him with rapaciousness and grievous oppressions, and his successor having demanded the like settlement upon him, they resolved to withstand the demand, and the spirit seemed to be as high there as in Massachusetts bay.

This had no small tendency to strengthen and confirm the resolution of the people here, who supposed their charter rather an additional privilege and security against this demand. There was a minor part, however, very desirous of an accommodation. The ill success of the controversy with Governor Shute was fresh in their minds. Many amiable qualities in Mr. Burnet caused them to wish he might continue their governor, and employ those powers and that attention which were now wholly engaged in this single point, in promoting the general welfare and prosperity of the province.

About a third part of the house of representatives and a major part of the council would have been content to have granted a salary for two, or perhaps three years. If we are to judge by his] [declarations, this would not have satisfied him, and it was far short of his instructions, but his friends were of opinion, that such a partial compliance would have produced a relaxation of the instruction and issued in lasting agreement and harmony.

The house made what they would have the governor think a small advance towards it. Instead of a grant for the salary, supposed, though not expressed, for half a year, they made a grant, September 20th, of 3000*l.* equal to 1000*l.* sterling, in order to enable him to manage the affairs of the province, and although it was not expressly mentioned, it was generally understood to be for a year. This was concurred in by the council, but he let it lie without signing his consent, which caused the house to make at least a seeming farther advance, for on the 24th of October they, by a message, entreated him to accept the grant, and added, "we cannot doubt but that succeeding assemblies, according to the ability of the province, will be very ready to grant as ample a support, and if they should not, your excellency will then have the opportunity of shewing your resentment." Still they had no effect, the governor knew how natural it would be for a future assembly to refuse being governed by the opinion of a former; besides, the reserve, "according to the ability of the province," left sufficient room for a further reason for reducing the sum whensoever a future assembly should think it proper.

A little before this message from the house, the governor had informed them that he was of opinion the act, which passed the last year, issuing 60,000*l.* in bills of credit by way of loan, would be disallowed, the lieutenant-governor having given his consent to it directly contrary to a royal instruction, and recommended to them, as the most likely way to obtain his Majesty's approbation, to apply the interest of the money arising from the loan towards the governor's salary. This was one of those acts which have their operation so far, before they are laid before his Majesty, that great confusion may arise from their disallowance. The house therefore had no great fears concerning it, but it would have been a sufficient reason to prevent their complying with the proposal, that it would be a fixing the salary so long as the loans continued, and for this reason they refused it.

The country in general, as we have observed, was averse to a compliance with the king's instruction, but no part more so than the town of Boston. Generally in the colonies, where there is a trading capital town, the inhabitants of it are the most zealous part of the colony in asserting their liberties when an opinion prevails that they are attacked. They follow the example of London, the capital of the nation. The governor had frequently said, that the members of the house could not act with freedom, being influenced by the inhabitants of the town. Besides, the town, at a general meeting of the inhabitants for that purpose, had passed a vote, which was called the unanimous declaration of the inhabitants of the town of Boston, against fixing a salary upon the governor, and this vote they ordered to be printed. The governor was in great wrath, and called it "an unnecessary forwardness, an attempt to give law to the country." This seems to have determined him to remove the court out of town, and on the 24th of October, he caused it to be adjourned to the 31st, then to meet at Salem in the county of Essex, "where prejudice had not taken root, and where of consequence his Majesty's service would in all probability be better answered." Jocosely, he said, "there might be a charm in the names of places, and that he was at a loss whether to carry them there or to Concord."

The house thought their being kept so long sitting at Boston a great grievance. In one of their messages they ask the governor, "Whether it has been customary that the knights, burgesses, and other freemen of the land, should be told that they are met to grant money in such a peculiar way and manner, and so they should be kept till they had done it, and this in order to gain their goodwill and assent?" In his reply he tells them he would consider their question in all its parts, 1st, "Whether freemen, &c. should be told they are met to grant money?" "I answer, the crown always tells them so." 2d, "In such a particular way and manner?" "I answer, if you mean the way and means of raising money, the crown leaves that to the commons; but if you mean the purpose for which it is to be granted, the crown always tells them what that is, whether it is for an honourable support, the defence of the kingdom, carrying on a war, or the like." 3dly, "And so they should be kept till they had done it." "The crown never tells the parliament so, that I know of, nor have I told you any thing like this as an expedient to get the thing done. I have given you a very different reason for not agreeing to a recess, altogether for your own sakes, lest I should thereby make your immediate regard to his Majesty's pleasure impossible," &c.

The house could not easily be persuaded they were kept so long together merely for their own sakes, and thought this part of the governor's answer evasive of the true reason, and considered them-]

[selves as under duresse whilst at Boston, and their removal to Salem to be a further hardship, and an earnest of what was still further to come, a removal from place to place until they were harassed into a compliance. The members of the general court lamented the measures which had driven away Governor Shute, who would have been easy with a salary of about 500*l*. sterling, granted from year to year. The same persons by whose influence his salary was reduced, were now pressing Mr. Burnet to accept 1000*l*. in the same way, and could not prevail.

The house met, according to the adjournment, but immediately complained of their removal from Boston as illegal or unconstitutional and a great grievance. The same and the only reason which was now given had been given before in the controversy with Governor Shute. The form of the writ for calling an assembly, directed by the province law, mentions its being to be held at the town house in Boston, but this had been determined by the king in council to be, as no doubt it was, mere matter of form or example only, and that it did not limit the power which the crown before had of summoning and holding assemblies at any other place. They prayed the governor, however, to adjourn them back to Boston, but without success.

They endeavoured to prevail upon the council to join with them, but the council declared they were of a different opinion, and urged the house to proceed upon business, which occasioned repeated messages upon the subject; but the whole stress of the argument on the part of the house lay upon the form of the writ for calling the assembly, which the board answered by saying, the house might as well insist that all precepts to the towns should go from the sheriff of Suffolk, because the form of the precept in the law has Suffolk ss.

The alteration of place had no effect upon the members of the house. Votes and messages passed, but no new arguments, the subject had been exhausted, nothing remained but a determined resolution on both sides to abide by their principles, and the house met and adjourned, day after day, without doing any business. The governor was the principal sufferer, not being allowed by the king to receive any thing towards his support, except in a way in which the assembly would not give it. The members of the court, in general, were as well accommodated at Salem as Boston, and the members of Boston, who had not been used to the expence and other inconvenience of absence from home, received a compensation from their town, over and above the ordinary wages of representatives. It was a time of peace without, and a cessation of public business for that reason was less felt.

The house, from an apprehension that their cause was just, and therefore that they were entitled to relief, resolved to make their humble application to his Majesty. Francis Wilks, a New England merchant in London, who had been friendly to Mr. Cooke in his agency, and who was universally esteemed for his great probity as well as his humane obliging disposition, was pitched upon for their agent.

Mr. Belcher, who had been several years of the council, always closely attached to Governor Shute, and in general, what was called a prerogative man, by some accident or other became, on a sudden, the favourite of the house, and he was thought the properest person to join with Mr. Wilks. At the last election he had been left out of the council, by what was called the country party, but now declared against the governor's measures, and became intimate with Mr. Cooke and other leading members of the house. Such instantaneous conversions are not uncommon. A grant was made by the house to defray the charges of the agency, but this was non-concurred by the council, because it was for the use of agents in whose appointment they had no voice. The want of money threatened a stop to the proceeding, but the public spirit of the town of Boston was displayed upon this occasion, and by a subscription of merchants and other principal inhabitants, a sum was raised which was thought sufficient for the purpose, the house voting them thanks, and promising their utmost endeavours that the sums advanced should be repaid in convenient time. The governor desired a copy of their address to the king, but they refused it.

The only argument or reason in the king's instruction for fixing a salary is, "that former assemblies have, from time to time, made such allowances and in such proportion as they themselves thought the governor deserved, in order to make him more dependent upon them." The house, in the first part of their memorial or address, declare they cannot in faithfulness settle or fix a salary, because, after that is done, the governor's particular interest will be very little affected by serving or deserving the interest of the people. This was shewing that they apprehended the reason given by his Majesty for settling a salary was insufficient, and that the governor ought to be paid, according to his services, in the judgment of those who paid him; but in the close of the address they say, "we doubt not succeeding assem-]

[blies, according to the ability of the province, will come into as ample and honourable support, from time to time, and should they not, we acknowledge, your Majesty will have just reason to shew your displeasure with them." It was remarked that, in order to make the last clause consist with the first, the ample and honourable support must be understood in proportion to the services of the governor in the judgment of the house, but in this sense, it was saying nothing and trifling with Majesty; for no case could happen, at any time, in which his Majesty would have just reason to shew his displeasure. It would always be enough to say that the house, in faithfulness to the people, had withheld part of the governor's support, because, in their judgment, he had neglected their interest and his duty.

Whether this remark was just or not, the house had great encouragement given them by Mr. Wilks, that their address would obtain for them the wished-for relief. He had been heard by counsel, Mr. Fazakerley and Dr. Sayes, before the board of trade, Mr. Belcher not being then arrived; but soon after they received letters from their joint agents, inclosing the report of the board of trade, highly disapproving the conduct of the house, and their agents let them know it was their opinion that, if the house should persist in their refusal to comply with the king's instruction, the affair might be carried before the parliament; but if this should be the case, they thought it better a salary should be fixed by the supreme legislature than by the legislature of the province, better the liberties of the people should be taken away from them than given up by their own act. The governor likewise communicated to the house his letters from the lords of trade approving his conduct. All hopes of success from the agents seemed to be over, and their business in England would have been very short if the governor had not given occasion for further application. His administration for many months, except in this affair of the salary, had been unexceptionable. Indeed the members of the house thought themselves aggrieved that he would not sign a warrant upon the treasury for their pay, and his reason for refusing it, viz. that one branch of the legislature might as well go without their wages as another, they thought insufficient. Being drove to straits, and obliged to his friends to assist him in the support of his family, he thought he might be justified in establishing a fee and perquisite which had never been known in the province before. At New York, all vessels took from the governor a let pass, for which there was no law, but the owners of vessels submitted to it, and it was said, *volenti non fit injuria.* Lord Coke perhaps would not have thought even this a justification.

The governor required all masters to take the same passes here, against their will, and demanded 6*s*. or 2*s*. sterling for every vessel bound a foreign voyage, and 4*s*. for coasters. The stated fee, by law, for registers was 6*s*. but the bills having depreciated more than one half in value since the law was made, he required 12*s*. This was a very different case from the other, and we do not know that it was exceptionable, but they were alike complained of as grievous and oppressive, and the governor's enemies were not displeased with the advantage he had given them against him; and upon a representation made by the agents, notwithstanding the hardship of being restrained from receiving a salary in any way except such as the assembly would not give it in, yet such was the regard to law and justice, that his conduct, so far as related to the let passes, was immediately disapproved. There were other matters besides that of the salary to be settled before Mr. Burnet could be easy in his government, but this grand affair caused the lesser to be kept off as much as possible. One was the appointment of an attorney-general. By the charter the election of the civil officers, except such as belong to the council and courts of justice, is in the general assembly. Until after Governor Dudley's time it had generally been allowed that the attorney-general was an officer of the courts of justice and included in the exception, but Lieutenant-governor Tailer, in the year 1716, consented to an election made by the two houses, and the choice had been annually made and approved ever since, not without notice from Mr. Shute of the irregularity of it, but he had so many other affairs upon his hands that he waved this.

Mr. Burnet was determined not to part with the right of nomination, and the council were of the opinion he ought not, and refused to join with the house in the election. There was some altercation between the two houses upon it, and both adhered to their principles.

Another affair of more extensive influence would have been more strenuously insisted upon.

In Governor Shute's administration, the house, after long disputes with the governor and with the council, carried the point as to the form of supply of the treasury, which differing, as we have already observed, from the former practice, and, as both governor and council insisted, from the rule prescribed by the charter, Mr. Burnet had determined to return to the first practice. The house passed a vote for supplying the treasury with]

[20,000*l.* which the council concurred, the practice having been the same for eight or nine years together, but the governor refused his consent, and assured them that he would agree to no supply of the treasury but such as was in practice before the year 1721. This declaration was made not long before his death. The settlement of the point in controversy remained for his successor.

(*Anno* 1729.)—The court was allowed a recess from the 20th of December to the 2d of April, and then sat until the 8th, at Salem again, without any disposition to comply.

The new assembly for the election of counsellors was held at the same place: There was a general expectation that a new set of counsellors would be chosen. The council of the last year had been of very different opinion from the house in many points. They had no doubt of the governor's power to call, adjourn, or prorogue the assembly to any part of the province he thought proper, and although they were not for a fixed salary according to the instruction, yet they would have willingly consented to settle it for longer term than a year, and some of them, during Mr. Burnet's administration, but the house were most offended with the non-concurrence of their grant of money to their agents. After all, only four new counsellors were elected. Immediately after the council was settled, the court was prorogued to the 25th of June, and having sat unto the 10th of July, he prorogued them again unto the 20th of August, having made no speech at either of the sessions, or taken any notice of any business he thought proper for them to do. The reason of this omission appeared at the session in August. He had waited the final determination of his Majesty in council, upon the report of the lords committee. This he now communicated to the house, whereby they perceived that his conduct was approved, that of the house condemned, and his Majesty advised to lay the case before the parliament. The house received a letter at the same time from their agents, who, it seems, had altered their opinions, and now intimated to the house, that notwithstanding the determination or advice of the privy council, it was not likely the affair would ever be brought before the parliament. This letter the house ordered to be printed. The governor, in one of his messages, calls it "an undeniable proof of their endeavours to keep the people in ignorance of the true state of their affairs." It seems to be preferring a present temporary convenience, in keeping up the spirit of the people and diffusing a favourable opinion of their representatives, to the future real advantage of the cause, for such a measure must weaken the hands of the agents in England, and tend to bring the matter before the parliament, when otherwise it might have been avoided.

The governor having held several sessions at Salem without any success, he adjourned the court, to meet the 21st of August at Cambridge. This widened the breach, and the house grew warmer in their votes and messages, and complained that they were to be compelled to measures against their judgment, by being harassed and drove from one p of the province to another. The governor's friends observed the effect the controversy had upon his spirits. In a few days he fell sick of a fever and died at Boston the 7th of September. Some attributed his illness to his taking cold, his carriage oversetting upon the causeway at Cambridge, the tide being high, and he falling into the water. The resentment which had been raised ceased with the people in general upon his death. Many amiable parts of his character revived in their minds. He had been steady and inflexible in his adherence to his instructions, but discovered nothing of a grasping avaricious mind, it was the mode more than the quantum of his salary upon which he insisted. The naval office had generally been a post for some relation or favourite of the governor, but Colonel Tailer having been lieutenant-governor, and in circumstances far from affluent, he generously gave the post to him, without any reserve of the issues or profits. The only instance of his undue exacting money, by some was thought to be palliated by the established custom of the government he had quitted. This did not justify it. In his disposal of public offices, he gave the preference to such as were disposed to favour his cause, and displaced some for not favouring it; and, in some instances, he went further than good policy would allow. He did not know the temper of the people of New England. They have a strong sense of liberty, and are more easily drawn than driven. He disobliged many of his friends by removing from his post Mr. Lynde, a gentleman of the house, esteemed by both sides for his integrity and other valuable qualities; and he acknowledged that he could assign no other reason except that the gentleman had not voted for a compliance with the instruction. However, an immoral or unfair character was a bar to office, and he gave his negative to an election of a counsellor, in one instance, upon that principle only. His superior talents and free and easy manner of communicating his sentiments made him the delight of men of sense and learning. His right of precedence in all com-]

[panics facilitated the exercise of his natural disposition to a great share in the conversation, and at the same time "caused it to appear more excusable." His own account of his genius was, that it was late before it budded, and that until he was near 20 years of age, his father despaired of his ever making any figure in life. This, perhaps, might proceed from the exact severe diseipline of the bishop's family, not calculated for every temper alike, and might damp and discourage his. To long and frequent religious services at home, in his youth, he would sometimes pleasantly attribute his indisposition to a very scrupulous exact attendance upon public worship; but this might really be owing to an abhorrence of ostentation and mere formality in religion, to avoid which, as most of the grave serious people of the province thought, he approached too near the other extreme. A little more caution and conformity to the different ages, manners, customs, and even prejudices of different companies, would have been more politic, but his open undisguised mind could not submit to it. Being asked to dine with an old charter senator who retained the custom of saying grace sitting, the grave gentleman desired to know which would be more agreeable to his excellency, that grace should be said standing or sitting, the governor replied, standing or sitting, any way or no way, just as you please. He sometimes wore a cloth coat lined with velvet. It was said to be expressive of his character. He was a firm believer of the truth of revealed religion, but a bigot to no particular profession among Christians, and laid little stress upon modes and forms. By a clause in his last will, he ordered his body to be buried, if he died at New York, by his wife, if in any other part of the world, in the nearest chureh-yard or burying-ground, all places being alike to God's all-seeing eye.

The assembly ordered a very honourable funeral at the public charge. A motion, at another time, was made in the house for a grant to a governor to bear the expence of his lady's funeral, a dry old representative objected to a grant for a governor's lady, had a motion been for a grant to bury the governor he should have thought the money well laid out.

Mr. Dummer reassumed the administration. He did not intend to enter into the controversy about the salary; no advantage could arise from it, no new arguments could be used, the king's instructions were to be his rule, and he would not depart from them by accepting any grant as lieutenant-governor; but the affair having been under consideration before his Majesty in council, and further proceedings expected, he would wait for further intelligence and directions. The house were not willing to admit that the instruction had any respect to the salary of a lieutenant-governor, but if it had they had given sufficient reasons against it, and were determined to come into no act for fixing a salary. Having continued the session at Cambridge until the 26th of September, he ordered an adjournment to the 29th of November, at Boston, which was a further indication that he did not intend to press the instruction; however, at their first coming together he recommended to them a compliance with it; and upon their assuring him, by a message, that although they could not settle a salary, yet they were ready to give him an ample and honourable support, he desired them to lose no time about it, for he would accept of no support unless it should be exactly conformable to his Majesty's instruction. The house, notwithstanding, made a grant of 750*l*. to enable him to manage the affairs of government. The council concurred with an amendment, adding "for the half year current;" but this being fixing a salary for half a year the house refused it.

Upon the news of Mr. Burnet's death Mr. Belcher applied with all his powers to obtain the commission for the government. Governor Shute might have returned, but he declined it, and generously gave his interest to Mr. Belcher, who, 14 years before, had given 500*l*. sterling, which was never repaid, to facilitate Colonel Shute's appointment. The controversy, which it was supposed a governor must be engaged in, caused fewer competitors, and the ministry were the more concerned to find a proper person. Lord Townshend asked Mr. Wilks, who had much of his confidence, whether he thought Mr. Belcher would be able to influence the people to a compliance with the king's instructions, he replied that he thought no man more likely. Their choosing him agent was a mark of their confidence in him, but it seemed natural to expect that they would be under stronger prejudices against him than against a person who had never engaged in their favour. Mr. Belcher's appointment occasioned the removal of Mr. Dummer from the place of lieutenant-governor. A young gentleman, with whose family Mr. Wilks was connected, Mr. Thornton, Mr. Belcher had engaged to provide for, and he had no post in his gift worth accepting besides the naval office. To make a vacancy there, Colonel Tailer was appointed lieutenant-governor. The pleasure, if there was any, in superseding Mr. Dummer, who had superseded him before, could be no equivalent for the difference between a post of] [naked honour and a post of profit which gave him a comfortable living. Mr. Dummer's administration has been justly well spoken of. His general aim was to do public service. He was compelled to some compliances which appeared to him the least of two evils. It lessened him in Mr. Burnet's esteem, who thought he should have shewn more fortitude; but he retired with honour, and, after some years, was elected into the council, where, from respect to his former commission, he took the place of president; but being thought too favourable to the prerogative, after two or three years he was left out. He seemed to lay this slight more to heart than the loss of his commission, and aimed at nothing more the rest of his life than *otium cum dignitate*, selecting for his friends and acquaintance men of sense, virtue, and religion, and enjoyed in life, for many years, that fame, which, for infinitely wise reasons, the great Creator has implanted in every generous breast a desire of, even after death.

Colonel Tailer's commission was received and published before Mr. Belcher's arrival, and it gave him an opportunity of doing a generous thing for Mr. Dummer. A vote had passed the two houses granting him 900*l.* which, from a regard to his instructions, he had not signed, nor had he expressly refused it, and the court having been adjourned only, not prorogued, the next meeting was considered as the same session, and Colonel Tailer ventured to sign it, not being a grant to himself and not against the letter of his instructions, and it was really saving money to Mr. Dummer; the grant being intended for services to come as well as past, would not have been renewed, or in part only.

Chap. VII.

From the arrival of Governor Belcher in 1730, *to the reimbursement of the charge of the expedition against Cape Breton, and the abolition of paper money, in* 1749.

(*Anno* 1730.)—Mr. Belcher arrived the beginning of August, in the Blandford man of war, Captain Prothero.

No governor had been received with a shew of greater joy. Both parties supposed they had an interest in him. For men to alter their principles and practice, according to their interest, was no new thing. A sketch of Mr. Belcher's life and character will in some measure account for his obtaining the government, for the principal events in his administration, and for the loss of his commission.

Being the only son of a wealthy father, he had high views from the beginning of life. After an academical education in his own country he travelled to Europe, was twice at Hanover, and was introduced to the court there, at the time when the Princess Sophia was the presumptive heiress to the British crown. The novelty of a British American, added to the gracefulness of his person, caused distinguishing notice to be taken of him, which tended to increase that aspiring turn of mind which was very natural to him. Some years after he made another voyage to England, being then engaged in mercantile affairs, which, after his return home, proved, in the general course of them, rather unsuccessful, and seem to have suppressed or abated the ruling passion; but being chosen agent for the house of representatives, it revived and was gratified to the utmost, by his appointment to the government of Massachusetts bay and New Hampshire, and discovered itself in every part of his administration. Before he was governor, except in one instance, he had always been a favourer of the prerogative, and afterwards he did not fail of acting up to his principles. A man of high principles cannot be too jealous of himself, upon a sudden advancement to a place of power. The council never enjoyed less freedom than in his time. He proposed matters for the sake of their sanction rather than advice, rarely failing of a majority to approve of his sentiments.

He lived elegantly in his family, was hospitable, made great shew in dress, equipage, &c. and although, by the depreciation of the currency, he was curtailed of his salary, yet he disdained any unwarrantable or mean ways of obtaining money to supply his expences. By great freedom in conversation, and an unreserved censure of persons whose principles or conduct he disapproved, he made himself many enemies. In a private person this may often pass with little notice, but from a governor it is very hardly forgot, and some never ceased pursuing revenge until they saw him displaced.

The general court met the 9th of September. The people waited with impatience the governor's first speech. Many flattered themselves that the instruction for a fixed salary was withdrawn others, that if it was continued, he would treat it rather as Dudley and Shute had done than as his immediate predecessor; others, who did not expect a relaxation, were, from curiosity, wishing to know how he would acquit himself with the people, who sent him to England to oppose the instruction. After premising, that the honour of the crown and interest of Great Britain are very] [compatible with the privileges and liberties of the plantations, he tells the two houses, that he had it in command from his royal master to communicate to them his 27th instruction, respecting the governor's support; that whilst he was in England he did every thing consistent with reason and justice for preserving and lengthening out the peace and welfare of the province; that they were no strangers to the steps taken by his Majesty with respect to the unhappy dispute between the late governor and them, and he hoped, after such a struggle, they would think it for the true interest of the province to do what might be perfectly acceptable; that nothing prevented this controversy, and several other matters of dangerous consequence, being laid before the parliament, but his Majesty's great lenity and goodness, which inclined him to give them one opportunity more of paying a due regard to what in his royal wisdom he thinks so just and reasonable. Had he stopped here, perhaps less could not have been expected from him; but he unfortunately attempted to shew the similitude between the case of Cato shut up in Utica, and the Massachusetts bay under the restraint of the royal instruction, commended the wisdom of Cato in making so brave a stand for the liberties of his country, but condemned his putting an end to his life, when affairs became desperate, rather than submit to a power he could no longer resist; which instance he brought as some illustration of the late controversy, though he would not allow it to run parallel, Cæsar being a tyrant, and the king the protector of the liberties of his subjects.

It was said, upon this occasion, that the governor must allow that the Massachusetts assembly had done wisely hitherto in defending their liberties, for, otherwise, he had brought an instance of a case in no one respect similar to theirs; and if they had done so, it was because the instruction was a mere exertion of power, and then the parallel would run farther than he was willing to allow.

The instruction was conceived in much stronger terms than that to Governor Burnet, and it is declared that in case the assembly refuses to conform to it, "his Majesty will find himself under a necessity of laying the undutiful behaviour of the province before the legislature of Great Britain not only in this single instance but in many others of the same nature and tendency, whereby it manifestly appears that this assembly, for some years last past, have attempted by unwarrantable practices to weaken, if not cast off, the obedience they owe to the crown, and the dependence which all colonies ought to have on their mother country." And in the close of the instruction his Majesty expects, "that they do forthwith comply with this proposal as the last signification of our royal pleasure to them upon this subject, and if the said assembly shall not think fit to comply therewith, it is our will and pleasure and you are required immediately to come over to this kingdom of Great Britain, in order to give us an exact account of all that shall have passed upon this subject, that we may lay the same before our parliament."

The house proceeded just as they had done with Governor Burnet. They made a grant to Mr. Belcher of 1000*l.* currency for defraying the expence of his voyage to New England and as a gratuity for services while in England; and some time after, they voted him a sum equal to 1000*l.* sterling, to enable him to manage the public affairs, &c. but would fix no time. The council concurred it with an amendment, viz. "and that the same sum be annually allowed for the governor's support." This, without a fund for the payment of it, was doing little more than the house had repeatedly done by their declarations, that they doubted not future assemblies would make the like honourable provision for the governor's support according to the ability of the province; the amendment, notwithstanding, was not agreed to, and the house adhered to their own vote. This produced a second amendment, viz. "that the same sum should be annually paid during his excellency's continuance in the government and residence here;" but this also was non-concurred. The two houses then conferred upon the subject, the governor being present, and before they parted he made a long speech, expressing the great pleasure the council had given him in the part they had taken, and his concern and surprise at the conduct of the house, in running the risk of the consequences of their refusal to comply with the instruction, reminded them of the vast expence which their former unsuccessful disputes with their governors had occasioned to the province, but used no arguments to convince them of the reasonableness of the demand and its compatibility with their rights and privileges.

The small-pox being in the town of Cambridge, where the court sat, the house desired to rise, but the governor let them know he would meet them in any other town, and the same day ordered an adjournment to Roxbury, where a bill passed both houses for the support of the governor, but, not coming up to the instruction, the governor could not consent to it. The country party in the house, as much a solecism as it is, were the most zealous for the prerogative, and except a few prerogative] [men, who were always willing to fix the salary, none went so great a length, at this time, towards fixing it as those who opposed any one step towards it under Mr. Burnet.

The people, in general, were well pleased with the governor. It is not improbable that he would have obtained the settlement of a salary during his administration, if it had not been, in effect, a settlement for his successors also, for such a precedent could not easily have been resisted. The two parties which had long subsisted in the government were vying, each with the other, in measures for an expedient or accommodation. The prerogative men were Mr. Belcher's old friends, who were pretty well satisfied that his going over to the other side was not from any real affection to the cause, and that he must, sooner or later, differ with those who adhered to it, and for this event they waited patiently. The other party, by whose interest he had been sent to England, adhered to him, expecting their reward. Accordingly, Mr. Cooke was soon appointed a justice of the common pleas for the county of Suffolk. To make way for him and another favourite, Colonel Byfield, to whom Mr. Belcher was allied, two gentlemen, Colonel Hutchinson and Colonel Dudley, were displaced. They were both in principle steady friends to government, and the first of them was a fast friend to the governor. Mr. Belcher would not have been able to advance so many of his friends as he did, if he had not persuaded the council, that upon the appointment of a new governor, it was necessary to renew all civil commissions. Having obtained this point, he took the most convenient time to settle the several counties. Before he settled the county of York, he recommended to the judges a person for clerk of the court. This officer the province law empowers the judges to appoint. Some of them sent their excuse, being well satisfied with the clerk they had, who was a faithful well-approved officer, but the governor let the judges know, if he could not appoint a clerk he could a judge, and accordingly removed those who were not for his purpose and appointed others in their stead. There was an inconsistency, in delaying appointments, with the principles he advanced. If new commissions were necessary, they were necessary immediately, and they might as well be delayed seven years as one.

(*Anno* 1731.)—Two or three sessions passed, when little more was done, on the governor's part, than repeating his demand for a fixed salary, and intimating that he should be obliged to go to England and render an account of their behaviour to the king. The major part of the house were very desirous of giving satisfaction to the governor and to their constituents both, but that could not be. Mr. Cooke's friends in the town of Boston began to be jealous of him. A bill was prepared, which sets forth in the preamble, that settling a salary would deprive the people of their rights as Englishmen. In the purview, after granting 3400*l.* which was about equal to 1000*l.* sterling, it is further enacted, that as his Majesty had been graciously pleased to appoint J. B. Esq. to be the governor, who was a native of the country, whose fortune was here, who, when a member of the council, as well as when in a private station, has always consulted the true interest of his country as well as the honour and dignity of the crown, therefore it is most solemnly promised and engaged to his most excellent Majesty that there shall be granted the like sum for the like purpose at the beginning of the sessions in May every year during the governor's continuance in the administration and residence within the province, provided this act shall not be pleaded as a precedent or binding on any future assembly for fixing a salary on any succeeding governor. The bill is in Mr. Cooke's hand-writing, and it is minuted at the bottom that the governor approved of it. The governor could not imagine so evasive a thing could be approved in England. He might hope to improve it as being a further advance than had been before made, and by using this argument, that it would be much more rational for the house to do what they now had fully in their power to do, than to make a solemn promise that another house should do the same thing, the performance of which promise they would not have in their own power. The scheme failed, the bill did not pass, and from that time Mr. Belcher, despairing of carrying his point, turned his thoughts to obtaining a relaxation of his instruction. Instead of applying himself, he advised to an address from the house, not for the withdraw of the instruction, but that the governor might have leave to receive the sum granted. This was allowed, but it was to be understood, that he was to insist upon a compliance with his instruction as much as ever. Leave for consent to particular grants was obtained two or three years, and at length, a general order of leave to receive such sums as should be granted. This was the issue of the controversy about a fixed salary. Until Mr. Belcher's arrival, Mr. Cooke had differed from most who, from time to time, have been recorded in history for popular men. Generally, to preserve the favour of the people, they must change with the popular air, and when we survey a course] of action it will not appear altogether consistent. He had the art of keeping the people steady in the applause of his measures. To be careful never to depart from the appearance of maintaining or enlarging rights, liberties, and privileges, was all he found necessary. As soon as he was defective in this respect, and tried to secure his interest both with the governor and town of Boston, he had like to have lost both. In the election of representatives for Boston, in 1733 or 1734, the governor's party appeared against him, he had lost many of the other party by what they called too great a compliance, and he had a majority, after several trials, of oue or two votes only in 6 or 700.

The dispute about the manner of issuing money out of the treasury was settled unfavourably for the house. The charter provides, that all money shall be issued by warrant from the governor with advice and consent of the council. Until the year 1720 the money was brought into the treasury, by a vote or act originating in the house, and destined to certain purposes, and drawn out for those purposes by warrant from the governor, with advice &c. but after that, the house not only destined the money when put into the treasury, but provided that none of it, except some trifling sums for expresses and the like, should be issued without a vote of the whole court for payment. After such a vote they were willing the governor should give his warrant. This appeared to the king to render his governor contemptible, and entirely to defeat the provision in the charter, and there was no prospect of any relaxation of the instruction to the governor. When the servants of the government had suffered a long time for want of their money, the house passed a bill which supplied the treasury in a way not materially differing from what had been in practice before 1720.

Mr. Belcher had another instruction not to consent to the issuing any bills of credit for a longer term than those were to remain current which had before been issued, none of which extended beyond the year 1741. It would have been but a small burden upon the inhabitants to have paid the charges of every year and the debt which lay upon such year besides, but, instead of that wise measure, they suffered one year after another to pass with light taxes, and laid heavy burdens upon distant years, and the last year, 1741, had more laid upon it than any four or five preceding years; and although even this was far short of what has been paid in some succeeding years, yet it was deemed an insupportable burden, and it was generally supposed the promises made by the acts of government to draw in the bills in that year would by some means or other be evaded or openly violated. Mr. Belcher seemed determined to adhere to his instruction, and there was an expectation of some great convulsion, which was prevented by his being superseded before that period arrived.

The project, of which we have already taken notice, for settling the *e.* country, Captain Coram pursued until he procured an order or instruction to Colonel Phillips, the governor of Nova Scotia, in 7130, to take possession of the land between St. Croix and Kennebeck, and 30 men with an officer were sent to the fort at Pemaquid, built by the Massachusetts. Colonel Dunbar, a gentleman out of employ, came over about the same time, took the command of the fort, and assumed the government of that part of the province. Mr. Belcher was applied to by the proprietors of the lands there, and the house of representatives asserted the right of the province. The governor, with advice of council, issued a proclamation requiring the inhabitants to remain in their obedience and due subjection to the laws and government of the province. This seems to have been all that in prudence he could do. Some were for taking further measures to remove Dunbar, which, as he had a royal commission, however liable to exceptions, Mr. Belcher thought by no means warrantable. The minds of the people were inflamed, and when Dunbar came up to Boston he persisted in his claim to the country, which, with reports of some not very decent expressions of the governor, raised the resentment of many. Persons of ill design perhaps might have been able to have caused a tumult. The lands indeed were claimed by a few particular persons, but it was spread abroad that when this country should be detached from the rest of the province the supplies of fuel to the sea-port towns would cease, or be burdened with heavy duties, and the poor oppressed.

(*Anno* 1732.)—It happened that Mr. Samuel Waldo, a gentleman of good capacity and who would not easily relinquish his right, undertook for the proprietors of the principal tract of the country claimed, and upon representation to his Majesty in council, the order to Phillips and the authority to Dunbar were revoked in 1732, and the government of the province afterwards thought it proper to place a garrison in their own pay at fort Frederick, the name given by Dunbar to the fort at Pemaquid.

We shall take notice of two or three only and those the most remarkable events during the rest of Mr. Belcher's administration.

(*Anno* 1733.)—In 1733, there was a general complaint throughout the four governments of New

[England of the unusual scarcity of money. There was as large a sum current in bills of credit as ever, but the bills having depreciated they answered the purposes of money so much less in proportion. The Massachusetts and New Hampshire were clogged with royal instructions. It was owing to them that those governments had not issued bills to as great an amount as Rhode Island. Connecticut, although under no restraint, yet consisting of more husbandmen and fewer traders than the rest, did not so much feel the want of money. The Massachusetts people were dissatisfied that Rhode Island should send their bills among them, and take away their substance and employ it in trade, and many people wished to see the bills of each government current within the limits of such government only. In the midst of this discontent, Rhode Island passed an act for issuing 100,000*l.* upon loan, for 20 years, to their own inhabitants, who would immediately have it in their power to add 100,000*l.* to their trading stock from the horses, sheep, lumber, fish, &c. of the Massachusetts inhabitants. The merchants of Boston therefore confederated, and mutually promised and engaged not to receive any bills of this new emission, but to provide a currency. A large number formed themselves into a company, entered into covenants, chose directors, &c. and issued 110,000*l.* redeemable in 10 years, in silver at 19*s.* per ounce, the then current rate, or gold in proportion, a tenth part annually. About the same time the Massachusetts treasury, which had been long shut, was opened, and the debts of two or three years were all paid at one time in bills of credit; to this was added the ordinary emissions of bills from New Hampshire and Connecticut, and some of the Boston merchants, tempted by an opportunity of selling their English goods, having broke through their engagements and received the Rhode Island bills, all the rest soon followed the example. All these emissions made a flood of money, silver rose from 19*s.* to 27*s.* the ounce, and exchange with all other countries consequently rose also, and every creditor was defrauded of about one third of his just dues. As soon as silver rose to 27*s.* the notes issued by the merchants payable at 19*s.* were boarded up and no longer answered the purposes of money. Although the currency was lessened by taking away the notes, yet what remained never increased in value, silver continuing several years about the same rate, until it took another large jump. Thus very great injustice was caused by this wretched paper currency and no relief of any sort obtained; for, by this sinking in value, though the nominal sum was higher than it had ever been before, yet the currency would produce no more sterling money than it would have done before the late emissions were made.

(*Anno* 1737.)—In 1737, a controversy which had long subsisted between the two governments of Massachusetts bay and New Hampshire was heard by commissioners for that purpose appointed by the crown. Various attempts had been made to settle this dispute, and it had been often recommended by the crown to the assemblies of the two provinces to agree upon arbitrators from neighbouring governments, and to pass acts which should bind each province to be subject to their determinations. Several such acts passed, but they were not exactly conformable one to the other, or the operation of them was by some means or other obstructed. The Massachusetts refused terms which afterwards they would gladly have accepted. They have done the like in other controversies. Long possession caused them to be loth to concede any part of the territory. New Hampshire took its name from the grants made by the council of Plymouth to Captain John Mason. Of these there had been four or five, all containing more or less of the same lands. Exceptions were taken to all of them, and that which was the least imperfect was dated after the grant of Massachusetts bay, so that the whole controversy turned upon the construction of the Massachusetts charters. The first charter made the *n.* boundary to be three miles to the *n.* of Merrimack river, or to the *n.* of any and every part thereof. After running *w.* about 80 miles from the sea the river alters its course and tends to the *n.*; or, to speak with more propriety, having run from its crotch or the meeting of Pemigewasset river and Winnepissiauke pond to the *s.* about 50 miles, it then tends to the *e.* about 30 miles, until it empties into the sea. It was urged by the advocates for Massachusetts colony that their boundary was to be three miles to the *n.* of the northernmost part of the river, and to extend *e.* and *w.* from the Atlantic to the S. sea. This swallowed all New Hampshire and the greatest part of the province of Maine. At a hearing before the king in council, in 1677, the agents for Massachusetts, by advice, disclaimed all right of jurisdiction beyond the three miles *n.* of the river according to the course, and it was determined they had a right as far as the river extended, but how far the river did extend was not then expressly mentioned. It seems, however, not to have been doubted, for although at the time of the grant of the first charter it does not appear that the course was known any great distance from the sea, yet, soon after the government was transferred from]

[Old England to New, it was as well known by the name of Merrimack as far as Penicook as it is at this day, and the tribe of Indians which dwelt there had a correspondence with the English, and in 1639 persons were employed by the government of Massachusetts to explore that part of the country, and there are still preserved the testimonies of divers persons, declaring that they before that time always understood the river to be called by the same name from the crotch to the mouth. If the first charter of the Massachusetts had continued, it is not probable any different construction would ever have been started; but in the new charter the boundary is thus expressed, "extending from the great river commonly called Monomack *alias* Merrimack on the *n.* part, and from three miles *n.* of the said river to the Atlantic or W. sea or ocean on the *s.* part, &c." The whole, however, of the old colony being included in the new province, many years passed without any thought of a different construction of bounds in the two charters, and the disputes between New Hampshire and the Massachusetts have been, principally, concerning the towns of Salisbury and Haverhill, which, when first granted by the Massachusetts, were made to extend more than three miles from the river, and the part beyond the three miles remained under the jurisdiction by which they had been granted, which New Hampshire complained of. A new line to begin three miles *n.* of the mouth of Merrimack and so run *w.* to the S. sea, is a modern construction. Some hints had been given of such a line before or about the year 1726, and it was supposed by New Hampshire that the Massachusetts were induced thereby to make grants of townships between Merrimack and Connecticut river, in order to strengthen their title by possession. Still there was a prospect of accommodation, and in the year 1731, the committees from the assemblies of the two provinces differed only upon the point of equivalents, the Massachusetts desiring to retain under their jurisdiction the whole of those towns which lay upon the river, and to give other lands as an equivalent for the property; but about the same time the gentlemen of New Hampshire, who had for many years before been at the helm, thinking, and perhaps justly, that they were not well treated by Mr. Belcher, determined to exert themselves to obtain a governor for that province, and to remain no longer under the same governor with the Massachusetts. They had but little chance for this unless they could enlarge their bounds. The very proposal of a distinct government, as it increased the number of officers of the crown, they thought would be a favourable circumstance in settling the controversy with Massachusetts.

The house of representatives of New Hampshire, October 7, 1731, by a vote appointed John Rindge, Esq. a merchant there who was bound to England, their agent, to solicit the settlement of the boundaries. But their main dependence was upon Mr. Thomlinson, a gentleman who had been in New Hampshire, and was then a merchant of note in London, and perhaps was as capable of conducting their cause as any person they could have pitched upon. He had the friendship of Colonel Bladen, who at that day had great weight in the board of trade, and had conceived very unfavourable sentiments of the Massachusetts in general, and did not like Mr. Belcher the governor. He employed a solicitor, Ferdinando Paris, one of the first rate, and who had a peculiar talent at slurring the characters of his antagonists. Many of his briefs have been known to abound in this way. The first step in consequence of Mr. Rindge's petition was a question sent by the lords of trade to the attorney and solicitor general for their opinion, "From what part of Merrimack river the three miles from whence the dividing line between the province of New Hampshire and the province of the Massachusetts bay is to begin, ought to be taken, according to the intent of the charter of William and Mary?" This was a plain intimation that if the point where to begin could be settled, nothing more was necessary, the *w.* line claimed by New Hampshire was to follow of course. The Massachusetts agent (Mr. Wilks), by his counsel, would say nothing upon the question, because it would not determine the matters in dispute. Report was made, however, that it ought to begin three miles *n.* of the mouth of Merrimack river. It was then proposed that commissioners should be appointed to settle this controversy. This the Massachusetts were averse to, unless they knew who they were to be. They were at the same time afraid of its being determined in England *ex parte*, if they should refuse to consent. A committee of the general court reported, that the agent should be instructed that the province would agree to commissioners to be appointed to settle the controversy here. This report was accepted, the house intending the commissioners should be agreed upon by the two governments, some of the committee intending the agent should understand his instructions, to consent to the appointment of commissioners provided they sat here or in one of the two governments. A comma after the word *appointed* and after the word *controversy* would give the sense of the house, the last comma left] [ont, it might be taken in the sense of the committee; but as it is most probable the letter had no regular pointing, their meaning was to be guessed at.

This was treating the agent ill, and he was censured by the house for not observing his instructions. This account of the affair was collected from some of the committee, who excused themselves for this equivocal report as being necessary for the public service, the house not being willing to consent to an explicit submission. It was made a condition of the submission that private property should not be affected. The ministry in later instances have not waited for an express submission, but have appointed commissioners upon application from one party only.

The commissioners were all such as the New Hampshire agent proposed, five counsellors from each of the governments of New York, Rhode Island, and Nova Scotia. With the two former governments the Massachusetts were then in controversy about lines. The latter, it was said, was disaffected to charter government. Connecticut, proposed by Massachusetts, was rejected because of a bias from their trade, religion, &c. which New Hampshire was afraid of. The place for the meeting of commissisoners was Hampton in New Hampsire, the first of August.

The commissioners from Nova Scotia, with some of Rhode Island, met at the time appointed, and were afterwards joined by Mr. Livingstone from New York, who presided. After many weeks spent in hearing the parties and examining their evidence, the only doubt in the commissioners minds was, whether the Massachusetts new charter comprehended the whole of the old colony. Not being able to satisfy themselves, and perhaps not being unwilling to avoid the determination, they agreed to make a special judgment or decree, the substance of which was, that, if the charter of William and Mary grants to the Massachusetts bay all the lands granted by the charter of Charles I. they then adjudge a curve line to begin three miles *n.* of the mouth of the river, and to keep the same distance from the river as far as the crotch or parting at Pemigewasset and Winnepissiauke, and then to run *w.* towards the S. sea until it meets with his Majesty's other governments; but if the charter of William and Mary did not contain, &c. then they adjudge a *w.* line to begin at the same place three miles *n.* of the mouth and to run to the S. sea. This point in doubt they submitted to his Majesty's royal pleasure.

The Massachusetts were sure of their cause. It was impossible, they thought, consistent with common sense, that the point in doubt shaould be determined against them. They thought it safest, however, to send to England a special agent, Edmund Quincy, Esq. one of the council, who had been one of the court's agents before the commissioners. He was joined with Mr. Wilks, and Mr. Belcher by his interest prevailed upon the assembly to add a third, his wife's brother, Richard Partridge. Exceptions, called an appeal, were offered to the judgment of the commissioners, Mr. Quincy died of the small pox by inoculation, soon after his arrival in London, the other two knew little or nothing of the controversy. The commissioners, however, had rendered it as difficult to determine a line against the Massachusetts as if they had given a general judgment in their favour. The New Hampshire agent and solicitor thought of no expedient. In their brief they pray the lords committee to report, "that all the lands lying to the *n.* of Merrimack river, which were granted by the charter of King Charles I. to the late colony of the Massachusetts bay, are not granted to the present province of the Massachusetts bay by the charter of King William and Queen Mary." This never could have been done. At the bearing, it was thought proper to lay aside all regard to the judgment of the commissioners, and to proceed upon an entirely new plan. No doubt was made, that the old colony was all included in the new province. The question was, what were the *n.* bounds of the colony of Massachusetts bay, which the council of Plymouth when they sold the territory to the patentees, and the king when he granted the jurisdiction, had in contemplation? This, it was said, must be a line three miles *n.* of a river not fully explored, but whose general course was supposed to be *e.* and *w.* So far therefore as it afterwards appeared that the river kept this course, so far it was equitable the line should continue; but, as on the one hand, if the river had altered its course and turned to the *s.* it would have been inequitable to have reduced the grant to a very small tract, so on the other hand, when it appeared to turn to the *n.* it was inequitable to extend the grant and make a very large territory, and therefore defeat other grants made about the same time. It was therefore determined that the *n.* boundaries of Massachusetts bay should be a line three miles from the river as far as Pantucket falls, then run *w.* 10° *n.* until it meets New York line.

The Massachusetts thought themselves aggrieved. They submitted the controversy to commissioners to be appointed by the crown, and had been fully heard. The whole proceedings of the commissioners were set aside, and, without any] [uetlee to the government, the controversy was determined by a committee of council upon a new point, on which their agent had never been instructed. And however there might be the appearance of equity in the principle upon which their lordships proceeded, yet the Massachusetts supposed, if their possession for 100 years, together with the determination of the king in council in 1677, and the acquiescence of all parties in this determination for about 50 years, had been urged and duly weighed, the balance, upon the sole principle of equity, would have been in their favour. It increased their mortification to find that they had lost by this new line several hundred thousand acres more than the utmost claim ever made by New Hampshire; for Merrimack river, from the mouth to Pantucket falls tending to the *s.* it made a difference of four or five miles in breadth, the whole length of the line, between a line to run *w.* from Pantucket falls and a line *w.* from the black rocks.

The dispute about the bounds of the province of Maine, which lies on the other side New Hampshire, was upon the construction of the word northwestward. The Massachusetts urged that it was the evident design of the granters of the province of Maine to describe a territory about 120 miles square. At that day this was probably the reputed distance from Newichawannock or Piscataqua river to Kennebeck, along the sea-coast, the general course of which was *n. e.* and *s. w.*; after going up the two rivers to the heads, the lines were to run north-westward until 120 miles were finished, and then a line back parallel to the line upon the sea. The agents for New Hampshire, at the court of commissioners, insisted that every body understood north-westward to be *n.* a little, perhaps less than a quarter of a point *w.* It not being possible to think of any reason for a line to run upon this course, the Massachusetts could scarce suppose the New Hampshire agents to be serious, and imagined the commissioners would need no other reply than that every body understood a line running *w.* to be a line from *e.* to *w.* and by the same rule of construction they supposed north-westward to be from *s. e.* to *n. w.*; that north-eastward being explained in the same grant to be as the coast lay, proved in fact to be from *s. w.* to *n. e.* They were, however, surprised with the determination of the commissioners, that north-westward intended *n.* 2° *w.* Why not 1° or 3° as well as 2°. From this part of the judgment the Massachusetts appealed. The agents in England obtained the celebrated Doctor Halley's opinion, in writing under his hand, that in the language and understanding of mathematicians a line to run north-westward is a line to run *n. w.* but this opinion did not prevail, and the judgment of the commissioners upon this point was confirmed by his Majesty in council.

It behoved Mr. Belcher, the governor of both provinces, to carry an even hand. It happened, that the general court of the Massachusetts, whilst it sat at Salisbury on the occasion of this controversy, made him a grant of 800*l.* currency, in consideration of the deficiency of their former grants, for his salary and his extraordinary expence and trouble in attending the court at a distance from his house and family. Soon after this grant, he adjourned the general courts of both provinces, in order to their determining whether to abide by the result of the commissioners or to appeal from it, but the court of New Hampshire was adjourned to a day or two after the Massachusetts court, and it was said they were prevented entering the appeal within the time limited. He did not care that either assembly should do any business when he was absent, and therefore intended first to finish the Massachusetts business, and immediately after proceed to New Hampshire.

This afforded matter of complaint from that province, which Mr. Belcher was called upon to answer, and it was determined the complaint was well founded; and it being urged that the 800*l.* was intended as a bribe to influence him to this measure, the Massachusetts thought their own honour concerned, and joined with him in his defence, which perhaps increased the suspicion of guilt and hastened his removal. That we may finish what relates to the controversy between the two provinces, we must take notice of the conduct of the Massachusetts upon the receiving his Majesty's order in council. The lines, by the order, were to be run by two surveyors, one on the part of each province; but if either province refused, the other was to proceed *ex parte*. New Hampshire, whose highest expectations were exceeded, proposed to join, but were refused by the Massachusetts, and thereupon appointed surveyors to run the lines of the Massachusetts and province of Maine *ex parte*. Both lines were complained of as being run favourably for New Hampshire: that of the province of Maine became a subject of new controversy, it having been suggested that the surveyor mistook the main branch of the river Newichewannock, which if he had pursued would have made five or six miles in breadth to the advantage of Massachusetts. This refusal to join proceeded from the feeble irresolute state of the minds of the house of representatives. Unwilling by any act of their own to express their submission to what they called an unequal decree, they ran the risk of its] [being carried into execution still more unequally, and yet succeeding houses, by a subsequent long continued passive submission, as effectually subjected the province, as if the same had been explicitly acknowledged at first.

(*Anno* 1738.)—After the controversy about the governor's salary and the supply of the treasury was finished, there seemed to be a general disposition to rest, and we hear little of a party in opposition to the governor for several years together. Whilst the controversy with New Hampshire was depending, all of every party engaged in defence of the right of the province. Besides, Mr. Cooke, who had been many years at the head of the popular party, was worn out with service, and having been some time in a declining state, died in the fall of 1737; and the town of Boston were so far from an apprehension of danger to their liberties that they chose in his stead Mr. Wheelwright, the commissiary general, who depended upon the governor every year for his approbation after being elected by the council and house, and in 1738 three of the representatives of the town had the character of friends to government; but towards the end of the year a great clamour arose against the governor for adhering to his instruction about paper money, and against the three representatives for their pernicious principles upon the subject of paper money; and at the town election for 1739 three others were chosen in their stead, two of them professedly disaffected to the governor and promoters of popular measures, the third, although of great integrity, and for that reason desirous of a fixed currency, yet in his judgment against reducing the paper money, and a favourer of schemes for preventing its depreciation.

(*Anno* 1739.)—Many country towns followed the example of Boston, and it appeared that a majority of the house were of the same principles with the town members. After Mr. Belcher's arrival, the house, as we have observed, had passed a vote for depositing 500*l.* sterling in the bank of England, to be used as they or their successors should think proper. This was concurred in council, and consented to by the governor. This money it was said could not be better applied than in soliciting a relaxation of the governor's instruction concerning paper money, and Mr. Kilby, one of the Boston representatives, was chosen agent for the house, and a petition was by him presented from the house to his Majesty in council, but it had no effect.

A general dread of drawing in all the paper money without a substitution of any other instrument of trade in the place of it, disposed a great part of the province to favour what was called the land bank or manufactory scheme, which was begun or rather revived in this year 1739, and produced such great and lasting mischiefs that a particular relation of the rise, progress, and overthrow of it, may be of use to discourage and prevent any attempts of the like nature in future ages. By a strange conduct in the general court they had been issuing bills of credit for 8 or 10 years annually for charges of government, and being willing to ease each present year, they had put off the redemption of the bills as far as they could, but the governor being restrained by his instruction from going beyond the year 1741, that year was unreasonably loaded with 30 or 40,000*l.* sterling taxes, which, according to the general opinion of the people, it was impossible to levy, not only on account of the large sum, but because all the bills in the province were but just sufficient to pay it, and there was very little silver or gold, which by an act of government was allowed to be paid for taxes as equivalent to the bills. A scheme was laid before the general court by a person of eminence, and then one of the representatives of Boston, in which it was proposed to borrow in England upon interest, and to import into the province, a sum in silver equal to all the bills then extant, and therewith to redeem them from the possessors and furnish a currency for the inhabitants, and to repay the silver at distant periods, which would render the burden of taxes tolerable by an equal division on a number of future years, and would prevent the distress of trade by the loss of the only instrument, the bills of credit, without another provided in its place. But this proposal was rejected. One great frailty of human nature, an inability or indisposition to compare a distant, though certain inconvenience or distress with a present convenience or delight, was said by some strangers, who came hither from Europe, to be prevalent in Americans, so as to make it one of their distinguishing characteristics. Be that as it may, it is certain that at this time a great number of private persons, alleging that the preceding general court having suffered the province to be brought into distress, from which it was not in the power of their successors to afford relief, the royal instruction being a bar to any future emissions of bills until all that were then extant should be redeemed, resolved to interpose. Royal instructions were no bar to the proceedings of private persons. The project of a bank in the year 1714 was revived.

(*Anno* 1740.)—The projector of that bank now put himself at the head of 7 or 800 persons,] [some few of rank and good estate, but generally of low condition among the plebeians and of small estate, and many of them perhaps insolvent. This notable company were to give credit to 150,000*l.* lawful money, to be issued in bills, each person being to mortgage a real estate in proportion to the sums he subscribed and took out, or to give bond with two sureties, but personal security was not to be taken for more than 100*l.* from any one person. Ten directors and a treasurer were to be chosen by the company. Every subscriber or partner was to pay three per cent. interest for the sum taken out, and five per cent. of the principal, and he that did not pay bills might pay the produce and manufacture of the province at such rates as the directors from time to time should set, and they should commonly pass in lawful money. The pretence was, that by thus furnishing a medium and instrument of trade, not only the inhabitants in general would be better able to procure the province bills of credit for their taxes, but trade, foreign and inland, would revive and flourish. The fate of the project was thought to depend upon the opinion which the general court should form of it. It was necessary therefore to have a house of representatives well disposed. Besides the 800 persons subscribers, the needy part of the province in general favoured the scheme. One of their votes will go as far in popular elections as one of the most opulent. The former are most numerous, and it appeared that by far the majority of the representatives for 1740 were subscribers to or favourers of the scheme, and they have ever since been distinguished by the name of the Land Bank House.

Men of estates; and the principal merchants in the province, abhorred the project and refused to receive the hills, but great numbers of shopkeepers who had lived for a long time before upon the fraud of a depreciating currency, and many small traders, gave credit to the bills. The directors, it was said, by a vote of the company, became traders, and issued just what bills they thought proper, without any fund or security for their ever being redeemed. They purchased every sort of commodity, ever so much a drug, for the sake of pushing off their bills, and by one means or other a large sum, perhaps 50 or 60,000*l.* was abroad. To lessen the temptation to receive the bills, a company of merchants agreed to issue their notes or bills, redeemable by silver and gold at distant periods, much like the scheme in 1733, and attended with no better effect. The governor exerted himself to blast this fraudulent undertaking, the land bank. Not only such civil and military officers as were directors or partners, but all who received or paid any of the bills, were displaced. The governor negatived the person chosen speaker of the house, being a director of the bank, and afterwards negatived 13 of the new-elected counsellors who were directors or partners in or reputed favourers of the scheme. But all was insufficient to suppress it. Perhaps the major part, in number, of the inhabitants of the province, openly or secretly, were well-wishers to it. One of the directors afterwards was said to acknowledge, that although he entered in the company with a view to the public interest, yet when he found what power and influence they had in all public concerns, he was convinced it was more than belonged to them, more than they could make a good use of, and therefore unwarrantable. Many of the most sensible discreet persons in the province saw a general confusion at hand. The authority of parliament to controul all public and private persons and proceedings in the colonies was, in that day, questioned by nobody. Application was therefore made to parliament for an act to suppress the company, which, notwithstanding the opposition made by their agent, was very easily obtained; and therein it was declared, that the act of the sixth of King George I. chap. 18, did, does, and shall extend to the colonies and plantations in America. It was said the act of George I. when it passed, had no relation to America, but another act 20 years after gave it a force even from the passing it, which it never could have had without. This was said to be an instance of the transcendent power of parliament. Although the company was dissolved, yet the act of parliament gave the possessors of the bills a right of action against every partner or director for the sums expressed, with interest. The company were in amaze. At a general meeting some, it was said, were for running all hazards, although the act subjected them to a præmunire; but the directors had more prudence, and advised them to declare that they considered themselves dissolved, and met only to consult upon some method of redeeming their bills from the possessors, which every man engaged to endeavour in proportion to his interest, and to pay into the directors, or some of them, to burn or destroy. Had the company issued their bills at the value expressed in the face of them, they would have had no reason to complain of being obliged to redeem them at the same rate, but as this was not the ease in general, and many of the possessors of the bills had acquired them for half their value,]

[being carried into execution still more
and yet succeeding houses, by a sul
continued passive submission, as effectu
ed the province, as if the same had b
acknowledged at first.

(*Anno* 1738.)—After the contro
governor's salary and the supply
was finished, there seemed to l
position to rest, and we hear li
opposition to the governor for se
Whilst the controversy with
depending, all of every party
of the right of the province.
who had been many year
popular party, was worn
having been some time in
the fall of 1737; and th
far from an apprehension
that they chose in his st
commissiary general,
governor every year
being elected by the
1738 three of the repr
the character of friends t
the end of the year a
the governor for adhe
paper money, and agai
for their pernicious prin
paper money; and at
three others were cho
them professedly disaff
promoters of popular m
of great integrity, and
a fixed currency, yet
reducing the paper m
schemes for preventing i

(*Anno* 1739.)—Many
the example of Boston,
majority of the house we
with the town member
arrival, the house, as we
a vote for depositing 50
England, to be used as
should think proper.
council, and consented t
money it was said could
in soliciting a relaxation
tion concerning paper m
of the Boston representa
the house, and a petiti
from the house to his Ma
no effect.

A general dread of
money without a sub
ment of trade in th

his
on-
loy-
c re-
icient
a plan-
t guile,
y have a
n.
er, was a
and had
rospect of a
nove to Bos-
ad resided six
al esteem; and
aid to be as ac-
r as any person
in London and
e collector's place
ld have preferred it
interest being made
Henry Frankland,
g for both, except by
r. Shirley.
in the first week in July.
idence in Rhode Island
the Massachusetts before
appointed to settle the
nments. As most of the
ords of that time are
ticular an account of
mmissioners as other-
It is certain that for
in controversy be-
s a small gore of
he Massachusetts
ice. A great part
ished it might be
tenacious men,
ces, influenced
ttlement made
nother settle-
confirmed in
l by the ill
controversy

h New Hampshire, applied to his Majesty to point commissioners to settle the line between e two governments. The consent or submission the Massachusetts to such appointment was not ought necessary, and, if they would not appear, he commissioners were to proceed *ex parte*. The Massachusetts assembly thought proper to appear by the committee, having no apprehensions the controversy would turn, in the judgment of the commissioners, upon a point never before relied upon, viz. that the colony of New Plymouth having no charter from the crown, Rhode Island charter must be the sole rule of determining the boundary, although the patent from the council of Plymouth to Bradford and associates was prior to it.

(*Ann.* 1741.)—The colony of New Plymouth was a government *de facto*, and considered by King Charles as such in his letters and orders to them before and after the grant of Rhode Island charter and when the incorporation was made of New Plymouth with Massachusetts, &c. the natural and legal construction of the province charter seems to be, that it should have relation to the time when the several governments incorporated respectively, in fact, became governments. A gentleman of the council of New York had great influence at the board of commissioners. The argument which had been made use of in former controversies, that Massachusetts was too extensive, and the other governments they were contending with, of which New York was one, were too contracted, was now revived. To the surprise of Massachusetts, a line was determined, which not only took from them the gore formerly in dispute, but the town of Bristol, Tiverton, and Little Compton, and great part of Swansey and Barrington. An appeal was claimed, and allowed to his Majesty in council, where, after lying four or five years, the decree of the court of commissioners was confirmed. In the prosecution and defence of this title it has been said, that some material evidence was never produced which would have supported the Massachusetts claim.

Mr. Shirley found the affairs of the province in a perplexed state. The treasury was shut, and could not be opened without some deviation from the royal instructions, the bills of credit were reduced, and nothing substituted as a currency in their stead the land bank party carried every point in the house, there seemed to be a necessity of securing them, the great art was to bring them over to his measures, and yet not give in to their measures so as to lose his interest with the rest of the province and with the ministry in England. Some of the principal of them, who knew their own [as expressed, equity could not be done; and so far as respected the company perhaps the parliament was not very anxious, the loss they sustained being but a just penalty for their unwarrantable undertaking, if it had been properly applied. Had not the parliament interposed, the province would have been in the utmost confusion, and the authority of government entirely in the land bank company.

Whilst Mr. Belcher, by his vigorous opposition to the land bank, was rendering himself obnoxious to one half the people of the province, measures were pursuing in England for his removal from the government. Besides the attempts which we have mentioned from New Hampshire, which had never been laid aside, there had always been a disaffected party in Massachusetts, who had been using what interest they had in England against him. Lord Wilmington, president of the council, the speaker of the house of commons, and Sir Charles Wager, first lord of the admiralty, all had a favourable opinion of Mr. Belcher, so had Mr. Holden, who was at the head of the dissenters in England, and all, upon one occasion or another, had appeared for him.

The most unfair and indirect measures were used with each of these persons to render Mr. Belcher obnoxious and odious to them. The first instance was several years before this time. A letter was sent to Sir Charles Wager in the name of five persons, whose hands were counterfeited, with an insinuation that Mr. Belcher encouraged the destruction of the pine-trees reserved for masts for the navy, and suffered them to be cut into logs for boards. Forgeries of this sort strike us with more horror than false insinuations in conversation, and perhaps are equally mischievous in their effects. The latter may appear the less criminal, because abundantly more common.

An anonymous letter was sent to Mr. Holden, but the contents of it declared, that it was the letter of many of the principal ministers of New England, who were afraid to publish their names lest Mr. Belcher should ruin them. The charge against him was, a secret undermining the Congregational interest, in concert with commissary Price and Dr. Cutler, whilst at the same time he pretended to Mr. Holden and the other dissenters in England to have it much at heart. To remove suspicion of fraud, the letter was superscribed in writing either in imitation of Dr. Colman's hand, a correspondent of Mr. Holden, or, which is more probable, a cover of one of his genuine letters had been taken off by a person of not an unblemished character, to whose care it was committed, and made use of to inclose the spurious one. Truth and right are more frequently, in a high degree, violated in political contests and animosities than upon any other occasion. It was well known that nothing would more readily induce a person of so great virtue as the speaker to give up Mr. Belcher than an instance of corruption and bribery. The New Hampshire agents therefore furnished him with the votes of the Massachusetts assembly, containing the grant of 800*l*. and evidence of the adjournment of New Hampshire assembly, alleged to be done in consequence; nor was he undeceived until it was too late.

Mr. Wilks, the Massachusetts agent, who was in great esteem with Lord Wilmington, and was really a person of a fair upright mind, had prevented any impressions to Mr. Belcher's prejudice, but it unluckily happened that the land bank company employed Richard Partridge, brother by marriage to Mr. Belcher, as their agent. He had been many years agent for his brother, which fact was well known to his lordship, but, from an expectation of obtaining the sole agency of the province by the interest of the prevailing party there, engaged zealously in opposing the petitions to the house of commons, and gave out bills at the door of the house. It was said that all Mr. Belcher's opposition to the scheme, in the province, was mere pretence; had he been in earnest, his agent in England would never venture to appear in support of it, and this was improved with Lord Wilmingtou to induce him to give up Mr. Belcher, and it succeeded. Still the removal was delayed one week after another, two gentlemen from the Massachusetts continually soliciting. At length, it being known that Lord Euston's election for Coventry was dubious, one of these gentlemen undertook to the Duke of Grafton to secure the election, provided Mr. Belcher might immediately be removed, and, to accomplish his design, he represented to Mr. Maltby, a large dealer in Coventry stuffs and a zealous dissenter, that Mr. Belcher was, with the Episcopal clergy, conspiring the ruin of the Congregational interest in New England, and unless he was immediately removed it would be irrecoverably lost; that the Duke of Grafton had promised, if Lord Euston's election could be secured, it should be done; that letters to his friends in Coventry would infallibly secure it; that he could not better employ his interest than in the cause of God and of religion. Maltby swallowed the bait, used all his interest for Lord Euston, the two gentlemen spent three weeks at Coventry, and] [having succeeded agreeable to the Duke's promise, Mr. Belcher was removed a day or two after their return. This account was received from Mr. Maltby himself, who lamented that he had suffered himself to be so easily imposed on.

A few weeks longer delay would have baffled all the schemes. The news arrived of his negativing 13 counsellors, and displacing a great number of officers concerned in the land bank, and his zeal and fortitude were highly applauded when it was too late. An American who was in London at this time, has given us some very full information concerning these facts. Certainly, in public employments no man ought to be condemned from the reports and accusations of a party, without sufficient opportunity given him to exculpate himself; a plantation governor especially, who, be he without guile, or a consummate politician, will infallibly have a greater or lesser number disaffected to him.

Mr. Shirley, successor to Mr. Belcher, was a gentleman of Sussex, bred in the law, and had been in office in the city, but having prospect of a numerous offspring, was advised to remove to Boston in the Massachusetts, where he had resided six or eight years and acquired a general esteem; and if there must be a change, it was said to be as acceptable to have it in his favour as any person whosoever. His lady was then in London and had obtained the promise of the collector's place for the port of Boston, and would have preferred it to the government, but a strong interest being made for Mr. Frankland, since Sir Henry Frankland, there was no way of providing for both, except by giving the government to Mr. Shirley.

The news came to Boston the first week in July. Mr. Shirley was at Providence in Rhode Island government, counsel for the Massachusetts before a court of commissioners appointed to settle the line between the two governments. As most of the public documents and records of that time are burnt, we cannot give so particular an account of the proceedings of those commissioners as otherwise we should have done. It is certain that for divers years past the only part in controversy between the two governments was a small gore of land between Attleborough in the Massachusetts and the old township of Providence. A great part of the Massachusetts assembly wished it might be ceded to Rhode Island, but a few tenacious men, who do not always regard consequences, influenced a majority against it. Besides a settlement made by commissioners in 1664 or 1665, another settlement had been made or the old one confirmed in 1708, but Rhode Island, encouraged by the ill success of the Massachusetts in the controversy with New Hampshire, applied to his Majesty to appoint commissioners to settle the line between the two governments. The consent or submission of the Massachusetts to such appointment was not thought necessary, and, if they would not appear, the commissioners were to proceed *ex parte*. The Massachusetts assembly thought proper to appear by their committee, having no apprehensions the controversy would turn, in the judgment of the commissioners, upon a point never before relied upon, viz. that the colony of New Plymouth having no charter from the crown, Rhode Island charter must be the sole rule of determining the boundary, although the patent from the council of Plymouth to Bradford and associates was prior to it.

(*Anno* 1741.)—The colony of New Plymouth was a government *de facto*, and considered by King Charles as such in his letters and orders to them before and after the grant of Rhode Island charter, and when the incorporation was made of New Plymouth with Massachusetts, &c. the natural and legal construction of the province charter seems to be, that it should have relation to the time when the several governments incorporated respectively, in fact, became governments. A gentleman of the council of New York had great influence at the board of commissioners. The argument which had been made use of in former controversies, that Massachusetts was too extensive, and the other governments they were contending with, of which New York was one, were too contracted, was now revived. To the surprise of Massachusetts, a line was determined, which not only took from them the gore formerly in dispute, but the towns of Bristol, Tiverton, and Little Compton, and great part of Swansey and Barrington. An appeal was claimed, and allowed to his Majesty in council, where, after lying four or five years, the decree of the court of commissioners was confirmed. In the prosecution and defence of this title it has been said, that some material evidence was never produced which would have supported the Massachusetts claim.

Mr. Shirley found the affairs of the province in a perplexed state. The treasury was shut, and could not be opened without some deviation from the royal instructions, the bills of credit were reduced, and nothing substituted as a currency in their stead, the land bank party carried every point in the house, there seemed to be a necessity of securing them, the great art was to bring them over to his measures, and yet not give in to their measures so as to lose his interest with the rest of the province and with the ministry in England. Some of the principal of them, who knew their own]

[importance, were willing to have some assurance of favour from him, at the same time they engaged to do every thing to serve him. The first step, on their part, was the advancement of the governor's salary to the full value of 1000*l*. sterling per annum. This had been most unjustifiably evaded all the latter part of Mr. Belcher's administration, by granting a sum in bills of credit without a due regard to their depreciation. Mr. Kilby, who had been very active for Mr. Shirley's interest and against Mr. Belcher, in England, was chosen agent for the province in England, and Mr. Wilks, who had been agent the whole of the last administration, was laid aside. Mr. Auchmuty, who had been one of the land bank directors, was joined with Mr. Kilby in the affair of the Rhode Island line. A grant of about 200*l*. sterling was made to John Sharpe, Esq. for his account of charge in defending Mr. Belcher against New Hampshire's complaint to the king in council. This had been repeatedly refused in Mr. Belcher's time, which gave great offence to Mr. Sharpe. It was thought extraordinary that Mr. Shirley should make it a point with the land bankers that this debt for his predecessor should be paid; but to take Mr. Sharpe off from Mr. Belcher and engage him for Mr. Shirley, the friends and solicitors for the latter in England had engaged that if he was appointed governor Mr. Sharpe's account should be paid.

But the grand affair to settle was that of the bills of credit. The instruction was express, not to consent to any act which should continue the bills beyond the time fixed for their being brought in. If this was complied with, a tax must have been made for the whole sum extant in that year, 1741. This it was said would be a burden that the people would never bear. Mr. Shirley was sensible that the intent of his instruction was the prevention of a depreciating currency. No matter how large a sum in bills was current if their value could be secured. If the spirit of the instruction could be preserved, an exact conformity to the letter would not be required. Every scheme for fixing the value of the bills had failed. A new project was reported by a committee of the house and accepted, and afterwards concurred by the council and consented to by the governor. This was a scheme to establish an ideal measure in all trade and dealings, let the instrument be what it would. The act which passed the court declared, that all contracts should be understood payable in silver at 6*s*. 8*d*. the ounce, or gold in proportion. Bills of a new form were issued, 20*s*. of which expressed in the face of the bill three ounces of silver, and they were to be received accordingly in all public and private payments, with this saving, that if they should depreciate in their value, an addition should be made to all debts, as much as the depreciation from the time of contract to the time of payment. How to ascertain the depreciation from time to time was the great difficulty in framing the act. To leave it to a common Jury would never do. There was some doubt whether a house of representatives would be wholly unbiassed. At length it was agreed that the eldest council in each county should meet once a year and ascertain the depreciation. This is said to have been the scheme of Colonel Stoddard of Northampton, a gentleman of good sense and great virtue, who probably saw the defects, but hoped to substitute a lesser evil in the place of a greater.

This at best must have been a very partial cure. It did not prevent the loss from the depreciation of the bills in those persons hands through which they were continually passing. All debts which were contracted and paid between the periods when the value of the bills were fixed annually, could not be affected by such fixing, and unless in debts of long standing, which the debtor could not pay without an action at law, demand was not ordinarily made for depreciation, and what rendered it of little effect in all other cases, the counsellors appointed to estimate the depreciation never had firmness enough in any instance to make the full allowance, but when silver and exchange had rose 20 per cent. or more, an addition was made of four or five only. The popular cry was against it, and one year, when Nathaniel Hubbard, Esq. the eldest counsellor for the county of Bristol, a gentleman of amiable character and who filled the several posts he sustained with applause, endeavoured to approach nearer to a just allowance than had been made in former years, he felt the resentment of the house, who left him out of the council the next election. In short, the act neither prevented the depreciation of the bills nor afforded relief in case of it, and was of no other service than to serve as a warning, when an act passed for the establishing a fixed currency a few years after, to leave nothing to be done by any person or bodies of men, or even future legislatures, to give the act its designed effect, but in the act itself to make full provision for its execution in every part.

Even this act which, with its fair appearance, justified Mr. Shirley in departing from his instruction, and afforded a supply of the treasury for the payment of debts and future support of government, could not have been obtained if he had] [not prevailed with the land bank party, contrary to the inclinations of many of them, to join in promoting it.

He made them return, by consenting to any new elections that were made of any of them into the council, by restoring now and then one and another to the posts they had been deprived of, which, though it was done by degrees, caused many who condemned the land bank and all who were concerned in it, to be very free in their censures upon it.

But the great favour they expected was relief from the severity of the act of parliament. This was to be touched with great tenderness and delicacy. Every person concerned was liable to the demands of the possessors of the bills. If large demands should be made upon any particular persons, it seemed but just that the rest should contribute their proportion ; but no demand was given by the act to one partner against another in such case. A bill was therefore prepared with a professed design to carry the act of parliament equitably into execution. Three commissioners were appointed by the bill, with power to tax all who had been concerned in the scheme in proportion to their interest in it, and with the moneys thus raised to redeem the company's bills from the possessors, and after the redemption of the bills to make an equitable adjustment between the members and the company. Great care was taken to avoid all opposition to the act of parliament; Mr. Shirley however did not think proper to sign the bill until he had sent a copy of it to England, and received directions concerning it. After it had passed both houses, to oblige the principal land bankers, he continued the session of the court by long repeated adjournments many months, and before the expiration of the year gave his consent to the bill. Having thus secured a considerable party in the government without losing those who had been in opposition to them, he rendered his administration easy, and generally obtained from the assembly such matters as he recommended to them.

From the Spanish war in 1740, a French war was expected every year to follow. Castle William, the key of the province, was not only effectually repaired, but a new battery of 20 forty-two pounders, which takes the name of Shirley battery, was added to the works, with a larger magazine than any before, and a large supply of powder, all at the expence of the province. The cannon, mortars, shot, and other stores, were the bounty of the crown. The forts upon the frontiers were also put into good order, and upon a representation from Mr. Mascarene, commander-in-chief at Annapolis in Nova Scotia, of the defenceless state of that province and the danger they were in from the enemy, Mr. Shirley, in 1744, prevailed upon the Massachusetts assembly to vote, pay, &c. for 200 men which were sent there, and who were the probable means of saving that country from falling into the enemy's hands.

(*Anno* 1744.)—But the great event in this administration was the siege and reduction of Louisburgh. Canso had been surprised and taken by 900 men under Duvivier from Louisburgh, before the war with France was known at Boston. With another party, Duvivier made an attempt the same summer upon Annapolis, but was disappointed. Many good vessels had been taken by the French men of war and privateers, and carried into Louisburgh. The fishermen had no intention to go upon their voyages the next summer, and every branch of trade, it was supposed, must be carried on by vessels under convoy. It was the general voice, in the fall of the year, that Louisburgh must be taken, but nobody supposed that the united force of the colonies could take it; application must be made to his Majesty for sea and land forces sufficient for the purpose. As winter approached, it began to be suggested that it was not improbable the place might be surprised or taken by a *coup de main*, the inhabitants and garrison being shut up within the walls. Some of the garrison at Canso, who had been prisoners and who professed to be well acquainted with the fortifications and garrison at Louisburgh, favoured this opinion, and declared that in winter the snow often lay in drifts or banks against a particular part of the wall, where there were no embrasures nor any cannon mounted; that the crust would bear a man's weight, and in that part at least, the walls might be sealed, and perhaps by the help of ladders it would not be difficult in other parts; that the grand battery, intended for defence in case of an attack by sea, would not be capable of long resisting if attacked by land. Mr. Vaughan, who had been a trader at Louisburgh, was very sanguine also that the place might be taken by surprise, and it was generally agreed that if they should be mistaken, yet it would not be possible for the enemy, who were scant of provisions, to stand a siege until the time the supplies usually arrive to them from France; and to prevent any chance vessels from entering, a sufficient naval force might be provided to cruise before the harbour.

Whilst this was the conversation abroad, Mr. Shirley was diligently inquiring of those persons who had been traders, and of others who had] [been prisoners there, into the condition of the place, the usual time for the arrival of supplies from Europe, the practicability of cruising off the harbour, &c. He had before wrote to the ministry and represented the necessity of a naval force early in the spring for the preservation of Annapolis. If this should arrive, he might be able to prevail with the commander to cover our forces with it. Commodore Warreu was with several ships at the Leeward islands. It was possible, when he was acquainted with the expedition, he would come with or send part of his force to strengthen it. These were the only chances for a naval strength sufficient to cope with a single capital French ship that might be bound to Louisburgh in the spring. The ministry indeed would, by express, be immediately acquainted with the expedition, if engaged in, but Europe was at too great distance to expect timely aid from thence. The plan of the expedition was, a land force of 4000 men in small transports to proceed to Canso, and the first favourable opportunity to land at Chapeaurouge bay, with cannon, mortars, ammunition, and warlike stores, and all other necessaries for carrying on a siege; and to prevent a supply of provision and stores to the enemy, several vessels were to cruise off the harbour of Louisburgh, as soon as the season of the year would permit. An estimate was made of all the naval force which could be procured in this and the neighbouring colonies, the largest vessel not exceeding 20 guns. With this land and sea force, it was said there was good chance for success, and if the men of war should arrive, which there was good reason to hope for, there was all imaginable grounds to depend upon the reduction of the place.

(*Anno* 1745.)—The general court being sitting the beginning of January, the governor sent a message to the two houses to let them know he had something to communicate to them of very great importance, but of such a nature that the publishing it, before they should come to any resolution upon it, might wholly defeat the design; he therefore desired they would lay themselves under an oath of secrecy for such time as each house should think proper. This they did, although it was the first instance in the house of representatives, without any scruple, and then be communicated to them his proposed plan of the expedition. Many of the members, who had heard little or nothing of the conversation upon the subject, were struck with amazement at the proposal. The undertaking was thought to be vastly too great, if there was a rational prospect of success. However, in deference to the recommendation of the governor, a committee of the two houses was appointed to consider the proposal. Here the proposal was for several days deliberated and weighed. Louisburgh, if left in the hands of the French, would infallibly prove the Dunkirk of New England; their trade had always been inconsiderable, their fishery was upon the decline, and for several years past they had bought fish of the English at Canso cheaper than they could catch and cure it themselves, both trade and fishery they might well lay aside, and by privateering enrich themselves with the spoils of New England; and to all these dangers was added that of losing Nova Scotia, which would cause an increase of 6 or 8000 enemies in an instant. The garrison of Louisburgh was disaffected, provisions were scant, the works mouldering and decayed, the governor an old man unskilled in the art of war; this therefore was the only time for success, another year the place would be impregnable. There was nothing to fear from the forces at Louisburgh, before additional strength could arrive from France they would be forced to surrender. There were, it must be owned, no ships of strength sufficient to match the French men of war, unless perhaps a single ship should fall in by herself, and in that case five or six of the state might be a match for her; but there was no probability of men of war so early, and it was very probable English men of war from Europe or the W. Indies would arrive before them. There was always uncertainty in war, a risk must be run, if the state failed it might be able to grapple with the disappointment, although it should bear the whole expence; but if it succeeded, not only the coasts of New England would be free from molestation, but so glorious an acquisition would be of the greatest importance to Great Britain, and might give peace to Europe, and it might depend upon a reimbursement of the whole charge it had been at.

On the other hand it was replied, that the state had better suffer in its trade, than by so expensive a measure deprive itself of all means of carrying on any future trade; that it was capable of annoying them in their fishery, as much as they could annoy it in its own, and in a short time, both sides would be willing to leave the fishery unmolested; that the accounts given of the works and the garrison at Louisburgh could not be depended upon, and it was not credible that any part of the walls should be unguarded and exposed to surprise; that instances of disaffection rising to mutiny were rare, and but few instances were to be met with in history where such expectation has not failed. The garrison at Louisburgh consisted of regular experienced troops, who, though unequal] [in number, would be more than a match in open field for all the raw unexperienced militia which could be sent from New England; that 20 cruizers at that season of the year would not prevent supplies going into the harbour, it being impossible to keep any station for any length of time, and the weather being frequently so thick, that a vessel was not to be discovered at a quarter of a mile's distance; that there was no room to expect any men of war for the cover of the troops; that if only one 60-gun ship should arrive from France, or the French islands, she would be more than a match for all the armed vessels that could be provided, the transports at Chapeaurouge bay would be every one destroyed, and the army upon Cape Breton obliged to submit to the mercy of the French; that the state would be condemned in England for engaging in such an affair without their direction or approbation, and that it would be nowhere pitied, its misfortunes proceeding from its own rash and wild measures. To these arguments were added the uncertainty of raising a sufficient number of men, or of being able to procure provisions, warlike stores, and transports, discouragement from the season of the year, when frequently, for many days together, no business could be done out of doors. Money indeed could be furnished, or bills of credit in lieu of it, but the infallible consequence would be the sinking the value of the whole currency, to what degree no man could determine, but probably in proportion to the sum issued; and finally, if the plan should succeed, a general national benefit would be the consequence, in which the state would be but small sharers, and far short of the vast expence of treasure and perhaps of lives in obtaining it, and if it failed, such a shock would be given to the province that half a century would not recover it to its former state. After mature deliberation, a majority of the committee disapproved the proposal, and their report was accepted, and for a few days all thoughts of the expedition with the members of the court were laid aside. In the mean time, the governor, who wished his proposal had been agreed to, but did not think it proper to press it any further by message or by privately urging the members, either directed or encouraged the carrying about a petition which was signed by many of the merchants in the town of Boston, but principally by those of Salem and Marblehead, directed to the house of representatives, or to the two houses, praying for reasons set forth, among others the saving the fishery from ruin, they would re-consider their vote and agree to the governor's proposal of an expedition against Louisburgh. A second committee, appointed upon this petition, reported in favour of it, and the 26th of January, their report came before the house, who spent the day in debating it, and at night a vote was carried in favour of it by a majority of one voice only. Never was any affair deliberated upon with greater calmness and moderation, the governor indeed laid the affair before the court, but left the members free to use their judgment without any solicitation, and there appeared no other division than what was caused by a real difference in opinion consulting the true interest of the province.

The point once settled, there was immediately a union of both parties in the necessary measures for carrying the design into execution, those who had opposed it before being employed upon committees, and exerting themselves with zeal equal to that of the principal promoters. An embargo was laid upon every harbour in the province, and messengers were immediately dispatched to the several governments, as far as Pennsylvania, to entreat an embargo on their ports, and that they would join in the expedition. All excused themselves from any share in the adventure, except Connecticut, who agreed to raise 500 men, New Hampshire 300, and Rhode Island 300. Connecticut and Rhode Island also consented their colony sloops should be employed as cruisers. A small privateer ship, about 200 tons, and a snow of less burden, belonging to Newport, were hired there by the Massachusetts; a new snow, Captain Rouse, a ship, Captain Snelling, were taken into the service at Boston, which, with a snow, Captain Smethurst, and a brig, Captain Fletcher, three sloops, Captains Sanders, Donahew, and Bosch, and a ship of 20 guns, purchased on the stocks, Captain Tyng the commodore, made the whole naval force.

From the day the vote passed until the place was reduced, a series of favourable incidents contributed to the general success. They will be obvious enough in the course of the narrative, and will not require being specially remarked. The time for preparing was short. The winter proved so favourable that all sorts of out-door business was carried on as well and with as great dispatch as at any other season of the year. In the appointment of a general officer one qualification was considered as essential, that he should be acceptable to the body of the people, the enlistment depended upon this circumstance. It was not easy to find a person, thus qualified, willing to accept the trust. Colonel Pepperell, having the offer from the governor, was rather pressed into the service than voluntarily engaged. Besides a very great landed interest, he was largely concerned in mercantile affairs, which must necessarily suffer by his absence;]

[and this being generally known had no small influence, from the example, with inferior officers and even private soldiers, to quit their lesser affairs for a season, for the service of their country. Many of the private soldiers were freeholders, and many more sons of wealthy farmers, who could have no other views in consenting to the enlistment of their children than the public interest.

Mr. Shirley had set his heart so much upon the expedition, that many points were conceded by him which he would not have given up at any other time, and the people of the province submitted to compulsory measures from the government, which at another time would have been grievous and not very patiently borne. Such officers were nominated by the governor as the people proposed or called for, because they were most likely to enlist men. Instead of a commissary-general, an officer appointed by the governor, a committee of war was chosen by the two houses out of their own members. Nothing further was heard of the royal instruction against bills of credit. Such sums as the service called for, and to be redeemed at such periods as the house thought proper, were consented to by the governor. It soon appeared that these sums would vastly exceed what had been computed, and many declared that had a right estimate been made they should never have voted for the expedition, but it was now too late to go back. It was found also, that transports and vessels of war could not be engaged unless the government would become insurers, which, although it occasioned no additional expence at first, yet, in case of ill success, would greatly increase the public debt and distress. The committee of war were likewise convinced that a sufficiency of provisions, clothing, and warlike stores, could not be procured within the province. Whosoever was possessed of any of these articles, by an act or order of government, his property was subjected to the committee, who set such price as they judged equitable, and upon refusal to deliver, entered warehouses, cellars, &c. by a warrant for that purpose to the sheriff, and took possession. In the course of the preparation, many vessels unexpectedly arrived with more or less of each of these articles, and after all, the army was poorly enough provided. Ten cannon, 18 pounders, were obtained upon loan, not without difficulty, from New York, otherwise Mr. Shirley himself seemed to doubt if the plan could proceed. Some dependence was placed upon cannon from the grand battery, but this was too manifest a disposal of the skin before the bear was caught. By force of a general exertion in all orders of men, the armament was ready, and the general, on board the Shirley snow, Captain Rouse, with the transports under her convoy, sailed from Nantasket the 24th of March, and arrived at Canso the 4th of April. The Massachusetts land forces consisted of 3250 men, exclusive of commission officers. The New Hampshire forces, 304, including officers, arrived four days before. Those of Connecticut, being 516, inclusive, did not arrive until the 25th. The deputy governor of the colony, Roger Walcot, Esq. had the command, and was the second officer in the army. Rhode Island waited until a better judgment could be made of the event, their 300 not arriving until after the place had surrendered. The 23d of March, an express boat, sent to Commodore Warren, in the W. Indies, returned to Boston.

As this was a provincial expedition, without orders from England, and as his small squadron had been weakened by the loss of the Weymouth, Mr. Warren excused himself from any concern in the affair. This answer must necessarily strike a damp into the governor as well as the general and Brigadier Waldo, then next in command, who were the only persons in the army made privy to it before the fleet sailed. Several of the cruising vessels had sailed the middle of March, but they could be no protection to the army against two capital ships; if they intercepted small vessels, it was the most that was expected. A blockhouse, with eight cannon, was built at Canso. Whether some good reason would not have been given for proceeding no further than Canso, if there had been a disappointment in the expected junction of men of war from the several quarters to which notice of the expedition had been sent, may well enough be made a question. Mr. Shirley hoped, if the reduction of Louisburgh was not effected, at least Canso would be regained, Nova Scotia preserved, the French fishery broke up, and the New England and Newfoundland fisheries restored. But on the 23d of April, to the great joy of the army, arrived at Canso, the Eltham, of 40 guns, from New England, by order from Mr. Warreu; and on the 23d the commodore himself, in the Superb, of 60 guns, with the Launceston and Mermaid of 40 each, arrived also. This gave great spirits to all who had the success of the expedition at heart, for although this was not a naval force to enter the harbour or annoy the forts, yet it was a cover to the army, and equal to any expected force from France. It seems that in two or three days after the express sailed from the W. Indies for Boston, the Hind sloop brought orders to Mr. Warren to repair to Boston with what ships could be spared, and to concert measures with Mr. Shirley for his]

[Majesty's general service in N. America. Upon the passage to Boston the commodore received intelligence that the fleet had sailed for Canso, and meeting with a schooner at sea he sent her to Boston to acquaint Mr. Shirley that he would proceed to Canso, and at the same time sent orders to any ships which might be in these seas to join him. The Eltham was actually under sail with the mast fleet when an express sent from Boston with the commodore's orders arrived at Portsmouth in New Hampshire, but being followed and overtaken by a boat, the captain ordered his convoy into port again, and sailed for Canso. After a short consultation with the general the men of war sailed to cruise before Louisburgh. The cruizers, before this, had intercepted several small vessels bound in there with W. India goods and provisions, and had engaged the Renommee, a French ship of 36 guns, sent from France with dispatches, and who kept a long running fight with the Massachusetts vessels, being able with ease to outsail them, and, after two or three attempts to enter the harbour, went back to France to give an account of what she had met with. She fell in with the Connecticut troops, under convoy of their own and the Rhode Island colony sloops, both which she had strength enough to have carried, but, after some damage to the Rhode Island sloop, she went her way. The forces landed at Chapeaurouge bay the 30th of April. The transports were discovered early in the morning from the town, which was the first knowledge of any design against them. The cruisers had been seen every fair day before the harbour, but these were supposed to be privateers in search after their trading and fishing vessels. The night before, it is said, there was a grand ball at the fort, and the company had scarce fallen asleep when they were called up by an alarm. Bouladric, a French officer, was sent with 150 men to oppose the landing, but the general making a feint of landing at one place, drew the detachment there, and this opportunity was taken for landing 100 men at another place without opposition, although they were soon after attacked by the detachment, six of which were killed on the spot, and about as many more, with Bouladric their leader, were taken prisoners: the rest fled to the town, or they would soon have fallen into the hands of the Massachusetts, who were landing fast one upon the back of another.

The next morning after they landed, 400 men marched round to the *n. e.* harbour, behind the hills, setting fire to all the houses and store-houses, until they came within a mile of the grand battery. Some of the store-houses having in them pitch, tar, and other combustible stuff, caused such a thick smoke, that the garrison were unable to discover an enemy, though but a few rods distant, and, expecting the body of the army upon them, they deserted the fort, having thrown their powder into a well, but leaving the cannon and shot for the service of the English. A small party of less than 20 English first came up to the battery, and discovering no signs of men, suspected a plot, and were afraid to enter; at length, it is said, a Cape Cod Indian went in alone and discovered the state of it to the rest of the party, just as some of the French were relanding in order to regain the possession of it.

The army found they had near two miles to transport their cannon, mortars, shot, &c. through a morass. This must be done by mere dint of labour. Such of the men as had been used to drawing piue-trees for masts, and those who had the hardiest and strongest bodies, were employed in this service. Horses and oxen would have been buried in mud and were of no use. Brigadier Waldo had the command of the grand battery. The French kept firing upon the battery from the town as well as from the island battery, but to little purpose, the town being near 2000 yards distant, and the island about 1600. A constant fire was kept from the grand battery upon the town with the 42 pounders. This greatly damaged the houses, but caused so great an expence of powder that it was thought advisable to stop and reserve it for the fascine batteries. Five of these were erected, the last the 20th of May, called Tidcomb's battery, with five 42 pounders, which did as great execution as any. The Massachusetts knew nothing of regular approaches, they took the advantage of the night, and when they heard Mr. Bastide's proposals for zigzags and epaulements, they made merry with the terms, and went on, void of art, in their own natural way. Captain Pierce, a brave officer, standing at one of these batteries, had his bowels shot away by a cannon ball, and lived just long enough to say, " it is hard to die."

Whilst the people of Massachusetts were thus busy ashore, the men of war and other vessels were cruising off the harbour whenever the weather would permit, and the 18th of May the Vigilant, a French man of war of 64 guns, having 560 men on board, and stores of all sorts for the garrison, was met with by the Mermaid, whom she attacked, but Captain Douglass, the commander, being of unequal force, suffered himself to be chased by her until he drew her under the command of the commodore and the other ships cruising with him, to whom, or, as some say, to the Mermaid, she struck, because she had first met with her. This capture gave great joy to the] [army, not so much for the addition made to the naval force, as for the disappointment to the enemy. A proposal had been made a few days before, that the men of war should anchor in Chapeaurouge bay, and that the marines, and as many sailors as could be spared, should land and join the army. The Vigilant would then have got in, and the siege would then have been given over. Affairs were now in such a state that the anxiety at Boston was much lessened. It was hoped the army might retreat with safety whenever it should be determined to give over the siege, for Bouladrie, who belonged to the town of Louisburgh, and the Marquis de la Maison Forte, commander of the Vigilant, who was well acquainted with the state of the place, when they came to Boston, were sanguine that it would hold out longer than the Massachusetts; and soon after we find the news of a fruitless, and perhaps a rash attempt upon the island battery by 400 men, 60 of whom were killed, and 116 taken prisoners. The Cæsar, Snelling, one of the ships in the provincial service, arrived at Boston with letters from the general, and an application for more men and a further supply of powder. The Massachusetts agreed, and actually did raise 400 men, and sent all the powder that could be purchased, and Connecticut raised 200 men, but there were neither men nor powder arrived when the siege was finished.

The Princess Mary, of 60, and the Hector, of 40 guns, unexpectedly had arrived at Boston from England, and were immediately sent to join the commodore, pursuant to his general orders, and arrived before Louisburgh the 22d of May. This increase of naval force occasioned conjectures, some being of opinion, that rather than the siege should be raised the ships would attempt to go in, but it was generally supposed the hazard would be too great. It was commonly reported that Colonel More, of the New Hampshire regiment, offered to go on board the Vigilant with his whole regiment and to lead the van, if, in case of success, he might be confirmed in the command of the ship. He had been an experienced sea captain, and had a very good character. It is certain, an attempt with the ships was not then thought advisable. A new battery, about this time, was erected upon the light-house point, which being well attended by Lieutenant-colonel Gridley, of the artillery, did great execution upon the island battery, silenced many of the guns, and it was expected it would not be long tenable. Soon after, viz. June 10th, arrived before Louisburgh the Chester, a 50-gun ship, in consequence of the dispatches from Mr. Shirley with an account of the expedition. The Canterbury and Sunderland, two 60-gun ships, sailed with her, and arrived the 12th. Here was now a fleet of 11 ships, and it is said to have been determined the ships should make an attack by sea the 18th, while the army did the same by land. It is not certain that when the day should come some sufficient reason would not have been found for a further delay. Those who give the most favourable accounts of the siege say, "the *w.* gate was entirely beat down, the wall adjoining very much battered, and a breach made ten feet from the bottom, the circular battery of 16 cannon, and the principal one against ships almost ruined, the *n. e.* battery of 17 cannon damaged, and the men drove from the guns, and the *w.* flank of the king's bastion almost demolished." Others say, "the *w.* gate was defaced, and the adjoining curtain, with the flank of the king's bastion, were much hurt, but no practicable breach." Whether a general storm was really intended upon the 18th or not, it seems the French expected it from the preparations on board the men of war, and did not incline to stand it; and on the 15th sent a flag of truce to the general, desiring a cessation that they might consider of articles to be proposed for a capitulation. Time was allowed for this purpose until the next morning, when such articles were offered as were rejected by the general and commodore, and others offered to the enemy in their stead, which they accepted of, and hostages were exchanged, and the next day, the 17th, the city was delivered up.

Many of the Massachusetts had taken colds and many fallen into dysenteries, so that 1500 were taken off from their duty at one time, but the weather proving remarkably fine during the 49 days siege they generally recovered. The day after the surrender the rains began, and continued ten days incessantly, which must have been fatal to many, they having nothing better than the wet ground to lodge on; and their tents, in general, being insufficient to secure them against a single shower, but in the city they found barracks to shelter them. Captain Bennet, in a schooner, was sent immediately to Boston, and arrived with the great news the 3d of July, about one in the morning. The bells of the town were ringing by break of day, and the day and night following were spent in rejoicing. The news flew through the continent. The colonies which declined any share in the expence and hazard were sensible they were greatly interested in the success. It was allowed every where, that if there had been no signal proof of bravery and courage in time of action, there having been only one sally from the town and a few skirmishes with the]

[French and Indians from the woods, in all which the men behaved well; yet here was the strongest evidence of a generous noble public spirit, which first induced to the undertaking, and of steadiness and firmness of mind in the prosecution of it, the labour, fatigue, and other hardships of the siege being without parallel in all preceding American affairs. A shade was thrown over the imprudence at first charged upon the New Englanders. Considerate persons among themselves could not, however, avoid gratefully admiring the favour of Divine Providence in so great a number of remarkable incidents which contributed to this success. The best use to be made by posterity seems to be, not to depend upon special interpositions of Providence, because their ancestors have experienced them, but to avoid the like imminent dangers, and to weigh the probability and improbability of succeeding in the ordinary course of events.

The commodore was willing to carry away a full share of the glory of this action. It was made a question whether the keys of the town should be delivered to him or to the general, and whether the sea or land forces should first enter. The officers of the army say they prevailed. The marines took possession of one or more of the batteries, and sometimes the commodore took the keys of the city gates. The command, however, until orders should arrive from England, was to be joint, and a dispute about precedence to be avoided as much as could be. The commodore dispatched Mr. Montague, in the Mermaid, to England, with intelligence, and the general, the day after, sent the Shirley galley, Captain Rouse. The Mermaid arrived first.

It was very happy that disputes arose to no height between the sea and land forces during the siege. This has often proved fatal. This expedition, having been begun and carried on under a commission from a provincial governor, seems to be distinguished from ordinary cases, and to leave less room for dispute. Whether the land or sea force had the greatest share in the acquisition may be judged from the relation of facts. Neither would have succeeded alone. The army, with infinite labour and fatigue to themselves, harassed and distressed the enemy, and, with perseverance a few weeks or days longer, must have compelled a surrender. It is very doubtful whether the ships could have lain long enough before the walls to have carried the place by storm, or whether, notwithstanding the appearance of a design to do it, they would have thought it advisable to attempt it; it is certain they prevented the arrival of the Vigilant, thus taking away all hopes of further supply and succour, and it is very probable the fears of a storm might accelerate the capitulation. The loss by the enemy and sickness did not exceed 101 men. The loss of the snow, Prince of Orange, belonging to the province, and supposed to be overset, was a heavy blow upon the town of Marblehead, the captain and most of the crew belonging to that town; and it is a rare thing for a Marblehead man to die without leaving a widow and a number of children surviving.

As it was a time of year to expect French vessels from all parts to Louisburgh, the French flag was kept flying to decoy them in. Two E. India and one S. sea ship, supposed to be all together of the value of 600,000*l.* sterling, were taken by the squadron at the mouth of the harbour, into which they would undoubtedly have entered. The army, at first, supposed they had acquired a right to the island of Cape Breton and its dependencies, and, until they were undeceived by Mr. Shirley, were for dividing the territory among the officers and men. With greater colour they might have claimed a share with the men of war in these rich prizes. Some of the officers expected a claim would have been laid in, but means were found to divert it, nor was any part decreed to the vessels of war in the province service, except a small sum to the brig Boston Packet, Captain Fletcher, who being chased by the S. sea ship, led her directly under the command of the guns of one of the men of war. It seemed to be conceded, that as this acquisition was made under the commission of the governor of Massachusetts bay, the exercise of government there appertained to him until his Majesty's pleasure should be known. We know of no precedent in the colonies except that of the conquest of Nova Scotia, in 1690. It was necessary then to admit this principle: the acquisition could not otherwise have been retained. Mr. Shirley made a voyage to Louisburgh, took the government upon him, prevailed upon a great part of the army to consent to remain in garrison over the winter, or until regiments which were expected arrived, engaged that their pay should be increased and clothing provided, and settled other matters to general satisfaction.

Duvivier had been sent to France the winter of 1744, to solicit a force, not to defend Cape Breton, but to conquer Nova Scotia, and accordingly sailed the beginning of July with seven ships of war for that purpose, who were to stop at Louisburgh. This fleet took a prize bound from Boston to London, on board which was Lieutenant-]

[governor Clark, of New York, and by this means they were informed of the conquest of Louisburgh, and the strong squadron there; otherwise some or all of them would also have probably fallen into the hands of the English. Upon this intelligence they went back to France. Thus Nova Scotia no doubt was saved by the Massachusetts expedition. There would not have been men of war in these seas sufficient to match this squadron.

(*Anno* 1746.)—The reduction of Louisburgh by a British colony must have been a surprise to Great Britain and to France. It caused very grand plans of American measures for the next year with both powers. Great Britain had in view the reduction of Canada, and the extirpation of the French from the *n.* continent. France intended the recovery of Louisburgh, the conquest of Nova Scotia, and the destruction of the English sea-coast from Nova Scotia to Georgia. Upon the English plan, eight battalions of regular troops, with the provincial forces to be raised in the four New England governments, were to rendezvous at Louisburgh, and, with a squadron under Admiral Warren, were to go up the river Saint Lawrence to Quebec; other provincials from Virginia and the colonies *n.* including New York, were to rendezvous at Albany, and go across the country to Montreal; the land forces to be under General St. Clair. No province had a certain number assigned, but it was expected there should be at least 5000 in the whole. The Massachusetts forces were ready to embark by the middle of July, about six weeks after the first notice. The preparations making at Brest for America were well known in England, and a squadron was ordered to block up that harbour. Notwithstanding all the caution used, the Brest squadron slipped out, and sailed to the *w.* and it is certain no English squadron followed. Whilst all were impatiently waiting for news of the arrival of the fleet at Louisburgh, a fisherman comes in some time in August, with an account of his being brought to by four French capital ships not far from Chibucto; that he was required to pilot them there; that as he lay under the stern of one of them he read the word Le Terrible, but a fog suddenly rising he made his escape. After that some days had passed without any further account, the fisherman's news was generally discredited. It appeared some months after that these were four ships under M. Conflans, who had escaped an English squadron from Jamaica, and were bound to Chibucto in order to join the Brest fleet; but after cruising some time, and meeting with storms and fogs upon a coast they were unacquainted with, they returned to France.

The beginning of September, vessels arrived at Boston from Hull and Liverpool, with advice that the Brest fleet had sailed, and it was supposed for N. America, and from the middle to the latter end of the month frequent accounts were brought of a great fleet seen to the *w.* of Newfoundland, which was supposed might be English as likely as French; but on the 28th, an express arrived from Louisburgh with certain advice that these ships were the French fleet, which it was affirmed consisted of 70 sail, 14 of which were capital ships, and that there were 20 smaller men of war, and the rest fire-ships, bombs, tenders, and transports for 8000 troops. The same day a vessel from Jamaica arrived, with advice that the four men of war who had engaged with Commodore Mitchell were intended to join the fleet, and it was now no longer doubted that these were the ships seen by the fishermen, and it was supposed soon after got into Chibucto. England was not more alarmed with the Spanish armada in 1588, than Boston and the other N. American sea-ports were with the arrival of this fleet in their neighbourhood. The firmest mind will bend upon the first advice of imminent danger to its country. Even the great De Witt swooned when he first opened a letter giving intelligence of England's confederating with France to enslave the Dutch, though the next moment he recovered his natural courage and vivacity.

Every practicable measure for defence was immediately pursued by the authority of the Massachusetts province, but the main dependence was upon a squadron from England sufficient, in conjunction with the ships then at Louisburgh, to overcome the French. It was impossible the ministry should be ignorant of the sailing of this fleet, and unless they were willing the colonies should be exposed to the ravages of the enemy, it was impossible an English squadron should not be soon after them. This was the general voice. But this dependence failed. However, as the probability of the arrival of the squadron was from day to day lessened, the apprehensions of danger from the enemy lessened in some proportion. At length was received such authentic account of the distresses of the French, that it was generally agreed that Admiral Townsend's ships at Louisburgh were more than a match for them, and it was every day expected to be heard they had sailed for Chibucto; but if that should prove otherwise, the utmost they would be able to effect by their] [grand plan would be the conquest of Annapolis, and the whole province of Nova Scotia. If the winter did not prevent a farther progress, it was agreed their strength was not sufficient for an attempt upon Boston.

The misfortunes of this grand armament are really very remarkable. The loss of Cape Breton filled the French with a spirit of revenge against the British colonies. The Duke D'Anville, a French nobleman, in whose courage and conduct great confidence was placed, was appointed to the command of the expedition. As early as the beginning of May the fleet was ready to sail, but was detained by contrary winds until the 22d of June, when it left Rochelle, and then consisted of 11 ships of the line, 30 smaller vessels from 10 to 30 guns, and transport ships with 3130 land forces, commanded by Monsieur Pommerit, a brigadier-general. The French of Nova Scotia, it was expected, would join them, and Ramsay, a French officer, with 1700 Canadians and Indians, were actually in arms there ready for their arrival. To this force Conflans, with four ships from the W. Indies, was to be added. It was the 3d of August before the fleet had passed the W. islands. The 24th, they were 300 leagues distant from Nova Scotia, and one of their ships complained so much that they burnt her. The 1st of September, in a violent storm, the Mars, a 64-gun ship, was so damaged in her masts and so leaky that she bore away for the W. Indies, and the Alcide, of 64 guns, which had also lost her topmast, was sent to accompany her. The 15th, the Argent, of 64 guns, most of her crew being sick, put back for Brest.

The Duke D'Anville, in the Northumberland, arrived at Chibucto the 12th of September, with only one ship of the line, the Renommee, and three or four of the transports. There he found only one of the fleet, which had been in three days, and after waiting three days and finding that only three more, and those transports, had arrived, the 16th in the morning he died, the French said of an apoplexy, the English that he poisoned himself. In the afternoon the vice-admiral, D'Estournelle, with three or four more of the line, came in. Monsieur de la Jonquiere, governor of Canada, was aboard the Northumberland, and had been declared a *chef d'escadre* after the fleet left France, and by this means was next in command to the vice-admiral. In a council of war, the 18th, the vice-admiral proposed returning to France. Four of the capital ships, the Ardent, Caribou, Mars, and Alcide, and the Argonaute fire-ship, they were deprived of; there was no news of Conflans and his ships, so that only seven ships of importance remained, more or less of the land forces were on board each of the missing ships, and what remained were in a very sickly condition. This motion was opposed for seven or eight hours by Jonquiere and others of the council, who supposed that at least they were in a condition to recover Annapolis and Nova Scotia, after which they might either winter securely at Casco bay, or at worst then return to France: The sick men, by the constant supply of fresh provisions from the Acadians, were daily recovering, and would be soon fit for service. The motion not prevailing, the vice-admiral's spirits were agitated to such a degree as to throw him into a fever attended with a delirium, in which he imagined himself among the English, and ran himself through the body. Jonquiere succeeded, who was a man experienced in war, and, although above 60, still more active than either of his predecessors, and the expectations of the fleet and army were much raised. From this time Annapolis seems to have been their chief object. An account, supposed to be authentic, having been received at Boston of the sailing of Admiral Lestock, Mr. Shirley sent an express to Louisburgh to carry the intelligence. The packet-boat was taken and carried into Chibucto, which accelerated the sailing of the fleet. Most of the sick had died at Chibucto, and but about one half their number remained alive. They sailed the 13th of October, and the 15th, being near cape Sables, they met with a violent cold storm, which, after some intermission, increased the 16th and 17th and separated the fleet, two of which only, a 50 and a 36 gun ship, were discovered from the fort at Annapolis, where the Chester man of war, Captain Spry, then lay with the Shirley frigate and a small vessel in the service of the board of ordnance, who being discovered by the French to be under sail they made off, and this was the last of the expedition. The news of the beginning of the misfortunes of the French having reached France by some of the returned vessels, two men of war were sent immediately, with orders, at all events, to take Annapolis, but the fleet had sailed three or four days before they arrived.

Pious men saw the immediate hand of Divine Providence in the protection or rather rescue of the British colonies this year, as they had done in the miraculous success of the Cape Breton expedition the former year.

When the summer had so far passed as to render it too late to prosecute the expedition against]

[Canada, if the fleet had arrived, Mr. Shirley's enterprising genius led him to project an attempt upon the French fort at Crown point, with part of the Massachusetts forces in conjunction with those of the other colonies, but the alarm of the French fleet prevented until it was judged, by some concerned, to be too late. Fifteen hundred of the Massachusetts men were intended for Nova Scotia, upon the news of Ramsay's appearing there, and 400 actually went there, convoyed by the Chester, and, late in the fall, an additional number were sent thither. Those posted at Minas were surprised, the 1st day of January, by a body of French and Indians commanded by Le Corne, a French officer, and after having 160 of their number killed, wounded, and taken prisoners, the rest capitulated, engaging not to bear arms against the French in Nova Scotia for the term of one year. De Ramsay with his troops soon after returned to Canada.

The troops raised for the Canada expedition continued in pay until September the next year, 1747. Some of them served for defence of the frontiers, the rest were inactive. The inactive prosecution of the war in Europe on both sides indicated peace to be near, which the next year was effected.

War had been declared in 1744 against the Cape Sable and St. John's Indians, and in 1745 against the Penobscots and Norridgewocks. The frontiers did not escape molestation. They suffered less than in any former wars. The Indians were lessened in number, and having withdrawn to the French frontiers were sometimes detained for their defence upon an apprehended invasion, and at other times engaged to be in readiness to join in the great designs against the English.

(*Anno* 1747.)—In 1747 (November 17th) happened a tumult in the town of Boston, equal to any which had preceded it, although far short of some that have happened since. Mr. Knowles was commodore of a number of men of war then in the harbour of Nantasket. Some of the sailors had deserted. These deserters generally fled to some of the neighbouring ports, where they were out of danger of discovery. The commodore thought it reasonable that Boston should supply him with as many men as he had lost, and sent his boats up to town early in the morning, and surprised not only as many seamen as could be found on board any of the ships, outward bound as well as others, but swept the wharfs also, taking some ship-carpenters apprentices and labouring land-men. However tolerable such a surprise might have been in London, it could not be borne here. The people had not been used to it, and men of all orders resented it, but the lower class were beyond measure enraged, and soon assembled with sticks, clubs, pitch-mops, &c. They first seized an innocent lieutenant who happened to be ashore upon other business. They had then formed no scheme, and the speaker of the house passing by and assuring them that he knew that the lieutenant had no hand in the press, they suffered him to be led off to a place of safety. The mob increasing, and having received intelligence that several of the commanders were at the governor's house, it was agreed to go and demand satisfaction. The house was soon surrounded, and the court or yard before the house filled, but many persons of discretion intruded themselves and prevailed so far as to prevent the mob from entering. Several of the officers had planted themselves at the head of the stair-way with loaded carbines, and seemed determined to preserve their liberty or lose their lives. A deputy sheriff attempting to exercise his authority, was seized by the mob and carried away in triumph, and set in the stocks, which afforded them diversion and tended to abate their rage, and disposed them to separate and go to dinner.

As soon as it was dusk, several thousand people assembled in King-street, below the town-house, where the general court was sitting. Stones and brickbats were thrown through the glass into the council chamber. The governor, however, with several gentlemen of the council and house, ventured into the balcony, and after silence was obtained, the governor, in a well-judged speech, expressed his great disapprobation of the impress, and promised his utmost endeavours to obtain the discharge of every one of the inhabitants, and at the same time gently reproved the irregular proceedings both of the forenoon and evening. Other gentlemen also attempted to persuade the people to disperse, and wait to see what steps the general court would take. All was to no purpose. The seizure and restraint of the commanders and other officers who were in town was insisted upon as the only effectual method to procure the release of the inhabitants aboard the ships.

It was thought advisable for the governor to withdraw to his house, many of the officers of the militia and other gentlemen attending him. A report was raised that a barge from one of the ships was come to a wharf in the town. The mob flew to seize it, but by mistake took a boat belonging to a Scotch ship, and dragged it with as much seeming ease through the streets as if it had been in the water, to the governor's house, and prepared] [to burn it before the house, but from a consideration of the danger of setting the town on fire, were diverted, and the boat was burnt in a place of less hazard. The nex day the governor ordered that the military officers of Boston should cause their companies to be mustered and to appear in arms, and that a military watch should be kept the succeeding night, but the drummers were interrupted and the militia refused to appear. The governor did not think it for his honour to remain in town another night, and privately withdrew to the castle. A number of gentlemen who had some intimation of his design, sent a message to him by Colonel Hutchinson, assuring him they would stand by him in maintaining the authority of government and restoring peace and order, but he did not think this sufficient.

The governor wrote to Mr. Knowles representing the confusions occasioned by this extravagant act of his officers, but he refused all terms of accommodation until the commanders and other officers on shore were suffered to go on board their ships, and he threatened to bring up his ships and bombard the town, and some of them being seen to weigh, caused different conjectures of his real intention. Captain Erskine of the Canterbury had been seized at the house of Colonel Brinley, in Roxbury, and given his parole not to go abroad, and divers inferior officers had been secured.

The 17th, 18th, and part of the 19th, the council and house of representatives, sitting in the town, went on with their ordinary business, not willing to interpose, lest they should encourage other commanders of the navy to future acts of the like nature, but towards noon of the 19th, some of the principal members of the house began to think more seriously of the dangerous consequence of leaving the governor without support when there was not the least ground of exception to his conduct. Some high spirits in the town began to question whether his retiring should be deemed a desertion or abdication. It was moved to appoint a committee of the two houses to consider what was proper to be done. This would take time and was excepted to, and the speaker was desired to draw up such resolves as it was thought necessary the house should immediately agree to, and they were passed by a considerable majority and made public.

"In the house of representatives, November 19th, 1747.

"Resolved, that there has been and still continues a tumultuous riotous assembling of armed seamen, servants, Negroes, and others, in the town of Boston, tending to the destruction of all government and order.

"Resolved, that it is incumbent on the civil and military officers in the province to exert themselves to the utmost, to discourage and suppress all such tumultuous riotous proceedings whensoever they may happen.

"Resolved, that this house will stand by and support, with their lives and estates, his excellency the governor, and the executive part of the government, in all endeavours for this purpose.

"Resolved, that this house will exert themselves by all ways and means possible in redressing such grievances as his Majesty's subjects are and have been under, which may have been the cause of the aforesaid tumultuous disorderly assembling together.

"T. Hutchinson, Speaker."

The council passed a vote ordering that Captain Erskine and all other officers belonging to his Majesty's ships should be forthwith set at liberty and protected by the government, which was concurred in by the house. As soon as these votes were known, the tumultuous spirit began to subside. The inhabitants of the town of Boston assembled in town meeting in the afternoon, having been notified to consider, in general, what was proper for them to do upon this occasion, and notwithstanding it was urged by many, that all measures to suppress the present spirit in the people would tend to encourage the like oppressive acts for the future, yet the contrary party prevailed, and the town, although they expressed their sense of the great insult and injury by the impress, condemned the tumultuous riotous acts of such as had insulted the governor and the other parts of the legislature, and committed many other heinous offences.

The governor, not expecting so favourable a turn, had wrote to the secretary to prepare orders for the colonels of the regiments of Cambridge, Roxbury, and Milton, and the regiment of horse, to have their officers and men ready to march at an hour's warning, to such place of rendezvous as he should direct, but the next day there was an uncommon appearance of the militia of the town of Boston, many persons taking their muskets who never carried one upon any other occasion, and the governor was conducted to his house with as great parade as when he first assumed the government.

The commodore dismissed most, if not all, of the inhabitants who had been impressed, and the squadron sailed to the joy of the rest of the town.

By the expedition to Louisburgh, the prepara-]

[tions for the reduction of Canada, and the several supplies of men for Nova Scotia, the province had issued an immense sum in bills of credit, between two and three millions, according to their denomination in the currency. The greatest part of this sum had been issued when between 500*l.* and 600*l.* was equal to 100*l.* sterling, and perhaps the real consideration the government received from the inhabitants who gave credit to them, was near 400,000*l.* sterling, but by thus multiplying the bills they had so much depreciated, that at the end of the war, 1000*l.* or 1200*l.* was not equal to more than 100*l.* sterling, and the whole debt of the province did not much exceed 200,000*l.* sterling. Thus the people had paid 200,000*l.* sterling in two or three years, besides a large sum raised by taxes each year, as much as it was supposed they were able to pay; but the effect was almost insensible, for however great and unequal might be the depreciation of the bills, yet as they were shifting hands every day, a possessor of a large sum for a few days could not perceive the difference in their value between the time when he received them, and the time when he parted with them. The apprehension of their depreciation tended to increase it, and occasioned a quick circulation, and for some time, even for English goods, which ordinarily sell for the longest credit, nobody pretended to ask credit. They were constantly, however, dying in somebody's hand, though nobody kept them long by them. Business was brisk, men in trade increased their figures, but were sinking the real value of their stock; and what is worse, by endeavours to shift the loss attending such a pernicious currency from one to another, fraudulent dispositions and habits are acquired, and the morals of the people depreciate with the currency.

The government was soliciting for the reimbursement of the charge in taking and securing Cape Breton, and by the address, assiduity, and fidelity of William Bollan, Esq. who was one of the agents of the province for that purpose, there was a hopeful prospect that the full sum, about 180,000*l.* sterling, would be obtained.

Some of the ministry thought it sufficient to grant such sum as would redeem the bills issued for the expedition, &c. at their depreciated value, and Mr. Kilby, the other agent, seemed to despair of obtaining more, but Mr. Bollan, who had an intimate knowledge of the public affairs, set the injustice of this proposal in a clear light, and made it evident that the depreciation of the bills was as effectually a charge borne by the people as if the same proportion of bills had been drawn in by taxes, and refused all proposals of accommodation, insisting upon the full value of the bills when issued. He certainly has great merit for this and other services.

Mr. Hutchinson, who was then speaker of the house of representatives, imagined this to be a most favourable opportunity for abolishing bills of credit, the source of so much iniquity, and for establishing a stable currency of silver and gold for the future. About 2,200,000*l.* would be outstanding in bills in the year 1749. One hundred and eighty thousand pounds sterling at eleven for one, which was the lowest rate of exchange with London for a year or two before, and perhaps the difference was really twelve for one, would redeem 1,980,000*l.* which would leave but 220,000*l.* outstanding: it was therefore proposed that the sum granted by parliament should be shipped to the province in Spanish milled dollars, and applied for the redemption of the bills as far it would serve for that purpose, and that the remainder of the bills should be drawn in by a tax on the year 1749. This would finish the bills. It was also proposed, silver of sterling alloy at 6*s.* 8*d.* the ounce, if payment should be made in bullion, or otherwise milled dollars at 6*s.* each, should be the lawful money of the province, and no person should receive or pay within the province, bills of credit of any of the other governments of New England. This proposal being made to the governor he approved of it, as founded in justice and tending to promote the real interest of the province, but he knew the attachment of the people to paper money and supposed it impracticable. The speaker, however, laid the proposal before the house, where it was received with a smile and generally thought to be an Utopian project, and, rather out of deference to the speaker, than from an apprehension of any effect, the house appointed a committee to consider of it. The committee treated it in the same manner, but reported that the speaker should be desired to bring in a bill for the consideration of the house. When this came to be known abroad, exceptions were taken and a clamour was raised from every quarter. The major part of the people, in number, were no sufferers by a depreciating currency, the number of debtors is always more than the number of creditors, and although debts on specialties had allowance made in judgments of court for depreciation of the bills, yet on simple contracts, of which there were ten to one specialty, no allowance was made. Those who were for a fixed currency were divided. Some supposed the bills might be] [reduced to so small a quantity as to be fixed and stable, and therefore were for redeeming as many by bills of exchange as should be thought superfluous; others were for putting an end to the bills, but in a gradual way, otherwise it was said a fatal shock would be given to trade. This last was the objection of many men of good sense. Douglass, who had wrote well upon paper currency and had been the oracle of the anti-paper party, was among them, and, as his manner was with all who differed from him, discovered as much rancour against the author and promoters of this new project as he had done against the fraudulent contrivers of paper money emissions.

The bills it was said had sunk gradually in their value from 6*s*. 10½*d*. to 60*s*. the ounce; by this means creditors had been defrauded, and it was but reasonable they should rise gradually that justice might be done. But the creditors and debtors would not be the same in one instance in a thousand, and where this was not the case the injury was the same, to oblige any one to pay more as to receive less than was justly due. Others were for exchanging the bills at a lower rate than the then current price of silver. The inhabitants had given credit to the government when silver was at 30*s*. the ounce, and ought to be paid accordingly. Two of the representatives of Boston urged their being exchanged at 30*s*. which would have given a most unreasonable profit to the present possessors, who had taken them at 55*s*. or 60*s*. To draw over some of this party, concessions were made and the bills were exchanged at 50*s*. the ounce instead of 55*s*. as was at first poposed.

Some of the directors and principal promoters of the land bank scheme, being at this time members of the general court, unexpectedly joined with the party who were for finishing paper money, but the opposition was so great, that after many weeks spent in debating and settling the several parts of the bill, and a whole day's debate at last in a committee of the whole house upon the expediency of passing the bill, as thus settled, it was rejected and the report of the committee accepted.

The house, although upon some occasions exceptions are taken to motions and proceedings which come before them as not being in parliamentary form, yet are not strict in conforming to some of the most useful rules of parliament. A bill or motion is not only referred from one session to another, but a bill, after rejecting upon a second or third reading, is sometimes taken up and passed suddenly the same session. They have an order of the house, that when any affair has been considered, it shall not be brought before the house again the same session unless there be as full a house as when it was passed upon. This, if observed, would still be liable to inconvenience, as any designing person might take an opportunity upon a change of faces, the number being as great as before, suddenly to carry any point, but even this rule, like many other of what are called standing orders, is too frequently by votes, on particular occasions, dispensed with, which lessens the dignity of the house.

(*Anno* 1749.)—It may be perhaps of no consequence to the prerogative whether the currency of a colony be silver or paper, but the royal instructions from time to time for preventing a depreciating currency, caused merely by a gracious regard to the interest of the people, had generally engaged what was called the country part in opposition to them and in favour of paper. It was the case at this time. However, the next morning, two of the members of the house, zealous adherers to this party and who had been strong opposers of the bill, came early to the house to wait the coming of the speaker, and in the lobby let him know, that although they were not satisfied with several parts of the bill, yet they were alarmed with the danger to the province from the schemes of those persons who were for a gradual reduction of the bills, and by that means for raising the value of the currency without any provision for the relief of debtors; and therefore they had changed their minds, and if the bill could be brought forward again, they would give their voice for it, and others who had opposed it would do the same. The speaker, who had looked upon any further attempt to be to no purpose, acquainted them that he did not think it proper to desire any of the favourers of the bill to move for a reconsideration of it, inasmuch as it had been understood and agreed in the house the day before, that if upon a full debate had, the bill should be rejected, no further motion should be made about it. As soon as the house met, upon a motion by one of these members, seconded by the other, the bill was again brought under consideration and passed the house, as it afterwards did the council, and had the governor's consent.

The provision made by this act for the exchange of the bills and for establishing a silver currency was altogether conditional, and depended upon a grant of parliament for reimbursement of the charge of the Cape Breton expedition. This being at a distance and not absolutely certain, the act had no sudden effect upon the minds of the people; but when the news of the grant arrived, the discontent appeared more visible, and upon] [the arrival of the money there were some beginnings of tumults, and the authors and promoters of the measure were threatened. The government passed an act with a severe penalty against riots, and appeared determined to carry the other act for exchanging the bills into execution. The apprehension of a shock to trade proved groundless; the bills being dispersed through every part of the province, the silver took place instead of them, a good currency was insensibly substituted in the room of a bad one, and every branch of business was carried on to greater advantage than before. The other governments, especially Connecticut and Rhode Island, who refused, upon being invited, to conform their currency to the Massachusetts, felt a shock in their trade which they did not for a long while recover. The latter had been the importers, for the Massachusetts, of W. India goods for many years, which ceased at once. New Hampshire, after some years, revived its business and increased their trade in English goods, which formerly they had been supplied with from the Massachusetts; and in this channel their successes have been considerable.

We shall finish this part of our history with a few remarks upon the trade of the province at this day, (1760), compared with its trade in 1692.

The other governments of New England, 60 or 70 years before this period, imported no English goods, or next to none, directly from England, they were supplied by the Massachusetts trader. Now, although the trade with Great Britain, upon the whole, was supposed to cause no addition to the wealth of Massachusetts, yet, at least so far as it was the channel for conveying supplies of goods to the other colonies for their consumption, a benefit undoubtedly accrued. New Hampshire, by their convenient situation, were induced to become their own importers in a great measure some years before the alteration of the currency. They made their returns by shipping lumber, &c. easier than Massachusetts. At this time they probably imported English goods equal to their consumption. Connecticut, until the state abolished its bills of credit and theirs with them, continued their trade with it for English goods, but soon after turned great part of their trade to New York, and some persons became importers from England. They soon discovered their error. The produce of New York was so much the same with that of Connecticut that the Massachusetts market was always the best. The importer found it more difficult to make his returns to England from Connecticut than from the Massachusetts. Connecticut trade therefore soon returned to the state it had formerly been in. Rhode Island, in part, became their own importers also, which they still continue.

As to the other colonies on the continent: between S. Carolina and the Massachusetts, there never has been any considerable trade. The chief benefit from that colony has been the affording freights for the Massachusetts ships in the European trade.

N. Carolina, Virginia, Maryland, the Jerseys, and Pennsylvania, until within 20 or 30 years, used to furnish Massachusetts with provisions, for which it paid them in W. India and sometimes English goods, and with its own produce and manufactures. Philadelphia has since become the mart for the grain of great part of Maryland, which they manufacture into flour and supply the Massachusetts, Rhode Island, and New Hampshire, and take little or no pay in return but money and bills of exchange. It seems agreed that the *s.* colonies as far as Virginia are designed by nature for grain countries.

The trade with the W. India islands was much more profitable to Massachusetts, from the beginning of King William's to the end of Queen Anne's war with France, than at any time to this period. Long since the peace of Utrecht it was continually growing worse. Barbadoes required then more *n.* produce than it does now. The other islands, except Jamaica, had very little increased their demand. From the growth of the *n.* colonies and the new methods of living, the produce of the islands was more than double the price it used to be. Perhaps tea and coffee, alone, caused as great consumption of sugar as all other uses to which it was applied did formerly. The produce of the *n.* colonies was as low in the islands as ever it was. Formerly their demand for *n.* produce not only afforded this state in return, rum, sugar, and molasses sufficient for the consumption of the state, but left a surplus which, in war time especially, every year gave freight to ships from Boston to England, and paid its debts there, or procured a supply of goods from thence, whereas, at this day, the whole supply of *n.* produce to the British islands would not pay for one half of the W. India goods consumed or used in the *n.* colonies. The trade to the Dutch colonies, it is true, had since increased, and Massachusetts goods from time to time have found their way into the French islands, sometimes through the Dutch, at other times, when French necessity called for them, by permission or other contrivances, and by this means was the state able to procure the W. India goods it wanted for its consumption over and above what it could obtain in pay for its produce from the British islands.]

[As to what relates to the subsequent history of Massachusetts, we refer our readers to the article United States; wherein a full and circumstantial detail of the revolution, and other interesting matter, relative alike to this as to the other states, will be found inserted.]

[Massachusetts Fort stands on the *n. w.* corner of the state of its name, in lat. 42° 41′ 30″ *n.* 19 miles *n. e.* by *n.* of Pittsfield, and 22 due *e.* of Lansinburgh city, in New York state.]

[Massachusetts Sound, on the *n. w.* coast of N. America, is situated on the *s.* side of the Quadras isles, and leads from the *w.* into Nootka sound along the *n.* side of Kenrick's island, and whose *e.* side forms, with point Breakfast, the mouth of Nootka sound.]

[MASSACRE River passes out of the straits of Magellan *s. w.* into the supposed channel of St. Barbara, which cuts through the island of Tierra del Fuego, through which, we are informed, Captain Marcanille of Marseilles passed in 1713 into the S. Pacific ocean.]

[Massacre River, on the *n.* side of the island of St. Domingo, falls into the Bay of Mancenilla; which see.]

[Massacre, a small island on the coast of W. Florida, two miles to the *e.* of Horn island, 10 miles from the mainland: all the way across there is from two to three fathoms, except the shoal called La Grand Bature, which stretches a league from the mainland, with two or three feet water on it, and in some places not so much. Behind it is a large bay called L'Ance de la Grand Bature, eight miles *e.* of Pascagoula bluff. See Dauphin Island.]

[MASSAFUERO, an island in the S. Pacific ocean, called by the Spaniards the Lesser Juan Fernandes, 22 leagues *w.* by *s.* of the Greater Juan Fernandes. It has always been represented by the Spaniards as a barren rock, without wood, water, or provisions; but Lord Anson found this to be a political falsity, asserted to prevent hostile vessels from touching there. There is anchorage on the *n.* side in deep water, where a single ship may be sheltered close under the shore, but is exposed to all winds except the *s.* According to Captain Magee, of the ship Jefferson, it is 78 miles to the *w.* of Juan Fernandes, and in about lat. 33° 47′ 30″ *s.* and long. 80° 41′ *w.* from Greenwich.]

[MASSANUTEN'S River, a *w.* branch of the Shenandoah.]

[MASSEDAN Bay, in the N. Pacific ocean, and *w.* coast of Mexico, is situated between Acapulco and Aquacara, a port near the cape of California, where Sir Thomas Cavendish lay after he had passed the straits of Magellan.]

[MASSEY'S Town, in the N. W. Territory, stands on the *n.* bank of Ohio river, between the rivers Little Miami and Sciota.]

[MASSY'S Cross Roads, in Kent county, Maryland, is *n. e.* of Newmarket, *s. e.* of George town, and *s.* by *w.* of Sassafras town, a little more than five miles from each.]

[MAST Bay, on the *n.* side of the island of Jamaica, in the *n. w.* part. It is *e.* of Montego bay, and near the shelf of rocks that lies from the shore, called Catlin's cliffs.]

MASTELES, a barbarous and ancient nation of Indians of the province of Popayán, towards the *w.* They were warlike, cruel, and cannibals, and lived in continual warfare with their neighbours the Abades; by whom they were at last extirpated.

MASTERKOUT, a small city of the county of Prince George, in the *w.* division of the province of Maryland.

MASTES, a barbarous and warlike nation of Indians, of the province and government of Santa Marta, to the *s. s. w.* and confederates of the Taironas.

[MASTIC Gut, on the *s. w.* side of the island of St. Christopher's in the W. Indies, is between Moline's gut on the *n. w.* and Godwin's gut on the *s. e.*]

[MASTIGON, a river which runs *w.* into lake Michigan, about 11 miles *n.* of La Grande Riviere. It is 150 yards wide at its mouth.]

MASTON, a settlement of the island of Barbadoes; situate in the *s.* part.

MATA, a large settlement of the province and *captainship* of Pernambuco in Brazil. It is well peopled through its commerce in Brazil-wood, which is cut in the neighbourhood in abundance. It has two convents of monks, and a very good parish; upon which is dependent another small settlement annexed to it. It is 11 leagues *s.* of its capital.

Mata, Cienega de, a settlement of the head settlement and *alcaldía mayor* of Sierra de Pinos in Nueva España. Five leagues to the *s.* of Ojo Caliente.

[Mata, Point, on the the *n.* side of the island of Cuba, and nine leagues *n. w.* of cape Maisy.]

MATACHIQUI, a settlement of the missions that were held by the regulars of the company, in the province of Taraumara, and kingdom of Nueva Vizcaya. It is 31 leagues from the town and *real* of the mines of San Felipe de Chiguagua.

MATA-FUEGO, some isles of the N. sea, near

the coast of the province and *captainship* of Rey in Brazil; situate *n.* of the island of Santa Catalina.

MATAGALPA, a province and *alcaldía mayor* of the kingdom of Guatemala in N. America.

MATAGEROS, some small islands of the N. sea, near the coast of the province and government of Honduras; opposite the entrance or mouth of the river Comecueros. They are many and desert.

MATAGORDA, or PESCADERO, a port of the coast of the province and *corregimiento* of Quillota in the kingdom of Chile, between the *quebrada* of Choapa and the river Quilimari.

MATAGUAYOS, or MATAGUAYES, a nation of barbarous Indians of the province of Chaco in Peru, who dwell in the *llanos* of Manso near the river Pilcomayo. The Fathers Ignacio de Medina and Andres de Lujan of the abolished order of Jesuits, endeavoured to reduce them to the Catholic faith in 1653, and entered their country, being conducted by the Cacique Nao, their chief, whom they had gained over by bribes and persuasions; but just as they had collected together a certain number of the Mataguayos, and were beginning to establish a settlément, these infidels rose of a sudden and retired to the mountains, determining to put to death the fathers; who, however, escaped their fury by the precaution of an Indian, their friend. In the following year they sent deputies to manifest their contrition for what had passed, and requesting that missionaries would again come amongst them to instruct them in the faith, and although their wishes were acceded to by the governor of Tucumán, and the provincial, a body of troops being also sent to secure the peace, yet did, at the same moment, one of the most terrible insurrections happen, which soon spread itself throughout the whole province, thereby putting aside all hope of reducing this nation to the bosom of the church.

MATAHUASI, a settlement of the province and *corregimiento* of Xanja in Peru.

MATAJES, a large river of the province and government of Esmeraldas in the kingdom of Quito. It runs from *s. e.* to *n. w.* in the country and territory of the Malaguas Indians, and enters the Pacific, in the bay of Sardinas, in latitude 1° 22′ *n.*

MATALIRON, a small island of the N. sea; one of the Lesser or Windward Antilles. It is desert.

MATAMEREQUE, a river of the province and *corregimiento* of Caxamarquilla in Peru.

MATAMUSKET, a settlement of Indians of N. Carolina and district of Hyde; situate on the coast.

[MATANCA, or MANANCES, a short and broad river of E. Florida, which falls into the ocean *s.* of St. Augustine.]

MATANCHEL, a town and port of the S. sea, on the coast of Nueva España, belonging to the jurisdiction and district of the *alcaldía mayor* of Tepic. It is on the *w.* part, and lying *e. w.* with the Marias isles. It is tolerably convenient and secure, and in 1745, a China ship anchored in it, having suffered much by a storm; also in 1746, a Dutch vessel which had been navigating those seas put in here. It is 65 miles from the capital, in lat. 20° 45′ *n.* and long. 105° 24′ *w.*

MATANCHEL, a settlement of the head settlement and *alcaldía mayor* of Compostela in Nueva España; situate on the coast of the S. sea, near the former port.

MATANE, a river of New France or Canada, the mouth of which is capable of receiving vessels of 200 tons. The whole of the sides of the river St. Lawrence, especially of that part of which we treat, abounds for more than 20 leagues in cod fish; and is capable of employing 500 fishing smacks at the same time. This fish is of an excellent quality, and much esteemed in España and in the *e.* parts. Here have also been seen many whales, which may be killed by the harpoon, and would bring ample profit.

MATANILLAS, a settlement of the island of Cuba; situate on the *n.* coast, between the settlement of Guanaja and Alivitas.

MATANZA, a settlement of the province and government of Buenos Ayres in Peru.

MATANZA, a river of the same province and government, which runs *n. n. e.* and enters the Paraná, between the Ramallos and that of Dos Hermanos.

MATANZAS, a city of the island of Cuba; situate on the *n.* coast, on the shore of a bay or port of the same name, and which is one of the most convenient, safest, and largest of any in America, having a good castle for its defence. The city is small, of poor commerce, and thinly inhabited. Formerly its commerce was great, since all the vessels that were going to Europe used to enter it. The Dutch admiral Pedro Hein routed and burnt in this bay the fleet of Nueva España in 1628; and in 1638, the admiral of the galleons, Don Carlos de Ibarra, had two smart combats with the Dutch privateer Pie de Palo. It is 30 leagues from the Cabeza de los Martires on the coast of

Florida, and 20 from the Havana, in lat. 23° 3′ *n.* and long. 81° 30′ *w.*

∴ Matanzas, Pan de, a very lofty mountain in the shape of a sugar-loaf; situate at the back of the former port, and serving as a land-mark or direction for entering the mouth of the canal of Bahama, putting the prow to the *n.* and leaving the mountain a-stern, in order to bear right a-head of the said entrance or mouth.

Matanzas, Pan de, an island of the N. sea, near the coast of Florida, beyond the canal of Bahama.

MATAOUACHIE, a river of the province and country of the Iroquees Indians in New France or Canada. It runs *n. e.* and enters the Otaivas.

MATAPE, a settlement of the province and government of Ostimuri in N. America.

MATAPI, a river of the province and government of La Guayana, in the part which was possessed by the Portuguese, according to the description and chart of the engineer Bellin; but not being able to find it in other charts, we may infer that he has mistaken the name for Matari.

MATAPONI, a river of the province and colony of Virginia in N. America. It runs *s. e.* and enters the river York.

MATAQUINO, or Mataquito, as others will have it, a large river of the province and district of Chanco in the kingdom of Chile. It runs many leagues to the *w.* and enters the sea between the rivers Maule and Boyeruca. On its shore by the *n.* side are two large estates, called Tilicura and Peralvillo. Near to this spot the Spaniards were engaged with the Araucanos Indians in a battle, called the battle of Mataquino, in which the brave Lautaro fell. On the *s.* side are two other estates, with the names of Huaquen and Naicuda.

MATARA, a settlement of the province and *corregimiento* of Caxatambo in Peru, where there is a spring of dark-coloured water, which flows into a river called De la Barranca, and which passing through the settlement acquires a deeper tint in proportion to its stagnation. This settlement is annexed to the curacy of Chiquian.

Matara, another settlement, in the province and *corregimiento* of Aimaraez in Peru.

Matara, another, in the province and *corregimiento* of Guarochiri in the same kingdom; annexed to the curacy of the settlement of Olleros.

Matara, another, in the province and *corregimiento* of Lucanas in the same kingdom; annexed to the curacy of the settlement of Paico.

Matara, another, called also Mopa, in the province and government of Tucumán in the same kingdom, of the district and jurisdiction of the city of Santiago del Estero, to the curacy of which it belongs.

MATARAS, or Mataranes, a nation of Indians of the province of Chaco in Peru. It is one of those which was happy enough to have employed in its conversion to the faith San Francisco Solano; but reverting, however, to its idolatrous state, it had lost nearly every trait of religion at the time that its conversion was again attempted by the Fathers Juan de Fonte, Juan Baptista Añasco, Francisco de Angulo, and Alonso de Barcena, of the abolished order of Jesuits. These succeeded in reducing the greater part of these Indians, who, having for many years been without any regular curate, again revolted to their pristine idolatry, when, in 1641, the true faith was again preached to them by the Fathers Pedro Pastor and Gaspar Cerqueira; and it was then that, becoming strict proselytes, they united together in a settlement of their name.

In the time of their gentilism these Indians had the custom of celebrating the anniversary of their relations death. A dead ostrich was brought in honour of each of the defunct, together with some articles of furniture, a ceremony with which even the guests were obliged to comply; and in the same manner all heirs, in refusing to keep up this practice, were considered as giving ample cause for open enmity. This festival used to last four days; at the end of these they would lament over the dead for one hour, and after this would succeed dances, convivialities, and the most excessive symptoms of mirth, so that they had the appearance of bacchanalians, in whom was a complication of all the disorderly habits, of which it was possible to conceive creatures, who had nothing but the form of rational beings, to be guilty.

[MATAVIA Bay, or Port Royal Bay, is situated within point Venus, near the *n.* part of the island of Otaheite, but open to the *n. w.* and in the S. Pacific ocean. The *e.* side of the bay has good anchorage in 14 and 16 fathoms. Lat. 17° 40′ *s.* Long. 149° 30′ *w.* and the variation of the compass 3° 34′ *e.*]

[MATCHADOCK Bay, in the easternmost part of lake Huron.]

[MATHANON Port, in the *s. e.* part of the island of Cuba, is one of those ports on that coast which afford good anchorage for ships, but without any use for want of them. It is between cape Cruz and cape Maizi, at the *e.* end of the island.]

[MATHEO River, in E. Florida, or St. John's; which see.]

[MATHEWS Fort stands on the *e.* side of

Oconee river, in the *s. w.* part of Franklin county, Georgia.]

[MATHEWS, a county of Virginia; bounded *w.* by Gloucester, from which it was taken since 1790; lying on the *w.* shore of the bay of Chesapeak. It is about 18 miles in length, and six in breadth.

MATIARE, a settlement of the province and government of Nicaragua, and kingdom of Guatemala, in the time of the gentilism of the Indians; but at present not existing. It was near the city of Leon.

MATIAS, S. a settlement of the head settlement of Tepalcaltepec, and *alcaldia mayor* of Nexapa, in Nueva España; situate on a mountain, the ascent to it being more than a league. It is of a cold temperature, and inhabited by 72 families of Indians.

MATIAS, S. a bay, also called SIN FONDO, on the coast which lies between the river La Plata and the straits of Magellan, to the *s.* between the bay Anegada and the port of Los Leones.

MATIAS, S. an island, the most *e.* of the two that were discovered by William Dampierre on the coast of New Britain of N. America. It is nine or 10 leagues long, is mountainous and full of woods, although not without certain spots and valleys with the appearance of cultivation.

[MATICALOC River, on the *w.* coast of New Mexico, is seven leagues from Catalta strand, or the port of Sansonate. It is much exposed to *n.* winds, and is known by some small but high hills that are opposite to it. There is another large river to the *w.* of it about four leagues, which has two fathoms upon the bar; and from thence to the bar of Estapa it is 15 leagues.]

[MATILDA, a village of Virginia; situated on the *s. w.* bank of Patowmac river, above Washington city, and near the great falls.]

MATILLA, a settlement of the province and *corregimiento* of Arica in Peru; annexed to the curacy of Pica.

MATINA, a district of the government and province of Costarica in the kingdom of Guatemala; situate between the district of Las Talamancas to the *e.* and the river San Juan to the *w.* on the coast of the N. sea. It has by the other part of the aforesaid river the coast of the Mosquitos Indians, who continually infest the same with their canoes; so that this spot is peopled only very thinly by Spaniards, who dwell in certain *rancherias* or temporary habitations, for the purpose of labourers, and who cultivate *cacao*, which both in and out of the kingdom is as highly esteemed as that of Soconusco: its crops are, however, much fallen off, and the produce is scarcely sufficient for the inhabitants of Nicaragua, who are eager in its monopoly. There was in this district a castle, which was demolished by the English in the war of 1740.

MATINA, a river of the province and government of Veragua in the kingdom of Tierra Firme, which serves as a limit, and divides this province from that of Costarica in Guatemala. It runs into the N. sea, and on its shores stood a castle, which was demolished by the English in 1744.

[MATINICUS Islands, on the coast of Maine. When you pass on the *w.* of these islands, the main passage from the sea to Penobscot bay lies about *n.* by *w.* They lie in lat. 43° 50′ *n.* Long. 68° 47′ *w.*]

MATITUI, a settlement of the province and government of Popayán in the Nuevo Reyno de Granada.

MATLACUCUEYE, a name which the Indians, in their idiom, gave to the *sierra* of Tlaxcala, and which served as a place of safety for their wives and children when Hernan Cortes entered that province with the Spaniards.

MATLAHUACALLAN, the spot where the Tultecas Indians settled for three years, the same being the first who peopled the Mexican empire, according to *Fr.* Juan de Torquemada.

MATLALCINCO, a large and rich province of the Mexican empire, in the time of the Indians, and conquered and united to it by the Emperor Axayacatl. It was in the valley of Tolucan, nine leagues from the city of this name. Its natives were of the most faithful of the allies of Hernan Cortes, and of the Spaniards; and on this account did the Otomies declare war against them, ravaging and destroying the whole of their territory. This oppressed people complaining to Cortes, had sent to their relief Gonzalo de Sondoval with 100 infantry and 18 cavalry, who chastised and reduced the Otomies to obedience. In gratitude for this succour, the natives, of whom we treat, assisted the Spaniards in a great measure in the siege and conquest of Mexico. They were extremely barbarous, superstitious, and vicious; but were greatly improved by *Fr.* Andres de Castro, of the order of San Francisco, who dedicated himself to their conversion and instruction for nearly 40 years, in which time he succeeded in reducing them to the Catholic faith, in which they have persisted, looking back upon his conduct as though he were an apostle of the most heroic virtues.

MATLALLAN, a settlement of the province and *alcaldia mayor* of Tlaxcala in Nueva España, in the time of the gentilism of the Indians.

MATLAPA, a settlement of the head settlement of Tamazunchale, and annexed to its curacy, in the *alcaldía mayor* of Valles and kingdom of Nueva España; situate on the bank of a beautiful stream, by which it is provided with water. It is of a hot and moist temperature, abounding in maize, French beans, seeds, and cotton. It contains 30 families of Indians, and is five leagues to the *n.* of its head settlement.

MATLATEPEC, a thinly peopled settlement of the head settlement of the district of Santa Ana, and *alcaldía mayor* of Zultepec, in Nueva España; united to the settlement of Hueztahualco, and being distant from it a short league.

MATLAZINCO, a large river of Nueva España. It rises near the settlement of Santiago, in the jurisdiction and *alcaldía mayor* of Lerma, from a small stream; and traversing various provinces for more than 500 leagues, in which it collects the waters of various other rivers, it takes the name of Rio Grande, and empties itself into the S. sea, opposite the island of California. In many parts its stream is so large as not to admit of being forded, especially after that it enters the sea Chapalico or lake of Chapala. To pass it near the city of Guadalajara, it has a singular bridge.

MATLICA, a settlement of the head settlement of Yautepec, and *alcaldía mayor* of Cuernavaca, in Nueva España.

MATO, a settlement of the province and *corregimiento* of Xauxa in Peru.

MATO, a river of the province and government of Moxos in the kingdom of Quito. It is also called COUITU; see this article.

MATO DENTRO, a settlement or village of the Portuguese, in the province and *captainship* of Espiritu Santo, and kingdom of Brazil; situate near the settlement of Castas Altas, close to the river Doce or Dulce.

MATO GROSO, a province and government of the Portuguese, in the country of Las Amazonas; bounded *n.* by the *captainship* of Para, *w.* by the viceroyalty of La Plata, *s.* by the *captainship* of S. Paul's, and *e.* by the *captainship* of Goias. It abounds in mines of the richest gold. It was desert, and the Portuguese, excited by its riches, made themselves masters of and established themselves in it, in 1761, placing in it a governor; his residence, which is also the capital, bearing the same name as the province. Its principal commerce is reduced to the working of the gold mines, which render well. It is of an hot and moist temperature, and scantily supplied with eatables, and these are consequently very dear. In 1765, it was attacked by the President Don Joseph Pestaña, by order of the viceroy of Peru, Don Manuel Amat, but he was obliged to desist from his enterprise, and to retire with his troops in 1766, owing to the valour manifested by the Portuguese, and the roughness of the territory.

MATOS, a small river of the province and government of Guayana or Nueva Andalucía. It runs *n.* and enters the Caura by the *w.* side, opposite the Yuruani.

MATOURI, QUARTEL DE, a settlement and parish of the English, in the island of Cayenne; situate in the vicinity of a mountainous tract of the same name, in the *w.* part of the island.

[MATTA DE BRAZIL, a town in the *captainship* of Pernambuco in Brazil, about nine leagues from Olinda. It is very populous, and quantities of Brazil-wood are sent from this country to Europe.]

[MATTAPONY, a navigable river of Virginia, which rises in Spottsylvania county, and running a *s. e.* course, joins Pamunky river below the town of De la War, and together form York river. This river will admit loaded floats to Downer's bridge, 70 miles above its mouth.]

[MATTES, or MATALINARES, a point on the *e.* coast of S. America, in the S. Atlantic ocean, is in lat. 45° 23′ *s.* and long. 67° 9′ *w.*]

[MATTHEO Island, ST. or ST. MATTHEW'S Island, in the S. Atlantic ocean. Lat. 1° 24′ *s.*]

[MATTHEW'S Bay, ST. in the gulf of Mexico, on the *w.* shore of Campeachy gulf, is more than 100 leagues to the *n.* of Tumbez.]

[MATTHEW'S, ST. or MATTHEO Bay, on the coast of Peru, on the N. Pacific ocean, is six leagues to the *n. e.* by *e.* from point Galera, and five or six leagues *s. s. w.* from the river St. Jago, between which there is anchorage all the way, if ships keep at least in six fathoms water. It is all high land with hollow red crags, and several points run out, forming good retreats for ships driven in by hard squalls and flaws from the hills, and by the seas running high, which often happen.]

MATUCANA, SAN JUAN DE, a settlement of the province and *corregimiento* of Guarochiri in Peru.

MATUMAGANTI, a river of the province and government of Darien, and kingdom of Tierra Firme. It rises in the mountains of the *n.* coast, and enters the Bayano.

MATUNA, BOCA DE, the entrance into the sea of the dike which communicates this river with that of the Magdalena, in the province and government of Cartagena, between the point of Barbacoa and that of San Bernardo.

MATURU, a settlement of the province and *captainship* of Pará in Brazil; situate on the shore of the river of Las Amazonas, near the town of Curupa.

[MAUGERVILLE, a township in Sunbury county, province of New Brunswick, situated on St. John's river, opposite St. Ann's, and 30 miles above Bellisle.]

MAULAS, Annoyo de, a small river of the province and government of Buenos Ayres in Peru. It runs *w.* and enters the Rio Negro, just before the mouth by which it enters the Uruguay.

MAULE, a province and *corregimiento* of the kingdom of Chile; bounded *e.* by the *cordillera*, *s.* by the province of Chillan, the river Maule serving as the limits, *w.* by the sea, and *n.* by the province of Colchagua. It is from *e.* to *w.* 130 miles, and about 60 from *n.* to *s.* It abounds in mines and in *lavaderos* or washing places of gold, and in the district of the curacy of the capital, near the ferry where the river Maule is crossed, is a mountain called Chivato, where there is a famous gold mine, which was formerly worked, and afterwards abandoned until the year 1767, when it was again worked: it always afforded a metal of very good quality.

In this province are cultivated all kinds of grain: there are good breeds of cattle of every species, especially goats, which are highly prized for their hides to make leather. Here are also vines more esteemed and of a better quality than are those of Peru, and some tobacco, the cultivation of which has ceased since that it became monopolized by the crown, and its sowing prohibited. In it also, as well as in the provinces of Quillota and Calchagua, are large forests of the coco nut tree. The rivers which water and fertilize this province, are Maule, Claro, and Lontue; this latter joins the Teno, divides this jurisdiction from that of Colchagua, and alter its union is known by the name of Mataquito. Here is also a mine of *brea* or tar, which is a kind of bituminous mud, and which, although it has been frequently clarified by fire and boiling, yet does it always retain a mixture of earth, so that its use will remain small until that a better means of preparing it be discovered.

Through some parts of this province passes the *cordillera* by the *e.* and in it are many muleteers, who carry on a traffic by carrying salt, which is found in small pieces of a very white quality in certain lakes. The inhabitants amount to 12,000, and the capital is Talca.

[The cheese made in this province is the best in Chile, and in no way inferior to that of Placentia or Holland. Its inhabitants, who are mostly the descendants of the ancient Promaucians, are courageous, robust, and warlike. The capital, Talca or St. Augustin, was built in the year 1742: it is situated among hills on the river Rioclaro, in latitude 35° 15′, and 71° 1″ of longitude. Its population is very considerable, owing, not only to rich mines of gold that are found in its mountains, but to the plentifulness of provisions, which are cheaper than in any other part of Chile. This latter circumstance has induced several noble families from St. Jago and Concepcion, whose finances had become diminished, to retire thither, an emigration which has been denominated, in derision, the bankrupt colony. In this province are also the towns of Curico, Canguenes, St. Saverio di Bella-isla, St. Antonio della Florida, Lora, and three or four other Indian villages. Curico, or St. Joseph of Buena Vista, was built in the year 1742, and is situated on a pleasant plain at the foot of a beautiful hill, in 34° 14′ of latitude. It contains a parish church, a convent of Meredarii, and another of strict Franciscans which is very large. Canguenas was built the same year, and lies in 35° 40′ of latitude, between the two small rivers Tutuben and Canguenes. Besides the parish church it has a convent of Franciscans. St. Saverio di Bella-isla and St. Antonio della Florida were founded in the year 1755; the first is in 35° 4′ of latitude, and the second in 35° 20′ of latitude. Loro, situate near the disemboguement of the river Mataquito, is a populous settlement of Promaucian Indians, and is governed by a *cacique* or *ulmen*.]

Maule, a river of which we have before spoken, in the former province, to the *s.* of the city of Santiago, and in the jurisdiction of La Concepcion. It rises in the mountains of the *cordillera*, and runs from *e.* to *w.* collecting the waters of several other rivers, of which are the Cauchenes, the Claro, and others of less note. As far as the last mentioned extended the conquests of the Inca Yupanqui, eleventh emperor of Peru.

This river runs to empty itself into the Pacific ocean, forming a very convenient bay, in lat. 34° 50′ *s.* on the borders of the jurisdiction and bishopric of Santiago. The Indians called this part *Promocaces*, which in their language signifies a place of dancing and delights, to denote the pleasantness of this territory. Before it enters the sea it has a celebrated dock, and here it spreads out and forms a very large sheet. The king provides a ferry for the passage of those who cross the river, and near to the ferry is a convent of Augustin monks, which serves as a parish to the Indians and Negroes, who dwell in different villages situate on its bank, and which are here called *estancias*.

[On the left bank of this river, at 400 paces distance from its mouth, is an insulated mass of white marble, consisting of a single piece, 75 feet in height, 224 in length, and 54 in breadth. This immense block, called from its appearance *The Church*, is excavated within like an arch, the third part of its height, and has on the outside three doors of a semicircular form, and proportionate height and breadth; through the one on the *w.* front the sea continually flows; the two others, which are on the *n.* and *s.* sides, and placed opposite, serve to admit those who wish to visit it at the ebb tide. This natural edifice, constantly washed by the sea, serves as a place of resort for the sea-wolves, who herd in great numbers in the lower part, and make the cavity re-echo with their lugubrious cries; while the upper is occupied by a species of sea-bird, very white, called *lili*, in figure and size resembling a house-pigeon.

In this river likewise is to be found a clay as white as snow, smooth and greasy to the touch, extremely fine, and sprinkled with brilliant specks. It is found on the borders of rivers and brooks in the province of Maule, in strata which run deep into the ground, and its surface, when seen at a distance, has the appearance of ground covered with snow, and is so unctuous and slippery that it is almost impossible to walk upon it without falling. It does not effervesce with acids, and instead of losing in the fire any portion of its shining whiteness, it acquires a slight degree of transparency. It is believed to be very analogous to the *kaolin* of the Chinese; and that, combined with fusible spar, of which there are great quantities in the same province, it would furnish an excellent porcelain.]

Maule, an island of the same province and kingdom, formed by the rivers Maule, Longomilla, and Putagan.

Maule, a *morro* or mountain of the same province; situate on the shore of the former river, from whence it takes its name.

[MAURA. See Society Islands.]

MAURE, a settlement of the province and *corregimiento* of Arica in Peru; annexed to the curacy of Tarata.

Maure, a river of the province and *corregimiento* of Pacajes in Peru. It rises at the foot of the *cordillera*, and runs nearly due *e.* until it enters the waste-water which runs into the lake of Chucuito.

[MAUREPAS, an island on the *n. e.* coast of lake Superior, and *n. e.* of Ponchartrain island.]

[Maurepas, a lake in W. Florida, which communicates *w.* with Mississippi river, through the gut of Ibberville, and *e.* with lake Ponchartrain. It is 10 miles long, seven broad, and has 10 or 12 feet water in it. The country round it is low, and covered with cypress, live oak, myrtle, &c. Two creeks fall into this lake, one from the *n.* side, called Nattabanie, the other from the peninsula of Orleans. From the Ibberville at its junction with Maurepas to the river Amit is 27 miles, and from thence, following the Ibberville to the Mississippi at the *w.* side of the peninsula of Orleans, 16 miles. From the Ibberville across the lake, it is seven miles to the passage leading to Ponchartrain. The length of this passage is seven miles, and only 300 yards in width, which is divided into two branches by an island that extends from Maurepas to about the distance of a mile from Ponchartrain. The *s.* channel is the deepest and shortest. The passage thence through lake Ponchartrain to the gulf of Mexico is above 46 miles.]

[Maurepas Island, on the coast of Cape Breton, the same as the Isle Madame, which see.]

MAURI, Quartel de, a settlement and parish of the French, in the island of Cayenne; situate in the *e.* part, and at the entrance of the river Ocuya.

[MAURICE Bay, on the *w.* side of Cape Farewell island, or *s.* extremity of E. Greenland, and the principal harbour of that sea.]

[Maurice River, the name of a place in Cumberland county, New Jersey.]

[Maurice River, in some maps called corruptly Morris, rises in Gloucester county, New Jersey, and runs *s.* about 30 miles, and empties into Delaware bay; is navigable for vessels of 100 tons 20 miles, and for small craft considerably further.]

MAURICIO, S. Bahia de, a bay in one of the islands which form the Land of Fuego, to the *e.* nearly opposite that of Los Estados on the *w.* side of Le Maire straits. It was discovered in 1616 by Jacob le Maire, who gave it this name in honour of Mauricio de Nassau, Prince of Orange. It is throughout its whole circumference of a rough, mountainous, and barren soil, and continually covered with snow. To the *n. e.* is a road called Verschoxsis, which is insecure, and on the *s.* is a bay of the name of Valentins, which is large, convenient, and capacious, but very open, having anchorage before it in about 12½ fathoms from the shore, over coral rocks. The cape or point which looks *s.* is called De Buen Suceso, and is in lat. 55° *s.* and long. 65° 18′ *w.*

MAURO, S. a settlement of the province and *captainship* of Todos Santos in Brazil; situate at the mouth of the river Yapetinga, and on the shore of the bay.

MAUTACONA, a small river of the province and government of Guayana or Nueva Andalucia. It rises to the *w.* of the lake Ipava, from whence the Orinoco takes its rise, according to the map of the geographer Crus, and enters into the last river near fort De S. Barbera.

MAXALTEPEC, a settlement of the *alcaldía mayor* of Nexapa in Nueva España; situate on the skirt of a mountain well covered with fine trees, and inhabited by 28 families of Indians.

[MAXANTALLA Island is near the port of Matanchel, on the *w.* coast of New Mexico, and on the N. Pacific ocean.]

MAXATES, a settlement of the province and government of Cartagena; situate near the *dique* or canal into which the sea enters to surround that piece of land.

MAXI, or MAXIS, a river of the province and country of Las Amazonas. It rises in some mountains much covered with wild *cacao*, in the territory of the Oreguatos and Curanaris Indians, runs *n.* and turning its course *w.* enters in a large body into the Madera, close to the great cataract which it forms.

MAXWELL'S, a city of the island of Barbadoes.

MAY, a small river of the province and colony of Georgia. It runs *e.* and enters the sea.

[MAY, Cape, the most *s.* point of land of the state of New Jersey, and the *n.* point of the entrance into Delaware bay and river, in lat. 38° 57′ *n.* and long. 74° 54′ *w.* The time of high water on spring-tide days is a quarter before nine o'clock.]

[MAY COUNTY, Cape. See CAPE.]

[MAY Point, on the *s.* side of Newfoundland island, a point of the peninsula between Fortune and Placentia bays.]

[MAY'S Lick, in Mason's county, Kentucky, a salt spring on a branch of Licking river, nine miles *s. s. w.* of Washington, on the *s.* bank of the Ohio, and 15 *n.* of the Blue licks.]

MAYA, SANTA ANA DE, a settlement of the head settlement of the district and *alcaldía mayor* of Cuiceo in Nueva España. It contains 12 families of Spaniards and *Mustees*, and 32 of Indians.

MAYA, a small port on the coast of the province and government of Venezuela, to the *w.* of the city of Caracas.

MAYA, or MAYAPAN. Such, in the time of the Indians, was the name of the province of Yucatán, which was then a kingdom separate from Nueva España, and which the Spaniards at first believed to be an island. It was afterwards discovered to be a peninsula surrounded by the N. sea, by this rhumb, as also by the *e.* and *w.* and united only by the *s.* and the kingdom of Guatemala. Moreover, between the two kingdoms laid an extremely rugged country, full of mountains and *sierras*, and inhabited by the infidels, the Itzaes, Petcnes, Lacandones, Cheaques, Mopanes, Choles, Chinamitas, Caboxes, Uchines, Ojoyes, Tirampies, and various other tribes; and these, ever after the conquest of the kingdom, so stood in the way of any connection between the two kingdoms, that this has been obliged to be effected by a very circuitous sea-voyage.

MAYACARE, a river of the province and government of Guayana. It runs *w.* and enters the sea in a bay near the N. cape.

MAYAGUANA, an island of the N. sea, one of the Lucayas, discovered by Christopher Columbus in his first voyage; situate *n.* of Hispañola or St. Domingo.

[MAYAGUANA is one of the uninhabited Bahama islands. Its *s. w.* point lies about 10 leagues in an *e. s. e.* direction from the *e.* point of the French Keys, and the *s. e.* point of it bears about 12 leagues *n. w.* from the *n. w.* point of the Blue Caicos. The *n. w.* point of the island (from whence a reef runs out about three miles) is situated about three leagues from the *s. w.* point; and there is an anchorage along the *s.* end of the island. Fresh water is to be found at the *n. w.* point. Along the *s.* side of the island, which is probably upwards of 30 miles in length, vessels may generally anchor, and about four leagues from the *s. w.* point there is a reef harbour, at Abraham's reef, for vessels drawing about 12 feet. There is fresh water at it. A dangerous reef runs out for several miles at the *e.* end of the island, and within it a harbour for very small vessels. A reef also runs all along the *n.* side of the island.]

MAYAGUES, a river of the island of Porto-rico. It rises from a mountain in the centre of the said island, runs *w.* and enters the sea between the river Guanagive and the bay of Añasco.

MAYANALA, SANTA MARIA DE, a settlement of the head settlement of the district and *alcaldía mayor* of Iguala in Nueva España. It contains 51 families of Indians.

MAYASQUER, a settlement of the province and government of Pastos, in the district of the jurisdiction of the audience of Quito.

MAYE, a mountain on the coast of the province and government of Guayana. It stands on the coast as it were an isolated platform, at a small distance from the river Cowanawini, and is covered with trees, serving as a landmark to those sailing for Cayenne. It takes its name from a nation of Indians who dwell in this part, in lat. 30° 15'.

MAYEN, an island situate to the *n. w.* of Spitzbergen, in lat. 71° 10' *n.* The sea, which washes its coasts, was formerly very abundant in whales, but these creatures having shifted to the *n.* the island was abandoned by such as used to occupy themselves in the fisheries. In its *n.* extremity is a very lofty mountain called Beerembergen or of the Bear; the same traverses the whole island, and is discovered at 30 miles distant. It has very good ports, and abounds in shell and other fish; but the great quantity of ice which forms in the sea around it, especially towards the *e.* renders it inaccessible in the spring.

[MAYES are Indians of N. America, who live on a large creek called St. Gabriel, on the bay of St. Bernard, near the mouth of Guadalupe river; are estimated at 200 men; never at peace with the Spaniards, towards whom they are said to possess a fixed hatred, but profess great friendship for the French, to whom they have been strongly attached since Mons. de Salle landed in their neighbourhood. The place where there is a talk of the Spaniards opening a new port, and making a settlement, is near them; where the party, with the governor of St. Antoine, who were there lately to examine it, say they found the remains of a French block-house: some of the cannon now at Labahie are said to have been brought from that place, and known by the engravings now to be seen on them.

The French speak highly of these Indians, for their extreme kindness and hospitality to all Frenchmen who have been amongst them: have a language of their own, but speak Attakapa, which is the language of their neighbours the Carankouas; they have likewise a way of conversing by signs.]

[MAYFIELD, a township in Montgomery county, New York, adjoining Broadabin on the *w.* taken from Caughnawaga, and incorporated in 1793. In 1796, 126 of its inhabitants were qualified electors.]

[MAYNAS, a government, formerly the *e.* limit of the jurisdiction of Quito in Peru, and joining on the *e.* to the governments of Quixos and Jaen de Bracamoros. It is separated from the possessions of the Portuguese by the famous line of demarcation, or the boundary of those countries belonging to Spain and Portugal. See Mainas.]

MAYO, Tablon de, a settlement of the province and government of Popayán in the Nuevo Reyno de Granada.

Mayo, another settlement, of the province and *corregimiento* of Canta in Peru; annexed to the curacy of Arabuay.

Mayo, an abundant river of the province of Ostimuri in Nueva España, which divides this province from that of Cinaloa, and enters the sea in the gulf of California, or Mar Roxo de Cortes.

Mayo, another river, of the province and colony of Virginia, in the county of Lunenburg. It runs *s. e.* and enters the Dan.

Mayo, a very lofty mountain, of the province and *captainship* of San Vicente in Brazil; situate on the shore of the river Tajai-Meri.

[Mayo, Santa Cruz de, a port at the mouth of the river Mayo, near the middle of the *w.* coast of the gulf of California. From this place the courier of Mexico goes to Loreto in California and Monterey, on the *w.* coast of N. America.]

MAYOBAMBA, a settlement of the province and *corregimiento* of Vilcas Huaman in Peru; annexed to the curacy of Hualla.

Mayobamba, another settlement, in the province and *corregimiento* of Lucanas in the same kingdom; annexed to the curacy of Chipán.

MAYOC, a settlement of the province and *corregimiento* of Guanta in Peru.

MAYOCMARCA. See Anco.

MAYORE, a lake of the province and government of Paraguay in Peru; formed in the territory of the Chiquitos Indians by a pool of waters of the river Paraguay, below the lake of Los Xareyes.

MAYORUNAS, a barbarous nation of Indians, who dwell in the woods *s.* of the Marañon, between the river Cusiquina to the *e.* and Ucayale to the *w.* bounded *s.* by the nations of the Cocamas and Cocamilas nations. Some of them have been reduced to the faith.

Mayorunas, with the dedicatory title of San Ignacio, a settlement of the above Indians, being a *reduccion* of the missions that were held by the regulars of the company, in the province of Mainas and kingdom of Quito.

MAYTOL, a river of the province and *alcaldia mayor* of Los Choles in the kingdom of Guatemala; discovered in 1675 by the Father Fraucisco Gallegos, who entered amongst those mountains to the reduction of the infidel Indians, accompanied by F. Joseph Delgado, both of the order of St. Domingo. This river is large, and on its shore is the mountain called Escurruchan, but which the Indians denominate God of the Mountains. On it is burning a continual fire, kept up by their sacrifices; and they are extremely particular to throw something into the flame as they pass, firmly believing that if they did not so they would soon die.

MAYURIAGA, or MARUACAS, a large river which laves the country of the Xibaros Indians, in the province and government of Mainas, of the kingdom of Quito. It runs *s. e.* and enters the Pastata near where this river enters the Marañon.

[MAYZI, the *e.* cape of the island of Cuba, and the *w.* point of the windward passage. Lat. 20° 14′ 30″ *n.* Long. 74° 1′ 30″ *w.* See MAISY.]

MAYAHUACAN, a settlement of the head settlement of the district and *alcaldía mayor* of Zochicoatlan in Nueva España. It contains 16 families of Indians.

[MAZALTAN, a province of Mexico or New Spain. It is well watered by the Alvarado, which discharges itself by three navigable mouths, at 30 miles distance from Vera Cruz.]

MAZAPIL, a *real* of the silver mines and settlement of the head settlement of the district and jurisdiction of the bishopric of Guadalaxara in Nueva España, which serves as a limit and division of the kingdom of Nueva Galicia with that of Nueva Vizcaya; the territories of the latter being on the *n.* It has always abounded in mines of good metal; and these at first were worked to great profit by slaves, owing to the intractability and savage state of the Indians; but since that these have become domesticated, they are not only employed in this labour, but also in the agricultural estates with which the territory is surrounded. The population amounts to more than 40 families of Spaniards and *Mustees*, and many more of Mulattoes, with a great number of Indians, there being also breeders of cattle of every kind. It is 220 miles *n. n. e.* of Guadalaxara, and in lat. 24″ 27′ *n.* Long. 101° 30′ *w.*

MAZARONI, MAZARUINI, or ATAPARAN, a large and abundant river of the province of Guayana, and government of Cumaná. It rises in the interior of the province, and runs nearly from *s.* to *n.* until it enters the Essequibo just close to where this runs into the sea. The Dutch, protected by the Caribes, navigate this river to pillage the Indians of the province, whom they make slaves to work in their estates; nor are there any stratagems which avarice and tyranny can invent that are not adopted for the purpose of entrapping those unhappy wretches. It is from this policy that the Dutch are in alliance and friendship with the Caribes.

MAZATAN, a settlement of the province and *alcaldía mayor* of Copala in Nueva España; situate near the coast of the S. sea. Its population is composed nearly altogether of Mulattoes, who are under the obligation of guarding its port, it having been invaded at various times by enemies. It is of a hot temperature, abounding in maize and French beans, and particularly in fish, which is caught in abundance in the large river of its name, and on the borders of which it is situate.

MAZATEPAN, S. MARTIN DE, a settlement and head settlement of the district of the *alcaldía mayor* of Tehuacan in Nueva España. It contains 150 families of Indians, and is 16 leagues to the *e.* ¼ to the *n. e.* of its capital.

MAZATEPEC, a settlement of the head settlement of the district of Huitepec, and *alcaldía mayor* of Cuernavaca, in Nueva España; it is situate on the top of a mountain so barren and ill-supplied with water that it is necessary to fetch this article from the neighbouring rivers; on its skirt, however, is a plain, which is the most woody of any part in the whole jurisdiction; and here there are many kinds of fruit and sugar-cane, of which a certain portion of sugar is made. It has a convent of the monks of San Francisco, and contains 57 families of Indians, and 27 of Spaniards and *Mustees*.

MAZATLAN, SAN JUAN DE, a settlement and head settlement of the district of the *alcaldía mayor* of Nexapa in Nueva España. It contains 111 families of Indians, who trade in cotton, cattle, and wood, which they cut. It is 25 leagues to the *n.* of its capital.

MAZATLAN, another settlement, in the head settlement of Zapotlan, and *alcaldía mayor* of Tepec, in the same kingdom. It contains 34 families of Indians, who traffic in seeds and cotton; and is a little more than 13 leagues between the *e.* and *s.* of its capital.

MAZATLAN, another, of the head settlement and *alcaldía mayor* of La Purificacion in the same kingdom; situate on the coast of the S. sea. Its inhabitants occupy themselves in keeping a look-out after vessels, and in giving intelligence to the *alcaldes mayores*. It is six leagues to the *s.* of its capital.

MAZATLAN, another, of the head settlement and *alcaldía mayor* of Compostela in the kingdom of Nueva Galicia.

Mazatlan, another, of the head settlement of the district of Tuzcacuesco, and *alcaldía mayor* of Amola, in Nueva España. It is of an hot temperature; situate between two rising grounds on the side of a small stream, with the waters of which the inhabitants irrigate their orchards and gardens. The population consists of 70 families of Indians, and it is four leagues to the *w.* of its head settlement.

MAZIBA, a large settlement of Indians of the Saliva nation, in the *llanos* of San Juan; bordering upon the river Sinaruco. The Caribes, united with the Dutch, took and pillaged it, with other settlements, in 1684.

MAZINGA, a settlement of the province and government of Santa Marta in the Nuevo Reyno de Granada; situate near the *serranía*. The English, commanded by William Gauson, sacked it in 1665.

MAZO, a settlement of the province and *corregimiento* of Chancay in Peru; annexed to the curacy of Huahura.

MAZOLA, Punta de, a point on the coast of the province and government of Santa Marta, and Nuevo Reyno de Granada. It is 22 leagues from the cape of Chichibacoa.

MAZOTECAS, settlements of Indians of the province and government of Honduras; discovered by Don Fernando Cortes, conqueror of Nueva España, in 1574, in the route which he made to the aforesaid province. The name of Mazotecas signifies, in the Indian language, Country of Deer, from the abundance of these animals, which were discovered to be so tame and domestic that they were not frightened at the appearance of the Spaniards. The Indians being asked the cause, they said that their god had appeared to them in the figure of one of these amimals, and that therefore it was unlawful to ill-treat or kill them.

MAZTITLAN, a settlement of the head settlement of the district of Ixtác, and *alcaldía mayor* of San Juan de los Llanos, in Nueva España.

MBOAPIARI, a river of the provinice and *captainship* of Rey in Brazil. It runs *s. w.* and enters the Rio Grande.

MBOCAE, a small river of the province and government of Paraguay. It runs *s.* and enters the Paraná near the mission of Itapua.

MBOCARIRAI, a river of the province and *captainship* of San Vicente in Brazil. It runs *w.* between the Tibiquari and the Tapiguy.

MBOERI, a river of the province and government of Paraguay. It runs *w.* between this river and the Paraná, and enters the former opposite the Rio Verde, in the country of the Zamucos Indians.

MBONGUIN, a river of the province and *captainship* of Rey in Brazil. It runs *n.* and enters the Rio Grande.

MBOTETEI, a large and copious river of the province and government of Paraguay. It runs *w.* and enters the Paraguay in the part which is called El Paso, according to Don Juan de Cruz. It runs *n. n. w.*

MBUTUAY, a river of the same province and government as the former. It also enters the Uruguay between the Spau and the Naumbi.

MBUTUI, a river of the same province and government as the former. It runs *w.* and enters the Uruguay between the Ibicuy and the settlement of the mission of San Borja.

MEAD, a settlement of the Indians of Barbadoes, in the district of the parish of St. Philip.

Mead, a river of Virginia, which runs *s. s. w.* and enters the Miamee the Great.

[MEADOWS, a small river which falls into Casco bay, in the district of Maine.]

[MEADS, a place situated on a fork of French creek; a branch of the Alleghany, in Pennsylvania. Lat. 41° 36′ *n.* and about 19 miles *n. w.* of fort Franklin, at the mouth of the creek.]

MEANA, a settlement of the province and *corregimiento* of Caxamarquilla in Peru.

MEARI, or Mari, a large river of the kingdom of Brazil. It flows down from the mountains of the *w.* part, and in its course receives the waters of the Ovaquczupi, Pinare, and Marañoa or Maracon. It runs into the gulf of Marañon, and forms the island Tatipera; is navigable for small vessels for upwards of 160 leagues, and its shores are covered with many cultivated estates and sugar-cane plantations, where there are some sugar-mills. Near its source dwell some barbarian Indians, called the Tapayos and Tapures.

MECA, a creek of the coast of the S. sea, in the province and *corregimiento* of Arica in Peru.

MECALAPA, a small settlement or ward of the *alcaldía mayor* of Guanchinango in Nueva España; annexed to the curacy of Pantepec.

MECAPACA, a settlement of the province and *corregimiento* of Sicasica in Peru.

MECAPALA, a settlement of the head settlement of the district and *alcaldía mayor* of Zochicoatlan in Nueva España. It contains 46 families of Indians, and is four leagues to the *w.* of its capital.

MECAPILLO, Nuestra Senora del Pilar de, a settlement of the province and government of Tucuman, in the territory of Chaco;

a *reduccion* of the Pasainas Indians made by the Jesuits, and at present under the care of the religious order of San Francisco.

MECATINA, GREAT, an island of the gulf of St. Lawrence, near the coast of the country of Labrador, near the islands of Channel and Little Mecatin, which is another isle of the name of which we treat, but smaller.

MECATLAN, a settlement of the jurisdiction and *alcaldia mayor* of Valles in Nueva España; situate on the shore of the river of the Desague of the lake of Mextitlan, and in which there is caught abundance of fish. It produces maize and other seeds, also cotton, of which are made various fabrics used by the natives for clothing. It is annexed to the curacy of Tamanzuchale, contains 87 families of Indians, and is 31 leagues from the capital.

MECATLAN, another settlement, of the head settlement of the district of Zaltocan, aud *alcaldia mayor* of Tepei, in the same kingdom. It contains 15 families of Indians, who exercise themselves in the cultivation of maize and many sorts of fruit. It has a convent of the religious order of San Francisco, and is 14 leagues to the *w.* of the capital.

MECHICOR, a river of Nova Scotia or Arcadia in N. America. It runs *s. e.* between those of Agoniche and St. Croix, and runs into the sea at the entrance of the bay of Fundy.

MECHISES, a port of the coast in the province of Sagadahoc, between the islands of Seal and Little Menan.

MECHISES, some islands of the aforesaid province; situate at the entrance of the said port.

MECHISES, a river of the same province. It is small, runs nearly due *e.* and enters the sea.

MECHOACAN, a province and bishopric of Nueva España in N. America, its name signifying a Place of Fishing. It is bounded *n.* by Nueva Galicia, *s.* by the Pacific sea, *e.* and *s. e.* by the province of Mexico, *w.* by that of Guadalaxara, *s. w.* by that of Xalisco. Its length *s. e. n. e.* is 30 leagues from the coast of the S. sea to the boundary of the jurisdiction of Valles, near the jurisdiction of the N. where the bishopric of Mexico is bordered by this bishopric and that of Guadalaxara. Its greatest width, following the coast of the S. sea, is 80 leagues, though in parts it narrows to 60, 40, and even 30.

This province was discovered by Cristoval de Olid, chieftain of the celebrated Hernan Cortés, and conquered and taken by him from its king Calzontzi. It is for the most part of a mild temperature, and so healthy that the Indians, in the time of their gentilism, when they were sick, used to make it a resort to establish their health, nor without succeeding in their object.

It is watered by many rivers and fountains of delicious waters, and of the latter are many that are hot and medicinal. It has also various lakes, in which are caught abundance of fish. The whole of the soil is rich and fertile in vegetable productions, and especially in wheat, maize, *chile* or pimiento, French beans, calabashes, and fruits of various sorts. It produces much honey, wax, cotton, of which very good woven fabrics are made, and silk, there being a large breed of silk-worms, and these constituting the principal profit of its commerce. In the woods are found abundance of the most excellent and most precious sorts of wood; and they are full of game, as well birds as hares and rabbits. Also in the estates are an infinite number of cattle, from the quantity of pasture. In different parts of this province are found some good saline earths, also a black stone which serves for sharpening razors, and another of a fine grain resembling jet.

This country is extremely rich in mines of gold, silver, copper, and tin; and, in 1725, a mine of silver was discovered so abundant that it was called De Morcillo; it produced very much metal, but a part of the mountain having fallen in, the mine became closed, and could not since be discovered.

The natives before the conquest by the Spaniards were most warlike, robust, handsome, and very dexterous marksmen with the arrow; and never were the Mexican emperors able to bring them under the yoke of the empire.

It is divided into five jurisdictions and *alcaldías mayores*, which are the following, the capital being Valladolid.

Pasquaro,	Cuiséo de la Laguna,
Chaco,	Chilchota,
Zelaya,	Zamora,
Salvatierra,	Colima,
San Miguel el Grande,	Tanzitaro,
Guanajuato,	Guimeo,
Leon,	Cinagua,
S. Luis de la Paz,	Motines,
S. Luis de Potosi,	Tinguindin,
S. Pedro Guadalcazar,	Xiquilpa,
Maravatio,	Tlasasalca,
Tlalpujagua,	Zacatula.
Jasso y Teremendo,	

Bishops who have presided in Mechoacán.

1. Don *Fr.* Luis de Fuensalida, a monk of the order of San Francisco, one of the 12 first who passed over to Nueva España: he found himself guardian in the convent of Tezcoco, when he was

presented to the bishopric of Mechoacán by the Emperor Charles V. in 1536, but he refused the office with profound humility.

2. Don Vasco de Quiroga, native of Madrigal; elected, for his virtue and literature, to be *oidor* of the audience of Mexico, and in this nominated through the above qualities to be visitor and pacificator of the tumults that had arisen amongst all the Indians of the province; and in this calling he acted with such skill, that on information being given to the emperor of the same, he was presented to this bishopric in 1537. He removed the episcopal see from the city of Tzinzunzan, where it had been erected and had remained for a year, to to the city of Pasquaro, on account of the bad climate of the former place. He passed over to Spain in 1547, and returned to his bishopric in 1554; dedicated his time to the making of the visitation, and died full of years and virtues in the settlement of Uruapán, in 1556.

3. Don Antonio Ruiz de Morales y Molina, native of Cordoba, knight of the order of Santiago, chanter of its holy church; presented to the bishopric of Mechoacán in 1557, and promoted to La Puebla de los Angeles in 1563.

4. Don *Fr.* Alonso de la Vera Cruz, of the order of San Agustin; presented by his Majesty Philip II. to this bishopric, which he renounced for weighty reasons; when at this moment the exemplary life and conduct of *Fr.* Diego de Chaves y Alvarado having excited the public attention, on him the bishopric was conferred; though he died before he received the bulls of his consecration.

5. Don *Fr.* Juan de Medina Rincon, of the order of San Agustin, native of Segovia; he passed over to Mexico, and received the habit, was an excellent theologist, and one of the most zealous evangelical missionaries; instructed himself in the Mexican and Otomian tongue to converse with the natives; was presented to the bishopric in 1572, and was obliged to accept after having first refused it; he defended the rights of his church, assisted at the third provincial council, removed the cathedral to the city of Valladolid, where it at present stands; and whilst on the visitation died, in 1580; his character being that of an apostle, poor in spirit, mild and charitable, and rich, zealous in the honour of God.

6. Don *Fr.* Alonso Guerra, of the order of Preachers; he passed from the bishopric of Paraguay to this in 1591, founded the convent of Santa Catalina de Sena of his own order, and that of the barefooted Carmelites; and died in 1596.

7. Don *Fr.* Domingo de Ulloa, of the same order as the former; descended from the illustrious house of the Marquises of la Mota; he took the habit in the convent of Nuestrá Señora de la Peña of France, was collegiate in the college of San Gregorio de Valladolid and its rector, prior of various convents, vicar-general of the province of Castilla; presented to the bishopric of Nicaragua, promoted to the church of Popayán, and afterwards to this of Mechoacán in 1596; he governed four years with great skill and applause, and died in 1600.

8. Don *Fr.* Andres de Ubilla, of the same order of Preachers, native of Guipuzcoa; he took the habit in Mexico, and having been professor of philosophy and theology, he took different prelacies until he became provincial; he then passed over to Spain on weighty matters concerning his religion, and was presented to the bishopric of Chiapa, and from thence removed to the mitre of Mechoacán in 1602, of which he did not take possession, having died before receiving the bulls.

9. Don Juan Fernandez Rosillo, dean of the holy church of Popayán and bishop of Vera Paz; from whence he was removed to this in 1605; he only governed a year and a half, and died in 1606.

10. Don *Fr.* Baltasar de Cobarrubias, native of Mexico, a monk of the order of San Agustin; presented through the fame of his virtue and literature to the bishopric of La Asuncion of Paraguay, afterwards to the church of Cazeres in Filipinas, from thence to Antequera, and lastly to Mechoacán in 1608; and having governed in every situation with zeal and edification, he died full of merits in 1622.

11. Don *Fr.* Alonso de Enriquez Toledo y Armendariz, of the order of Nuestra Señora de la Merced, native of Sevilla; he passed to the kingdom of Peru in quality of vicar-general in those provinces, and at his return to Spain was elected bishop of Cuba in 1622, and from thence promoted to the church of Mechoacán, where he governed with great skill; and died in the settlement of Irumbo, in 1628.

12. Don *Fr.* Francisco de Rivera, of the same order as the former, native of Alcalá de Henares; after having finished his studies and filled the professorships of philosophy and theology with great approbation, he was nominated vicar-general of his religion in Nueva España; made the division of the provinces of Mexico and Guatemala; passed over to Spain, where he was nominated vicar-general of Aragon, Cataluña, and Castilla, removed to the bishopric of Guadalaxara, and from thence to Mechoacan in 1629; he visited the whole of his diocese, endowed the festival of the Most Holy Trinity, and died in 1637.

13. Don *Fr.* Marcos Ramirez de Prado, of the order of San Francisco, native of Madrid; he studied in the university of Salamanca; nominated vice-commissary-general of the Indies, when he was appointed by Philip IV. to be bishop of Chiapa, of which dignity he took possession in 1634; and was removed, in 1639, to this church of Mechoacán; in 1648 nominated by his Majesty visitor of the tribunal of Crusade, and afterwards archbishop of Mexico, where he died before he received the pall.

14. Don *Fr.* Payo Enriquez de Rivera, of the order of San Agustin, native of Sevilla; he passed from the bishopric of Guatemala to this, and on his road received advice of his promotion to the archbishopric of Mexico.

15. Don *Fr.* Francisco Sarmiento y Luna, of the same order as the former; in which, after having filled many situations as well as prelacies, he was presented to this bishopric in 1668; governed five years with great peace and quietness, and in 1674 embarked for Spain, being promoted to the bishopric of Almeria.

16. Don Francisco Berdin de Molina, promoted to the bishopric of Guadalaxara in 1674; and he died 11 months after.

17. Don Francisco de Aguiar Seixas y Villoa, native of Betanzos; he was archbishop of Santiago, studied philosophy and theology with much profit, and led such an edifying life, that this prelacy only prognosticated his future greatness; he visited La Beca of the college of Fonseca, and passed to the university of Salamanca; was magisterial canon of Astorga and penitentiary in the church of Santiago; passed to the mitre of Guadalaxara, and from thence to this of Mechoacán, and afterwards to the archbishopric of Mexico; where he died, with general regret, in 1698.

18. Don *Fr.* Antonio Monroy, of the order of Preachers; he died elected bishop of this church before he took possession.

19. Don Juan de Ortega Montañes, promoted from the church of Guatemala to this; which he enriched with the costly silver throne in which the sacrament is deposited; made endowments for female children, built the episcopal palace, in which he laid out above 80,000 dollars, passed to the vice-royalty of Mexico, and afterwards to its bishopric.

20. Don Garcia de Lagaspi, of the church of San Luis de Potosi, canon and arch-deacon of the cathedral of Mexico; presented to the bishopric of Guadiana, and promoted to this, which he held until 1704; passing then to La Puebla de los Angeles.

21. Don Manuel de Escalante Colombres y Mendoza, native of Peru; he passed over to Mexico with his father, who went as fiscal of that audience, studied in the university there, obtained the professorship of rhetoric, was morning and evening lecturer, chanter in the cathedral, commissary in the tribunal of the holy crusade; elected bishop of Durango in 1703, and the following year promoted to this, which he entered in 1706; he was so charitable that he devoted the most of his episcopal life to the relieving of other's necessities, and died in the city of Salvatierra in 1708.

22. Don Felipe Ignacio de Truxillo y Guerrero, native of Cadiz; collegiate *mayor* of San Bartolomé el Viego of the univerity of Salamanca, fiscal of the tribunal of the holy office of the city of Barcelona, inquisitor *mayor* of the tribunal of Palermo, judge in ordinary of the royal tribunal of Nápoles, abbot of Santa Maria de Terrana, governor of the sacred religion of San Juan, fiscal regent in the royal and supreme council of Italy, deputy of the kingdom of Sicily, and being nominated in the general parliament of the Cortes, he was presented to the bishopric of Mechoacán in 1711; he governed for nine years with great skill, and died in 1720.

23. Don *Fr.* Francisco de la Cuesta, of the order of San Geronimo, native of Colmenar of Oreja, in the archbishopric of Toledo; he took the habit in the royal monastery of San Lorenzo del Escorial, where he lived 32 years, and King Charles II. presented him with the archbishopric of Manila, which he governed 18 years, being at the same time president and captain-general; he was promoted to this church of Mechoacán, for which place he embarked, but the labours of the voyage so harassed him, that he was obliged to be carried from Acapulco in a litter; he governed only a month and 13 days, dying in the year 1724.

24. Don *Fr.* Angel Maldonado, native of Ocaña, a Basilican monk; presented to the bishopric of Mechoacán at the time that he was serving in that of Antequera; but he refused the offiee.

25. Don Francisco Garzeron, inquisitor of Mexico, and visitor-general of the kingdom, presented through the renunciation of the former; but he died before he received the cedule.

26. Don Juan Joseph de Escalona and Calatayud, native of the town of Quer in La Rioja; collegiate in San Gerónimo de Alcalá de Henares, and of the *mayor* of San Bartolomé el Viejo of Salamanca, penitentiary canon of the cathedral of Calahorra, and chaplain major of the royal con-

vent of La Encarnacion de Madrid; from whence he was elected bishop of Caracas, from thence promoted to Mechoacán in 1729, governed eight years with great virtue, and his sanctity was borne testimony to by his blood remaining uncorrupt for seven years after his death; this took place in the estate of the Rincon in 1737.

27. Don Francisco Pablo Matos Coronado, born in the Canaries: after that his letters had gained the general approbation in the universities of Sevilla and Salamanca, he was presented to the church of Yucatán, and from thence removed to this of Mechoacán, which he governed with extreme tranquillity; he was much beloved for his talents and charity; he passed over to Mexico for the recovery of his health, and died there in 1744.

28. Don Martin de Elizacochea, originally of the town of Azpilcueta in Navarra; he studied in the university of Alcalá, where he graduated as doctor in theology, was canon of the holy church of Mexico, where he had the dignified titles of school-master and dean, was chancellor of that university, commissary apostolic, subdelegate of the tribunal of the holy crusade, bishop of Cuba, afterwards of Durango, until 1745, when he was promoted to that of Mechoacán, where he erected the sumptuous temple of Santa Rosa, endowing the collegiates, constructed the episcopal prisons, and did innumerable acts of charity; he died in 1756.

29. Don Pedro Anselmo Sanchez de Tagle, originally of Santillana, bishop of Santander; he studied in the universities of Valladolid and Salamanca; was collegiate of the *mayor* of San Bartolomé el Viejo, graduated as licentiate of canons for the chapel of Santa Barbara, was rector and deputy of his university, fiscal and inquisitor of Mexico, bishop of Durango; and from thence removed to this church of Mechoacán in 1757; he died in 1772.

30. Don Luis Fernando de Hoyos y Mier, elected in the aforesaid year, 1772; he died in 1776.

31. Don Juan Ignacio de la Rocha, elected in 1777; he died in 1783.

32. Don *Fr.* Antonio de San Miguel, promoted from the bishopric of Comayagua to that of Mechoacán in 1784.

[MECKLENBURG, a county of Virginia, bounded *n.* by the state of N. Carolina. It contains 14,733 inhabitants, of whom 6762 are slaves.]

[MECKLENBURG, a county of N. Carolina, in Salisbury district, bounded *s.* by the state of S. Carolina. It contains 11,395 inhabitants, of whom 1603 are slaves. Chief town, Charlotte.]

[MECOWBANISH, a lake in N. America, in lat. 49° *n.*]

MEDANO, a settlement of the province and government of Buenos Ayres in Peru; situate at the source of the river of Las Conchas, to the *s.* of the capital.

MEDELLIN, a settlement of the jurisdiction and government of Vera Cruz in Nueva España; founded by Hernan Cortes, who gave it this name in honour of his native place in Estremadura. It is of a hot temperature, and has at two leagues distance the river called Enmedio, where its jurisdiction terminates in that part. It is three league *n. w.* of the capital.

MEDELLIN, a river of the former jurisdiction and kingdom. It rises near the city of San Ildefonso de los Zapotecas, runs nearly from *e.* to *w.* and runs into the sea at the gulf of Campeche, opposite the island of Sacrificios. It was discovered by Hernan Cortes, who named it as well as the above settlement.

MEDELLIN, a town of the province and government of Antioquia in the Nuevo Reyno de Granada; situate on the shore of the river San Bartolomé, which empties itself into the Magdalena.

[MEDFIELD, a township in Norfolk county, Massachusetts, 20 miles *s. w.* of Boston. It was incorporated in 1650, and contains 731 inhabitants.]

[MEDFORD, a pleasant, thriving, compact town in Middlesex county, Massachusetts, four miles *n.* of Boston; situated on Mystic river, three miles from its mouth. Here are several distilleries and brick-works, which give employment to a considerable number of people. The river is navigable for small vessels to this place. The township was incorporated in 1630, and contains 1029 inhabitants, who are noted for their industry.]

MEDINA, a settlement of the province and government of Los Llanos in the Nuevo Reyno de Granada; founded by *Fr.* Alonso Ronquillo, of the order of St. Domingo, in 1670. It is of an hot temperature, but abounding in wild *cacao*, and other vegetable productions, such as maize, *yucas*, and dates. It has a vein of black virgin salt in a very strong rock, which is carried to every part of the province for the cattle.

MEUINA, a river of the kingdom of Nueva Galicia in N. America, which runs into the sea at the bay of San Joseph, of the bay of Mexico.

MEDIO, RIO DEL, a river in the island of St. Domingo, which rises in the *sierras* of the mines of

Ciboo. It runs *s. s. w.* making a curve, and enters the Jacques, a little before this runs into the Neiba.

MEDIO, another river, in the province and government of Buenos Ayres. It runs *n. n. e.* and enters the Paraná, between those of Pabon and Ramallos.

[MEDOCTU, a settlement in New Brunswick; situated on the *w.* side of St. John's river, 37 miles above St. Ann's.]

[MEDOROSTA, a lake in the *n.* part of the district of Maine, whose *n.* point is within eight miles of the Canada line, in lat. 47° 56′ and long. 68° 22′ *w.* It gives rise to Spey river, which runs *s. s. e.* into St. Jobu's river.]

MEDOUSA, a lake of Nova Scotia or Acadia, in N. America. It is formed by the river Pistoles, and empties itself into the San Juan.

[MEDUNCOOK, a plantation in Lincoln county, district of Maine, 230 miles from Boston, having 322 inhabitants.]

[MEDWAY, a township in Norfolk county, Massachusetts, bounded *e.* and *s.* by Charles river, which separates it from Medfield, and of which it was a part until 1713. It has two parishes of Congregationalists, and contains 1035 inhabitants. It is 23 miles *s. w.* of Boston, on the middle post-road from thence to Hartford.]

[MEDWAY, or MIDWAY, a settlement in Liberty county, Georgia, formed by emigrants from Dorchester in S. Carolina, about the year 1750, and whose ancestors migrated from Dorchester and the vicinity of Boston about the year 1700. A handsome Congregational meeting-house, belonging to this settlement, was burnt by the British during the war, and the settlement was destroyed. It has since recovered, in a considerable degree, its former importance. Medway is 26 miles *s. w.* of Savannah, and nine *w.* of Sunbury.]

MEGENA, a small river of the province and government of Guayana or Nueva Andalucia, one of those which enter the Orinoco by the *e.* side.

MEGUIN, a river of the district of Guadalabquen in the kingdom of Chile. It runs *w.* and enters the sea, between the point of Quenli and the Morro Bonifacio.

MEHANECK, a settlement of the English, in the territory and country of the Iroquees Indians, and on the confines of Pennsylvania; situate on the shore of the river Ohio.

[MEHERRIN, a principal branch of Chowan river, in N. Carolina, which rises in Charlotte county, Virginia; and running an *e.* by *s.* course, unites with the Nottaway about seven miles *s.* of the Virginia line. See CHOWAN River.]

MEJARI, or MEARIM, a river of the province and *captainship* of Marañan in Brazil. It rises in a lake in the mountains of the Topinambes Indians, runs nearly due *n.* and enters the sea in the bay formed by the mouth of the river Marañan.

MEJORADA. See COTUY.

[MELA. See MALA.]

[MELAQUE Port, on the *w.* coast of New Mexico, is to the *n. w.* of port Natividad or Nativity, and near three leagues at *s. e.* from a row of four or five rocks or naked islands above water, in the direction of *n. w.* This port is land-locked against all winds from the *n. w.* to the *s. w.*]

[MELAWASKA, a French settlement of about 70 families, secluded in a singular manner from the rest of mankind, in the *n. e.* part of the district of Maine. These people are Roman Catholics, and are industrious, humane, and hospitable.]

[MELETECUNK River, in Monmouth county, New Jersey, falls *e.* into Beaver Dam, which is at the head of the bay which is *n.* of Cranberry New Inlet.]

[MELFORD's Place, on Tallapoosee river, in the *w.* part of Georgia, is separated from some Indian towns by that river, a considerable distance from its mouth.]

MELGAR, SAN JUAN DE, a settlement of the jurisdiction of Tocaima, and government of Mariquita, in the Nuevo Reyno de Granada. It is extremely hot, and abounding in fruits of this climate, such as maize, *yucas*, dates, and sugarcane.

MELILLA, a city in the island of Jamaica, one of the first that were founded there by the Spaniards, and abandoned shortly after from the badness of the territory. It stood towards the *w.* and it was removed, with the name of Sevilla, towards the *n.*

[MELINCUE, a parish of the province and government of Buenos Ayres; situate on a plain between the Tercero and Saladillo rivers, in lat. 33° 44′ 30″ *s.* Long. 61° 49′ 56″ *w.*]

MELINQUE, a settlement of the province and government of Tucumán in Peru; situate in the extremity of the Pampas, where this jurisdiction is divided from that of Buenos Ayres.

MELIPILLA, a province and *corregimiento* of the kingdom of Chile; bounded *e.* by the jurisdiction of St. Iago, the river Mapocho serving as the limits; *w.* by the coast, and *s.* by the province of Rancagua, the river Meipo running between. In the *w.* part is a convent of the religious order of San Francisco, called Del Monte. The extent of this province is very limited; its vegetable productions are barley, wheat, and other seeds; it has a good quantity of cattle and wine of excellent quality. On the coast is found much fish, especially on the coast of the mouth of the river Maipo and the port of San Antonio. The inhabitants, who amount to 3000, live for the most part in estates in the country, so that the villages or towns are but few. Through it passes the road which is traversed by the carts going from St. Iago to Valparaiso.

This province is of small extent upon the sea, but is about 25 leagues from e. to *w.* Its rivers are the Mapocho and Poangue, and it abounds, as before observed, with wine and grain. Melipilla, or St. Joseph de Logioño, situated not far from the Maypo, in lat. 32° 32′, is the capital. Although the situation of this place is beautiful, and the land near it very fertile, yet from its vicinity to St. Iago, where the greater part of the proprietors reside, it is but thinly peopled. Notwithstanding, besides a parish church, the Augustins and the Mercedarii have establishments there, and the Jesuits had also a college. Near the river Mapocho is the town of St. Francis del Monte, so called from an ancient convent of Franciscans, around which a number of poor families having collected, formed the population of this place. In its vicinity are several country houses belonging to some of the principal inhabitants of St. Iago. Not far from the mouth of the river Maypo is the port of St. Antonio, which was much frequented at an early period of the Spanish settlement; but since the trade has been transferred to Valparaiso, few or no vessels continue to load there.

MELIRUPU, a settlement of Indians of the kingdom of Chile; situate at the source of the river Cauchupil.

MELLAQUI, a small river of New France or Canada. It runs *s.* between lake Kitchigamin and that of Mitchigan, and turning *e.* enters the latter.

MELLO, a port of the coast of the N. sea, in the province and *captainship* of Seara in Brazil. It lies between the river Omoneses and the Salinas Grandes.

MELO, a town of the province and government of Buenos Ayres; situate at the foot of the *sierra* De S. Paulo, on a small branch of the river Taquari. Lat. 32° 23′ 14″ *s.* Long. 54° 17′ 24″ *w.*

MELONES, a small and desert island of the S. sea, in the bay and gulf of Panamá.

MEMAROBKE, a small lake of the province and country of the Iroquees Indians; *s.* of the river St. Lawrence, on the confines of New England.

MEMBRILLAR, a settlement of Indians of the district of Puchacay in the kingdom of Chile; situate on the shore of the river Itata, in the part where the ferry is.

MEME, a river of the province and government of Esmeraldas in the kingdom of Quito, flowing down from the mountain of Guanas. It runs *n. w.* and enters by the *e.* shore into the Toachi, in lat. 13° 34′ *s.*

MEMKECHKAOUCK, a small island near the coast of the province of Nova Scotia or Acadia, near cape Sable. It is one of those which the French call Loups Marins.

MEMNISTE, a bay of the *s.* coast of the straits of Magellan, five leagues from the bay of Mauricio towards cape Deseado. It was discovered by the English in 1600, and the pilot of the vessel, who was a Menmistan Anabaptist, gave it this name.

[MEMORONCOK, a stream a little *w.* of Byram river. Douglass says the partition line between New York and Connecticut, as settled Dec. 1, 1664, ran from the mouth of this river *n. n. w.* and was the ancient limits of New York, until Nov. 23, 1683, when the line was run nearly the same as it is now settled.]

[MEMORY Rocks, amongst the Bahama islands; situate 19 miles to the *n.* of Great Bahama island, in lat. 27° 4′ *n.* and long. 78° 49′ *w.*]

[MEMPHREMAGOG, a lake chiefly in the province of Canada, 19 miles in length from *n.* to *s.* and two or three wide from *e.* to *w.* The *n.* line of Vermont state passes over the *s.* part of the lake, in lat. 45° *n.* Memphremagog, which has communication by the river St. Francis with St. Lawrence river, is the reservoir of three considerable streams, viz. Black, Barton, and Clyde rivers, which rise in Vermont. The soil on its banks is rich, and the country round it is level. See VERMONT, &c.]

[MEMRAMCOOK River has been recommended as the most proper boundary between the province of New Brunswick and Nova Scotia. It lies a little to the *e.* of Petitcodiak, and takes a *n. e.* direction.]

MEMRUNCOOK, a settlement of Indians of the province of Nova Scotia or Acadia, in N. America; situate at the mouth of the river Patcolyeak.

[MENADOU Bay, or PANADOU, is two leagues from port Baleine, or port Neuf, on the coast of Cape Breton island, at the *s.* part of the gulf of St. Lawrence, having the island of Scatari, heretofore called Little Cape Breton, opposite to it.]

MENA-HERMOSA, Santo Domingo de, a settlement of the province and government of Tarma in Peru, with a small fort for its defence; as it is situate in the frontier of the rebellious Chunchos Indians. It was founded by Don Joseph de Llamas, Marquis of Mena-hermosa, who gave it his name. He was also general of Callao and of the armies in Peru, in 1744, when he passed, by a commission of the viceroy, into that kingdom, in order to settle the disturbances in this province, this settlement being made the military rendezvous.

MENAN, the name of two islands, the one larger than the other, and distinguished by Menan Great, and Menan Small, near the coast of the province and colony of Sagadahook, between the island of Pascamadie and the port of Mechises.

MENCHIXEQUE, or Menchiquijo, as others call it, a settlement of the province and government of Cartagena, in the district and division of the town of Mompox; situate on the shore of the river Magdalena, to the *s.* of that town.

MENCOPA, a settlement in the province and government of Tucumán in Peru, of the district and jurisdiction of the capital; situate *s. s. e.* of the same.

MENDAN, a settlement and *asiento* of the silver mines of the province and *corregimiento* of Chachapoyas in Peru; annexed to the curacy of Quillay.

[MENDHAM, a township in Morris county, New Jersey, three miles *n. w.* of Veal town, and six *w.* of Morristown.]

[MENDOCIN, a cape on the *n. w.* coast of America, and N. Pacific ocean. Lat. 40° 25′ *n.* Long. 124° 25′ *w.*]

[MENDON, a post-town in Worcester county, Massachusetts, 32 miles *s. w.* of Boston, and 24 *n. e.* of Pomfret in Connecticut. This township, called Quanshipauge by the Indians, was incorporated in 1667, and contains two Congregational parishes, a society of Friends, and 1555 inhabitants. It is bounded on the *s.* by the state of Rhode Island. It is watered by Charles and Mill rivers, and other small streams, which serve five grist-mills, two saw-mills, two clothier's works, and a forge. There are three hills here, viz. Caleb's, Wigwam, and Miskee, from either of which may be seen, in a clear day, the four New England states.]

MENDOZA, a city, the capital of the province and *corregimiento* of Cuyo, bearing also this latter name, in the kingdom of Chile: founded by Don Garcia Hurtado de Mendoza, Marquis of Cañete, he being the president, and giving it his name in 1559, and not in 1593, as is asserted by the ex-jesuit Coleti. Its situation is pleasant and beautiful, on the *e.* part of the *cordillera* of the Andes, and on a level plain. It is of a good size, and has some fine buildings, all of the houses having their respective orchard or garden: also to irrigate it there are some reservoirs formed from the river of its name, the which cause it to be extremely fertile and abundant in fruit and vegetables. The climate is mild and pleasant. It has a very good parish church, and some convents of the orders of San Francisco, and St. Domingo, San Augustin, La Merced, an hospital of Bethlemites, a church with the dedicatory title of Santa Barbara, destined for the establishment of a monastery of nuns; four chapels of ease; of the names of Nuestra Señora de Nieva, San Vicente, San Joseph, and Nuestra Señora de Buen Viage; and a college which belonged to the Jesuits. Its population is composed of about 300 families, the half Spaniards and whites, and the other half of *Mustees*, Mulattoes, and Negroes. It lies in the direct road to Peru, and is 95 miles to the *e.* of the city of Santiago, but the journey from thence is very rough. It lies on the shore of a river of its name, and *e. s. e.* of the volcano of Santiago. On the *n.* part it has various estates, such as those of Cienega and El Agua del Chayado; and between this city and that of San Luis de la Punta dwell the Plainches and Pehuenches Indians, who, mixed with the Hubliches and Moluches, descendants of the ancient Aucas, with many outlaw Creoles and Mulattoes, have various settlements and dwellings; from whence they sally forth to plunder and destroy the inhabitants of the jurisdictions of Mendoza and of Cordoba in the province of Tucumán. This city is in lat. 32° 52′ 30″ *n.* and long. 68° 58′ *w.*

Mendoza, a large and abundant river of the same province as the former city. It rises in the *cordillera*, and running *e.* collects in its course the waters of all the streams which flow down from those mountains, and shortly after forms the lakes of Huanacache or Guanacha, which run 20 leagues, forming various others; and from the last of these lakes this river issues into the river of Tunuyán by a wide trench called the Desaguadero, near the settlement of Corocorto. It also forms many islands, and empties itself by different mouths. It meets in one of its courses a mountain which it has washed completely through, forming a bridge over which three carts may pass abreast. Its arch is adorned with figures and points, being an efflorescence of stone, resembling the concretion of salt particles, forming an enchanting appearance, and such as could never be equalled by art. Below this natural bridge, which is called *of the Inca*, is a fine tablet of stone, which serves as a pavement, and from which gush out boiling five streams of salt water. This river enters the sea with the name of Colorado. See Chile, Chap. IV. Sect. 29.

MENDOZINO, Card, an extremity of the coast of California, which looks to the S. sea. It is near White cape, in lat. 41° and long. 122°.

MENEMOCKACI, a small river of the province and colony of Virginia in N. America. It runs *w.* and enters the Ohio between the rivers Scalp and Molcochinecon.

MENEARO, a settlement of the province and *corregimiento* of Caxamarquilla in Peru.

[MENICHLICK Lake, in the *n. w.* part of N. America; *n.* of this is lake Dohount.]

[MENIOLAGOMEKAH, a Moravian settlement *e.* of the Great swamp, at the head of Lehigh river in Pennsylvania, about 33 miles *n. w.* by *n.* of Bethlehem.]

[MENOLOPEN, a wealthy and pleasant farming settlement in Monmouth county, New Jersey; making a part of a rich glade of land extending from the sea *w.* to Delaware river. It is 18 miles *s. e.* of Princeton.]

MENORES, a settlement of the province and government of Santa Marta in the Nuevo Reyno de Granada, of the district and division or jurisdiction of the Rio del Hacha; situate near this city, in the road which leads to Maracaibo.

MENTOS, a settlement of Indians, of the province and government of Luisiana, where the French have built a fort. It is situate on the shore of the river Akansas.

MENZABE, a settlement of the *alcaldía mayor* of Natá in the province and government of Tierra Firme; situate on the shore of a river on the coast of the S. sea, opposite the isle called Farallon de Guararé.

MENZAPA, San Francisco de, a settlement of the head settlement of the district of Tenantitlan, and *alcaldía mayor* of Acayuca, in Nueva España. It contains 73 families of Indians, and is eight leagues to the *e.* one quarter to the *n. e.* of its head settlement.

[MERASHEEN Island, in Placentia bay, Newfoundland island.]

MERASSI, an island or neck of land of the Atlantic sea, near the coast of Guayana, opposite the mouth or entrance of the river Surinam. It is about 70 miles long from *e.* to *w.* and by this part covers and defends the bay of Surinam. The river Ceneca or Cottica divides this island from the continent: the climate here is very hot, and it is little cultivated. In lat. 6° *n.*

MERCADEN, or Mercaderes, a small settlement of the province and government of Popayán; situate on the *s.* of the rivers Mayo and San Jorge, in the high road which leads to Quito; and it is the first settlement that is met with in leaving this kingdom, and from whence you proceed to La Herradura, in lat. 1° 46′ *n.*

MERCALO, Punta de, a point on the coast of the S. sea, and province and government of Veragua in the kingdom of Tierra Firme. It lies between the mountain of Puercos and the Punta Blanca.

MERCED, La, a settlement of the province and government of Sonora in Nueva España.

Merced, La, another settlement, of the district of Chanco in the kingdom of Chile; situate between the river Tinguiririca and the Estero of Chimbarongo. It is of the *corregimiento* of Maule, and lies at the source of the river Rapél.

MERCEDES, a settlement of the province and *corregimiento* of Cuyo in the kingdom of Chile; situate on the shore of the river Concaray.

[MERCER, a county of Kentucky, adjoining Woodford, Shelby, and Madison counties. Harodsburgh is the chief town.]

[Mercer's Creek, in the *n. e.* part of the island of Antigua in the W. Indies, is a pretty deep inlet of the coast, the entrance to which is between the islands of Codrington, Crumps, or Pelican. Lavicount's island is a small island, also within it, towards the *s.* shore; and in the *s. w.* part of it is Farley's bay, at the mouth of a river.]

[MERCERSBOROUGH, a village of Pennsylvania, *s. e.* of N. mountain, and about 13 miles *s. w.* of Chambersburgh.]

[MERCHANT'S Careening Place, within the harbour of Port Royal in Jamaica, on the *n.* side of the long peninsula. Along this narrow slip of beach is the only way to pass by land to Port Royal, for nine or 10 miles, the careening place being almost at midway, but somewhat nearer to the *e.* end of the peninsula.]

[MERCY, Cape of God's, the most *s.* point of Cumberland's island, on the *n.* side of Cumberland straits, and has cape Walsingham on its *n. e.* and Exeter sound on its *n.*]

MERE, La, or The Mother, a small island near the coast of the province and government of Guayana, in the part possessed by the English. It is opposite the mouth of the river Orapu.

[MEREDITH, Cape, among the Falkland islands in the S. Atlantic ocean, is between port Stephen's and cape Orford.]

[Meredith, a township in Strafford county, New Hampshire; situated on the *w.* side of lake Winnipiseogee, 15 miles *n.* of Gilmantown, nine *s. e.* of Plymouth, and 52 *n. w.* of Portsmouth. It was incorporated in 1768. In 1775, it contained 259, and in 1790, 881 inhabitants. It was first called New Salem.]

MERI, an arm of the river Orinoco, in the province and government of Guayana. It forms a large island opposite the coast of the Hovaroux Indians.

MERIDA, a city of the Nuevo Reyno de Granada, the capital of the government, in the province of Venezuela; founded in 1558 by Captain Juan Rodriguez Suarez, with the name of Santiago de los Caballeros, in the province of Las Sierras Nevadas, the surname of Merida being added to it (and by which it is at present known) in memory of his native place in Estremadura. This is as the case really is, and not so, that it was founded by Jmm de los Pinos in 1562, as according to the ex-jesuit Coleti. In the year after its foundation it was removed by Captain Juan de Maldonado to the spot where it now stands; this being a pleasant plain surrounded by three rivers, the first called Chama, which is the largest, and could not in the time of the Indians be forded, being now passed *en taravita*, or by cords; the second Mucusin; and the third Albarregas, which surrounds the city, and the water of which is the best, being that which is used both for drinking and washing. At the distance of three leagues these rivers unite, and are entered by another smaller river, which passes at uo great distance from the city. This enjoys the peculiarity of experiencing every day the four seasons of the year; since there are 12 hours of cold, in proportion to its climate, from six in the evening till six in the morning; five hours of spring, from six to ten o'clock, when the heat of the sun begins to shed a warmth over the Sierra Nevada; and from ten till six in the evening the heat is very great. It is surrounded by mountains, and in its vicinity is one in which there is a lake, and which is called Mountain de las Flores, (of Flowers), from the variety which it produces, together with laurels and other trees and plants, which cover it and render it pleasing to behold, its charms being heightened by a great variety of birds. This city is of a very healthy temperature, fertile, and abounding in wheat, maize, *papas*, *arracaches*, cabbages, exquisite *cacao*, in which consists its principal commerce, and which is highly esteemed, much cotton, delicate fruits, such as apples, peaches, quinces, pomegranates, dates, and other fruits of an hot and cold climate. It has a beautiful parish church, and at its entrance four chapels or hermitages, some convents of the religious orders of St. Francis, St. Domingo, St. Agustin, an hospital, a college which belonged to the Jesuits, and a monastery of nuns of the order of Santa Clara. The buildings are very good, and the streets wide; the inhabitants should amount to 400 housekeepers, and it would be far more populous and rich did not the party spirit and bickering between the Serradas and Gavirias, two classes of its first settlers, continue to perpetuate animosity amongst their descendants, causing many murders and losses both of fortunes and estates. It has suffered much by earthquakes, and more particularly in those which happened successively in 1644, and which left it nearly desolate. It is, at the present day, the head of a bishopric, erected in 1782, suffragan of St. Domingo, and afterwards of the archbishopric of Santa Fé, its first bishop having been Don *Fr.* Juan Marcos de Lora, of the order of San Francisco. It has gold mines which are not worked.

[The city of Merida, which was founded in 1558 by John Rodriguez Suarez, under the name of Santiago de los Caballeros, is situate in a valley of about three leagues long, and in the widest part about three quarters of a league broad. It is surrounded by three rivers: the first is named Mucujun, and has its source to the *n.* in what is called Los Paramos de los Conejos, (or the Rabbit Heaths); it flows from it to *s.* and runs through the *e.* part of the town. The second is called Albarregas, and rises to the *n. e.* and running to the *w.* of the town. The third is the Chama; it flows from the *e.* to the *n.* round the *s.* side of the town until it falls into the lake of Maracaibo. It receives the two first-mentioned rivers at a little distance from Merida, and by the junction also of a multitude of other rivers it at length acquires an immense size. There are wooden bridges for foot-passengers and horses over these three rivers, which are constructed so substantially as to stand throughout all the seasons. Not one of them is navigable, on account of the rapidity of the current and the obstacles of rocks, as well as of the mountains, which, by making the beds of the rivers narrow in certain places, form cascades too powerful for any vessel whatever to stem.

Another strong reason why these difficulties are not overcome, is the excessive insalubrity of the part of the lake of Maracaibo where it receives the river Chama. It is indeed impossible to pass two hours at this place without catching a fever, the malignity of which generally proves fatal. The soil is all that benefits by the rivers; and, to the

2

[praise of the inhabitants they have, by their activity, taken advantage of these favours of Nature. At some distance from the town are plantations of sugar, cocoa, and coffee of a very superior quality to what is cultivated throughout the rest of the province.

The environs of Merida are covered with the provisions of the country, fruits, limes, beans and pease of every kind, potatoes, wheat, barley, &c. These are all consumed by the people of Merida, and are so abundant that the poorest have always more than sufficient food. Their butchers meat comes from Varinas and Pedraza, it is very good and exceedingly cheap.

The climate is very changeable, almost every day exhibits (as already observed) the four seasons of the year. The people, however, assert that they never feel inconvenience from cold or heat, and can wear indifferently silk or worsted throughout the year; but it is certain that the transitions are so rapid and severe as to cause frequently disorders. The *w.* wind is especially dreaded; it never blows without leaving the effects of its malignity. The rains fall at all seasons, and are very heavy, but they are most violent between March and November: they have intervals between the rains.

Merida is the seat of a bishop and has a chapter. It has a college and a seminary in which the catholic clergy and all other professions are instructed. There are masters for reading, writing, and arithmetic, and professors of theology, philosophy, ethics, and the civil and canon law. These schools are all under the management and inspection of a governor and vice-governor, and are immediately under the authority of the bishop.

The sciences had made so much progress at Merida, that they resolved to obtain a university, which would relieve them from going to Santa Fé or Caracas to obtain their degrees. They sent in 1801 the vice-governor of the college to the university of Caracas to entreat them to approve of the demand they wished to make of his Catholic Majesty for the establishment of a university. This question was scrupulously examined, and, in spite of the talents and personal qualities of the delegate, the decision was against the views of the applicants. This refusal, more likely to irritate than quell their desire, did not repulse the partisans of the new university. Their demand has been transmitted to the king. It certainly will not be easily complied with, for the system of the government is not to multiply this sort of establishments.

Independently of the cathedral, there are at least as many chapels in Merida as are proportionate to the number of the inhabitants. There are three convents of the order of St. Dominic, St. Augustin, and St. Claire. A church of a suppressed convent of Cordeliers is supported with care. That of the hospital is remarkable; after these succeed the chapels of Millay, Mucugun del Espego, and De Uano; and lastly, the chapel of Mercy.

The number of inhabitants of all descriptions amount to 11,500. The slaves are the least numerons class. The whites have for a long time been split into two factions; the Serradas and Guavirias, the names of the two principal founders of the city, who had vowed a hatred against each other, and which has been perpetuated by their descendants, so that the feud cannot be considered as extinct, although the violence, formerly so frequent, has not latterly broken out. But for this the population would be greater, and the state of cultivation more flourishing.

A frankness, a spirit of justice, and a love of literature, are observable among these people. No class despises labour. The cultivation of the soil, the rearing of cattle, or the pulpit, are the employments of the whites. The people of colour exercise useful employments. Manufactories of cotton and wool are carried on here, and the different articles are so cheap as to give them a preference over those of Europe. Among these manufactures are carpets made of the wool of the country, an ell long and more than a half of an ell wide, ornamented with flowers and worked with the plants of the country: the red, green, blue, and yellow, are as bright and as permanent as the tints of the most famous manufactories of Europe. It is 112 miles *n.e.* of Pamplona, in lat. 8° 11′ *n.* Long. 70° 58′ *w.*

Merida, another city, the capital of the province and government of Yucatán in N. America, and kingdom of Nueva España; founded by Captain Francisco de Montejo in 1542. It is well situate, and has beautiful buildings, and streets wide and running straight from *e.* to *w.* and divided by others which intersect them and form certain equal squares; the chief square is also handsome and very large, and is entered by eight streets, the front of which is that looking to the *e.* and in which is the cathedral church, one of the handsomest of any in Nueva España; another, which looks *n.* containing the governor's house, and another looking *s.* composed of houses which were built with much magnificence by the founder. The territory is level, that the streams can scarcely run down the streets, and on this account there are many wells for holding the water. It is the head of a bishopric, erected in 1518, has too curacies for the administration of the sacraments, one for the Spaniards, another for the Indians, being also used by five settlements or wards of the city, called Santiago, Santa Ana, Santa Lucia, Santa Catalina, and San Sebastian; two convents of the religious order of San Francisco, one called the Grande, and is magnificent, the other the church of Nuestra Señora de la Mejorada, built in imitation of that of Doña Maria de Aragon in the court of Madrid, and close to this the parish church of San Cristoval for the Indians, an hospital of San Juan de Dios, a college which belonged to the abolished order of Jesuits, and a monastery of nuns. Its population is composed of only 400 housekeepers, having much diminished through an epidemic disorder experienced in 1548; but the number of *Mustees*, Mulattoes, and Negroes, is very great. It is 28 miles from the sea-coast, in lat. 20° 50′ *n*. Long. 89° 30′ *w*.

[MERIDA, Intendancy of. This intendancy, concerning which valuable information has been furnished to us by M. Gilbert, comprehends the great peninsula of Yucatán, situated between the bays of Campeche and Honduras. It is at cape Catoche, 51 leagues distant from the calcareous hills of cape St. Antony, that Mexico appears, before the irruption of the ocean, to have been joined to the island of Cuba.

The province of Merida is bounded on the *s*. by the kingdom of Guatemala, on the *e*. by the intendancy of Vera Cruz, from which it is separated by the Rio Baraderas, called also the River of Crocodiles, (*Lagartos*), and on the *w*. by the English establishments which extend from the mouth of the Rio Hondo to the *n*. of the bay of Hanover, opposite the island of Ubero, (Ambergris key). In this quarter, Salamanca, or the small fort of San Felipe de Bacalar, is the most *s*. point inhabited by the Spaniards.

The peninsula of Yucatán, of which the *n*. coast from cape Catoche, near the island of Contoy, to the Punta de Piedras, (a length of 81 leagues), follows exactly the direction of the current of rotation, is a vast plain intersected in its interior from *n. w.* to *s. w.* by a chain of hills of small elevation. The country which extends *e*. from these hills towards the bays of the Ascension and Santo Spirito appears to be the most fertile, and was earliest inhabited. The ruins of European edifices discoverable in the island Cosumel, in the midst of a grove of palm trees, indicate that this island, which is now uninhabited, was at the commencement of the conquest peopled by Spanish colonists. Since the settlement of the English between Omo and Rio Hondo, the government, to diminish the contraband trade, concentrated the Spanish and Indian population in the part of the peninsula *w*. from the mountains of Yucatán. Colonists are not permitted to settle on the *e*. coast on the banks of the Rio Bacalar and Rio Hondo. All this vast country remains uninhabited, with the exception of the military post (*presidio*) of Salamanca.

The intendancy of Merida is one of the warmest and yet one of the healthiest of Equinoctial America. This salubrity ought undoubtedly to be attributed, in Yucatán as well as at Coro, Cumana, and the island of Marguerite, to the extreme dryness of the soil and atmosphere. On the whole coast from Campeche, or from the mouth of the Rio de San Francisco to cape Catoche, the navigator does not find a single spring of fresh water. Near this cape Nature has repeated the same phenomenon which appears in the island of Cuba, in the bay of Xagua, described by us in another place. On the *n*. coast of Yucatán, at the mouth of the Rio Lagartos, 400 metres from the shore, (1312 feet), springs of fresh water spout up from amidst the salt water. These remarkable springs are called the Mouths (*Boccas*) *de Coni*. It is probable, that from some strong hydrostatical pression, the fresh water, after bursting through the banks of calcareous rock, between the clefts of which it had flowed, rises above the level of the salt water.

The Indians of this intendancy speak the Maya language, which is extremely guttural, and of which there are four tolerably complete dictionaries, by Pedro Beltan, Andres de Avendano, Fray Antonio de Ciudad Real, and Luis de Villalpando. The peninsula of Yucatán was never subject to the Mexican or Aztec kings. However, the first conquerors, Bernal Diaz, Hernandez de Cordova, and the valorous Juan de Grixalva, were struck with the advanced civilization of the inhabitants of this peninsula. They found houses built of stone cemented with lime, pyramidal edifices (*teocallis*) which they compared to Moorish mosques, fields enclosed with hedges, and the people clothed, civilized, and very different from the natives of the island of Cuba. Many ruins, particularly of sepulchral monuments, (*guacas*), are still to be discovered to the *e*. of the small central chain of mountains. Several Indian tribes have preserved their independence in the *s*. part of this hilly district, which is almost inaccessible from thick forests and the luxuriance of the vegetation.

The province of Merida, like all the countries of the torrid zone, of which the surface does not rise more than 1300 metres (4264 feet)] [above the level of the sea, yields only for the sustenance of the inhabitants maize, *jatropha*, and *dioscorea* roots, but no European grain. The trees which furnish the famous Campeche wood (*hæmatoxylon Campechianum L.*) grow in abundance in several districts of this intendancy. The cutting (*cortes de palo Campeche*) takes place annually on the banks of the Rio Champoton, the month of which is *s.* from the town of Campeche, within four leagues of the small village of Lerma. It is only with an extraordinary permission from the intendant of Merida, who bears the title of Governor Captain-general, that the merchant can from time to time cut down Campeche wood to the *e.* of the mountains near the bays of Ascension, Todos los Santos, and El Espirito Santo. In these creeks of the *e.* coast the English carry on an extensive and lucrative contraband trade. The Campeche wood, after being cut down, must dry for a year before it can be sent to Vera Cruz, the Havanah, or Cadiz. The quintal of this dried wood (*palo de tinta*) is sold at Campeche for two piastres to two piastres and a half (from 8*s.* 9*d.* to 10*s.* 11*d.*) The hæmotoxylon, so abundant in Yucatán and the Honduras coast, is also to be found scattered throughout all the forces of Equinoctial America, wherever the mean temperature of the air is not below 22° of the centigrade thermometer (71° of Fahrenheit.) The coast of Paria, in the province of New Andalusia, may one day carry on a considerable trade in Campeche and Brazil (*cæsalpinia*) wood, which it produces in great abundance.

The most remarkable places of the intendancy of Merida are, the capitol of this name, Campeche, and Valladolid. Population in 1803, 465,800; extent of surface in square leagues, 5977; number of inhabitants to the square league, 81.]

In Merida, the capital, which is 10 leagues in the interior of the country, and has been already described, is the small port called Sizal, to the *w.* of Chaboana, opposite a sand bank nearly 12 leagues in length. The population of the capital is 10,000.

[MERIDIONAL. See South.]

[MERIM, a large lake in Paraguay in S. America, very near the coast of the S. Atlantic ocean, where the land is very flat. Fort St. Miguel stands at the *s.* end, and fort Mangaveira at its *n. e.* extremity. There is a very narrow lake parallel to lake Merim, between it and the ocean, and nearly as long. The forts command the extremities of the peninsula.]

[MERIMEG, or Marameg, a large river of Louisiana, which empties into the Mississippi, below the mouth of the Missouri, and 22 miles above the settlement of Genevieve. Fine meadows lie between this and the Missouri.]

[MERION, Upper and Lower, two townships in Montgomery county, Pennsylvania.]

[MERISCHI, a settlement of the province and government of Sonora in N. America; situate on the shore of a river to the *e.* of Santa Maria Magdalena.

MERLO, a settlement of the province and government of Buenos Ayres in Peru; situate on the shore of the river La Plata, to the *n. w.* of its capital.

MERO, Punta de, a point on the coast of the S. sea, of the province and *corregimiento* of Piura in the bay of Tumbez, one of the two which form the same bay. It is low and covered with sand, and at ebb-tide a passage is open to the shore of the road which leads from Tumbez to Piura. It is extremely barren.

[Meno District, in the state of Tennessee, on the banks of Cumberland river. It comprehends the counties of Davidson, Sumner, and Tennessee. In 1790, it contained 7042 inhabitants, including 1151 slaves. By the state census of 1795 there were 14,390, of which number 2466 were slaves.]

[MERRIMACK River has its course *s.* through the state of New Hampshire, till it enters Massachusetts; it then turns *e.* and passes into the ocean at Newbury port. This river is formed by the confluence of Pemigewasset and Winnipiscogee rivers, in about latitude 43° 26′. This river is navigable for vessels of burden about 15 miles from its mouth, where it is obstructed by the first falls or rapids, called Mitchell's Eddy, between Bradford and Haverhill. Vast quantities of ship-timber, and various kinds of lumber, are brought down in rafts, so constructed as to pass all the falls in the river except those of Amuskeag and Pawtucket. In the spring and summer, considerable quantities of salmon, shad, and alewives are caught, which are either used as bait in the cod-fishery, or pickled, and shipped to the W. Indies. As many as six or seven bridges have been thrown over this fine river, at different distances, from New Concord, downwards; the most elegant and expensive are the one two miles above Newbury port, and the one at Haverhill. A canal is now in process to open a communication between the waters of the Merrimack at Chelmsford and the harbour of Boston, through Mystic river. See Middlesex Canal. The bar across the mouth of this river is a very great incumbrance to navigation, and is especially terrible to strangers. There are 16 feet of water upon it at common tides. There are two light-houses of wood removable at pleasure, according to the shifting of the bar. The lights now bear *e.* one-half *n.* and *w.* one-half *s.* Bringing both the light-houses to bear into one, until you are abreast of the lower one, will bring you in over the bar in the deepest water, where is a bold shore and good anchoring ground. The *n.* point of Plumb island, which forms the *s.* side of the entrance into the river, lies in lat. 42° 47′ 40″.]

[MERRIMACK, a township in Hillsborough county, New Hampshire; situated on the *s.* side of Souhegan river, which runs *e.* into the Merrimack. It is 35 miles *w.* of Portsmouth, was incorporated in 1746, and contains 819 inhabitants.]

[MERRIMICHI River, falls into the head of a bay of that name on the *n. e.* coast of the province of New Brunswick. A little above its confluence with the bay, it forms into two branches, and runs through a fertile tract of choice intervale land; and the land is, in general, well clothed with timber of all kinds. From this river there is a communication with St. John's, partly by land, but principally by water carriage in canoes. The salmon fishery is carried on with success, and the cod fishery is improving near the entrance of the bay.]

[MERRYCONEAG. See HARPSWELL.]

[MERRY-MEETING Bay, in Strafford county, New Hampshire, is the *s. e.* arm of lake Winnipiscogee. Mount Major stands on its *w.* side.]

[MERRY-MEETING Bay, in the district of Maine, is formed by the junction of Androscoggin and Kennebeck rivers, opposite to the town of Woolwich, 20 miles from the sea. Formerly from this bay to the sea, the confluent stream was called Sagadahoc. The lands here are good. Steven's river heads within a mile of the bay, and a canal has lately been opened which unites these waters. A company has been incorporated to build a bridge over Androscoggin river, at its entrance into the bay, to connect the towns of Brunswick and Topsham; the former on its *s.* side, the latter on its *n.* side.]

[MERTEQUE, a town in the province of Honduras in New Spain, which produces the cochineal.]

MESCA, a settlement and head settlement of the district of the *alcaldía mayor* of Senticpac in Nueva España. It is of an hot temperature, contains 60 families of Indians, whose trade is fishing for prawns and other fish, being close to the sea, and it is five leagues *w.* of its capital.

[MESCALA, Village of. Humboldt found its lat. by the culmination of Antares, 17° 56′ 4″, and the long. by the chronometer, 6° 47′ 16″, supposing Acapulco 6° 48′ 24″. The city of Chilpanzingo, from angles taken at Mescala, appears to be 17° 36′ of lat. and 6° 46′ 53″ of long.]

MESLE, a bay on the *s.* coast and in the French possessons of the island of St. Domingo, opposite the Isla Vaca.

MESPA, a head settlement of the district of the *alcaldía mayor* of Xala in Nueva España. It contains 20 families of Indians, who occupy themselves only in the barter of seeds; and it is somewhat less than a quarter of a league *s. w.* of its capital.

MESQUIC, a settlement of the *alcaldía mayor* of Chalco in Nueva España; situate on the shore of the lake, and with the greater part of it within the same. It is fertile and of an agreeable temperature: by it pass the canoes loaded with vegetable productions, such as sugar, sugar-canes, honey, and fruit, which render its prospect very pleasing; and, as we before observed, it is necessary to go from one part to another, in a great degree, by water. It contains 197 families of Indians, and a convent of the monks of St. Augustin. It is four leagues *s. s. e.* of Chalco.

MESQUIQUEJOS, a small and poor settlement of the Nuevo Reyno de Granada, taking the name of the nation thus called, and of whom it was first composed. It is situate *n.* of Chilloa, and *s.* of the town of Mompox, on the *w.* shore of the river Magdalena. Its climate is very hot and unhealthy, and in it is produced the most delicate plantains of any in the kingdom. Lat. 9° 10′ *n.*

MESQUITAL, a settlement and *real* of the gold mines of the *alcaldía mayor* of Juchipila in Nueva Galicia. It contains 200 families of Spaniards, *Mustees* and Mulattoes, and many Indians who live by the labour and commerce of its mines; they are very rich, and the gold is excellent for its fine colour, ductility, and superior quality. This settlement is eight leagues from the capital, and 18 to the *n.* one-fourth to the *n. e.* of Guadalaxara.

MESQUITAL, another settlement of the missions held by the religious order of San Francisco in the kingdom of Nueva Vizcaya in N. America.

MESQUITAL, another, with the surname of La Sierra; situate opposite that of Tulazinco, 60 leagues from Mexico, having a beautiful plain of more than six leagues in length. In one of its mountains are found described on a rock a cross and other characters, which, as the tradition goes, were done by the apostle St. Thomas when he preached in that province.

MESQUITIC, San Miguel de, a settlement of the *alcaldía mayor* of San Luis de Potosi, and bishopric of Mechoacán, in Nueva Espana. It contains 80 families of Indians, and a convent of monks of the order of San Francisco. It is a boundary of division between the aforesaid bishopric and that of Guadalaxara ; and in it begins the *sierra* of Pinos of La Nueva Galicia. It is five leagues to the *n. w.* of its capital.

MESQUITLAN, a settlement of the head settlement of the district and *alcaldía mayor* of Chilapa in Nueva España. It contains 22 families of Indians, and is three leagues to the *n.* of its capital.

MESQUITULA, a settlement of the head settlement of the district and *alcaldía mayor* of Juchipila in Nueva España; four leagues to the *s.* of the said head settlement.

MESSA, a settlement of the government and jurisdiction of Merida in the Nuevo Reyno de Granada. It is of a mild, healthy, and pleasant climate, produces maize, *yucas*, plantains, many *yuamas*, and other fruits of a warm climate. It has mills for expressing the sugar from the canes; and its population is composed of 50 Indians, and of about 30 or 40 housekeepers. It is four leagues from Merida.

Messa, another settlement, with the dedicatory title of San Marcelo, in the province and government of Veragua, and kingdom of Tierra Firme ; situate on the top of a mountain called the Mesa de Tabaraba, and abounding in vegetable productions and swine. It is five leagues from the capital.

Messa, another, with the surname of Tonati, of the missions that were held by the regulars of the company of Jesuits in the province of Nayarith, and kingdom of Nueva Galicia, in N. America.

Méssa, another, with the addition of Grande, in the province and government of Neiba, of the Nuevo Reyno de Granada ; situate on the shore of the Rio Grande de la Magdalena.

Messa, another, a small settlement and ward of the *alcaldía mayor* of Guauchinango in Nueva España :. annexed to the curacy of Pantepec.

[MESSASAGUES, Indians inhabiting between lakes Superior and Huron. They have about 1500 warriors.]

MESSIETA, a settlement of Indians of the province and *corregimiento* of Maule in the kingdom of Chile ; situate on the bank and at the head of the river Carampangue.

[MESSILLONES, Mixillones, (by the Spaniards), or Muscle Bay, on the coast of Chile in S. America. It is 33 miles *n. e.* of Morro Jorge, and 74 *s. s. w.* of Atacama, and is so deep on the *s.* side that there are no soundings; but at the entrance or anchoring-place it is moderate, and ships may ride in 15 fathoms, clean ground, and secured from most winds.]

MESTITLAN, a settlement of the head settlement of the district and *alcaldía mayor* of Cuquio in Nueva España; nine leagues distant to the *n.* of the same head settlement.

MESTIZOS, Punta de, a point on the coast of the province and government of Cartagena, and Nuevo Reyno de Granada ; one of those which form the bay of Zipato.

[MESTRE Bay, Little, on the *n. e.* part of Newfoundlad island, *s.* of St. Julian, and *n.* by *w.* of the islands Gros and Belle.]

[MESUCKAMA Lake, in the *n.* part of N. America. Lat. 50° 10′ *n.* Long. 80° *w.*

META, a large, abundant, and navigable river of the Nuevo Reyno de Granada, which gives its name to the *llanos* of San Juan, through which it runs. It rises in the mountains which lie between Santa Fé and Tunja, in the *paramo* of Albarracin, thus called from an inn and estate of this name which are upon it. This river runs *e.* and after collecting the waters of many others, enters, united with the Pato, by the *w.* shore, into the Orinoco. Passing the valley of Turmeque in the Nuevo Reyno de Granada, it takes this name, and being increased by other streams which it collects in the different valleys of that broken *serranía*, enters with a large body into the *llanos* of San Juan, with the name of Upia, turns its course *n. w.* and receives the Cusiana, which has its origin in the *paramos* of Toquilla, not far distant from Tunja : shortly afterwards it collects the waters of the Cravo, at the month of which stands the colony of La Concepcion de Achagua, is then entered by the Guirripa, near the mission of San Miguel of Salivas Indians, below the Guanapo, four leagues from Paulo, and lastly by the rich streams of the Casanare and Elne, and being swoln to an immense size by them, and after running 300 leagues, enters, as we have before said, into the Orinoco; forming, however, first such a level body of water as that its current is scarce susceptible, and forming a beautiful appearance before the settlement of the mission of Santa Teresa of Salivas Indians, the same being near its mouth or entrance.

In its vicinity dwell some barbarian nations of Indians, spread through the spacious *llanuras* of Cazanare and Meta or San Juan, where the regulars of the abolished order of Jesuits of the province of Santa Fé had some flourishing missions, by which they had reduced to the Catholic faith an infinite number of barbarians, and which, since 1767, have been under the care of the monks of the barefooted order of San Francisco. The mouth of this river is in lat. 6° 10′ 30″ *n.* Long. 67° 45′ *w.*

[The source of the Meta, observes Depons, is 150 leagues *s. w.* of its junction with the Orinoco. Several rivers of Santa Fé fail into it. It is navigable as far as Macuco, near the plains of Santiago de las Atalayas, 40 leagues from the capital of the kingdom. Its banks are still deserted or inhabited only by Indians, who have an equal aversion to civilized life and to labour. They are intractable without being fierce, and for this reason more adapted to attack than to defence; they, therefore, owe their independence to flight. The navigator can traverse their country without danger. Seventy-five leagues before the Meta falls into the Orinoco it receives the river Casanare. Its course is silent and majestic, and it may be distinguished from the other rivers that are received into the Orinoco by the silence with which it forms the junction.]

META, with the addition of Incognita, a piece of territory which was taken possession of for Queen Elizabeth of England, in 1578, by Martin Forbisher, in his third voyage to discover a *n. w.* pass; but which has not been met with or seen by any navigator since.

METALTEPEC, a settlement of the head settlement of the district of Atitlan, and *alcaldía mayor* of Villalta, in Nueva España. It contains 39 families of Indians, and is 14 leagues from its capital.

METATI, a river of the province and government of Darien and kingdom of Tierra Frime. It rises near the *e.* coast, and following a course to this rhumb, enters the grand river Atrato near its mouth.

METATLAN, a settlement and head settlement of the district of the *alcaldía mayor* of Papantla in Nueva España. It contains 70 families of Indians.

METAU, a small river of the province and government of Tucumán in Peru, and of the jurisdiction of the city of Salta. It runs *e.* and enters the Pasage between the Concha and Yatasco.

METCHIGAMIA, a lake of the province and government of Louisiana in N. America, on the shore of the river San Francisco, and from which it is formed, just where this river enters into the Mississippi.

METCHIGAMIA, a settlement of the former province and colony, founded by the French on the shore of the river Mississippi.

METENQUEN, a settlement of Indians of the province and *corregimiento* of Maule in the kingdom of Chile.

METEPEC, a small settlement or ward of the head settlement of the district of the *alcaldía mayor* of Tetela del Volcan in Nueva España. It is close to its capital.

METEPEC, another settlement, in the head settlement of the district of Ocotepec, and *alcaldía mayor* of Villalta, in the same kingdom. It contains 33 families of Indians, and is six leagues *s. w.* of its capital.

METEPEC, a jurisdiction and *alcaldía mayor* of Nueva España, and one of the largest there, extending more than 20 leagues from *n.* to *s.* and 12 from *e.* to *w.* divided into two other districts, which are Ixtlahuaca and Tianguistengo, and in which the *alcalde mayor* nominates two lieutenants, for the better and more ready administration of justice. It is very fertile in maize, barley, beans, and lentils, these being the vegetable productions in which it trades; also in a good quantity of swine, which are bred for the supply of Mexico, and by which the inhabitants make great profit. The population consists of 36 principal settlements, which are heads of districts; and to these are subject above 300 small settlements or wards, reduced to 13 curacies.

S. Miguel Temazcalzingo,
S. Pedro de Tultepec,
S. Francisco Chichicocuitla,
S. Francisco Xonacatlan,
S. Juan Xiquipilco,
Santiago Temoya,
S. Lorenzo Huitzizlapa,
S. Geronimo Amanalco,
S. Pedro Atlapulco,
Ixtlahuaca,
S. Felipe,
Asuncion de Malacatepec,
S. Miguel Almololoya,
S. Miguel Zinacantepec,
S. Mateo Texcalicaque,
Tepemaxalco,
Calimaya,
S. Mateo Mexicalzingo,
S. Miguel Chapultepec,
S. Miguel Mimialpa,
Asuncion Yalatlaco,
S. Mateo Tlachichilpa,
Santiago Tlacotepec,
Santa Maria Atlamulco,
S. Mateo Atengo,
S. Gaspar,
Santiago Tianguistengo,
Santa Ana Xilotzingo,
S. Bartolomé Otzolotepeque.
Xocotitlán,
S. Bartolomé Capuluaque,
Asuncion de Tepexoyuca,
S. Martin Ocuyoacaque,
S. Juan Guapanoya.

The capital is of the same name, with the dedicatory title of San Juan, situate in the spacious valley of Tolucá, at the foot of a small mountain. Its population consists of 62 families of Spaniards and *Mustees*, and nearly 900 of Indians, including those of the wards of its district. It has a convent of the religious order of San Francisco, and is 33 miles to the *s. w.* of Mexico, in long. 99° 20′ *w.* Lat. 19° 20′ *n.*

METETA, a river of the province and government of San Juan de los Llanos in the Nuevo Reyno de Granada. It rises between those of Meta and Doma, runs *e.* and enters the Orinoco near the settlement of San Francisco de Borja, of the missions that were held there by the regulars of the company of Jesuits.

[METHUEN, the *n.* westernmost township in Essex county, Massachusetts; situate on the *e.* bank of Merrimack river, between Dracut and Haverhill. It contains two parishes and 1297 inhabitants. It was incorporated in 1725. Husbandry and the cutting and selling lumber divide the attention of the inhabitants.]

METINACAS, some islands of the N. sea; situate near the coast of the province and colony of Sagadahock. They are many, all small, and lying opposite the bay of Penobscot.

METLATONO, a settlement of the head settlement of the district and *alcaldía mayor* of Tlapa, in the same kingdom as the former. It contains 55 families of Mexican Indians, employed in the cultivation of seeds, cochineal, and cotton.

METOS, a small river of the province and government of Buenos Ayres in Peru. It runs *n. n. w.* and enters the Santa Lucia Grande.

METRANO, a settlement of the province and *corregimiento* of Xacamarquilla in Peru.

METWAY, a port of the *s.* coast of Nova Scotia or Acadia, between those of Senior and La Heve.

METZTITLAN, an ancient province of Nueva España in the time of the Indians, and, when the Spaniards entered, bounded by the province at the present day called Panuco. It was peopled at that time by an infinite number of the Chiehimecas Indians, a ferocious people and cannibals, and who, together with the Aculhuas, were the first inhabitants. They rebelled against the Emperor Tlaltecatzin, who fought them in a battle which, according to Torquemada, lasted 40 days successively, and in which they lost the greater part of their people; so that they were obliged to surrender, obtaining pardon, with the exception of some few of their ringleaders who were severely chastised. In the time of Techotlalatzin, the successor of the former emperor, they returned to their state of rebellion, at least such of them as lived in the *sierra;* and these are called at the present day by the title of those of La Misteca Alta.

MEUIS, a name which the English give to one of the Antilles isles. See NIEVES.

[MEW Islands, on the coast of the Spanish main in the W. Indies, between cape Cameron and cape Gracias a Dios, lie across the entrance into the bay of Cotroe or Crotoe. They are surrounded with rocks, and are very dangerous, especially in case of hard gusts from the *n.* and *n. e.*]

MEXICAL, a river of the island of St. Domingo, which rises near the *n.* coast, runs to this rhumb, and enters the Jacques.

MEXICALTZINCO, a jurisdiction and *corregimiento* of Nueva España, very fertile in maize, beans, barley, French beans, and garden herbs. It is as it were the principal key of the lake of Chalco. Carried in vessels through this lake, pass not only all the vegetable productions of the province of Chalco, but also of all the other neighbouring provinces of a warm climate; and by the channel called the Real are introduced honey, sugar, fruit, grain, and other effects, which being warehoused in Chalco, avoid the duties of freightage, and are conveyed by the lake up to the bridge of the palace of Mexico. This *corregimiento* contains three settlements which are head settlements of districts, and subject to these are five others, as follows:

The capital,

Ixtapalapan, which was the court of the King Cuitlahuatzin, exalted to the empire of Mexico through the death of Motezuma, and where there remains a beautiful pool for breeding fish, as also vestiges of royal fabrics of the gentiles.

Santa Maria Aztahuacan,

Santa Marta,

Couhuacán,

Santiago Chahualtepeque,

San Lorenzo,

San Mateo Huitzolopozteo, Noviciado of the Dieguinos.

The capital is of the same name, beautifully situate, and very pleasant and fertile, from the waters of the lake which are close to it, the richness of foliage, and as being the direct road to Mexico either by water or by land, the latter lying along the causeway of the *llano* of Santa Marta, and much frequented by the drovers of other jurisdictions. It is two leagues to the *s.* a quarter to the *s. e.* of Mexico, and contains 61 families of Indians. Near this settlement, and in the middle of the lake going to Mexico, at the distance of about two leagues from this capital, is a stream of water which is called the Estrilla, and which is easily to be distinguished by its pellucid course in the lake. Not far from the same spot are some flower gardens, and *peonies* and other flowers are to be seen mixed amongst the brush-wood and reeds of the lake. Here the Indians have their dwellings and estates, changing them as often as they think proper.

MEXICALZINCO, a settlement of the head settlement of the district of the *alcaldía mayor* of Analco, in the kingdom of Nueva Galicia and bishopric of Guadalaxara. It contains 15 families of Indians, who occupy themselves in carrying for sale to this city wheat, maize, seeds, and fruits. It lies a little to the *n.* of its head settlement.

Mexicalzinco, another, with the dedicatory title of San Mateo, the head settlement of the district of the *alcaldía mayor* of Metepec in Nueva España. It contains 415 families of Indians.

MEXICANOS, Barrio de los, a settlement of the province and *alcaldía mayor* of Sonsonate in the kingdom of Guatemala.

Mexicanos, a nation of Indians of Nueva España, and one of the principal and most numerous of all in the new world. They are descendants of the Aztecas, one of the first nations that left the kingdom called Aztán, taking the name of Mexicos from their idol, and who, being led by Huitziton and Tecpatzin, great idolaters and soothsayers, wandered about for above 50 years without establishing themselves in any place, until that, as they say, directed by their god Huitzilopuchtli, they collected and fixed themselves on the lake, which takes its name from its floating inhabitants, having been before called Tenuititlan, meaning Stone of Tuna.

Having collected themselves here, from the reasons we have just mentioned, as also from their having lost their two leaders in the privations, sickness, and disasters they had undergone, their families began to increase, and their population to multiply, until they at last nominated a king, and formed the powerful Mexican empire. These Indians are of a darker colour than the rest, of a quick genius, and lived in civil and political order previous to the arrival of the Spaniards. They were idolatrous, and had an infinite number of gods and deities, to whom they made sacrifices of human blood, in order to draw down mercies in their necessities; and as they had different deities as tutelar to different circumstances, and as they all were thought to require sacrifice, the number of captives that were slain before these monstrous representations of a divinity were exceedingly great.

Their government was monarchical, and organized with singular skill and harmony. In the court they had a council of revenue, which took cognizance of the expences of the royal palace, and to which the collectors of the tributes of the different provinces rendered in their accounts; a court of justice, in which was vested the nomination of inferior tribunals; a council of war, which took charge of the formation and supplies of the army; another of state, which frequently deliberated in the king's presence; also judges of commeree and of supplies, and other ministers of the police. The judgments were summary and verbal, the plaintiff and defendant with their witnesses told their story, and the cause was finished. They had no written laws, but they were governed by traditional ordinances, save in cases where the will of the prince interfered. They were regardful of rewards and punishments, holding as capital crimes, theft, homicide, and adultery; also whatever was committed against the king or religion, however trifling, although other crimes were easily pardoned. Any fault of integrity amongst the ministers was to be paid by the life of the person offending; pardon in this instance was never granted, and indeed such was the diligence used in scanning the faults of such persons, that even the emperor's friends and confidants were obliged to silence the investigators by bribes.

They took singular care of the education of their youth in public schools and colleges for the nobility, taught them their mode of writing, which consisted of certain characters and figures, and made them learn by heart certain historical songs, which contained an account of the exploits of their ancestors, and which were in praise of their deities; after these they instilled modesty, courteousness, decent deportment, and when they became old enough, exercised them by trying their strength by carrying weights, running, and wrestling, in the use of arms, to endure hunger and thirst, and to combat the inclemencies of the weather; and thus they were, according to the report of their different masters, destined either for political government, to the army, or to the priesthood, which were the three roads to nobility. On the other hand, the girls of rank were brought up by certain matrons, who lived in other colleges, and who were dedicated to the care of the temples; they were kept closely confined from their earliest infancy, nor suffered to leave their mistresses but by the approbation of their parents and the kingly consent; nor were they ill-skilled in all those arts and occupations which render the female character useful and interesting.

The Mexicanos constituted all their happiness in war, a profession which their kings looked upon as what principally constituted their power, and all the vassals as the peculiar attribute of their nation: these considerations made them naturally valorous, eager of gaining the prizes that were offered as rewards to bravery, and historians recount that Mocthezuma had no less than 30 vassals, so powerful that each of them could bring into the field 30,000 men at the first order. This mixture of their barbarian and savage customs with those of civilized life would deserve a more extensive description than we can admit in this article, but an account may be seen in Herrera, Gomara, Acosta, Torquemada, Solis, Garcia, and others, such as will give every satisfactory information that may be further required.

The empire of the Mexicanos was conquered by the incomparable Hernan Cortes in 1521, nor was it less than a miracle that with an army of little more than 300 Spaniards he should succeed in subjecting so many millions of men, whose patriotism might be put in competition with that of the Romans.

MEXICAPA, San Martin de, a settlement of the head settlement of the district of the town of Marquesado del Valle, and of the *alcaldía mayor* of Quatro Villas, in Nueva Espana. It contains 30 families of Indians, who live by cultivating and trading in wheat, maize, cochineal, and fruit, and in collecting woods, coal, salt, and fuel. It is one mile to the *w.* with an inclination to the *s. w.* of its capital.

MEXICO, a province and *corregimiento* of Nueva España in N. America. It is 313 miles long in a direct line from *s. w.* to *n. e.* from the port of San Diego de Acapulco in the S. sea to the bay or lake of Paunco near the *n.* It is bounded by the province and bishopric of Mechoacán on one side, and on the other by the province of Tlaxcala, a line being drawn through these from *e. s. e.* to *w.* Its width is 37 leagues, forming an irregular figure; for being narrow in the strip on the coast of the S. sea, it continues widening as it runs *n.*

It enjoys different climates, is for the most part mountainous, but not without many fertile valleys, watered by different rivers, which render it a country extremely productive of grain, fruit, seeds, and herbs. It produces also cochineal, which is cultivated in many parts, and an incredible multitude of cattle of every species. The woods are thronged with birds notorious for their plumage and their song; but, as this forms a part of Nueva España, we reserve a further description for that article. The extent of this province, which is as far as its archbishopric reaches, comprehends many *alcaldías mayores* and *corregimientos*, in which are counted 235 enracies and 23 missions, viz.

S. Agustin de las Cuebas,
S. Angel,
Ascapotzalco,
San Bartolomé,
Naucalpan,
S. Christóval Ecatepec,
Culhuacán,
Cuyuacán,
Churubusco,
Santa Fé,
Ntra. Señora de Guadalupe,
Ixtapalapan,
Ixtacalco,
Mexicalzinco,
Mixcoac,
Tlalneplantla,
Tacuba,
Tacubaya,
Xochimilco,
Atitalaquia,
Actopam,
Alfaxayuca,
Acolman,
Acapuxco,
Apán,
Ayozinco,
Ayapango,
Amecamecan,
Atzompan,
Achichipico,
Real de Atotonilco Chico,
Atotonilco el Grande,
Acatlan,
Aculco,
Acambay,
Atlacomulco,
Asuncion de Malacatepec,
Amanalco,
Almoloyán,
Acamistla,
Acapetlahuayán,
Amatepec Tlatlaya,
Alahuistlan,
Apaxtla,
Atlatlauca,
Ayacapixtla,
Acapulco,
Amealco,
Ayahualicán,
Santa Ana Tianguistenco,
San Bartolomé Ozolotepec,
Chapantongo,
Chilcuautla,
Cardonal,
Chautla,
Coatlinchan,
Chimalhuacán Atenco,
Coatepec,
Chimalhuacán Chalco,
Chalco,
Chiconcuautla,
Chapa de Mota,
Capuluac,
Calimaya,
Cacalotenango,
Coatepec de los Costales,
Coyuca,
Cuernavaca,
Casas Viejas,
Cañada,
Cadereita,
Calnale,
Coscatlan,
Real del Doctor,
Escanela,
Ecazinco,
Epazoyucán,
S. Felipe Ixtlahuaca,
Ixtapán,
Ixcatiopán,
Yautepec,
Iguala,
S. Juan Teotihuacan,
S. Joseph Malacatepec,
S. Juan del Rio,
Landa,
Lerma,
Lolotlan,
Misquiahuala,
Misquic,
Milpa Alta,
Real del Monte,
S. Martin Ozoloapan.
Metepec,
Malinalco,
Malinaltenanco,
Matzatepec,
Mestitlan,
Molanco,

Otumba,
Ocuituco,
Real de Omitlan,
Ocoyoacan,
Ocuila,
Oapán,
Pacula y Xiliapan,
Real de Pachuca,
Pilcayán,
Pueblito,
Panuco,
Quantitlan,
Quautla Amilpas,
Singuilucan,
Sinacantepec,
Santiago de Queretaro,
S. Sebastian de Queretaro,
Sochicoatlan,
Tultitlan,
Tepotzotlan,
Teoloyucan,
Tequisquiac,
Tetepanco,
Tepexi del Rio,
Tula,
Tepititlan,
Tasquillo,
Tepeapulco,
Tepatlastoc,
Tezcnco,
Tlahuac,
Tenango Tepopula,
Temamatlac,
Tlalmanalco,
Tetela del Volcan,
Tochimilco,
Tecama,
Tizayucán,
Tolcayucan,
Tetzontepec,
Tulancinco,
Tlaola,
Temoaya,
Temascalcinco,
Temascaltepec del Valle,
Real de Temascaltepec,
Texupilco,
Toluca,
Tenango del Valle,
Tescaliacac,
Tenancinco,
Tecualoyan,
Real de Tezicapán,
Tetipac,
Real de Tasco,
Teloloapan,
Tetela del Rio,
Tlayacapan,
Tlalneplanta,
Totolapán,
Tlaltizapán,
Tlalquitenanco,
Tenango del Rio,
Tepoxtlan,
Tepequacuilco,
Tolimancjo,
Toliman,
Tequisquiapan,
Tecoutzatla,
Tlalchichilco,
Tlalchinol,
Tepehuacan,
Tamazunchale,
Tampamolon,
Tancanhuitz,
Tempoal,
Tantoyuca,
Tantima,
Villanueva de la Peña de Francia.
Xacala,
Xalpan,
Xuchitepec,
Xumiltepec,
Xaltenco,
Xiquipilco,
Xocotitlan,
Xilotepec,
Xulatlaco,
Xantetelco,
Xonacatepec,
Xiutepec,
Xochitepec,
Xichu de Indios,
Real de Xichu,
Real de Zimapan,
Zempoala,
Zumpango de la Laguna,
Ziautchutla,
Real de Zultepec,
Zumpahuacán,
Real de Zacualpan,
Zacualpan del Rio,
Zaqualtipan,
Zontecomatlan,
Zoquiatipan.

The Missions.

Tampico,
Ozuluama,
Taralon,
Tanquayalal,
Tanlaxac,
Altamira,
Huehuetlan,
Tamaquichimin,
Tamapach,
Tamitas,
Villa del Valles,
Tampasquil,
Palma, or Salto del Agua,
Tamuya,
Huayabos,
Sauz,
Tanlacrin,
Tanlacun,
Santa Maria Acapulco,
Santa Barbara,
Escandon,
Horcasitas,
Las Palmas.

The capital is the city of the same name, with the dedicatory title of San Hipolito, in memory of the day on which the Spaniards took it from the Indians. It is the court and metropolis of the Mexican empire, or of Los Aculhuas, an archbishop's see, and the residence of the viceroy, governor, and captain-general, and of the audience and royal chancery of Nueva España, erected in 1527, and the jurisdiction of which extends from the cape of Honduras to that of Florida in the N. sea, and by the S. sea from the point where the jurisdiction of the audience of Guatemala terminates to where that of Guadalaxara or Nueva Galicia begins, and to the provinces of Yucatán, Cozumel, and Tabasco. It is the most beautiful, grand, and sumptuous city of the whole of the Spanish monarchy.

It was founded by the Indians in 1327, with the name of Tenochitlan, upon a lake in the midst of a valley, 14 leagues long, seven wide, and 40 in circumference. It is surrounded by 90 leagues of mountains and *serranías*, fruitful in cedars, trees, both rare and common, gums, drugs, salts, and metallic productions of all classes, marbles and precious stones, plain and vari-coloured. As well in the mountains as in the valley are beautiful settlements, farms, granges, and estates, in which the European fruits and those peculiar to the country are cultivated. The fields abound in herbage, which render the prospect beautiful, and afford pasturage for innumerable herds of cattle of every kind. The rivers and lakes fertilize the territory, and by them are carried to the capital all the most delicate fruits, during the different seasons of the year. Here are all kinds of vegetables, garden herbs, and grain, ducks, geese, widgeons, quails, fish, eels, and other productions, birds much valued for their song and plumage; and above all, the soil is extremely fertile in flax, hemp, cotton, tobacco, indigo, sugar, and *magueyes* or *pita*, of the branches of which is made a great commerce, and such as greatly enriches the royal exchequer.

In the time of the gentilism of the Indians there

were 140,000 houses, and these were divided into four quarters or wards, an infinite number of temples dedicated to their idols, the most celebrated of which was that erected to Huitzilopochitli, the god of war, built upon a pyramidical square table-land of 40 feet high, and to which there was an ascent by a staircase of 140 steps, wide enough to receive eight men abreast; the whole forming an edifice so magnificent as to strike the Spaniards with astonishment; though not less so did the great *plaza* or market-place of Tlateluco, of the which now not even the vestiges remain.

The Spaniards, led on by the celebrated Hernan Cortés, gained this city on the 13th day of August 1521. It was dedicated to the martyr San Hipolito, its sworn patron, and in memory of whom the pennant is taken down every year by the royal Alferez, and carried in splendid procession by the viceroy, the audience, the tribunals, the city council, and the nobility, to the church of the said saint, to the vespers, and to the mass, said by way of gratitude and thanks.

The plant of the city is square; its diameter within the gates is 4340 Spanish yards from *n.* to *s.* and 3640 from *e.* to *w.*; the ground is level, the streets straight, and drawn at right lines, being a little more than 14 yards wide. The town is surrounded with a wall of uncemented stones, and the channels which lead from the lake disperse their waters in various small canals, which flow through some beautiful streets, and are covered with craft and canoes, which every day appear loaded with supplies of fruit, flowers, &c. and make their way up as far as the walls of the palace of the viceroy, which is situate in the *plaza mayor.* The buildings are magnificent, and some of them of the most beautiful architecture. There are different markets, where there is a regular supply of every thing that the public can require. This city is entered by seven stone causeways, which are, Guadalupe to the *n.* Tacuba to the *w.* San Antonio to the *s.* built by the Indians, and the others by the Spaniards, their names being La Piedad, Ascapuscalo, Tacuba, Santiago, and Chapultepec. The whole of the city is paved, the principal streets with free-stone, and some of these are arched, so that the filth is carried off.

Here are some beautiful fountains, the waters of which come from various parts; but that which has the best and sweetest water is that which is brought from the settlement of Santa Fé, for more than two leagues upon an aqueduct of more than 900 arches, each of eight yards diameter, above three high, and a quarter wide: this canal is open at top, and has a rail-way of half a yard deep on each side. Another aqueduct similar to this comes from the pool of Chapultepec, about a league's distance, and formerly there was one towards the *s.* through Churubusco, of which nothing but the vestiges remain. Here are some beautiful promenades, both in the country and by the water side, the principal of which is that which was made in the time of the viceroy Don Antonio Bucareli; also a theatre for plays, a cock-pit, a tennis-court, &c.; many gambling places, billiard tables, inns, and taverns; 42 *pulquerias*, in which are sold daily 100,000 quarts of *pulque*, this being the liquor or wine of the Magueyes. There are different palaces and public buildings, such as that of the viceroy, where reside the tribunals, the secretaries, the officers of the treasury and of the royal revenues. There is also a mint, in which in some years have been coined upwards of 23,000,000 of dollars; also the inquisition, together with the office for its ministers, and the prison for the culprits.

Here is a royal and pontifical university, erected by the Emperor Charles V. in 1551, with the same privileges as that of Salamanca; its cloister being composed of more than 225 doctors and masters, with 22 professors of all the sciences, with a grand library; a most ancient royal college of San Ildefonso, which superb edifice contains within it two other colleges, namely, those of Filotos and that of Rosario, containing above 300 students; likewise the Real de Christo, incorporated on the day of St. Peter and St. Paul; the college called of Los Comendadores de San Ramon, for the natives of Valladolid and the Havana; that of Santiago Tlateluco, for the Indians of rank, the same having been founded by the Emperor Charles V.; the Real de San Juan de Letrán, the Semenario Tridentino de San Gregorio, for the Indians, and the seminary of Los Infantes; not to mention various other colleges for religious students, and, besides the university, public professorships, amounting altogether to the number of 43. Besides these there are some free schools and academies for the instruction of youth, especially those under the care of the religious order of the Betlemitas, the royal college for the instruction of miners, a royal academy of the three noble arts of painting, sculpture, and architecture, erected in the reign of Don Carlos III. four colleges for educating female children, with the names of Santa Maria de la Caridad, or De las Niñas, San Ignacio, San Miguel de Bellén, and of Guadalupe, for the Indian nobility; the Misericordia, being a refuge for married women, the Beaterio de San Lorenzo, the Casa de Magdalena, or house of repentants, for the chastisement of bad women; the Cuna, or a house for foundlings; and a general house of entertainment for poor invalids and beggars; 13 hospitals, which are the General and the *Real* for the Indians, and those of San Andres, Espiritu Santo, Santissima Trinidad or San Pedro, San Juan de Dios, San Lazaro, San Anton, Amor de Dios, De la Concepcion or Jesus Nazareno, Betlemitas, San Hipolito, El Salvador de Sayago, and La Tercera Orden de San Francisco.

Its archbishopric extends 180 leagues from the port of Acapulco, in the S. sea, to the bay of San Esteban, of the port of Panuco, in the *n.* upon a line which runs from *s. w.* to *n. e.* and bounded *w.* by the province and bishopric of Mechoacán, and *e.* by that of Tlaxcala. It was first only a bishopric, and the church was called Carolense, with the dedicatory title of La Asuncion de Maria Santisima, in 1525, and afterwards erected into a bishopric by Pope Clement VII. during the reign of the Emperor Charles V. in 1534. It has for suffragans the bishops of La Puebla, Tlaxcala, Mechoacán, Guadalaxara, Oaxaca, Yucatán, Durango, Nuevo Reyno de Leon, &c. Its cathedral is a sumptuous and magnificent temple, 94 years old, and ornamented by the piety and generosity of the Kings Philip II. III. and IV. and Charles II. and in which 18 viceroys have seemed to shew their competition in benefiting the sacred cause of religion. It is 133¼ Spanish yards in length, 74 in width; of Ionic order, and with 74 windows: divided into five naves, with three doors on the *s.* side, two on the *e.* and *w.* and two on the *n.* In it is venerated two images of Maria Santisima, the one of the Assumption, of the most pure gold, weighing 6984 *castellanos*, and adorned with precious stones, and the other of silver, concerning the mystery of the Conception, and weighing 38 marks; in short, the grandeur, riches, and majesty with which the worship is executed in this temple by the archbishop and venerable *cabildo*, exceeds all description. The other sumptuous temples, and at which a numerous clergy officiate, amount to above 100. Here are 14 parishes, with the titles of Sagrario, San Miguel, Santa Catalina, Martir, La Santa Vera Cruz, San Joseph, Santa Ana, Santa Cruz, San Sebastian, Santa Maria de Redonda, San Pablo, Santa Cruz Acatlan, Nuestra Señora del Salto del Agua, Santo Tomas, and San Antonio de las Huertas; four convents of Dominican monks, which are, the Convento Grande, Porta Celi, La Piedad, and San Jacinto; five convents of the order of San Francisco, namely, Convento Grande, Santiago Tlatelolco, Recolleccion de San Cosme, San Fernando de Misoneros de Propoganda Fide, and of Los Descalzos de San Diego; four of the order of San Agustin, the Convento Grande, San Pablo, and the house of entertainment of San Nicolás de Recoletos y Santo Tomas; three of La Merced, the Convento Grande, San Pedro Pasqual de Belen y la Merced de las Huertas; one of the Carmelites; two of San Juan de Dios; two of the order of La Caridad, of which the one is San Hipolito, where resides the general of the religion, and the other the Espiritu Santo; one of the religious Betlemitas, the residence of the general of its order; the Casa de San Antonio Abad, for the regular canons; the Oratorio de San Felipe Neri, the house of entertainment of Monges Benitos; a college of the Padres Agonizantes; besides other chapels and churches of the following orders, viz. that of San Francisco of La Orden Tercera, that of Santa Escuela, the parish of Indians of San Joseph, Nuestra Señora de Aranzazu, Santo Christo de Burgos, and Nuestra Señora de Balvanera; 20 monasteries of nuns, entitled the Convento Real de la Concepcion, Regina, Balvanera, Real de Jesus Maria, Real de San Geronímo, of the same order, Lorenzo el Real de Escorial, where the Mexican poetess Sor Juana Inez de la Cruz flourished; La Incarnacion, San Lorenzo, Santa Ines, San Joseph de Garcia, San Bernardo, Santa Teresa la Antigua, Santa Teresa la Nueva, Capuchinas, Santa Brigida, La Enseñanza or Salesas, Santa Catalina de Sena, Santa Clara, San Juan de la Penitencia, where flourished the venerable Sebastiana; Santa Isabel and Corpus Christi de Señoras Capuchinas Indias. It also had a possessed house of the regulars of the company, a Mexican college of San Pedro y San Pablo, an house of noviciates of the title of San Andres, and a college of San Gregorio; and under its charge the Real de San Ildefonso, the Chico del Rosario, and that of Guadalupe de Indias.

What further tends to ornament this city are the royal audience and court for the judgment of crimes, composed of the viceroy, the regent, 10 *oidors*, five *alcaldes*, three *fiscals*, an *alguacil mayor*, a chancellor, four secretaries of the chamber, and corresponding subordinate officers; the general court of justice for those beyond the seas, or who have died intestate; the royal tribunal, *mayor*, and audit of accounts; the *real caxa matrix*, or general treasury; that of land and water; that of the general of the Indians; that of the half-yearly revenue and of the spear-armed soldiery; and the superintendance of the *azogues* or quick-silver; of the sealed paper; that of the general direction of the *alcababas* and *pulques*; of the *real* stores; the mint; the tribunal of La Santa

Hermandad; the offiee that takes cognizance of forbidden drinks; the royal junta of tobacco, and the general direction of this revenue; a general post-office, and the office for the revenues arising from powder, cards, &c.; the royal depôts of dies, colours, leather and snow; the tribunals of the inquisition, crusade, temporalities, missions of California, for the direction of the lottery; the sacred and royal place called the Monte de Piedad, for the erection of which the Count of Regla most beneficently and patriotically advanced a sum of 315,000 dollars; also the royal tribunal of the *consulado*, of the mines, the college of physicians, and the tribunal of the Estado and Marquesado del Valle; and lastly, the most illustrious *cabildo*, and the other offices of justice, &c.

Independently of the title of most noble, most loyal, renowned, and imperial city, there was conceded to it by the Emperor Charles V. in 1523, the title of Cabeza y Corte del Reyno, (Head and Court of the Kingdom), who also granted to it the liberty of using the arms which it had in the time of its gentilism, the which were, a shield, with a castle of three towers, an eagle upon a *tunal* tree, with a snake in its beak; at the foot of the tree ran some waters; on the side without the shield were two lions, and upon the top a crown: also by a cedule of 4th of July, of the same year, there were further conceded for the arms of its corporation and of the city, a blue shield of the colour of water, to represent the lake, a gold castle in the middle, and three bridges of stone leading to it, the two side bridges not quite touching the castle, and upon each a lion standing, and having his feet upon the bridge and his talons on the castle; and within the orle were ten green *tuna* leaves, and above all the imperial crown.

In 1530, the same emperor granted to this city the title and privileges of Burgos Cabeza de Castilla, and in 1548, the titles of most noble, most loyal and illustrious city: and again the Señor Don Felipe V. confirming the above ordinances, granted them to be perpetual, in 1728.

It also enjoys the privilege and pre-eminence of being called the Grande de España, and the Señor Don Carlos III. granted in 1773, to the persons belonging to the chapter, the use of gold embroidery to their dress and undress uniforms, declaring that they should be preferred before all the tribunals and bodies, with the exception of the royal audience and the tribunal of accounts.

The principal patroness of this city is Maria Santisima de Guadalupe, [the most holy Maria of Guadalupe], who was solemnly shown as such in 1737, and from thence she has become almost universal patroness in America. There is a representation of her here, which was found to be miraculously painted on the cloak of the Indian Juan Diego, in the presence of the first archbishop, in 1531.

The temperature of this place is most benign, and it enjoys a continual spring, neither the heat nor the cold being at any time troublesome; and although, indeed, the four seasons are perceptible, yet do they all abound almost equally in the production of flowers and fruits. The whole of the city is surrounded by estates, orchards, and gardens, so as to pour in a continual supply of flowers, fruits, vegetables, and garden herbs, for the use and luxury of the inhabitants, save on the *e.* side, which is barren, owing to the great lake of Tezcoco, the vapours arising from which, and the particles of saltpetre surrounding it, give a complete check to all kinds of vegetation.

The population consists of more than 350,000 souls of all classes and casts; and amongst these are counted many noble families, descendants from the conquerors and settlers; and for the subsistence of this population there are consumed annually 300,000 sheep, 15,500 cows and calves, 30,000 swine, near 2,000,000 *arrobas* of flour, and 170,000 bushels of maize. The natives are clever, and have a great disposition for the liberal arts; and in these some have excelled very much, as is proved by the paintings of Cabrera, Enriquez, Vallejo, Pelaez, and Don Juan Patricio. The general character of the Mexicans is that of being liberal, courteous, affable, and charitable. Mexico, in its university and colleges, has produced many characters noted for their virtue, science, and arts, in all times; it has had 84 archbishops and bishops, many viceroys, captains-general, ambassadors, generals of religious, counsellors; *oidors*, dignitaries, and magistrates, who have done honour to the tribunals, ecclesiastic and secular; to enumerate all of whom would form a catalogue too extensive for the limits of this article; we shall therefore confine ourselves to the mention of such only as have rendered themselves pre-eminently conspicuous, viz.

San Felipe de Jesus, sworn patron of the city after the Virgin of Guadalupe: he was baptized in the cathedral church, in the chapel of which the font is still preserved, suffered martyrdom in Japon, his mother having the glory of assisting at his beatification in 1629, and his father, Alonso de las Casas, that of declaring in his will that he had two sons who had died martyrs to the holy faith.

The venerable mother, Sebastiana, of the Most Holy Trinity, a nun in the monastery of San Juan de la Penitencia.

The venerable Don Alonso de las Culvas Davalos, descended of the most illustrious families, whose virtues and learning raised him to the dignity of archbishop of his native place, where he died, as it is supposed, a saint.

The venerable *Fr.* Bartolomé Gutierrez, burnt at Japon on account of the faith.

Don *Fr.* Antonio Monrroy, general of St. Domingo; a most religious character, and archbishop of Santiago.

Don Nicolas Gomez de Cervantes, of the ancient houses of the noble founders of Mexico, imitator of St. Tomas de Villanueva, archbishop of Guatemala.

Don Manuel de Ocio y Villafañe, collegiate of the *real* De San Ildefonso, and of that of San Ramon, doctor and dean of Manila, bishop of Zebú; in the troublesome visitation of which he died from a venomous bite.

Don *Fr.* Ignacio de Padilla y Estrada, bishop of Yucatán, archbishop of St. Domingo; the honour of his family, and in his religion of the order of Preachers.

Don Miguel Poblete, archbishop of Manila; a model of wise and holy prelates.

Don Manuel Antonio Roxo del Rio y Vieira, collegiate *real* of San Ildefonso, rector of Salamanca, archbishop, governor, and captain-general of Manila: the calamities which were suffered here at the time of the assault made by the English, preyed so much on his kind heart that he died a victim to his feelings for the interest of his country and his king.

The Father Christobal de Villafañe, a victim to his charity; having been put to death whilst visiting the prisons in the city of Guatemala, where he was prefect of prisons.

Don Joaquin Antonio de Ribadeneira, collegiate *mayor* of Santa Maria de los Santos, advocate, fiscal, and *oidor* of the royal audience; his learning is borne testimony to by his works.

Don Francisco Xavia de Gamboa, collegiate of the *real* De San Alfonso, *alcalde del crimen*, *oidor*, and actual regent of that audience; noted both in Europe and America for his talents, and for his celebrated Treatise on Mining which he published at Madrid.

Don Antonio de Villaurrutia, collegiate of the college of Todos los Santos, *oidor* of the audience of St. Domingo, deacon of that of his native place, and regent in that of Guadalaxara; a minister most exemplary for his probity, justice, and humility.

Don Baltasar Ladron de Guevara, who, following the career of jurisprudence, was invited, from his literature and talents, to accept the viceroyalty; he was fiscal-agent, *alcalde del crimen*, and *oidor* in his native place, manifesting a continual zeal and ability in the service of God and of his king.

The Doctor Don Juan Joseph de Eguiara, magistrate and dignitary of the church of Mexico, his native place; elected bishop of Yucatán, which office he renounced; wrote more than 40 volumes, and died a victim to fatigue and study; not more than one of the volumes of his excellent work, the Mexican bible, being published at the time.

We here finish this list, and observe that Mexico is 167 miles *s. w.* of Acapulco in the S. sea, and about the same distance from the N. sea, or from the port of Vera Cruz. In lat. 19° 26′ 53″, and long. 274° 10′. [But the long. taken by Humboldt, at the convent of St. Augustin, makes it in 101° 25′ 30″ or 99° 5′ 30″ *w.* from Greenwich; the lat. being 19° 25′ 45″.

Some further and very interesting accounts of this city, extracted from the above author, will be found interspersed amongst the new matter inserted under the head MEXICO, Intendancy of, Chap. I. the whole of which new matter we have inserted before Alçedo's chronological series of the Indian emperors.]

[INDEX TO NEW MATTER RESPECTING MEXICO.

[CHAP. VIII. *Table of the geographical positions of the kingdom of Nueva España; for which see the end of the general preface to this work.*
CHAP. IX. *Table of the most remarkable elevations measured in the interior of Nueva España.*
CHAP. X. *General considerations on the possibility of uniting the South sea and Atlantic ocean, viz. By the rivers of Peace and Tacoutche—Tesse—Sources of the Rio Bravo and Rio Colorado—Isthmus of Tehuantepec—Lake of Nicaragua—Isthmus of Panáma—Bay of Cupica—Canal of Choco—Rio Guallaga—Gulf of St. George.*
CHAP. XI. *Glance on the coast of the Great ocean, which extends from the port of San Francisco and from cape Mendocino to the Russian establishments in Prince William's sound.*

CHAP. I.

Mexico, Intendancy of.---Population, extent, and boundaries.---Physical appearance.---Teocalli and other edifices.---Lake of Tezcuco.---Grandeur of the city.---Its monuments and antiquities.---Its population.---Its consumption of provisions.--Floating islands.--Valley of Tenochtitlan.---Hill of Chapultepec.---Hydrographical view of the valley of Mexico.---Remarkable cities, towns, and mines of this intendancy.

The population of this intendancy, in 1803, amounted to 1,511,800 souls; the extent of surface in square leagues to 5927; making the number of inhabitants to the square league 255.

The whole of this intendancy is situated under the torrid zone. It extends from the 16° 34′ to the 21° 57′ of *n.* lat. It is bounded on the *n.* by the intendancy of San Luis Potosi, on the *w.* by the intendancies of Guanaxuato and Valladolid, and on the *e.* by those of Vera Cruz and La Puebla de los Angeles. It is washed towards the *s.* by the S. sea, or Great ocean, for a length of coast of 82 leagues from Acapulco to Zacatula.

Its greatest length from Zacatula to the mines of the Doctor is 136 leagues; and its greatest breadth from Zacatula to the mountains situated to the *e.* of Chilpansingo is 92 leagues. In its *n.* part, towards the celebrated mines of Zimapan and the Doctor, it is separated by a narrow strip from the gulph of Mexico. Near Mextitlan this strip is only nine leagues in breadth.

More than two-thirds of the intendancy of Mexico are mountainous, in which there are immense plains, elevated from 2000 to 2300 metres (6561 and 7545 feet) above the level of the ocean. From Chalco to Queretaro are almost uninterrupted plains of 50 leagues in length and eight or ten in breadth. In the neighbourhood of the *w.* coast the climate is burning and very unhealthy. One summit only, the Nevado de Toluca, situated in a fertile plain of 2700 metres (or 8857 feet) in height, enters the region of perpetual snow. Yet the porphyritical summit of this old volcano, whose form bears a strong resemblance to that of Pichincha near Quito, and which appears to have been formerly extremely elevated, is uncovered with snow in the rainy months of September and October. The elevation of the Pico del Fraile, or the highest summit of the Nevado de Toluca, is 4520 metres (2370 toises, or 15,156 feet.) No mountain in this intendancy equals the height of Mount Blane.

The valley of Mexico, or Tenochtitlan, of which M. Humboldt has published a very minute map, is situated in the centre of the *cordillera* of Anahuac, on the ridge of the *porphyritical* and *basaltic amygdaloid* mountains, which run from the *s. s. e.* to the *n. n. w.* This valley is of an oval form. According to his observations, and those of a distinguished mineralogist, M. Don Luis Martin, it contains, from the entry of the Rio Tenango into the lake of Chalco, to the foot of the Cerro de Sincoque, near the *desague real* of Huehuetoca, 18¼ leagues in length, and from S. Gabriel, near the small town of Tezcuco, to the sources of the Rio de Escapusalco, near Guisquiluca, 12½ leagues in breadth. The territorial extent of the valley is 244½ square leagues, of which only 22 square leagues are occupied by the lakes, which is less than a tenth of the whole surface.

The circumference of the valley, reckoning from the crest of the mountains, which surround it like a circular wall, is 67 leagues. This crest is most elevated on the *s.* particularly on the *s. e.* where the great volcanoes of La Puebla, the Popocatepetl, and Iztaccihuatl, bound the valley. One of the roads which lead from the valley of Tenochtitlan to that of Cholula and La Puebla passes even between the two volcanoes, by Tlamanalco, Ameca, La Cumbre, and La Cruz del Coreo. The small army of Cortes passed by this road on his first iavasion.

Six great roads cross the *cordillera* which incloses the valley, of which the medium height is 3000 metres (or 9842 feet) above the level of the ocean. 1. The road from Acapulco to Guchilaque and Cuervaracca by the high summit called La Cruz del Marques (alluding to Hernan Cortes, Marques del Valle de Oaxaca.) 2. The road of Toluca by Tianguillo and Lerma, a magnificent causeway, not sufficiently to be admired, constructed with great art, partly over arches. 3. The road of Queretaro, Guanaxuato, and Durango el] [camino de tierra adentro, which passes by Guautitlan, Huehuetoca, and the Puerto de Reyes, near Bata, through hills scarcely 80 (or 262 feet) metres above the pavement of the great square of Mexico. 4. The road of Pachuco, which leads to the celebrated mines of *real* del Monte, by the Cerro Ventoso, covered with oak, cypress, and rose trees, almost continually in flower. 5. The old road of La Puebla, by S. Bonaventura and the Llanos de Apan. And, 6. the new road of La Puebla by Rio Frio and Tesmelucos, *s. e.* from the Cerro del Telapon, of which the distance from the Sierra Nevada, as well as that from the Sierra Nevada (Iztaccihuatl) to the great volcano (Popocatepetl), served for bases to the trigonometrical operations of MM. Velasquez and Costanzo.

From being long accustomed to hear the capital of Mexico spoken of as a city built in the midst of a lake, and connected with the continent merely by dikes, those who look at Humboldt's map will be no doubt astonished on seeing that the centre of the present city is 4500 metres (or 14,763 feet) distant from the lake of Tezcuco, and more than 9000 metres (or 29,527 feet) from the lake of Chalco. They will be inclined, therefore, either to doubt the accuracy of the descriptions in the history of the discoveries of the new world, or they will believe that the capital of Mexico does not stand on the same ground with the old residence of Montezuma, or Muteczuma; but the city has certainly not changed its place, for the cathedral of Mexico occupies exactly the ground where the temple of Huitzilopochtli stood, and the present street of Tacuba is the old street of Tlacopan, through which Cortes made his famous retreat in the melancholy night of the 1st of July 1520, which goes by the name of *noche triste*. The difference of situation between the old maps and those published by Humboldt, arises solely from the diminution of water of the lake of Tezcuco.

It may be useful in this place to lay before the readers a passage from a letter addressed by Cortes to the Emperor Charles V. dated 30th October 1520, in which he gives the description of the valley of Mexico. This passage, written with great simplicity of style, gives us at the same time a very good idea of the sort of police which prevailed in the old Tenochtitlan. "The province in which the residence of this great lord Muteczuma is situated," says Cortes, "is circularly surrounded with elevated mountains, and intersected with precipices. The plain contains near 70 leagues in circumference, and in this plain are two lakes, which fill nearly the whole valley; for the inhabitants sail in canoes for more than 50 leagues round." (We must observe that the General speaks only of two lakes, because he knew but imperfectly those of Zumpango and Xaltocan, between which he hastily passed in his flight from Mexico to Tlascala, before the battle of Otumba.) "Of the two great lakes of the valley of Mexico, the one is fresh and the other salt water. They are separated by a small range of mountain (the conical and insulated hills near Iztapalapan); these mountains rise in the middle of the plain, and the waters of the lake mingle together in a strait between the hills and the high *cordillera* (undoubtedly the *e.* declivity of Cerros de Santa Fe.) The numerous towns and villages constructed in both of the two lakes carry on their commerce by canoes, without touching the continent. The great city of Temixtitan or Tenochtitlan is situated in the midst of the salt-water lake, which has its tides like the sea; and from the city to the continent there are two leagues, whichever way we wish to enter. Four dikes lead to the city: they are made by the hand of man, and are of the breadth of two lances. The city is as large as Seville or Cordova. The streets, I merely speak of the principal ones, are very narrow and very large; some are half dry and half occupied by navigable canals, furnished with very well constructed wooden bridges, broad enough for 10 men on horseback to pass at the same time. The market-place, twice as large as that of Seville, is surrounded with an immense portico, under which are exposed for sale all sorts of merchandize, eatables, ornaments made of gold, silver, lead, pewter, precious stones, bones, shells, and feathers; delft ware, leather, and spun cotton. We find hewn stones, tiles, and timber fit for building. There are lanes for game, others for roots and garden fruits; there are houses where barbers shave the head (with razors made of obsidian); and there are houses resembling our apothecary shops, where prepared medicines, unguents, and plasters are sold. There are houses where drink is sold. The market abounds with so many things, that I am unable to name them all to your highness. To avoid confusion, every species of merchandize is sold in a separate lane; every thing is sold by the yard, but nothing has hitherto been seen to be weighed in the market. In the midst of the great square is a house, which I shall call *l'audiencia*, in which 10 or 12 persons sit constantly for determining any disputes which may arise respecting the sale of goods. There are other persons who mix continually with the crowd, to see that a just price is asked. We have seen them break the false measures which they had seized from the merchants."

Such was the state of Tenochtitlan in 1520,]

[according to the description of Cortes himself. Humboldt sought in vain in the archives of his family, preserved at Mexico in the Casa del Estado, for the plan which this great captain ordered to be drawn up of the environs of the capital, and which he sent to the emperor, as he says, in his third letter published by Cardinal Lorenzana. The Abbe Clavigero has ventured to give a plan of the lake of Tezcuco, such as he supposes it to have been in the sixteenth century. This sketch is very inaccurate, though much preferable to that given by Robertson, and other European authors, equally unskilled in the geography of Mexico. Humboldt has drawn on the map of the valley of Tenochtitlan the old extent of the salt-water lake, such as he conceived it from the historical account of Cortes, and some of his contemporaries. In 1520, and long after, the villages of Iztapalapan, Coyohuacan (improperly called Cuyacan), Tacubaja, and Tacuba, were quite near the banks of the lake of Tezcuco. Cortes says expressly, (*Lorenzana*, p. 229, 195, 102), that the most part of the houses of Coyohuacan, Culuacan, Chulubuzco, Mexicaltzingo, Iztapalapan, Cuitaguaca, and Mizqueque, were built in the water on piles, so that frequently the canoes could enter by an under-door. The small hill of Chapultepec, on which the viceroy Count Galvez constructed a castle, was no longer an island in the lake of Tezcuco in the time of Cortes. On this side, the continent approached to within about 3000 metres (or 9812 feet) of the city of Tenochtitlan, consequently the distance of two leagues indicated by Cortes in his letter to Charles V. is not altogether accurate: he ought to have retrenched the one half of this, excepting, however, the part of the *w.* side at the small porphyritical hill of Chapultepec. We may well believe, however, that this hill was, some centuries before, also a small island, like the Peñol del Marques, or the Peñol de los Baños. It appears extremely probable, from geological observations, that the lakes had been on the decrease long before the arrival of the Spaniards, and before the construction of the canal of Huehuetoca.

The Aztecs, or Mexicans, before founding on a group of islands, in 1325, the capital which yet subsists, had already inhabited for 52 years another part of the lake farther to the *s.* of which the Indians have not been lately able to point out the site. The Mexicans left Aztlan towards the year 1160, and only arrived, after a migration of 56 years, in the valley of Tenochtitlan, by Malinalco, in the *cordillera* of Toluca, and by Tula. They established themselves first at Zumpango, then on the *s.* declivity of the mountains of Tepeyac, where the magnificent temple, dedicated to Our Lady of Guadaloupe, is situated. In the year 1245 (according to the chronology of the Abbe Clavigero), they arrived at Chapultepec. Harassed by the petty princes of Zaltocan, whom the Spanish historians honour with the title of kings, the Aztecs, to preserve their independence, withdrew to a group of small islands called Acoculco, situated towards the *s.* extremity of the lake of Tezcuco. There they lived for half a century in great want, compelled to feed on roots of aquatic plants, insects, and a problematical reptile called *axolotl*, which Mr. Cuvier looks upon to be the nympha of an unknown salamander. Having been reduced to slavery by the kings of Tezcuco or Acolhuacan, the Mexicans were forced to abandon their village in the midst of the lake, and to take refuge on the continent at Tizapan. The services which they rendered to their masters in a war against the inhabitants of Xochimilco again procured them liberty. They established themselves first at Acatzitzintlan, which they called Mexicalzing, from the name of Mexitli, or Huitzilopochtli, their god of war, and next at Iztacalco. They removed from Iztacalco to the little islands which then appeared to the *e. n. e.* of the hill of Chapultepec, in the *w.* part of the lake of Tezcuco, in obedience to an order of the oracle of Aztlan. An ancient tradition was preserved among this horde, that the fatal term of their migration was to be a place where they should find an eagle sitting on the top of a *nopal*, of which the roots penetrated the crevices of a rock. This *nopal* (cactus), alluded to in the oracle, was seen by the Aztecs in the year 1325, which is the second *calli* of the Mexican era, on a small island, which served for foundation to the Teocalli, or Teopan, *i. e.* the house of God, afterwards called by the Spaniards the Great Temple of Mexitli.

The first *Teocalli*, around which the new city was built, was of wood, like the most ancient Grecian temple, that of Apollo at Delphi, described by Pausanias. The stone edifice, of which Cortes and Bernal Diaz admired the symmetry, was constructed on the same spot by King Ahuitzotl in the year 1486. It was a pyramidal monument of 37 metres, or 121 feet, in height, situated in the middle of a vast inclosure of walls, and consisted of five stories, like several pyramids of Sacara, and particularly that of Mehedun. The Teocalli of Tenochtitlan, very accurately laid out, like all the Egyptian, Asiatic, and Mexican pyramids, contained 97 metres, or 318 feet, of base, and formed so truncated a pyramid, that when seen from a distance the monument appeared an enormous] [cube, with small altars, covered with wooden cupolas on the top. The point where these cupolas terminated was 54 metres, or 177 feet, elevated above the base of the edifice or the pavement of the inclosure. We may see from these details that the Teocalli bore a strong resemblance in form to the ancient monument of Babylon, called by Strabo the Mausoleum of Belus, which was only a pyramid dedicated to Jupiter Belus. (*Zoega de Obeliscis*, p. 50.) Neither the Teocalli nor the Babylonian edifice were temples, in the sense which we attach to the word, according to the ideas derived by us from the Greeks and Romans. All edifices consecrated to Mexican divinities formed truncated pyramids. The great monuments of Teotihuacan, Cholula, and Papantla, still in preservation, confirm this idea, and indicate what the more inconsiderable temples were in the cities of Tenochtitlan and Tezcuco. Covered altars were placed on the top of the Teocallis; and these edifices must hence be classed with the pyramidal monuments of Asia, of which traces were anciently found even in Arcadia; for the conical mausoleum of Callistus (*Pausanias*, lib. viii. c. 35.) was a true *tumulus*, covered with fruit trees, and served for a base to a small temple consecrated to Diana.

We know not of what materials the Teocalli of Tenochtitlan was constructed. The historians merely relate, that it was covered with a hard and smooth stone. The enormous fragments which are from time to time discovered around the present cathedral are of porphyry, with a base of grunstein filled with amphibolos and vitreous feldspath. When the square round the cathedral was recently paved, carved stones were found at a depth of 10 and 12 metres, or 32 and 38 feet. Few nations have moved such great masses as were moved by the Mexicans. The calendar stone and the sacrifice stone, exposed to public view in the great square, contain from eight to 10 cubic metres, or from 282 to 353 cubic feet. The colossal statue of Teoyaomiqui, covered with hieroglyphics, lying in one of the vestibules of the university, is three metres, or $9\frac{4}{5}$ feet, in breadth. M. Gamboa, one of the canons, assured Humboldt, that on digging opposite the chapel of the Sagrario, a carved rock was found among an immense quantity of idols belonging to the Teocalli, which was seven metres in length, six in breadth, and three in height, viz. $22\frac{7}{8}$, $19\frac{2}{3}$, and $9\frac{4}{5}$ feet. They endeavoured in vain to remove it.

The Teocalli was in ruins a few years after the siege of Tenochtitlan, which, like that of Troy, ended in an almost entire destruction of the city. We are therefore inclined to believe that the exterior of the truncated pyramid was clay, covered with porous amygdaloid called *tetzontli*. In fact, a short time before the construction of the temple, under the reign of King Ahuitzotl, the quarries of this cellular and spongy rock began to be worked. Now nothing could be easier destroyed than edifices constructed of porous and light materials, like pumice-stone. Notwithstanding the coincidence of a great number of accounts, it is not impossible that the dimensions attributed to the Teocalli are somewhat exaggerated; but the pyramidal form of this Mexican edifice, and its great analogy to the most ancient monuments of Asia, ought to interest us much more than its mass and size.

The old city of Mexico communicated with the continent by the three great dikes of Tepejacac, (Guadalupe), Tlacopan, (Tacuba), and Iztapalapan. Cortes mentions four dikes, because he reckoned, without doubt, the causeway which led to Chapultepec. The Calzada of Iztapalapan had a branch which united Coyohuacan to the small fort Xaloe, the same in which the Spaniards were entertained at their first entry by the Mexican nobility. Robertson speaks of a dike which led to Tezcuco, but such a dike never existed, on account of the distance of the place, and the great depth of the *e.* part of the lake.

In 1338, 17 years after the foundation of Tenochtitlan, a part of the inhabitants, in a civil dissension, separated from the rest: they established themselves in the small islands to the *n. w.* of the temple of Mexitli. The new city, which at first bore the name of Xaltilolco, and afterwards Tlateloleo, was governed by a king independent of Tenochtitlan. In the centre of Anahuac, as well as in the Peloponnesus, Latium, and wherever the civilization of the human species was merely commencing, every city, for a long time, constituted a separate state. The Mexican king Axajacatl (*Clavigero*, i. p. 251. Axajacall reigned from 1464 to 1477, iv. p. 58.) conquered Tlatelolco, which was thenceforth united by bridges to the city of Tenochtitlan. Humboldt discovered in the hieroglyphical manuscripts of the ancient Mexicans, preserved in the palace of the viceroy, a curious painting, which represents the last king of Tlatelolco, called Moquihuix, as killed on the top of a *house of God*, or truncated pyramid, and then thrown down the stairs which led to the stone of the sacrifices. Since this catastrophe, the great market of the Mexicans, formerly held near the Teocalli of Mexitli, was transferred to Tlatelolco. The description of the Mexican market, which we have given from Cortes, relates to the market of Tlatelolco.]

[What is now called the Barrio of Santiago composes but a part of the ancient Tlatelolco. We proceed for more than an hour on the road to Tanepantla and Ahuahuetes, among the ruins of the old city. We perceive there, as well as on the road to Tacuba and Iztapalapan, how much the Mexico rebuilt by Cortes is smaller than Tenochtitlan under the last of the Montezumas. The enormous magnitude of the market-place of Tlatelolco, of which the boundaries are still discernible, proves the great population of the ancient city. The Indians show in this same market-place an elevation surrounded by walls. It was one of the Mexican theatres, the same on which Cortes, a few days before the end of the siege, erected his famous Catapulta, *trabuco de palo*, (*Lorenzana*, p. 289.) the appearance of which alone terrified the besieged; for the machine was incapable of being used from the awkwardness of the artillery-men. This elevation is now included in the porch of the chapel of Santiago.

The city of Tenochtitlan was divided into four quarters, called Teopan, or Xochimilca, Atzacualco, Moyotla, and Tlaguechiuchan, or Cuepopan. The old division is still preserved in the limits assigned to the quarters of St. Paul, St. Sebastian, St. John, and St. Mary; and the present streets have for the most part the same direction as the old ones, nearly from *n.* to *s.* and from *e.* to *w.* though more properly from the *s.* 16° *w.* to *n.* 74° *e.* at least towards the convent of St. Augustin, where Humboldt took his azimuths. The direction of the old streets was undoubtedly determined by that of the principal dikes. Now, from the position of the places where these dikes appear to have terminated, it is very improbable that they represented exactly meridians and parallels. But what gives the new city, as we have already observed, a peculiar and distinctive character, is that it is situated entirely on the continent, between the extremities of the two lakes of Tezcuco and Xochimilco, and that it only receives, by means of navigable canals, the fresh water of the Xochimilco.

Many circumstances have contributed to this new order of things. The part of the salt-water lake between the *s.* and *w.* dikes was always the shallowest. Cortes complained that his flotilla, the brigantines which he constructed at Tezcuco, could not, notwithstanding the openings in the dikes, make the circuit of the besieged city. Sheets of water of small depth became insensibly marshes, which, when intersected with trenches or small defluous canals, were converted into *chinampas* and arable land. The lake of Tezcuco, which Valmont de Bomarc (in his *Dictionnaire d'Histoire Naturelle*, article *Lac*) supposed to communicate with the ocean, though it is at an elevation of 2277 metres, or 7468 feet, has no particular sources, like the lake of Chalco. When we consider, on the one hand, the small volume of water with which in dry seasons this lake is furnished by very inconsiderable rivers, and on the other, the enormous rapidity of evaporation in the table-land of Mexico, of which Humboldt has made repeated experiments, we must admit, what geological observations appear also to confirm, that for centuries the want of equilibrium between the water lost by evaporation, and the mass of water flowing in, has progressively circumscribed the lake of Tezcuco within more narrow limits. We learn from the Mexican annals, (viz. paintings preserved in the Vatican, and testimony of Father Acosta), that in the reign of King Ahuizotl, this salt-water lake experienced such a want of water as to interrupt navigation; and that to obviate this evil, and to increase its supplies, an aqueduct was constructed from Coyohuacan to Tenochtitlan. This aqueduct brought the sources of Huitzilopochco to several canals of the city which were dried up.

This diminution of water, experienced before the arrival of the Spaniards, would no doubt have been very slow and very insensible, if the hand of man, since the period of the conquest, had not contributed to reverse the order of nature. Those who have travelled in the peninsula know how much, even in Europe, the Spaniards hate all plantations which yield a shade round towns or villages. It would appear that the first conquerors wished the beautiful valley of Tenochtitlan to resemble the Castilian soil, which is dry and destitute of vegetation. Since the 16th century they have inconsiderately cut, not only the trees of the plain in which the capital is situated, but those on the mountains which surround it. The construction of the new city, begun in 1524, required a great quantity of timber for building and piles. They destroyed, and they daily destroy, without planting any thing in its stead, except around the capital, where the last viceroys have perpetuated their memory by promenades, (namely, by the *Paseos de Buccarelli, de Revillagigedo, de Galvez, de Asanza*), which bear their names. The want of vegetation exposes the soil to the direct influence of the solar rays; and the humidity which is not lost by filtration through the amygdaloid, basaltic, and spongy rock, is rapidly evaporated and dissolved in air, wherever the foliage of the trees] [or a luxuriant verdure does not defend the soil from the influence of the sun and the dry winds of the *s.*

As the same cause operates throughout the whole valley, the abundance and circulation of water has sensibly diminished. The lake of Tezcuco, the finest of the five lakes, which Cortes in his letters habitually calls an interior sea, receives much less water from infiltration than in the 16th century. Every where the clearing and destruction of forests have produced the same effects. General Andreossi, in his classical work on the *Canal du Midi*, has proved that the springs have diminished around the reservoir of St. Fencol, merely through a false system introduced in the management of the forests. In the province of Caracas, the picturesque lake of Tacarigua has been drying gradually up ever since the sun darted his rays without interposition on the naked and defenceless soil of the valleys of Aragua.

But the circumstance which has contributed the most to the diminution of the lake of Tezcuco is the famous open drain, known by the name of the *Desague real de Huehuetoca*, which we shall afterwards discuss. This cut in the mountain, first begun in 1607 in the form of a subterranean tunnel, has not only reduced within very narrow limits the two lakes in the *n.* part of the valley, *i. e.* the lakes of Zumpango (Tzompango) and San Christobal; but has also prevented their waters in the rainy season from flowing into the basin of the lake of Tezcnco. These waters formerly inundated the plains, and purified a soil strongly covered with carbonate and muriate of soda. At present, without settling into pools, and thereby increasing the humidity of the Mexican atmosphere, they are drawn off by an artificial canal into the river of Panuco, which flows into the Atlantic ocean.

This state of things has been brought about from the desire of converting the ancient city of Mexico into a capital better adapted for carriages, and less exposed to the danger of inundation. The water and vegetation have in fact diminished with the same rapidity with which the *tequesquite* (or carbonate of soda) has increased. In the time of Montezuma, and long afterwards, the suburb of Tlatelolco, the *barrios* of San Sebastian, San Juan, and Santa Cruz, were celebrated for the beautiful verdure of their gardens; but these places now, and especially the plains of San Lazaro, exhibit nothing but a crust of efflorescent salts. The fertility of the plain, though yet considerable in the *s.* part, is by no means what it was when the city was surrounded by the lake. A wise distribution of water, particularly by means of small canals of irrigation, might restore the ancient fertility of the soil, and re-enrich a valley which nature appears to have destined for the capital of a great empire.

The actual bounds of the lake of Tezcnco are not very well determined, the soil being so argillaceous and smooth that the difference of level for a mile is not more than two decimetres, or 7874 inches. When the *e.* winds blow with any violence, the water withdraws towards the *w.* bank of the lake, and sometimes leaves an extent of more than 600 metres, or 1968 feet, dry. Perhaps the periodical operation of these winds suggested to Cortes the idea of regular tides, (See *Journal de Savans* for the year 1676, p. 34. The lake of Geneva manifests also a regular motion, which Saussure attributes to periodical winds.) of which the existence has not been confirmed by late observations. The lake of Tezcuco is in general only from three to five metres, or 9¾ to 16⅓ feet, in depth, and in some places even less than one. Hence the commerce of the inhabitants of the small town of Tezcuco suffers much in the very dry months of January and February; for the want of water prevents them from going in canoes to the capital. The lake of Xochimilco is free from this inconvenience; for from Chalco, Mesquie, and Tlahuac, the navigation is never once interrupted, and Mexico receives daily, by the canal of Iztapalapan, roots, fruits, and flowers in abundance.

Of the five lakes of the valley of Mexico, the lake of Tezcuco is most impregnated with muriate and carbonate of soda. The nitrate of barytes proves that this water contains no sulphate in dissolution. The most pure and limpid water is that of the lake of Xochimilco, the specific weight of which Humboldt found to be 1.0009, when that of water distilled at the temperature of 18° centigrade, or 54° Fahrenheit, was 1.000, and when water from the lake of Tezcuco was 1.0215. The water of this last lake is consequently heavier than that of the Baltic sea, and not so heavy as that of the ocean, which, under different latitudes, has been found between 1.0269 and 1.0285. The quantity of sulphuretted hydrogen which is detached from the surface of all the Mexican lakes, and which the acetite of lead indicates in great abundance in the lakes of Tezcuco and Chalco, undoubtedly contributes in certain seasons to the unhealthiness of the air of the valley. However, and the fact is curious, intermittent fevers are very rare on the banks of these very lakes, of which the sur-]

[face is partly concealed by rushes and aquatic herbs.

Adorned with numerous Teocallis, like so many Mahometan steeples, surrounded with water and dikes founded on islands covered with verdure, and receiving hourly in its streets thousands of boats, which vivified the lake, the ancient Tenochtitlan, according to the accounts of the first conquerors, must have resembled some of the cities of Holland, China, or the Delta of Lower Egypt. The capital, reconstructed by the Spaniards, exhibits, perhaps, a less vivid, though a more august and majestic appearance. Mexico is undoubtedly one of the finest cities ever built by Europeans in either hemisphere. With the exception of Petersburgh, Berlin, Philadelphia, and some quarters of Westminster, there does not exist a city of the same extent which can be compared to the capital of New Spain, for the uniform level of the ground on which it stands, for the regularity and breadth of the streets, and the extent of the public places. The architecture is generally of a very pure style, and there are even edifices of very beautiful structure. The exterior of the houses is not loaded with ornaments. Two sorts of hewn stone, the porous amygdaloid called *tetzontli*, and especially a porphyry of vitreous feld-spath without any quartz, give to the Mexican buildings an air of solidity, and sometimes even magnificence. There are none of those wooden balconies and galleries to be seen which disfigure so much all the European cities in both the Indies. The balustrades and gates are all of Biscay iron, ornamented with bronze, and the houses, instead of roofs, have terraces, like those in Italy and other s. countries.

Mexico has been very much embellished since the residence of the Abbe Chappe there in 1769. The edifice destined to the school of mines, for which the richest individuals of the country furnished a sum of more than 3,000,000 of francs, or 124,800*l*. sterling, would adorn the principal places of Paris or London. Two great palaces, were recently constructed by Mexican artists. pupils of the academy of fine arts of the capital, One of these palaces, in the quarter Della Traspana, exhibits in the interior of the court a very beautiful oval peristyle of coupled columns. The traveller justly admires a vast circumference paved with porphyry flags, and inclosed with an iron railing, richly ornamented with bronze, containing an equestrian statue of King Charles IV. placed on a pedestal of Mexican marble, in the midst of the *plaza major* of Mexico, opposite the cathedral and the viceroy's palace. This colossal statue was executed at the expence of the Marquis de Branciforte, formerly viceroy of Mexico, brother-in-law of the Prince of Peace. It weighs 450 quintals, and was modelled, founded, and placed by the same artist, M. Tolsa, whose name deserves a distinguished place in the history of Spanish sculpture. The merits of this man of genius can only be appreciated by those who know the difficulties with which the execution of these great works of art are attended even in civilized Europe. However, it must be agreed, that notwithstanding the progress of the arts within these last 30 years, it is much less from the grandeur and beauty of the monuments, than from the breadth and straightness of the streets, and much less from its edifices than from its uniform regularity, its extent and position, that the capital of New Spain attracts the admiration of Europeans. Humboldt, who, from a singular concurrence of circumstances, had seen successively, within a very short space of time, Lima, Mexico, Philadelphia, Washington, Paris, Rome, Naples, and the largest cities of Germany; and notwithstanding such unavoidable comparisons, of which several, one would think, must have proved disadvantageous for the capital of Mexico, remained nevertheless smitten with a recollection of the grandeur of this latter city, a circumstance which he attributes principally to the majestic character of its situation and the surrounding scenery.

In fact, nothing can present a more rich and varied appearance than the valley, when, in a fine summer morning, the sky without a cloud, and of that deep azure which is peculiar to the dry and rarefied air of high mountains, we transport ourselves to the top of one of the towers of the cathedral of Mexico, or ascend the hill of Chapultepec. A beautiful vegetation surrounds this hill. Old cypress trunks, (namely, the Ahuahuetes—Cupressus disticha Lin.), of more than 15 and 16 metres, (49 and 52 feet), in circumference, raise their naked heads above those of the *schinus*, which resemble in their appearance the weeping willows of the east. From the centre of this solitude, the summit of the porphyritical rock of Chapultepec, the eye sweeps over a vast plain of carefully cultivated fields, which extend to the very feet of the colossal mountains covered with perpetual snow. The city appears as if washed by the waters of the lake of Tezcuco, whose basin, surrounded with villages and hamlets, brings to mind the most beautiful lakes of the mountains of Switzerland. Large avenues of elms and poplars lead in every direction to the capital; and two aqueducts, constructed over arches of very great elevation, cross the plain, and exhibit an appearance equally] [agreeable and interesting. The magnificent convent of Nuestra Señora de Guadalupe appears joined to the mountains of Tepeyacac, among ravines which shelter a few date and young *yuca* trees. Towards the *s.* the whole tract between San Angel, Tacabaya, and San Agustin de las Cuevas, appears an immense garden of orange, peach, apple, cherry, and other European fruit trees. This beautiful cultivation forms a singular contrast with the wild appearance of the naked mountains which inclose the valley, among which the famous volcanoes of La Puebla, Popocatepetl, and Iztaccicibuatl, are the most distinguished. The first of these forms an enormous cone, of which the crater, continually inflamed and throwing up smoke and ashes, opens in the midst of eternal snows.

The city of Mexico is also remarkable for its excellent police. The most part of the streets have very broad pavements; and they are clean and well lighted. These advantages are the fruits of the activity of the Count de Revillagigedo, who on his arrival found the capital extremely dirty.

Water is every where to be had in the soil of Mexico, a very short way below the surface, but it is brackish, like the water of the lake of Tezcuco. The two aqueducts already mentioned, by which the city receives fresh water, are monuments of modern construction worthy of the traveller's attention. The springs of potable water are situated to the *e.* of the town, one in the insulated hill of Chapultepec, and the other in the *cerros* of Santa Fé, near the *cordillera*, which separates the valley of Tenochtitlan from that of Lerma and Toluca. The arches of the aqueduct of Chapultepec occupy a length of more than 3300 metres, or 10,826 feet. The water of Chapultepec enters by the *s.* part of the city, at the Salto del Agua. It is not the most pure, and is only drank in the suburbs of Mexico. The water which is least impregnated with carbonate of lime is that of the aqueduct of Santa Fé, which runs along Almeda, and terminates at La Traspana, at the bridge De la Marescalla. This aqueduct is nearly 10,200 metres, or 33,464 feet, in length; but the declivity of the ground is such, that for not more than a third of this space the water can be conducted over arches. The old city of Tenochtitlan had aqueducts no less considerable. (*Clavigero*, iii. p. 195; *Solis*, i. p. 406.) In the beginning of the siege, the two captains Alvarado and Olid destroyed that of Chapultepec. Cortes, in his first letter to Charles V. speaks also of the spring of Amilco, near Churubusco, of which the waters were brought to the city by pipes of burnt earth. This spring is near to that of Santa Fé. We still perceive the remains of this great aqueduct, which was constructed with double pipes, one of which received the water, while they were employed in cleaning the other. This water was sold in canoes, which traversed the streets of Tenochtitlan. The sources of San Agustin de las Cuevas are the finest and purest; and Humboldt imagined he discovered on the road leading from this charming village to Mexico traces of an ancient aqueduct. The largest and finest construction, however, of the Indians in this way is the aqueduct of the city of Tezcuco. We still admire the traces of a great mound which was constructed to heighten the level of the water. How must we admire the industry and activity displayed in general by the ancient Mexicans and Peruvians in the irrigation of arid lands! In the maritime part of Peru Humboldt saw the remains of walls, along which water was conducted for a space of from 5 to 6000 metres, (from 16,404 to 19,685 feet), from the foot of the *cordillera* to the coast. The conquerors of the 16th century destroyed these aqueducts, and that part of Peru is become, like Persia, a desert destitute of vegetation. Such is the civilization carried by the Europeans among the people whom they are pleased to call barbarous.

How much it is to be regretted that Robertson gives usually such general descriptions, that we have a difficulty in forming any thing like a distinct conception of the subjects of them! He says of the Peru canals of irrigation, "By means of artificial canals, conducted with much patience and considerable art from the torrents that poured across their country, they conveyed a regular supply of moisture to their fields."—Would it have been beneath the dignity of a historian to have specified that art and that patience to his readers, for which he did not want materials?

We have already named the three principal dikes by which the old city was connected with the Tierra Firme. These dikes partly still exist, and the number has been even increased. They form at present great paved causeways across marshy grounds; and as they are very elevated, they possess the double advantage of admitting the passage of carriages, and containing the overflowings of the lake. The Calzada of Astapalapan is founded on the very same old dike on which Cortes performed such prodigies of valour in his encounters with the besieged. The Calzada of San Anton is still distinguished in our days for the great number of small bridges which the Spaniards and Tlascaltecs found there, when Sandoval,]

[Cortes's companion in arms, was wounded near Coyohuacan. These Calzadas of San Antonio Abad, of La Piedad, of San Christobal, and of Guadalupe, (anciently called the Dike of Tepeyacac), were newly reconstructed after the great inundation of 1604, under the viceroy Don Juan de Mendoza y Lima, Marquis De Montesclaros. The only *savans* of that time, Fathers Torquemada and Geronimo de Sarate, executed the survey and marking out of the causeways. At this period the city of Mexico was paved for the first time; for before the Count De Revillagigedo, no other viceroy had employed himself more successfully in effecting a good police than the Marquis de Montesclaros.

The objects which generally attract the attention of the traveller are, 1. The cathedral, of which a small part is in the style vulgarly called Gothic: the principal edifice, which has two towers ornamented with pilasters and statues, is of very beautiful symmetry and very recent construction. 2. The treasury, adjoining to the palace of the viceroys, a building from which, since the beginning of the 16th century, more than 6,500,000,000, or 270,855,000*l.* sterling, in gold and silver coin, have been issued. 3. The convents, among which the convent of St. Francis is particularly distinguished, which from alms alone possesses an annual revenue of half a million of francs, or 20,835*l.* sterling. This vast edifice was at first intended to be constructed on the ruins of the temple of Huitzilopochtli; but these ruins having been destined for the foundation of the cathedral, the convent was begun in 1531 in its actual situation. It owes its existence to the great activity of a serving brother or lay monk, *Fray* Pedro de Gante, an extraordinary man, who was said to have been the natural son of the Emperor Charles V. and who was a great benefactor of the Indians, to whom he was the first who taught the most useful mechanical arts of Europe. 4. The hospital, or rather the two united hospitals, of which the one maintains 600, and the other 800 children and old people. This establishment, in which both order and cleanliness may be seen, but little industry, has a revenue of 250,000 francs, or 10,470*l.* sterling. A rich merchant lately bequeathed to it by his testament 6,000,000 of francs, or 250,020*l.* sterling, which the royal treasury laid hold of, on the promise of paying five per cent. for it. 5. The Acordada, a fine edifice, of which the prisons are generally spacious and well aired. They reckon in this house, and in the other prisons of the Acordada which depend on it, more than 1200 individuals, among whom are a great number of smugglers, and the unfortunate Indian prisoners dragged to Mexico from the *provincias internas* (Indios Mecos.) 6. The school of mines, the newly begun edifice, and the old provisory establishment, with its fine collections in physics, mechanics, and mineralogy. 7. The botanical garden, in one of the courts of the viceroy's palace. It is very small, but extremely rich in vegetable productions, either rare or interesting for commerce. 8. The edifices of the university and the public library, which is very unworthy of so great and ancient an establishment. 9. The academy of fine arts, with a collection of ancient casts. 10. The equestrian statue of King Charles IV. in the *plaza mayor* and the sepulchral monument which the Duke de Monteleone consecrated to the great Cortes, in a chapel of the Hospital de los Naturales. It is a simple family monument, adorned with a bust in bronze, representing the hero in the prime of life, executed by M. Tolsa. Wherever we traverse Spanish America, from Buenos Ayres to Monteroy, and from Trinidad and Portorico to Panáma and Veragua, we no where meet with a national monument erected by the public gratitude to the glory of Christopher Columbus and Hernan Cortes!

Those who are addicted to the study of history, and who love to investigate American antiquities, will not find in this capital those great remains of works which are to be seen in Peru, in the environs of Cusco and Guamachuco, at Pachacamac near Lima, or at Mansiche near Truxillo; at Canar and Cayo in the province of Qnito; and in Mexico, near Mitla and Cholula, in the intendancies of Oaxaca and Puebla. It appears that the Teocallis (of which we have already attempted to describe the strange form) were the sole monuments of the Aztecs. Now the invading Spaniards were not only highly interested in their destruction, but the very safety of the conqueror rendered such a destruction necessary. It was partly effected during the siege; for those truncated pyramids, rising up by layers, served for refuge to the combatants, like the temple of Baal-Berith to the people of Canaan. They were so many castles from which it was necessary to dislodge the enemy.

As to the houses of individuals, which the Spanish historians describe as very low, we are not to be surprised to find merely their foundations or low ruins, such as we discover in the Barrio de Tlatelolco, and towards the canal of Istacalco. Even in the most part of our European cities, how small is the number of houses of which the construction goes so far back as the beginning of the]

[16th century! However, the edifices of Mexico are not fallen into ruins through age. Animated by the same spirit of destruction which the Romans displayed at Syracuse, Carthage, and in Greece, the Spanish conquerors believed that the siege of a Mexican city never was finished till they had razed every building in it. Cortes, in his third letter (*Lorenzana*, p. 278.) to the Emperor Charles V. discloses himself the fearful system which he followed in his military operations. "Notwithstanding all these advantages," says he, "which we have gained, I saw clearly that the inhabitants of the city of Temixtitlan (Tenochtitlan) were so rebellious and obstinate that they wished rather to perish than surrender. I knew not what means to employ to spare so many dangers and hardships, and to avoid completing the entire ruin of the capital, which was the most beautiful thing in the world, (*a la ciudad, porque era la mas hermosa cosa del mundo.*) It was in vain to tell them that I would never raise my camp, nor withdraw my flotilla of brigantines; and that I would never cease to carry on the war by land and water till I was master of Temixtitlan; and it was in vain I observed to them that they could expect no assistance, and that there was not a nook of land from which they could hope to draw maize, meat, fruits, and water. The more we made these exhortations to them, the more they showed us that they were far from being discouraged. They had no other desire but that of fighting. In this state of things, considering that more than 40 or 50 days had already elapsed since we began to invest the place, I resolved at last to adopt means, by which, in providing for our own security, we should be able to press our enemies more closely. *I formed the design of demolishing on all sides all the houses in proportion as we became masters of the streets, so that we should not advance a foot without having destroyed and cleared down whatever was behind us, converting into firm ground whatever was water, however slow the operation might be, and notwithstanding the delay to which we should expose ourselves.*—(Accordè de tomar un medio para nuestra seguridad y para poder mas estrechar a los enemigos; y fue que como fuessemos ganando por las calles de la ciudad, que fuessen derocando todas las casas de ellas, de un lado y del otro; por manera que no fuessemos un passo adelante sin la dejar todo asolado y que lo que era agua hacerlo tierra firme; aunque hubiesse todo la dilacion que se pudiesse seguir. *Lorenzana*, No. xxxiv.) For this purpose I assembled the lords and chiefs of our allies; and I explained to them the resolution which I had formed. I engaged them to send a great number of labourers with their *coas*, which are somewhat like the hoes which are used in Spain for excavations; and our allies and friends approved my project, for they hoped that the city would be laid in complete ruins, which they had ardently desired for a long time. Three or four days passed without fighting, for we waited the arrival of the people from the country, who were to aid us in demolishing."

After reading the naif recital of this commander in chief to his sovereign, we are not to be surprised at finding almost no vestige of the ancient Mexican edifices. Cortes relates that the Indians, to revenge themselves for the oppressions which they had suffered from the Aztec kings, flocked in great numbers, even from the remotest provinces, whenever they learned that the destruction of the capital was going on. The rubbish of the demolished houses served to fill up the canals. The streets were made dry to allow the Spanish cavalry to act. The low houses, like those of Pekin and China, were partly constructed of wood and partly of *tetzontli* a spongy stone, light, and easily broken. "More than 50,000 Indians assisted us," says Cortes, "that day, when, marching over heaps of carcases, we at length gained the great street of Tacuba, and burned the house of King Guatimucin." The true name of this unfortunate king, the last of the Aztec dynasty, was Quauhtemotzin. He is the same to whom Cortes caused the soles of the feet to be gradually burned, after having soaked them in oil. This torment, however, did not induce the king to declare in what place his treasures were concealed. His end was the same as that of the king of Acolhuacan (Tezcuco), and of Tetlepanguetzaltzin, king of Tlacopan (Tacuba.) These three princes were hung on the same tree, and, as Humboldt saw in a hieroglyphical picture possessed by Father Pichardo (in the convent of San Felipe Neri), they were hung by the feet to lengthen out their torments. This act of cruelty in Cortes, which recent historians have the meanness to describe as the effect of a far-sighted policy, excited murmurs in the very army. "The death of the young king," says Bernal Diaz del Castillo (an old soldier full of honour and of naivety of expression), "was a very unjust thing; and it was accordingly blamed by us all, so long as we were in the suite of the captain, in his march to Comajahua.")

The Abbe Clavigero observes, on what authority we know not, that this cruelty made Cortes very melancholy, and gave him a few sleepless] nights, *una gran malinconia, ed alcune vegghie.* Well indeed it might; but whether we are indebted for these *vegghie* to the native suggestions of his own conscience, or to the murmurs of his army, is not so easy to be determined; for heroes consciences are made of stern stuff, as many can witness, who have known several of them perform certain actions in a certain neighbouring country, and neither eat nor sleep the worse for it; at the bare recital of which other people's cheeks turn either pale or flushed as their different temperaments dispose them. We must not think that the Spaniards monopolized cruelty in foreign settlements. Mr. Orme, in his excellent History of Hindostan, celebrates some feats of our own countrymen, and those the bravest of our countrymen, which yield very little to any thing in the Mexican annals. Three or four hundred of the brave grenadiers who long distinguished themselves so gallantly on the plains of Trichinopoly, and who, rushing on certain destruction, swore, in their energetic way, "they would follow their leader to hell," on taking possession of a fortified town in Arcot put every soul in it to death, man, woman, and child, for no other reason than that the place had been gallantly defended. Heroes are nearly the same all the world over.

But, to be sure, the poor Mexican kings were better off. Juan de Varillas, a friar of the order of Nuestra Señora de la Merced, confessed them, and comforted them in their sufferings, that they were good Christians, and that they died in good preparation, seeing they were baptized: *li confessò e confortò nel supplicio: ch'eglino erano buoni Cristiani, e che morirono ben disposti: ond' è manifesto ch' erano stato battezzati.* (Clavigero, iii. p. 233, note.)

But to continue the observations of Cortes:—"No other thing accordingly," observes he, "was done than burn and demolish houses. Those of the city said to our allies, that they did wrong in assisting us to destroy, because one day they would have to re-construct with their hands the very same edifices, either for the besieged if they were to conquer, or for us Spaniards, who, in reality, now compel them to rebuild what was demolished." (*Lorenzana*, p. 286.) In going over the Libro del Cabildo, Humboldt found a manuscript containing the history of the new city of Mexico, from the year 1524 to 1529, in all the pages of which there was nothing but names of people who appeared before the *alguazils* "to demand the situation (*solar*) on which formerly stood the house of such or such a Mexican lord." Even at present they are occupied in filling and drying up the old canals which run through the capital. The number of these canals has diminished in a particular manner since the government of the Count de Galvez, though, on account of the great breadth of the streets of Mexico, the canals are less inimical to the passage of carriages than in the most part of the cities of Holland.

We may reckon among the small remains of Mexican antiquities which interest the intelligent traveller, either in the bounds of the city of Mexico, or in its environs, the ruins of the Aztec dikes (*albaradones*) and aqueducts; the stone of the sacrifices, adorned with a relievo which represents the triumph of a Mexican king; the great calendar monument (exposed with the foregoing at the *plaza mayor*;) the colossal statue of the goddess Teoyaomiqui, stretched out in one of the galleries of the edifice of the university, and habitually covered with three or four inches of earth; the Aztec manuscripts, or hieroglyphical pictures, painted on agave paper, on stag skins and cotton cloth, (a valuable collection unjustly taken away from the Chevalier Boturini, (the author of the ingenious work, *Ydea de una nueva Historia general de la America Septentrional*), very ill preserved in the archives of the palace of the viceroys, displaying in every figure the extravagant imagination of a people who delighted to see the palpitating heart of human victims offered up to gigantic and monstrous idols; the foundations of the palace of the kings of Alcolhuacan at Tezcuco; the colossal relievo traced on the *w.* face of the porphyritical rock, called the Peñol de los Baños; as well as several other objects which recal to the intelligent observer the institutions and works of people of the Mongol race, of which descriptions and drawings are promised to be given by Humboldt, in the historical account of his travels to the equinoctial regions of the new continent.

The only ancient monuments in the Mexican valley which from their size or their masses can strike the eyes of an European, are the remains of the two pyramids of San Juan de Teotihuacan, situated to the *n. e.* of the lake of Tezcuco, consecrated to the sun and moon, which the Indians called Tonatiuh Ytzaqual, House of the Sun, and Meztli Ytzaqual, House of the Moon. According to the measurements made in 1803, by a young Mexican *savant*, Dr. Oteyza, the first pyramid, which is the most *s.* has in its present state a base of 208 metres (682 feet English) in length, and 55 metres (66 Mexican vara, containing exactly 31 inches of the old *pied du roi* of Paris), or 180 feet, of perpendicular elevation. The second, the pyramid of the moon, is eleven metres, or 36 feet, lower, and its base is much less. These monu-

[ments, according to the accounts of the first travellers, and from the form which they yet exhibit, were the models of the Aztec Teocallis. The nations whom the Spaniards found settled in New Spain attributed the pyramids of Teotihuacan to the Toultec nation; consequently their construction goes as far back as the eighth or ninth century; for the kingdom of Tolula lasted from 667 to 1031. Siguenza, however, in his manuscript notes, believes them to be the work of the Olmec nation, which dwelt round the Sierra de Tlascala, called Matlacueje. If this hypothesis, of which we are unacquainted with the historical foundations, be true, these monuments would be still more ancient. For the Olmecs belong to the first nations mentioned in the Aztec chronology as existing in New Spain. It is even pretended that the Olmecs are the only nation of which the migration took place, not from the *n.* and *n. w.* (viz. Asia), but from the *e.* (Europe.) The faces of the above edifices are to within 52′ exactly placed from *n.* to *s.* and from *e.* to *w.* Their interior is clay mixed with small stones. This kernel is covered with a thick wall of porous amygdaloid. We perceive, besides, traces of a bed of lime which covers the stones (the *tetzontli*) on the outside. Several authors of the 16th century pretend, according to an Indian tradition, that the interior of these pyramids is hollow. Boturini says, that Siguenza, the Mexican geometrician, in vain endeavoured to pierce these edifices by a gallery. They formed four layers, of which three are only now perceivable, the injuries of time and the vegetation of the cactus and agaves having exercised their destructive influence on the exterior of these monuments. A stair of large hewn stones formerly led to their tops, where, according to the accounts of the first travellers, were statues covered with very thin laminæ of gold. Each of the four principal layers was subdivided into small gradations of a metre, or three feet three inches, in height, of which the edges are still distinguishable, which were covered with fragments of obsidian, that were undoubtedly the edge instruments with which the Toultec and Aztec priests in their barbarous sacrifices (*Papahua Tlemacazque* or *Teopixqui*) opened the chest of the human victims. We know that the obsidian (*itztli*) was the object of the great mining undertakings, of which we still see the traces in an innumerable quantity of pits between the mines of Moran and the village of Atotonilco el Grande, in the porphyry mountains of Oyamel and the Jacal, a region called by the Spaniards the Mountain of Knives, El Cerro de las Navajas.

It would be undoubtedly desirable to have the question resolved, whether these curious edifices, of which the one (*the Tonatiuh Ytzaqual*), according to the accurate measurement of M. Oteyza, has a mass of 128,970 cubic toises, or 33,743,201 cubic feet, were entirely constructed by the hand of man, or whether the Toultecs took advantage of some natural hill which they covered over with stone and lime? This very question has been recently agitated with respect to several pyramids of Giza and Sacara; and it has become doubly interesting from the fantastical hypotheses which M. Witte has thrown out as to the origin of the monuments of colossal form in Egypt, Persepolis, and Palmyra. As neither the pyramids of Teotihuacan, nor that of Cholula, have been diametrically pierced, it is impossible to speak with certainty of their interior structure. The Indian traditions, from which they are believed to be hollow, are vague and contradictory. Their situation in plains where no other hill is to be found, renders it extremely probable that no natural rock serves for a kernel to these monuments. What is also very remarkable (especially if we call to mind the assertions of Pococke, as to the symmetrical position of the lesser pyramids of Egypt) is, that around the Houses of the Sun and Moon of Teotihuacan we find a group, we may say a system, of pyramids, of scarcely nine or 10 metres of elevation, or 29 or 32 feet. These monuments, of which there are several hundreds, are disposed in very large streets, which follow exactly the direction of the parallels and of the meridians, and which terminate in the four faces of the two great pyramids. The lesser pyramids are more frequent towards the *s.* side of the Temple of the Moon than towards the Temple of the Sun; and, according to the tradition of the country, they were dedicated to the stars. It appears certain enough that they served as burying-places for the chiefs of tribes. All the plain which the Spaniards, from a word of the language of the island of Cuba, call *Llano de los Cues*, bore formerly in the Aztec and Toultec languages the name of *Micaotl*, or Road of the Dead.

Another ancient monument, worthy of the traveller's attention, is the military entrenchment of Xochicalco, situated to the *s. s. w.* of the town of Cuernavaca, near Tetlama, belonging to the parish of Xochitepeque. It is an insulated hill of 117 metres of elevation, surrounded with ditches or trenches, and divided by the hand of man into five terraces covered with masonry. The whole forms a truncated pyramid, of which the four faces are exactly laid down according to the four cardinal] [points. The porphyry stones with basaltic bases are of a very regular cut, and are adorned with hieroglyphical figures, among which are to be seen crocodiles spouting up water, and, what is very curious, men sitting cross-legged in the Asiatic manner. The platform of this extraordinary monument contains more than 9000 square metres, or 96,825 square feet, and exhibits the ruins of a small square edifice, which undoubtedly served for a last retreat to the besieged.

We shall conclude this rapid view of the Aztec antiquities with pointing out a few places which may be called classical, on account of the interest they excite in those who have studied the history of the Spanish conquest of Mexico.

The palace of Motezuma occupied the very same site on which at present stands the hotel of the Duke de Monteleone, vulgarly called Casa del Estado, in the *plaza mayor*, *s. w.* from the cathedral. This palace, like those of the emperor of China, of which we have accurate descriptions from Sir George Staunton and M. Barrow, was composed of a great number of spacious but very low houses. They occupied the whole extent of ground between the Empedradillo, the great street of Tacuba, and the convent De la Professa. Cortes, after the taking of the city, fixed his abode opposite to the ruins of the palace of the Aztec kings, where the palace of the viceroy is now situated. But it was soon thought that the house of Cortes was more suitable for the assemblies of the *audiencia*, and the government consequently made the family of Cortes resign the Casa del Estado, or the old hotel belonging to them. This family, which bears the title of the Marquesado del Valle de Oaxaca, received in exchange the situation of the ancient palace of Motezuma, and they there constructed the fine edifice in which the archives Del Estado are kept, and which descended, with the rest of the heritage, to the Neapolitan Duke de Monteleone.

At the first entry of Cortes into Tenochtitlan, on the 8th November 1519, he and his small army were lodged, not in the palace of Motezuma, but in an edifice formerly possessed by King Axajacatl. It was in this edifice that the Spaniards and the Tlascaltecs, their allies, sustained the assault of the Mexicans; it was there that the unfortunate King Motezuma perished of the consequences of a wound which he received in haranguing his people. We still perceive inconsiderable remains of these quarters of the Spaniards in the ruins behind the convent of Santa Teresa, at the corner of the streets of Tacuba and Del Indio Triste.

A small bridge near Bonavista preserves the name of Alvarado's Leap (Salto de Alvarado), in memory of the prodigious leap of the valorous Don Pedro de Alvarado, when in the famous melancholy night, (*noche triste*, July 1, 1520), the dike of Tlacopan having been cut in several places by the Mexicans, the Spaniards withdrew from the city to the mountains of Tepayacac. It appears that even in the time of Cortes the historical truth of this fact was disputed, which, from popular tradition, is familiar to every class of the inhabitants of Mexico. Bernal Diaz considers the history of the leap as a mere boast of his companion in arms, of whose courage and presence of mind he, however, elsewhere makes honourable mention. He affirms that the ditch was much too broad to be passed at a leap. We have, however, to observe, that this anecdote is very minutely related in the manuscript of a noble Mestizoe of the republic of Tlascala, Diego Muñoz Camargo, which Humboldt consulted at the convent of San Felipe Neri, and of which Father Torquemada appears also to have had some knowledge. This Mestizoe historian was the contemporary of Hernan Cortes. He relates the history of Alvarado's leap with much simplicity, without any appearance of exaggeration, and without mentioning the breadth of the ditch. We imagine we perceive in his naif recital one of the heroes of antiquity, who, with his shoulder and arm supported on his lance, takes an enormous leap to escape from the hands of his enemies. Camargo adds, that other Spaniards wished to follow the example of Alvarado, but that, having less agility than he had, they fell into the ditch (*azequia.*)

Strangers are shown the bridge of Clerigo, near the *plaza mayor* de Tlatelolco, as the memorable place where the last Aztec King Quauhtemotzin, nephew of his predecessor King Cuitlahuatzin, and son-in-law of Motezuma II. was taken. But the result of the most careful researches which Humboldt and the Father Pichardo could make, was, that the young king fell into the hands of Garci Holguin, in a great basin of water which was formerly between the Garita del Peralvillo, the square of Santiago de Tlatelolco, and the bridge of Amaxaca. This event happened on the 31st of August 1521, the 75th day of the siege of Tenochtitlan, and Saint Hyppolitus's day. The same day is still celebrated every year by a tour round the city by the viceroy and *oidores* on horseback, following the standard. Cortes was upon the terrace of a house of Tlatelolco when the young king was brought a prisoner to him. « I made [him sit down," says the conqueror in his third letter to the Emperor Charles V. "and I treated him with confidence; but the young man put his hand on the poniard which I wore at my side and exhorted me to kill him, because, since he had done all that his duty to himself and his people demanded of him, he had no other desire but death." This trait is worthy of the best days of Greece and Rome. Under every zone, and whatever be the colour of men, the language of energetic minds struggling with misfortune is the same.

After the entire destruction of the ancient Tenochtitlan, Cortes remained with his people for four or five months at Cojohuacan, a place for which he constantly displayed a great predilection. He was at first uncertain whether he should reconstruct the capital on some other spot around the lakes. He at last determined on the old situation, "because the city of Temixtitlan had acquired celebrity, because its position was delightful, and because in all times it had been considered as the head of the Mexican provinces," (*como principal y señora de todas estas provincias.*) It cannot, however, admit of a doubt, that, on account of the frequent inundations suffered by Old and New Mexico, it would have been better to have rebuilt the city to the e. of Tezcuco, or on the heights between Tacuba and Tacubaya. The capital was, in fact, about to be transferred to these heights by a formal edict of King Philip III. at the period of the great inundation in 1607. The *ajuntamiento*, or magistracy of the city, represented to the court that the value of the houses condemned to destruction amounted 105,000,000 of francs, or 4,375,350*l.* sterling. They appeared to be ignorant at Madrid that the capital of a kingdom, constructed for more than 88 years, is not a flying camp, which may be changed at will. It is, however, to be confessed, that the most part of the great cities of the Spanish colonies, however new their appearance may be, are in disagreeable situations. We do not here speak of the site of Caraças, Quito, Pasto, and several other cities of S. America, but merely of the Mexican cities; for example, Valladolid, which might have been built in the beautiful valley of Tepare; Guadalaxara, which is quite near the delightful plain of the Rio Chiconahuatenco, or San Pedro; Pazcuaro, which we cannot help wishing to have been built at Tzintzontza. One would say that every where the new colonists of two adjoining places have uniformly chosen either the one most mountainous, or most exposed to inundations. But indeed the Spaniards have constructed almost no new cities; they merely inhabited or enlarged those which were already founded by the Indians.

It is impossible to determine with any certainty the number of inhabitants of old Tenochtitlan. Were we to judge from the fragments of ruined houses, and the recital of the first conquerors, and especially from the number of the combatants whom the kings Cuitlahuatzin and Quauhtimotzin opposed to the Tlascaltecs and Spaniards, we should pronounce the population of Tenochtitlan three times greater than that of Mexico in our days. Cortes asserts, that after the siege, the concourse of Mexican artisans who wrought for the Spaniards, as carpenters, masons, weavers, and founders, was so enormous, that in 1524 the new city of Mexico already numbered 30,000 inhabitants. Modern authors have thrown out the most contradictory ideas regarding the population of this capital. The Abbe Clavigero, in his excellent work on the ancient history of New Spain, proves that these estimations vary from 60,000 to 1,500,000 of inhabitants. (*Clavigero*, iv. p. 278, note p.) We ought not to be astonished at these contradictions, when we consider how new statistical researches are even in the most cultivated parts of Europe.

According to the most recent and least uncertain data, the actual population of the capital of Mexico appears to be (including the troops) from 135 to 140,000 souls. The enumeration in 1790, by orders of the Count de Revillagigedo, gave a result of only 112,926 inhabitants for the city; but we know that this result is one-sixth below the truth. The regular troops and militia in garrison in the capital are composed of from 5 to 6000 men in arms. We may admit with great probability that the actual population consists of

2,500 white Europeans.
65,000 white Creoles.
33,000 indigenous (copper-coloured.)
26,500 Mestizoes, mixture of whites and Indians.
10,000 Mulattoes.

137,000 inhabitants.

There are consequently in Mexico 69,500 men of colour, and 67,500 whites; but a great number of the Mestizoes are almost as white as the Europeans and Spanish Creoles!

In the 23 male convents which the capital contains there are nearly 1200 individuals, of whom 580 are priests and choristers. In the 15 female convents there are 2100 individuals, of whom nearly 900 are professed *religieuses*.

The clergy of the city of Mexico is extremely]

numerous, though less numerous by one-fourth than at Madrid. The enumeration of 1790 gives

		Individuals.
In the convents of monks.	573 priests and choristers. 59 novices. 235 lay brothers.	867
In the convents of nuns.	888 professed *religieuses* 35 novices.	923
Prebendaries		26
Parish priests (*curés*)		16
Curates		43
Secular ecclesiastics		517
	Total	2392

and without including lay-brothers and novices, 2068. The clergy of Madrid, according to the work of M. de Laborde, (which received several contributions from M. de Humboldt), is composed of 3470 persons: consequently the clergy is to the whole population of Mexico as 1¼ to 100, and at Madrid as 2 to 100.

The archbishop of Mexico possesses a revenue of 682,500 livres, or 18,420*l.* sterling. This sum is somewhat less than the revenue of the convent of Jeronimites of the Escurial. An archbishop of Mexico is, consequently, much poorer than the archbishops of Toledo, Valencia, Seville, and Santiago. The first of these possesses a revenue of 3,000,000 of livres, or, 105,000*l.* sterling. M. de Laborde has proved, and the fact is by no means generally known, that the clergy of France, before the revolution, was more numerous, compared to the total population, and richer as a body than the Spanish clergy. The revenues of the tribunal of inquisition of Mexico, a tribunal which extends over the whole kingdom of New Spain, Guatemala, and the Philippine islands, amount to 200,000 livres, or 8334*l.* sterling.

The number of births at Mexico, for a mean term of 100 years, is 5930; and the number of deaths 5050. In the year 1802 there were even 6155 births, and 5166 deaths, which would give, supposing a population of 137,000 souls, for every 22¼ individuals, one birth, and for every 26¼ one death. They reckon in general, in the country parts, in Nueva España, the relation of the births to the population as one to 17; and the relation of the deaths to the population as one to 30. There is consequently, in appearance, a very great mortality and a very small number of births in the capital. The conflux of patients to the city is considerable, not only of the most indigent class of the people, who seek assistance in the hospitals, of which the number of beds amounts to 1100, but also of persons in easy circumstances, who are brought to Mexico because neither advice nor remedies can be procured in the country. This circumstance accounts for the great number of deaths on the parish registers. On the other hand, the convents, the celibacy of the clergy, progress of luxury, the militia, and the indigence of the Saragates Indians, who live like the Lazaroni of Naples, in idleness, are the principal causes which influence the disadvantageous relation of the births to the population.

MM. Alzate and Clavigero, from a comparison of the parish registers of Mexico with those of several European cities, have endeavoured to prove that the capital of Nueva España must contain more than 200,000 inhabitants; but how can we suppose in the enumeration of 1790 an error of 87,000 souls, more than two-fifths of the whole population? Besides, the comparisons of these two learned Mexicans can, from their nature, lead to no certain results, because the cities of which they exhibit the bills of mortality are situated in very different elevations and climates, and because the state of civilization and comfort of the great mass of their inhabitants afford the most striking contrasts. At Madrid the births are one in 34, and at Berlin one in 28. The one of these proportions can no more, however, than the other, be applicable to calculations regarding the population of the cities of Equinoctial America. Yet the difference between these proportions is so great, that it would alone, on an annual number of 6000 births, augment or diminish to the extent of 36,000 souls the population of the city of Mexico. The number of deaths or births is, perhaps, the best of all means for determining the number of the inhabitants of a district, when the numbers which express the relations of the births and deaths to the whole population in a given country have been carefully ascertained; but these numbers, the result of a long induction, can never be applied to countries whose physical and moral situation are totally different. They denote the medium state of prosperity of a mass of population, of which the greatest part dwell in the country; and we cannot, therefore, avail ourselves of these proportions to ascertain the number of inhabitants of a capital.

Mexico is the most populous city of the new continent. It contains nearly 40,000 inhabitants fewer than Madrid; and as it forms a great square, of which each side is nearly 2750 metres, or 9021 feet, its population is spread over a great extent of ground. Its greatest length is nearly 3900 metres (12,794 English feet); of Paris 8000 [metres (26,216 English feet.) The streets being very spacious, they in general appear rather deserted. They are so much the more so, as in a climate considered as cold by the inhabitants of the tropics, people expose themselves less to the free air than in the cities at the foot of the *cordillera*. Hence the latter (*ciudades de tierra caliente*) appear uniformly more populous than the cities of the temperate or cold regions (*ciudades de tierra fria.*) If Mexico contains more inhabitants than any of the cities of Great Britain and France, with the exception of London, Dublin, and Paris; on the other hand, its population is much less than that of the great cities of the Levant and East Indies. Calcutta, Surat, Madras, Aleppo, and Damascus, contain all of them from 2 to 4 and even 600,000 inhabitants.

The Count de Revillagigedo set on foot accurate researches into the consumption of Mexico. The following table, drawn up in 1791, may be interesting to those who have a knowledge of the important operations of MM. Lavoisier and Arnould, relative to the consumption of Paris and all France.

CONSUMPTION OF MEXICO.

1. *Eatables.*

Beeves	16,300
Calves	450
Sheep	278,923
Hogs	50,676
Kids and rabbits	24,000
Fowls	1,255,340
Ducks	125,000
Turkeys	205,000
Pigeons	65,300
Partridges	140,000

2. *Grain.*

Maize or Turkey wheat, cargas of three fanegas	117,224
Barley, cargas	40,219

3. *Liquid Measure.*

Wheat flour, cargas of 12 arrobas	130,000
Pulque, the fermented juice of the agava, cargas	294,790
Wine and vinegar, barrels of 4½ arrobas	4,507
Brandy, barrels	12,000
Spanish oil, arrobas of 25 pounds	5,585

Supposing, with M. Puchet, the population of Paris to be four times greater than that of Mexico, we shall find that the consumption of beef is nearly proportional to the number of inhabitants of the two cities, but that that of mutton and pork is infinitely more at Mexico. The difference is as follows:

	Consumption Of Mexico.	Consumption Of Paris.	Quadruple of the Consumption of Mexico.
Beeves	16,300	70,000	65,200
Sheep	273,000	350,000	1,116,000
Hogs	50,100	35,000	200,400

M. Lavoisier found by his calculations that the inhabitants of Paris consumed annually in his time 90,000,000 of pounds of animal food of all sorts, which amounts to 163 pounds ($79\frac{7}{10}$ kilogrammes, or $175\frac{9}{10}$ pounds averd.) per individual. In estimating the animal food yielded by the animals designated in the preceding table, according to the principles of Lavoisier, modified according to the localities, the consumption of Mexico in every sort of meat is 26,000,000 of pounds, or 189 pounds, or 204 pounds averd. per individual. This difference is so much the more remarkable as the population of Mexico includes 33,000 Indians, who consume very little animal food.

The consumption of wine has greatly increased since 1791, especially since the introduction of the Brownonian system in the practice of the Mexican physicians. The enthusiasm with which this system was received in a country where asthenical or debilitating remedies had been employed to an excess for ages, produced, according to the testimony of all merchants of Vera Cruz, the most remarkable effect on the trade in luscious Spanish wines (*vins liquoreux.*) These wines, however, are only drunk by the wealthy class of the inhabitants. The Indians, Mestizoes, Mulattoes, and even the greatest number of white Creoles, prefer the fermented juice of the agave, called *pulque*, of which there is annually consumed the enormous quantity of 44,000,000 of bottles, containing 48 French cubic inches each, or 58.141 cubic inches English. The immense population of Paris only consumed annually in the time of M. Lavisier 281,000 muids of wine, brandy, cyder, and beer, equal to 80,928,000 bottles.

The consumption of bread at Mexico is equal to that of the cities of Europe. This fact is so much the more remarkable, as at Caraças, at Cumana, and Carthagena de las Indias, and in all the cities of America situated under the torrid zone, but on a level with the ocean, or very little above it, the Creole inhabitants live on almost nothing but maize bread, and the *jatropha manihot*. If we suppose, with M. Arnould, that 325 pounds of flour yield 416 pounds of bread, we shall find that the 130,000 loads of flour consumed at Mexico yield 49,900,000] [pounds of bread, which amounts to $391\frac{1}{10}$ pounds averd. per individual of every age. Estimating the habitual population of Paris at 547,000 inhabitants, and the consumption of bread at 206,788,000 pounds, we shall find the consumption of each individual in Paris 377 pounds French, or $406\frac{2}{10}$ pounds averd. At Mexico the consumption of maize is almost equal to that of wheat. The Turkish corn is the food most in request among the Indians. We may apply to it the denomination which Pliny gives to barley (the κρι of Homer) *antiquissimum frumentum;* for the zea maize was the only farinaceous gramen cultivated by the Americans before the arrival of the Europeans.

The market of Mexico is richly supplied with eatables, particularly with roots and fruits of every sort. It is a most interesting spectacle, which may be enjoyed every morning at sun-rise, to see these provisions, and a great quantity of flowers, brought in by Indians in boats, descending the canals of Istacalco and Chalco. The greater part of these roots is cultivated on the *chinampas*, called by the Europeans floating gardens. There are two sorts of them, of which the one is moveable and driven about by the winds, and the other fixed and attached to the shore. The first alone merit the denomination of floating gardens, but their number is daily diminishing.

The ingenious invention of *chinampas* appears to go back to the end of the 14th century. It had its origin in the extraordinary situation of a people surrounded with enemies, and compelled to live in the midst of a lake little abounding in fish, who were forced to fall upon every means of procuring subsistence. It is even probable that Nature herself suggested to the Aztecs the first idea of floating gardens. On the marshy banks of the lakes of Xochimilco and Chalco, the agitated water in the time of the great rises carries away pieces of earth covered with herbs, and bound together by roots. These floating about for a long time as they are driven by the wind, sometimes unite into small islands. A tribe of men, too weak to defend themselves on the continent, would take advantage of these portions of ground, which accident put within their reach, and of which no enemy disputed the property. The oldest *chinampas* were merely bits of ground joined together artificially, and dug and sown upon by the Aztecs. These floating islands are to be met with in all the zones. Humboldt saw them in the kingdom of Quito, on the river Guayaquil, of eight or nine metres (or 26 or 29 feet) in length, floating in the midst of the current, and bearing young shoots of *bambusa*, *pistia stratiotes*, *pontederia*, and a number of other vegetables, of which the roots are easily interlaced. He found also in Italy, in the small *lago di aqua solfa* of Tivoli, near the hot baths of Agrippa, small islands formed of sulphur, carbonate of lime, and the leaves of the *ulva thermalis*, which change their place with the smallest breath of wind. Floating gardens are, as is well known, also to be met with in the rivers and canals of China, where an excessive population compels the inhabitants to have recourse to every shift for increasing the means of subsistence.

Simple lumps of earth, carried away from the banks, have given rise to the invention of *chinampas;* but the industry of the Aztec nation gradually carried this system of cultivation to perfection. The floating gardens, of which very many were found by the Spaniards, and of which many still exist in the lake of Chalco, were rafts formed of reeds (*totora*), rushes, roots, and branches of brushwood. The Indians cover these light and well-connected materials with black mould, naturally impregnated with muriate of soda. The soil is gradually purified from this salt by washing it with the water of the lake; and the ground becomes so much the more fertile as this lixiviation is annually repeated. This process succeeds even with the salt water of the lake of Tezcuco, because this water, by no means at the point of its saturation, is still capable of dissolving salt as it filtrates through the mould. The *chinampas* sometimes contain even the cottage of the Indian who acts as guard for a group of floating gardens. They are towed or pushed with long poles when wished to be removed from one side of the banks to the other.

In proportion as the fresh-water lake has become more distant from the salt-water lake, the moveable chinampas have been fixed. We see this last class all along the canal De la Viga, in the marshy ground between the lake of Chalco and the lake of Tezcuco. Every chinampa forms a parallelogram of 100 metres in length, and from five to six metres (or 328 by 16 or 19 feet) in breadth. Narrow ditches, communicating symmetrically between them, separate these squares. The mould fit for cultivation, purified from salt by frequent irrigations, rises nearly a metre, or 3.28 feet above the surface of the surrounding water. On these chinampas are cultivated beans, small peas, pimento (*chile*, *capsicum*), potatoes, artichokes, cauliflowers, and a great variety of other vegetables. The edges of these squares are generally ornamented with flowers, and sometimes even with a hedge of rose bushes. The promenade in boats around the chinampas of Istacalco is one of the most agreeable that can be] [enjoyed in the environs of Mexico. The vegetation is extremely vigorous on a soil continually refreshed with water.

The valley of Tenochtitlan offers to the examination of naturalists two sources of mineral water, that of Nuestra Señora de Guadalupe, and that of the Peñon de los Baños. These sources contain carbonic acid, sulphate of lime and soda, and muriate of soda. Baths have been established there in a manner equally salutary and convenient. The Indians manufacture their salt near the Peñon de los Baños. They wash clayey lands full of muriate of soda, and concentrate water which has only 12 or 13 to the 100 of salt. Their caldrons, which are very ill constructed, have only six square feet of surface, and from two to three inches of depth. No other combustible is employed but the mules and cow dung. The fire is so ill managed, that to produce twelve pounds of salt, which sells at 35 sous, or 1*s.* 5¼*d.* they consume 12 sous, or 5¼*d.* worth of combustibles. This salt-pit existed in the time of Motezuma, and no change has taken place in the technical process, but the substitution of caldrons of beaten copper to the old earthen vats.

The hill of Chapultepec was chosen by the young viceroy Galvez as the site of a villa (Chateau de Plaisance) for himself and his successors. The castle has been finished externally, but the apartments are not yet furnished. This building cost the king nearly 1,500,000 livres, or 62,505*l.* sterling. The court of Madrid disapproved of the expence, but, as usual, after it was laid out. The plan of this edifice is very singular. It is fortified on the side of the city of Mexico. We perceive salient walls and parapets adapted for cannon, though these parts have all the appearance of mere architectural ornaments. Towards the *n.* there are fosses and vast vaults capable of containing provisions for several months. The common opinion at Mexico is, that the house of the viceroy at Chapultepec is a disguised fortress. Count Bernardo de Galvez was accused of having conceived the project of rendering Nueva España independent of the peninsula; and it was supposed that the rock of Chapultepec was destined for an asylum and defence to him in case of attack from the European troops. Men of respectability in the first situations are said to have entertained this suspicion against the young viceroy. It is the duty of a historian, however, not to yield too easy an acquiescence to accusations of so grave a nature. The Count de Galves belonged to a family that King Charles III. had suddenly raised to an extraordinary degree of wealth and power. Young, amiable, and addicted to pleasures and magnificence, he had obtained from the munificence of his sovereign one of the first places to which an individual could be exalted; and without more undeniable proofs of conviction, we cannot but consider it as highly unnatural that he should have endeavoured to break the ties which for three centuries had united the colonies to the mother country. The Count de Galves, notwithstanding his conduct was well calculated to gain the favour of the populace of Mexico, and notwithstanding the influence of the Countess de Galvez, as beautiful as she was generally beloved, would have experienced the fate of every European viceroy who aims at independence. In a great revolutionary commotion, it would never have been forgiven him that he was not born an American.

The castle of Chapultepec should be sold for the advantage of the government. As in every country it is difficult to find individuals fond of purchasing strong places, several of the ministers of the Real Hacienda have begun, by selling to the highest bidder the glass and sashes of the windows. This vandalism, which passes by the name of economy, has already much contributed to degrade an edifice on an elevation of 2325 metres, or 7626 feet, and which, in a climate so rude, is exposed to all the impetuosity of the winds. It would, perhaps, be prudent to preserve this castle as the only place in which the archives, bars of silver, and coin, could be placed, and the person of the viceroy could be in safety, in the first moments of a popular commotion. The commotions of the 12th February 1608, 15th January 1624 and 1692, are still in remembrance at Mexico. In the last of these, the Indians, from want of maize, burned the palace of the viceroy Don Gaspar de Sandoval, Count of Galves, who took refuge in the garden of the convent of St. Francis. But it was only in those times that the protection of the monks was equivalent to the security of a fortified castle.

To terminate the description of the valley of Mexico, it remains for us to give a rapid hydrographical view of this country, so intersected with lakes and small rivers. This view, we flatter ourselves, will be equally interesting to the naturalist and the civil engineer. We have already said, that the surface of the four principal lakes occupies nearly a tenth of the valley, or 22 square leagues. The lake of Xochimilco (and Chalco) contains 6¼, the lake of Tezcuco $10\frac{1}{10}$, San Christobal $3\frac{6}{10}$, and Zumpango $1\frac{3}{10}$ square leagues (of 25 to the equatorial degree.) The valley of Te-]

[nochtitlan, or Mexico, is a basin surrounded by a circular wall of porphyry mountains of great elevation. This basin, of which the bottom is elevated 2277 metres, or 7468 feet, above the level of the sea, resembles, on a small scale, the vast basin of Bohemia, and (if the comparison is not too bold) the valleys of the mountains of the moon, described by MM. Herschel and Schroeter. All the humidity furnished by the *cordilleras* which surround the plain of Tenochtitlan is collected in the valley. No river issues out of it, if we except the small brook (*arroyo*) of Tequisquiac, which, in a ravine of small breadth, traverses the *n.* chain of the mountains, to throw itself into the Rio de Tula, or Moteuczoma.

The principal supplies of the lakes of the valley of Tenochtitlan are, 1. The rivers of Papalotla, Tezcuco, Teotihuacan, and Tepeyacac (Gnadalope), which pour their waters into the lake of Tezcuco; 2. The rivers of Pachuca and Guautitlan (*Quauhtitlan*), which flow into the lake of Zumpango. The latter of these rivers (the Rio de Guautitlan) has the longest course; and its volume of water is more considerable than that of all the other supplies put together.

The Mexican lakes, which are so many natural recipients, in which the torrents deposit the waters of the surrounding mountains, rise by stages, in proportion to their distance from the centre of the valley, or the site of the capital. After the lake of Tezcuco, the city of Mexico is the least elevated point of the whole valley. According to the very accurate survey of MM. Velasquez and Castera, the *plaza mayor* of Mexico, at the *s.* corner of the viceroy's palace, is one Mexican *vara*, (the Mexican *vara* being equal to 0^{m}.839), one foot and one inch higher than the mean level of the lake of Tezcuco, which again is four *varas* and eight inches lower than the lake of San Christobal, whereof the *n.* part is called the lake of Xaltocan. The elevation of the *plaza mayor*, therefore, above Tezcuco is 47.245 inches, and that of San Christobal 8.863 inches. In the aforesaid *n.* part, on two small islands, the villages of Xaltocan and Tonanitla are situated. The lake of San Christobal, properly so called, is separated from that of Xaltocan by a very ancient dike which leads to the villages of San Pablo and San Tomas de Chiconautla. The most *n.* lake of the valley of Mexico, Zumpango (Tzompango), is 10 *varas* 1 foot 6 inches, or 29 feet 1 inch 888 English, higher than the mean level of the lake of Tezcuco. A dike (la Calzada de la Cruz del Rey) divides the lake of Zumpango into two basins, of which the most *w.* bears the name of Laguna de Zitlaltepec, and the most *e.* the name of Laguna de Coyotepec. The lake of Chalco is at the *s.* extremity of the valley. It contains the pretty little village of Xico, founded on a small island; and it is separated from the lake of Xochimilco by the Calzada de San Pedro de Tlahua, a narrow dike which runs from Tuliagualco to San Francisco Tlaltengo. The level of the fresh-water lakes of Chalco and Xochimilco is only 1 *vara* 11 inches, or 3 feet 9 inches, higher than the *plaza mayor* of the capital. Humboldt has given these details as thinking they might be interesting to civil engineers wishing to form an exact idea of the great canal (*desague*) of Huehuetoca.

The difference of elevation of the four great reservoirs of water of the valley of Tenochtitlan was sensibly felt in the great inundations to which the city of Mexico for a long series of ages has been exposed. In all of them the sequence of the phenomena has been uniformly the same. The lake of Zumpango, swelled by the extraordinary increases of the Rio de Guautitlan, and the influxes from Pachuca, flows over into the lake of San Christobal, with which the *cienegas* of Tepejuelo and Tlapanahuiloya communicate. The lake of San Christobal bursts the dike which separates it from the lake of Tezcuco. Lastly, the water of this last basin rises in level from the accumulated influx more than a metre, or 39.371 inches, and traversing the saline grounds of San Lazaro, flows with impetuosity into the streets of Mexico. Such is the general progress of the inundations: they proceed from the *n.* and the *n. w.* The drain or canal called the Desague Real de Huehuetoca is destined to prevent any danger from them; but it is certain, however, that from a coincidence of several circumstances, the inundations of the *s.* (*avenidas del sur*), on which, unfortunately, the *desague* has no influence, may be equally disastrous to the capital. The lakes of Chalco and Xochimilco would overflow, if in a strong eruption of the volcano Popocatepetl, this colossal mountain should suddenly be stripped of its snows. While Humboldt was at Guayaquil, on the coast of the province of Quito, in 1802, the cone of Cotopaxi was heated to such a degree by the effect of the volcanic fire, that almost in one night it lost the enormous mass of snow with which it is covered. In the new continent eruptions and great earthquakes are often followed with heavy showers, which last for whole months. With what dangers would not the capital be threatened were these phenomena to take place in] [the valley of Mexico, under a zone, where, in years by no means humid, the rain which falls amounts to 15 decimetres, or 59 inches.

The inhabitants of Nueva España think that they can perceive something like a constant period in the number of years which intervene between the great inundations. Experience has proved that the extraordinary inundations in the valley of Mexico have followed nearly at intervals of 25 years. Toaldo pretends to be able to deduce from a great number of observations, that the very rainy years, and consequently the great inundations, return every 19 years, according to the terms of the cycle of Saros. Since the arrival of the Spaniards the city has experienced five great inundations, viz. in 1553, under the viceroy Don Luis de Velasco (el Viejo), constable of Castile; in 1580, under the viceroy Don Martin Enriquez de Alamanza; in 1604, under the viceroy Montesclaros; in 1607, under the viceroy Don Luis de Velasco (el Segundo), Marquis de Salinas; and in 1629, under the viceroy Marquis de Ceralvo. This last inundation is the only one which has taken place since the opening of the canal of Huehuetoca; and we shall see hereafter what were the circumstances which produced it. Since the year 1629 there have still been, however, several very alarming swellings of the waters, but the city was preserved by the *desague*. These seven very rainy years were 1648, 1675, 1707, 1732, 1748, 1772, 1795. Comparing together the foregoing 11 epochas, we shall find for the period of the fatal recurrence the numbers of 27, 24, 3, 26, 19, 27, 32, 25, 16, 24, and 23; a series which undoubtedly denotes somewhat more regularity than what is observed at Lima in the return of the great earthquakes.

The situation of the capital of Mexico is so much the more dangerous, that the difference of level between the surface of the lake of Tezcuco and the ground on which the houses are built is every year diminishing. This ground is a fixed plane, particularly since all the streets of Mexico were paved under the government of the Count de Revillagigedo; but the bed of the lake of Tezcuco is progressively rising, from the mud brought down by the small torrents, which is deposited in the reservoirs into which they flow. To avoid a similar inconvenience, the Venetians turned from their *lagunas* the Brenta, the Piave, the Livenza, and other rivers, which formed deposits in them. If we could rely on the results of a survey executed in the 16th century, we should no doubt find that the *plaza mayor* of Mexico was formerly more than 11 decimetres, or $43\frac{3}{10}$ feet, elevated above the level of the lake of Tezcuco, and that the mean level of the lake varies from year to year. If, on the one hand, the humidity of the atmosphere and the sources have diminished in the mountains surrounding the valley, from the destruction of the forests; on the other hand, the cultivation of the land has increased the depositions and the rapidity of the inundations. General Andreossi, in his excellent work on the Canal of Languedoc, has insisted a great deal on these causes, which are common to all climates. Waters which glide over declivities covered with sward, carry much less of the soil along with them than those which run over loose soil. Now the sward, whether formed from gramina, as in Europe, or small alpine plants, as in Mexico, is only to be preserved in the shade of a forest. The shrubs and underwood oppose also powerful obstacles to the melted snow which runs down the declivities of the mountains. When these declivities are stripped of their vegetation, the streams are less opposed, and more easily unite with the torrents which swell the lakes in the neighbourhood of Mexico.

It is natural enough, that in the order of hydraulical operations undertaken to preserve the capital from the danger of inundation, the system of dikes preceded that of evacuating canals or drains. When the city of Tenochtitlan was inundated to such a degree in 1446, that none of the streets remained dry, Motezuma I. (*Huehue Moteuczoma*), by advice of Nezahualcojotl, king of Tezcuco, ordered a dike to be constructed of more than 12,000 metres in length, and 20 in breadth, viz. 395,369 by 65.6 feet. This dike, partly constructed in the lake, consisted of a wall of stones and clay, supported on each side by a range of palisadoes, of which considerable remains are yet to be seen in the plains of San Lazaro. This dike of Motezuma I. was enlarged and repaired after the great inundation in 1498, occasioned by the imprudence of King Ahuitzotl. This prince, as we have already observed, ordered the abundant sources of Huitzilopochco to be conducted into the lake of Tezcuco. He forgot that the lake of Tezcuco, however destitute of water in time of drought, becomes so much the more dangerous in the rainy season, as the number of its supplies is increased. Ahuitzotl ordered Tzotzomatzin, citizen of Coyohuacan, to be put to death, because he had courage enough to predict the danger to which the new aqueduct of Huitzilopochco would expose the capital. Shortly afterwards the young Mexican King very narrowly escaped drowning in his] [palace. The water increased with such rapidity, that the prince was grievously wounded in the head, while saving himself, by a door which led from the lower apartments to the street.

The Aztecs had thus constructed the dikes (*calzadas*) of Tlahua and Mexicaltzingo, and l'Albaradon, which extends from Iztapalapan to Tepeyaeae (Guadalupe), and of which the ruins at present are still very useful to the city of Mexico. This system of dikes, which the Spaniards continued to follow till the commencement of the 17th century, afforded means of defence, which, if not quite secure, were at least nearly adequate, at a period when the inhabitants of Tenochtitlan, sailing in canoes, were more indifferent to the effects of the more trifling inundations. The abundance of forests and plantations afforded them great facilities for constructions on piles. The produce of the floating gardens (*chinampas*) was adequate to the wants of a frugal nation. A very small portion of ground fit for cultivation was all that the people required. The overflow of the lake of Tezcuco was less alarming to men who lived in houses, many of which could be traversed by canoes.

When the new city, rebuilt by Hernan Cortes, experienced the first inundation in 1553, the viceroy Velasco I. caused the Albaradon de San Lazaro to be constructed. This work, executed after the model of the Indian dikes, suffered a great deal from the second inundation of 1580. In the third of 1604 it had to be wholly rebuilt. The viceroy Montesclaros then added, for the safety of the capital, the Presa d'Oculma, and the three *calzadas* of Nuestra Señora de Guadalupe, San Christobal, and San Antonio Abad.

These great constructions were scarcely finished, when, from a concurrence of extraordinary circumstances, the capital was again inundated in 1607. Two inundations had never before followed so closely upon one another; and the fatal period of these calamities has never since been divided by a lapse of more than 16 or 17 years. Tired of constructing dikes (*albaradones*) which the water periodically destroyed, they discovered at last that it was time to abandon the old hydraulical system of the Indians, and to adopt that of canals of evacuation. This change appeared so much the more necessary, as the city inhabited by the Spaniards had no resemblance in the least to the capital of the Aztec empire. The lower part of the houses was now inhabited; few streets could be passed through in boats; and the inconveniences and real losses occasioned by the inundations were consequently much greater than what they had been in the time of Motezuma.

The extraordinary rise of the river Guautitlan and its tributary streams being looked upon as the principal cause of the inundations, the idea naturally occurred of preventing this river from discharging itself into the lake of Zumpango, the mean level of the surface of which is $7\frac{1}{2}$ metres, or $24\frac{6}{10}$ feet, higher than the *plaza mayor* of Mexico. In a valley circularly surrounded by high mountains, it was only possible to find a vent for the Rio de Guautitlan through a subterraneous gallery, or an open canal through these very mountains. In fact, in 1580, at the epoch of the great inundation, two intelligent men, the *Licenciado* Obregon, and the *Maestro* Arciniega, proposed to government to have a gallery pierced between the Cerro de Sincoque and the Loma of Nochistongo. This was the point which more than any other was likely to fix the attention of those who had studied the configuration of the Mexican ground. It was nearest to the Rio de Guautitlan, justly considered the most dangerous enemy of the capital. No where the mountains surrounding the valley are less elevated, and present a smaller mass than to the *n. n. w.* of Huehuetoca, near the hills of Nochistongo. One would say, on examining attentively the marl soil, of which the horizontal strata fill a porphyritical defile, that the valley of Tenochtitlan formerly communicated at that place with the valley of Tula.

In 1607, the Marquis de Salinas, viceroy, employed Enrico Martinez to carry through the artificial evacuation of the Mexican lakes. It is generally believed in New Spain that this celebrated engineer, the author of the *Desague de Huehuetoca*, was a Dutchman or a German. His name undoubtedly denotes that he was of foreign descent; but he appears, however, to have received his education in Spain. The king conferred on him the title of cosmographer; and there is a treatise of his on trigonometry, printed at Mexico, which is now become very scarce. Enrico Martinez, Alonso Martinez, Damian Davila, and Juan de Ysla, made an exact survey of the valley, of which the accuracy was ascertained by the operations of the learned geometrician Don Joaquim Velasquez in 1774. The royal cosmographer, Enrico Martinez, presented two plans of canals, the one to evacuate the three lakes of Tezcuco, Zumpango, and San Christobal, and the other the lake of Zumpango alone; and, agreeably to both projects, the evacuation of the water was to take place through the subterraneons gallery of Nochis-]

[tongo, proposed in 1580 by Obregon and Arciniega. But the distance of the lakes of Tezcuco from the month of the Rio de Guautitlan being nearly 32,000 metres, or 104,987 feet, the government confined themselves to the canal of Zumpango. This canal was so constructed as to receive at the same time the waters of the lake, and those of the river of Guautitlan; and it is consequently not true that the *desague* projected by Martinez was negative in its principle, that is to say, that it merely prevented the Rio de Guautitlan from discharging itself into the lake of Zumpango. The branch of the canal which conducted the water from the lake to the gallery was filled up by depositions of mud, and the *desague* was only useful then for the Rio de Guautitlan, which was turned from its course; so that when M. Mier recently undertook the direct evacuation of the lakes of San Christobal and Zumpango, it was hardly remembered at Mexico that 188 years before the same work had already been carried into execution with respect to the latter of these great basins.

The famous subterraneous gallery of Nochistongo was commenced on the 28th November 1607. The viceroy, in presence of the *audiencia*, applied the first pick-axe: 15,000 Indians were employed at this work, which was terminated with extraordinary celerity, because the work was carried on in a number of pits at the same time. The unfortunate Indians were treated with the greatest severity. The use of the pick-axe and shovel was sufficient to pierce such loose and crumbling earth. After eleven months of continued labour, the gallery (*el socabon*) was completed. Its length was more than 6600 metres, or 21,653 feet, (1.48 common leagues, of 25 to the sexagesimal degree, 4443 metres each); its breadth $3^m.5$, or 11.482 feet; and its height $4^m.2$, or 13.779 feet. In the month of December 1608, the viceroy and archbishop of Mexico were invited by Martinez to repair to Huehuetoca, to see the water flow from the lake of Zumpango and the Rio de Guautitlan through the gallery. It actually began to flow for the first time on the 17th September 1608. The Marquis de Salinas, the viceroy, according to Zepeda's account, entered more than 2000 metres, or 6561 feet, on horseback into this subterraneous passage. On the opposite side of the hill of Nochistongo is the Rio de Moctezuma (or Tula), which runs into the Rio de Panuco. From the *n.* extremity of the *socabon*, called the Boca de San Gregorio, Martinez carried on an open trench for a direct distance of 8600 metres, or 28,214 feet, which conducted the water from the gallery to the small cascade (*salto*) of the Rio de Tula. From this cascade the water has yet to descend, according to Humboldt's measurement, before it reaches the gulf of Mexico, near the bar of Tampico, nearly 2153 metres, or 7056 feet, which gives for a length of 323,000 metres, or 1,059,714 feet, a mean fall of $6\frac{1}{4}$ metres in the 1000.

A subterraneous passage serving for a canal of evacuation, of 6600 metres in length, and an aperture of 14.7 square metres in section (corresponding to 158 square feet), finished in less than a year, is a hydraulical operation which in our times, even in Europe, would draw the attention of engineers. It is only, in fact, since the end of the 17th century, from the example set by the illustrious Francis Andreossi in the canal of Languedoc, that these subterraneous apertures have become common. The canal which joins the Thames with the Severn, passes near Sapperton, for a length of more than 4000 metres, or 13,123 feet, through a chain of very elevated mountains. The great subterraneous canal of Bridgewater, which, near Worsley, in the neighbourhood of Manchester, serves for the carriage of coals, has an extent, including its different ramifications, of 19,200 metres, or 62,991 feet, (or $4\frac{3}{10}$ common leagues.) The canal of Picardy, which is at present going on, ought, according to the first plan, to have a subterraneous navigable passage of 13,700 metres, or 45,300 feet, in length; seven metres, or 26.246 feet, in breadth; and eight metres, or 26.965 feet, in height.

Scarcely had a part of the water of the valley of Mexico began to flow towards the Atlantic ocean, when Enrico Martinez was reproached with having dug a gallery neither broad nor durable, nor deep enough to admit the water of the great swellings. The chief engineer (Maestro del Desague) replied, that he had presented several plans, but that the government had chosen the remedy of most prompt execution. In fact, the filtrations and erosions occasioned by the alternate states of humidity and aridity caused the loose earth frequently to crumble down. They were soon compelled to support the roof, which was only composed of alternate strata of marl and a stiff clay called *tepetate*. They made use at first of wood, by throwing planks across, which rested on pillars; but as resinous wood was not very plentiful in that part of the valley, Martinez substituted masonry in its place. This masonry, if we judge of it from the remains discovered in the] [*obra del consulado*,] was very well executed; but it was conducted on an erroneous principle. The engineer, in place of fortifying the gallery from top to bóttom with a complete vault of an elliptical form (as is done in mines whenever a gallery is cut through loose sand), merely constructed arches, which had no sufficient foundation to rest on. The water, to which too great a fall was given, gradually undermined the lateral walls, and deposited an enormous quantity of earth and gravel in the water-course of the gallery, because no means were taken to filtrate it, as might have been effected by making it previously pass, for example, through reticulations of *petate*, executed by the Indians with filaments of the shoots of palm trees. To obviate these inconveniences, Martinez constructed in the gallery at intervals a species of small sluices, which, in opening rapidly, were to clear the passage. This means, however, proved insufficient, and the gallery was stopt up by the perpetual falling in of earth.

From the year 1608 the Mexican engineers began to dispute whether it was proper to enlarge the *socabon* of Nochistongo, or to finish the walling, or to make an uncovered aperture by taking off the upper part of the vault, or to commence a new gallery farther down, capable of also receiving, besides the waters of the Rio de Guautitlan, and the lake of Zumpango, those of the lake of Tezcuco. The archbishop Don Garcia Guerra, a Dominican, then viceroy, ordered new surveys to be made in 1611 by Alonso de Arias, superintendant of the royal arsenal *(armero mayor)*, and inspector of fortifications *(maestro mayor de fortificaciones)*, a man of probity, who then enjoyed great reputation. Arias seemed to approve of the operations of Martinez, but the viceroy could not fix on any definitive resolution. The court of Madrid, wearied out with these disputes of the engineers, sent to Mexico in 1614 Adrian Boot, a Dutchman, whose knowledge of hydraulic architecture is extolled in the memoirs of those times preserved in the archives of the viceroyalty. This stranger recommended to Philip III. by his ambassador at the court of France, held forth again in favour of the Indian system; and he advised the construction of great dikes and well protected mounds of earth around the capital. He was unable, however, to bring about the entire relinquishment of the gallery of Nochistongo till the year 1623. A new viceroy, the Marquis de Guelves, had recently arrived at Mexico; and he had consequently never witnessed the inundations produced by the overflow of the river of Guautitlan. He had the temerity, however, to order Martinez to stop up the subterraneous passage, and make the water of Zumpango and San Christobal return to the lake of Tezcuco, that he might see if the danger was, in fact, so great as it had been represented to him. This last lake swelled in an extraordinary manner, and the orders were recalled. Martinez recommenced his operations in the gallery, which he continued till the 20th June (though, according to some manuscript memoirs, the 20th September) 1629, when an event occurred, the true causes of which have ever remained secret.

The rains had been very abundant; and the engineer stopt up the subterraneous passage. The city of Mexico was in the morning inundated to the height of a metre, or 3½ feet. The *plaza mayor*, La Plaza del Volador, and the suburb of Tlatelolco, alone remained dry. Boats went up and down the other streets. Martinez was committed to prison. It was pretended that he had shut up the gallery to give the incredulous a manifest and negative proof of the utility of his work; but the engineer declared, that seeing the mass of water was too considerable to be received into his narrow gallery, he preferred exposing the capital to the temporary danger of an inundation, to seeing destroyed in one day, by the impetuosity of the water, the labours of so many years. Contrary to every expectation, Mexico remained inundated for five years, from 1629 to 1634. (Several memoirs, however, state that the inundation only lasted till 1631, but that it broke out afresh towards the end of the year 1633.) Be this as it may, the streets were passed in boats, as had been done before the conquest in the old Tenochtitlan. Wooden bridges were constructed along the sides of the houses for the convenience of foot passengers.

In this interval four different projects were presented and discussed by the Marquis de Ceralvo, the viceroy. An inhabitant of Valladolid, Simon Mendez, affirmed in a memoir, that the ground of the valley of Tenochtitlan rose considerably on the *n. w.* side towards Huehuetoca and the hill of Nochistongo; that the point where Martinez had opened the chain of mountains which circularly shuts in the valley corresponds to the mean level of the most elevated lake (Zumpango), and not to the level of the lowest (Tezcuco); and that the ground of the valley falls considerably to the *n.* of the village of Carpio, *e.* from the lakes of Zumpango and San Christobal. Mendez proposed to draw off the water of the lake of Tezcuco by a gallery which should pass between Xaltocan and Santa Lucia, and open into the brook (*arroyo*) of Tequisquiac, which, as has been already ob-]

[served, falls into the Rio de Moctezuma or Tula. Mendez began this *desague*, projected at the lowest point; and four pits of ventilation (*lumbreras*) were already completed, when the government, perpetually irresolute and vacillating, abandoned the undertaking as being too long and too expensive. Another desiccation of the valley was projected in 1630 by Antonio Roman and Juan Alvarez de Toledo, at an intermediate point, by the lake of San Christobal, the waters of which were proposed to be conducted to the ravine (*barranca*) of Huiputztla, *n.* of the village of San Mateo, and four leagues *w.* from the small town of Pachuca. The viceroy and *audiencia* paid as little attention to this project as to another of the mayor of Oculma, Christobal de Padilla, who, having discovered three perpendicular caverns, or natural gulfs (*boquerones*), even in the interior of the small town of Oculma, wished to avail himself of these holes for drawing off the water of the lakes. The small river of Teotihuacan is lost in these *boquerones.* Padilla proposed to turn also the water of the lake of Tezcuco into them, by bringing it to Oculma through the farm of Tezquititlan.

This idea of availing themselves of the natural caverns formed in the strata of porous amygdaloid gave rise to an analogous and equally gigantic project, in the head of Francisco Calderon the Jesuit. This monk pretended that at the bottom of the lake of Tezcuco, near the Peñol de los Baños, there was a hole (*sumideró*), which, on being enlarged, would swallow up all the water. He endeavoured to support this assertion by the testimony of the most intelligent Indians, and by old Indian maps. The viceroy commissioned the prelates of all the religious orders to examine this project. The monks and Jesuit kept sounding in vain for three months, from September till December 1635; but no *sumideró* was ever found, though, even yet, many Indians believe as firmly in its existence as Father Calderon. Whatever geological opinion may be formed of the volcanic or neptunian origin of the porous amygdaloid (*blasiger Mandelstein*) of the valley of Mexico, it is very improbable that this problematical rock contains hollows of dimension enough to receive the water of the lake of Tezcuco, which even in time of drought ought to be estimated at more than 251,700,000 cubic metres. It is only in secondary strata of gypsum, as in Thuringia, where we can sometimes venture to conduct inconsiderable masses of water into natural caverns (*gypsschlotten*), where galleries of discharge opened from the interior of a mine of coppery schistus are allowed to terminate, without any concern about the ulterior direction taken by the waters which impede the metallic operations. But how is it possible to employ this local measure in the case of a great hydraulical operation?

During the inundation of Mexico, which lasted five successive years, the wretchedness of the lower orders was singularly increased. Commerce was at a stand, many houses tumbled down, and others were rendered uninhabitable. In these unfortunate times the Archbishop Francisco Manzo y Zuniga distinguished himself by his beneficence. He went about daily in his canoe distributing bread among the poor. The court of Madrid gave orders a second time to transfer the city into the plains between Tacuba and Tacubaya; but the magistracy (*cabildo*) represented that the value of the edifices (*fincas*) which, in 1607, amounted to 150,000,000 of livres, now amounted to more than 200,000,000, or 8,334,000*l.* sterling. In the midst of these calamities the viceroy ordered the image of the Holy Virgin of Guadalupe to be brought to Mexico. The waters, as we have before observed, retired in 1634, when from very strong and very frequent earthquakes the ground of the valley opened, a phenomenon which was imputed in no small degree to the protecting influence of the Virgin.

The Marquis de Ceralvo, viceroy, set the engineer Martinez at liberty. He constructed the *calzada* (dike) of San Christobal, such nearly as we now see it. Sluices (*compertuas*) admit the communication of the lake of San Christobal with the lake of Tezonco, of which the level is generally from 30 to 32 decimetres, or from 118 to 125 inches, lower. Martinez had already begun, in 1609, to convert a small part of the subterraneous gallery of Nochistongo into an open trench. After the inundation in 1634, he was ordered to abandon this work as too tedious and expensive, and to finish the desague by enlarging his old gallery. The produce of a particular impost on the consumption of commodities (*derecho de sisas*) was destined by the Marquis de Salinas for the expences of the hydraulical operations of Martinez. The Marquis de Cadereyta increased the revenues of the desague by a new imposition of 25 *piastres* on the importation of every pipe of Spanish wine. These duties still subsist, though but a small part of them is applied to the desague. In the beginning of the 18th century the court destined the half of the excise on wines to keep up the great fortifications of the castle of San Juan d'Ulua. Since 1779 the chest of the hydraulical operations of the valley of Mexico does not draw more than five francs of the duties levied on each barrel of wine from Europe imported at Vera Cruz.]

[The operations of the desague were carried on with very little energy from 1634 to 1637, when the Marquis de Villena (Duke d'Escalona), viceroy, gave the charge of it to Father Luis Flores, commissary general of the order of St. Francis. The activity of this monk is much extolled, under whose administration the system of desiccation was changed for the third time. It was definitively resolved to abandon the gallery (*socabon*), to take off the top of the vault, and to make an immense cut through the mountain (*tajo abierto*), of which the old subterraneous passage was merely to be the water-course.

The monks of St. Francis contrived to retain the direction of hydraulical operations. It was so much the easier for them to do this, as at that epocha, (viz. from 9th June 1641, to 13th December 1673), the viceroyalty was almost consecutively in the hands of Palafox, a bishop of Puebla, Torres, a bishop of Yucatán, a Count de Baños, who ended his brilliant career by becoming a barefooted Carmelite, and Enriquez de Ribera, a monk of St. Augustin, archbishop of Mexico. Wearied with the monastical ignorance and delay, a lawyer, the fiscal Martin de Solis, obtained from the court of Madrid, in 1675, the administration of the desague. He undertook to finish the cut through the chain of the mountains in two months; and his undertaking succeeded so well, that 80 years were hardly sufficient to repair the mischief which he did in a few days. The fiscal, by advice of the engineer Francisco Posuelo de Espinosa, caused more earth to be thrown at one time into the water-course than the shock of the water could carry along. The passage was stopt up. In 1760, remains of what had fallen in by the imprudence of Solis were still perceptible. The Count de Monclova, viceroy, very justly thought that the tardiness of the monks of St. Francis was still preferable to the rash activity of the jurisconsult. Father *Fr.* Manuel Cabrera was reinstated in 1687 in his place of superintendant (*superintendente de la real obra del desague de Huehuetoca*). He took his revenge of the fiscal, by publishing a book which bears the strange title of "Truth cleared up and impostures put to flight, by which a powerful and envenomed pen endeavoured to prove, in an absurd report, that the work of the desague was completed in 1675." (*Verdad aclarada y desvanecidas imposturas, con que lo ardiente y envenenado de una pluma poderosa en esta Nueva España, en un dictamen mal instruido, quisò persuadir averse acabado y perfeccionao el año de 1675, la fabrica del Real Desague de Mexico.*)

The subterraneous passage had been opened and walled in a few years. It required two centuries to complete the open cut in a loose earth, and in sections of from 80 to 100 metres, or from 262 to 328 feet in breadth, and from 40 to 50, or from 131 to 164 feet, in perpendicular depth. The work was neglected in years of drought; but it was renewed with extraordinary energy for a few months after any great swelling or any overflow of the river of Guautitlan. The inundation with which the capital was threatened in 1747 induced the Count de Guemes to think of the desague. But a new delay took place till 1762, when after a very rainy winter there were strong appearances of inundation. There were still at the *n.* extremity of the subterraneous opening of Martinez 2310 Mexican *varas*, viz. 1938 metres, or 6356 feet, which had never been converted into an open trench (*tajo abierto.*) This gallery being too narrow, it frequently happened that the waters of the valley had not a free passage towards the Salto de Tula.

At length, in 1767, under the administration of a Flemish viceroy, the Marquis de Croix, the body of merchants of Mexico, forming the tribunal of the *consulado* of the capital, undertook to finish the desague, provided they were allowed to levy the duties of *sisa* and the duty on wine, as an indemnification for their advances. The work was estimated by the engineers at 6,000,000 of francs, or 250,020*l.* sterling. The *consulado* executed it at an expence of 4,000,000 of francs, or 166,680*l.* sterling; but in place of completing it in five years (as had been stipulated), and in place of giving a breadth of eight metres, or 26¼ feet, to the water-course, the canal was only completed in 1789 of the old breadth of the gallery of Martinez. Since that period they have been incessantly endeavouring to improve the work by enlarging the cut, and especially by rendering the slope more gentle. However, the canal is yet far from being in such a state that fallings in are no more to be apprehended, which are so much the more dangerous as lateral erosions increase in the proportion of the obstacles which impede the course of the water.

On studying in the archives of Mexico the history of the hydraulical operations of Nochistongo, we perceive a continual irresolution on the part of the governors, and a fluctuation of ideas, calculated to increase the danger instead of removing it. We find visits made by the viceroy, accompanied by the *audiencia* and canons; papers drawn up by the fiscal and other lawyers; advices given by the monks of St. Francis; an active impetuosity every 15 or 20 years, when the lakes threatened an over-]

[flow;. and a tardiness and culpable indifference whenever the danger was past. Twenty-five millions of livres, or 1,041,750*l.* sterling, were expended, because they never had courage to follow the same plan, and because they kept hesitating for two centuries between the Indian system of dikes and that of canals, between the subterraneous gallery, (*socabon*), and the open cut through the mountain (*tajo abierto.*) The gallery of Martinez was suffered to be choked up, because a large and deeper one was wished; and the cut (*tajo*) of Nochistongo was neglected to be finished, while they were disputing about the project of a canal of Tezcuco, which was never executed.

The *desague* in its actual state is undoubtedly one of the most gigantic hydraulical operations ever executed by man. We look upon it with a species of admiration, particularly when we consider the nature of the ground, and the enormous breadth, depth, and length of the aperture. If this cut were filled with water to the depth of 10 metres, or 32.8 feet, the largest vessels of war could pass through the range of mountains which bound the plain of Mexico to the *n. e.* The admiration which this work inspires is mingled, however, with the most afflicting ideas. We call to mind at the sight of the cut of Nochistongo the number of Indians who perished there, either from the ignorance of the engineers, or the excess of the fatigues to which they were exposed in ages of barbarity and cruelty. We examine if such slow and costly means were necessary to carry off from a valley inclosed in on all sides so inconsiderable a mass of water; and we regret that so much collective strength was not employed in some greater and more useful object; in opening, for example, not a canal, but a passage through some isthmus which impedes navigation.

The project of Henry Martinez was wisely conceived, and executed with astonishing rapidity. The nature of the ground and the form of the valley necessarily prescribed such a subterraneous opening. The problem would have been resolved in a complete and durable manner, 1. If the gallery had been commenced in a lower point, that is to say, corresponding to the level of the inferior lake; and, 2. If this gallery had been pierced in an elliptical form, and wholly protected by a solid wall equally elliptically vaulted. The subterraneous passage executed by Martinez contained only 15 square metres, or 161 square feet, in section, as we have already observed. To judge of the dimensions necessary for a gallery of this nature, we must know exactly the mass of water carried along by the river of Guautitlan and the lake of Zumpango at their greatest rise. Humboldt could find no estimation in the memoirs drawn up by Zepeda, Cabrera, Velasquez, and by M. Castera. But from the researches which he himself made on the spot, in the part of the cut of the mountain (*el corte o tajo*) called La Obra del Consulado, it appeared to him that at the period of the ordinary rains the waters afford a section of from eight to ten square metres, or from 86 to 107½ square feet, and that this quantity increases in the extraordinary swellings of the river Guautitlan to 30 or 40 square metres, or from 322¼ to 430⅓ square feet. The Indians assured him, that in this last case, the water-course which forms the bottom of the *tajo* is filled to such a degree, that the ruins of the old vault of Martinez are completely concealed under water. Had the engineers found great difficulties in the execution of an elliptical gallery of more than from four to five metres, or from 13 to 16 feet, in breadth, it would have been better to have supported the vault by a pillar in the centre, or to have opened two galleries at once, than to have made an open trench. These trenches are only advantageous when the hills are of a small elevation and small breadth, and when they contain strata less subject to falling down. To pass a volume of water of a section in general of eight metres, or 86 square feet, and sometimes from 15 to 20 square metres, or from 161 to 215 square feet, it has been judged expedient to open a trench, of which the section for considerable distances is from 1800 to 3000 square metres, or from 19,365 to 32,275 square feet.

In its present state the canal of derivation (*desague*) of Huehuetoca, according to the measurements of M. Velasquez, in his "*Informe y exposicion de las operaciones hechas para examinar la possibilidad del desague general de la laguna de Mexico y otros fines a el conducientes,* 1774, (manuscript memoir, folio 5.)," contains,

	Mex. varas.	Metres.
From the sluice of Vertideros to the bridge of Huehuetoca -	4870 or	4087
From the bridge of Huehuetoca to the sluice of Santa Maria	2660	2232
From the Compuerta de Santa Maria to the sluice of Valderas	1400	1175
From the Compuerta de Valderas to La Boveda Real - -	3290	2761
From La Boveda Real to the remains of the old subterraneous gallery called Techo Basso	650	545
From Techo Basso to the gallery of the viceroys - -	1270	1066
Carry over -	14,140	11,866

	Mex.varas.	Metres.
Brought over -	14,140	11,866
From the Cañon de los Vireyes to La Bocca de San Gregorio	610	512
From the Bocca de San Gregorio to the demolished sluice -	1400	1175
From La Presa Demolida to the cascade bridge - -	7950	6671
From La Puente del Salto to the cascade itself (Salto del Rio de Tula) - - - - -	430	361
Length of the canal from Vertideros to the Salto	24,530 or	20,585
		equal to 67,535 feet.

In this length of 4¾ common leagues, the chain of the hills of Nochistongo (to the *e.* of the Cerro de Sincoque), constituting a fourth part of it, has been cut to an extraordinary depth. At the point where the ridge is highest near the old well of Don Juan Garcia, for more than a length of 800 metres, or 2624 feet, the cut in the mountains is from 45 to 60 metres, or from 147 to 196 feet, in perpendicular depth. From the one side to the other, the breadth at top is from 85 to 110 metres, or from 278 to 360 feet. To have a clearer idea of the enormous breadth of this trench in the Obra del Consulado, we have only to recollect that the breadth of the Seine at Paris is at Port Bonaparte 102 metres, (334 English feet), at Pont-Royal 136 metres, (446 feet), and at the Pont d'Austerlitz, near the botanical garden, 175 metres, (574 feet). The depth of the above cut is from 30 to 50 metres, or from 98 to 131 feet, for a length of more than 3500 metres, or 11,482 feet. The water-course is generally only from three to four metres, or from 9.84 to 13.1 feet, in breadth; but in a great part of the desague the breadth of the cut is by no means in proportion to its depth, so that the sides in place of having a slope of 40° or 50° are much too rapid, and are perpetually falling in. It is in the Obra del Consulado where we principally see the enormous accumulations of moveable earth which nature has deposited on the porphyries of the valley of Mexico. Humboldt reckoned, in descending the stair of the viceroys, 25 strata of hardened clay, with as many alternate strata of marl, containing fibrous calcareous balls of a cellular surface. It was in digging the trench of the desague that he discovered some elephant bones, mentioned in his "*Recueil des Observations de Zoologie et d'Anatomie comparée.*"

On both sides of the cut we see considerable hills formed of the rubbish, which are gradually beginning to be covered with vegetation. The extraction of the rubbish having been an infinitely laborious and tedious operation, the method of Enrico Martinez was at last resorted to. They raised the level of the water by small sluices, so that the force of the current carried along the rubbish thrown into the water-course. During this operation, from 20 to 30 Indians have sometimes perished at a time. Cords were fastened round them, by which they were kept suspended in the current for the sake of collecting the rubbish into the middle of it; and it frequently happened that the impetuosity of the stream dashed them against detached masses of rock, which crushed them to death.

We have already observed that from the year 1643, the branch of Martinez's canal, directed towards the lake of Zumpango, had filled up, and that by that means (to use the expression of the Mexican engineers of the present day) the desague had become simply negative; that is to say, it prevented the river of Guautitlan to discharge itself into the lake. At the period of the great rises the disadvantages resulting from this state of things were sensibly felt in the city of Mexico. The Rio de Guautitlan, in overflowing, poured part of its water into the basin of Zumpango, which, swelled by the additional confluents of San Mateo and Pachuca, formed a junction with the lake of San Christobal. It would have been very expensive to enlarge the bed of the Rio de Guautitlan, to cut its sinuosities, and rectify its course; and even this remedy would not have wholly removed the danger of inundation. The very wise resolution was therefore adopted, at the end of the last century, under the direction of Don Cosme de Mier y Trespalacios, superintendant-general of the desague, of opening two canals to conduct the water from the lakes of Zumpango and San Christobal to the cut in the mountain at Nochistongo. The first of these canals was begun in 1796, and the second in 1798. The one is 8900 metres, or 29,228 feet, and the other 13,000 metres, or 42,650 feet, in length. The canal of San Christobal joins that of Zumpango to the *s. e.* of Huehoetoca, at 5000 metres, or 16,404 feet, distance from its entry into the desague of Martinez. These two works cost more than 1,000,000 of livres, or 41,670*l.* sterling. They are water-courses, in which the level of the water is from eight to 12 metres, or from 26 to 39 feet, lower than the neighbouring ground; and they have the same defects on a small scale with the great trench of Nochistongo. Their slopes are much too rapid; in several places they are almost perpendicular. Hence the loose earth falls so frequently in, that it [requires from 16,000 to 20,000 francs, equal to from 666*l.* to 833*l.* sterling, annually to keep these two canals of M. Mier in a proper condition. When the viceroys go to inspect (*hacer la visita*) the desague (a two days journey, which formerly brought them in a present of 3000 double piastres, or 656*l.* sterling), they embarked near their palace from the *s.* bank of the lake of San Christobal, and went even farther than Huehuetoca by water, a distance of seven common leagues. The above *Palacio de los Vireyes*, from which there is a magnificent view of the lake of Tezcnco, and the volcano of Popocatepec, covered with eternal snow, bears more resemblance to a great farm-house than to a palace.

It appears from a manuscript memoir of Don Ignacio Castera, present inspector (*maestro mayor*) of hydraulical operations in the valley of Mexico, that the desague cost, including the repairs of the dikes (*albaradones*), between 1607 and 1789, the sum of 5,547,670 double piastres. If we add to this enormous sum from 6 to 700,000 piastres expended in the 15 following years, we shall find that the whole of these operations (the cut through the mountains of Nochistongo, the dikes, and the two canals from the upper lakes) have not cost less than 31,000,000 of livres, or 1,291,770*l.* sterling. The estimate of the expence of the canal Du Midi, of which the length is 238,648 metres, or 782,966 feet, (notwithstanding the construction of 62 locks, and the magnificent reservoir of St. Ferreol) was only 4,897,000 francs, or 204,057*l.* sterling; but it has cost from 1686 to 1791 the sum of 22,999,000 of francs, or 958,368*l.* sterling, to keep this canal in order, (Andreossi, *Histoire du Canal du Midi*, p. 289.)

Resuming what we have been stating relative to the hydraulical operations carried on in the plains of Mexico, we see that the safety of the capital actually depends, 1. On the stone dikes which prevent the water of the lake of Zumpango from flowing over into the lake of San Christobal, and San Christobal from flowing into the lake of Tezcuco; 2. On the dikes and sluices of Tlahuac and Mexicaltzingo, which prevent the lakes of Chalco and Xochimilco from overflowing; 3. On the desague of Enrico Martinez, by which the Rio de Guautitlan makes its way through the mountains into the valley of Tula; and, 4. On the two canals of M. Mier, by which the two lakes of Zumpango and San Christobal may be thrown dry at pleasure.

However, all these multiplied means do not secure the capital against inundations proceeding from the *n.* and *n. w.* Notwithstanding all the expence which has been laid out, the city will continue exposed to very great risks till a canal shall be immediately opened from the lake of Tezcuco. The waters of this lake may rise, without those of San Christobal bursting the dike which confines them. The great inundation of Mexico under the reign of Ahuitzotl was solely occasioned by frequent rains, and the overflowing of the most *s.* lakes, Chalco and Xochimilco. The water rose to five or six metres, or 16 and 19 feet, above the level of the streets. In 1763, and the beginning of 1764, the capital was from a similar cause in the greatest danger. Inundated in every quarter, it formed an island for several months, without a single drop from the Rio de Guautitlan entering the lake of Tezcnco. This overflow was merely occasioned by small confluents from the *e. w.* and *s.* Water was every where seen to spring up, undoubtedly from the hydrostatical pression which it experienced in filtration in the surrounding mountains. On the 6th of September 1772, there fell so sudden and abundant a shower in the valley of Mexico, that it had all the appearance of a water spout, (*manga de agua.*) Fortunately, however, this phenomenon took place only in the *n.* and *n. w.* part of the valley. The canal of Huehuetoca was then productive of the most beneficial effects, though a great portion of ground between San Christobal, Ecatepec, San Mateo, Santa Iñes, and Guautitlan, were inundated to such a degree that many edifices became entire ruins. If this deluge had burst above the basin of the lake of Tezcuco, the capital would have been exposed to the most imminent danger. These circumstances, and several others which we have already adverted to, sufficiently prove how indispensable a duty it becomes for the government to take in hand the draining the lakes which are nearest to the city of Mexico. This necessity is daily increasing, because the bottoms of the basins of Tezcuco and Chalco are continually becoming more elevated, from the depositions which they receive.

In fact, while Humboldt was at Huehuetoca in the month of January 1804, the viceroy Iturrigaray gave orders for the construction of the canal of Tezcuco, formerly projected by Martinez, and more recently surveyed by Velasquez. This canal, the estimate of the expence of which amounts to 3,000,000 of livres tournois, or 125,010*l.* sterling, is to commence at the *n. w.* extremity of the lake of Tezcuco, in a point situated at a distance of 4593 metres, or 15,067 feet, *s.* 36° *e.* from the first sluice of the Calzada de San Christobal. It is to pass, first, through the great arid plain containing the] [insulated mountains of Las Cruces de Ecatepec and Chiconautla, and it will then take the direction of the farm of Santa Iñes towards the canal of Huehuetoca; the former of those summits, according to the geodesical measurements of M. Velasquez, being 404, and the latter 378 Mexican varas (339 and 317 metres) above the mean level of the lake of Tezcuco. Its total length to the sluice of Vertideros will be 37,978 Mexican varas, (viz. 31,901 metres, or 104,660 feet); but what will render the execution of this plan the most expensive, is the necessity of deepening the course of the old desague all the way from Vertideros to beyond the Boveda Real; the first of these two points being 9m.078 above, and the second 9m.181, (viz. 357.108 inches, and 361.464 inches), lower than the mean level of the lake of Tezcuco. To complete the description of this great hydraulical undertaking, we shall here insert the principal results of M. Velazquez's survey. These results, on correcting the error of the refraction, and reducing the apparent to the true level, coincide well enough with those obtained by Enrico Martinez and Arias in the commencement of the 17th century; but they prove the erroneousness of the surveys executed in 1764 by Don Ildefonso Yniesta, according to which the draining of the lake of Tezcuco appeared a much more difficult problem to resolve than it is in realty. We shall designate by + the points which are more elevated, and by — the points which are less elevated than the mean level of the water of Tezcuco, in 1773 and 1774, or the signal placed near its bank, at the distance of 5475 Mexican varas, *s.* 36° *e.* from the first sluice of the Calzada de San Christobal.

		Varas.	Palmos.	Dedos.	Granos.
The channel of the Rio de Guautitlan near the sluice of Vertideros	+	10	3	2	3
The channel of the desague under the port of Huehuetoca	+	8	0	2	1
Id. near the sluice of Santa Maria	+	4	3	8	3
Id. below the sluice of Valderas	+	2	1	11	2
The channel of the desague below the Boveda Real	—	10	3	9	3
Id. below the Boveda de Techo Baxo	—	15	0	6	1
Id. below the Bocca de San Gregorio	—	23	1	11	2
Id. above the Salto del Rio	—	90	1	9	0
The channel of the desague below the Salto del Rio	—	107	2	9	0

It is to be observed, that the vara is divided into four palmos, 48 dedos, and 192 granos; that a toise is equal to 2.32258 Mexican varas, and that a Mexican vara is .839169 metres, according to the experiments made on a vara preserved in the Casa del Cabildo of Mexico since the time of King Philip II.

Thus then a toise being equivalent to 2.32258 Mexican varas, a vara being equal to .839169 of a metre, 2.32258 varas correspond to 1.949 metres = 6.394 English feet = 1 toise.

But, to return to the plan of the canal, the distance from the aforesaid points, Vertederos to beyond the Boveda Real, is almost 10,200 metres (33,464 feet English.) To avoid deepening the bed of the present desague for a still more considerable length, it is proposed to give to the new canal a fall of only 0m.2 in 1000 metres. The plan of the engineer Martinez was rejected in 1607, purely because it was supposed that a current ought to have a fall of half a metre in the hundred. Alonso de Arias then proved on the authority of Vitruvius (L. VIII. C. 7.), that to convey the water of the lake of Tezcuco into the Rio de Tula a prodigious depth would be requisite for the new canal, and that even at the foot of the cascade near the Hacienda del Salto, the level of its water would be 200 metres, or 656 feet, below the river. Martinez could not stand against the power of prejudices and the authority of the ancients!

When we take into consideration the expence of the excavations required in the Rio del Desague, from the sluice of Vertideros or that of Valderas to the Boveda Real, we are tempted to believe that it would be, perhaps, easier to secure the capital from the dangers with which it is still threatened by the lake of Tezcuco, by recurring to the project attempted to be carried into execution by Simon Mendez during the great inundation from 1629 to 1634. M. Velasquez examined this project in 1774. After surveying the ground, that geometrician affirmed that 28 pits of ventilation, and a subterraneous gallery of 13,000 metres, or 42,650 feet, in length, for bringing the water of Tezcuco across the mountain of Citlaltepec towards the river of Tequixquiac, could be sooner finished, and at less expence, than the enlarging the bed of the desague, deepening it for a course of more than 9000 metres, or 29,527 feet, and cutting a canal from the lake of Tezcuco to the sluice of Vertideros near Huehuetoca. Humboldt was present at the consultations which took place] [in 1804, before deciding that the water of Tezcuco should pass through the old cut of Nochistongo. The advantages and disadvantages of Mendez's project were never discussed in these conferences.

It is to be hoped that in digging the new canal of Tezcuco more attention will be paid to the situation of the Indians than has hitherto been done, even so late as 1796 and 1798, when the courses of Zumpango and San Christobal were executed. The Indians entertain the most bitter hatred against the desague of Huehuetoca. A hydraulical operation is looked upon by them in the light of a public calamity, not only because a great number of individuals have perished by unfortunate accidents in Martinez's operations, but especially because they were compelled to labour to the neglect of their own domestic affairs, so that they fell into the greatest indigence while the desiccation was going on. Many thousands of Indian labourers have been almost constantly occupied in the desague for two centuries; and it may be considered as a principal cause of the poverty of the Indians in the valley of Mexico. The great humidity to which they were exposed in the trench of Nochistongo gave rise to the most fatal maladies among them. Only a very few years ago the Indians were cruelly bound with ropes, and forced to work like galley slaves, even when sick, till they expired on the spot. From an abuse of law, and especially from an abuse of the principles introduced since the organization of intendancies, the work at the desague of Huehuetoca is looked upon as an extraordinary *corvée*. It is a personal service exigible from the Indian, a remain of the *mita*, which we should not expect in a country where the working of the mines is perfectly voluntary, and where the Indian enjoys more personal liberty than in the *n. e.* part of Europe. The Indian is paid at the desague at the rate of two reals of *plata*, or 25 sous per day (=1*s.* 0½*d.*) In Martinez's time, in the 17th century, the Indians were only paid at the rate of five reals or three francs per week (=2*s.* 6*d.*), but they also received a certain quantity of maize for their maintenance.

Amongst other proofs of the light in which these hydraulical operations were considered, there are numerous testimonies contained in the *Informe de Zepeda*. In every passage of it we read, "that the desague has diminished the population and prosperity of the Indians, and that such or such a hydraulical project dare not be carried into execution, because the engineers have no longer so great a number of labourers at their disposal as in the time of the viceroy Don Luis de Velasco II." It is consoling, however, to observe, as we have elsewhere endeavoured to explain, that this progressive depopulation has only taken place in the central part of the old Anahuac, and ought therefore by no means to be considered general.

In all the hydraulical operations of the valley of Mexico, water has been always regarded as an enemy, against which it was necessary to be defended either by dikes or drains. We have already proved that this mode of proceeding, especially the European method of artificial desiccation, has destroyed the germ of fertility in a great part of the plain of Tenochtitlan. Efflorescences of carbonate of soda (*tequesquite*) have increased in proportion as the masses of running water have diminished. Fine *savannas* have gradually assumed the appearance of arid *steppes*. For great spaces the soil of the valley appears merely a crust of hardened clay (*tepetate*), destitute of vegetation, and cracked by contact with the air. It would have been easy, however, to profit by the natural advantages of the ground, in applying the same canals for the drawing of water from the lakes for watering of the arid plains, and for interior navigation. Large basins of water ranged as it were in stages above one another facilitate the execution of canals of irrigation. To the *s. e.* of Huehuetoca are three sluices, called Los Vertideros, which are opened when the Rio de Guautitlan is wished to be discharged into the lake of Zumpango, and the Rio del Desague to be thrown dry for the sake of cleaning or deepening the course. The channel of the old mouth of the Rio de Guautitlan, that which existed in 1607, having become gradually obliterated, a new canal has been cut from Vertideros to the lake of Zumpango. In place of continually drawing the water from this lake, and from San Christobal, out of the valley towards the Atlantic ocean, in the interval of 18 or 20 years, during which no extraordinary rise takes place, the water of the desague might have been distributed to the great advantage of agriculture in the lower parts of the valley. Reservoirs of water might have been constructed for seasons of drought. It was thought preferable, however, blindly to follow the order issued from Madrid, which bears, "that not a drop of water ought to enter into the lake of Tezcuco from the lake of San Christobal, unless once a year, when the sluices (Las Compuertas de la Calzada) are opened for the sake of fishing in the basin of San Christobal." This fishing is a grand rural festival for the inhabitants of the capital. The Indians construct huts on the banks of the lake of San Christobal, which is thrown] [almost dry during the fishing. This bears some resemblance to the fishing which Herodotus relates the Egyptians carried on twice a year in the lake Moeris, on opening the sluices of irrigation.

The trade of the Indians of Tezcnco languishes for whole months from the want of water in the salt lake which separates them from the capital; and districts of ground lie below the mean level of the water of Guautitlan and of the *n.* lakes; and yet no idea has ever been entertained for ages of supplying the wants of agriculture and interior navigation. From a remote period there was a small canal (*sanja*) from the lake of Tezcuco to the lake of San Christobal. A lock of four metres, or 13 feet, of fall would have admitted canoes from the capital to the latter of these lakes; and the canals of M. Micr would have even conducted them to the village of Huehuetoca. In this manner a communication would have been established from the *s.* bank of the lake of Chalco to the *n.* bounds of the valley, for an extent of more than 80,000 metres, or 262,468 feet. Men of the best information, animated with the noblest patriotic zeal, have had the courage to propose these measures, (M. Velasquez, for example, at the end of his *Informe sobre el Desague*, MS.); but the government, by rejecting the best conceived projects for such a length of time, seems to be resolved to consider the water of the Mexican lakes merely as a destructive element, from which the environs of the capital must be freed, and to which no other course ought to be permitted than that towards the Atlantic ocean.

Now that the canal of Tezcuco, by order of the viceroy Don Josef de Iturrigary, is to be opened, there will remain no obstacle to a free navigation through the large and beautiful valley of Tenochtitlan. Corn and the other productions of the districts of Tula and Guautitlan will come by water to the capital. The carriage of a mule load, estimated at 300 pounds weight, costs from Huehuetoca to Mexico five reals, or 3*s.* 4*d.* It is computed that when the navigation will be set on foot, the freight of an Indian canoe of 15,000 pounds burden will not be more than four or five piastres, or 1*l.* 1*s.* 10*d.* sterling; so that the carriage of 300 pounds (which make a *carga*) will only cost nine sous, or 4½*d.* Mexico, for example, will get lime at six or seven piastres, or 1*l.* 10*s.* 7*d.* the cart load (*carretada*), while the present price is from 10 to 12 piastres, or from 2*l.* 3*s.* 9*d.* to 2*l.* 12*s.* 6*d.*

But the most beneficial effect of a navigable canal from Chalco to Huehuetoca will be experienced in the commerce of the interior of Nueva España, known by the name of Comercio de Tierra Adentro, which goes in a straight line from the capital to Durango, Chihuhua, and Sante Fé, in New Mexico. Huehuetoca may hereafter become the emporium of this important trade, in which from 50 to 60,000 beasts of burden (*requas*) are constantly employed. The muleteers (*arrieros*) of New Biscay and Santa Fé fear nothing so much in the whole road of 500 leagues as the journey from Huehuetoca to Mexico. The roads in the *n. w.* part of the valley, where the basaltic amygdaloid is covered with a large stratum of clay, are almost impassable in the rainy season. Many mules perish in them. Those which stand out cannot recover from their fatigues in the environs of the capital, where there is no good pasturage and no large commons (*exidos*), which Huehuetoca would easily supply. It is only by remaining some length of time in countries where all commerce is carried on by caravans, either of camels or mules, that we can correctly appreciate the influence of the objects under discussion on the prosperisy and comfort of the inhabitants.

The lakes situated in the *s.* part of the valley of Tenochtitlan throw off from their surface miasmata of sulphuretted hydrogen, which become sensible in the streets of Mexico every time the *s.* wind blows. This wind is therefore considered in the country as extremely unhealthy. The Aztecs in their hieroglyphical writing represented it by a death's head. The lake of Xochimilco is partly filled with plants of the family of the junci and cyperoides, which vegetate at a small depth under a bed of stagnating water. It has been recently proposed to the government to cut a navigable canal in a straight line from the small town of Chalco to Mexico, a canal which would be shorter by a third than the present one; and it has at the same time been projected to drain the basins of the lakes of Xochimilco and Chalco, and sell the ground, which from having been for centuries washed with fresh water is uncommonly fertile. The centre of the lake of Chalco being somewhat deeper than the lake of Tezcnco, its water will never be completely drawn off. Agriculture and the salubrity of the air will be equally improved by the execution of M. Castera's project; for the *s.* extremity of the valley possesses in general the soil best adapted for cultivation. The carbonate and muriate of soda are less abundant, from the continual filtrations occasioned by the numerous rills which descend from the Cerro d'Axusco, the Guarda, and the volca-]

[noes. It must not, however, be forgotten that the draining of the two lakes will have a tendency to increase still farther the dryness of the atmosphere in a valley where the hygrometer of Deluc frequently descends to 15. This evil is inevitable, if no attempt is made to connect these hydraulical operations with some general system; the multiplying at the same time canals of irrigation, forming reservoirs of water for times of draught, and constructing sluices for the sake of counteracting the different pressures of the inequality of levels, and for receiving and withholding the increases of the rivers. These reservoirs of water distributed at suitable elevations might be employed at the same time in cleaning and working periodically the streets of the capital.

In the epocha of a nascent civilization, gigantic projects are much more seductive than more simple ideas of easier execution. Thus, in place of establishing a system of small canals for the interior navigation of the valley, the minds of the inhabitants have been bewildered since the time of the viceroy Count Revillagigedo with vague speculations on the possibility of a communication by water between the capital and the port of Tampico. Seeing the water of the lakes descend by the mountains of Nochistongo into the Rio de Tula (called also Rio de Moctezuma), and by the Rio de Panuco into the gulf of Mexico, they entertain the hope of opening the same route to the commerce of Vera Cruz. Goods to the value of more than 100,000,000 of livres, or 4,167,000*l.* sterling, are annually transported on mules from the Atlantic coast over the interior table-land, while the flour, hides, and metals descend from the central table-land to Vera Cruz. The capital is the emporium of this immense commerce. The road, which, if no canal is attempted, is to be carried from the coast to Perote, will cost several millions of piastres. Hitherto the air of the port of Tampico has appeared not so prejudicial to the health of Europeans and the inhabitants of the cold regions of Mexico as the climate of Vera Cruz. Although the bar of Tampico prevents the entry of vessels into the port drawing more than from 45 to 60 decimetres, or from 14¾ feet to 19 feet 8 inches, water, it would still be preferable to the dangerous anchorage among the shallows of Vera Cruz. From these circumstances a navigation from the capital to Tampico would be desirable, whatever expence might be requisite for the execution of so bold an undertaking.

But it is not the expence which is to be feared in a country where a private individual, the Count de la Valenciana, dug in a single mine, near Guanaxuato, three pits at an expence of 8,500,000 of francs, or 354,195*l.* sterling. Nor can we deny the possibility of carrying a canal into execution from the valley of Tenochtitlan to Tampico. In the present state of hydraulical architecture, boats may be made to pass over elevated chains of mountains, wherever nature offers points of separation which communicate with two principal recipients. Many of these points have been indicated by General Andreossi in the Vosges and other parts of France (*Andreossi sur le Canal du Midi*). M. de Piony made a calculation of the time that a boat would take to pass the Alps, if by means of the lakes situated near the hospital of mount Cenis a communication were established by water between Lans-le-bourg and the valley of Suze. This illustrious engineer proved by his calculation how much, in that particular case, land carriage was to be preferred to the tediousness of locks. The inclined planes, invented by Reynolds, and carried to perfection by Fulton, and the locks of MM. Huddleston and Betancourt, two conceptions equally applicable to the system of small canals, have greatly multiplied the means of navigation in mountainous countries. But however great the economy of water and time at which we can arrive, there is a certain maximum of height, in the predominant point, beyond which water is no longer preferable to land carriage. The water of the lake of Tezcuco, *e.* from the capital of Mexico, is more than 2276 metres, or 7465 feet, elevated above the level of the sea, near the port of Tampico! Two hundred locks would be requisite to carry boats to so enormous a height. If on the Mexican canal the levels were to be distributed, as in the canal du Midi, the highest point of which (at Naurouse) has only a perpendicular elevation of 189 metres, or 620 feet, the number of locks would amount to 330 or 340. We know nothing of the bed of the Rio de Moctezuma beyond the valley of Tula (the ancient Tollan); and we are ignorant of its partial fall from the vicinity of Zimapan and the Doctor. It is observed, however, that in the great rivers of S. America canoes ascend without locks for distances of 180 leagues, against the current, either by towing or rowing, to elevations of 300 metres, or 984 feet; but notwithstanding this analogy, and that of the great works executed in Europe, we can hardly persuade ourselves that a navigable canal from the plain of Anahuac to the Atlantic coast is a hydraulical work, the execution of which is anywise advisable.]

[The following are the remarkable cities and towns of the intendancy of Mexico.

Mexico, the capital of the kingdom of New Spain, height 2277 metres, or 7470 feet, population 137,000;

Tezcuco,	Zacatula,
Tacubaya,	Lerma,
Cuyoacan,	Tolaca,
Tacuba,	Pachuca,
Cuernavaca,	Caderelta,
Chilpansingo,	San Juan del Rio,
Tasco,	Queretaro.
Acapulco,	

The most important mines of this intendancy, considering them only in the relation of their present wealth, are:

La Veta Biscaina de Real del Monte, near Pachuca; Zimapan, El Doctor, and Tehulilotepec, near Tasco.

Chap. II.

Recent mediation between Spain and her colonies, being concise particulars of the secret sittings of the Cortes on that subject.

It is already known that the commissioners appointed to go out to Spanish America to mediate between them and the mother country, have returned to England without proceeding to fulfil the objects of their intended mission. The reason was the obstinate refusal of the Cortes to give them the powers which were necessary to success; for they would not consent to include Mexico in the commission, or permit them to go thither at all. It had been considered to be in vain to proceed to the execution of the trust under these circumstances; and the measure was abandoned. Such is still the conduct of the Cortes; and we lament to say they came to this decision after the arrival of the Duke de l'Infantado at Cadiz. But that a more correct opinion may be formed of this mediation; and that a more specific idea may be had of the grounds on which England has entered on this business, we have collected the following concise particulars of the secret sittings in the Cortes, in which the mediation was discussed.

On the proposals made by the British government for the ground-work of their interference, being laid before the Cortes, a committee was by them chosen to take cognizance of the affair, and report thereon. The persons named were Messrs. Morales Gallego, Gutierrez de la Huerta, Navarro, Cea, Alcour, Mexia, and Jauregui. The four first Europeans, and the other three Americans. The votes of the committee were equal; that is, three were of opinion that the mediation ought to be accepted, and three that it ought to be rejected, the remaining vote, which was that of Cea, being withheld, and not given on either side.

On the 10th of July 1812, secret sittings were held in the Cortes on this question, when the report of the committee was read, after which Senor Villa Gomez proposed the reading of the opinion of the regency. Senor Morales Gallego answered, that the opinion of the regency was expressed in the answers of the ministers of foreign relations to the notes of the English ambassador, and that he proposed the reading of the whole correspondence that had passed between both parties since the affair was first agitated.

Senor Asnarez was of opinion, that the council of state ought to be consulted on this affair, to which Senor Arguelles objected, by saying, that the council of state being recently installed, was not in any manner informed thereon, as it had been in agitation for more than a year, adding, that notwithstanding the regency, in conformity to the constitution, might listen to the opinion of the council of state, the Cortes were not under any such obligation, much less, when the members thereof (Cortes) were better informed on the affair, from having had it before them since its commencement. The president then observed, that as it was then too late to read the whole of the aforesaid correspondence, it might be done next day in the sittings which were to commence at 12 o'clock precisely.

Sitting of the 11th July 1812.

In the secret sittings of this day, which lasted from 12 till two and a quarter *p. m.*—the greatest part of the said notes from the English ambassador, and the answers of the minister of foreign relations, were read.

Sitting of the 12th.

The sitting of this day commenced at 11 o'clock, and the reading of the remaining part of the said correspondence was concluded; which done, a profound silence ensued for some time in the Cortes, which was broken by Senor Arguelles (European), who observed, that in an affair of such importance to the nation, he had resolved to give his opinion in writing, when he read a paper containing the same, the purport of which was, to shew the state of the revolution in America, the conduct which had been observed by the Spanish government, in employing pacific measures and conciliatory means to regain the ill-affected provinces; the conduct observed by the English government in receiving the rebels, and in holding correspondence with them; and lastly, he argued that the nature of the revolution in New Spain was] [entirely different from those of the other points of America; after which statements, he concluded, that English mediation ought not to be extended to the said kingdom of Mexico.

Senor Mexia (American) retorted by observing, that the causes of the revolutions in America, in their beginning, had been a wish for the removal of the authorities which governed therein despotically, and were inclined to deliver them up to the French; for which reason the inhabitants considered it necessary to establish local governments under the dependence of Ferdinand VII. which just and necessary measure of precaution on the part of the Americans alarmed the Spanish government, who, considering it as an act of rebellion, in concert with the mercantile junta of Cadiz, declared war against Caracas; which violence, together with other acts of a similar nature, had progressively continued to exasperate the minds of the Americans, driving at length some sections to the extreme of declaring their independence; and that the Spanish government, as far as it had been able, had used nothing but force against America, even resorting to the impolitical measure of availing themselves of the Portuguese against Buenos Ayres; that in New Spain the acts of violence used by the military chiefs against the revolutionary parties were notorious; that their complaints were yet unheard; and that they had been assassinated in the very act of parleying under a flag of truce. In short, he supported with most solid reasons the opinion of the Americans of the committee, in which state of the argument the president closed the sitting.

Sitting of the 13th.

At 12 this day the secret sitting commenced, when Senor Villa Gomez (European) rose and observed, that New Spain was not a dissentient province, even in the opinion of the English; because Captain Flemming had exhorted some of the provinces of S. America to follow the example of Mexico in sending their deputies to the Cortes. Senor Vegas (European) read a sound discourse, in which he retorts against the report of the European members of the committee, as well as the allegations of Senor Arguelles, concluding by demanding that the opinion of the three American members should be followed. (Reference is here made to a certain singular correspondence which Captain Flemming addressed to the government of Chile, in which, in the name of his government, he opposes the establishment of the new governments in those regions; which officious interference has been the cause of so much animosity to the English. The date of this memorable correspondence is 27th July, 2d August, and 3d October, 1811.)

Senor Gutiernez de la Huerta (European) endeavoured to sustain the opinion he had given as a member of the committee, in a heated and declamatory style. Amongst other things he said, that in an English club it had been asserted, that the felicity of the English nation depended on the independence of Spanish America. He treated the views of the English in the mediation in a most mysterious manner, giving to understand that this nation was interested in the disturbances of America. Senor Ribera (European) answered him with great warmth, and clearly demonstrated the futility of his arguments. The sitting then ended.

Sitting of the 14th.

This secret sitting commenced by the reading of an address of Vigodet, governor of Monte Video, in which he observed, that notwithstanding his repeated remonstrances, the Spanish government did not aid him with the necessary succours, and that if 4000 men at least were not sent out to him, he could not answer for the holding out of the fortress, which it would be necessary either to deliver up to the Portuguese, or to the insurgents. In consequence of which, Senor Mexia (American) observed, that the passage just read proved the certainty of what he had already stated in the Cortes; that is, that a great number of European troops were necessary to pacify the different sections of America; that it was not an ephemeral or partial movement, but a general and well-organized rising on the part of the natives; and that as it was impossible for the peninsula to send such forces in the present situation of things, there resulted the absolute necessity of acceding to the proposed mediation on the part of the British.

Senor Ramos Arispe (American) answered and denied that part of Senor Arguelles' speech, in which he asserted, that since the mediation was first agitated in the Cortes, the regency had abstained from taking active and hostile measures to subject the provinces of America; adding, that the Cortes had not hindered the regency from employing the means in its power to preserve the union of the American provinces; that this authority had never been considered as belonging to the Cortes, but to the executive power; and finally, that the Cortes, by virtue of a proposition made by Senor Del Monte, and approved, had urged the regency to send troops to quell the revolutions. The said Senor Ramos Arispe then proceeded to shew the necessity of English mediation in the kingdom of Mexico, founding his argu-]

[ment on the fact of the constituted authorities in that country having openly refused to treat with the insurgents, violating in this manner every principle of reason, equity, and prudence; that up to the present time, neither the Cortes, the regency, nor any one else, had sufficient knowledge of the causes and motives of the revolutions of America; and that it was not contrary to the decorum of the Spanish nation to treat with the insurgents, in order to accord with them, and settle matters, in like manner as Charles III. had capitulated with the insurgents of Madrid, and as the Cortes themselves had treated with the people of Cadiz on the 25th October 1811, when the latter, in opposition to the sovereignty of the nation and the inviolability of a deputy of the Cortes, demanded the head of Senor Valiente, contrary to every sentiment of justice.

Senor Golfin (European) remarked, that there was a deviation from the subject in question, for the point in agitation was, whether the Cortes ought, or ought not, to take cognizance of the matter? El Senor Conde de Torreno (European) endeavoured to support that part of the discourse of Senor Arguelles which had been answered by Senor Ramos de Arispe, by saying, that the views of the English in pretending the mediation for Mexico were too well known; that the notes of the English ambassador manifested that in fact there was a wish to acknowledge the independence of the American provinces, and make of them states federated with the peninsula, which was not only contrary to the constitution, but also to the treaty made with England, who had contracted to support the integrity of the Spanish monarchy. The sitting then ended.

Sitting of the 15th.

In the secret sitting of this day, Senor Alcocer (American) rose and said, that the mediation affair exclusively belonged to the Cortes, in conformity to several articles of the constitution, which he quoted; that of consequence it was there that it ought to be discussed, whether or not the same was to be extended to Mexico? He proved, by the most solid arguments, that it was not only advisable, but even absolutely necessary, to adopt the proposed plan of mediation; that without it there remained not even the most distant hope of tranquillizing those provinces; and that besides the Spanish government was exposed to incur the displeasure of the British, which might be attended with the most fatal consequences; that every possible measure ought to be adopted to spare the effusion of blood; that the means of reconciliation ought to be preferred to force and rigour, even when the latter measures had the appearance of better answering the end proposed; that, in short, Spain was not able to oppose to the insurgents a force capable of reducing them, because the insurrection every day became greater and more general, as well in numbers as from the increasing discipline of the troops which sustain the cause; ending his discourse by addressing himself to the feelings of the Cortes, in favour of the American provinces.

Senor Garcia Herreros (European) observed, that he was surprised to hear the American deputies speak so strongly in favour of the bandittis of New Spain, and that they should forget those who remained there faithful to the Spanish government; that the blood of the Europeans and other faithful subjects spilt by the insurgents, ought to be more interesting to the Cortes than that of the latter. He asserted, at the same time, that the insurrection had considerably increased in consequence of the means of rigour not having been sufficiently resorted to; but that with regard to the proposed mediation, the views of the English were sinister, as they had been proved in the last sitting by El Senor Torreno; that every thing possibly might be settled by means of a commercial treaty with the English, the only object they had in view; that in the last note of the English ambassador he observed, that the latter ungenerously reproached Spain with the succours which the British had expended, rather for their own interest, in order to sustain the war against the common enemy of Europe; and that Spain would still, at all times, be grateful for these services, and would recompence them with liberality, even more than the ambition of the English could expect. The sitting then closed.

Sitting of the 16th.

In the secret sitting of this day, El Senor Morales (European) observed, that the question of mediation belonged exclusively to the regency, and not to the Cortes. Senor Perez (American) read a long speech, contradicting what had been alleged by the American deputies. He confessed that the insurgents in New Spain had a form of government, or junta; but in order to turn the same into ridicule, he read a decree which, he said, was issued by the same junta, giving power to a curate to dispense in a case of a marriage. He added, that the insurgents had been heard, which he proved by a private letter from Mexico, mentioning, that the bishop of La Puebla had sent two curates to treat with Rayon, and that they had returned without having been able to do any thing; that it was false that pacific measures had] not been adopted towards the insurgents; for the said bishop had published a pastoral letter, in which he offered pardon to all those who should return to the obedience of the mother country. He ended by observing, that he was of opinion that the mediation of the English ought not to be accepted; that what the Cortes had to do was to strengthen the regency by placing at its head a royal personage, and that, in the mean time, all possible troops ought to be sent out in order to act offensively and defensively against the insurgents.

Senor Jauregui (American) read a discourse, in which he asserted, that since he had heard the opinion of the minister of foreign relations, in a meeting of the committee to which he was called, no doubt had been lett on his mind of the necessity of the mediation in the kingdom of Mexico, not only because the government was unacquainted with the forces the insurgents had there collected, and the progress they were likely to make hereafter, but because it was of importance to the whole nation to have their allies impressed with favourable sentiments, and not to give them any cause of complaint, as this might be injurious to the general cause.

Senor Lespergues (American) read a discourse proving the necessity of the mediation.

Finally, Senor Felice (American) also read an eloquent discourse in favour of the mediation.

The question being then declared to be sufficiently discussed, a small altercation took place respecting which of the two reports ought to be proposed for voting, when it was resolved, that the vote was first to be taken on the report of the three European members of the committee, which approved the refusal of the regency to adopt the mediation.

The question being then put to the vote, it resulted that this proposal was approved by 101 votes against 46. Of the first, two only were Americans, viz. Senor Perez and Senor Manian, both from New Spain; the rest were all Europeans; of the minority, six were Europeans, and the remaining 40 all Americans. Thus terminated the famous mediation affair, which would appear to seal the independence of Spanish America.

Amongst all the disturbances which have shaken the power of Spain in her Transatlantic possessions, that of Mexico is the most interesting, as well from its importance in being the head seat of the deputed government, as from the deeply stained traces of bloodshed which have, even at this early period, marked its career. It were impossible to give an adequate idea of the causes which have led to these revolutions, without an impartial retrospect of events relating alike to S. America and to the mother country. Similar causes of discontent appear to have operated in Mexico, Caracas, and Buenos Ayres, in short in the whole S. American possessions; but the effects have necessarily varied according to circumstances, which will, in their proper places, be treated of with a minuter attention. The former part, therefore, of the facts we are about to communicate, may be considered as attaching, in a great measure, to the whole of the Spanish colonies, whilst the latter contain specific accounts of the revolution of Mexico alone.

Chap. III.

Account of the present revolution.

The population of the Spanish colonies may be considered as divided into five classes; 1st, Spaniards born in Old Spain; 2dly, The descendants of Europeans, without any mixture of African or Indian blood, called Creoles; 3dly, The different races of Mulattoes and Mestizoes, or the issue of the crossings of the European, Indian, and African blood; 4thly, The Indians or Aborigines; 5thly, The imported African slaves. The first two classes, from their political importance, chiefly deserve our attention.

What the old Spaniards are, when transplanted to their American colonies, or what peculiar turn their national character takes in that particular situation, would not be a difficult point for conjecture, even if we were deprived of facts and observations. Prejudices are strong in proportion to their range, and evidently derive activity from the numbers which adopt them. Family prejudices are more tenacious than those of individuals, and national prejudices exceed both, in violence and duration. Those, especially, which are grounded on pretensions to superiority over a particular set or nation, are so early imbibed by all classes of the state, so indissolubly blended with every individual feeling, that their conjoint or national effects are astonishing, even when culture has scarcely left any visible traces of them in the common intercourse of life.

We may conceive what the national prejudices of the Spaniards, with respect to their colonies, now are, from the manner in which their ancestors took possession of them, and the authority which the descendants of those conquerors have enjoyed there during four centuries. The Spanish adventurers who flocked to America, immediately after the discovery of those countries, considered them in the light of a wilderness occupied merely by four and two footed game, of which they might dispose at their pleasure. The avowed and infinite cruelties which they committed without the least feeling of remorse, would demonstrate, if other proofs were wanting, the general opinion which prevailed for some time among them, of the irrationality of the Indians.

It will be easily conceived that the overbearing pride of the first conquerors, swelled with the destruction or submission of the Indians, was transmitted in full force to the adventurers whom the thirst of gold, and the desire of living freely at an immense distance from the seat of government, allured to those fertile regions. Those whose haughty and turbulent character was scarcely to be curbed by the authority of a powerful sovereign, must have exerted a dreadful sway over the conquered Indians. Every Spaniard thought himself a sovereign from the moment that he set his foot on the shores of America; and the kings of Spain would have soon lost their newly acquired dominions, but for the uncontroulable pride of the adventurers, which operated as a check on their mutual ambition.

The first generation of Creoles, though born upon the soil of America, naturally considered themselves as true Spaniards, since they could boast no other title to the superiority which they claimed over the natives; and it is probable that many years elapsed before any degree of national interest was felt by those new natives of the American continent. But when they began to multiply, and the ties of parentage between them and the European Spaniards were successively weakened; when, in the course of centuries, the natural connections which arise from a native soil, made the Creoles consider themselves as a people, seeds of jealousy against the mother country sprung up, the growth of which nothing could check but a system of equity and moderation, seldom, if ever, observed by any government with respect to colonies or conquered countries: by none less than the despotic and tyrannical court of Madrid.

The government of the Spanish colonies was entirely confided to the hands of viceroys and captains-general, who had under them several military governors and intendants; the adminstration of justice being committed to the *audiencias* or tribunals, which resided in the capitals, and were presided over by the respective viceroys and captains-general. The people, though nominally represented by the *cabildos*, or town corporations, had, in fact, no check upon the authority of their governors. The members of the *audiencias* were old Spaniards, and partook of the haughty spirit which considered the Creoles as inferior to their own countrymen. With respect to the town corporations, nothing could be more insignificant. The seats were, for the most part, filled up by the court of Spain; several were the property of particular families, and all of them were considered as empty honours, with which the timid ambition of some wealthy Creoles might be amused.

The viceroy was, in fact, as absolute as the monarch whom he represented; and, although by law responsible for his conduct to the council of Indies resident at Madrid, on the expiration of his commission, the same laws declared that the viceroy was to be obeyed as the king in person. It would be needless to expatiate upon the futility of such responsibily. The hope of redress is but a feeble consolation for actual oppression, even when the redresser is at hand. Let those, then, who are not blind to every abuse of power, and know how easily it is made the instrument of oppression when not checked by some effective restraint, consider what sort of government the Spanish colonists must have enjoyed, under nine European Spaniards, who had nothing to dread but an examination of their conduct at 2000 leagues distance from the theatre of their injustice.

The consequences of this system were sufficiently apparent. Prosperity and its foundation, security, were only to be found in interest and favour. The crowds of flatterers who thronged the palace of the Spanish monarch fell infinitely short of those which surrounded the viceroy of Mexico. His secretary was generally the favourite, the mediator through whom petitions reached the idol; and the grants descended to those who could enforce them with the most suitable offerings. Dreadful as the corruption of the late court of Madrid was, it must have appeared pure and exemplary when compared with the venality of the viceroyal courts of Spanish America. That honourable exceptions are to be found among the Spanish viceroys, we are far from bringing into question; but how cruelly must that people be oppressed, whose moments of happiness are to be counted by exceptions!

Oppression can never bear equally upon all classes, and especially when the community is divided into casts, as in Spanish America. Without speaking of those which are constitutionally degraded, as the Indians and Mestizoes, we shall merely point out the effect which the unlimited powers of the Spanish governors naturally produced on that numerous and powerful class, the Spanish Creoles. We shall not enter into a separate discussion about the state of opinion among [the Indians, for this poor degraded race have none at all. But we do not pretend to say that this state of mental degradation renders them insignificant in the present contest. On the contrary, we reckon them a most powerful tool. Their number, in Spanish America, is about 7,000,000, which forms more than one half the population of the country. Enjoying very little or no property, they are ready to follow any leaders who will conduct them to war against the Spaniards.

Those who are thoroughly acquainted with the character and circumstances of the two rival parties, the old Spaniards and Creoles, in Spanish America, will rather feel inclined to wonder at the extraordinary forbearance of the latter, than at the war which they are now waging against the former. Let it be considered that the number of Spaniards in the colonies, bears no proportion to the Creole population; that these Creoles, being the descendants of Spanish merchants, enjoy considerable wealth, and an education far superior to that of which their fathers could boast; while, on the other hand, very few of their rivals have the least title, from birth, education, or any other circumstance, to that superiority which they claim. Exclusively of those who are employed in the higher situations of government, the Spaniards who resort to the colonies to acquire a fortune, are, with few exceptions, a low, plodding set of people, who would never have risen from the humblest situations had they remained in the peninsula, and who generally commence their operations in America in the same way. Biscay, Asturias, Galicia, and Catalonia, have constantly sent out swarms of adventurers, among whom, those who expected to begin their career behind a counter in one of the shops of Vera Cruz or Mexico, thought too highly of themselves to associate with the rest of their companions. But the means of making a fortune are so easy in Spanish America, for those who object to no sort of occupation, that there is hardly one of these adventurers who, in the course of a few years, is not enabled to vie in riches with the old families of the country. At first they limit their pride to that superiority which Spaniards of all ranks claim in the colonies, and to the privilege of *hidalguia* or nobility, which is to be found even among Spanish beggars: but no sooner have they acquired property, than a part of it is destined to purchase honours at the court of Madrid. The wealthy drudge enjoys them behind his counter; and nothing is more common than to see people of this description, in their tawdry uniforms of captains or colonels, with a badge of one of the orders of Spain on their breasts, sitting in their shops, and occasionally helping their clerks to dispatch the customers who come for a yard of cloth or calico.

While the proud pretensions of this gross uneducated party, supported by the Spaniards in power, naturally excite dissatisfaction in the Creole gentry, the oppressive measures which they promote against the interest of the land, cannot fail to produce hatred, and an eager thirst for revenge. The Spanish merchants of America consider themselves exclusively entitled to the profits of trade,—trade, not grounded upon the mutual advantages of buyer and seller, but rather an oppressive monopoly, by which they oblige a whole population to take whatever they import from the mother country, extorting the most extravagant prices, by all the means which a market that excludes competition can afford.

The Spanish merchants were not, however, the only monopolists in the colonies. The government which supported them was the first to derive a paltry profit from shackling the industry of the Americans. The well known simile of the savage, who cut down the tree in order to pluck its fruit, (used by Montesquieu to exemplify the effects of despotism), was literally applicable to the Spanish colonial system. A Spanish colonist could not enjoy the advantages so lavishly bestowed on those beautiful countries. The eyes of a suspicious and oppressive government were constantly watching the progress of his industry. To sow or plant, he was not to consult the nature of the soil, but the government. Vines and olives, the two great blessings of temperate countries, were forbidden to grow in his fields, by proclamation. Some individuals had planted vineyards in Mexico. Whether the viceroy winked at this infraction of the colonial regulations, or was ignorant of it, we cannot say; the Spanish merchants, however, who were quicker sighted, gave the alarm to their correspondents at Cadiz. Complaint was instantly made to the court of Madrid, whence an order issued for rooting up the vines, in pursuance of the right enjoyed by the Cadiz merchants of administering to the wants of the American people at their own discretion.

It would be endless to enumerate the grievances which the colonies suffered, from the combined action of tyranny and monopoly. Mr. Walton's account of this system of exclusion on the part of Old Spain, appears more than sufficient to account for the state of habitual discontent, to which the Creoles were imperceptibly brought, not less by this palpable injustice, than by the civilization which the natural progress of human societies must] [always increase, in spite of the trammels imposed by the blindest of governments.

While the Creoles conceived that their security against the Indians, the Negro slaves, and the mixed casts, depended on the union of the whole European race, the Spaniards could oppress them with impunity. From this principle, Humboldt very satisfactorily accounts for the passive state of the Spanish colonies, during the succession-war in Spain. But the Indians have been so completely subdued, and the Creole population has so much increased since that period, that the same tranquillity and passiveness could not be looked for, when the late shock of the Spanish throne awakened them to the hopes of bettering their condition.

There was a period, when the whole mass of native population entertained such an opinion of the knowledge and power of the mother country, that they would have shut their eyes, in reverential awe, to whatever injustice she might commit; but the political events of our own times have destroyed all traces of this powerful illusion. The American war, in which Spain engaged with the most unaccountable degree of folly, could not but excite the attention of the Spanish Creoles. They must have compared their own situation with that of their neighbours, and perceived how much more galling were their own grievances, than those which produced the successful resistance of the English colonies. They must have reflected on the inconsistency and injustice of the Spanish government, who with one hand was helping English subjects to throw off their allegiance, and with the other binding its own in the most intolerable chains ever devised by oppression. About that period, the works of the French philosophers found their way into Spanish America, in despite of the terrors of the inquisition. This circumstance, which was scarcely noticed at the time, proved momentous in the highest degree, and amidst silence and obscurity, operated with fearful effect in undermining the fabric of despotism.

Reading is one of those pleasures which a certain degree of ease and comfort will never fail to generate among all sorts of people. The higher classes in the Spanish colonies had long arrived at that state, in consequence of their wealth, and books were an article not a little in request among them. Books, of course, were always put in the assortment of those cargoes of trash of all kinds, which were constantly sent out from Spain to the colonies. The glass beads which the first adventurers battered for gold with the simple tribes of Indians, were real treasures in comparison of the literary filth which the Spaniards exported to the colonies, with the certainty of selling it at the most extravagant price.

With the inconsistency peculiar to despotic governments, universities had been established at Mexico and Lima, to which even professors of mathematics were appointed. Thus, while they exalted the thirst for knowledge, they foolishly expected that the American youth would be still contented to seek it in those ponds of ignorance which had been prescribed to them.

The consequences of such a system may be easily guessed. No sooner had the works of the French philosophers found their way into the colonies, than they were read with an avidity beyond expression. The facility with which their general principles are seized, the common-place knowledge with which they enable young people to shine in conversation, the contempt and hatred which they breathe against what they denominate oppression, occasioned them to be looked on as invaluable treasures. The danger which attended their perusal, naturally enhanced the interest which they excited. There are instances of people who retired from all sorts of business into the country, to devote themselves wholly to the study of the French political and moral writers.

We, who have witnessed the effect of their doctrines in this free and happy country, during the ferment of the French revolution, when they threatened to overthrow the majestic fabric of our constitution, may easily conceive how they must have operated where every civil institution tends to countenance the bold assertions of those artful apostles of anarchy and atheism.

It would be difficult, without these premises, to account for the contrast which Humboldt observed between the people of the interior provinces of Mexico, and the enlightened classes of the capital. This part of his work deserves the attention of our readers, as it will be a clue to the knowledge of the character and principles of the present disturbances, of which we now hasten to give a passing sketch.

"The words European and Spaniard (says Humboldt) are become synonymous in Mexico and Peru. The inhabitants of the remote provinces have therefore a difficulty in conceiving, that there can be Europeans who do not speak their language; and they consider this ignorance as a mark of low extraction, because every where around them, all, except the very lowest class of the people, speak Spanish. Better acquainted with the history of the sixteenth century, than with that of our own times, they imagine] [that Spain continues to possess a decided preponderance over the rest of Europe. To them, the peninsula appears the very centre of European civilization:—It is otherwise with the Americans of the capital. Those of them who are acquainted with French or English literature, fall easily into a contrary extreme, and have a still more unfavourable opinion of the mother country than the French had, at a time when communication was less frequent between Spain and the rest of Europe. They prefer strangers from other countries to the Spaniards; and they flatter themselves with the idea, that intellectual cultivation has made more rapid progress in the colonies, than in the peninsula."

The public opinion being thus divided with respect to the mother country, it is evident that if the first class lost their enthusiasm for Spain, they might easily be led into rebellion by that more enlightened part of the community, who dispised and hated her government.

The news of the invasion of the French, together with that of the captivity of the king, and the resignations of Bayonne, produced a kind of stupor, which pervaded the whole population of Spanish America; but this was soon followed by a general enthusiasm in favour of the mother country. The prevailing sentiments were abhorrence of the French, and desire to support the Spaniards against their tyranny and injustice. If we wanted arguments to confirm the correctness of Humboldt's description, we should find a very strong one in the confidence with which the Americans looked for a speedy and successful issue to the Spanish cause. If there were any who doubted of that success, they were to be found among the higher classes, and even among the Spanish authorities. Those who, according to Humboldt, considered Spain just as if only a day had passed since the battle of Pavia, hourly expected to hear of the patriotic armies having reached Paris, and of Buonaparte being a prisoner at Madrid.

Few examples can be found of such an attachment, between what might be called two nations, as that which was evinced by the American population towards the mother country. The opinion in favour of supporting Spain was so general and decided, that not a single voice was heard from the discontented Creoles, who had been long meditating a revolution. Had the Spanish governmeated acted wisely, the French invasion would have strengthened the ties of union between Spain and her colonies; and what force had at first established, friendship, gratitude, and compassion would have sanctioned and confirmed for centuries.

The news of the general insurrection of Spain reached Mexico on the 29th July 1808. The enthusiasm which it had produced was still in full force, when the arrival of two deputies from the junta of Seville was announced, who were come to claim the sovereign command of Spanish America for that corporation, which had assumed the title of Supreme Gubernative Junta of Spain and the Indies. Such was the general disposition in favour of the peninsula, that it appears probable, from existing documents, that the Mexicans would have acceded to the demands of the junta, if, during the deliberation of a meeting of the public authorities, which the viceroy had convened, dispatches had not arrived from London, in which the deputies of the junta of Asturias announced their installation, and warned the Mexicans expressly against the pretension of the Andalusian junta. We may easily conceive how this declared rivalship must have affected the opinion which the Mexicans had formed of the spirit of the Spanish revolution.

The resignations of the royal family produced no diminution of American loyalty. The acclamations of "Ferdinand VII." were as sincere as they were general: but the blind submission which the old Spaniards demanded for whoever called himself his representative in the peninsula, was not so readily accorded. In Mexico the *cabildo*, or town corporation, had suggested the propriety of forming a junta which should govern that kingdom in the name of the captive sovereign. The viceroy appeared inclined to the measure, and the old Spaniards were in consequence determined to depose him. Had this chief made use of his power, and ordered to the capital the troops which, to the number of 1200, were stationed between Mexico and Vera Cruz, the country would probably have been spared the horrors which are now laying it waste. But the viceroy had no fixed plan: he was old, and wanted vigour: he was besides afraid of exciting suspicions against his loyalty, and had even proposed to resign his authority.

This weakness was soon perceived by the Spaniards. One of the wealthiest merchants among them, a personal enemy of the viceroy, was placed at the head of the conspiracy. The officers who were to command the guard on the appointed day were bribed; and this person, followed by about 200 Spaniards taken from the shops of Mexico, entered the palace of the viceroy at midnight, without resistance, and seizing him and his lady, committed the latter to a nunnery, and the former to the prison of the inquisition.

The *audiencia*, or supreme court of justice, had]

[secretly supported this measure, and the imprisonment of the viceroy was announced to the public, together with the circumstance of their having taken upon themselves to nominate a new viceroy. Though no disturbance followed this act of violence, the Creoles were by no means pleased or satisfied with it: not that they had any particular fondness for the deposed viceroy, but because the power which the Spaniards were assuming was now become intolerable to them.

When the news of this event reached the peninsula, the central junta was still in the full enjoyment of that tranquil slumber at Seville, during which the French, trembling for their safety, and hopeless of succour, on account of the Austrian war, found leisure to recover their spirits, and recruit their armies. On hearing that the viceroy of Mexico had been brought a prisoner to Spain upon suspicion of treachery, the joy of the junta was unbounded. It never occurred to them to examine the grounds of accusation; nor did they once condescend to reflect how greatly the ties of subordination must be relaxed, when a handful of persons, under no legitimate authority, could force the seat of government, and seize the chief magistrate with impunity. The junta was weak, and of course suspicious: a denunciation therefore, in any shape, was welcome to them.

Meanwhile advices of the ferment, which was rapidly spreading through the colonies, arrived by every packet. The declarations of their attachment had been sincere; but some time had now elapsed, and as the first impressions of sympathy grew fainter, the colonists began to reflect upon their situation, and to grow weary of the protracted hopes of that amelioration which had been promised to them in the most positive terms. The central junta conceived that the repetition of these promises would be sufficient to lull them again into apathy; and a pompous proclamation was issued, in which the colonies were declared equal to the mother country, and the Spanish Americans told, in direct terms, that "they belonged to nobody; and that they were masters of their own fate."

What this fate would have been, had the cause of Spain been crowned with the early successes which was anticipated, it is needless now to conjecture. In justice, however, to the Americans, we must say, that from the sentiments which they constantly manifested with regard to Spain, there is every reason to conclude that they would have continued faithful to her, if the unhappy course of events in the peninsula, and the more unhappy system of the central government, had not obliged them to take those steps which have progressively conducted them to a state of open rebellion.

Two years had elapsed since the Spanish Americans had heard of the victories of Baylen, Valencia, &c. and of the unprincipled invader of their mother country being driven to collect his scattered forces behind the Ebro. A supreme government had been created, and every blessing was hoped from the political principles which its members had ostensibly adopted. But while the distance of the scene raised the expectations of the Spanish Americans to the highest pitch, and they were daily expecting to hear of the restoration of Ferdinand VII. news arrived that Buonaparte was master of Madrid; that the central junta had with difficulty escaped to Andalusia; that several generals had been massacred by their troops on a suspicion of disaffection; that others, among whom was Morla, had openly betrayed their country; and that the public opinion had scarcely any one in whom it could venture to repose the slightest confidence. Though the disappointment of the Americans must have been proportioned to the exaltation of their hopes, not a symptom of commotion appeared through the whole extent of the Spanish colonies. Supplies were regularly dispatched to the mother country; subscriptions raised among all classes of people; and it seemed as if their loyalty had increased with the misfortunes of their European brethren. These misfortunes were attributed to treason, and the opinion of the Spanish superiority remained unshaken.

The Austrian war restored them to the plenitude of their first hopes, and the news of the victory of Talavera came in time to confirm them. But, alas! this was but a passing gleam of sunshine—a long period of gloom rapidly followed:—the Spanish armies completely defeated; the juntas of Seville and Valencia protesting against the central government; the brave Romana publishing a manifesto, in which the power of the supreme government was declared illegal! All this regularly dispatched, and carefully spread through the colonies by the discontented parties of the peninsula, naturally weakened their confidence, and gave the first shock to their enthusiasm.

The decisive blow was now impending: The French had dispersed the whole Spanish army at Ocaña, and nothing could stop them in their way to Andalusia. The boasted works of Sierra Morena were found to be a deception on the people, and the French entered Seville without the loss of a man, while the members of the central junta, dispersed and insulted in their flight, could scarcely escape the popular fury. These men, publicly] [proclaimed as traitors, assembled in the isle of Leon, and still trembling at the death with which they had been threatened, hastened to deposit their powers in the hands of a regency, chosen by themselves.

A government thus formed was little calculated to re-establish the confidence of the colonies: so conscious, indeed, were the members of their weakness, that they did not dare to communicate their installation to them, before they had been countenanced by a manifesto of the merchants of Cadiz; a species of support which, while it ensured them the attachment of the Spanish factors in the colonies, was certain to produce the contempt and abhorrenee of the rest of the people.

The Spaniards themselves must have foreseen the consequences of these events. Caracas was the first province where the news arrived, and the first also to effect a revolution. A month after, the information reached Buenos Ayres, and a similar event took place. The fermentation now began to spread through the *s.* continent: the alarm of the old Spaniards was general, but instead of inspiring them with a spirit of moderation, it seemed to embitter their animosities against the natives. The governor of the province of Socorro, in the kingdom of Santa Fé, ordered the military to fire on the unarmed people, who had assembled to petition him. An immense multitude flocked from the neighbouring country to revenge this act of cruelty; the governor took refuge in a convent, where he was surrounded and taken. Another insult from an European had a similar effect in the capital of Santa Fé. Quito became a scene of carnage. Carthagena formed a junta, which deprived the governor of his command. Lima was threatened with an insurrection; and every thing announced that a general explosion was at hand.

That these commotions were the effect of some general causes, and not of partial intrigues, is evident from the simultaneous movements in provinces which have scarcely any communication, such as those of Caracas and Buenos Ayres. These two provinces knew nothing of each other's revolution till some months after it was effected. Had both been the consequence of the same plan, the leaders would not have failed to cheer the public expectation with the hopes at least of having partners in their enterprise.

But although, wherever the insurrection broke out, the mass of the Creole population had eagerly declared in its favour, they were far from intending a total separation from the mother country. The motives alleged at the same moment in the most distant provinces, bear an extraordinary similarity, and shew that they were the gennine expression of the public opinion. "The supreme government of the peninsula (they said) has been declared infamous and treacherous: the members of it are even accused by the people of Spain, of having betrayed the country into the hands of the enemy. Can we then trust to the suspicions offspring of such a corrupted stock? Shall we wait till they choose to make their peace with Buonaparte, by betraying us into his hands? It was owing to our decided determination that the orders sent from Bayonne by the French ruler were not put into execution by our European governors. They were then ready to submit to his treachery. They will scarcely be less so now, when they have lost all hopes of succeeding in the peninsula. But setting all this aside, how can the ephemeral governments of Spain pretend to rule us, when they are manifestly incompetent to direct the people among whom they dwell! If they represent Ferdinand VII. let them exercise their power over those who have elected them—we will do the same in our own country—we will create a government in the name of our beloved sovereign, and that we will obey. Our brethren of the peninsula shall have our aid, our friendship, and our good wishes."

Such is the tenor of all the early proclamations of the insurgents of Spanish America. We do not pretend to say that they contained the genuine sentiments of the leaders; but they evidently were a correct statement of the prevailing sentiments of the people. The difference of opinion which divided the Creoles with respect to the mother country, and which we have noticed from Humboldt, was certainly the cause of this forbearance in the chiefs of the revolution. They hated the Spanish government, and were for the most part ardent and enthusiastic admirers of the metaphysical principles of liberty, which they had imbibed from the French publications; but they were obliged to yield to the more general opinion of their countrymen, who were heartily attached to Ferdinand VII. and had a great regard for Spain, which the misconduct of her revolutionary governments had only weakened, after two years of perpetual disappointment. Instead of fostering this excellent disposition, the Spanish government listened only to the dictates of wounded pride, and adopted every measure that was calculated to alienate the well disposed, and strengthen the party of their inveterate enemies.

The first step of the regency, upon hearing of the occurrence of Caracas, was to declare their proceedings rebellious, and to blockade their ports. The declaration itself was conceived in the most] violent and outrageous terms; the governors of the surrounding districts were ordered to stop all communication with the insurgent provinces, and to intercept their supplies. The effect of that unfeeling and insulting decree was to increase the contempt of a government, which, while it was obliged to court the protection of a handful of merchants in the peninsula, was thundering vengeance against 2,000,000 of people, who had the Atlantic between them and their pretended masters. In fact, the regency was the mere tool of the Cadiz merchants, and the orders—the dictates of their alarmed avarice. A singular fact, which we have it in our power to state, made this sufficiently evident in the eyes of the Spanish Americans.

So strong was the persuasion of the enlightened part of the Spanish people, that the news of the dispersion of the central junta would excite commotions in America, that the regency, in spite of its short-sighted policy, found it necessary to do something in favour of the colonies, which might reconcile them to their government, and preserve their union with Spain. The measure of granting them a free trade was proposed by the minister of the Indies, and ardently seconded by his under-secretary, a man distinguished in the revolution of Spain for his zeal and patriotism. This was a few days after the installation of the regency, when the new government, though timid and irresolute, had not entirely submitted to the yoke of the mercantile junta of Cadiz. The measure was put in practice after the pitiful intriguing manner of the old court. The order was signed by the minister and secretly printed; precautions were then taken to send it with the same secrecy to the colonies, that when the merchants came to the knowledge of it, it might be too late to repeal it. The whole transaction, however, transpired; and the rage of the mercantile junta knew no bounds. The regents were intimidated, and submitted to the disgrace of charging their minister and his under-secretary with having forged the order. Both of them were arrested; a counter order was issued, and the two prisoners were then set at liberty, without any farther inquiry.

But the most lamentable part of the American revolution was now at hand. The kingdom of Mexico had enjoyed an apparent tranquillity since the conspiracy of the Spaniards against the viceroy. The central junta had given the civil command of that kingdom to the archbishop, who, though an European by birth, was beloved by the Creoles for his moderation. The Spanish government had happily stumbled on one good measure; the rest, however, were calculated to increase the disaffection.

The viceroy had been deposed merely because he appeared favourable to the plan of erecting a junta for the government of Mexico, when Spain was without a supreme power. The Spaniards of the capital, who had defeated this plan, were already become unpopular from the intoxication of success; when intelligence arrived that the central junta had lavished on them its highest honours. The state of the Creoles became intolerable, when in addition to the insults which they had borne, their friend the archbishop was removed from the command; and the high court of justice, whom they considered as their most violent enemies, made temporary governors of the kingdom, until the arrival of the viceroy Venegas, nominated by the new regency of Cadiz.

Although the regular forces of Mexico had checked the spirit of insurrection, those who know the state of civilization at which that kingdom has arrived, and which puts it, according to Humboldt, at the head of the Spanish colonies in every respect, will easily suppose that discontented and enterprising individuals could not be wanting, who would watch every opportunity of shaking off the Spanish yoke. In fact, several of this description were to be found among the military and clergy, and even among the monks of New Spain. The most conspicuous was a country vicar of the name of Hidalgo, who enjoyed a valuable living in Dolores, a considerable town in the province of Valladolid Mechoacán. Hidalgo was a man of no vulgar talents, and of a knowledge far superior to that of the clergy of New Spain; this, as was commonly the case, had excited suspicions of his orthodoxy. We find that he had been accused to the inquisition, but had the good fortune or the art to remove their jealousy. He had thoroughly gained the affections of the Indians, whom he had taken great pains to enlighten. Several manufactories had risen by his care, and he had even established a foundry of cannon, alleging the immense advantage which might accrue to the crown from it, there being some rich copper mines in the neighbourhood of his parish.

When the viceroy was deposed by the Spaniards of Mexico, the troops constantly stationed, in times of war, between that capital and Vera Cruz, to prevent any attempt which our cruisers might make on that coast, were ordered into the interior. The regiment of cavalry De la Reyna was sent to San Miguel el Grande, a populous town in the vicinity of Dolores. Three captains of the [names of Allende, Aldama, and Abasolo, who served in that regiment, were natives of the place, and friends of the vicar Hidalgo, whom they readily joined. Their activity was extraordinary in disseminating discontent, and pourtraying, with the darkest colours, whatever tended to alienate the minds of the natives, in the actual circumstances of Spain.

Allende was sent to Queretaro, one of the most considerable towns in the kingdom of Mexico, where he recruited a great number of partisans. The Spaniards perceived that something was in agitation among the Creoles, and their suspicions fell upon the mayor or *corregidor* of the town. Information was sent to some of the *acuerdo*, or corporation, which was, at that time, split into two parties. Those who received it concealed it from the rest, and privately advised the Spaniards of Queretaro to act, with respect to the *corregidor*, as those of the capital had with the viceroy. The *corregidor* was accordingly seized and sent to Mexico. This second instance of insubordination, and contempt of the law, this trampling upon all authority in the person of a magistrate who proved to be innocent of the crime imputed to him, furnished a new pretence to the chiefs of the insurrection for instigating the Creoles against that handful of Spaniards who considered themselves superior to all established authority.

Venegas was now arrived at Vera Cruz, and the report of his bringing new honours for the enemies of the late viceroy, Yturrigaray, inflamed the whole Creole population. Hidalgo and his associates, indignant at this fresh outrage, and dreading the discovèry of their plan, determined to hasten its execution. On the 17th of September 1810, the vicar assembled the Indians to a sermon, the drift of which was to point out the tyranny of the Europeans, the state to which the treachery of the Spaniards had brought the peninsula, and the danger of being delivered up to the French or the English, who would assuredly extirpate the holy catholic religion.

Nothing could more strongly affect the minds of the poor Indians, who have ever submitted to be implicitly governed at the nod of a priest. Hidalgo ended his discourse with calling his Indians to arms; and to arms they flew with incredible fury. Allende appeared at the side of Hidalgo, and they led the mob to the town of St. Miguel el Grande, where the houses of the Spaniards were pillaged. No sooner was the insurrection at Dolores known, than the mass of the inhabitants of the extensive kingdom of Mechoacán acknowledged the authority of Hidalgo. Three regiments of veterans joined his standard, the towns of Salamanca and Valladolid fell into his hands. Wherever he appeared, crowds of Indians flocked to his army. The wealthy town of Guanaxuato, in the vicinity of which lay the richest mine of Mexico, supplied him with 5,000,000 of dollars. The insurgents possessed every thing but discipline and good leaders.

Meanwhile Venegas, who had now taken possession of his command at Mexico, was not wanting to himself. He secured the town of Queretaro, which may be considered as the key to Mexico. He awed into submission the Creoles of the capital by forming a camp with his troops without the walls. The governors of St. Luis Potosi and Guadalaxara armed the militia of the country; and even the wealthy Creoles of the principal towns supported the cause of the Spaniards, in order to avoid suspicion.

The insurgents, instead of falling immediately upon Mexico, marched to Valladolid, which they entered on the 20th of October, amidst the shouts of the Indian and Creole population. The greatest marks of honour were bestowed upon Hidalgo by the corporations of the town, and 1,500,000 of dollars were emptied into his military chest from the royal treasury. Two regiments of veteran cavalry joined him at this place. The province of Guadalaxara and the city of Zacatecas were also at his devotion. His army being now extremely large, he flattered himself that the viceroy would not hazard an action, and that the capital contained such a number of disaffected, as would oblige him to surrender it as soon as the insurgents came in sight. In this belief he marched to Toluca, while the troops of the viceroy fell back on Lerma.

While Hidalgo was advancing towards Mexico, another corps pushed through Ajusco to Cuernavaca, to take possession of the neighbouring part of the coast of the Pacific ocean. The main body of the vice-royal troops had gone too far to the *n.* and nothing was known of it in the capital.

Mexico was in imminent danger. The populace and a considerable part of the higher classes hated the Spaniards. Venegas had but a handful of men on whom he could rely. In this critical moment he resorted to an expedient which, however ridiculous it may appear in the eyes of many, was assuredly the only thing that saved him. He applied to the archbishop and the inquisition for a sentence of excommunication against Hidalgo, and all his troops and abettors. The Mexicans] [were struck with terror; and the whole town remained quiet, as if every inhabitant had been put in shackles.

But the dreadful sentence made no impression in the insurgent camp, where Hidalgo succeeded in persuading his Indians that the excommunication would fall upon the archbishop. The army had now advanced to the mount of Las Cruzes, a few miles from Mexico, where a division of the Spanish troops defended the pass. The insurgents dispersed them without difficulty, and presented themselves before the capital. But Hidalgo wanted decision. He summoned the town when he should have stormed it. The summons was answered with contempt, and the next morning his troops were seen retiring without any further effort.

Hidalgo's natural moderation and horror of bloodshed were reported to be the causes of this apparent timidity. It is well known that he alleviated the evils of war as much as possible, and that he sometimes ordered the artillery to fire upon his troops, when he had no other means to prevent pillage and devastation. His summons to the viceroy is said to have been very moderate; for he declared that his only desire was to see a junta established for the government of the kingdom; and that it was his intention to send immediate supplies of money to the peninsula. That Hidalgo's proposals were calculated to conciliate the public opinion, we are at liberty to conjecture from the care which the viceroy employed to conceal them from the inhabitants of Mexico. The true cause of Hildalgo's retreat, however, was the information he received of the advantages which the main corps of the vice-royal troops had gained in his rear. General Callejas, who commanded then, had taken the town of Dolores, where the revolution began, and massacred all the inhabitants. Hidalgo wanted skill to secure his retreat and watch the movements of the Spaniards; and he was now obliged to fall back in confusion. Callejas met the insurgents at Aculco, and completely defeated them. He then directed his march to Guanaxuato, which he entered on the 25th of November, taking a dreadful revenge on the inhabitants. Another corps of Spaniards, under General Cruz, entered the town of Irapurato, repeating the same cruelties and horrors.

The catastrophe of Hidalgo was now at hand. He had just reached the *provincias internas* with a considerable body of forces, which still followed his fortunes, when the governor of that part of the kingdom offered him his alliance. Hidalgo and his companions trusted to his faith, and incautiously presented themselves for a conference, when they were seized, and immediately put to death, as if the Spaniards were afraid of having them rescued out of their hands.

The insurrection, however, was far from being terminated by the death of its authors. The whole Creole and Indian population had now risen and formed detached corps in every part of the kingdom. The system of guerillas has been adopted by the Mexican insurgents, who improve every hour in boldness and dexterity. There are even large organized corps commanded by more skilful leaders than Hidalgo. One Rayon, a lawyer, had established an insurgent government at Zitaquaro. When that town was in danger of falling into the viceroy's hands, Rayon and his partisans made good their escape, and joined another numerous band of insurgents under the priest Morelos. This chief has lately obtained considerable advantages; and we find by accounts as late as the 7th of April (1812,) that he is master of the whole coast to the s. and that his comrade, Sanchez, with 30,000 men, preserves his authority in the plains of Puebla, and throughout the mountainous districts of Orezava.

We also find that the city of Orezava itself is in the hands of the insurgents, and that Vera Cruz is in alarm, its communication by Xalapa having been entirely cut off. But it would be an endless task to trace the actual state of the country from the confused and partial accounts of the viceroy, the only official information which is allowed to reach Europe. Suffice it to say, that, according to the last letters from Mexico, all the roads from the interior were occupied within a few days march of the capital, the fate of which depended on the resistance of an inconsiderable body of troops, which, as its losses could not be supplied, must finally perish by the effects of its own victories. Trade was at a stand; and the mines were totally abandoned, with the exception of one which an insurgent chief had been working for eight or 10 months, and with the produce of which he had been able to support his army. Several persons of the first rank had quitted the city, and gone over to the insurgents; from which it was naturally concluded that the chances of ultimate success began to appear in their favour.

Chap. IV.

Distances from Mexico to Acapulco.

It will be useful, for a minute acquaintance with the country, to add the distances which the natives, particularly the muleteers, who travel as it were in caravans to the great fair of Acapulco,] [reckon from one village to another. The true distance from the capital to the port being known, and supposing a third more for windings in a road both straight and of easy access, we shall find the value of the leagues in use in these countries. This datum is interesting for geographers, who in remote regions must avail themselves of simple itineraries. It is evident that the people shorten the leagues as the road becomes more difficult. However, under equal circumstances, we may have some confidence in the judgments formed by the muleteers of comparative distances; they may not know whether their beasts of burden go 2 or 3000 metres, or 6561 or 9842 feet English, in the space of an hour, but they learn from long habit if one distance be the third or fourth or the double of another.

The Mexican muleteers estimate the road from Acapulco to Mexico at 110 leagues. They reckon from Acapulco to the Passo d'Aguacatillo, four leagues; El Limon, three leagues; Los dos Aroyos, five; Alto de Camaron, four; La Guarita de los dos Caminos, three; La Moxonera, one-half; Quaxiniquilapa, two and a half; Acaguisotla, four; Masatlan, four; Chilpansingo, four; Sampango, three; Sapilote, four; Venta Vieja, four; Mescala, four; Estola, five; Palula, one and a half; La Tranca del Conexo, one and a half; Cuagolotal, one; Tuspa, or Pueblo Nuevo, four; Los Amates, three; Tepetlalapa, five; Punte de Istla, four; Alpuyeco, six; Xuchitepeque, two; Cuernavaca, two; S. Maria, three-fourths; Guchilaque, two and a half; Sacapisca, two; La Cruz del Marques, two; El Garda, two; Axusco, two; San Augustin de las Cuevas, three; Mexico, four. In this itinerary the numbers indicate how many leagues one place is distant from the one which immediately precedes it. Other itineraries, which are distributed to travellers who come by the S. sea, estimate the total distance at 104 or 106 leagues. Now, according to Humboldt's observations, it is in a straight line 151,766 toises. Adding a quarter for windings, we shall have 189,708 toises, or 1725 toises, or 11,040 feet, for the league of the country.

Chap. V.

General considerations on the extent and physical aspect of the kingdom of Nueva España.

In bestowing a rapid glance on the extent and population of the Spanish possessions in the two Americas, we must generalize our ideas, and consider each colony in its relations with the neighbouring colonies and with the mother country, if we would obtain accurate results, and assign to the country described the place to which it is entitled from its territorial wealth.

The Spanish possessions of the new continent occupy the immense extent of territory comprised between lat. 41° 43′ *s.* and lat. 37° 48′ *n.* This space of 79 degrees equals not only the length of all Africa, but it even much surpasses the breadth of the Russian empire, which includes about 167 dedrees of longitude, under a parallel of which the degrees are not more than half the degrees of the equator.

The most *s.* point of the new continent inhabited by the Spaniards is fort Maullin, near the small village of Carelmapu, on the coast of Chile, opposite to the *n.* extremity of the island of Chiloe. A road is opening from Valdivia to this fort of Maullin; a bold but useful undertaking, as a stormy sea prevents navigators for a great part of the year from landing on so dangerous a coast. On the *s.* and *s. e.* of fort Maullin, in the gulfs of Ancud and Reloncavi, by which we reach the great lakes of Nahuelhapi and Todos los Santos, there are no Spanish establishments; but we meet with them in the islands near the *e.* coast of Chiloe, even in lat. 43° 34′ *s.* where the island Caylin (opposite the lofty summit of the Corcobado) is inhabited by several families of Spanish origin.

The most *n.* point of the Spanish colonies is the mission of San Francisco, on the coast of New California, seven leagues to the *n. w.* of Santa Cruz. The Spanish language is thus spread over an extent of more than 1900 leagues in length. Under the wise administration of Count Florida Blanca, a regular communication of posts was established from Paraguay to the *n. w.* coast of N. America; and a monk in the mission of the Guaranis Indians can maintain a correspondence with another missionary inhabiting New Mexico, or the countries in the neighbourhood of cape Mendocin, without their letters ever passing at any great distance from the continent of Spanish America.

The dominions of the king of Spain in America exceed in extent the vast regions possessed by the Russian empire or Great Britain in Asia.

The Spanish possessions in America are divided into nine great governments, which may be regarded as independent of one another. Of these nine governments, five, viz. the viceroyalties of Peru and of New Granada, *capitanias generales* of Guatemala, of Portorico, and of Caracas, are wholly comprised in the torrid zone; the four other divisions, viz. the viceroyalties of Mexico and Buenos Ayres, the *capitanias generales* of Chile and Havannah, including the Floridas, are] [composed of countries of which a great part is situated without the tropics, that is to say, in the temperate zone. We shall afterwards see that this position alone does not determine the nature of the productions of these fine regions. The union of several physical causes, such as the great height of the *cordilleras*, their enormous masses, the number of plains, elevated more than from 2 to 3000 metres, or from 6561 to 9842 feet, above the level of the ocean, give to a part of the equinoctial regions a temperature adapted to the cultivation of the wheat and fruit trees of Europe. The geographical latitude has small influence on the fertility of a country, where, on the ridge and declivity of the mountains, nature exhibits a union of every climate.

Among the colonies subject to the king of Spain, Mexico occupies at present the first rank, both on account of its territorial wealth, and on account of its favourable position for commerce with Europe and Asia. We speak here merely of the political value of the country, considering it in its actual state of civilization, which is very superior to that of the other Spanish possessions. Many branches of agriculture have undoubtedly attained a higher degree of perfection in the province of Caraças than in New Spain. The fewer mines a colony has, the more the industry of the inhabitants is turned towards the productions of the vegetable kingdom. The fertility of the soil is greater in the provinces of Cumana, of New Barcelona, and Venezuela; and it is greater on the banks of the Lower Orinoco, and in the *n.* part of New Granada, than in the kingdom of Mexico, of which several regions are barren, destitute of water, and incapable of vegetation. But on considering the greatness of the population of Mexico, the number of considerable cities in the proximity of one another, the enormous value of the metallic produce, and its influence on the commerce of Europe and Asia; in short, on examining the imperfect state of cultivation observable in the rest of Spanish America we are tempted to justify the preference which the court of Madrid has long manifested for Mexico above its other colonies.

The denomination of New Spain designates, in general, the vast extent of country over which the viceroy of Mexico exercises his power. Using the word in this sense, we are to consider as *n.* and *s.* limits the parallels of the 38th and 10th degrees of latitude. But the captain-general of Guatemala, considered as administrator, depends very little on the viceroy of New Spain. The kingdom of Guatemala contains, according to its political division, the governments of Costa Rica and of Nicaragua. It is conterminous with the kingdom of New Granada, to which Darien and the isthmus of Panama belong. Whenever in this part of the work we use the denominations of New Spain and Mexico, we exclude the *captania-general* of Guatemala, a country extremely fertile, well peopled, compared with the rest of the Spanish possessions, and so much the better cultivated as the soil, convulsed by volcanoes, contains almost no metallic mines. We consider the intendancies of Merida and Oaxaca as the most *s.* and at the same time the most *e.* parts of New Spain. The confines which separate Mexico from the kingdom of Guatemala are washed by the great ocean to the *e.* of the port of Tehuantepec, near La Barra de Tonala. They terminate on the shore of the Atlantic, near the bay of Honduras.

We are tempted to compare together the extent and population of Mexico, and that of two empires with which this fine colony is in relations of unity and rivalry. Spain is five times smaller than Mexico. Should no unforeseen misfortune occur, we may reckon that in less than a century the population of New Spain will equal that of the mother country. The United States of N. America since the cession of Louisiana, and since they recognise no other boundary than the Rio Bravo del Norte, contain 240,000 square leagues. Their population is not much greater than that of Mexico, as we shall afterwards see on examining carefully the population and the area of New Spain.

If the political force of two states depended solely on the space which they occupy on the globe, and on the number of their inhabitants; if the nature of the soil, the configuration of the coast; and if the climate, the energy of the nation, and above all the degree of perfection of its social institutions, were not the principal elements of this grand dynamical calculation, the kingdom of New Spain might, at present, be placed in opposition to the confederation of the American republics. Both labour under the inconvenience of an unequally distributed population; but that of the United States, though in a soil and climate less favoured by nature, augments with an infinitely greater rapidity. Neither does it comprehend, like the Mexican population, nearly 2,500,000 of aborigines. These Indians, degraded by the despotism of the ancient Aztec sovereigns, and by the vexations of the first conquerors, though protected by the Spanish laws, wise and humane in general, enjoy very little, however, of this protection, from the great distance of the supreme].

[authority. The kingdom of New Spain has one decided advantage over the United States. The number of slaves there, either Africans or of mixed race, is almost nothing; an advantage which the European colonists have only begun rightly to appreciate since the tragical events of the revolution of St. Domingo. So true it is, that the fear of physical evils acts more powerfully than moral considerations on the true interests of society, or the principles of philanthropy and of justice, so often the theme of the parliament, the constituent assembly, and the works of the philosophers.

The number of African slaves in the United States amounts to more than 1,000,000, and constitute a sixth part of the whole population. The *s.* states, whose influence is increased since the acquisition of Louisiana, very inconsiderately increase the annual importation of these Negroes. It has not yet been in the power of congress, nor in that of the chief of the confederation, the present president, or even the former, to oppose this augmentation, and to spare by that means much distress to the generations to come.

In taking a general view of the whole surface of Mexico, we see that one-half is situated under the burning sky of the tropics, and the other belongs to the temperate zone. The latter contains 60,000 square leagues, and comprehends the *provincias internas*, both those which are under the immediate administration of the viceroy of Mexico (for example, the new kingdom of Leon, and the province of New Santander), and those governed by a particular commandant-general. The influence of this commandant extends over the intendancies of Durango and Sonora, and the provinces of Cohahuila, Texas, and New Mexico, regions thinly inhabited, which go all under the designation of *provincias internas de la commandancia general*, to distinguish them from the *provincias internas del vireynato*.

The *n.* provinces of Sonora and New Santander stretch as far *n.* as 38°, and part of the *s.* intendancies of Guadalaxara, Zacatecas, and S. Luis de Potosi, lie *s.* of the tropic of Cancer. We know, however, that the physical climate of a country does not altogether depend on its distance from the pole, but also on its elevation above the level of the sea, proximity to the ocean, configuration, and a great number of other local circumstances. Hence, of the 50,000 square leagues situated in the torrid zone, more than three-fifths enjoy rather a cold or temperate than a burning climate. The whole interior of the viceroyalty of Mexico, especially the interior of the countries comprised under the ancient denominations of Anahuac and Mechoacán, probably even all New Biscay, form an immense plain elevated 2000 or 2500 metres, or 6561. and 8201 feet, above the level of the neighbouring seas.

There is scarcely a point on the globe where the mountains exhibit so extraordinary a construction as in New Spain. In Europe, Switzerland, Savoy, and the Tyrol, are considered very elevated countries; but this opinion is merely founded on the aspect of the groups of a great number of summits perpetually covered with snow, and disposed in parallel chains to the great central chain. Thus the summits of the Alps rise to 3900 and even 4700 metres, or 12,794 and 15,419 feet, while the neighbouring plains in the canton of Berne are not more than from 1312 and 1968 feet in height. The former of these numbers (1312), a very moderate elevation, may be considered as that of the most part of plains of any considerable extent in Suabia, Bavaria, and New Silesia, near the sources of the Wartha and Piliza. In Spain, the two Castilles are elevated more than 580 metres, or 1902 feet. The highest level in France is Auvergne, on which the Mont d'Or, the Cantal, and the Puy de Dôme repose. The elevation of this level, according to the observations of M. de Buch, is 720 metres, or 2360 feet. These examples serve to prove that in general the elevated surfaces of Europe which exhibit the aspect of plains, are seldom more than from 400 to 800 metres, or from 1312 to 2624 feet, higher than the level of the ocean.

In Africa, perhaps, near the sources of the Nile, and in Asia, under lat. 31° and 37° *n.* there are plains analogous to those of Mexico; but the travellers who have visited Asia have left us completely ignorant of the elevation of Thibet. The elevation of the great desert of Cobi, to the *n. w.* of China, exceeds, according to Father Duhalde, 1400 metres, or 5511 feet. Colonel Gordon assured M. Labillardiere, that from the cape of Good Hope to lat. 21° *s.* the soil of Africa rose gradually to 2000 metres, or 6561 feet, of elevation. (*Labillardiere*, t. i. p. 89.) This fact, as new as it is curious, has not been confirmed by other naturalists.

The chain of mountains which form the vast plain of Mexico is the same with what, under the name of the Andes, runs through all S. America; but the construction, we may say the skeleton (*charpente*) of this chain, varies to the *s.* and *n.* of the equator. In the *s.* hemisphere, the *cordillera* is every where torn and interrupted by crevices like open furrows, not filled with heterogenous sub-]

[stances. If there are plains elevated from 2700 to 3000 metres, or from 10,629 to 11,811 feet, as in the kingdom of Quito, and farther *n.* in the province of Los Pastos, they are not to be compared in extent with those of New Spain, and are rather to be considered as longitudinal valleys bounded by two branches of the great *cordillera* of the Andes; while in Mexico it is the very ridge of the mountains which forms the plain, and it is the direction of the plain which designates, as it were, that of the whole chain. In Peru, the most elevated summits constitute the crest of the Andes; but in Mexico these same summits, less colossal it is true, but still from 4900 to 5400 metres, or from 16,075 to 17,715 feet, in height, are either dispersed on the plain, or ranged in lines which bear no relation of parallelism with the direction of the *cordillera*. Peru and the kingdom of New Granada contain transversal valleys, of which the perpendicular depth is sometimes 1400 metres, or 4854 feet. The existence of these valleys prevents the inhabitants from travelling except on horseback, a-foot, or carried on the shoulders of Indians (called *cargadores*); but in the kingdom of New Spain carriages roll on to Santa Fé in the province of New Mexico, for a length of more than 1000 kilometres, or 500 leagues. On the whole of this road there were few difficulties for art to surmount.

The table-land of Mexico is in general so little interrupted by valleys, and its declivity is so gentle, that as far as the city of Durango in New Biscay, 140 leagues from Mexico, the surface is continually elevated from 1700 to 2700 metres, or from 5576 to 8856 feet, above the level of the neighbouring ocean. This is equal to the height of mount Cenis, St. Gothard, or the great St. Bernard. Humboldt, that he might examine this geological phenomenon with the attention which it deserved, executed five barometrical surveys:— The first was across the kingdom of New Spain, from the S. sea to the Mexican gulf, from Acapulco to Mexico, and from Mexico to Vera Cruz. The second survey extended from Mexico by Tula, Queretaro, and Salamanca, to Guanaxuato. The third comprehended the intendancy of Valladolid, from Guanaxuato to the volcano of Jorullo at Pascuaro. The fourth extended from Valladolid to Toluca, and from thence to Mexico. Lastly, the fifth included the environs of Moran and Actopan. The number of points of which he determined the height, either barometrically or trigonometrically, amounts to 208; and they are all distributed over a surface comprehended between lat. 16° 50′ and 21° 0′ *n.* and long. 102° 8′ and 98° 28′ *w.* from Paris. Beyond these limits but one place was accurately ascertained, and that is the city of Durango, elevated, according to a deduction from a mean barometrical altitude, 2000 metres, or 6561 feet, above the level of the sea. Thus the table-land of Mexico preserves its extraordinary elevation much farther *n.* than the tropic of Cancer.

These measurements of heights, with the astronomical observations which Humboldt made on the same extent of ground, enabled him to construct the physical maps which accompany his work. They contain a series of vertical sections. In the statistics of the kingdom of New Spain, we must confine ourselves to plans likely to attract interest from views of political economy. The physiognomy of a country, grouping of mountains, extent of plains, elevation which determines its temperature; in short, whatever constitutes the construction of the globe, has the most essential influence on the progress of population and welfare of the inhabitants. It influences the state of agriculture, which must vary with the difference of climate, the means of internal commerce, the communications which depend on the nature of the territory, and the military defence, on which the external security of the colony depends. In these relations alone extensive geological views can interest the statesman, when he calculates the force and territorial wealth of a nation.

In S. America, the *cordillera* of the Andes exhibits at immense heights plains completely level. Such is the plain of 2565 metres, or 8413 feet, elevation, on which the city of Santa Fé de Bogota is built. Wheat, potatoes, and *chenopodium quinoa*, are there carefully cultivated. Such is also the plain of Caxamarca in Peru, the ancient residence of the unfortunate Atahualpa, of 2750 metres, or 9021 feet, elevation. The great plains of Antisana, in the middle of which rises the part of the volcano which penetrates the region of perpetual snow, are 4100 metres, or 13,451 feet, higher than the level of the ocean. These plains exceed in length the summit of the Pic of Teneriffe by 389 metres, or 1541 feet; and yet they are so level, that at the aspect of their natal soil, those who inhabit these countries have no suspicion of the extraordinary situation in which Nature has placed them. But all the plains of New Granada, Quito, or Peru, do not exceed 40 square leagues. Of difficult access, and separate from one another by profound valleys, they are very unfavourable for the transport of goods and internal commerce. Crowning insulated summits, they form as it were small islands in the middle of the aerial ocean.]

[Those who inhabit these frozen plains remain concentrated there, and dread to descend into the neighbouring regions, where a suffocating heat prevails, prejudicial to the primitive inhabitants of the higher Andes.

In Mexico, however, the soil assumes a different aspect. Plains of a great extent, but of a surface no less uniform, are so approximated to one another, that they form but a single plain on the lengthened ridge of the *cordillera;* such is the plain which runs from lat. 18° to 40° *n.* Its length is equal to the distance from Lyons to the tropic of Cancer, which traverses the great African desert. This extraordinary plain appears to decline insensibly towards the *n.* No measurement, as we have already remarked, was ever made in New Spain beyond the city of Durango; but travellers observe, that the ground lowers visibly towards New Mexico, and towards the sources of the Rio Colorado. The three sections aecompanying Humboldt's essay, show at a glance the difficulty which the extraordinary configuration of the country opposes to the transport of productions from the interior to the commercial cities of the coast.

In travelling from the capital of Mexico to the great mines of Guanaxuato, we remain at first for 10 leagues in the valley of Tenochtitlan, elevated 2277 metres, or 7468 feet, above the level of the sea. The level of this beautiful valley is so uniform, that the village of Gueguetoque, situated at the foot of the mountain of Sincoque, is only 10 metres, or 328 feet, higher than Mexico. The bill of Barientos is merely a promontory which stretches into the valley. From Gueguetoque we ascend near Botas to Puerto de los Reyes, and from thence descend into the valley of Tula, which is 115 metres, or 376 feet, lower than the valley of Tenochtitlan, and across which the great canal of evacuation of the lakes San Christoval and Zumpango passes to the Rio de Moctezuma and the gulf of Mexico. To arrive at the bottom of the valley of Tula, in the great plain of Queretaro, we must pass the mountain of Calpulalpan, which is only 1379 metres, or 4522 feet, above the level of the sea, and is consequently less elevated than the city of Quito, though it appears the highest point of the whole road from Mexico to Chihuahua. To the *n.* of this mountainous country the vast plains of S. Juan del Rio, Queretaro, and Zelaya begin, plains covered with villages and considerable cities. Their mean height equals Puy de Dôme in Auvergne, and they are near 30 leagues in length, extending to the foot of the metalliferous mountains of Guanaxuato. Those who have travelled into New Mexico assert, that the rest of the way consists of immense plains, appearing like so many basins of old dried-up lakes, following one another, and only separated by hills which hardly rise 200 or 250 metres (656 or 820 feet) at most above the bottom of these basins. The four plains surrounding the valley of Mexico are as follows, viz. the first, which comprehends the valley of Toluca, 2600 metres, or 8529 feet; the second, or the valley of Tenochtitlan, 2274 metres, or 7459 feet; the third, or the valley of Actopan, 1966 metres, or 6447 feet; and the fourth, the valley of Istla, 981 metres, or 3247 feet, of elevation. These four basins differ as much in their climate as in their elevation above the level of the sea; each exhibits a different cultivation: the first, and least elevated, is adapted for the cultivation of sugar; the second, cotton; the third, for European grain; and the fourth, for agava plantations, which may be considered as the vineyards of the Aztec Indians.

The barometrical survey which Humboldt executed from Mexico to Guanaxuato proves how much the configuration of the soil is favourable in New Spain for the transport of goods, navigation, and even the construction of canals. It is different in the transversal sections from the Atlantic to the S. sea. These sections show the difficulties opposed by nature to the communication between the interior of the kingdom and the coast. They every where exhibit an enormous difference of level and temperature, while from Mexico to New Biscay the plain preserves an equal elevation, and consequently a climate rather cold than temperate. From the capital of Mexico to Vera Cruz, the descent is shorter and more rapid than from the same point to Acapulco. We might almost say, that the country has a better military defence from nature against the people of Europe than against the attack of an Asiatic enemy; but the constancy of the trade-winds, and the great current of rotation which never ceases between the tropics, almost annihilate every political influence which China, Japan, or Asiatic Russia, in the succession of ages might wish to exercise over the new continent.

Taking our direction from the capital of Mexico towards the *e.* in the road to Vera Cruz, we must advance 60 marine leagues before arriving at a valley, of which the bottom is less than 1000 metres, or 3280 feet, higher than the level of the sea, and in which, consequently, oaks cease to grow. In the Acapulco road, descending from Mexico towards the S. sea, we arrive at the same temperate regions in less than 17 leagues. The *e.* de-]

[clivity of the *cordillera* is so rapid, that when once we begin to descend from the great central plain, we continue the descent till we arrive at the *e.* coast.

The *w.* coast is furrowed by four very remarkable longitudinal valleys, so regularly disposed, that those which are nearest the ocean are even deeper than those more remote from it. Casting our eyes on the section drawn up by Humboldt from exact measurements, we shall observe, that from the plain of Tenochtitlan the traveller first descends into the valley of Istla, then into that of Mascala, then into that of Papagayo, and lastly into the valley of Peregrino. The bottom of these four basins rise 981, 514, 170, and 158 metres (3217, 1685, 557, and 518 feet) above the level of the ocean. The deepest are also the narrowest. A curve drawn over the mountains which separate these valleys, over the Pic of the Marquis (the old camp of Cortes), the summits of Tasco, Chilpansingo, and Posquelitos, would preserve an equally regular progress. We might even be tempted to believe that this regularity is conformable to the type generally followed by nature in the construction of mountains; but the aspect of the Andes of S. America will soon destroy these systematic delusions. Many geological considerations prove to us, that at the formation of mountains, causes apparently very trivial have determined the accumulation of matter in colossal summits, sometimes towards the centre, and sometimes on the edges of the *cordilleras*.

Thus the Asiatic road differs very much from the European. For the space of 72.5 leagues, the distance in a straight line from Mexico to Acapulco, we continually ascend and descend, and arrive every instant from a cold climate in regions excessively hot. Yet the road of Acapulco may be made fit for carriages. On the contrary, of the 84.5 leagues from the capital to the port of Vera Cruz, one-fourth belongs to the great plain of Anahuac. The rest of the road is a laborious and continued descent, particularly from the small fortress of Perote to the city of Xalapa, and from this site, one of the most beautiful and picturesque in the known world, to La Rinconada. It is the difficulty of this descent which raises the carriage of flour from Mexico to Vera Cruz, and prevents it to this day from competing in Europe with the flour of Philadelphia. There is actually at present constructing a superb causeway along this *e.* descent of the *cordillera*. This work, due to the great and praiseworthy activity of the merchants of Vera Cruz, will have the most decided influence on the prosperity of the inhabitants of the whole kingdom of New Spain. The places of thousands of mules will be supplied by carriages fit to transport merchandises from sea to sea, which will connect, as it were, the Asiatic commerce of Acapulco with the European commerce of Vera Cruz.

We have already stated that in the Mexican provinces situated in the torrid zone, a space of 23,000 square leagues enjoys a cold, rather than a temperate climate. All this great extent of country is traversed by the *cordillera* of Mexico, a chain of colossal mountains which may be considered as a prolongation of the Andes of Peru. Notwithstanding their lowness in Choco, and the province of Darien, the Andes traverse the isthmus of Panama, and recover a considerable height in the kingdom of Guatemala. Sometimes their crest approaches the Pacific ocean, at other times it occupies the centre of the country, and sometimes it approaches the gulf of Mexico. In the kingdom of Guatemala, for example, this crest, jagged with volcanic cones, runs along the *w.* coast from the lake of Nicaragua towards the bay of Tehuantepec; but in the province of Oaxaca, between the sources of the rivers Chimalapa and Guasacualco, it occupies the centre of the Mexican isthmus. From lat. 18°½ to the 21°, in the intendancies of La Puebla and Mexico, from Misteca to the mines of Zimapan, the *cordillera* stretches from *s.* to *n.* and approaches the *e.* coast.

In this part of the great plain of Anahuac, between the capital of Mexico, and the small cities of Xalapa and Cordoba, a group of mountains appears which rivals the most elevated summits of the new continent. It is enough to name four of these colossi, whose heights were unknown before Humboldt's expedition; Popocatepetl, 5400 metres, or 17,716 feet; Iztaccihuatl, or the White Woman, 4768 metres, or 15,700 feet; Citlaltepetl, or the Pic d'Orizaba, 5295 metres, or 17,371 feet; and Nauhcampatepetl, or the Cofre de Perote, 4089 metres, or 13,314 feet. This group of volcanic mountains bears a strong analogy with that of the kingdom of Quito. If the height attributed to mount St. Elie be exact, we may admit that it is only under the 19° and 60° of lat. that mountains in the *n.* hemisphere reach the enormous elevation of 5400 metres above the level of the ocean.

Farther to the *n.* of the parallel of 19°, near the celebrated mines of Zimapan and the Doctor, situated in the intendancy of Mexico, the *cordillera* takes the name of Sierra Madre; and then leaving the *e.* part of the kingdom it runs to the *n. w.* towards the cities of San Miguel el Grande and Gua-]

[naxuato. To the *n.* of this last city, considered as the Potosi of Mexico, the Sierra Madre becomes of an extraordinary breadth. It divides immediately into three branches, of which the most *e.* runs in the direction of Charcas and the Real de Catorce, and loses itself in the new kingdom of Leon. The *w.* branch occupies a part of the intendancy of Guadalaxara. After passing Balaños it sinks rapidly, and stretches by Culiacan and Arispe, in the intendancy of Sonora, to the banks of the Rio Gila. However, it acquires again a considerable degree of height under the 30° of lat. in Tarahumara, near the gulf of California, where it forms the mountains De la Primeria Alta, celebrated for the gold washed down from them. The third branch of the Sierra Madre, which may be considered as the central chain of the Mexican Andes, occupies the whole extent of the intendancy of Zacatecas. We may follow it through Durango and the Parral in New Biscay, to the Sierra de los Mimbres (situated to the *w.* of the Rio Grande del Norté). From thence it traverses New Mexico, and joins the Crane mountains (Montagnes de la Grue) and the Sierra Verde. This mountainous country, situated under the 40° of lat. was examined in 1777 by Fathers Escalante and Font. The Rio Gila rises here, of which the sources are near those of the Rio del Norte. It is the crest of this central branch of the Sierra Madre which divides the waters between the Pacific and Atlantic ocean. It was a continuation of this branch which Fidler and the intrepid Mackenzie examined under the 50° and 55° of *n.* lat.

We have thus sketched a view of the *cordilleras* of New Spain. We have remarked that the coasts alone of this vast kingdom possess a warm climate adapted for the productions of the West Indies. The intendancy of Vera Cruz, with the exception of the plain which extends from Perote to the Pic d'Orizaba, Yucatán, the coast of Oaxaca, the maritime provinces of New Santander and Texas, the new kingdom of Leon, the province of Cohahuila, the uncultivated country called Bolson de Mapimi, the coast of California, the *w.* part of Sonora, Cinaloa, and New Galicia, the *s.* regions of the intendancies of Valladolid, Mexico, and La Puebla, are low grounds intersected with very inconsiderable hills. The mean temperature of these plains, of those at least situated within the tropics, and whose elevation above the level of the sea does not exceed 300 metres, or 984 feet, is from 25° to 26° of the centigrade thermometer, or 77° of Fahrenheit's; that is to say, from 8° to 9° of the centigrade, or from 14° to 16° of Fahrenheit, greater than the mean heat of Naples.

These fertile regions, which the natives call *tierras calientes*, produce in abundance sugar, indigo, cotton, and *bananas*. But when Europeans, not seasoned to the climate, remain in these countries for any time, particularly in populous cities, they become the abode of the yellow fever, known by the name of black vomiting, or *vomito prieto*. The port of Acapulco, and the valleys of Papagayo and Peregrino, are among the hottest and unhealthiest places of the earth. On the *e.* coast of New Spain, the great heats are occasionally interrupted by strata of cold air, brought by the winds from Hudson's bay towards the parallels of the Havannah and Vera Cruz. These impetuous winds blow from October to March; they are announced by the extraordinary manner in which they disturb the regular recurrence of the small atmospherical tides, or horary variations of the barometer; (see this phenomenon explained in the first volume of Humboldt's Travels, *Physique Generale*, p. 92, 94); and they frequently cool the air to such a degree, that at Havannah the centigrade thermometer descends to 0°, or 32° of Fahrenheit, and at Vera Cruz to 16°, or 60° of Fahrenheit; a prodigious fall for countries in the torrid zone.

On the declivity of the *cordillera*, at the elevation of 1200 or 1500 metres, or from 3936 to 4920 feet, there reigns perpetually a soft spring temperature, which never varies more than four or five degrees (seven or nine of Fahrenheit). The extremes of heat and cold are there equally unknown. The natives give to this region the name of *tierras templadas*, in which the mean heat of the whole year is from 20° to 21°, or from 68° to 70° of Fahrenheit. Such is the fine climate of Xalapa, Tasco, and Chilpansingo, three cities celebrated for their great salubrity, and the abundance of fruit trees which grow in their neighbourhood. Unfortunately, this mean height of 1300 metres, or 4264 feet, is the height to which the clouds ascend above the plains adjoining to the sea; from which circumstance these temperate regions, situated on the declivity (for example, the environs of the city of Xalapa), are frequently enveloped in thick fogs.

It remains for us to speak of the third zone, known by the denomination of *tierras frias*. It comprehends the plains elevated more than 2200 metres, or 7217 feet, above the level of the ocean, of which the mean temperature is under 17°, or 62° of Fahrenheit. In the capital of Mexico, the centigrade thermometer has been known to fall several degrees below the freezing point; but this is a very rare phenomenon; and the winters are usually as] [mild there as at Naples. In the coldest season, the mean heat of the day is from 13° to 14°, from 55° to 70° of Fahrenheit. In summer the thermometer never rises in the shade above 24°, or 75° of Fahrenheit. The mean temperature of the whole table-land of Mexico is in general 17°, or 62° of Fahrenheit, which is equal to the temperature of Rome. Yet this same table-land, according to the classification of the natives, belongs, as we have already stated, to the *tierras frias;* from which we may see that the expressions, hot or cold, have no absolute value. At Guayaquil, under a burning sky, the people of colour complain of excessive cold, when the centigrade thermometer suddenly sinks to 24°, (75° of Fahrenheit), while it remains the rest of the day at 30°, (86° of Fahrenheit).

But the plains more elevated than the valley of Mexico, for example, those whose absolute height exceeds 2500 metres, or 8201 feet, possess, within the tropics, a rude and disagreeable climate, even to an inhabitant of the *n.* Such are the plains of Toluca, and the heights of Guchilaque, where, during a great part of the day, the air never heats to more than 6° or 8°, (43° or 46° of Fahrenheit), and the olive tree bears no fruit, though it is cultivated successfully a few hundred metres lower in the valley of Mexico.

All those regions called *cold* enjoy a mean temperature of from 11° to 13°, or from 51° to 55° of Fahrenheit, equal to that of France and Lombardy. Yet the vegetation is less vigorous, and the European plants do not grow with the same rapidity as in their natal soil. The winters, at an elevation of 2500 metres, are not extremely rude; but the sun has not sufficient power in summer over the rarefied air of these plains to accelerate the development of flowers, and to bring fruits to perfect maturity. This constant equality, this want of a strong ephemeral heat, imprints a peculiar character on the climate of the higher equinoctial regions. Thus the cultivation of several vegetables succeeds worse on the ridge of the Mexican *cordilleras* than in plains situated to the *n.* of the tropic, though frequently the mean heat of these plains is less than that of the plains between the 19° and 22° of lat.

These general considerations on the physical division of New Spain are extremely interesting in a political view. In France, even in the greatest part of Europe, the employment of the soil depends almost entirely on geographical latitude; but in the equinoctial regions of Peru, New Grenada, and Mexico, the climate, productions, aspect, we may say physiognomy, of the country, are solely modified by the elevation of the soil above the level of the sea. The influence of geographical position is absorbed in the effect of this elevation. Lines of cultivation similar to those drawn by Arthur Young and M. Decandolle on the horizontal projections of France can only be indicated on sections of New Spain. Under the 19° and 22° of lat. with some few exceptions, sugar, cotton, particularly *cacao*, and indigo, are only produced abundantly at an elevation of from 6 to 800 metres, or from 1968 to 2624 feet. The wheat of Europe occupies a zone on the declivity of the mountains, which generally commences at 1400 metres, or 4592 feet, and ends at 3000 metres, or 9842 feet. The banana tree (*musa paradisiaca*), the fruit of which constitutes the principal nourishment of all the inhabitants of the tropics, bears almost no fruit above 1550 metres, or 5084 feet; the oaks of Mexico grow only between 800 and 3000 metres, (2624 and 9842 feet); and the pines never descend towards the coast of Vera Cruz farther down than 1850, or 6068 feet, nor rise near the region of perpetual snow to an elevation of more than 4000 metres, or 13,123 feet.

The provinces called *internas*, situated in the temperate zone (particularly those included between the 30° and 38° of lat.) enjoy, like the rest of N. America, a climate essentially different from that of the same parallels in the old continent. A remarkable inequality prevails between the temperature of the different seasons. German winters succeed to Neapolitan and Sicilian summers. It would be superfluous to assign here other causes for this phenomenon than the great breadth of the continent and its prolongation towards the *n.* pole. This subject has been discussed by enlightened natural philosophers, particularly by M. Volney, in his excellent work on the soil and climate of the United States, with all the care which it deserves. We shall merely observe, that the difference of temperature observable between the same latitudes of Europe and America, is much less remarkable in those parts of the new continent bordering on the Pacific ocean than in the *e.* parts. M. Barton has proved, from the state of agriculture and the natural distribution of vegetables, that the Atlantic provinces are much colder than the extensive plains situated to the *w.* of the Alleghany mountains.

A remarkable advantage for the progress of national industry arises from the height at which nature, in New Spain, has deposited the precious metals. In Peru the most considerable silver mines, those of Potosi, Pasco, and Chota, are immensely elevated very near the region of per-]

[petual snow. In working them, men, provisions, and cattle must all be brought from a distance. Cities situated in plains, where water freezes the whole year round, and where trees never vegetate, can hardly be an attractive abode. Nothing can determine a free-man to abandon the delicious climate of the valleys to insulate himself on the top of the Andes but the hope of amassing wealth. But in Mexico, the richest seams of silver, those of Guanaxuato, Zacatecas, Tasco, and Real del Monte, are in moderate elevations of from 1700 to 2000 metres, (5576 to 6561 feet). The mines are surrounded with cultivated fields, towns, and villages; the neighbouring summits are crowned with forests; and every thing facilitates the acquisition of this subterraneous wealth.

In the midst of so many advantages bestowed by nature on the kingdom of New Spain, it suffers in general, like old Spain, from the want of water and navigable rivers. The great river of the *n.* (Rio Bravo del Norte) and the Rio Colorado, are the only rivers worthy of fixing the attention of travellers, either for the length of their course, or the mass of water which they pour into the ocean. The Rio del Norte, from the mountains of the Sierra Verde (to the *e.* of the lake of Timpanogos) to its mouth in the province of New Santander, has a course of 512 leagues. The course of the Rio Colorado is 250. But these two rivers, situated in the most uncultivated part of the kingdom, can never be interesting for commerce, till great changes in the social order, and other favourable events, introduce colonization into these fertile and temperate regions. These changes are not perhaps very distant. The banks of the Ohio were even in 1797 so thinly inhabited, (*Voyage de Michaux a l'Ouest des Monts Alleghanys*, p. 115), that 30 families could hardly be found in a space of 130 leagues, while the habitations are now so multiplied that they are never more than one or two leagues distant from one another.

In the whole equinoctial part of Mexico there are only small rivers, the mouths of which are of considerable size. The narrow form of the continent prevents the collection of a great mass of water. The rapid declivity of the *cordillera* abounds more properly with torrents than rivers. Mexico is in the same state with Peru, where the Andes approach so near to the coast as to occasion the aridity of the neighbouring plains. Among the small number of rivers in the *s.* part of New Spain, the only ones which may in time become interesting for interior commerce are, 1. The Rio Guasacualco, and the Rio Alvarado, both to the *s. e.* of Vera Cruz, and adapted for facilitating the communication with the kingdom of Guatemala; 2. The Rio de Moctezuma, which carries the waters of the lakes and valley of Tenochtitlan to the Rio de Panuco, and by which, forgetting that Mexico is 2277 metres, or 7468 feet, elevated above the level of the sea, a navigation has been projected between the capital and the *w.* coast; 3. The Rio de Zacatula; 4. The great river of Santiago, formed by the junction of the rivers Lerma and Las Laxas, which might carry the flour of Salamanca, Zelaya, and perhaps the whole intendancy of Guadalaxara, to the port of San Blas, or the coast of the Pacific ocean.

The lakes with which Mexico abounds, and of which the most part appear annually on the decline, are merely the remains of immense basins of water, which appear to have formerly existed on the high and extensive plains of the *cordillera.* We shall merely mention in this physical view the great lake of Chapala in New Galicia, of nearly 160 square leagues, double the size of the lake of Constance; the lakes of the valley of Mexico, which include a fourth part of its surface; the lake of Patzcuaro, in the intendancy of Valladolid, one of the most picturesque situations found in either continent; and the lakes of Mextitlan and Parras in New Biscay.

The interior of New Spain, especially a great part of the high table-land of Anahuac, is destitute of vegetation: its arid aspect brings to mind in some places the plains of the two Castiles. Several causes concur to produce this extraordinary effect. The evaporation which takes place on great plains is sensibly increased by the great elevation of the Mexican *cordillera.* On the other hand, the country is not of sufficient elevation for a great number of summits to penetrate the region of perpetual snow. This region commences under the equator at 4800 metres, or 15,747 feet, and under the 45° of lat. at 2550 metres, or 8365 feet, above the level of the sea. In Mexico the eternal snows commence in the 19° and 20° of lat. at 4600 metres, or 15,091 feet, of elevation. Hence, of six colossal mountains which nature has ranged in the same line, between the parallels of 19° and 19°¼, only four, the Pic d'Orizaba, Popocatepetl, Iztaccihuatl, and the Nevado de Toluca, are covered with perpetual snow, while the two others, the Cofre de Perote, and the Volcan de Colima, remain uncovered the greatest part of the year. To the *n.* and *s.* of this parallel of great elevations, beyond this singular zone, in which the new Volcan de Jorullo is also ranged, there are no mountains which exhibit the phenomenon of perpetual snow.]

[These snows, at the period of their minimum, in the month of September, never descend in the parallel of Mexico below 4500 metres, or 14,763 feet. But in the month of January they fall as low as 3700 metres, or 12,138 feet: this is the period of their maximum. The *oscillation* of the limits of perpetual snow is, consequently, under the lat. of 19°, from one season to the other, 800 metres, or 2624 feet; while under the equator it never exceeds 60 or 70 metres, (196 or 229 feet). We must not confound these eternal snows with the snows which in winter accidentally fall in much lower regions. Even this phenomenon, like every other in nature, is subject to immutable laws worthy the investigation of philosophers. This ephemeral snow is never observed under the equator below 3800 or 3900 metres, (12,466 to 12,794 feet); but in Mexico, under the lat. of 18° and 22° it is commonly seen at an elevation of 3000 metres, or 9842 feet. Snow has even been seen in the streets of the capital of Mexico at 2277 metres, or 7468 feet, and 400 metres, or 6156 feet, lower in the city of Valladolid.

In general, in the equinoctial regions of New Spain, the soil, climate, physiognomy of vegetables, all assume the character of the temperate zones. The proximity of Canada, the great breadth of the new continent towards the *n.* the mass of snows with which it is covered, occasion in the Mexican atmosphere frigorifications by no means to be expected in these regions.

If the table-land of New Spain is singularly cold in winter, its temperature is, on the other hand, much higher in summer than what was found by the thermometrical observations of Bouguer and La Condamine in the Andes of Peru. The great mass of the *cordillera* of Mexico, and the immense extent of its plains, produce a reverberation of the solar rays, never observed in mountainous countries of greater inequality. This heat, and other local causes, produce the aridity of these fine regions.

To the *n.* of 20°, from the 22° to the 30° of lat. the rains, which only fall in the months of June, July, August, and September, are very unfrequent in the interior of the country. We have already observed that the great height of this table-land, and the small barometrical pressure of the rarefied air, accelerate the evaporation. The ascending current or column of warm air which rises from the plains prevents the clouds from precipitating in rain to water a land, dry, saline, and destitute of vegetation. The springs are rare in mountains composed principally of porous amygdaloid and fendilated porphyry. The filtrated water, in place of collecting in small subterraneous basins, is lost in the crevices which old volcanic revolutions have opened, and only issues forth at the bottom of the *cordillera*. It forms a great number of rivers on the coast, of which the course is very short, on account of the configuration of the country.

The aridity of the central plain, the want of trees, occasioned, perhaps, in a good measure by the length of time the great valleys have remained covered with water, obstruct very much the working of the mines. These disadvantages have augmented since the arrival of Europeans in Mexico, who have not only destroyed without planting, but in draining great extents of ground have occasioned another more important evil. Muriate of soda and lime, nitrate of potash, and other saline substances, cover the surface of the soil, and spread with a rapidity very difficult to be explained. Through this abundance of salt, and these efflorecences, hostile to cultivation, the table-land of Mexico bears a great resemblance in many places to Thibet and the saline *steppes* of central Asia. In the valley of Tenochtitlan, particularly, the sterility and want of vigorous vegetation have been sensibly augmenting since the Spanish conquest; for this valley was adorned with beautiful verdure when the lake occupied more ground, and the clayey soil was washed by more frequent inundations.

Happily, however, this aridity of soil, of which we have been indicating the principal physical causes, is only to be found in the most elevated plains. A great part of the vast kingdom of New Spain belongs to the most fertile regions of the earth. The declivity of the *cordillera* is exposed to humid winds and frequent fogs; and the vegetation nourished with these aqueous vapours exhibits an uncommon beauty and strength. The humidity of the coasts, assisting the putrefaction of a great mass of organic substances, gives rise to maladies, to which Europeans and others not seasoned to the climate are alone exposed; for under the burning sun of the tropics the unhealthiness of the air almost always indicates extraordinary fertility of soil. Thus at Vera Cruz the quantity of rain in a year amounts to 1^{m}.62, equal to 63.780 inches, while in France it scarcely amounts to 0^{m}.80, or 37.496 inches. Yet with the exception of a few sea-ports and deep valleys, where the natives suffer from intermittent fevers, New Spain ought to be considered as a country remarkably salubrious.

The inhabitants of Mexico are less disturbed by earthquakes and volcanic explosions than the in-]

[habitants of Quito, and the provinces of Guatemala and Cumaná. There are only five burning volcanoes in all New Spain, Orizaba, Popocatepetl, and the mountains of Tustla, Jorullo, and Colima. Earthquakes, however, are by no means rare on the coast of the Pacific ocean, and in the environs of the capital; but they never produce such desolating effects as have been witnessed in the cities of Lima, Riobamba, Guatemala, and Cumaná. On the 14th of September 1759, a horrible catastrophe took place: the volcanoes of Jorullo burst, and were seen surrounded with an innumerable multitude of small smoking cones. Subterraneous noises, so much the more alarming as they were followed by no phenomenon, were heard at Guanaxuato in the month of January 1784. All these phenomena seem to prove, that the country between the parallels of 18° and 22° contains an active internal fire, which pierces, from time to time, through the crust of the globe, even at great distances from the sea shore.

The physical situation of the city of Mexico possesses inestimable advantages, if we consider it in the relation of its communication with the rest of the civilized world. Placed on an isthmus, washed by the S. sea and Atlantic ocean, Mexico appears destined to possess a powerful influence over the political events which agitate the two continents. A king of Spain resident in the capital of Mexico, might transmit his orders in five weeks to the peninsula in Europe, and in six weeks to the Philippine islands in Asia. The vast kingdom of New Spain, under a careful cultivation, would alone produce all that commerce collects together from the rest of the globe, sugar, cochineal, *cacao*, cotton, coffee, wheat, hemp, flax, silk, oils, and wine. It would furnish every metal without even the exception of mercury. Superb timber and an abundance of iron and copper would favour the progress of Mexican navigation; but the state of the coasts and the want of ports from the month of the Rio Alvarado to the month of the Rio Bravo, oppose obstacles in this respect which would be difficult to overcome.

These obstacles, it is true, do not exist on the coast of the Pacific ocean. San Francisco in New California, San Blas in the intendancy of Guadalaxara, near the mouth of the river Santiago, and especially Acapulco, are magnificent ports. The last, probably formed by a violent earthquake, is one of the most admirable basins in the whole world. In the S. sea there is only Coquimbo on the coast of Chile which can be compared with Acapulco; yet in winter, during great hurricanes, the sea becomes very rough in Acapulco. Farther *s.* we find the port of Rialexo, in the kingdom of Guatemala, formed, like Guayaquil, by a large and beautiful river. Sonsonate is very much frequented during the fine season, but it is merely an open road like Tehuantepec, and is consequently very dangerous in winter.

When we examine the *e.* coast of New Spain, we see that it does not possess the same advantages as the *w.* coast. We have already observed, that, properly speaking, it possesses no port; for Vera Cruz, by which an annual commerce of 50 or 60,000,000 of piastres is carried on, is merely a bad anchorage between the shallows of La Caleta, La Gallega, and La Lavandera. The physical cause of this disadvantage is easily discovered. The coast of Mexico, along the Mexican gulf, may be considered as a dike against which the trade winds, and perpetual motion of the waves from *e.* to *w.* throw up the sands which the agitated ocean carries along. This current of rotation runs along S. America from Cumaná to the isthmus of Darien; it ascends towards cape Catoche, and after whirling a long time in the Mexican gulf, issues through the canal of Florida, and flows towards the banks of Newfoundland. The sands heaped up by the vortices of the waters, from the peninsula of Yucatán to the mouths of the Rio del Norte and the Mississippi, insensibly contract the the basin of the Mexican gulf. Geological facts of a very remarkable nature prove this increase of the continent; we see the ocean every where retiring. M. Ferrer found near Sotto la Marina, to the *e.* of the small town of New Santander, 10 leagues in the interior of the country, moving sands filled with sea shells. Humboldt observed the same thing in the environs of Antigua and New Vera Cruz. The rivers which descend from the Sierra Madre and enter the Atlantic ocean have in no small degree contributed to increase the sand banks. It is curious to observe that the *e.* coasts of Old and New Spain are equally disadvantageous for navigation. The coast of New Spain, from the 18° to the 26° of lat. abounds with bars; and vessels which draw more than 32 centimetres, or 12½ inches, of water, cannot pass over any of these bars, without danger of grounding. Yet obstacles like these, so unfavourable for commerce, would at the same time facilitate the defence of the country against the ambitious projects of a European conqueror.

The inhabitants of Mexico, discontented with the port of Vera Cruz, if we may give the name of port to the most dangerous of all anchorages, entertain the hope of finding out surer channels] for the commerce with the mother country. We shall merely name the months of the rivers Alvarado and Guasacualco to the *s.* of Vera Cruz; and to the *n.* of that city the Rio Tampico, and especially the village of Sotto la Marina, near the bar of Santander. These four points have long fixed the attention of the government; but even there, however advantageous in other respects, the sand-banks prevent the entry of large vessels. These ports would require to be artificially corrected; but it becomes necessary in the first place to inquire if the localities are such as to warrant a belief that this expensive remedy would be durable in its effects. It is to be observed, however, that we still know too little of the coasts of New Santander and Texas, particularly that part to the *n.* of the lake of S. Bernard or Carbonera, to be able to assert that in the whole of this extent nature presents the same obstacles and the same bars. Two Spanish officers of distinguished zeal and astronomical knowledge, MM. Cevallos and Herrera, have engaged in this interesting and useful investigation. At present Mexico is in a military dependeuce on the Havannah, which is the only neighbouring port capable of receiving squadrons, and the most important point for the defence of the *e.* coast of New Spain. Accordingly, the government, since the last taking of the Havannah by the English, has been at enormous expences in increasing the fortifications of the place. Sensible of its true interests, the court of Madrid has wisely laid it down as a principle, that the dominion of the island of Cuba is essential for the preservation of New Spain.

A very serious inconvenience is common to the *e.* coast, and to the coast washed by the Great ocean, falsely called the Pacific ocean. They are rendered inaccessible for several months by violent tempests, which effectually prevent all navigation. The *n.* winds (*los nortes*), which are *n. w.* winds, blow in the gulf of Mexico from the autumnal to the spring equinox. These winds are generally moderate in the months of September and October; their greatest fury is in the month of March; and they sometimes last to April. Those navigators who have long frequented the port of Vera Cruz know the symptoms of the coming tempest as a physician knows the symptoms of an acute malady. According to the excellent observations of M. Orta, a great change in the barometer, and a sudden interruption in the regular recurrence of the horary variations of that instrument, are the sure forerunners of the tempest. It is accompanied by the following phenomena. At first a small land wind (*terral*) blows from the *w. n. w.*; and to this *terral* succeeds a breeze, first from the *n. e.* and then from the *s.* During all this time a most suffocating heat prevails; and the water dissolved in the air is precipitated on the brick walls, the pavement, and iron or wooden balustrades. The summits of the Pic d'Orizaba and the Cofre de Perote, and the mountains of Villa Rica, particularly the Sierra de San Martin, which extends from Tustla to Guasacualco, appear uncovered with clouds, while their bases are concealed under a veil of demi-transparent vapours. These *cordilleras* appear projected on a fine azure ground. In this state of the atmosphere the tempest commences, and sometimes with such impetuosity, that before the lapse of a quarter of an hour it would be dangerons to remain on the mole in the port of Vera Cruz. All communication between the city and the castle of S. Juan d'Ulua is thenceforth interrupted. These *n.* wind hurricanes generally remain for three or four days, and sometimes for 10 or 12. If the *n.* wind change into a *s.* breeze the latter is very inconstant, and it is then probable that the tempest will recommence; but if the *n.* veers to *e.* by the *n. e.* then the breeze or fine weather is durable. During winter we may reckon on the breeze continuing for three or four successive days, an interval more than sufficient for allowing any vessel leaving Vera Cruz to get out to sea and escape the sand-banks adjoining to the coast. Sometimes even in the months of May, June, July, and August, very strong hurricanes are felt in the gulf of Mexico. They are called *nortes de hueso colorado;* but fortunately they are not very common. The periods in which the black vomiting (yellow fever) and tempests from the *n.* prevail at Vera Cruz do not coincide; consequently the European who arrives in Mexico, and the Mexican whose affairs compel him to embark, or to descend from the table-land of New Spain to the coast, have both to make their election between the danger of navigation and a mortal disease.

The *w.* coast of Mexico is of very dangerous navigation during the months of July and August, when terrible hurricanes blow from the *s. w.* At that time, and even in September and October, the ports of San Blas and Acapulco are of very difficult access. Even in the fine season, from the month of October to the month of May (*verano de la mar del sur*), the tranquillity of the Pacific ocean is interrupted on this coast by impetuous winds from the *n. e.* and the *n. n. e.* known by the names of *papagallo* and *tehuantepec.*

In illustration of this phenomenon, we might be led to believe that the equilibrium of the atmo-

[sphere being disturbed in the months of January and February on the coast of the Atlantic, the agitated air flows back with impetuosity towards the Great ocean. The *tehuantepec* should seem therefore to be merely the effect, or rather the continuation, of the *n.* wind of the Mexican gulf and the *brisottes* of St. Martha. It renders the coast of Solinas and La Ventosa almost as inaccessible as that of Nicaragua and Guatemala, where violent *s. w.* winds prevail during the months of August and September, known by the name of *tapayaguas*.

These *s. w.* winds are accompanied with thunder and excessive rains, while the *tehuantepec* and *papagallos*, which blow particularly from cape Blane de Nicoya (lat. 9° 30′) to L'Ensenada de S. Catharina (lat. 10° 45′), exert their violence during a clear and azure sky. Thus at certain periods almost all the coasts of New Spain are dangerons for navigators.

Chap. VI.

Particular statistical account of the intendancies of Nueva España.

Before giving the table which contains a particular statistical account of the intendancies of New Spain, we shall discuss the principles on which the new territorial divisions are founded. These divisions have been, till lately, entirely unknown to the most modern geographers; and it was M. Humboldt who first afforded a general map of New Spain, in which were contained the limits of the intendancies established since 1776.

Before the introduction of the new administration by Count Don Jose de Galvez, minister of the Indies, New Spain contained, 1. El Reyno de Mexico; 2. El Reyno de Nueva Galicia; 3. El Nuevo Reyno de Leon; 4. La Colonia del Nuevo Santander; 5. La Provincia de Texas; 6. La Provincia de Cohahuila; 7. La Provincia de Nueva Biscaya; 8. La Provincia de la Sonora; 9. La Provincia de Nuevo Mexico; and, 10. Ambas Californias, or Las Provincias de la Vieja y Nueva California. These old divisions are still very frequently used in the country. The limits which separate La Nueva Galicia from El Reyno de Mexico, to which a part of the old kingdom of Mechoacán belongs, are also the line of demarcation between the jurisdiction of the two audiences of Mexico and Guadalaxara. This line begins on the coast of the gulf of Mexico, 10 leagues to the *n.* of the Rio de Panuco and the city of Altamira near Bara Ciega, and runs through the intendancy of S. Luis Potosi to the mines of Potosi and Bernalejo; from thence passing along the *s.* extremity of the intendancy of Zacatecas, and the *w.* limits of the intendancy of Guanaxuato, it traverses the intendancy of Guadalaxara between Zapotlan and Sayula, between Ayotitlan and the Ciudad de la Purificacion, to Guatlan, one of the ports of the S. sea. All *n.* of this line belongs to the *audiencia* of Guadalaxara; and all *s.* of it to the *audiencia* of Mexico.

In its present state New Spain is divided into 12 intendancies, to which we must add three other districts, very remote from the capital, which have preserved the simple denomination of provinces. These fifteen divisions are,

I. Under the Temperate Zone, 82,000 leagues, with 677,000 souls, or eight inhabitants to the square league.

A. *Region of the North*, an interior region.

1. Provincia de Nuevo Mexico, along the Rio del Norte to the *n.* of the parallel of 31°.
2. Intendencia de Nueva Biscaya, to the *s. w.* of the Rio del Norte, on the central table-land which declines rapidly from Durango towards Chihuahua.

B. *Region of the North-west*, in the vicinity of the Great ocean.

3. Provincia de la Nueva California, or *n. w.* coast of N. America, possessed by the Spaniards.
4. Provincia de la Antigua California. Its *s.* extremity ends the torrid zone.
5. Intendencia de la Sonora. The most *s.* part of Cinaloa, in which the celebrated mines of Copala and Rosario are situated, also passes the tropic of cancer.

C. *Region of the North-east*, adjoining the gulf of Mexico.

6. Intendencia de San Luis Potosi. It comprehends the provinces of Texas, La Colonia de Nuevo Santander and Cohahuila, El Nuevo Reyno de Leon, and the districts of Charcas, Altamira, Catorce, and Ramos. These last districts compose the intendancy of San Luis properly so called. The *s.* part, which extends to the *s.* of the Barra de Santander and the *real* de Catorce, belongs to the torrid zone.

II. Under the Torrid Zone, 36,500 square leagues, with 5,160,000 souls, or 141 inhabitants to the square league.

D. *Central Region.*

7. Intendencia de Zacatecas, excepting the part which extends to the *n.* of the mines of Fresnillo.
8. Intendencia de Guadalaxara.
9. Intendencia de Guanaxuato.]

[10. Intendencia de Valladolid.
11. Intendencia de Mexico.
12. Intendencia de la Puebla.
13. Intendencia de Vera Cruz.
E. *Region of the South-west.*
14. Intendencia de Oaxaca.
15. Intendencia de Merida.

The divisions in this table are founded on the physical state of the country. We see that nearly seven-eighths of the inhabitants live under the torrid zone. The population becomes thinner as we advance towards Durango and Chihuahua. In this respect New Spain bears a striking analogy to Hindostan, which in its *n.* parts is bounded by regions almost uncultivated and uninhabited. Of 5,000,000 who inhabit the equinoctial part of Mexico, four-fifths live on the ridge of the *cordillera* or table-lands, whose elevation above the level of the sea equals that of the passage of mount Cenis.

New Spain, considering its provinces according to their commercial relations, or the situation of the coasts, is divided into three regions.

I. Provinces of the Interior, which do not extend to the ocean.
1. Nuevo Mexico.
2. Nueva Biscaya.
3. Zacatecas.
4. Guanaxuato.

II. Maritime Provinces of the *e.* coast opposite to Europe.
5. San Luis Potosi.
6. Vera Cruz.
7. Merida, or Yucatán.

III. Maritime Provinces of the *w.* coast opposite to Asia.
8. New California.
9. Old California.
10. Sonora.
11. Guadalaxara.
12. Valladolid.
13. Mexico.
14. Puebla.
15. Oaxaca.

These divisions may, as Humboldt observes, one day possess great political interest, when the cultivation of Mexico shall be less concentrated on the central table-land or ridge of the *cordillera*, and when the coasts shall become more populous. The maritime provinces of the *w.* will send their vessels to Nootka, to China, and the E. Indies. The Sandwich islands, inhabited by a ferocious, but industrious and enterprising people, appear more likely destined to receive Mexican than European colonists. They afford an important stage to the nations who carry on commerce in the Great ocean. The inhabitants of New Spain and Peru have never yet been able to profit by their advantageous position on a coast opposite Asia and New Holland. They do not even know the productions of the S. sea islands. What efforts have not been made by the United States of North America, within the last 10 years, to open a communication with the *w.* coast, with the same coast on which the Mexicans possess the finest ports, but without activity and without commerce!

According to the ancient division of the country, the Reyno de Nueva Galicia contained more than 14,000 square leagues, and nearly a million of inhabitants: it included the intendancies of Zacatecas and Guadalaxara, (with the exception of the most *s.* part, which contains the volcano of Colima and the village of Ayotitau,) as well as a small part of that of San Luis Potosi. The regions now known by the denomination of the seven intendancies of Guanaxuato, Valladolid or Mechoacán, Mexico, Puebla, Vera Cruz, Oaxaca, and Merida, formed along with a small portion of the intendancy of San Luis Potosi, (the most *s.* part through which the river of Panuco runs), the Reyno de Mexico, properly so called. This kingdom consequently contained more than 27,000 square leagues, and nearly 4,500,000 of inhabitants.

Another division of New Spain, equally ancient and less vague, is that which distinguishes New Spain, properly so called, from the *provincias internas.* To the latter belongs all to the *n.* and *n. w.* of the kingdom of Nueva Galicia, with the exception of the two Californias; consequently, 1. The small kingdom of Leon; 2. The colony of New Santander; 3. Texas; 4. New Biscay; 5. Sonora; 6. Cohahuila; and, 7. New Mexico. The *provincias internas del Vireynato*, which contain 7814 square leagues, are distinguished from the *provincias internas de la comandancia* (of Chihuahua), erected into a *capitania general* in 1779, which contain 59,375 square leagues. Of the twelve new intendancies, three are situated in the *provincias internas*, Durango, Sonora, and San Luis Potosi. We must not, however, forget that the intendant of San Luis is only under the direct authority of the viceroy for Leon, Santander, and the districts near his residence, those of Charcas, Catorce, and Altamira. The governments of Cohahuila and Texas make also part of the intendancy of San Luis Potosi, but they belong directly to the comandancia general de Chihuahua. The following tables will throw some light on these very] [complicated territorial divisions. Let us divide all New Spain into,

A. *Provincias sujetas al Virey de Nueva España*, 59,103 square leagues, with 5,479,095 souls: the 10 intendancies of Mexico, Puebla, Vera Cruz, Oaxaca, Merida, Valladolid, Guadalaxara, Zacatecas, Guanaxuato, and San Luis Potosi (without including Cohahuila and Texas.)
The two Californias.

B. *Provincias sujetas al comandante general de provincias internas*, 59,375 square leagues, with 359,200 inhabitants.
The two intendancies of Durango and Sonora.
The province of Nuevo Mexico.
Cohahuila and Texas.

The whole of New Spain, 118,478 square leagues, with 5,837,100 inhabitants.

These tables exhibit the surface of the provinces, calculated in square leagues of 25 to the degree, according to the general map accompanying Humboldt's work. The first calculations were made at Mexico in the end of 1803, by M. Oteyza and Humboldt. His geographical labours having since that period attained to greater perfection, M. Oltmanns took the pains to recalculate the whole territorial surfaces. He executed this operation with the precision which characterises whatever he undertakes, having formed squares of which the sides did not contain more than three minutes.

The population indicated in the following tables is what may be supposed to have existed in 1803. In all times the population of Asia has been exaggerated, and that of the Spanish possessions in America lowered. We forget that with a fine climate and fertile soil, population makes rapid advances even in countries the worst administered; and we also forget that men scattered over an immense territory suffer less from the imperfections of the social state than when the population is very concentrated.

We are uncertain as to the limits which ought to be assigned to New Spain to the *n.* and *e.* It is not enough that a country has been run over by a missionary monk, or that a coast has been seen by a vessel of war, to consider it as belonging to the Spanish colonies of America. Cardinal Lorenzana printed at Mexico, even in 1770, that New Spain, through the bishopric of Durango, bordered perhaps on Tartary and Greenland! We are now too well instructed in geography to yield ourselves up to such vague suppositions. A viceroy of Mexico caused the American colonies of the Russians on the peninsula of Alaska to be visited from San Blas. The attention of the Mexican government was for a long time turned to the *n. w.* coast, especially since the establishment at Nootka, which the court of Madrid was compelled to abandon to avoid a war with England. The inhabitants of the United States carry their civilization towards the Missoury. They gradually approach the coast of the Great ocean, to which the fur trade invites them. The period approaches when, through the rapid progress of human cultivation, the boundaries of New Spain will join those of the Russian empire, and the great confederation of American republics. At present, however, the Mexican government extends no farther along the *w.* coast than the mission of St. Francis, to the *s.* of cape Mendocin, and the village of Taos in New Mexico. The boundaries of the intendancy of San Luis Potosi on the *e.* towards the state of Louisiana, are not very well determined; the congress of Washington endeavoured to confine them to the right bank of the Rio Bravo del Norte, while the Spaniards comprehend under the denomination of province of Texas, the savanas which extend to the Rio Mexicano or Mermentas, to the *e.* of the Rio Sabina.

The following table exhibits the surface and population of the greatest political associations of Europe and Asia. It will furnish curious comparisons with the present state of Mexico.]

GREAT

Great political Associations in 1808.	Square leag. of 25 to the degree.	Total population.	Inhab. to the square league.
Russian empire - - - - - -	942,452	40,000,000	42
1. European part - - - - - -	215,809	36,400,000	169
2. Asiatic part - - - - - -	726,644	3,597,000	5
The single government of Irkutzk - - -	350,000	680,600	2
The single government of Tobolsk - - -	200,000	72,547	1
All Europe - - - - - - - -	476,111	182,599,000	383
The united States of North America, viz.			
1. With Louisiana - - - - -	260,340	6,800,000	22
2. Without Louisiana - - - - - -	117,478	6,715,000	43
3. Without Louisiana and the Indian territory (in Georgia and Western Waters) - - - -	78,120	6,655,000	85
Hindostan on this side (*en-deça*) the Ganges* - -	162,827		
English territory, of which the East India company possess the sovereignty - - - - - -	48,299	23,806,000	493
Allies and tributaries of the English company - -	32,647	16,900,000	518
Turkish empire in Europe, Asia, and Africa - - -	136,110	25,330,000	186
Austrian monarchy - - - - - - -	33,258	25,588,000	769
France, according to M. Peuchet - - - - -	32,000	35,000,000	1094
Spain, according to M. Laborde - - - -	25,147	10,409,000	413
New Spain,			
1. With the provincias internas - - - -	118,378	5,837,100	49
2. Without the provincias internas - - -	51,289	5,413,900	105

* According to Arrowsmith's beautiful map of India, 1804. (Journal Astronomique de MM. Zach et Lindenau, 1807, p. 361.) The rest of the date from the classical work of M. Hassel, Statistical View of the States of Europe, No. I. (1805,) in German.

We see from this table, which may suggest very curious considerations as to the disproportion of European cultivation, that New Spain is almost four times larger than the French empire, with a population which till this day is seven times smaller. We also see that the points of analogy in a comparison of the United States with Mexico are very striking, especially if we consider Louisiana and the *w.* territory as the *provincias internas* of the great confederation of American republics.

The state of the *provincias internas* are described as it was when Humboldt left Mexico. A considerable change has since taken place in the military government of these vast provinces, of which the surface almost doubles that of the French empire. In 1807, two *commandantes generales*, brigadier generals, by name Don Nemesio Salcedo and Don Pedro Grimarest, governed these *n.* provinces. The following is the present division of the *gobierno militar*, which is now no longer in the hands of the governor of Chihuahua alone:

Provincias Internas Del Reyno De Nueva Espana.

A. *Provincias internas occidentales.*

1. Sonora.
2. Durango o Nueva Biscaya.
3. Nuevo Mexico.
4. Californias.

B. *Provincias internas orientales.*

1. Cohahuila.
2. Texas.
3. Colonia del Nuevo Santander.
4. Nuevo Reyno de Leon.

The new *commandantes generales* of the internal provinces, as well as the old, are considered as at the head of the administration of finances in the two intendancies of Sonora and Durango, in the province of Nuevo Mexico, and in that part of the intendancy of San Luis Potosi which comprehends Texas and Cohahuila. As to the small kingdom of Leon and New Santander, they are only subject to the commandant in a military point of view.]

Territorial Divisions.	Surface in sq. leagues of 25 to the degree.	Population reduced to the epocha of 1803.	No. of inhab. to the sq. league.
New Spain, (extent of the whole viceroyalty without including the kingdom of Guatemala.)	118,478	5,837,100	49
A. Provincias internas - - - - - -	67,189	423,200	6
a. Immediately subject to the viceroy, (provincias internas del Vireynato) - - - - - -	7,814	64,000	8
1. Nuevo Reyno de Leon - - - - -	2,621	29,000	10
2. Nuevo Santander - - - - - -	5,193	38,000	7
b. Subject to the governnor of Chihuahua (provincias internas de la comandancia general) - - - - -	59,375	359,200	6
1. Intendencia de la Nueva Biseaya o Durango - -	16,873	159.700	10
2. Intendencia de la Sonora - - - - -	19,143	121,400	6
3. Cohahuila - - - - - - -	6,702	16,900	2
4. Texas - - - - - - -	10,948	21,000	2
5. Nuevo Mexico - - - - - -	5,709	40,200	7
B. New Spain, properly so called, immediately subject to the viceroy, comprehending los Reynos de Mexico, Mechoacán y Nneva Galicia, and the two Californias	51,289	5,413,900	105
1. Intendencia de Mexico - - - - -	5,927	1,511,900	255
2. Intendencia de Puebla - - - - -	2,696	813,300	301
3. Intendencia de Vera Cruz - - - - -	4,141	156,000	38
4. Intendencia de Oaxaca - - - - -	4,447	534,800	120
5. Intendencia de Merida, or Yucatán - - -	5,977	465,800	81
6. Intendencia de Valladolid - - - - -	3,446	476,400	273
7. Intendencia de Guadalaxara - - - - -	9,612	630,500	66
8. Intendencia de Zacatecas - - - - -	2,355	153,300	65
9. Intendencia de Guanaxuato - - - - -	911	517,300	568
10. Intendencia de San Luis Potosi, (without including New Santander, Texas, Cohahuila, and the kingdom of Leon) - - - - - - -	2,357	230,000	98
11. Old California, (Antigua California) - - -	7,295	9,000	1
12. New California, (Nueva California) - - -	2,125	15,600	7

This statistical table proves the imperfection of the territorial division. It appears that in confiding to intendants the administration of police and finances, the object was to divide the Mexican soil on principles analogous to those followed by the French government on the division of the kingdom into generalities. In New Spain every intendancy comprehends several sub-delegations. In the same manner the generalities in France were governed by sub-delegates, who exercised their functions under the orders of the intendant. But in the formation of the Mexican intendancies, little regard has been paid to the extent of territory or the greater or less degree of concentration of the population. This new division indeed took place at a time when the ministers of the colonies, the council of the Indies, and the viceroys, were unfurnished with the necessary materials for so important an undertaking. How is it possible to possess the detail of the administration of a country of which there has never been any map, and regarding which the most simple calculations of political arithmetic have never been attempted?

Comparing the extent of surface of the Mexican intendancies, we find several of them 10, 20, even 30 times larger than others. The intendancy of San Luis Potosi, for example, is more extensive than all European Spain, while the intendancy of Guanaxuato does not exceed in size two or three of the departments of France. The following is an exact table of the extraordinary disproportion among the several Mexican intendancies in their territorial extent; we have arranged them in the order of their extent:]

[Intendancy of	San Luis Potosi,	27,821 sq. leag.
	Sonora, - -	19,143
	Durango, -	16,873
	Guadalaxara,	9,612
	Merida, - -	5,977
	Mexico, - -	5,927
	Oaxaca, - -	4,447
	Vera Cruz, -	4,141
	Valladolid, -	3,447
	Puebla, - -	2,696
	Zacatecas, -	2,355
	Guanaxuato,	911

With the exception of the three intendancies of San Luis Potosi, Sonora, and Durango, of which each ocenpies more ground than the whole empire of Great Britain, the other intendancies contain a mean surface of 3 or 4000 square leagues. We may compare them for extent to the kingdom of Naples, or that of Bohemia. We can conceive that the less populous a country is, the less its administration requires small divisions. In France no department exceeds the extent of 550 square leagues: the mean extent of the departments is 300. But in European Russia and Mexico the governments and intendancies are 10 times more extensive.

In France, the heads of departments, the prefeets, watch over the wants of a population which rarely exceeds 450,000 souls, and which on an average we may estimate at 300,000. The governments into which the Russian empire is divided, as well as the Mexican intendancies, comprehend, notwithstanding their very different states of civilization, a greater number of inhabitants. The following table will show the disproportion of population among the territorial divisions of New Spain. It begins with the most populous intendancy, and ends with the one most thinly inhabited.

Intendancy of	Mexico, -	1,511,800 inhab.
	Puebla, -	813,300
	Guadalaxara, -	630,500
	Oaxaca, -	534,800
	Guanaxuato,	517,300
	Valladolid, -	476,400
	Merida, -	465,700
	San Luis Potosi,	331,900
	Durango, -	159,700
	Vera Cruz -	156,000
	Zacatecas, -	153,000
	Sonora, -	121,400

It is in comparing together the tables of the population of the 12 intendancies, and the extent of their surface, that we are particularly struck with the inequality of the distribution of the Mexican population, even in the most civilized part of the kingdom. The intendancy of Puebla, which in the second table occupies one of the first places, is almost at the end of the first table. Yet no principle ought more to guide those who chalk out territorial divisions than the proportion of the population to the extent expressed in square leagues or myriametres. A third table exhibits the state of the population, which may be called *relative*. To arrive at numerical results which indicate the proportion between the number of inhabitants and extent of inhabited soil, we must divide the absolute population by the territory of the intendancies. The following are the results of this operation:

Intendancy of	Guanaxuato, -	568 inhab. to the sq. leag.
	Puebla, - -	301
	Valladolid, -	273
	Mexico, - -	255
	Oaxaca, - -	120
	Merida, - -	81
	Guadalaxara,	66
	Zacatecas, -	65
	Vera Cruz, -	38
	San Luis Potosi,	12
	Durango, - -	10
	Sonora, - -	6

This last table proves that in the intendancies where the cultivation of the soil has made least progress, the *relative* population is from 50 to 90 times less than the old civilized regions adjacent to the capital. This extraordinary difference in the distribution of the population is also to be found in the *n.* and *n. e.* of Europe. In Lapland we scarcely find one inhabitant to the square league, while in other parts of Sweden, in Gothland, for example, there are more than 248. In the states subject to the king of Denmark, the island of Zealand contains 944, and Iceland 11 inhabitants, to the square league. In European Russia, the governments of Archangel, Olonez, Kalonga, and Moscow, differ so much in their relative population to the extent of the territory, that the two former of these governments contain six and 26, and the two last 812 and 974 souls to the square league. These enormous differences indicate that one province is 160 times better inhabited than another.

In France, where the whole of the population gives 1094 inhabitants to the square league, the best peopled departments, those of L'Escaut, Le Nord, and La Lys, afford a relative population of 3869, 2786, and 2274. The worst peopled department, that of the Hautes-Alpes, composed of a part of old Dauphiny, contains only 471 inhabitants to the square league. Hence the extremes are in]

[France in the relation of 8 : 1; so that the intendancy of Mexico in which the population is the most concentrated, that of Guanaxuato, is scarcely so well inhabited as the worst peopled department of continental France.

The three tables which have been given of the extent, absolute population, and relative population of the intendancies of New Spain, will sufficiently prove the great imperfection of the present territorial division. A country in which the population is dispersed over a vast extent requires that the provincial administration be restricted to smaller portions of ground than those of the Mexican intendancies. Whenever a population is under 100 inhabitants to the square league, the administration of an intendancy or a department should not extend over more than 100,000 inhabitants. We may assign a double or triple number to regions in which the population is more concentrated.

It is on this concentration that the degree of industry, the activity of commerce, and the number of affairs consequently demanding the attention of government, undoubtedly depend. In this point of view the small intendancy of Guanaxuato gives more occupation to an administrator than the provinces of Texas, Cohahuila, and New Mexico, which are six times more extensive. But, on the other hand, how is it possible for an intendant of San Luis Potosi ever to know the wants of a province of 28,000 square leagues in extent? How can he, even while he devotes himself with the most patriotic zeal to the duties of his place, superintend the sub-delegates, and protect the Indian from the oppressions which are exercised in the villages?

This point of administrative organization cannot be too carefully discussed. A reforming government ought, before every other object, to set about changing the present limits of the intendancies. This political change ought to be founded on the exact knowledge of the physical state, and the state of cultivation of the provinces which constitute the kingdom of Nueva España.

Chap. VII.

Minuter details of the state of the agriculture of Nueva España, and of its metallic mines, viz. Of the vegetable productions of the Mexican territory.—Progress of the cultivation of the soil.—Influence of the mines on cultivation.—Plants which contribute to the nourishment of man.

We have run over the immense extent of territory comprehended under the denomination of Nueva España. We have rapidly described the limits of each province, the physical aspect of the country, its temperature, its natural fertility, and the progress of a nascent population. It is now time to enter more minutely into the state of agriculture and territorial wealth of Mexico.

An empire extending from lat. 16° to 37° affords us, from its geometrical position, all the modifications of climate to be found on transporting ourselves from the banks of the Senegal to Spain, or from the Malabar coast to the *steppes* of the Great Bucharia. The variety of climate is also augmented by the geological constitution of the country, by the mass and extraordinary form of the Mexican mountains. On the ridge and declivity of the *cordilleras* the temperature of each table-land varies as it is more or less elevated; not merely insulated peaks, of which the summits approach the region of perpetual snow, are covered with oaks and pines, but whole provinces spontaneously produce alpine plants; and the cultivator inhabiting the torrid zone frequently loses the hopes of his harvest from the effects of frost or the abundance of snow.

Such is the admirable distribution of heat on the globe, that in the aerial ocean we meet with colder strata in proportion as we ascend, while in the depth of the sea the temperature diminishes as we leave the surface of the water. In the two elements the same latitude unites, as it were, every climate. At unequal distances from the surface of the ocean, but in the same vertical plane, we find strata of air and strata of water of the same temperature. Hence, under the tropics, on the declivity of the *cordilleras*, and in the abyss of the ocean, the plants of Lapland, as well as the marine animals in the vicinity of the pole, find the degree of heat necessary to their organic development.

From this order of things, established by nature, we may conceive that, in a mountainous and extensive country like Mexico, the variety of indigenous productions must be immense, and that there hardly exists a plant in the rest of the globe which is not capable of being cultivated in some part of Nueva España. Notwithstanding the laborious researches of three distinguished botanists, MM. Sesse, Mociño, and Cervantes, employed by the court in examining the vegetable riches of Mexico, we are far from yet being able to flatter ourselves that we know any thing like all the plants scattered over the insulated summits, or crowded together in the vast forests at the foot of the *cordilleras*. If we still daily discover new herbaceous species on the central table-land, and even] [in the vicinity of the city of Mexico, how many arborescent plants have never yet been discovered by botanists in the humid and warm region along the *e.* coast, from the province of Tobasco, and the fertile banks of the Guasacualco, to Colipa and Papantla, and along the *w.* coast from the port of San Blas and Sonora to the plains of the province of Oaxaca? Hitherto no species of quinquina (*cinchona*), none even of the small group, of which the stamina are longer than the corolla, which form the genus exostema, has been discovered in the equinoctial part of Nueva España. It is probable, however, that this precious discovery will one day be made on the declivity of the *cordilleras*, where arborescent ferns abound, and where the region of the true febrifuge quinquina with very short stamina and downy corollæ commences.

We do not propose here to describe the innumerable variety of vegetables with which nature has enriched the vast extent of Nueva España, and of which the useful properties will become better known when civilization shall have made farther progress in the country. We mean merely to speak of the different kinds of cultivation which an enlightened government might introduce with success; and we shall confine ourselves to an examination of the indigenous productions which at this moment furnish objects of exportation, and which form the principal basis of the Mexican agriculture.

Under the tropics, especially in the W. Indies, which have become the centre of the commercial activity of the Europeans, the word agriculture is understood in a very different sense from what it receives in Europe. When we hear at Jamaica or Cuba of the flourishing state of agriculture, this expression does not offer to the imagination the idea of harvests which serve for the nourishment of man, but of ground which produces objects of commercial exchange, and rude materials for manufacturing industry. Moreover, whatever be the riches or fertility of the country, in the valley De los Gnines, for example, to the *s. e.* of the Havanah, one of the most delicious situations of the New World, we see only plains carefully planted with sugar-cane and coffee; and these plains are watered with the sweat of African slaves! Rural life loses its charms when it is inseparable from the aspect of the sufferings of our species.

But in the interior of Mexico, the word agriculture suggests ideas of a less afflicting nature. The Indian cultivator is poor, but he is free. His state is even greatly preferable to that of the peasantry in a great part of the *n.* of Europe. There are neither corvées nor villanage in Nueva España; and the number of slaves is extremely small. Sugar is chiefly the produce of free hands. There the principal objects of agriculture are not the productions to which European luxury has assigned a variable and arbitrary value, but cereal gramina, nutritive roots, and the agave, the vine of the Indians. The appearance of the country proclaims to the traveller that the soil nourishes him who cultivates it, and that the true prosperity of the Mexican people neither depends on the accidents of foreign commerce, nor on the unruly politics of Europe.

Those who only know the interior of the Spanish colonies from the vague and uncertain notions hitherto published, will have some difficulty in believing that the principal sources of the Mexican riches are by no means the mines, but an agriculture which has been gradually ameliorating since the end of the last century. Without reflecting on the immense extent of the country, and especially the great number of provinces which appear totally destitute of precious metals, we generally imagine that all the activity of the Mexican population is directed to the working of mines. Because agriculture has made a very considerable progress in the *capitania-general* of Caracas, in the kingdom of Guatemala, the island of Cuba, and wherever the mountains are accounted poor in mineral productions, it has been inferred that it is to the working of the mines that we are to attribute the small care bestowed on the cultivation of the soil in other parts of the Spanish colonies. This reasoning is just when applied to small portions of territory. No doubt, in the provinces of Choco and Antioquia, and the coast of Barbacoas, the inhabitants are fonder of seeking for the gold washed down into the brooks and ravines than of cultivating a virgin and fertile soil; and in the beginning of the conquest, the Spaniards who abandoned the peninsula or Canary islands to settle in Peru and Mexico, had no other view but the discovery of the precious metals. "*Auri rabida sitis à cultura Hispanos divertit*," says a writer of those times, Pedro Martyr, in his work on the discovery of Yucatán and the colonization of the Antilles. But this reasoning cannot now explain why in countries of three or four times the extent of France agriculture is in a state of languor. The same physical and moral causes which fetter the progress of national industry in the Spanish colonies have been inimical to a better cultivation of the soil. It cannot be doubted that under improved social institutions the countries which most abound with mineral productions will be as well, it] [not better, cultivated than those in which no such productions are to be found. But the desire natural to man of simplifying the causes of every thing has introduced into works of political economy a species of reasoning, which is perpetuated, because it flatters the mental indolence of the multitude. The depopulation of Spanish America, the state of neglect in which the most fertile lands are found, and the want of manufacturing industry, are attributed to the metallic wealth, to the abundance of gold and silver; as, according to the same logic, all the evils of Spain are to be attributed to the discovery of America, or the wandering race of the Merinos, or the religious intolerance of the clergy.

We do not observe that agriculture is more neglected in Peru than in the province of Cumana or Guayana, in which, however, there are no mines worked. In Mexico the best cultivated fields, those which recal to the mind of the traveller the beautiful plains of France, are those which extend from Salamanca towards Silao, Guanaxuato, and the Villa de Leon, and which surround the richest mines of the known world. Wherever metallic seams have been discovered in the most uncultivated parts of the *cordilleras*, on the insulated and desert table-lands, the working of mines, far from impeding the cultivation of the soil, has been singularly favourable to it. Travelling along the ridge of the Andes, or the mountainous part of Mexico, we every where see the most striking examples of the beneficial influence of the mines on agriculture. Were it not for the establishments formed for the working of the mines, how many places would have remained desert? how many districts uncultivated in the four intendancies of Guanaxuato, Zacatecas, San Luis Potosi, and Durango, between the parallels of 21° and 25°, where the most considerable metallic wealth of Nueva España is to be found? If the town is placed on the arid side or the crest of the *cordilleras*, the new colonists can only draw from a distance the means of their subsistence, and the maintenance of the great number of cattle employed in drawing-off the water, and raising and amalgamating the mineral produce. Want soon awakens industry. The soil begins to be cultivated in the ravines and declivities of the neighbouring mountains, wherever the rock is covered with earth. Farms are established in the neighbourhood of the mine. The high price of provision, from the competition of the purchasers, indemnifies the cultivator for the privations to which he is exposed from the hard life of the mountains. Thus from the hope of gain alone, and the motives of mutual interest, which are the most powerful bonds of society, and without any interference on the part of government in colonization, a mine, which at first appeared insulated in the midst of wild and desert mountains, becomes in a short time connected with the lands which have been long under cultivation.

Moreover, this influence of the mines on the progressive cultivation of the country is more durable than they are themselves. When the seams are exhausted, and the subterraneous operations are abandoned, the population of the canton undoubtedly diminishes, because the miners emigrate elsewhere; but the colonist is retained by his attachment for the spot where he received his birth, and which his fathers cultivated with their hands. The more lonely the cottage is, the more it has charms for the inhabitant of the mountains. It is with the beginning of civilization as with its decline: man appears to repent of the constraint which he has imposed on himself by entering into society; and he loves solitude because it restores to him his former freedom. This moral tendency, this desire for solitude, is particularly manifested by the copper-coloured indigenous, whom a long and sad experience has disgusted with social life, and more especially with the neighbourhood of the whites. Like the Arcadians, the Aztec people love to inhabit the summits and brows of the steepest mountains. This peculiar trait in their disposition contributes very much to extend population in the mountainous regions of Mexico. What a pleasure it is for the traveller to follow these peaceful conquests of agriculture, and to contemplate the numerous Indian cottages dispersed in the wildest ravines and necks of cultivated ground advancing into a desert country between naked and arid rocks!

The plants cultivated in these elevated and solitary regions differ essentially from those cultivated on the plains below, on the declivity and at the foot of the *cordilleras*. The height requisite for the different kinds of cultivation depends, in general, on the latitude of the places; but such is the flexibility of organization in cultivated plants, that with the assistance of the care of man they frequently break through the limits assigned to them by the naturalist.

Under the equator, the meteorological phenomena, such as those of the geography of plants and animals, are subject to laws which are immutable and easily to be perceived. The climate there is only modified by the height of the place, and the temperature is nearly constant, notwithstanding the difference of seasons. As we leave] the equator, especially between the 15th degree and the tropic, the climate depends on a great number of local circumstances, and varies at the same absolute height, and under the same geographical latitude. This influence of localities, of which the study is of such importance to the cultivator, is still much more manifest in the *n.* than the *s.* hemisphere. The great breadth of the new continent, the proximity of Canada, the winds which blow from the *n.* and other causes already developed, give the equinoctial region of Mexico and the island of Cuba a particular character. One would say that in these regions the temperate zone, the zone of variable climates, increases towards the *s.* and passes the tropic of Cancer. It is sufficient here to state that in the environs of the Havannah (lat. 23° 8′) the thermometer has been seen to descend to the freezing point at the small elevation of 80 metres, or 262 feet, above the level of the ocean, and that snow has fallen near Valladolid (lat. 19° 42′), at an absolute elevation of 1900 metres, or 6232 feet, while under the equator this last phenomenon is only observable at the double of the elevation.

These considerations prove to us that towards the tropic, where the torrid zone approaches the temperate zone, the plants under cultivation are not subject to fixed and invariable heights. We might be led to distribute them according to the mean temperature of the places in which they vegetate. We observe, in fact, that in Europe the minimum of the mean temperature which a proper cultivation requires, is, for the sugar-cane, from 19° to 20°; for coffee 18°; for the orange 17°; for the olive 13° 5′ to 14°; and for the vine yielding wine fit to be drunk from 10° to 11° of the centigrade thermometer, viz. from 66° to 68°; 64°; 62°; from 56°.3 to 57°; and from 50° to 51°.8 of Fahrenheit. This thermometrical agricultural scale is accurate enough when we embrace the phenomena in their greatest generality. But numerous exceptions occur when we consider countries of which the mean annual heat is the same, while the mean temperatures of the months differ very much from one another. It is the unequal division of the heat among the different seasons of the year which has the greatest influence on the kind of cultivation proper to such or such a latitude, as has been very well proved by M. Decandole. Several annual plants, especially gramina with farinaceous seed, are very little affected by the rigour of winter, but, like fruit-trees and the vine, require a considerable heat during summer. In part of Maryland, and especially Virginia, the mean temperature of the year is equal and perhaps even superior to that of Lombardy; yet the severity of winter will not allow the same vegetables to be there cultivated with which the plains of the Milanese are adorned. In the equinoctial region of Peru or Mexico, rye, and especially wheat, attain to no maturity in plains of 3500 or 4000 metres, or 11,482 and 13,123 feet, of elevation, though the mean heat of these alpine regions exceeds that of the parts of Norway and Siberia in which cerealia are successfully cultivated. But for about 30 days the obliquity of the sphere and the short duration of the nights render the summer heats very considerable in the countries in the vicinity of the pole, while under the tropics or the table-land of the *cordilleras* the thermometer never remains a whole day above 10 or 12 centigrade degrees.

To avoid mixing ideas of a theoretical nature and hardly susceptible of rigorous accuracy with facts, the certainty of which has been ascertained, we shall neither divide the cultivated plants in Nueva España according to the heat of the soil in which they vegetate most abundantly, nor according to the degrees of mean temperature which they appear to require for their development: but we shall arrange them in the order of their utility to society. We shall begin with the vegetables which form the principal support of the Mexican people; we shall afterwards treat of the cultivation of the plants which afford materials to manufacturing industry; and we shall conclude with a description of the vegetable productions which are the subject of an important commerce with the mother country.

The *banana* is for all the inhabitants of the torrid zone what the cereal gamina, wheat, barley, and rye, are for W. Asia and for Europe, and what the numerous varieties of rice are for the countries beyond the Indus, especially for Bengal and China. In the two continents, in the islands throughout the immense extent of the equinoctial seas, wherever the mean heat of the year exceeds 24 centigrade degrees, or 75° of Fahrenheit, the fruit of the banana is one of the most interesting objects of cultivation for the subsistence of man. The celebrated traveller George Forster, and other naturalists after him, pretended that this valuable plant did not exist in America before the arrival of the Spaniards, but that it was imported from the Canary islands in the beginning of the 16th century. In fact, Oviedo, who in his Natural History of the Indies, very carefully distinguishes the indigenous vegetables from those which were introduced there, positively says that the first bananas were planted in 1516 in the island of St. Domingo, by Thomas [de Berlangas, a monk of the order of preaching friars. He affirms that he himself saw the musa cultivated in Spain, near the town of Armeria in Grenada, and in the convent of Franciscans at the island of La Gran Canaria, where Berlangas procured suckers, which were transported to Hispaniola, and from thence successively to the other islands and to the continent. In support of M. Forster's opinion it may also be stated, that in the first accounts of the voyages of Columbus, Alonzo Negro, Penzon, Vespucci, and Cortes, there is frequent mention of maize, the papayer, the jatropha manihot, and the agave, but never of the banana. However, the silence of these first travellers only proves the little attention which they paid to the natural productions of the American soil. Hernandez, who, besides medical plants, describes a great number of other Mexican vegetables, makes no mention of the musa. Now this botanist lived half a century after Oviedo, and those who consider the musa as foreign to the new continent cannot doubt that its cultivation was general in Mexico towards the end of the 16th century, at an epocha when a crowd of vegetables of less utility to man had already been carried there from Spain, the Canary islands, and Peru. The silence of authors is not a sufficient proof in favour of M. Forster's opinion.

It is, perhaps, with the true country of the bananas as with that of the pear and cherry-trees. The prunus avium, for example, is indigenous in Germany and France, and has existed from the most remote antiquity in French forests, like the robur and the linden-tree; while other species of cherry-trees which are considered as varieties become permanent, and of which the fruits are more savoury than the prunus avium, have originally come through the Romans from Asia Minor, and particularly from the kingdom of Pontus. In the same manner, under the name of banana, a great number of plants, which differ essentially in the form of their fruits, and which, perhaps, constitute true species, are cultivated in the equinoctial regions, and even to the parallel of 33 or 34 degrees. If it is an opinion not yet proved, that all the pear-trees which are cultivated descend from the wild pear-tree as a common stock, we are still more entitled to doubt whether the great number of constant varieties of the banana descend from the musa troglodytarum, cultivated in the Molucca islands, which itself, according to Gaertner, is not perhaps a musa, but a species of the genus ravenala of Adanson.

The musæ, or *pisangs*, described by Rumphius and Rheede, are not all known in the Spanish colonics. Three species, however, are there distinguished, still very imperfectly determined by botanists, the true *platano* or *arton* (*musa paradisiaca Lin.?*); the *camburi* (*M. sapientum Lin.?*); and the *dominico* (*M. regia Rumph.?*) There is also a fourth species of very exquisite taste cultivated in Peru, the *meiya* of the S. sea, which is called in the market of Lima the *platano de taiti*, because the first roots of it were brought in the frigate Aguila from the island of Otaheite. Now it is a constant tradition in Mexico and all the continent of S. America, that the platano arton and the dominico were cultivated there long before the arrival of the Spaniards, but that the *guinco*, a variety of the camburi, as its name proves, came from the coast of Africa. The author who has most carefully marked the different epochas at which American agriculture was enriched with foreign productions, the Peruvian Garcillasso de la Vega, expressly says, "that in the time of the Incas the maize, *quinoa*, potatoes, and in the warm and temperate regions bananas, constituted the basis of the nourishment of the natives." He describes the musa of the valleys of the Antis, and he even distinguishes the most rare species with small sugary and aromatic fruit, the dominico, from the common or arton banana. Father Acosta also affirms, (*Historia natural de Indias*, 1608, p. 250), though not so positively, that the musa was cultivated by the Americans before the arrival of the Spaniards. "The banana," says he, "is a fruit to be found in all the Indies, though there are people who pretend that it is a native of Ethiopia, and that it came from thence into America." On the banks of the Orinoco, the Cassiquiare, or the Beni, among the mountains De l'Esmeralda and the sources of the river Carony, in the midst of the thickest forests, wherever we discover Indian tribes who have had no connections with European establishments, we find plantations of manioc and bananas.

Father Thomas de Berlangas could not transport from the Canary Islands to St. Domingo any other species but the one which is there cultivated, the camburi (*caule nigrescente striato fructu minore ovato-elongato*), and not the platano arton or *zapalote* of the Mexicans, (*caule albovirescente lævi, fructu longiore apicem versus subarcuato acute trigono*). The first of these species only grows in temperate climates, in the Canary islands, at Tunis, Algiers, and the coast of Malaga. In the valley of Caracas also, placed under lat. 10° 30′ but at 900 metres, or 2952 feet, of absolute elevation, we find only the camburi and the dominico (*caule albo-virescente, fructu minimo obsolete trigono*), and not the platano arton, of which the] [fruit only ripens under the influence of a very high temperature. From these numerous proofs we cannot doubt that the banana, which several travellers pretend to have found wild at Amboina, at Gilolo, and the Mariana islands, was cultivated in America long before the arrival of the Spaniards, who merely augmented the number of the indigenous species. However, we are not to be astonished that there was no musa seen in the island of St. Domingo before 1516. Like the animals around them, savages generally draw their nourishment from one species of plant. The forests of Guayana afford numerous examples of tribes whose plantations (*conucos*) contain *manihot*, *arum* or *dioscorea*, and not a single banana.

Notwithstanding the great extent of the Mexican table-land, and the height of the mountains in the neighbourhood of the coast, the space of which the temperature is favourable for the cultivation of the musa is more than 50,000 square leagues, and inhabited by nearly a million and a half of inhabitants. In the warm and humid valleys of the intendancy of Vera Cruz, at the foot of the *cordillera* of Orizaba, the fruit of the platano arton sometimes exceeds three decimetres, or 11.8 inches, and often from 20 to 22 centimetres, or 7.87 to 8.66 inches, in length. In these fertile regions, especially in the environs of Acapulco, San Blas, and the Rio Guasacualco, a cluster (*regime*) of bananas contains from 160 to 180 fruits, and weighs from 30 to 40 *kilogrammes*, or from 66 to 88lb. avoird.

We doubt whether there is another plant on the globe which on so small a space of ground can produce so considerable a mass of nutritive substance. Eight or nine months after the sueker has been planted, the banana commences to develop its clusters; and the fruit may be collected in the tenth or eleventh month. When the stalk is cut, we find constantly among the numerous shoots which have put forth roots a sprout (*pimpollo*), which having two-thirds of the height of the mother plant, bears fruit three months later. In this manner a plantation of musa, called in the Spanish colonies *platanar*, is perpetuated without any other care being bestowed by man than to cut the stalks of which the fruit has ripened, and to give the earth once or twice a year a slight dressing, by digging round the roots. A spot of ground of 100 square metres, or 1076 square feet, of surface, may contain at least from 30 to 40 banana plants. In the space of a year, this same ground, reckoning only the weight of a cluster at from 15 to 20 kilogrammes, or from 33 to 44lb. avoird. yields more than 2000 kilogrammes, or 4414lb. avoird. or 4000 pounds of nutritive substance. What a difference between this produce and that of the cereal *gramina* in the most fertile parts of Europe! Wheat, supposing it sown and not planted in the Chinese manner, and calculating on the basis of a decuple harvest, does not produce on 100 square metres more than 15 kilogrammes, (33lb. avoird.) or 30 pounds of grain. In France, for example, the *demi-hectare*, or legal *arpent*, of 1344¼ square toises, or 54,995 square feet, of good land, is sown (*à la volée*) with 160 pounds of grain; and if the land is not so good or absolutely bad, with 200 or 220 pounds. The produce varies from 1000 to 2500 pounds per acre. The potato, according to M. Tessie, yields in Europe on 100 square acres of well cultivated and well manured ground a produce of 45 kilogrammes, or 99lb. avoird. of roots. We reckon from 4 to 6000 pounds to the legal arpent. The produce of bananas is consequently to that of wheat as 133 : 1, and to that of potatoes as 44 : 1.

Those who in Europe have tasted bananas ripened in hot-houses have a difficulty in conceiving that a fruit, which from its great mildness has some resemblance to a dried fig, can be the principal nourishment of many millions of men in both Indies. We seem to forget that in the act of vegetation the same elements form very different chemical mixtures according as they combine or separate. How should we even discover in the lacteous mucilage, which the grains of gramina contain before the ripening of the ear, the farinaecous *perisperma* of the *cerealia*, which nourishes the majority of the nations of the temperate zone? In the musa, the formation of the amylaceous matter precedes the epocha of maturity. We must distinguish between the banana fruit collected when green, and what is allowed to grow yellow on the plant. In the second the sugar is quite formed; it is mixed with the pulp, and in such abundance that if the sugar-cane was not cultivated in the banana region, we might extract sugar from this fruit to greater advantage than is done in Europe from red beet and the grape. The banana, when gathered green, contains the same nutritive principle which is observed in grain, rice, the tuberose roots, and the *sagou*, namely the amylaceous sediment united with a very small portion of vegetable gluten. By kneading with water meal of bananas dried in the sun, Humboldt could only obtain a few atoms of this ductile and viscous mass, which resides in abundance in the perisperma, and especially in the embryo of the cerealia. If, on the one hand, the gluten, which has so much analogy to animal matter, and which] [swells with heat, is of great use in the making of bread; on the other hand, it is not indispensable to render a root or fruit nutritive. M. Proust discovered gluten in beans, apples, and quinces; but he could not discover any in the meal of potatoes. Gums, for example, that of the *mimosa nilotica* (*acacia vera Willd.*), which serves for nourishment to several African tribes in their passages through the desert, prove that a vegetable substance may be a nutritive aliment without containing either gluten or amylaceous matter.

It would be difficult to describe the numerous preparations by which the Americans render the fruit of the musa, both before and after its maturity, a wholesome and agreeable diet. Humboldt frequently observed in ascending rivers, that the natives, after the greatest fatigues, make a complete dinner on a very small portion of manioc and three bananas (platano arton) of the large kind. In the time of Alexander, if we are to credit the ancients, the philosophers of Hindostan were still more sober. "Arbori nomen *palæ* pomo arienæ, quo sapientes Indorum vivunt. Fructus admirabilis succi dulcedine ut uno quaternos satiet." (*Plin.* xii. 12). In warm countries the people in general not only consider sugary substances as a food which satisfies for the moment, but as truly nutritive. Humboldt has frequently observed, that the mule-drivers who carry the baggage on the coast of Caracas give the preference to unprepared sugar (*papelon*) over fresh animal food.

Physiologists have not yet determined with precision what characterises a substance eminently nutritive. To appease the appetite by stimulating the nerves of the gastric system, and to furnish matter to the body which may easily assimilate with it, are modes of action very different. Tobacco, the leaves of the *erythroxylon* cocca mixed with quick lime, the opium which the natives of Bengal have frequently used for whole months in times of scarcity, will appease the violence of hunger; but these substances act in a very different manner from wheaten bread, the root of the jatropha, gum-arabic, the lichen of Iceland, or the putrid fish which is the principal food of several tribes of African Negroes. There can be no doubt, the bulk being equal, superazoted matter, or animals, are more nutritive than vegetable matter; and it appears that, among vegetables, gluten is more nutritive than starch, and starch more than mucilage; but we must beware of attributing to these insulated principles what depends, in the action of the aliment on living bodies, on the varied mixture of hydrogen, carbonate, and oxygen. Hence a matter becomes eminently nutritive if it contains, like the bean of the cocoa-tree (*theobroma cacao*), besides the amylaceous matter, an aromatic principle which excites and fortifies the nervous system.

These considerations, to which we cannot give more development here, will serve to throw some light on the comparisons which we have already made of the produce of different modes of cultivation. If we draw from the same space of ground three times as many potatoes as wheat in weight, we must not therefore conclude that the cultivation of tuberous plants will on an equal surface maintain three times as many individuals as the cultivation of cereal *gramina*. The potato is reduced to the fourth part of its weight when dried by a gentle heat; and the dry starch that can be separated from 2300 kilogrammes, the produce of half a *hectare* of ground, would hardly equal the quantity furnished by 800 kilogrammes of wheat. It is the same with the fruit of the banana, which before its maturity, even in the state in which it is very farinaceous, contains much more water and sugary pulp than the seeds of gramina. We have seen that the same extent of ground in a favourable climate will yield 106,000 kilogrammes of bananas, 2400 kilogrammes of tuberous roots, and 800 kilogrammes of wheat. These quantities bear no proportion to the number of individuals which can be maintained by these different kinds of cultivation on the same extent of ground. The aqueous mucilage which the banana contains, and the tuberous root of the *solanum*, possess undoubted nutritive properties. The farinaceous pulp, such as is presented by nature, yields undoubtedly more aliment than the starch which is separated from it by art. But the weights alone do not indicate the absolute quantities of nutritive matter; and to shew the amount of the aliment which the cultivation of the musa yields on the same space of ground to man more than the cultivation of wheat, we ought rather to calculate according to the mass of vegetable substance necessary to satisfy a full-grown person. According to this last principle, and the fact is very curious, we find that in a very fertile country a demi-hectare, or legal arpent (54,998 square feet), cultivated with bananas of the large species (platano arton), is capable of maintaining 50 individuals; when the same arpent in Europe would only yield annually, supposing the eighth grain 576 kilogrammes, or 1271 lb. avoird. of flour, a quantity not equal to the subsistence of two individuals. Accordingly, a European newly arrived in the torrid zone is struck with nothing so much as the extreme smallness of the spots under cultivation] [round a cabin which contains a numerous family of Indians.

The ripe fruit of the musa, when exposed to the sun, is preserved like our figs. The skin becomes black and takes a particular odour, which resembles that of smoked ham. The fruit in this state is called *patano pasado*, and becomes an object of commerce in the province of Mechoacán. This dry banana is an aliment of an agreeable taste, and extremely healthy. But those Europeans who newly arrive consider the ripe fruit of the platano arton, newly gathered, as very ill to digest. This opinion is very ancient, for Pliny relates that Alexander gave orders to his soldiers to touch none of the bananas which grow on the banks of the Hyphasus. Meal is extracted from the musa by cutting the green fruit into slices, drying it in the sun on a slope, and pounding it when it becomes friable. This flour, less used in Mexico than in the islands, may serve for the same use as flour from rice or maize.

The facility with which the banana is reproduced from its roots gives it an extraordinary advantage over fruit trees, and even over the bread-fruit tree, which for eight months in the year is loaded with farinaceous fruit. When tribes are at war with one another and destroy the trees, the disaster is felt for a long time. A plantation of bananas is renewed by suckers in the space of a few months.

We hear it frequently repeated in the Spanish colonies, that the inhabitants of the warm region (*tierra caliente*) will never awake from the state of apathy in which for centuries they have been plunged, till a royal *cedula* shall order the destruction of the banana plantations (*platanares*). The remedy is violent, and those who propose it with so much warmth do not in general display more activity than the lower people, whom they would force to work by augmenting the number of their wants. It is to be hoped that industry will make progress among the Mexicans without recurring to means of destruction. When we consider, however, the facility with which our species can be maintained in a climate where bananas are produced, we are not to be astonished that in the equinoctial region of the new continent civilization first commenced on the mountains, in a soil of inferior fertility, and under a sky less favourable to the development of organized beings, in whom necessity even awakes industry. At the foot of the *cordillera*, in the humid valleys of the intendancies of Vera Cruz, Valladolid, and Guadalaxara, a man who merely employs two days in the week in a work by no means laborious may procure subsistence for a whole family. Yet such is the love of his native soil, that the inhabitant of the mountains, whom the frost of a single night frequently deprives of the whole hopes of his harvest, never thinks of descending into the fertile but thinly inhabited plains, where Nature showers in vain her blessings and her treasures.

The same region in which the banana is cultivated produces also the valuable plant of which the root affords the flour of *manioc*, or *magnoc*. The green fruit of the musa is eaten dressed, like the bread fruit, or the tuberous root of the potato; but the flour of the manioc is converted into bread, and furnishes to the inhabitants of warm countries what the Spanish colonists call *pan de tierra caliente*. The maize, as we shall afterwards see, affords the great advantage of being cultivated under the tropics, from the level of the ocean to elevations which equal those of the highest summits of the Pyrenees. It possesses that extraordinary flexibility of organization for which the vegetables of the family of the *gramina* are characterised; and it even possesses it in a higher degree than the *cerealia* of the old continent, which suffer under a burning sun, while the maize vegetates vigorously in the warmest regions of the earth. The plant whose root yields the nutritive flour of the manioc takes its name from *juca*, a word of the language of Haity, or St. Domingo. It is only successfully cultivated within the tropics; and the cultivation of it in the mountainous part of Mexico never rises above the absolute height of 6 or 800 metres, or 1968 and 2624 feet. This height is much surpassed by that of the camburi, or banana of the Canaries, a plant which grows nearer the central table-land of the *cordilleras*.

The Mexicans, like the natives of all equinoctial America, have cultivated, from the remotest antiquity, two kinds of juca, which the botanists, in their inventory of species, have united under the name of *jatropha manihot*. They distinguish, in the Spanish colony, the sweet (*dulce*) from the tart or bitter (*amarga*) juca. The root of the former, which bears the name of *camagnoc* at Cayenne, may be eaten without danger, while the other is a very active poison. The two may be made into bread; however, the root of the bitter juca is generally used for this purpose, the poisonous juice of which is carefully separated from the *fecula* before making the bread of the manioc, called *cazavi*, or cassave. This separation is operated by compressing the root after being grated down in the *cibucan*, which is a species of long sack. It appears from a passage of Oviedo, (lib. vii. c. 2)] [that the juca dulce, which he calls *boniata*, and which is the *huacamote* of the Mexicans, was not found originally in the W. India islands, and that it was transplanted from the neighbouring continent. "The boniata," says Oviedo, "is like that of the continent; it is not poisonous, and may be eaten with its juice either raw or prepared." The natives carefully separate in their fields (*conucos*) the two species of jatropha.

It is very remarkable that plants, of which the chemical properties are so very different, are yet so very difficult to distinguish from their exterior characters. Brown, in his Natural History of Jamaica, imagined he found these characters in dissecting the leaves. He calls the sweet juca, sweet cassava, *jatropha foliis palmatis lobis incertis;* and the bitter or tart juca, common cassava, *jatropha foliis palmatis pentadactylibus.* Humboldt, having examined many plantations of manihot, found that the two species of jatropha, like all cultivated plants with lobed or palmated leaves, vary prodigiously in their aspect. He also observed that the natives distinguish the sweet from the poisonous manioc, not so much from the superior whiteness of the stalk and the reddish colour of the leaves as from the taste of the root, which is not tart or bitter. It is with the cultivated jatropha as with the sweet orange-tree, which botanists cannot distinguish from the bitter orange-tree, but which, however, according to the beautiful experiments of M.Galesio, is a primitive species, propagated from the grain, as well as the bitter orange-tree. Several naturalists, from the example of Dr. Wright of Jamaica, have taken the sweet juca for the true *jatropha janipha* of Linnæus, or the *jatropha frutescens* of Loffling. (*Reza til Spanska Loenderna*, 1758, p. 309). But this last species, which is the *jatropha Carthaginensis* of Jacquin, differs from it essentially by the form of the leaves (*lobis utrinque sinuatis*), which resemble those of the *papayer.* We very much doubt whether the jatropha can be transformed by cultivation into the jatropha manihot. It appears equally improbable that the sweet juca is a poisonous jatropha, which, by the care of man, or the effect of a long cultivation, has gradually lost the acidity of its juices. The juca amarga of the American fields has remained the same for centuries, though planted and cultivated like the juca dulce. Nothing is more mysterious than this difference of interior organization in cultivated vegetables, of which the exterior forms are nearly the same.

Raynal (*Histoire Philosophique*, tom. iii. p. 212—214) has advanced that the manioc was transplanted from Africa to America to serve for the maintenance of the Negroes, and that if it existed on the continent before the arrival of the Spaniards, it was not, however, known by the natives of the W. Indies in the time of Columbus. We are afraid that this celebrated author has confounded the manioc with the ignames; that is to say, the jatropha with a species of *dioscorea.* We should wish to know by what authority we can prove that the manioc was cultivated in Guinea from the remotest period. Several travellers have also pretended that the maize grew wild in this part of Africa, and yet it is certain that it was transported there by the Portuguese in the 16th century. Nothing is more difficult to resolve than the problem of the migration of the plants useful to man, especially since communications have become so frequent between all continents. Fernandez de Oviedo, who went in 1513 to the island of Hispaniola, or St. Domingo, and who for more than 20 years inhabited different parts of the new continent, speaks of the manioc as of a very ancient cultivation, and peculiar to America. If, however, the Negro slaves introduced the manioc, Oviedo would himself have seen the commencement of this important branch of tropical agriculture. If he had believed that the jatropha was not indigenous in America, he would have cited the epocha at which the first maniocs were planted, as he relates in the greatest detail the first introduction of the sugar-cane, the banana of the Canaries, the olive, and the date. Amerigo Vespucci relates in his letter addressed to the Doke of Loraine, (*Grynæus*, p. 215), that he saw bread made of the manioc on the coast of Paria in 1497. "The natives," says this adventurer, in other respects by no means accurate in his recital, "know nothing of our corn and our farinaceous grains; they draw their principal subsistence from a root which they reduce into meal, which some of them call *jucha*, others *chambi*, and others *igname.*" It is easy to discover the word *jucca* in *jucha.* As to the word *igname*, it now means the root of the *dioscorea alata*, which Columbus describes under the name of *ages*, and of which we shall afterwards speak. The natives of Spanish Guayana who do not acknowledge the dominion of the Europeans have cultivated the manioc from the remotest antiquity. Running out of provisions in repassing the rapids of the Orinoco, Humboldt, on his return from the Rio Negro, applied to the tribe of Piraoas Indians, who dwell to the *e.* of the Maypures, and they supplied him with jatropha bread. There can therefore remain no doubt that the manioc is a plant of which the cultivation is of a much] [earlier date than the arrival of the Europeans and Africans into America.

The manioc bread is very nutritive, perhaps on account of the sugar which it contains, and a viscous matter which unites the farinaceous molecules of the cassava. This matter appears to have some analogy with the *caoutchouc*, which is so common in all the plants of the group of the *tithymaloides*. They give to the cassava a circular form. The disks, which are called *turtas*, or *xauxau* in the old language of Haity, have a diameter of from five to six decimetres, or from 19.685 inches to 23.622 inches, of thickness. The natives, who are much more sober than the whites, generally eat less than half a kilogramme, or about a pound, of manioc per day. The want of gluten mixed with the amylaceous matter, and the thinness of the bread, render it extremely brittle and difficult of transportation. This inconvenience is particularly felt in long navigations. The fecula of manioc grated, dried, and smoked, is almost unalterable. Insects and worms never attack it, and every traveller knows in equinoctial America the advantages of the *couaque*.

It is not only the fecula of the juca amarga which serves for nourishment to the Indians, they use also the juice of the root, which in its natural state is an active poison. This juice is decomposed by fire. When kept for a long time in ebullition it loses its poisonous properties gradually as it is skimmed. It is used without danger as a sauce, and Humboldt himself frequently used this brownish juice, which resembles a very nutritive *bouillon*. At Cayenne, (*Aublet, Hist. des Plantes de la Guyane Françoise*, tom. ii. p. 72), it is thickened to make *cabiou*, which is analogous to the *souy* brought from China, and which serves to season dishes. From time to time very serious accidents happen when the juice has not been long enough exposed to the heat. It is a fact very well known in the islands, that formerly a great number of the natives of Haity killed themselves voluntarily by the raw juice of the root of the juca amarga. Oviedo relates, as an eye-witness, that these unhappy wretches, who, like many African tribes, preferred death to involuntary labour, united together by fifties to swallow at once the poisonous juice of the jatropha. This extraordinary contempt of life characterises the savage in the most remote parts of the globe.

Reflecting on the union of accidental circumstances which have determined nations to this or that species of cultivation, we are astonished to see the Americans, in the midst of the richness of their country, seek in the poisonous root of a *tithymaloid* the same amylaceous substance which other nations have found in the family of gramina, in bananas, asparagus (dioscorea alata), aroides (arum macrorrhizen, dracontium polyphillum), solana, lizerons (convolvulus batatas, c. chrysorbizus), narcissi (tacca pinnatifida), polygonoi (p. fagopyrum), urticæ (artocarpus), legumens and arborescent ferns (cycas circinnalis). We ask why the savage who discovered the jatropha manihot did not reject a root of the poisonous qualities of which a sad experience must have convinced him before he could discover its nutritive properties? But the cultivation of the juca dulce, of which the juice is not deleterious, preceded perhaps that of the juca amarga, from which the manioc is now taken. Perhaps also the same people who first ventured to feed on the root of the jatropha manihot had formerly cultivated plants analogous to the *arum* and the *dracontium*, of which the juice is acrid, without being poisonous. It was easy to remark, that the fecula extracted from the root of an aroid is of a taste so much the more agreeable, as it is carefully washed to deprive it of its milky juice. This very simple consideration would naturally lead to the idea of expressing the fecula, and preparing it in the same manner as the manioc. We can conceive that a people who knew how to dulcify the roots of an aroid could undertake to nourish themselves on a plant of the group of the *euphorbia*. The transition is easy, though the danger is continually augmenting. In fact, the natives of the Society and Molucca islands, who are unacquainted with the jatropha manihot, cultivate the *arum macrorrhizon* and the *tacca pinnatifida*. The root of this last plant requires the same precaution as the manioc, and yet the *tacca* bread competes in the market of Banda with the *sagou* bread.

The cultivation of the manioc requires more care than that of the banana. It resembles that of potatoes, and the harvest takes place only from seven to eight months after the slips have been planted. The people who can plant the jatropha have already made great advances towards civilization. There are even varieties of the manioc, for example, those which are called at Cayenne *manioc bois blanc*, and *manioc mai-pourri-rouge*, of which the roots can only be pulled up at the end of 15 months. The savage of New Zealand would not certainly have the patience to wait for so tardy a harvest.

Plantations of jatropha manihot are now found along the coast from the mouth of the river of Guasacualco to the *n.* of Santander, and from Tehuantepec to San Blas and Sinaloa, in the low] [and warm regions of the intendancies of Vera Cruz, Oaxaca, Puebla, Mexico, Valladolid, and Guadalaxara. M. Aublet, a judicious botanist, who, happily, has not disdained in his travels to inquire into the agriculture of the tropics, says very justly, "that the manioc is one of the finest and most useful productions of the American soil, and that with this plant the inhabitants of the torrid zone could dispense with rice and every sort of wheat, as well as all the roots and fruits which serve as nourishment to the human species."

Maize occupies the same region as the banana and the manioc; but its cultivation is still more important and more extensive, especially than that of the two plants which we have been describing. Advancing towards the central table-land we meet with fields of maize all the way from the coast to the valley of Toluca, which is more than 2800 metres, or 9185 feet, above the level of the ocean. The year in which the maize harvest fails is a year of famine and misery for the inhabitants of Mexico.

It is no longer doubted among botanists, that maize, or Turkey corn, is a true American grain, and that the old continent received it from the new. It appears also that the cultivation of this plant in Spain long preceded that of potatoes. Oviedo (*Rerum Medicarum Novæ Hispanæ Thesaurus*, 1651, lib. vii. c. 40, p. 247), whose first essay on the natural history of the Indies was printed at Toledo in 1525, says that he saw maize cultivated in Andalusia, near the chapel of Atocha, in the environs of Madrid. This assertion is so much the more remarkable, as from a passage of Hernandez (book vii. chap. 40) we might believe that maize was still unknown in Spain in the time of Philip II. towards the end of the 16th century.

On the discovery of America by the Europeans, the zea maize (*tlaolli* in the Aztec language, *mahiz* in the Haitian, and *cara* in the Quichua) was cultivated from the most *s.* part of Chile to Pennsylvania. According to a tradition of the Aztec people, the Toultecs, in the 7th century of our era, were the first who introduced into Mexico the cultivation of maize, cotton, and pimento. It might happen, however, that these different branches of agriculture existed before the Toultecs, and that this nation, the great civilization of which has been celebrated by all the historians, merely extended them successfully. Hernandez informs us, that the Otamites even, who were only a wandering and barbarous people, planted maize. The cultivation of this grain consequently extended beyond the Rio Grande de Santiago, formerly called Tololotlan.

The maize introduced into the *n.* of Europe suffers from the cold wherever the mean temperature does not reach 7° or 8° of the centigrade thermometer, or 44° or 46° of Fahrenheit. We therefore see rye, and especially barley, vegetate vigorously on the ridge of the *cordilleras*, at heights where, on account of the roughness of the climate, the cultivation of maize would be attended with no success. But, on the other hand, the latter descends to the warmest regions of the torrid zone, even to plains where wheat, barley, and rye cannot develop themselves. Hence on the scale of the different kinds of cultivation, the maize, at present, occupies a much greater extent in the equinoctial part of America than the cerealia of the old continent. The maize, also, of all the grains useful to man, is the one whose farinaceous perisperma has the greatest volume.

It is commonly believed that this plant is the only species of grain known by the Americans before the arrival of the Europeans. It appears, however, certain enough, that in Chile, in the 15th century, and even long before, besides the zea maize and the zea curagua, two gramina called *magu* and *tuca* were cultivated, of which, according to the Abbe Molina, the first was a species of rye, and the second a species of barley. The bread of this araucan bread went by the name of *covque*, a word which afterwards was applied to the bread made of European corn. (*Molina, Histoire naturelle du Chile*, p. 101). Hernandez even pretends to have found among the Indians of Mechoacán a species of wheat, (p. vii. 43. *Clavigero*, i. p. 56, note F.), which, according to his very succinct description, resembles the corn-of-abundance, (*triticum compositum*), which is believed to be a native of Egypt. Notwithstanding every information which Humboldt procured during his stay in the intendancy of Valladolid, it was impossible for him to clear up this important point in the history of cerealia. Nobody there knew any thing of a wheat peculiar to the country, and he suspected that Hernandez gave the name of *triticum Michuacanense* to some variety of European grain become wild and growing in a very fertile soil.

The fecundity of the *tlaolli*, or Mexican maize, is beyond any thing that can be imagined in Europe. The plant, favoured by strong heats and much humidity, acquires a height of from two to three metres, or from 6½ to 9$\frac{8}{10}$ feet. In the beautiful plains which extend from San Juan del Rio to Queretaro, for example, in the lands of the great] [plantation of L'Esperanza, one *fanega* of maize produces sometimes 800. Fertile lands yield, *communibus annis*, from 3 to 400. In the environs of Valladolid a harvest is reckoned bad which yields only the seed 130 or 150 fold. Where the soil is even most sterile it still returns from 60 to 80 grains for one. It is believed that we may estimate the produce of maize in general, in the equinoctial region of the kingdom of Nueva España, at 150 for one. The valley of Toluca alone yields annually more than 600,000 *fanegas*, or 66,210,600 lbs. on an extent of 30 square leagues, of which a great part is cultivated in agave. Between the parallels of 18° and 22° the frosts and cold winds render this cultivation by no means lucrative on plains whose height exceeds 3000 metres, or 9842 feet. The annual produce of maize in the intendancy of Guadalaxara is, as we have already observed, more than 80,000,000 of kilogrammes, or 176,562,400 lbs. avoirdupois.

Under the temperate zone, between lat. 33° and 38°, in New California for example, maize produces in general only, *communibus annis*, from 70 to 80 for one. By comparing the manuscript memoirs of Father Fermin Lassuen with the statistical tables published in the historical account of the voyage of M. de Galeano, we should be enabled to indicate village by village the quantities of maize sown and reaped. We find that in 1791, 12 missions of New California reaped 7625 fanegas on a piece of ground sown with 96. In 1801 the harvest of 16 missions was 4661 fanegas, while the quantity sown only amounted to 66. Hence for the former year the produce was 79, and for the latter 70 for one. This coast in general appears better adapted for the cultivation of the cerealia of Europe. However, it is proved by the same tables, that in some parts of New California, for example in the fields belonging to the villages of San Buenaventura and Capistrano, the maize has frequently yielded from 180 to 200 for one.

Although a great quantity of other grain is cultivated in Mexico, the maize must be considered as the principal food of the people, as also of the most part of the domestic animals. The price of this commodity modifies that of all the others, of which it is, as it were, the natural measure. When the harvest is poor, either from the want of rain or from premature frost, the famine is general, and produces the most fatal consequences. Fowls, turkies, and even the larger cattle, equally suffer from it. A traveller who passes through a country in which the maize has been frost-bitten finds neither egg nor poultry, nor *arepa* bread, nor meal for the *atolli*, which is a nutritive and agreeable soup. The dearth of provisions is especially felt in the environs of the Mexican mines; in those of Guanaxuato, for example, where 14,000 mules, which are necessary in the process of amalgamation, annually consume an enormous quantity of maize. We have already mentioned the influence which dearths have periodically had on the progress of population in Nueva España. The frightful dearth of 1784 was the consequence of a strong frost, which was felt at an epocha when it was least to be expected in the torrid zone, the 28th August, and at the inconsiderable height of 1800 metres (5904 feet) above the level of the ocean.

Of all the gramina cultivated by man none is so unequal in its produce. This produce varies in the same field, according to the changes of humidity and the mean temperature of the year, from 40 to 200 or 300 for one. If the harvest is good, the colonist makes his fortune more rapidly with maize than with wheat; and we may say that this cultivation participates in both the advantages and disadvantages of the vine. The price of maize varies from 2 livres 10 sons to 25 livres the fanega. The mean price is five livres in the interior of the country; but it is increased so mnch by the carriage, that during Humboldt's stay in the intendancy of Guanaxuato, the fanega cost at Salamanca 9, at Queretaro 12, and at San Luis Potosi 22 livres. In a country where there are no magazines, and where the natives merely live from hand to mouth, the people suffer terribly whenever the maize remains for any length of time at 2 piastres, or 10 livres, the fanega. The natives then feed on unripe fruit, on cactus berries, and on roots. This insufficient food occasions diseases among them; and it is observed that famines are usually accompanied with a great mortality among the children.

In warm and very humid regions the maize will yield from two to three harvests annually; but generally only one is taken. It is sown from the middle of June till near the end of August. Among the numerous varieties of this gramen there is one of which the ear ripens two months after the grain has been sown. This precious variety is well known in Hungary, and M. Parmentier has endeavoured to introduce the cultivation of it into France. The Mexicans who inhabit the shores of the S. sea give the preference to another, which Oviedo (lib. vii. c. 1, p. 103) affirms he saw in his time, in the province of Nicaragua, and which is reaped in between 30 and 40 days. Humboldt also observed it near Tomependa, on the banks of the river of the Amazons; but all these varieties of maize, of which the vegetation is so] [rapid, appear to be of a less farinaceous grain, and almost as small as the zea curagua of Chile.

The utility which the Americans draw from maize is too well known to make it necessary for us to dwell on it. The use of rice is not more various in China and the E. Indies. The ear is eaten boiled or roasted. The grain when beat yields a nutritive bread (*arepa*), though not fermented and ill baked, on account of the small quantity of gluten mixed with the amylaceous fecula. The meal is employed like gruel in the boullies, which the Mexicans call *atlolli*, in which they mix sugar, honey, and sometimes even ground potatoes. The botanist Hernandez (lib. vii. c. 40, p. 244) describes 16 species of attolis which were made in his time.

A chemist would have some difficulty in preparing the innumerable variety of spirituous, acid, or sugary beverages, which the Indians display a particular address in making, by infusing the grain of maize, in which the sugary matter begins to develop itself by germination. These beverages, generally known by the name of *chicha*, have some of them a resemblance to beer, and others to cider. Under the monastic government of the Incas it was not permitted in Peru to manufacture intoxicating liquors, especially those which are called *vinapu* and *sora*. — (*Garcilasso*, lib. viii. c. 9, tom. i. p. 277. *Acosta*, lib. iv. c. 16, p. 238.)— The Mexican despots were less interested in the public and private morals; and drunkenness was very common among the Indians of the times of the Aztec dynasty. But the Europeans have multiplied the enjoyments of the lower people, by the introduction of the sugar-cane. At present in every elevation the Indian has his particular drinks. The plains in the vicinity of the coast furnish him with spirit from the sugar-cane, (*guarapo*, or *aguardiente de caña*), and the *chicha de manioc*. The *chicha de mais* abounds on the declivity of the *cordilleras*. The central table-land is the country of the Mexican vines, the agave plantations, which supply the favourite drink of the natives, the *pulque de maguey*. The Indian in easy circumstances adds to these productions of the American soil a liquor still dearer and rarer, grape brandy (*aguardiente de Castilla*), partly furnished by European commerce, and partly distilled in the country. Such are the numerous resources of a people who love intoxicating liquors to excess.

Before the arrival of the Europeans, the Mexicans and Peruvians pressed out the juice of the maize-stalk to make sugar from it. They not only concentrated this juice by evaporation; they knew also to prepare the rough sugar by cooling the thickened syrup. Cortes, describing to the Emperor Charles V. all the commodities sold in the great market of Tlatelolco, on his entry into Tenochtitlan, expressly names the Mexican sugar. "There is sold," says he, "honey of bees and wax, (honey from the stalks of maize), which are as sweet as sugar-cane, and honey from a shrub called by the people *maguey*. The natives make sugar of these plants, and this sugar they also sell." The stalk of all the gramina contains sugary matter, especially near the knots. The quantity of the sugar that maize can furnish in the temperate zone appears, however, to be very inconsiderable; but under the tropics its fistulous stalk is so sugary that the Indians have been frequently seen sucking it, as the sugar-cane is sucked by the Negroes. In the valley of Toluca the stalk of the maize is squeezed between cylinders, and then is prepared from its fermented juice a spirituous liquor, called *pulque de mahis*, or *tlaolli*, a liquor which becomes a very important object of commerce.

From the statistical tables drawn up in the intendancy of Guadalaxara, of which the population is more than 500,000 of inhabitants, it appears extremely probable that, *communibus annis*, the actual produce of maize in all Nueva España amounts to more than 17,000,000 of fanegas, or more than 800,000,000 of kilogrammes, or 1765¼ millions of pounds avoirdupois, of weight. This grain will keep in Mexico, in the temperate climates, for three years, and in the valley of Toluca, and all the levels of which the mean temperature is below 14 centigrade degrees, or 57° of Fahrenheit, for five or six years, especially if the dry stalk is not cut before the ripe grain has been somewhat struck with the frost.

In good years the kingdom of Nueva España produces much more maize than it can consume. As the country unites in a small space a great variety of climates, and as the maize almost never succeeds at the same time in the warm region (*tierras calientes*) and on the central table-land, in the *terras frias*, the interior commerce is singularly vivified by the transport of this grain. Maize compared with European grain has the disadvantage of containing a smaller quantity of nutritive substance in a greater volume. This circumstance, and the difficulty of the roads on the declivities of the mountains, present obstacles to its exportation, which will be less frequent when the construction of the fine causeway from Vera Cruz to Xalapa and Perote shall be finished. The islands in general, and especially the island of Cuba, consume an enormous quantity of maize.]

[These islands are frequently in want of it, because the interest of their inhabitants is almost exclusively fixed on the cultivation of sugar and coffee; although it has been long observed by well informed agriculturists, that in the district contained between the Havanah, the port of Batabano, and Matanzas, fields cultivated with maize and by free hands yield a greater nett revenue than a sugar plantation, for which enormous advances are necessary in the purchase and maintenance of slaves and the construction of edifices.

If it is probable that in Chile formerly, besides maize, there were two other gramina with farinaceous seed sown, which belonged to the same genus as our barley and wheat, it is no less certain that before the arrival of the Spaniards in America none of the cerealia of the old continent were known there. Could we suppose that all mankind were descended from the same stock, we might be tempted to admit that the Americans, like the Atlantes, (see the opinion of Diodorus Siculus, *Bibl.* lib. iii. p. 186. *Rhodom.*), separated from the rest of the human race before the cultivation of wheat on the central plains of Asia. But are we to lose ourselves in fabulous times to explain the ancient communications which appear to have existed between the two continents? In the time of Herodotus all the *n.* part of Africa presented no other agricultural nations but the Egyptians and the Carthaginians. (*Heeren über Africa*, p. 41). In the interior of Asia the tribes of the Mongol race, the Hiong-nu, the Burattes, the Kalkas, and the Sifanes, have constantly lived as wandering shepherds. Now, if the people of central Asia, or if the Lybians of Africa, could have passed into the new continent, neither of them would have introduced the cultivation of cerealia. The want of these gramina then proves nothing either against the Asiatic origin of the Americans, or against the possibility of a very recent transmigration.

The introduction of European grain having had the most beneficial influence on the prosperity of the natives of Mexico, it becomes interesting to relate at what epocha this new branch of agriculture commenced. A Negro slave of Cortes found three or four grains of wheat among the rice which served to maintain the Spanish army. These grains were sown, as it appears, before the year 1530. History has brought down to us the name of a Spanish lady, Maria d'Escobar, the wife of Diego de Chaves, who first carried a few grains of wheat into the city of Lima, then called Rimac. The produce of the harvest which she obtained from these grains was distributed for three years among the new colonists, so that each farmer received 20 or 30 grains. Garcilasso already complained of the ingratitude of his countrymen, who hardly knew the name of Maria d'Escobar. We are ignorant of the epocha at which the cultivation of cerealia commenced in Peru, but it is certain that in 1547 wheaten bread was hardly known in the city of Cuzco. (*Commentarios Reales*, ix. 24, t. ii. p. 332). At Quito the first European grain was sown near the convent of St. Francis by Father Josse Rixi, a native of Gand in Flanders. The monks still show there with enthusiasm the earthen vase in which the first wheat came from Europe, which they look upon as a precious relic. (See *Humboldt's Tableaux de la Nature*, t. ii. p. 166). Why have not every where the names of those been preserved, who, in place of ravaging the earth, have enriched it with plants useful to the human race?

The temperate region, especially the climate where the mean heat of the year does not exceed from 18 to 19 centigrade degrees, or 64° and 66° of Fahr. appears most favourable to the cultivation of cerealia, embracing under this denomination only the nutritive gramina known to the ancients, namely, wheat, spelt, barley, oats, and rye. In fact, in the equinoctial part of Mexico, the cerealia of Europe are no where cultivated in plains of which the elevation is under from 8 to 900 metres, or from 2629 to 2952 feet; and we have already observed, that on the declivity of the *cordilleras* between Vera Cruz and Acapulco, we generally see only the commencement of this cultivation at an elevation of 1200 or 1300 metres, or 3936 and 4264 feet. A long experience has proved to the inhabitants of Xalapa that the wheat sown around their city vegetates vigorously, but never produces a single ear. It is cultivated because its straw and its succulent leaves serve for forage (*zacate*) to cattle. It is very certain, however, that in the kingdom of Guatemala, and consequently nearer the equator, grain ripens at smaller elevations than that of the town of Xalapa. A particular exposure, the cool winds which blow in the direction of the *n.* and other local causes, may modify the influence of the climate. In the province of Caracas the finest harvests of wheat near Victoria (lat. 10° 13′) are found at 500 or 600 metres (1640 or 1968 feet) of absolute elevation; and it appears that the wheaten fields which surround the Quatro Villas in the island of Cuba (lat. 21° 58′) have still a smaller elevation. At the isle of France (lat. 20° 10′) wheat is cultivated on a soil almost level with the ocean.

The European colonists have not sufficiently]

[varied their experiments to know what is the *minimum* of height at which cerealia grow in the equinoctial region of Mexico. The absolute want of rain during the summer months is so much the more unfavourable to the wheat as the heat of the climate is greater. It is true that the droughts and heats are also very considerable in Syria and Egypt; but this last country, which abounds so much in grain, has a climate which differs essentially from that of the torrid zone, and the soil preserves a certain degree of humidity from the beneficent inundations of the Nile. However, the vegetables, which are of the same kind with our cerealia, grow only wild in temperate climates, and even in those only of the old continent. With the exception of a few gigantic arundinaceous, which are social plants, the gramina appear in general infinitely rarer in the torrid zone than in the temperate zone, where they have the ascendancy, as it were, over the other vegetables. We ought not, then, to be astonished that the cerealia, notwithstanding the great flexibility of organization attributed to them, and which is common to them with the domestic animals, thrive better on the central table-land of Mexico, in the hilly region, where they find the climate of Rome and Milan, than in the plains in the vicinity of the equinoctial ocean.

Were the soil of Nueva España watered by more frequent rains, it would be one of the most fertile countries cultivated by man in the two hemispheres. The hero who, in the midst of a bloody war, had his eyes continually fixed on every branch of national industry, Hernan Cortes, wrote to his sovereign shortly after the siege of Teñochtitlan: "All the plants of Spain thrive admirably in this land. We shall not proceed here as we have done in the isles, where we have neglected cultivation and destroyed the inhabitants. A sad experience ought to render us more prudent. I beseech your Majesty to give orders to the Casa de Contratacion of Seville, that no vessel set sail for this country without a certain quantity of plants and grain." The great fertility of the Mexican soil is incontrovertible, but the want of water frequently diminishes the abundance of the harvests.

There are only two seasons known in the equinoctial region of Mexico, even as far as the 28° of *n.* lat.: the rainy season (*estacion de las aguas*), which begins in the month of June or July, and ends in the month of September or October, and the dry season (*el estio*), which lasts eight months, from October to the end of May. The first rains generally commence on the *e.* declivity of the *cordillera*. The formation of the clouds and the precipitation of the water dissolved in the air commence on the coast of Vera Cruz. These phenomena are accompanied with strong electrical explosions, which take place successively at Mexico, Guadalaxara, and on the *w.* coast. The chemical action is propagated from *e.* to *w.* in the direction of the trade-winds, and the rains begin 15 or 20 days sooner at Vera Cruz than on the central table-land. Sometimes we see in the mountain, even below 2000 metres, or 6561 feet, of absolute height, rain mixed with rime (*gresil*) and snow in the months of November, December, and January; but these rains are very short, and only last from four to five days; and however cold they may be, they are considered as very useful for the vegetation of wheat and the pasturages. In Mexico in general, as in Europe, the rains are most frequent in the mountainous regions, especially in that part of the *cordilleras* which extends from the Pic d'Orizaba by Guanaxuato, Sierra de Pinos, Zacatecas, and Bolaños, to the mines of Guarisamey and the Rosario.

The prosperity of Nueva España depends on the proportion established between the duration of two seasons of rain and drought. The agriculturist has seldom to complain of too great a humidity, and if sometimes the maize and the cerealia of Europe are exposed to partial inundations in the plains, of which several form circular basins shut in by the mountains, the grain sown on the slopes of the hills vegetates with so much the greater vigour. From the parallel of 24° to that of 30° the rains are seldomer, and of short duration. Happily the snow, of which there is great abundance from the 26° of lat. supplies the want of rain.

The extreme drought to which Nueva España is exposed from the month of June to the month of September, compels the inhabitants in a great part of this vast country to have recourse to artificial irrigations. The harvests of wheat are rich in proportion to the water taken from the rivers by means of canals of irrigation. This system is particularly followed in the fine plains which border the river Santiago, called Rio Grande, and in those between Salamanca, Irapuato, and the Villa de Leon. Canals of irrigation (*acequias*), reservoirs of water (*presas*), and the hydraulical machines called *norias*, are objects of the greatest importance for Mexican agriculture. Like Persia and the lower part of Peru, the interior of Nueva España is infinitely productive in nutritive gramina wherever the industry of man has diminished the natural dryness of the soil and the air.]

[No where does the proprietor of a large farm more frequently feel the necessity of employing engineers skilled in surveying ground and the principles of hydraulic constructions. However, at Mexico, as elsewhere, those arts have been preferred which please the imagination to those which are indispensable to the wants of domestic life. They possess architects, who judge learnedly of the beauty and symmetry of an edifice; but nothing is still so rare there as to find persons capable of constructing machines, dikes, and canals. Fotunately the feeling of their want has excited the national industry, and a certain sagacity peenliar to all mountainous people supplies in some sort the want of instruction.

In the places which are not artificially watered the Mexican soil yields only pasturage to the months of March and April. At this period, when the *s. w.* wind, which is dry and warm, (*viento de la Misteca*) frequently blows, all verdure disappears, and the gramina and other herbaceous plants gradually dry up. This change is more sensibly felt when the rains of the preceding year have been less abundant and the summer has been warmer. The wheat then, especially in the month of May, suffers much if it is not artificially watered. The rain only excites the vegetation in the month of June; with the first falls the fields become covered with verdure; the foliage of the trees is renewed; and the European, who recals to his mind incessantly the climate of his native country, enjoys doubly this season of the rains, because it presents to him the image of spring.

In indicating the dry and rainy months we have described the course which the meteorological phenomena commonly follow. For several years, however, these phenomena appear to have deviated from the general law, and the exceptions have unfortunately been to the disadvantage of agriculture. The rains have become more rare, and especially more tardy. It is observed in Mexico that the maize, which suffers much more than the wheat from the frosts in autumn, has the advantage of recovering more easily after long droughts. In the intendancy of Valladolid, between Salamanca and the lake of Cuizeo, are seen fields of maize which were believed to be destroyed, vegetate with an astonishing vigour after two or three days of rain. The great breadth of the leaves undoubtedly contributes greatly to the nutrition and vegetative force of this American gramen.

In the farms (*haciendas de trigo*) in which the system of irrigation is well established, in those of Silao and Irapuato, for example, near Leon, the wheat is twice watered; first, when the young plant springs up in the month of January; and the second time in the beginning of March, when the ear is on the point of developing itself. Sometimes even the whole field is inundated before sowing. It is observed, that in allowing the water to remain for several weeks, the soil is so impregnated with humidity that the wheat resists more easily the long droughts. They scatter the seed (*semer a la volée*) at the moment when the waters begin to flow from the opening of the canals. This method brings to mind the cultivation of wheat in Lower Egypt, and these prolonged inundations diminish at the same time the abundance of the parasitical herbs which mix with the harvest at reaping, and of which a part has unfortunately past into America with the European grain.

The riches of the harvests are surprising in lands carefully cultivated, especially in those which are watered or properly separated by different courses of labour. The most fertile part of the table-land is that which extends from Queretaro to the town of Leon. These elevated plains are 30 leagues in length by eight or ten in breadth. The wheat harvest is 35 and 40 for one, and several great farms can even reckon on 50 or 60 to one. An equal fertility is found in the fields which extend from the village of Santiago to Yurirapundaro in the intendancy of Valladolid. In the environs of Puebla, Atlisco, and Zelaya, in a great part of the bishoprics of Mechoacán and Guadalaxara, the produce is from 20 to 30 for one. A field is considered there as far from fertile when a fanega of wheat yields only, *communibus annis*, 16 tanegas. At Cholula the common harvest is from 30 to 40, but it frequently exceeds from 70 to 80 for one. In the valley of Mexico the maize yields 200, and the wheat 18 or 20. We have to observe, that the numbers which we here give have all the accuracy which can be desired in so important an object for the knowledge of territorial riches. Being eagerly desirous of knowing the produce of agriculture under the tropics, Humboldt procured all the information on the very spots; and compared it with the data with which he was furnished by intelligent colonists, who inhabited provinces at a distance from one another. He was induced to be so much the more precise in this operation, as from having been born in a country where grain scarcely produces four or five for one, he was naturally more apt than another to be disposed to suspect the exaggerations of agriculturists, exaggerations which are the same in Mexico, China, and wherever the vanity of the inhabitants wishes to take advantage of the credulity of travellers.]

[The same author was aware that on account of the great inequality with which different countries sow, it would have been better to compare the produce of the harvest with the extent of ground sown up. But the agrarian measures are so inexact, and there are so few farms in Mexico in which we know with precision the number of square toises or varas which they contain, that he was obliged to confine himself to the simple comparison between the wheat reaped and the wheat sown. The researches to which he applied himself during his stay in Mexico gave him for result, *communibus annis*, the mean produce of all the country at 22 or 25 for one. When he returned to Europe he began again to entertain doubts as to the precision of this important result, and he asserts, he should perhaps have hesitated to publish it, if had not had it in his power to consult on this subject quite recently, and in Paris even, a respectable and enlightened person who has inhabited the Spanish colonies these 30 years, and who applied himself with great success to agriculture. M. Abad, a canon of the metropolitan church of Valladolid de Mechoacán, assured him, that from his calculations the mean produce of the Mexican wheat, far from being below 22 grains, is probably from 25 to 30, which, according to the calculations of Lavoisier and Neckar, exceeds from five to six times the mean produce of France.

Near Zelaya the agriculturists shewed him the enormous difference of produce between the lands artificially watered and those which are not. The former, which receive the water of the Rio Grande, distributed by drains into several pools, yield from 40 to 50 for one; while the latter, which do not enjoy the benefit of irrigation, only yield 15 or 20. The same fault prevails here of which agricultural writers complain in almost every country of Europe, that of employing too much seed, so that the grain chokes itself. Were it not for this the produce of the harvests would still appear greater than what we have stated.

It may be of use to insert here an observation made near Zelaya by a person worthy of confidence, and very much accustomed to researches of this nature. M. Abad took at random, in a fine field of wheat of several acres in extent, 40 wheaten plants (*triticum hybernum*); he put the roots in water to clear them of all earth, and he found that every grain had produced 40, 60, and even 70 stalks. The ears were almost all equally well furnished. The number of grains which they contained was reckoned, and it was found that this number frequently exceeded 100 and even 120. The mean term appeared 90. Some ears even contained 160 grains. What an astonishing example of fertility! It is remarked in general that wheat divides enormously in the Mexican fields, that from a single grain a great number of stalks shoot up, and that each plant has extremely long and bushy roots. The Spanish colonists call this effect of the vigour of vegetation *el macollar del trigo*.

To the *n.* of this very fertile district of Zelaya, Salamanca, and Leon, the country is arid in the extreme, without rivers, without springs, and presenting vast extents of crusts of hardened clay (*tepetate*), which the cultivators call *hard* and *cold* lands, and through which the roots of the herbaceous plants with difficulty penetrate. These beds of clay, which are also found in the kingdom of Quito, resemble at a distance banks of rock destitute of every sort of vegetation. They belong to the *trappish formation*, and constantly accompany on the ridge of the Andes of Peru and Mexico the basaltes, the grünstein, the amygdaloid, and the amphibolic porphyry. But in other parts of Nueva España, in the beautiful valley of Santiago, and to the *s.* of the town of Valladolid, the decomposed basaltes and amygdaloids have formed in the succession of ages a black and very productive earth. The fertile fields which surround the Alberca of Santiago bring to mind the basaltic districts of the Mittelgebirge of Bohemia.

All the table-land which extends from Sombrerete to the Saltillo, and from thence towards La Punta de Lampazos, is a naked and arid plain, in which cactus and other prickly plants only vegetate. The sole vestige of cultivation is on some points, where, as around the town of Saltillo, the industry of man has procured a little water for the watering of the fields. We have also traced under its proper head a view of Old California, of which the soil is a rock both destitute of earth and water. All these considerations concur to prove, that on account of its extreme dryness a considerable part of Nueva España situate to the *n.* of the tropic is not susceptible of a great population. Hence, what a remarkable contrast between the physiognomy of two neighbouring countries, between Mexico and the United States of N. America! In the latter the soil is one vast forest, intersected by a great number of rivers, which flow into spacious gulfs; while Mexico presents from *e.* to *w.* a wooded shore, and in its centre an enormous mass of colossal mountains, on the ridge of which stretch out plains destitute of wood, and so much the more arid, as the temperature of the ambient air is augmented by the reverberation of the solar rays. In the *n.* of Nueva España, as in Thibet, Persia, and all] [the mountainous regions, a part of the country will never be adapted for the cultivation of cerealia till a concentrated and highly civilized population shall have vanquished the obstacles opposed by nature to the progress of rural economy. But this aridity, we repeat it, is not general; and it is compensated for by the extreme fertility observable in the *s.* countries, even in that part of the *provincias internas* in the neighbourhood of rivers, in the basins of the Rio del Norte, the Gila, the Hiaqui, the Mayo, the Culiacan, the Rio del Rosario, the Rio de Conchos, the Rio de Santander, the Tigre, and the numerous torrents of the province of Texas.

In the most *n.* extremity of the kingdom, on the coast of New California, the produce of wheat is from 16 to 17 for one, taking the mean term among the harvest of 18 villages for two years. We believe that agriculturists will peruse with pleasure the detail of these harvests in a country situated under the same parallel as Algiers, Tunis, and Palestine, between the 32° 39′ and 37° 48′ of lat.

Names of the villages of New California.	1791. Fanegas of wheat.		1802. Fanegas of wheat.		Harvest considered as multiple of the grain sown.	
	Sown.	Reaped.	Sown.	Reaped.	1791.	1802.
San Diego	60	3021	....	....	$50\frac{3}{10}$	....
San Luis Rey de Francia	....	....	100	1200	....	12
San Juan Capistrano	80	1586	103	2908	$19\frac{8}{10}$	$28\frac{2}{10}$
San Gabriel	178	3700	282	3800	$20\frac{7}{10}$	$13\frac{4}{10}$
San Fernando	....	....	100	2800	....	28
San Buenaventura	44	259	96	3500	$5\frac{8}{10}$	$36\frac{4}{10}$
Santa Barbara	65	1500	113	2876	23	$25\frac{4}{10}$
La Purissima Concepcion	76	800	96	3500	$10\frac{5}{10}$	$36\frac{4}{10}$
San Luis Obispo	86	1078	161	4000	$12\frac{5}{10}$	$25\frac{4}{10}$
San Miguel	....	....	70	1600	....	$22\frac{8}{10}$
Soledad	....	....	78	500	....	$6\frac{4}{10}$
San Antonio de Padua	90	952	139	1200	$10\frac{5}{10}$	$8\frac{7}{10}$
San Carlos	71	221	60	240	$3\frac{1}{10}$	4
San Juan Baptista	....	....	52	1200	....	$23\frac{1}{10}$
Santa Cruz	....	....	60	550	....	$9\frac{1}{10}$
Santa Clara	64	1400	129	2000	$21\frac{8}{10}$	$15\frac{5}{10}$
San Jose	....	....	84	1200	....	$14\frac{3}{10}$
San Francisco	60	680	233	2322	$11\frac{3}{10}$	$9\frac{9}{10}$
	874	15,197	1956	35,396	$17\frac{4}{10}$	$17\frac{2}{10}$

It appears that the most *n.* part of this coast is less favourable to the cultivation of wheat than that which extends from San Diego to San Miguel. However, in newly cultivated grounds the produce of the soil is more unequal than in lands which have been long under cultivation, though we observe in no part of Nueva España that progressive diminution of fertility which is so distressing to new colonists wherever forests have been converted into arable land.

Those who have seriously reflected on the riches of the Mexican soil know that by means of a more careful cultivation, and without supposing any extraordinary labour in the irrigation of the soil, the portion of ground already under cultivation might furnish subsistence for a population eight or ten times more numerous. If the fertile plains of Atlixco, Cholula, and Puebla, do not produce very abundant harvests, the principal cause ought to be sought for in the want of consumers, and in the obstacles opposed by the inequality of the soil to the interior commerce of grain, especially to its carriage towards the Atlantic coast. We shall afterwards return to this interesting subject when we come to treat of the exportation from Vera Cruz.]

[What is actually the produce of the grain harvest in the whole of Nueva España? We can conceive how difficult must be the resolution of this problem in a country where the government, since the death of the Count de Revillagigedo, has been very unfavourable to statistical researches. In France, even the estimations of Quesnay, Lavoisier, and Arthur Young, vary from 45 and 50 to 75 millions of septiers of 117 kilogrammes (11,620, 12,911, and 19,366 millions of pounds avoird.) in weight. We have no positive data as to the quantity of rye and oats reaped in Mexico, but we conceive ourselves enabled to calculate approximately the mean produce of wheat. The most sure estimate in Europe is the computed consumption of each individual. This method was successfully employed by MM. Lavoisier and Arnould; but it is a method which cannot be followed in the case of a population composed of very heterogeneous elements. The Indian and Mestizo, the inhabitants of the country, are only fed on maize and manioc bread. The white Creoles who live in great cities consume much more wheaten bread than those who habitually live on their farms. The capital, which includes more than 33,000 Indians, requires annually 19,000,000 of kilogrammes of flour. This consumption is almost the same as that of the cities of Europe of an equal population; and if, according to this basis, we were to calculate the consumption of the whole kingdom of Nneva España, we should attain to a result which would be five times too high.

From these considerations we prefer the method which is founded on partial estimations. The quantity of wheat reaped in 1802 in the intendancy of Guadalaxara was, according to the statistical table communicated by the intendant of this province to the chamber of commerce at Vera Cruz, 43,000 *cargas*, or 645,000 kilogrammes. Now the population of Guadalaxara is nearly a ninth of the total population. In this part of Mexico there is a great number of Indians who eat maize bread, and there are few populous cities inhabited by whites in easy circumstances. According to the analogy of this partial harvest, the general harvest of Nueva España would only be 59,000,000 kilogrammes. But if we add 36,000,000 of kilogrammes on account of the beneficial influence of the consumption of the cities of Mexico, Puebla, and Guanaxuato, on the cultivation of circumjacent districts, and on account of the *provincias internas*, of which the inhabitants live almost exclusively on wheaten bread, we find for the whole kingdom nearly 10,000,000 of myriagrammes, or upwards of 220,500,000 of pounds avoird. This estimate gives too small a result, because in the above calculation we have not suitably separated the *n.* provinces from the equinoctial region. This separation is dictated, however, by the very nature of the population.

In the *provincias internas* the greatest number of the inhabitants are either white or reputed white; and they are calculated at 400,000. Supposing their consumption of wheat equal to that of the city of Puebla, we shall find 6,000,000 of myriagrammes. We may admit, calculating according to the annual harvest of the intendancy of Guadalaxara, that in the *s.* regions of Nueva España, of which the mixed population is estimated at 5,437,000, the consumption of wheat in the country amounts to 5,800,000 myriagrammes. If we add 3,600,000 myriagrammes for the consumption of the great interior cities of Mexico, Puebla, and Guanaxuato, we shall find the total consumption of Nueva España above 15,000,000 of myriagrammes, or 331,000,000 of pounds avoird.

We might be astonished to find from this calculation that the *provincias internas*, of which the population is only a fourteenth of the whole population, consume more than the third of the harvest of Mexico. But we must not forget that in these *n.* provinces the number of whites is to the total mass of Spaniards, (Creoles and Europeans), as one to three, and that it is principally this cast by which the wheaten flour is consumed. Of the 800,000 whites who inhabit the equinoctial region of Nueva España, nearly 150,000 live in an excessively warm climate in the plains adjacent to the coast, and feed on manioc and bananas. These results, we repeat, are merely simple approximations.

In France the whole grain harvest, that is to say, wheat, rye, and barley, was, according to Lavoisier, before the revolution, and consequently at a period when the population of the kingdom amounted to 25,000,000 of inhabitants, 58,000,000 of setiers, or 6,786,000,000 of kilogrammes. Now, according to the authors of the *Feuille du Cultivateur*, the wheat reaped in France is to the whole mass of grain as 5:17. Hence the produce of wheat alone was, previous to 1789, 17,000,000 of setiers, which, taking merely absolute quantities, and without considering the populations of the two empires, is nearly 13 times more than the produce of wheat in Mexico. This comparison agrees very well with the bases of the anterior estimation. For the number of inhabitants of Nueva España who habitually live on wheaten bread does not exceed 1,300,000; and it is well known that the French consume more bread than the Spa-]

[nish race, especially those who inhabit America.

But on account of the extreme fertility of the soil, the 15,000,000 of myriagrammes annually produced by Nueva España are reaped on an extent of ground four or five times smaller than would be requisite for the same harvest in France. We may expect, it is true, as the Mexican population shall increase, that this *fertility*, which may be called *medium*, and which indicates a total produce of 24 for one, will decrease. Every where men begin with the cultivation of the least arid lands, and the mean produce must naturally diminish when agriculture embraces a greater extent, and consequently a greater variety of ground. But in a vast empire like Mexico this effect can only be very tardy in its manifestation, and the industry of the inhabitants increases with the population and the number of increasing wants.

We shall collect into one table the knowledge which we have acquired as to the mean produce of the cerealia in the two continents. We are not here adducing examples of an extraordinary fertility observable in a small extent of ground, nor of grain sown according to the Chinese method. The produce would nearly be the same in every zone, if, in choosing our ground, we were to bestow the same care on cerealia which we bestow on our garden plants. But in treating of agriculture in general, we speak merely of extensive results, of calculations, in which the total harvest of a country is considered as the *multiple* of the quantity of wheat sown. It will be found that this multiple, which may be considered as one of the first elements of the prosperity of nations, varies in the following manner:

5 to 6 grains for 1, in France, according to Lavoisier and Neckar. We estimate, with M. Penchet, that 4,100,000 *arpens* sown with wheat yield annually 5,280,000,000 of pounds, which amounts to 1173 kilogrammes per hectare, (2588 lb. avoird. per 107,639 square feet). This is also the mean produce in the *n.* of Germany, Poland, and, according to M. Rühs, in Sweden. They reckon in France in some remarkably fertile districts of the departments of L'Escant and Le Nord 15 for 1; in the good land of Picardy and the isle of France from 8 to 10 for 1; and in the lands of less fertility from 4 to 5 for 1. (*Penchet, Statistique*, p. 290).

8 to 10 grains for 1 in Hungary, Croatia, and Sclavonia, according to the researches of M. Swartner.

12 grains for 1 in the Reyno de la Plata, especially in the environs of Montevideo, according to Don Felix Azara. Near the city of Buenos Ayres they reckon even 16. In Paraguay the cultivation of cerealia does not extend farther *n.* than the parallel of 24°, (*Voyage d'Azara*, t. i. p. 140).

17 grains for 1 in the *n.* part of Mexico, and at the same distance from the equator as Paraguay and Buenos Ayres.

24 grains for 1 in the equinoctial region of Mexico at 2 or 3000 metres of elevation above the level of the ocean. They reckon 5000 kilogrammes per hectare, or 11,035 lb. avoird. per 107,639 square feet. In the province of Pasto of the kingdom of Santa Fe, the plains of La Vega de San Lorenzo, Pansitara, and Almaguer, lat. 1° 54′ *n.* commonly produce 25, in very fertile years 35, and in cold and dry years 12 for 1. In Peru, in the beautiful plain of Caxamarca, lat. 7° 8′ *n.* watered by the rivers Mascon and Utusco, and celebrated from the defeat of the Inca Atahualpa, wheat yields from 18 to 20 for 1.

The Mexican flour enters into competition at the Havanah market with that of the United States. When the road which is constructing from the table-land of Perote to Vera Cruz shall be completely finished, the grain of Nueva España will be exported for Bourdeaux, Hamburgh, and Bremen. The Mexicans will then possess a double advantage over the inhabitants of the United States, that of a greater fertility of territory, and that of a lower price of labour. It would be very interesting in this point of view could we compare here the mean produce of the different provinces of the American confederation with the results which we have obtained for Mexico. But the fertility of the soil and the industry of the inhabitants vary so much in different provinces, that it becomes difficult to find the mean term which corresponds to the total harvest. What a difference between the excellent cultivation of the environs of Lancaster and several parts of New England and that of N. Carolina! "An English farmer," says the immortal Washington in one of his letters to Arthur Young, "ought to have a horrid idea of the state of our agriculture, or the nature of our soil, when he is informed that an acre with us only produces eight or 10 bushels. But it must be kept in mind that in all countries where land is cheap and labour dear, men are fonder of cultivating much than cultivating well. Much ground has been scratched over, and none cultivated as it ought to have been." According to the recent researches of M. Blodget,] [which may be regarded as sufficiently exact, we find the following results:

In the Atlantic provinces to the e. of the Alleghany mountains.	per acre.	per hectare.
	bushels.	kilogrammes
In rich lands - - - -	32	2372
In common lands - - -	9	667
In the w. territory between the Alleghany and the Mississipi.		
In rich lands - - - -	40	2965
In common lands - - -	25	1853

We see from these *data*, that in the Mexican intendancies of Puebla and Guanaxuato, where on the ridge of the *cordillera* the climate of Rome and Naples prevails, the territory is more rich and productive than the most fertile parts of the United States; the comparative fertility, taking the highest of the American produce, being 5000 : 2965.

As since the death of General Washington the progress of agriculture has been very considerable in the *w.* territory, especially in Kentucky, Tennessee, and Louisiana, we believe we may consider from 13 to 14 bushels as the mean term of the annual produce, which, however, only amounts to 700 kilogrammes (less than 13 bushels) per hectare, or less than four for one. In England the wheat harvest is generally estimated at from 19 to 20 bushels per acre, which gives 1100 kilogrammes per hectare. This comparison, we have to repeat, does not announce a greater fertility of the soil of Great Britain. Far from giving us an unfavourable idea of the sterlity of the Atlantic provinces of the United States, it proves only that whenever the colonist is master of a vast extent of ground, the art of cultivating the soil comes extremely slow to perfection. The Memoirs of the Agricultural Society of Philadelphia furnish us with different examples of harvests exceeding 38 and 40 bushels per acre, whenever the fields have been laboured in Philadelphia with the same care as in Ireland and Flanders.

After comparing the mean produce of the lands in Mexico and Buenos Ayres with those in the United States and France, let us bestow a rapid glance at the price of labour in these different countries. In Mexico it amounts to two reals de plata (1*s*. 1*d*.) per day in the cold regions, and to two reals and a half (1*s*. 4½*d*.) per day in the warm regions, where there is a want of hands, and where the inhabitants in general are very lazy. This price of labour ought to appear moderate enough when we consider the metallic wealth of the country, and the quantity of money constantly in circulation. In the United States, where the whites have pushed the Indian population beyond the Ohio and the Mississipi, the price of labour varies from 3 livres 10 sols, to 4 francs, (from 2*s*. 11*d*. to 3*s*. 4*d*.) In France we may estimate it from 30 to 40 sols, (from 1*s*. 3*d*. to 1*s*. 8*d*.) and in Bengal, according to M. Titzing, at six sols, or 3*d*. Hence, notwithstanding the enormous difference of freight, the E. India sugar is cheaper at Philadelphia than that of Jamaica. From these data it follows, that the present price of labour in Mexico is to the price of labour,

In France - - - = 12 : 6.
In the United States = 26 : 13.
In Bengal - - - = 2 : 1.

The mean price of wheat is in Nueva España from four to five piastres, or from 20 to 25 francs, the carga, which weighs 150 kilogrammes, or from 17*s*. 6*d*. to 21*s*. 10*d*. This is the price at which it is purchased in the country, even from the firmers. At Paris, for several years, 150 kilogrammes of wheat cost 30 francs. In the city of Mexico the high price of carriage adds so much to the price of the grain, that it generally sells there at 9 and 10 piastres the carga, (that is to say, from 1*l*. 17*s*. 6*d*. to 2*l*. 3*s*. 4*d*.) The extremes, at the periods of the greatest or least fertility, are 8 and 14 piastres. It is easy to foresee that the price of Mexican grain will suffer a considerable fall when the roads shall be constructed on the declivity of the *cordilleras*, and the progress of agriculture shall be favoured by greater commercial freedom.

The Mexican wheat is of the very best quality; and it may be compared with the finest Andalusian grain. It is superior to that of Monte Video, which, according to M. Azara, has the grain smaller by one half than the Spanish grain. In Mexico the grain is very large, very white, and very nutritive, especially in farms where watering is employed. It is observed that the wheat of the mountains (*trigo de sierra*), that it is to say, that which grows at very great elevations on the ridge of the *cordillera*, has its grain covered with a thicker husk, while the grain of the temperate regions abounds in glutinous matter. The quality of the flour depends principally on the proportion which exists between the gluten and starch, and it appears natural that, under a climate favourable] [to the vegetation of gramina, the embryo and the cellular reticulation of the albumen should become more voluminous.

In Mexico grain is with difficulty preserved for more than two or three years, especially in the temperate climates, and the causes of this phenomenon have never been sufficiently attended to. It would be advisable to establish magazines in the coldest parts of the country. We find, however, a prejudice spread through several parts of Spanish America, that the flour of the *cordillera* does not preserve so long as the flour of the United States. The cause of this prejudice, which has been of particular detriment to the agriculture of New Granada, is easily to be discovered. The merchants who inhabit the coasts opposite to the W. Indies, and who find themselves constrained by commercial prohibitions, particularly the merchants of Carthagena for example, have the greatest interest in maintaining a connection with the United States. The custom-house officers are sometimes indulgent enough to take a Jamaica vessel for a vessel of the United States.

Rye, and especially barley, resists cold better than wheat. They are cultivated on the highest regions. Barley yields abundant harvests at heights where the thermometer rarely keeps up during the day beyond 14°, or 57° of Fahrenheit. In New California, taking the term of the harvests of 13 villages, the barley produced in 1791, 24, and in 1802, 18 for 1.

Oats are very little cultivated in Mexico. They are even very seldom seen in Spain, where the horses are fed on barley, as in the times of the Greeks and Romans. The rye and barley are seldom attacked by a disease called by the Mexicans *chaquistle*, which frequently destroys the finest wheat harvests when the spring and the beginning of the summer have been very warm, and when storms are frequent. It is generally believed that this disease is occasioned by small insects, which fill the interior of the stalk, and hinder the nutritive juice from mounting up to the ear.

A plant of a nutritive root, which belongs originally to America, the potato (*solanum tuberosum*), appears to have been introduced into Mexico nearly at the same period as the cerealia of the old continent. We shall not take upon ourselves to decide whether the *papas* (the old Peruvian name by which potatoes are now known in all the Spanish colonies) came to Mexico along with the schinus molle of Peru, and consequently by the S. sea; or whether the first conquerors brought them from the mountains of New Granada. However this may be, it is certain that they were not known in the time of Montezuma; and this fact is the more important, because it is one of those in which the history of the migrations of a plant is connected with the history of the migrations of nations.

The predilections manifested by certain tribes for the cultivation of certain plants, indicates most frequently either an identity of race, or ancient communications between men who live under different climates. In this view the vegetables, like the languages and physiognomy of nations, may become historical monuments. Not merely pastoral tribes, or those who live solely on the chase, undertake long voyages, instigated by an unquiet and warlike spirit; the hordes of Germanic origin, the swarm of people who transported themselves from the interior of Asia to the banks of the Borysthenes and the Danube, and the savages of Guayana, afford numerous examples of tribes, who, fixing themselves for a few years, cultivate small pieces of ground, on which they sow the grain reaped by them elsewhere, and abandon these imperfect cultivations when a bad year, or any other accident, disgusts them with the situation. It is thus that the people of the Mongol race have transported themselves from the wall which separates China from Tartary to the very centre of Europe; and it is thus that, from the *n.* of California and the banks of the Rio Gila, the American tribes poured even into the *s.* hemisphere. We every where see torrents of wandering and warlike hordes pave a way for themselves through the midst of peaceable and agricultural nations. Immoveable as the shore, the latter collect and carefully preserve the nutritive plants and domestic animals which accompanied the wandering tribes in these distant courses. Frequently the cultivation of a small number of vegetables, as well as the foreign words mingled with languages of a different origin, serve to point out the route by which a nation has passed from one extremity of the continent to the other.

These considerations are sufficient to prove how important it is for the history of our species to know with precision how far the primitive dominion of certain vegetables extended before the spirit of colonization among the Europeans collected together the productions of the most distant climates. If the cerealia, if the rice of the E. Indies, were unknown to the first inhabitants of America, on the other hand, maize, the potato, and the quinoa, were neither cultivated in *e.* Asia, nor in the] [islands of the S. sea. Maize was introduced into Japan by the Chinese, who, according to the assertion of some authors, ought to have known it from the remotest period. (*Thunberg, Flora Japonica*, p. 37.) This assertion, if it was founded, would throw light on the ancient communications supposed to have taken place between the inhabitants of the two continents. But where are the monuments which attest that maize was cultivated in Asia before the 16th century? According to the learned researches of Father Gaubil, (see astronomical MS. of the Jesuits preserved in the *Bureau des Longitudes* at Paris,) it appears even doubtful whether, a thousand years before that period, the Chinese ever visited the *w.* coast of America, as was advanced by a justly celebrated historian, M. de Guignes. We persist in believing that the maize was not transported from the table-land of Tartary to that of Mexico, and that it is equally improbable that, before the discovery of America by the Europeans, this precious gramen was transported from the new continent into Asia.

The potato presents us with another very curious problem, when we consider it in a historical point of view. It appears certain, as we have already advanced, that this plant, of which the cultivation has had the greatest influence on the progress of population in Europe, was not known in Mexico before the arrival of the Spaniards. It was cultivated at this epocha in Chile, Peru, Quito, in the kingdom of New Granada, on all the *cordillera* of the Andes, from lat. 40° *s.* to 50° *n.* It is supposed by botanists that it grows spontaneously in the mountainous part of Peru. On the other hand, the learned who have inquired into the introduction of potatoes into Europe, affirm that the potato was found in Virginia by the first settlers sent there by Sir Walter Raleigh in 1584. Now, how can we conceive that a plant, said to belong originally to the *s.* hemisphere, was found under cultivation at the foot of the Alleghany mountains, while it was unknown in Mexico and the mountainous and temperate regions of the W. Indies? Is it probable that Peruvian tribes may have penetrated *n.* to the banks of the Rapahannoc in Virginia; or have potatoes first come from *n.* to *s.* like the nations who from the 7th century have successively appeared on the table-land of Anahuac? In either of these hypotheses, how came this cultivation not to be introduced or preserved in Mexico? These are questions which have hitherto been very little agitated, but which, nevertheless, deserve to fix the attention of the naturalist, who, in embracing at one view the influence of man on nature, and the re-action of the physical world on man, appears to read in the distribution of the vegetables the history of the first migrations of our species.

We have first to observe, stating here only what facts are to be relied on, that the potato is not indigenous in Peru, and that it is nowhere to be found wild in the part of the *cordilleras* situated under the tropics. Humboldt and Bonpland herborized on the back and on the declivity of the Andes from 5° *n.* to 12° *s.*; they informed themselves from persons who have examined this chain of colossal mountains as far as La Pas and Oruro, and are certain that in this vast extent of ground no species of solanum with nutritive root vegetates spontaneously. It is true that there are places not very accessible, and very cold, which the natives call *Paramos de las Papas*, (desert potato plains); but these denominations, of which it is difficult to conjecture the origin, by no means indicate that these great elevations produce the plant of which they bear the name.

Passing further *s.* beyond the tropic, we find it, according to Molina, in all the fields of Chile, (*Hist. Nat. de Chile*, p. 102.) The natives distinguish the wild potato, of which the tubercles are small and somewhat bitter, from that which has been cultivated for a long series of ages. The first of these plants bears the name of *maglia*, and the second that of *pogny*. Another species of solanum is also cultivated in Chile, which belongs to the same group, with pennated and not prickly leaves, and which has a very sweet root of a cylindrical form. This is the *solanum cari*, which is still unknown, not only in Europe, but also in Quito and Mexico.

We might ask if these useful plants are truly natives of Chile, or if, from the effect of a long cultivation, they have become wild there. The same question has been put to the travellers who have found cerealia growing spontaneously in the mountains of India and Caucasus. MM. Ruiz and Pavon, whose authority is of so great weight, affirm that they found the potato in cultivated grounds, *in cultis*, and not in forests, and on the ridges of the mountains. But we are to observe, that among us the solanum and the different kinds of grain do not propagate of themselves in a durable manner, when the birds transport the grains into meadows and woods. Wherever these plants appear to become wild under our eyes, far from multiplying like the erigeron Canadense, the oenothera biennis, and other colonists of the vegetable kingdom, they disappear in a very short] [space of time. Are not the maglia of Chile, the grain of the banks of the Terek, and the wheat of the mountains (hill-wheat) of Boutan, which M. Banks (*Bibl. Britt.* 1809, n. 322, p. 86) has recently made known, more likely to be the primitive type of the solanum and cultivated cerealia?

It is probable that from the mountains of Chile the cultivation of potatoes gradually advanced *n.* by Peru and the kingdom of Quito to the table-land of Bogota, the ancient Cundinamarca. This is also the course followed by the Incas in their conquests. We can easily conceive why long before the arrival of Manco Capac, in those remote times when the province of Collao and the plains of Tiahuanacu were the centre of the first civilization of mankind, (*Pedro Cieca de Leon*, c. 105. *Garcilasso*, iii. 1), the migrations of the S. American nations would rather be from *s.* to *n.* than in an opposite direction. Every where in the two hemispheres the people of the mountains have manifested a desire to approach the equator, or at least the torrid zone, which, at great elevations, affords the mildness of climate and the other advantages of the temperate zone. Following the direction of the *cordilleras*, either from the banks of the Gila to the centre of Mexico, or from Chile to the beautiful valleys of Quito, the natives found in the same elevations, and without descending towards the plains, a more vigorous vegetation, less premature frosts, and less abundance of snow. The plains of Tiahuanacu (lat. 17° 10′ *s.*) covered with ruins of an august grandeur, and the banks of the lake of Chucuito, a basin which resembles a small interior sea, are the Himala and Thibet of S. America. These men, under the government of laws, and collected together on a soil of no great fertility, first applied themselves to agriculture. From this remarkable plain, situated between the cities of Cuzco and La Paz, descended numerous and powerful tribes, who carried their arms, language, and arts even to the *n.* hemisphere.

The vegetables which were the object of the agriculture of the Andes, must have been carried *n.* in two ways; either by the conquests of the Incas, who were followed by the establishment of Peruvian colonies in the conquered countries, or by the slow but peaceable communications which always take place between neighbouring nations. The sovereigns of Cuzco did not extend their conquests beyond the river of Mayo (lat. 1° 34′ *n.*) of which the course is *n.* from the town of Pasto. The potatoes which the Spaniards found under cultivation among the Muysca tribes in the king- of the *zaque* of Bogota (lat. 4° 6′ *n.*) could only have been transported there from Peru by means of the relations which are gradually established even among mountainous tribes separated from one another by deserts covered with snow, or impassable valleys. The *cordilleras*, which preserve a formidable height from Chile to the province of Antioquia, fall suddenly near the sources of the great Rio Atracto. Choco and Darien present merely a group of hills, which, in the isthmus of Panama, are only a few hundred toises in height. The cultivation of the potato succeeds well in the tropics only on very elevated grounds in a cold and foggy climate. The Indian of the warm regions gives the preference to maize, the manioc, and banana. Besides Choco, Darien, and the isthmus, covered with thick forests, have always been inhabited by hordes of savages and hunters, enemies to every sort of cultivation. We are not, therefore, to be astonished that both physical and moral causes have prevented the potato from penetrating into Mexico.

We know not a single fact by which the history of S. America is connected with that of N. America. In Nueva España, as we have already several times observed, the flux of nations was from *n.* to *s.* A great analogy of manners and civilization has been thought to be perceived between the Toultecs, driven by a pestilence from the table-land of Anahuac in the middle of the 12th century, and the Peruvians under the government of Manco Capac. It might, no doubt, have happened, that people from Aztlan advanced beyond the isthmus or gulf of Panama; but it is very improbable that by migrations from *s.* to *n.* the productions of Peru, Quito, and New Granada, ever passed to Mexico and Canada.

From all these considerations it follows, that if the colonists sent out by Raleigh really found potatoes among the Indians of Virginia, we can hardly refuse our assent to the idea that this plant was originally wild in some country of the *n.* hemisphere, as it was in Chile. The interesting researches carried on by MM. Beckman, Banks, and Dryander, (*Beckmann's Grundsätze der Teutschen Landwirthschaft*, 1806, p. 289. *Sir Joseph Banks's Attempt to ascertain the Time of the Introduction of Potatoes*, 1808,) prove that vessels which returned from the bay of Albemarle in 1586, first carried potatoes into Ireland, and that Thomas Harriot, more celebrated as a mathematician than as a navigator, described this nutritive root by the name of *openawk*. Gerard, in his Herbal, published in 1597, calls it Virginian patatate, or *norembega*. We might be tempted to believe that the English colonists received it from Spanish America.]

[Their establishment had been in existence from the month of July 1584. The navigators of those times were not in the habit of steering straight *w.* to reach the coast of N. America; they were still in the practice of following the tract indicated by Columbus, and profiting by the trade winds of the torrid zone. This passage facilitated communication with the W. India islands, which were the centre of the Spanish commerce. Sir Francis Drake, who had been navigating among these islands, and along the coast of Tierra Firme, put in at Roanoke, in Virginia. It appears then natural enough to suppose, that the English themselves brought potatoes from S. America or from Mexico into Virginia. At the time when they were brought from Virginia into England they were common both in Spain and Italy. We are not then to be astonished that a production which had past from one continent to the other, could in America pass from the Spanish to the English colonies. The very name by which Harriot describes the potato seems to prove its Virginian origin. Were the savages to have a word for a foreign plant, and would not Harriot have known the name *papa?*

The plants which are cultivated in the highest and coldest part of the Andes and Mexican *cordilleras* are the potato, the *tropæolum esculentum*, and the *chenopodium quinoa*, of which the grain is an aliment equally agreeable and healthy. In Nueva España the first of these becomes an object of cultivation, of so much greater importance from its extent, as it does not require any great humidity of soil. The Mexicans, like the Peruvians, can preserve potatoes for whole years by exposing them to the frost and drying them in the sun. The root, when hardened and deprived of its water, is called *chunu*, from a word of the Quichua language. It would be undoubtedly very useful to imitate this preparation in Europe, where a commencement of germination frequently destroys the winter's provisions; but it would be still of greater importance to procure the grain of the potatoes cultivated at Quito and on the plain of Santa Fé. These roots have been seen of a spherical form of more than three decimetres (11 inches) in diameter, and of a much better taste than any in our continent. We know that certain herbaceous plants which have been long multiplied from the roots, degenerate in the end, especially when the bad custom is followed of cutting the roots into several pieces. It has been proved by experience in several parts of Germany, that, of all the potatoes, those which grow from the seed are the most savoury. We may ameliorate the species by collecting the seed in its native country, and by choosing on the *cordillera* of the Andes the varieties which are most recommendable from their volume and the savour of their roots. We have long possessed in Europe a potato which is known by agricultural writers under the name of red potato of Bedfordshire, and of which the tubercles weigh more than a kilogramme, or $2\frac{1}{10}$ lb. avoird. but this variety (*conglomerated potato*) is of an insipid taste, and can almost be applied only to feed cattle, while the *papad chogota*, which contains less water, is is very farinaceous, contains very little sugar, and is of an extremely agreeable taste.

Amongst the great number of useful productions which the migrations of nations and distant navigations have made known, no plant since the discovery of cerealia, that is to say from time immemorial, has had so decided an influence on the prosperity of mankind as the potato. This root, according to the calculations of Sir John Sinclair, can maintain nine individuals per acre of 5368 square metres, or 55,356 square feet. It has become common in New Zealand, (see *John Savage's Account of New Zealand*, 1807, p. 18), in Japan, in the island of Java, in the Boutan, and in Bengal, where, according to the testimony of M. Bockford, potatoes are considered as more useful than the bread-fruit tree introduced at Madras. Their cultivation extends from the extremity of Africa to Labrador, Iceland, and Lapland. It is a very interesting spectacle to see a plant descended from the mountains under the equator advance towards the pole, and resist better than the cereal gramina all the colds of the *n.*

We have successively examined the vegetable productions which are the basis of the food of the Mexican population, the banana, the manioc, the maize, and the cerealia; and we have endeavoured to throw some interest into this subject by comparing the agriculture of the equinoctial regions with that of the temperate climate of Europe, and by connecting the history of the migration of the vegetables with the events which have brought the human race from one part of the globe to the other. Without entering into botanical details, which would be foreign to the aim of this part of the work, we shall terminate this chapter by a succinct indication of the other alimentary plants which are cultivated in Mexico.

A great number of these plants has been introduced since the 16th century. The inhabitants of *w.* Europe have deposited in America what they had been receiving for 2000 years by their communications with the Greeks and Romans, by the irruption of the hordes of central Asia, by the] [conquests of the Arabs, by the crusades, and by the navigations of the Portuguese. All these vegetable treasures, accumulated in an extremity of the old continent by the continual flux of nations towards the *w.* and preserved under the happy influence of a perpetually increasing civilization, have become almost at once the inheritance of Mexico and Peru. We see them afterwards augmented by the productions of America, pass farther still to the islands of the S. sea, and to the establishments which a powerful nation has formed on the coast of New Holland. In this way the smallest corner of the earth, if it become the domain of European colonists, and especially if it abounds with a great variety of climates, attests the activity which our species has been for centuries displaying. A colony collects in a small space every thing most valuable which wandering man has discovered over the whole surface of the globe.

America is extremely rich in vegetables with nutritive roots. After the manioc and the papas, or potatoes, there are none more useful for the subsistence of the common people than the oca (*oxalis tuberosa*), the batate, and the igname. The first of these productions only grows in the cold and temperate climates, on the summit and declivity of the *cordilleras;* and the two others belong to the warm region of Mexico. The Spanish historians, who have described the discovery of America, confound the words *axes* and *batates*, though the one means a plant of the group of asparagus, and the other a convolvulus.

The igname, or *dioscorea alata*, like the banana, appears proper to all the equinoctial regions of the globe. The account of the voyage of Aloysio Cadamusto (*Cadamusti Navigatio ad Terras incognitas. Grynæus Orb. nov. p.* 47) informs us that this root was known by the Arabs. Its American name may even throw some light on a very important fact in the history of geographical discoveries, which never appears hitherto to have fixed the attention of the learned. Cadamusto relates, that the king of Portugal sent in 1500 a fleet of 12 vessels round the cape of Good Hope to Calcutta, under the command of Pedro Aliares. This admiral, after having seen the Cape Verd islands, discovered a great unknown land, which he took for a continent. He found there naked men, swarthy, painted red, with very long hair, who plucked out their beards, pierced their chins, slept in hammocks, and were entirely ignorant of the use of metals. From these traits we easily recognise the natives of America. But what renders it extremely probable that Aliares either landed on the coast of Paria or on that of Guayana, is, that he said he found in cultivation there a species of millet (maize), and a root of which bread is made, and which bears the name of *igname*. Vespucci had heard the same word three years before pronounced by the inhabitants of the coast of Paria. The Haitian name of the dioscorea alata is *axes* or *ajes*. It is under this denomination that Columbus describes the igname in the account of his first voyage; and it is also that which it had in the times of Garcilasso, Acosta, and Oviedo, who have very well indicated the characters by which the *axes* are distinguished from batates. (See *Christophori Columbi Navigatio*, c. lxxxix. *Comentarios Reales*, t. i. p. 278. *Historia natural de Indas*, p. 242. *Oviedo*, libro vii. c. 3.)

The first roots of the dioscorea were introduced into Portugal in 1596, from the small island of St. Thomas, situated near the coast of Africa, almost under the equator, (*Clusii Rariorum Plantarum Hist.* lib. iv. p. 77). A vessel which brought slaves to Lisbon had embarked these ignames to serve for food to the Negroes in their passage. From similar circumstances several alimentary plants of Guinea have been introduced into the W. Indies. They have been carefully propagated, for the sake of furnishing the slaves with a diet to which they have been accustomed in their native country. It is observed that the melancholy of these unfortunate beings diminishes sensibly when they discover the plants familiar to them in their infancy.

In the warm regions of the Spanish colonies the inhabitants distinguish the *axe* from the *ñamas* of Guinea. The latter came from the coast of Africa to the W. Indies, and the name of *igname* has gradually prevailed there over *axe*. These two plants are only, perhaps, varieties of the dioscorea alata, although Brown has endeavoured to elevate them to the rank of species, forgetting that the form of the leaves of the ignames undergoes a singular change by cultivation. We have no where discovered the plant called by Linnæus *dioscorea sativa;* neither does it exist in the islands of the S. sea, where the root of the dioscorea alata, mixed with the white of cocoa-nuts and the pulp of the banana, is the favourite dish of the Otaheitans. The root of the igname acquires an enormous volume when it grows in a fertile soil. In the valley of Aragua, in the province of Caracas, it has been seen to weigh from 25 to 30 kilogrammes (from 55 to 66 lb. avoird).

The *batates* go in Peru by the name of *apichu*, and in Mexico by that of *camotes*, which is a corruption of the Aztec word *cacamotic*. Several va-]

rieties are cultivated with white and yellow roots; those of Queretaro, which grow in a climate analogons to that of Andalusia, are the most in request. We doubt very much if these batates were ever found wild by the Spanish navigators, though it has been advanced by Clusius. It would be so much the more interesting to know whether the batates cultivated in Peru, and those which Cook found in Easter island (*ile de Paques*), are the same, as from the position of that island and the monuments which have been there discovered, several of the learned have been led to suspect the existence of ancient communications between the Peruvians and the inhabitants of the island discovered by Roggeween.

Gomara relates that Columbus, after his return to Spain, when he first made his appearance before Queen Isabella, brought to her grains of maize, igname roots, and batates. Hence the cultivation of the last of these must have been already common in the *s.* part of Spain towards the middle of the 16th century. In 1591 they were even sold in the market of London. (*Clusius*, iii. c. 51). It is generally believed that the celebrated Drake, or Sir John Hawkins, made them known in England, where they were long thought to be endowed with the mysterious properties for which the Greeks recommended the onions of Megara. The cultivation of batates succeeds very well in the *s.* of France. It requires less heat than the igname, which, otherwise, on account of the enormous mass of nutritive matter furnished by its roots, would be much preferable to the potato, if it could be successfully cultivated in countries of which the mean temperature is under 18 centigrade degrees, (64° of Fahrenheit).

We must also reckon among the useful plants proper to Mexico the *cacomite*, or *oceloxochitl*, a species of tigridia, of which the root yielded a nutritive flour to the inhabitants of the valley of Mexico; the numerous varieties of love-apples, or *tomatl* (*solanum lycopersicum*), which was formerly sown along with maize; the earth-pistachio, or *mani* (*arachis hypogea*), of which the root is concealed in the earth, and which appears to have existed in Cochin China (see *Loureiro*, *Flora Cochinchinensis*, p. 522) long before the discovery of America; lastly, the different species of pimento (*capsicum baccatum*, *c. annuum*, and *c. frutescens*), called by the Mexicans *chilli*, and the Peruvians *uchu*, of which the fruit is as indispensably necessary to the natives as salt to the whites. The Spaniards call pimento *chile* or *axi* (*ahi*). The first word is derived from *quauh-chilli*, the second is a Haitian word that we must not confound with *axe*, which, as we have already observed, designates the dioscorea alata.

The *topinambours* (*helianthus tuberosus*), which, according to M. Correa, are not even to be found in the Brazils, are not known to be cultivated elsewhere on this continent, though in all our works on botany they are said to be natives of the country of the Brazilian Topinambas. The *chimalatl*, or sun with large flowers (*helianthus annuus*), came from Peru to Nueva España. It was formerly sown in several parts of Spanish America, not only to extract oil from its seed, but also for the sake of roasting it and making it into a very nutritive bread.

Rice (*oryza sativa*) was unknown to the people of the new continent, as well as the inhabitants of the S. sea islands. Whenever the old historians use the expression small Peruvian rice (*arroz pequeño*), they mean the *chenopodium quinoa*, which is found very common in Peru and the beautiful valley of Bogota. The cultivation of rice, introduced by the Arabs into Europe, and by the Spaniards into America, is of very little importance in Nneva España. The great drought which prevails in the interior of the country seems hostile to its cultivation. At Mexico they are not agreed as to the utility with which the introduction of the mountain rice might be attended, which is common to China, Japan, and known to all the Spaniards who have lived in the Philippine islands. It is certain that the mountain rice, so much extolled of late, only grows on the slopes of hills, which are watered either by natural torrents or by canals of irrigation cut at very great elevations. On the coast of Mexico, especially to the *s. e.* of Vera Cruz, in the fertile and marshy grounds situated between the mouths of the rivers Alvarado and Goasacualco, the cultivation of the common rice may one day become as important as it has long been for the province of Guayaquil, for Louisiana, and the *s.* part of the United States.

It is so much the more to be desired that this branch of agriculture should be followed with ardour, as from the great droughts and premature frosts the grain and maize harvests frequently fail in the mountainous region, and the Mexican people suffer periodically from the fatal effects of a general famine. The rice contains a great deal of alimentary substance in a very small volume. In Bengal, where 40 kilogrammes may be purchased for three francs, (viz. 88lb. avoird. for 2*s.* 6*d.*) the daily consumption of a family of five individuals consists of two kilogrammes of rice, two of pease, ($4\frac{4}{10}$ lb. rice and $4\frac{4}{10}$ lb. pease), and two ounces of salt. (*Bockford's Indian Recrea-*

[*tions. Calcutta*, 1807, p. 18. The frugality of the indigenous Aztec is almost equal to that of the Hindoo; and the frequent scarcities in Mexico might be avoided by multiplying the objects of cultivation, and directing the industry to vegetable productions easier to be preserved and transported than maize and farinaceous roots. At Louisiana, in the basin of the Mississippi, they compute that an acre of land commonly produces in rice 18 barrels, in wheat and oats 8, in maize 20, and in potatoes 26. In Virginia they reckon, according to M. Blodget, that an acre yields from 20 to 30 bushels of rice, while wheat only yields from 15 to 16. We are aware that in Europe rice grounds are considered very pernicious to the health of the inhabitants; but the long experience of *e.* Asia seems to prove that the effect is not the same in every climate. However this may be, there is little room to fear that the irrigation of the rice grounds will add to the insalubrity of a country already filled with marshes and *paletuviers* (*rhizophora mangle*), which forms a true delta between the rivers Alvarado, San Juan, and Goasacualco.

The Mexicans now possess all the garden-stuffs and fruit-trees of Europe. It is not easy to indicate which of the former existed in the new continent before the arrival of the Spaniards. The same uncertainty prevails among botanists as to the species of turnips, sallads, and cabbage cultivated by the Greeks and Romans. We know with certainty that the Americans were always acquainted with onions (in Mexican *xonacatl*), haricots (in Mexican *ayacotli*, in the Peruvian or Quichua language *purutu*), gourds (in Peruvian *capallu*), and several varieties of cicer. Cortes, speaking of the eatables which were daily sold in the market of the ancient Tenochtitlan, expressly says, that every kind of garden-stuff (*legume*) was to be found there, particularly onions, leeks, garlic, garden and water cresses (*mastuerzo y berro*), borrage, sorrel, and artichokes (*cardo y tagarninas*). It appears that no species of cabbage or turnip (*brassica et raphanus*) was cultivated in America, although the indigenous are very fond of dressed herbs. They mixed together all sorts of leaves, and even flowers, and they called this dish *iraca*. It appears that the Mexicans had originally no pease; and this fact is so much the more remarkable, as our *pisum sativum* is believed to grow wild on the *n. w.* coast of America.

In general, if we consider the garden-stuffs of the Aztecs, and the great number of farinaceous roots cultivated in Mexico and Peru, we see that America was by no means so poor in alimentary plants as has been advanced by some learned men from a false spirit of system, who were only acquainted with the New World through the works of Herrera and Solis. The degree of civilization of a people has no relation with the variety of productions which are the objects of its agriculture or gardening. This variety is greater or less, as the communications between remote regions have been more or less frequent, or as nations separated from the rest of the human race in very distant periods have been in a situation of greater or less insulation. We must not be astonished at not finding among the Mexicans of the 16th century the vegetable stores now contained in our gardens. The Greeks and Romans even neither knew spinach nor cauliflowers, nor scorzoneras, nor artichokes, nor a great number of other kitchen vegetables.

The central table-land of Nueva España produces in the greatest abundance cherries, prunes, peaches, apricots, figs, grapes, melons, apples, and pears. In the environs of Mexico, the villages of San Augustin de las Cuevas and Tacubaya, the famous garden of the convent of Carmelites at San Angel, and that of the family of Fagoaga at Tanepantla, yield in the months of June, July, and August, an immense quantity of fruit, for the most part of an exquisite taste, although the trees are in general very ill taken care of. The traveller is astonished to see in Mexico, Peru, and New Granada, the tables of the wealthy inhabitants loaded at once with the fruits of temperate Europe, ananas, different species of *passiflora* and tacsonia, sapotes, mameis, goyavas, anonas, chilimoyas, and other valuable productions of the torrid zone. This variety of fruits is to be found in almost all the country from Guatemala to New California. In studying the history of the conquest, we admire the extraordinary rapidity with which the Spaniards of the 16th century spread the cultivation of the European vegetables along the ridge of the *cordilleras*, from one extremity of the continent to the other. The ecclesiastics, and especially the religions missionaries, contributed greatly to the rapidity of this progress. The gardens of the convents and of the secular priests were so many nurseries, from which the recently imported vegetables were diffused over the country. The *conquistadores* even, all of whom we ought by no means to regard as warlike barbarians, addicted themselves in their old age to a rural life. These simple men, surrounded by Indians, of whose language they were ignorant, cultivated in preference, as if to console them in their solitude, the plants which recalled to them the plains of Estramadura and the Castilles. The epocha at which an Eu-]

[ropean fruit ripened for the first time was distinguished by a family festival. It is impossible to read without being warmly affected what is related by the Inca Garcilasso as to the manner of living of these first colonists. He relates, with an exquisite *naiveté*, how his father, the valorous Andres de la Vega, collected together all his old companions in arms to share with them three asparaguses, the first which ever grew on the table-land of Cuzco.

Before the arrival of the Spaniards, Mexico and the *cordilleras* of S. America produced several fruits, which bear great analogy to those of the temperate climates of the old continent. The physiognomy of vegetables bears always a great mutual resemblance where the temperature and humidity are the same. The mountainous part of S. America has a cherry (*padus capuli*), nut, apple, mulberry, strawberry, rubus, and gooseberry, which are peculiar to it. Cortes relates that he saw, on his arrival at Mexico, besides the indigenous cherries, which are very acid, prunes, *ciruelas*. He adds, that they entirely resemble those of Spain. We doubt the existence of these Mexican prunes, although the Abbe Clavigero also mentions them. Perhaps the first Spaniards took the fruit of the *spondias*, which is a *drupa ovoide*, for European prunes.

Although the *w.* coast of Nueva España be washed by the Great ocean, and although Mendana, Gaetano, Quiros, and other Spanish navigators, were the first who visited the islands situated between America and Asia, the most useful productions of these countries, the bread-fruit, the flax of New Zealand (*phormium tenax*), and the sugar-cane of Otaheite, remained unknown to the inhabitants of Mexico. These vegetables, after travelling round the globe, will reach them gradually from the W. India islands. They were left by Captain Bligh at Jamaica, and they have propagated rapidly in the island of Cuba, Trinidad, and on the coast of Caracas. The bread-fruit (*artocarpus incisa*), of which are to be seen considerable plantations in Spanish Guayana, would vegetate vigorously on the humid and warm coasts of Tabasco, Tustla, and San Blas. It is very improbable that this cultivation will ever supersede among the natives that of bananas, which, on the same extent of ground, furnish more nutritive substance. It is true that the artocarpus, for eight months in the year, is continually loaded with fruits, and that three trees are sufficient to nourish an adult individual. (*Georg Forster vom Brodbaume*, 1784, s. xxiii). But an arpent or demi-hectare of ground can only contain from 35 to 40 bread-fruit trees; for when they are planted too near one another, and when their roots meet, they do not bear so great a quantity of fruit.

The extreme slowness of the passage from the Philippine islands and Mariana to Acapulco, and the necessity in which the Manilla galleons are under of ascending to higher latitudes to get the *n. w.* winds, render the introduction of vegetables from oriental Asia extremely difficult. Hence, on the *w.* coast of Mexico we find no plant of China or the Philippine islands, except the *triphasia aurantiola* (*limonia trifoliata*), an elegant shrub, of which the fruits are dressed, and which, according to Loureiro, is identical with the *citrus trifoliata*, or *karatats-banna* of Kämpfer. As to the orange and citron trees, which in the *s.* of Europe support, without any bad consequences, a cold for five or six days below 0, (32° of Fahrenheit), they are now cultivated throughout all Nueva España, even on the central table-land. It has frequently been discussed, if these trees existed in the Spanish colonies before the discovery of America, or if they were introduced by the Europeans from the Canary islands, the island of St. Thomas, or the coast of Africa. It is certain that there is an orange-tree, of a small and bitter fruit, and a very prickly citron, yielding a green, round fruit, with a singularly oily bark, which is frequently hardly of the size of a large nut, growing wild in the island of Cuba and on the coast of Tierra Firme. But Humboldt, notwithstanding all his researches, could never discover a single individual in the interior of the forests of Guayana, between the Orinoco, the Cassiquiare, and the frontiers of Brazil. Perhaps the small green citron (*limoncito verde*) was anciently cultivated by the natives; and perhaps it has only grown wild when the population, and consequently the extent of cultivated territory, were most considerable. We are inclined to believe that only the citron-tree, with large yellow fruit (*limon sutil*), and the sweet orange, were introduced by the Portuguese and Spaniards. Humboldt saw them on the banks of the Orinoco, where the Jesuits had established their missions. The orange, on the discovery of America, had only existed for a few centuries even in Europe. If there had been any ancient communication between the new continent and the islands of the S. sea, the true *citrus aurantium* might have arrived in Peru or Mexico by the way of the *w.*; for this tree was found by M. Forster in the Hebrides islands, where it was seen by Quiros long before him.

The great analogy between the climate of the table-land of Nueva España and that of Italy, Greece, and the *s.* of France, ought to invite the Mexi-]

[cans to the cultivation of the olive. This cultivation was successfully attempted at the beginning of the conquest, but the government, from an unjust policy, far from favouring, endeavoured rather indirectly to frustrate it. As far as we know, there exists no formal prohibition; but the colonists have never ventured on a branch of national industry which would have immediately excited the jealousy of the mother country. The court of Madrid has always seen with an unfavourable eye the cultivation of the olive and the mulberry, hemp, flax, and the vine, in the new continent; and if the commerce of wines and indigenous oils has been tolerated in Peru and Chile, it is only because those colonies, situated beyond cape Horn, are frequently ill provisioned from Europe, and the effect of vexations measures is dreaded in provinces so remote. A system of the most odious prohibitions has been obstinately followed in all the colonies of which the coast is washed by the Atlantic ocean. During Humboldt's stay at Mexico the viceroy received orders from the court to pull up the vines (*arancar las cepas*) in the *n.* provinces of Mexico, because the merchants of Cadiz complained of a diminution in the consumption of Spanish wines. Happily this order, like many others given by the ministers, was never executed. It was judged that, notwithstanding the extreme patience of the Mexican people, it might be dangerous to drive them to despair by laying waste their properties and forcing them to purchase from the monopolists of Europe what the bounty of nature produces on the Mexican soil.

The olive-tree is very rare in all Nueva España; and there exists but a single olive plantation, the beautiful one of the archbishop of Mexico, situated two leagues *s. e.* from the capital. This *olivar del Arzobispo* annually produces 200 *arrobas*, 5500 lb. avoird. of an oil of a very good quality. We have already spoken of the olive cultivated by the missionaries of New California, especially near the village of San Diego. The Mexican, when at complete liberty in the cultivation of his soil, will in time dispense with the oil, wine, hemp, and flax of Europe. The Andalusian olive introduced by Cortes sometimes suffers from the cold of the central table-land; for although the frosts are not strong, they are frequent and of long duration. It might be useful to plant the Corsican olive in Mexico, which is more than any other calculated to resist the severity of the climate.

In terminating the list of alimentary plants, we shall give a rapid survey of the plants which furnish beverages to the Mexicans. We shall see that in this point of view the history of the Aztec agriculture presents us with a trait so much the more curious, as we find nothing analogous among a great number of nations much more advanced in civilization than the ancient inhabitants of Anahuac.

There hardly exists a tribe of savages on the face of the earth who cannot prepare some kind of beverage from the vegetable kingdom. The miserable hordes who wander in the forests of Guayana make as agreeable emulsions from the different palm-tree fruits as the barley water prepared in Europe. The inhabitants of Easter island, exiled on a mass of arid rocks without springs, besides the sea-water, drink the juice of the sugar-cane. The most part of civilized nations draw their drinks from the same plants which constitute the basis of their nourishment, and of which the roots or seeds contain the sugary principle united with the amylaceous substance. Rice in *s.* and *e.* Asia, in Africa the igname root with a few arums, and in the *n.* of Europe cerealia, furnish fermented liquors. There are few nations who cultivate certain plants merely with a view to prepare beverages from them. The old continent affords us no instance of vine plantations but to the *w.* of the Indus. In the better days of Greece this cultivation was even confined to the countries situated between the Oxus and Euphrates, to Asia Minor, and *w.* Europe. On the rest of the globe nature produces species of wild vitis; but nowhere else did man endeavour to collect them round him to ameliorate them by cultivation.

But in the new continent we have the example of a people who not only extracted liquors from the amylaceous and sugary substance of the maize, the manioc, and bananas, or from the pulp of several species of *mimosa*, but who cultivated expressly a plant of the family of the ananas, to convert its juice into a spirituous liquor. On the interior table-land, in the intendancy of Puebla, and in that of Mexico, we run over vast extents of country, where the eye reposes only on fields planted with *pittes or maguey*. This plant, of a coriaceous and prickly leaf, which with the *cactus opuntia* has become wild since the sixteenth century throughout all the *s.* of Europe, the Canary islands, and the coast of Africa, gives a particular character to the Mexican landscape. What a contrast of vegetable forms between a field of grain, a plantation of agava, and a group of bananas, of which the glossy leaves are constantly of a tender and delicate green! Under every zone, man, by multiplying certain vegetable productions, modifies at will the aspect of the country under cultivation.]

[In the Spanish colonies there are several species of *maguey* which deserve a careful examination, and of which several, on account of the division of their corolla, the length of their stamina, and the form of their stigmata, appear to belong to different genus! The maguey or *metl* cultivated in Mexico are numerous varieties of the *agave Americana*, which has become so common in our gardens, with yellow fasciculated and straight leaves, and stamina twice as long as the pinking of the corolla. We must not confound this metl with the *agave Cubensis* of Jacquen, (*floribus ex albo virentibus, longe paniculatis, pendulis, staminibus corolla duplo brevioribus*), called by M. Lamarck *a. Mexicana*, and which has been believed by some botanists, for what reason we know not, the principal object of the Mexican cultivation.

The plantations of the *maguey de pulque* extend as far as the Aztec language. The people of the Otomite, Totonac, and Mistec race, are not addicted to the *octli*, which the Spaniards call *pulque*. On the central plain we hardly find the maguey cultivated to the *n.* of Salamanca. The finest cultivations are in the valley of Toluca and on the plains of Cholula. The agaves are there planted in rows at a distance of 15 decimetres, or 58 inches, from one another. The plants only begin to yield the juice, which goes by the name of *honey*, on account of the sugary principle with which it abounds, when the *hampe* is on the point of its development. It is on this account of the greatest importance for the cultivator to know exactly the period of efflorescence. Its proximity is announced by the direction of the radical leaves, which are observed by the Indians with much attention. These leaves, which are till then inclined towards the earth, rise all of a sudden; and they endeavour to form a junction to cover the hampe which is on the point of formation. The bundle of central leaves (*el corazon*) becomes at the same time of a clearer green, and lengthens perceptibly. It is said by the Indians that it is difficult to be deceived in these signs, but that there are others of no less importance which cannot be precisely described, because they have merely a reference to the carriage of the plant. The cultivator goes daily through his agave plantations to mark those plants which approach efflorescence. If he has any doubt, he applies to the experts of the village, old Indians, who, from long experience, have a judgment, or rather tact, more securely to be relied on.

Near Cholula, and between Toluca and Cacanumaean, a maguey of eight years old gives already signs of the development of its hampe. They then begin to collect the juice, of which the pulque is made. They cut the corazon, or bundle of central leaves, and enlarge insensibly the wound, and cover it with lateral leaves, which they raise up by drawing them close, and tying them to the extremities. In this wound the vessels appear to deposit all the juice which would have formed the colossal hampe loaded with flowers. This is a true vegetable spring, which keeps running for two or three months, and from which the Indian draws three or four times a day. We may judge of the quickness or slowness of the motion of the juice by the quantity of honey extracted from the maguey at different times of the day. A foot commonly yields, in 24 hours, four cubic decimetres, or 200 cubic inches, (242 cubic inches English), equal to eight *quartillos*. Of this total quantity they obtain three quartillos at sun-rise, two at mid-day, and three at six in the evening. A very vigorous plant sometimes yields 15 quartillos, or 375 cubic inches (454 cubic inches English), per day, for from four to five months, which amounts to the enormous volume of more than 1100 cubic decimetres, or 67,130 cubic inches. This abundance of juice produced by a maguey of scarcely a metre and a half in height, or $4\frac{9}{10}$ feet, is so much the more astonishing, as the agave plantations are in the most arid grounds, and frequently on banks of rocks hardly covered with vegetable earth. The value of a maguey plant near its efflorescence is at Pachuca five piastres, or *1l. 2s. 4d.* In a barren soil the Indian calculates the produce of each maguey at 150 bottles, and the value of the pulque furnished in a day at from 10 to 12 sols. The produce is unequal, like that of the vine, which varies very much in its quantity of grapes.

The cultivation of the agave has real advantages over the cultivation of maize, grain, and potatoes. This plant, with firm and vigorous leaves, is neither affected by drought nor hail, nor the excessive cold which prevails in winter on the higher *cordilleras* of Mexico. The stalk perishes after efflorescence. If we deprive it of the central leaves, it withers, after the juice which nature appears to have destined to the increase of the hampe is entirely exhausted. An infinity of shoots then spring from the root of the decayed plant; for no plant multiplies with greater facility. An arpent of ground contains from 12 to 1300 maguey plants. If the field is of old cultivation, we may calculate that a 12th or 14th of these plants yields honey annually. A proprietor who plants from 30 to 40,000 maguey is sure to establish the fortune of his children; but it requires patience and courage to follow a species of cultivation which only begins] [to grow lucrative at the end of 15 years. In a good soil the agave enters on its efflorescence at the end of five years; and in a poor soil no harvest can be expected in less than 18 years. Although the rapidity of the vegetation is of the utmost consequence for the Mexican cultivators, they never attempt artificially to accelerate the development of the hampe by mutilating the roots or watering them with warm water. It has been discovered that by these means, which weaken the plant, the confluence of juice towards the centre is sensibly diminished. A maguey plant is destroyed, if, misled by false appearances, the Indian makes the incision long before the flowers would have naturally developed themselves.

The honey or juice of the agave is of a very agreeable sour taste. It easily ferments, on account of the sogar and mucilage which it contains. To accelerate this fermentation they add, however, a little old and acid pulque. The operation is terminated in three or four days. The vinous beverage, which resembles cider, has an odour of putrid meat extremely disagreeable; but the Europeans who have been able to get over the aversion which this fœtid odour inspires, prefer the pulque to every other liquor. They consider it as stomachic, strengthening, and especially as very nutritive; and it is recommended to lean persons. Whites also have been known, like the Mexican Indians, totally to have abstained from water, beer, and wine, and to have drunk no other liquor than the juice of the agave. The connoisseurs speak with enthusiasm of the pulque prepared in the village of Hocotitlan, situated to the *n.* of Toluca, at the foot of a mountain almost as elevated as the Nevado of this name. They affirm that the excellent quality of this pulque does not altogether depend on the art with which the liquor is prepared, but also on a taste of the soil communicated to the juice, according to the fields in which the plant is cultivated. There are plantations of maguey near Hocotitlan (*haciendas de pulque*) which bring in annually more than 40,000 livres, or 1666*l.* sterling. The inhabitants of the country differ very much in their opinions as to the true cause of the fetid odour of the pulque. It is generally affirmed that this odour, which is analogous to that of animal matter, is to be ascribed to the skins in which the first juice of the agave is poured. But several well informed individuals pretend that the pulque when prepared in vessels has the same odour, and that if it is not found in that of Toluca, it is because the great cold there modifies the process of fermentation. Perhaps this odour proceeds from the decomposition of a vegeto-animal matter, analogous to the gluten contained in the juice of the agave. .

The cultivation of the maguey is an object of such importance for the revenue, that the entry duties paid in the three cities of Mexico, Toluca, and Puebla, amounted, in 1793, to the sum of 817,739 piastres, or 178,880*l.* sterling. The expences of perception were then 56,608 piastres, or 12,383*l.* sterling; so that the government drew from the agave juice a nett revenue of 761,131 piastres, or 166,497*l.* or more than 3,800,000 francs. The desire of increasing the revenues of the crown occasioned latterly a heavy tax on the fabrication of pulque, equally vexatious and inconsiderate. It is time to change the system in this respect, otherwise it is to be presumed that this cultivation, one of the most ancient and lucrative, will insensibly decline, notwithstanding the decided predilection of the people for the fermented juice of the agave.

A very intoxicating brandy is formed from the pulque, which is called *mexical*, or *aguardiente de maguey*. We have been assured that the plant cultivated for distillation differs essentially from the common maguey, or *maguey de pulque*. The sugar-cane has also a particular variety, with a violet stalk, which came from the coast of Africa (*caño de Guinea*), and which is preferred in the province of Caracas for the fabrication of rum to the sugar-cane of Otaheite. The Spanish government, and particularly the *real hacienda*, has been long very severe against the *mexical*, which is strictly prohibited, because the use of it is prejudicial to the Spanish brandy trade. An enormous quantity, however, of this maguey brandy is manufactured in the intendances of Valladolid, Mexico, and Durango, especially in the new kingdom of Leon. We may judge of the value of this illicit traffic by considering the disproportion between the population of Mexico and the annual importation of European brandy into Vera Cruz. The whole importation only amounts to 32,000 barrels! In several parts of the kingdom, for example in the *provincias internas* and the district of Tuxpan, belonging to the intendancy of Guadalaxara, for some time past the mexical has been publicly sold on payment of a small duty. This measure, which ought to be general, has been both profitable to the revenue, and has put an end to the complaints of the inhabitants.

But the maguey is not only the vine of the Aztecs, it can also supply the place of the hemp of Asia, and the papyrus (*cyperus papyrus*) of the Egyptians. The paper on which the ancient Mexicans painted their hieroglyphical figures was] [made of the fibres of agave leaves, macerated in water, and disposed in layers like the fibres of the Egyptian cyperus, and the mulberry *(broussonetia)* of the S. sea islands. Humboldt brought with him several fragments of Aztec manuscripts written on maguey paper, of a thickness so different that some of them resemble pasteboard, while others resemble Chinese paper. These fragments were so much the more interesting, as the only hieroglyphics which exist at Vienna, Rome, and Veletri, are on Mexican stag-skins. The thread which is obtained from the maguey is known in Europe by the name of pite thread, and it is preferred by naturalists to every other, because it is less subject to twist. It does not, however, resist so well as that prepared from the fibres of the phormium. The juice (*xugo de cocuyza*) which the agave yields when it is still far from the period of efflorescence is very acrid, and is successfully employed as a caustic in the cleaning of wounds. The prickles which terminate the leaves served formerly, like those of the cactus, for pins and nails to the Indians. The Mexican priests pierced their arms and breast with them in their acts of expiation, analogous to those of the buddists of Hindostan.

We may conclude from all that we have related respecting the use of the different parts of the maguey, that next to the maize and potato, this plant is the most useful of all the productions with which nature has supplied the mountaineers of equinoctial America.

When the fetters which the government has hitherto put on several branches of the national industry shall be removed, when the Mexican agriculture shall be no longer restrained by a system of administration, which, while it impoverishes the colonies, does not enrich the mother country, the maguey plantations will be gradually succeeded by vineyards. The cultivation of the vine will augment with the number of the whites, who consume a great quantity of the wines of Spain, France, Madeira, and the Canary islands. But in the present state of things, the vine can hardly be included in the territorial riches of Mexico, the harvest of it being so inconsiderable. The grape of the best quality is that of Zapotitlan, in the intendancy of Oaxaca. There are also vineyards near Dolores and San Luis de la Paz to the *n.* of Guanaxuato, and in the *provincias internas* near Parras, and the Passo del Norte. The wine of the Passo is in great estimation, especially that of the estate of the Marquis de San Miguel, which keeps for a great number of years, although very little care is bestowed on the making of it. They complain in the country that the must of the table-land ferments with difficulty; and they add *arope* to the juice of the grape, that is to say, a small quantity of wine in which sugar has been infused, and which by means of dressing has been reduced into a syrup. This process gives to the Mexican wines a flavour of must, which they would lose if the making of wine was more studied among them. When in the course of ages the new continent, jealous of its independence, shall wish to dispense with the productions of the old, the mountainous and temperate parts of Mexico, Guatemala, New Granada, and Caracas, will supply wine to the whole of N. America; and they will then become to that country what France, Italy, and Spain have long been to the *n.* of Europe.

Chap. VIII.

Table of the geographical positions of the kingdom of Nueva España; for which see the end of the general preface to this work.

Chap. IX.

Table of the most remarkable elevations measured in the interior of Nueva España.

The work published with the title of *Nivellement barometrique fait dans les Regions Equinoxiales du Nouveau Continent*, in 1799—1804, contains more than 200 points in the interior of Nueva España, of which Mr. Humboldt determined the elevation above the level of the sea, either by the barometer, or by trigonometrical methods. We have merely inserted in the following table the absolute heights of the most remarkable mountains and cities. The points marked with an asterisk are doubtful. The *Recueil d'Observations astronomiques et de Mesures barometriques* of Mr. Humboldt, edited by M. Oltmanns, may also be consulted, (vol. i. pages 318 to 334).]

Names of places of observation.	Height above the level of the sea, according to the formula of M. Laplace.		
	In metres.	In toises.	In Eng. feet.
Volcan de Popocatepetl, volcan grande de Mexico ó de Puebla	5400	2771	17716
Pic d'Orizaba or Citlaltepetl - - -	5295	2717	17371
Nevado d'Iztaccihuatl, Sierra Nevada of Mexico - -	4786	2456	15700
Nevado de Toluca, at the rock of Frailes - - -	4621	2372	15159
Coffre de Perote or Nauhcampatepetl - - -	4089	2098	13514
Cerro de Axusco, six leagues to the *s. s. w.* of Mexico -	3674*	1885*	12052
Pic de Tancitaro - - - - - - -	3200*	1642	10498
El Jacal, summit of the Cerro de las Nabajas - -	3124	1603	10249
Mamanchota or Organos d'Actopan, *n. e.* from Mexico -	2977	1527	9766
Volcan de Colima - - - - - -	2800*	1437	9186
Volcan de Jorullo, in the intendancy of Valladolid - -	1301	667	4267
Mexico, at the convent of St. Augustin - - -	2277	1168	7470
Pachuca - - - - - - - - -	2484	1274	8149
Moran, mine near the Real del Monte - - -	2595	1331	8513
Real del Monte, mine - - - - - -	2781	1427	9057
Tula, city - - - - - - - - -	2053	1053	6735
Toluca, city - - - - - - - -	2688	1379	8818
Cuernavaca, city - - - - - - - -	1656	849	5433
Tasco, city - - - - - - - - -	1784	915	5852
Chilpansingo, city - - - - - -	1380	708	4527
Puebla de los Angeles, city - - - - - -	2194	1126	7198
Perote, town - - - - - - - -	2354	1208	7723
Xalapa, city - - - - - - - -	1321	678	4333
Valladolid, city - - - - - - - -	1952	1001	6404
Pazcuaro, city - - - - - - - -	2202	1130	7224
Charo, city - - - - - - - -	1907	978	6256
Villa de Islahuaca, in the intendancy of Valladolid - -	2585	1326	8481
San Juan del Rio, town - - - - - -	1978	1015	6489
Queretaro, city - - - - - - - -	1940	995	6364
Celaya, city - - - - - - - -	1835	941	6020
Salamanca, city - - - - - - - -	1757	902	5763
Guanaxuato, city - - - - - - - -	2084	1069	6836
Mine de la Valenciana - - - - - -	2328	1194	7637
Durango, city - - - - - - - -	2087*	1071	6847

CHAP. X.

General considerations on the possibility of uniting the South sea and Atlantic ocean, viz. By the rivers of Peace and Tacoutche Tesse—Sources of the Rio Bravo and Rio Colorado—Isthmus of Tehuantepec—Lake of Nicaragua—Isthmus of Panama—Bay of Cupica—Canal of Choco—Rio Guallaga—Gulf of St. George.

THE part of Mexico in which the two oceans, the Atlantic and the S. sea, approach the nearest to one another, is unfortunately not that part which contains the two ports of Acapulco and Vera Cruz, and the capital of Mexico. There are, according to Mr. Humboldt's astronomical observations, from Acapulco to Mexico an oblique distance of 2° 40′ 19″, or 155,885 toises, or 997,664 feet; from Mexico to Vera Cruz 2° 57′ 9″, or 158,572 toises, or 1,014,860 feet; and from the port of Acapulco to the port of Vera Cruz, in a direct line, 4° 10′ 7″. It is in these distances that the old maps are most faulty. From the ob.]

[servations published by M. de Cassini, in the account of the voyage of Chappe, the distance from Mexico to Vera Cruz appears 5° 10′ of long. instead of 2° 57′, the real distance between these two great cities. In adopting for Vera Cruz the longitude given by Chappe, and for Acapulco that of the map of the Depôt drawn up in 1784, the breadth of the Mexican isthmus betwixt the two ports would be 175 leagues, 75 leagues beyond the truth.

The isthmus of Tehuantepec, to the *s. e.* of the port of Vera Cruz, is the point of Nueva España in which the continent is narrowest. From the Atlantic ocean to the S. sea the distance is 45 leagues. The approximation of the sources of the rivers Huasacualo and Chimalapa seems to favour the project of a canal for interior navigation; a project with which the Count of Revillagigedo, one of the most zealous viceroys for the public good, has been for a long time occupied. When we come to speak of the intendancy of Oaxaça, we shall return to this object, so important to all civilized Europe. We must confine ourselves here to the *problem of the communication between the two seas*, in all the generality of which it is susceptible; and although it may appear that the nature of the question of which we are about to treat, does not exclusively apply to the article Mexico, yet as it is in this kingdom that the two oceans, the Atlantic and the S. sea, as we have before observed, approach the nearest to each other; and as it is consequently to this point that the eyes of the inquirer will be naturally bent; we shall endeavour to present in one view nine points, several of which are not sufficiently known in Europe, and all offering a greater or less probability either of canals or interior river communications. At a time when the new continent, profiting by the misfortunes and perpetual dissensions of Europe, advances rapidly towards civilization; and when the commerce of China, and the *n. w.* coast of America, becomes yearly of greater importance, the subject which we here summarily discuss is of the greatest interest for the commerce and political preponderancy of nations.

These nine points, which at different times have fixed the attention of statesmen and merchants in the colonies, present very different advantages. We shall range them according to their geographical position, beginning with the most *n.* part of the new continent, and following the coasts to the *s.* of the island of Chiloe. It can only be after having examined all the projects hitherto formed for the communication of the two seas, that the government can decide which of them merits the preference. Before this examination, exact materials for which are not yet collected, it would be imprudent to cut canals in the isthmuses of Guasacualco or Panama.

1. Under the 54° 37′ of *n.* lat. in the parallel of Queen Charlotte's island, the sources of the river of Peace, or Ounigigah, approach to within seven leagues of the sources of the Tacoutche Tesse, supposed the same with the river of Colombia. The first of these rivers discharges itself into the N. ocean, after having mingled its waters with those of the Slave lake, and the river Mackenzie. The second river, Colombia, enters the Pacific ocean, near cape Disappointment, to the *s.* of Nootka sound, according to the celebrated voyager Vancouver, under the 46° 19′ of lat. The *cordillera*, or chain of the stony mountains, abounding in coal, was found by M. Fiedler to be elevated in some places 3520 English feet, or 550 toises, above the neighbouring plains. It separates the sources of the rivers of Peace and Colombia. According to Mackenzie's account, who passed this *cordillera* in the month of August 1793, it is practicable enough for carriages, and the mountains appear of no very great elevation. To avoid the great winding of the Colombia, another communication still shorter might be opened from the sources of the Tacoutche Tesse to the Salmon river, the mouth of which is to the *e.* of the Princess Royal islands, in the 52° 26′ of lat. Mackenzie rightly observes, that the government which should open this communication between the two oceans, by forming regular establishments in the interior of the country, and at the extremities of the rivers, would get possession of the whole fur trade of N. America, from the 48° of lat. to the pole, excepting a part of the coast which has been long included in Russian America. Canada, from the multitude and course of its rivers, presents facilities for internal commerce similar to those of oriental Siberia. The mouth of the river Colombia seems to invite Europeans to found a fine colony there; for its banks afford fertile land in abundance, covered with superb timber. It must be allowed, however, that notwithstanding the examination by Mr. Broughton, we still know but a very small part of Colombia, which, like the Severn and the Thames, appears of a disproportionate contraction as it leaves the coast. Every geographer who carefully compares Mackenzie's maps with Vancouver's, will be astonished that the Colombia, in descending from these stony mountains, which we cannot help considering as a prolongation of the Andes of Mexico, should traverse the chain of mountains which approach the]

[shore of the Great ocean, whose principal summits are mount St. Helen and mount Rainier. But M. Malte-Brun has started important doubts concerning the identity of the Tacoutche Tesse and the Rio Colombia. He even presumes, (as may be seen in the *Geogr. Mathem.* vol. xv. p. 117), that the former discharges itself into the gulf of California; a bold supposition, which would give to the Tacoutche Tesse a course of an enormous length. It must be allowed that all that part of the *w.* of N. America is still but very imperfectly known.

In the 50° of lat. the Nelson river, the Saskashawan, and the Missoury, which may be regarded as one of the principal branches of the Mississippi, furnish equal facilities of communication with the Pacific ocean. All these rivers take their rise at the foot of the stony mountains. But we have not yet sufficient acquaintance with the nature of the ground through which the communication is proposed to be established, to pronounce upon the utility of these projects. The journey of Captain Lewis, at the expence of the Anglo-American government, on the Mississippi and the Missoury, has thrown considerable light on this interesting problem.

2. The next projection is through the Rio del Norte, or Rio Bravo; the sources of which are only separated from the sources of the Rio Colorado by a mountainous tract of from 12 to 13 leagues in breadth. See article BRAVO.

3. The isthmus of Tehuantepec comprises, under the 16° of lat. the sources of the Rio Huasacualco, which is discharged into the golf of Mexico, and the sources of the Rio de Chimalapa. The waters of this last river mix with those of the Pacific ocean near the Barra de S. Francisco. We consider here the Rio del Passo as the principal source of the river Huasacualco, although the latter only takes its name at the Paso de la Fabrica, after one of its arms, which comes from the mountains De los Mexes, unites with the Rio del Passo. We shall examine afterwards the possibility of cutting a canal, of from six to seven leagues, in the forests of Tarifa. We shall merely observe here, that since, in 1798, a road has been opened which leads by land from the port of Tehuantepec, to the Embarcadero de la Cruz, the same road having been completed in 1800; the Rio Huasacualco forms, in reality, a commercial communication between the two oceans. During the course of the war with the English, the indigo of Guatemala, the most precious of all known indigos, came by the way of this isthmus to the port of Vera Cruz, and from thence to Europe.

4. The great lake of Nicaragua communicates not only with the lake of Leon, but also on the *e.* by the river of San Juan, with the sea of the Antilles. The communication with the Pacific ocean would be effected in cutting a canal across the isthmus which separates the lake from the gulf of Papagayo. On this strait isthmus are to be found the volcanic and isolated summits of Bombacho, at 11° 7′ of lat., of Granada, and of the Papagayo, at 10° 50′ of lat. The old maps point out a communication by water as existing across the isthmus from the lake to the Great ocean. Other maps, somewhat newer, represent a river under the name of Rio Partido, which gives one of its branches to the Pacific ocean, and the other to the lake of Nicaragua; but this divided stream does not appear on the last maps published by the Spaniards.

There are in the archives of Madrid several French and English memoirs on the possibility of the junction of the lake of Nicaragua with the Pacific ocean: viz. *Memoire sur le passage de la mer du Sud a la mer du Nord, par M. la Bastide,* en 1791. *Voyage de Marchand,* vol. i. p. 565. *Mapa del Golfo de Mexico por Thomas Lopez y Juan de la Cruz,* 1755. The commerce carried on by the English on the coast of Mosquitos has greatly contributed to give celebrity to this project of communication between the two seas. In none of the memoirs which have come to our knowledge is the principal point, the height of the ground in the isthmus, sufficiently cleared up.

From the kingdom of New Granada to the environs of the capital of Mexico, there is not a single mountain, a single level, a single city, of which we know the elevation above the level of the sea. Does there exist an uninterrupted chain of mountains in the provinces of Veragua and Nicaragua? Has this *cordillera,* which is supposed to unite the Andes of Peru to the mountains of Mexico, its central chain to the *w.* or the *e.* of the lake of Nicaragua? Would not the isthmus of Papagayo rather present a hilly tract than a continued *cordillera?* These are problems whose solution is equally interesting to the statesman and the geographical naturalist!

There is no spot on the globe so full of volcanoes as this part of America, from lat. 11° or 13°; but do not these conical summits form groups which, separately from one another, rise from the plain itself? We ought not to be astonished that we are ignorant of these very important facts; we shall soon see that even the height of the mountains which traverse the isthmus of Panama is not yet known. Perhaps the communication of the lake of Nicaragua with the Pacific ocean could be carried on by the lake of Leon, by means of the river]

[Tosta, which, on the road from Leon to Realexo, descends from the volcano of Telica. In fact, the ground appears there very little elevated. The account of the voyage of Dampier leads us even to suppose that there exists no chain of mountains between the lake of Nicaragua and the S. sea. "The coast of Nicoya," says this great navigator, "is low, and covered at full tide. To arrive from Realexo to Leon, we must go 20 miles across a country flat and covered with mangle-trees." The city of Leon itself is situated in a savanna. There is a small river which, passing near Realexo, might facilitate the communication between the latter port and that of Leon. (See *Collection of Dampier's and Wafer's Voyages*, vol. i. p. 113, 119, 218). From the *w.* bank of the lake of Nicaragua there are only four marine leagues to the bottom of the gulf of Papagayo, and seven to that of Nicoyo, which navigators call La Caldera. Dampier says expressly that the ground between La Caldera and the lake is a little hilly, but for the greatest part level and like a savanna.

The coast of Nicaragua is almost inaccessible in the months of August, September, and October, on account of the terrible storms and rains; in January and February, on account of the furious *n. e.* and *e. n. e.* winds called Papagayos. This circumstance is exceedingly inconvenient for navigation. The port of Tehuantepec, on the isthmus of Guasacualco, is not more favoured by nature; it gives its name to the hurricanes which blow from the *n. w.* and which frighten vessels from landing at the small ports of Sabinas and Ventosa.

5. The isthmus of Panama was crossed for the first time by Vasco Nuñez de Balboa, in 1513. Since this memorable epocha in the history of geographical discoveries, the project of a canal has occupied every mind; and yet at this day, after the lapse of 300 years, there neither exists a survey of the ground, nor an exact determination of the positions of Panama and Portobello. The longitude of the first of these two ports has been found with relation to Carthagena; the longitude of the second has been fixed from Guayaquil. The operations of Fidalgo and Malaspina are undoubtedly deserving of very great confidence; but errors are insensibly multiplied, when by chronometrical operations from the isle of Trinidad to Portobello, and from Lima to Panama, one position becomes dependent on another. It would be important to carry the time directly from Panama to Portobello, and thus to connect the operations in the S. sea with those which the Spanish government has carried on in the Atlantic ocean. Perhaps MM. Fidalgo, Tiscar, and Noguera, may one day advance with their instruments to the *s.* coast of the isthmus, while MM. Colmenares, Irasvirivill, and Quartara, shall carry their operations to the *n.* coast. The expedition of Fidalgo was destined for the coast situated between the isle of Trinidad and Portobello, the expedition of Colomenares for the coast of Chile, and the expedition of Moraleda and Quartara for the part between Guayaquil and Realexo. To form an idea of the uncertainty which still prevails as to the form and breadth of the isthmus (for example towards Nata), we have only to compare the maps of Lopez with those of Arrowsmith, and with the more recent ones of the Deposito Hydrografico of Madrid. The river Chagre, which flows into the sea of the Antilles to the *w.* of Portobello, presents, notwithstanding its sinuosities and its rapids, great facility for commerce; its breadth is 120 toises at its mouth, and 20 toises near Cruces, where it begins to be navigable. It requires four or five days at present to ascend the Rio Chagre from its mouth to Cruces. If the waters are very high, the current must be struggled with for 10 or 12 days. From Cruces to Panama merchandizes are transported on the backs of mules, for a space of five small leagues. The barometrical heights related in the travels of Ulloa, in his *Observations Astronomiques*, p. 97, lead us to suppose that there exists in the Rio Chagre, from the sea of the Antilles to the Embarcadero, or Venta de Cruces, a difference of level of from 35 to 40 toises. This must appear a very small difference to those who have ascended the Rio Chagre; they forget that the force of the current depends as much on a great accumulation of water near the sources, as on the general descent of the river; that is to say, of the descent of the Rio Chagre above Cruces. On comparing the barometrical survey of Ulloa with that made by Humboldt in the river of Magdalena, we perceive that the elevation of Cruces above the ocean, far from being small, is, on the contrary, very considerable. The fall of the Rio de la Magdelena from Honda to the dike of Mahates, near Barrancas, is nearly 170 toises, or 1088 feet; and this distance, nevertheless, is not, as we might suppose, four times, but eight times, greater than that of Cruces, at the fort of Chagre.

The engineers in proposing to the court of Madrid that the river Chagre should serve for establishing a communication between the two oceans, have projected a canal from the Venta de Cruces to Panama. This canal would have to pass through a hilly tract, of the height of which we are completely ignorant. We only know that from Cruces the ascent is at first rapid, and that there is then a descent for several hours towards the S. sea. It is very astonishing, that in cross-]

[ing the isthmus neither La Condamine nor Don George Juan and Ulloa had the curiosity to observe their barometer, for the sake of informing us what is the height of the most elevated point on the route of the castle of Chagre at Panama. These illustrious *savans* sojourned three months in that interesting region for the commercial world; but their stay has added little to the old observations which we owe to Dampier and to Wafer. However, it appears beyond a doubt that we find the principal *cordillera*, or rather a range of bills that may be regarded as a prolongation of the Andes of New Granada, towards the S. sea, between Cruces and Panama. It is from thence that the two oceans are said to be discernible at the same time, which would only require an absolute height of 290 metres, or 947 English feet. However, Lionel Wafer complains that he could not enjoy this interesting spectacle. He assures us, moreover, that the hills which form the central chain are separated from one another by valleys which allow free course for passage of the rivers. (See *Description of the Isthmus of America*, 1729, p. 297.) Ulloa also asserts, vol. i. p. 101, that near the town of Panama, a little to the *n.* of the port, is the mountain of L'Ancon, which, according to a geometrical measurement, is 101 toises (646 feet) in height. If these assertions be well founded, we might believe in the possibility of a canal from Cruces to Panama, of which the navigation would only be interrupted by a very few locks.

There are other points where, according to memoirs drawn up in 1528, the isthmus has been proposed to be cut; for example, in joining the sources of the rivers called Caimito and Rio Grande, with the Rio Trinidad. The *e.* part of the isthmus is the narrowest, but the ground appears to be also most elevated there. This is at least what has been remarked in the frightful road travelled by the courier from Portobello to Panama, a two days journey, which goes by the village of Pequeni, and is full of the greatest difficulties.

In every age and climate, of two neighbouring seas, the one has been considered as more elevated than the other. Traces of this common opinion are to be found among the ancients. Strabo relates, that in his time the gulf of Corinth near Lechæum was believed to be above the level of the sea of Cenchreæ. He is of opinion, (Lib. i. *ed. Siebenkees*, v. i. p. 146.) that it would be very dangerous to cut the isthmus of the Peloponnesus in the place where the Corinthians, by means of particular machines, had established a portage. In America, the S. sea is generally supposed to be higher at the isthmus of Panama than the Atlantic ocean. After a struggle of several days against the current of the Rio Chagre, we naturally believe the ascent to be greater than the descent from the hills near Cruces to Panama. Nothing, in fact, can be more treacherous than the estimates which we are apt to form of the difference of level on a long and easy descent. Humboldt could hardly believe his own eyes at Peru, when he found, by means of a barometrical measurement, that the city of Lima was 91 toises, or 582 feet, higher than the port of Callao. An earthquake must cover entirely the rock of the isle San Lorenzo with water before the ocean can reach the capital of Peru. The idea of a difference of level between the Atlantic and S. sea has been combated by Don George Juan, who found the height of the column of mercury the same at the mouth of the Chagre and at Panama.

The imperfection of the meteorological instruments then in use, and the want of every sort of thermometrical correction of the calculation of heights, might also give rise to doubts. These doubts have acquired additional force since the French engineers, in the expedition to Egypt, found the Red sea six toises, or 38 feet, higher than the Mediterranean. Till a geometrical survey be executed in the isthmus itself, we can only have recourse to barometrical measurements. Those made by Humboldt at the mouth of the Rio Sinu in the Atlantic sea, and on the coast of the S. sea in Peru, prove, with every allowance for temperature, that if there is a difference of level between the two seas, it cannot exceed six or seven metres, or 19 or 22 feet.

When we consider the effect of the current of rotation, (that is to say, the general motion from *e.* to *w.* observed in the part of the ocean comprised in the torrid zone), which carries the waters from *e.* to *w.* and accumulates them towards the coast of Costa Rica and Veragua, we are tempted to admit, contrary to the received opinion, that the Atlantic is a little higher than the S. sea. Trivial causes of a local nature, such as the configuration of the coast, currents and winds (as in the straits of Babelmandel), may trouble the equilibrium which ought necessarily to exist between all the parts of the ocean. As the tides rise at Portobello to a third part of a metre, or 13 inches, and at Panama to four or five metres, or 13 or 16 feet, the levels of the two neighbouring seas ought to vary with the different establishments of the ports. But these trivial inequalities, far from obstructing hydraulical operations, would even be favourable for sluices.

We cannot doubt that if the isthmus of Panama were once burst by some similar catastrophe to] [that which opened the columns of Hercules, (see Diodorus Siculus, lib. iv. p. 226. lib. xvii. p. 533. edit. Rhodom.) the current of rotation in place of ascending towards the gulf of Mexico, and issuing through the canal of Bahama, would follow the same parallel from the coast of Paria to the Philippine islands. The effect of this opening, or new strait, would extend much beyond the banks of Newfoundland, and would either occasion the disappearance or diminish the celerity of the Hotwater river, known by the name of Gulf Stream, which leaving Florida on the *n. e.* flows in the 43° of latitude to the *e.* and especially the *s. e.* towards the coast of Africa. Such would be the effects of an inundation analogous to that of which the memory has been preserved in the traditions of the Samothracians. But shall we dare to compare the pitiful works of man with canals cut by Nature herself, with straits like the Hellespont and the Dardanelles!

Strabo, (Strabo, ed. Siebenkees, t. i. p. 156), appears inclined to believe that the sea will one day open the isthmus of Suez. No such catastrophe can be expected in the isthmus of Panama, unless enormous volcanic convulsions, very improbable in the actual state of repose of our planet, should occasion extraordinary revolutions. A tongue of land lengthened out from *e.* to *w.* in a direction almost parallel to that of the current of rotation escapes, as it were, the shock of the waves. The isthmus of Panama would be seriously threatened, if it extended from *s.* to *n.* and was situated between the port of Carthago and the mouth of the Rio San Juan, if the narrowest part of the new continent lay between the 10° and the 11° of latitude.

The navigation of the river Chagre is difficult, both on account of its sinuosities and the celerity of the current, frequently from one to two metres per second, or from 3.28 to 6.56 feet. These sinuosities, however, afford a counter current, by means of which the small vessels called *bongos*, and *chatas*, ascend the river, either with oars, poles, or towing. Were these sinuosities to be cut, and the old bed of the river to be dried up, this advantage would cease, and it would be infinitely difficult to arrive from the N. sea to Cruces.

From all the information which Humboldt could procure relating to this isthmus, while he remained at Carthagena and Guayaquil, it appeared to him that the expectation of a canal of seven metres, or 22 feet 11 inches, in depth, and from 22 to 28 metres, or from 72 feet 2 inches, to 91 feet 10 inches, in breadth, which, like a pass or a strait, should go from sea to sea, and admit the vessels which sail from Europe to the East Indies, ought to be completely abandoned. The elevation of the ground would force the engineer to have recourse either to subterraneous galleries, or to the system of sluices; and the merchandizes destined to pass the isthmus of Panama could only, therefore, be transported in flat-bottomed boats unable to keep the sea. Entrepots at Panama and Portobello would be requisite. Every nation which wished to trade in this way would be dependent on the masters of the isthmus and canal; and this would be a very great inconvenience for the vessels dispatched from Europe. Supposing then that this canal were cut, the greatest number of these vessels would probably continue their voyage round cape Horn. We see that the passage of the sound is still frequented, notwithstanding the existence of the Eyder canal, which connects the ocean with the Baltic sea.

It would be otherwise with the productions of *w.* America, or the goods sent from Europe to the coast of the Pacific ocean. These goods would cross the isthmus at less expence, and with less danger, particularly in time of war, than in doubling the *s.* extremity of the new continent. In the present state of things, the carriage of three quintals on mule-back from Panama to Portobello costs from three to four piastres (from 12*s.* 6*d.* to 16*s.* 8*d.*) But the uncultivated state in which the government allows the isthmus to remain is such, that the carriage of the copper of Chile, the *quinquina* of Peru, and the 60 or 70,000 *vanegas* of *cacao* (the vanega weighing 110 Castilian pounds), annually exported by Guayaquil, across this neck of land, requires many more beasts of burden than can be procured, so that the slow and expensive navigation round cape Horn is preferred.

In 1802 and 1803, when the Spanish commerce was every where harassed by the English cruisers, a great part of the cacao was carried across the kingdom of Nueva España, and embarked at Vera Cruz for Cadiz. They preferred the passage from Guayaquil to Acapulco and a land journey of 100 leagues from Acapulco to Vera Cruz, to the danger of a long navigation by cape Horn, and the difficulty of struggling with the current along the coasts of Peru and Chile. This example proves, that, if the construction of a canal across the isthmus of Panama, or that of Guasacualco, abounds with too many difficulties from the multiplicity of sluices, the commerce of America would gain the most important advantages from good causeways, carried from Tehuantepec to the Embarcadero de la Cruz, and from Panama to Portobello. It is true that in the isthmus, the pasturage to this day] [is very unfavourable to the nourishment and multiplication of cattle; but it is no less true that the assertion of Raynal (t. iv. p. 150), that domestic animals transported to Portobello lose their fecundity, should be considered as totally destitute of truth. The fact is, that it would be easy, in so fertile a soil, to form savannas by cutting down forests, or to cultivate the *paspalum purpureum*, the *milium nigricans*, and particularly the *medicago sativa*, which grows abundantly in Peru in the warmest districts. The introducion of camels would be still a surer means of diminishing the expence of carriage. These land-ships, as they are called by the orientals, hitherto exist only in the province of Caracas, and were brought there from the Canary islands by the Marquis de Toro.

Moreover, no political consideration should oppose the progress of population, agriculture, commerce, and civilization, in the isthmus of Panama. The more this neck of land shall be cultivated, the more resistance will it oppose to the enemies of the Spanish government. The events which took place at Buenos Ayres prove the advantages of a concentrated population in the case of an invasion. If any enterprising nation wished to become possessed of the isthmus, it could do so with the greatest ease at present, when good and numerous fortifications are destitute of arms to defend them. The unhealthiness of the climate, though now much diminished at Potobello, would alone oppose great obstacles to any military undertaking in the isthmus. It is from St. Charles de Chiloe, and not from Panama, that Peru can be attacked. It requires from three to five months to ascend from Panama to Lima. But the whale and *cachalot* fishery, which in 1803 drew 60 English vessels to the S. sea, and the facilities for the Chinese commerce and the furs of Nootka sound, are baits of a very seductive nature. They will draw, sooner or later, the masters of the ocean to a point of the globe destined by nature to change the face of the commercial system of nations.

6. To the *s. e.* of Panama, following the coast of the Pacific ocean, from cape S. Miguel to cape Corientes, we find the small port and bay of Cupica, which has acquired celebrity, on account of a new p a of communication between the two seas. See this article, (Cupica).

7. In the interior of the province of Choco, the small ravine (*quebrada*) De la Raspadura, unites the neighbouring sources of the Rio de Noanama, called also Rio San Juan, and the small river Quito. The latter, the Rio Andageda, and the Rio Zitara, form the Rio d'Atrato, which discharges itself into the Atlantic ocean, while the Rio San Juan flows into the S. sea. A monk of great activity, *curé* of the village of Novita, employed his parishioners to dig a small canal in the ravine De la Raspadura, by means of which, when the rains are abundant, canoes loaded with cacao pass from sea to sea. This interior communication has existed since 1788, unknown in Europe. The small canal of Raspadura unites, on the coasts of the two oceans, two points 75 leagues distant from one another.

8. In lat. 10° *s.* two or three days journey from Lima, we reach the banks of the Rio Guallaga (or Huallaga), by which we may without doubling cape Horn arrive at the banks of the Grand Para in Brazil. The sources even of the Rio Huanuco, which runs into the Guallaga, are only four or five leagues distant from the source of the Rio Huaura, which flows into the Pacific ocean. The Rio Xauxa, also, which contributes to form the Apuremac and the Ucayale, has its rise near the source of the Rio Rimae. The height of the *cordillera*, and the nature of the ground, render the execution of a canal impossible; but the construetion of a commodious road, from the capital of Peru to the Rio de Huanuco, would facilitate the transport of goods to Europe. The great rivers Ucayale and Guallaga would carry in five or six weeks the productions of Peru to the mouth of the Amazons, and to the neighbouring coasts of Europe, while a passage of four months is requisite to convey the same goods to the same point, in doubling cape Horn. The cultivation of the fine regions situated on the *e.* declivity of the Andes, and the prosperity and wealth of their inhabitants, depend on a free navigation of the river of the Amazons. This liberty, denied by the court of Portugal to the Spaniards, might have been acquired in the sequel to the events which preceded the peace of 1801.

9. Before the coast of the Patagonians was sufficiently known, the gulf of St. George, situated between the 45° and the 47° of *s.* lat. was supposed to enter so far into the interior of the country, as to communicate with the arms of the sea which interrupt the continuity of the *w.* coast, that is to say, with the coast opposite to the archipelago of Chayamapu. Were this supposition founded on solid bases, the vessels destined for the S. sea might cross S. America 7° to the *n.* of the straits of Magellan, and shorten their route more than 700 leagues. In this way, navigators might avoid the dangers which, notwithstanding the perfection of nautical science, still accompany the voyage round cape Horn and along the Patagonian coast, from cape Pilares to the parallel of the Chonos islands. These ideas, in 1790, occupied the attention of]

[the court of Madrid. M. Gil Lemos, viceroy of Peru, an upright and zealous administrator, equipped a small expedition under the orders of M. Moraleda, to examine the *s.* coast of Chile. The aforesaid person visited the archipelagos of Chiloe and Chonos, and the *w.* coast of the Patagonians, from 1787 down to 1796. Two very interesting manuscripts, drawn up by M. Moraleda, are to be found in the archives of the viceroyalty of Lima: the title of the one is, *Viage al Reconocimiento de las Islas de Chiloe*, 1786; the other comprehends the *Reconocimiento del Archipelago de los Chonos y Costa occidental Patagonica*, 1792—1796. Curious and interesting extracts might be published from these journals, which contain details regarding the cities De los Cesares and De l'Arguello, which are said to have been founded in 1554, and are placed by apocryphal accounts between 42° and 49° of *s.* lat. Humboldt saw the instructions the above person received at Lima, which recommended to him the greatest secrecy in case he should be happy enough to discover a communication between the two seas. But M. Moraleda discovered in 1793, that the Estero de Aysen, visited before him in 1763 by the Jesuits, Fathers Jose Garcia and Juan Vicuña, was, of all the arms of the sea, that in which the waters of the ocean advance the farthest towards the *e.* Yet it is but eight leagues in length, and terminates at the isle De la Cruz, where it receives a small river, near a hot spring. Hence the canal of Aysen, situated in the 45° 28′ of lat. is still 88 leagues distant from the gulf of St. George. This gulf was exactly surveyed by the expedition of Malaspina. In the year 1746 a communication was, in the same manner, suspected in Europe between the bay of St. Julien (lat. 50° 53′) and the Great ocean.

M. Humboldt has sketched in one plate the nine points which appear to afford means of communication between the two oceans, by the junction of neighbouring rivers, either by canals or carriage-roads between the places where the rivers become navigable. These sketches are not of equal accuracy, astronomically considered; but he wished to save the reader the labour of seeking in several maps what may be contained in one; and it is the duty of the government which possesses the finest and most fertile part of the globe to perfect what he has merely hinted at in this discussion. Two Spanish engineers, MM. Le Maur, drew up superb plans of the canal De los Guines, projected for traversing the whole island of Cuba, from Batabano to the Havanah. A similar survey of the isthmus of Guasacualco, the lake Nicaragua, of the country between Cruces and Panama, and between Cupica and the Rio Naipi, would direct the statesman in his choice, and enable him to decide, if it is at Mexico or Darien that this undertaking should be executed; an undertaking calculated to immortalize a government occupied with the true interests of humanity.

The long circumnavigation of S. America would then be less frequent; and a communication would be opened for the goods which pass from the Atlantic ocean to the S. sea. The time is past (as observes M. de Fleurieu, in his learned notes on the Voyage de Marchand, t. i. p. 566) "when Spain, through a jealous policy, refused to other nations a thoroughfare through the possessions of which she so long kept the world in ignorance." Those who are at present at the head of the government are enlightened enough to give a favourable reception to the liberal ideas proposed to them; and the presence of a stranger is no longer regarded as a danger for the country.

Should a canal of communication be opened between the two oceans, the productions of Nootka sound and of China will be brought more than 2000 leagues nearer to Europe and the United States. Then only can any great changes be effected in the political state of *e.* Asia, for this neck of land, the barrier against the waves of the Atlantic ocean, has been for many ages the bulwark of the independence of China and Japan.

CHAP. XI.

Glance on the coast of the Great ocean, which extends from the port of San Francisco and from cape Mendocino to the Russian establishments in Prince William's sound.

THE whole of this coast has been visited since the end of the 16th century by Spanish navigators; but it has only been carefully examined by order of the viceroys of New Spain since 1774. Numerous expeditions of discovery have followed one another up to 1792. The colony attempted to be established by the Spaniards at Nootka fixed for some time the attention of all the maritime powers of Europe. A few sheds erected on the coast, and a miserable bastion defended by swivel guns, and a few cabbages planted within an enclosure, were very near exciting a bloody war between Spain and England; and it was only by the destruction of the establishment founded at the island of Quadra and of Vancouver, that Macuina, the *tays* or prince of Nootka, was enabled to preserve his independence. Several nations of Europe have frequented these latitudes since 1786, for the sake of the trade in sea otter skins; but their rivalry has had the most disadvantageous] [consequences both for themselves and the natives of the country. The price of the skins as they rose on the coast of America fell enormously in China. Corruption of manners has increased among the Indians; and by following the same policy by which the African coasts have been laid waste, the Europeans endeavoured to take advantage of the discord among the *tays*. Several of the most debauched sailors deserted their ships to settle among the natives of the country. At Nootka, as well as at the Sandwich islands, the most fearful mixture of primitive barbarity with the vices of polished Europe is to be observed. It is difficult to conceive that the few species of roots of the old continent transplanted into these fertile regions by voyagers, which figure in the list of the benefits that the Europeans boast of having bestowed on the inhabitants of the S. sea islands, have proved any thing like a compensation for the real evils which they introduced among them.

At the glorious epocha in the 16th century, when the Spanish nation, favoured by a combination of singular circumstances, freely displayed the resources of their genius and the force of their character; the problem of a passage to the *n. w.* and a direct road to the E. Indies, occupied the minds of the Castilians with the same ardour displayed by some other nations within these 30 or 40 years. We do not allude to the apocryphal voyages of Ferrer Maldonado, Juan de Fuca, and Bartolome Fonte, to which for a long time only too much importance was given. The most part of the impostures published under the names of these three navigators were destroyed by the laborious and learned discussion of several officers of the Spanish marine. (See Memoirs of Don Ciriaco Cevallos. Researches into the Archives of Seville, by Don Augustin Cean. Historical Introduction to the Voyage of Galiano and Valdes, p. xlix. lvi. and lxxvi. lxxxiii.) In place of bringing forward names nearly fabulous, and losing ourselves in the uncertainty of hypotheses, we shall confine ourselves to indicate here what is incontestibly proved by historical documents. The following notices, partly drawn from the manuscript memoirs of Don Antonio Bonilla and M. Casasola, preserved in the archives of the viceroyalty of Mexico, present facts which, combined together, deserve the attention of the reader. These notices displaying, as it were, the varying picture of the national activity, sometimes excited and sometimes palsied, will no doubt be interesting.

The names of Cabrillo and Gali are less celebrated than Fuca and Fonte. The true recital of a modest navigator has neither the charm nor the power which accompany deception. Juan Rodriguez Cabrillo visited the coast of New California to the 37° 10′, or the Punta del Año Nuevo, to the *n.* of Monterey. He perished (on the 3d January 1543) at the island of San Bernardo, near the channel of Santa Barbara, according to the manuscript preserved in the Archivo-general de Indias at Madrid. But Bartolome Ferrelo, his pilot, continued his discoveries *n.* to the 43° of lat. when he saw the coast of cape Blane, called by Vancouver cape Orford.

Francisco Gali, in his voyage from Macao to Acapulco, discovered in 1582 the *n. w.* coast of America under the 57° 30′. He admired, like all those who since his time have visited New Cornwall, the beauty of those colossal mountains, of which the summit is covered with perpetual snow, while their bottom is covered with the most beautiful vegetation. On correcting the old observations by the new, in places of which the identity is ascertained, we find that Gali coasted part of the archipelago of the Prince of Wales, or that of King George. Sir Francis Drake only went as far as the 48° of lat. to the *n.* of cape Grenville in New Georgia.

Of the two expeditions undertaken by Sebastian Viscayno in 1596 and 1602, the last only was directed to the coast of New California. Thirty-two maps, drawn up at Mexico by the cosmographer Henry Martinez, prove that Viscayno surveyed those coasts with more care and more intelligence than was ever done by any pilot before him. The diseases of his crew, the want of provision, and the extreme rigour of the season, prevented him, however, from ascending higher than cape S. Sebastian, situated under the 42° of lat. a little to the *n.* of the bay of the Trinity. One vessel of Viscayno's expedition, the frigate commanded by Antonio Florez, alone passed cape Mendocino. This frigate reached the mouth of a river in the 43° of lat. which appears to have been already discovered by Cabrillo in 1543, and which was believed by Martin de Aguilar to be the *w.* extremity of the straits of Anian. We must not confound this entry or river of Aguilar, which could not be found again in our times, with the mouth of the Rio Colombia (lat. 46° 15′) celebrated from the voyage of Vancouver, Gray, and Captain Lewis.

The brilliant epocha of the discoveries made anciently by the Spaniards on the *n. w.* coast of America ended with Gali and Viscayno. The history of the navigations of the 17th century, and the first half of the 18th, offers us no expedition directed from the coast of Mexico to the immense shore from cape Mendocino to the confines of *e.* Asia. In place of the Spanish the Russian flag] [was alone seen to float in these latitudes, waving on the vessels commanded by two intrepid navigators, Bering and Tschiricow.

At length, after an interruption of nearly 170 years, the court of Madrid again turned its attention to the coast of the Great ocean. But it was not alone the desire of discoveries useful to science which roused the government from its lethargy. It was rather the fear of being attacked in its most *n.* possessions in New Spain; it was the dread of seeing European establishments in the neighbourhood of those of California. Of all the Spanish expeditions undertaken between 1774 and 1792, the two last alone bear the true character of expeditions of discovery. They were commanded by officers whose labours display an intimate acquaintance with nautical astronomy. The names of Alexander Malaspina, Galiano, Espinosa, Valdes, and Vernaci, will ever hold an honourable place in the list of the intelligent and intrepid navigators to whom we owe an exact knowledge of the *n. w.* coast of the new continent. If their predecessors could not give the same perfection to their operations, it was because, setting out from San Blas or Monterey, they were unprovided with instruments and the other means furnished by civilized Europe.

The first important expedition made after the voyage of Viscayno was that of Juan Perez, who commanded the corvette Santiago, formerly called La Nueva Galicia. As neither Cook nor Barrington, nor M. de Fleurieu, appear to have had any knowledge of this important voyage, we shall here insert several facts extracted from a manuscript journal, which was kept by two monks, *Fray* Juan Crespi, and *Fray* Tomas de la Peña, and for which Humboldt was indebted to the kindness of M. Don Guillermo Aguirre, a member of the *audiencia* of Mexico. Perez and his pilot, Estevan Jose Martinez, left the port of San Blas on the 24th January 1774. They were ordered to examine all the coast from the port of San Carlos de Monterey to the 60° of latitude. After touching at Monterey they set sail again on the 7th June. They discovered on the 20th July the island De la Marguerite (which is the *n. w.* point of Queen Charlotte's island), and the strait which separates this island from that of the Prince of Wales. On the 9th August they anchored, the first of all the European navigators, in Nootka road, which they called the port of San Lorenzo, and which the illustrious Cook four years afterwards called King George's sound. They carried on barter with the natives, among whom they saw iron and copper. They gave them axes and knives for skins and otter furs. Perez could not land on account of the rough weather and high seas. His sloop was even on the point of being lost in attempting to land; and the corvette was obliged to cut its cables and to abandon its anchors to get into the open sea. The Indians stole several articles belonging to M. Perez and his crew; and this circumstance, related in the journal of Father Crespi, may serve to resolve the famous difficulty attending the European silver spoons found there by Captain Cook in 1778 in the possession of the Indians of Nootka. The corvette Santiago returned to Monterey on the 27th of August 1774, after a cruise of eight months.

In the following year a second expedition set out from San Blas, under the command of Don Bruno Heceta, Don Juan de Ayala, and Don Juan de la Bodega y Quadra. This voyage, which singularly advanced the discovery of the *n. w.* coast, is known from the journal of the pilot Maurelle, published by M. Barrington and joined to the instructions of the unfortunate La Perouse. Quadra discovered the mouth of the Rio Colombia, called Entrada de Heceta, the Pic of San Jacinto (mount Edgecumbe), near Norfolk bay, and the fine port of Bucareli (lat. 55° 24′), which from the researches of Vancouver we know to belong to the *w.* coast of the great island of the archipelago of the Prince of Wales. This port is surrounded by seven volcanoes, of which the summits, covered with perpetual snow, throw up flames and ashes. M. Quadra found there a great number of dogs which the Indians use for hunting. Humboldt states to have in his possession two very curious small maps, engraved in 1788, in the city of Mexico, which give the bearings of the coast from the 17° to the 58° of latitude, as they were discovered in the expedition of Quadra. One of these maps is entitled, "Carta geografica de la costa occidental de la California, situada al Norte de la linea sobre el mar Asiatico que se discubrio en los años de 1769 y 1775, por el Teniente de Navio, Don Juan Francisco de Bodega y Quadra y por el Alferez de Fragata, Don Jose Cañizares desde los 17 hasta los 58 grados." On this map the coast appears almost without *entradas* and without islands. In this we remark L'Ensenada de Ezeta (Rio Colombia) and L'Entrada de Juan Perez, but under the name of the port of San Lorenzo (Nootka), seen by the same Perez in 1774. The other is called "Plan del gran puerto de San Francisco discubierto por Don Jose de Cañizares en el mar Asiatico." Vancouver distinguishes the ports of St. Francis, Sir Francis Drake, and Bodega, as three different ports. M.]

[de Fleurieu considers them as identical. Voyage de Marchand, vol. i. p. 54. Quadra believes, as we have already observed, that Drake anchored at the port De la Bogeda.

The court of Madrid gave orders in 1776 to the viceroy of Mexico, to prepare a new expedition to examine the coast of America to the 70° of *n.* latitude. For this purpose two corvettes were built, La Princesa and La Favorita; but this building experienced such delay, that the expedition, commanded by Quadra and Don Ignacio Arteaga, could not set sail from the port of San Blas till the 11th February 1779. During this interval Cook visited the same coast. Quadra and the pilot Don Francisco Maurelle carefully examined the port De Bucareli, the Mont Sant Elie, and the island De la Magdalena, called by Vancouver Hinchinbrook island (lat. 60° 25′), situated at the entry of Prince William's bay, and the island of Regla, one of the most sterile islands in Cook river. The expedition returned to San Blas on the 21st November 1779. We find from a manuscript procured at Mexico, that the schistous rocks in the vicinity of the port of Bucareli in Prince of Wales's island contain metalliferous seams.

The memorable war which gave liberty to a great part of N. America, prevented the viceroys of Mexico from pursuing expeditions of discovery to the *n.* of Mendocino. The court of Madrid gave orders to suspend the expeditions so long as the hostilities should endure between Spain and England. This interruption continued even long after the peace of Versailles; and it was not till 1788 that two Spanish vessels, the frigate La Princesa, and the packet-boat San Carlos, commanded by Don Esteban Martinez and Don Gonzalo Lopez de Haro, left the port of San Blas with the design of examining the position and state of the Russian establishments on the *n. w.* coast of America. The existence of these establishments, of which it appears that the court of Madrid had no knowledge till after the publication of the third voyage of the illustrious Cook, gave the greatest uneasiness to the Spanish government. It saw with chagrin that the fur trade drew numerous English, French, and American vessels towards a coast which, before the return of Lieutenant King to London, had been as little frequented by Europeans as the land of the Nuyts, or that of Endracht in New Holland.

The expedition of Martinez and Haro lasted from the 8th March to the 5th of December 1788. These navigators made the direct route from San Blas to the entry of Prince William, called by the Russians the gulf Tschugatskaja. They visited Cook river, the Kichtak (Kodiak) islands, Schumagin, Unimak, and Unalaschka (Onalaska).—They were very friendly treated in the different factories which they found established in Cook river and Unalaschka, and they even received communication of several maps drawn up by the Russians of these latitudes. Humboldt found in the archives of the viceroyalty of Mexico a large volume in folio, bearing the title of *Reconocimiento de los quatros establecimientos Russos al Norte de la California, hecho en* 1788. The historical account of the voyage of Martinez contained in this manuscript furnishes, however, very few data relative to the Russian colonies in the new continent. No person in the crew understanding a word of the Russian language, they could only make themselves understood by signs. They forgot, before undertaking this distant expedition, to bring an interpreter from Europe. The evil was without remedy. However, M. Martinez would have had as great difficulty in finding a Russian in the whole extent of Spanish America as Sir George Staunton had to discover a Chinese in England or France.

Since the voyages of Cook, Dixon, Portlock, Mears, and Duncan, the Europeans began to consider the port of Nootka as the principal fur market of the *n. w.* coast of N. America. This consideration induced the court of Madrid to do in 1789 what it could easier have done 15 years sooner, immediately after the voyage of Juan Perez. M. Martinez, who had been visiting the Russian factories, received orders to make a solid establishment at Nootka, and to examine carefully that part of the coast comprised between the 50° and the 55° of latitude, which Captain Cook could not survey in the course of his navigation.

The port of Nootka is on the *e.* coast of an island, which, according to the survey in 1791 by MM. Espinosa and Cevallos, is 20 marine miles in breadth, and which is separated by the channel of Tasis from the great island, now called the island of Quadra and Vancouver. It is therefore equally false to assert that the port of Nootka, called by the natives Yucuatl, belongs to the great island of Quadra, as it is inaccurate to say that cape Horn is the extremity of Tierra del Fuego. It was an extraordinary misconception in the illustrious Cook in converting the name of Yucuatl into Nootka, this last word being unknown to the natives of the country, and having no analogy to any of the words of their language excepting *noutchi*, which signifies mountain. It would appear, however, from what is said of Captain Cook] [by Mr. King, that his ear was by no means very accurate in distinguishing sounds.

Don Esteban Martinez, commanding the frigate La Princesa, and the packet-boat San Carlos, anchored in the port of Nootka on the 5th May 1789. He was received in a very friendly manner by the chief Macuina, who recollected very well having seen him with M. Perez in 1774, and who even shewed the beautiful Monterey shells which were then presented to him. Maçuina, the *tays* of the island of Yucuatl, has an absolute authority; he is the Montezuma of these countries; and his name has become celebrated among all the nations who carry on the sea-otter skin trade. We know not if Macuina yet lives; but it was said at Mexico in the end of 1803, that, more jealous of his independence than the king of the Sandwich islands, who has declared himself the vassal of England, he was endeavouring to procure fire-arms and powder to protect himself from the insults to which he was frequently exposed by European navigators. See Nootka.

Martinez did not carry his researches beyond the 50° of latitude. Two months after his entry into the port of Nootka he saw the arrival of an English vessel, the Argonaut, commanded by James Colnet, known by his observations at the Galapagos islands. Colnet showed the Spanish navigator the orders which he had received from his government to establish a factory at Nootka, to construct a frigate and a cutter, and to prevent every other European nation from interfering with the fur trade. It was in vain Martinez replied, that long before Cook, Juan Perez had anchored on the same coast. The dispute which arose between the commanders of the Argonaut and the Princesa was on the point of occasioning a rupture between the courts of London and Madrid. Martinez, to establish the priority of his rights, made use of a violent and very illegal measure: he arrested Colnet, and sent him by San Blas to the city of Mexico. The true proprietor of the Nootka country, the Tays Macuina, declared himself prudently for the vanquishing party; but the viceroy, who deemed it proper to hasten the recal of Martinez, sent out three other armed vessels in the commencement of the year 1790 to the *n. w.* coast of America.

Don Francisco Elisa, and Don Salvador Fidalgo, the brother of the astronomer who surveyed the coast of S. America, from the mouth of the Dragon to Portobello, commanded this new expedition. M. Fidalgo visited Cook creek and Prince William's sound, and he completed the examination of that coast, which was only afterwards examined by the intrepid Vancouver. Under the 60° 54′ of latitude, at the *n.* extremity of Prince William's sound, M. Fidalgo was witness of a phenomenon, probably volcanic, of a most extraordinary nature. The Indians conducted him into a plain covered with snow, where he saw great masses of ice and stone thrown up to prodigious heights in the air with a dreadful noise. Don Francisco Elisa remained at Nootka to enlarge and fortify the establishment founded by Martinez in the preceding year. It was not yet known in this part of the world, that by a treaty signed at the Escurial on the 28th October 1790, Spain had desisted from her pretensions to Nootka and Cox channel in favour of the court of London. The frigate Dedalus, which brought orders to Vancouver to watch over the execution of this treaty, only arrived at the port of Nootka in the mouth of August 1792, at an epocha when Fidalgo was employed in forming a second Spanish establishment to the *s. e.* of the island of Quadra on the continent, at the port of Nuñez Gaoua, or Quinacamet, situated under the 48° 20′ of latitude, at the creek of Juan de Fuca.

The expedition of Captain Elisa was followed by two others, which, for the importance of their astronomical operations, and the excellence of the instruments with which they were provided, may be compared with the expeditions of Cook, La Perouse, and Vancouver. We mean the voyage of the illustrious Malaspina in 1791, and that of Galiano and Valdes in 1792.

The operations of Malaspina and the officers under him embrace an immense extent of coast from the mouth of the Rio de la Plata to Prince William's sound. But this able navigator is still more celebrated for his misfortunes than his discoveries. After examining both hemispheres, and escaping all the dangers of the ocean, he had still greater to suffer from his court; and he dragged out six years in a dungeon, the victim of a political intrigue. He obtained his liberty from the French government, and returned to his native country; and he enjoys in solitude on the banks of the Arno the profound impressions which the contemplation of nature and the study of man under so many different climates have left on a mind of great sensibility, tried in the school of adversity.

The labours of Malaspina remain buried in the archives, not because the government dreaded the disclosure of secrets, the concealment of which might be deemed useful, but that the name of this intrepid navigator might be doomed to eternal oblivion. Fortunately the directors of the Deposito Hydrografico of Madrid, (established by a royal order] [on the 6th August 1797), have communicated to the public the principal results of the astronomical observations of Malaspina's expedition. The charts which have appeared at Madrid since 1799 are founded in a great measure on those important results; but instead of the name of the chief, we merely find the names of the corvettes La Desenbierta and La Atrevida, which were commanded by Malaspina.

His expedition, which set out from Cadiz on the 30th July 1789, only arrived at the port of Acapulco on the 2d February 1791. At this period the court of Madrid again turned its attention to a subject which had been under dispute in the beginning of the 17th century, the pretended straits by which Lorenzo Ferrer Maldonado passed in 1588 from the Labrador coast to the Great ocean. A memoir read by M. Buache at the Academy of Sciences revived the hope of the existence of such a passage; and the corvettes La Descubierta and L'Atrevida received orders to ascend to high latitudes on the *n. w.* coast of America, and to examine all the passages and creeks which interrupt the continuity of the shore between the 53° and 60° of latitude. Malaspina, accompanied by the botanists Haenke and Nee, set sail from Acapulco on the 1st May 1791. After a navigation of three weeks he reached cape S. Bartholomew, which had already been ascertained by Quadra in 1775, by Cook in 1778, and in 1786 by Dixon. He surveyed the coast from the mountain of San Jacinto, near cape Edgecumbe (Cabo Engano), lat. 57° 1′ 30″ to Montagu island, opposite the entrance of Prince William's sound. During the course of this expedition, the length of the pendulum and the inclination and declination of the magnetic needle were determined on several points of the coast. The elevation of S. Elie and mount Fairweather (or Cerro de Buen Tempo), which are the principal summits of the *cordillera* of New Norfolk, were very carefully measured: the height of the former is 17,850, and of the second 14,992 feet. The knowledge of their height and position may be of great assistance to navigators when they are prevented by unfavourable weather from seeing the sun for whole weeks; for by seeing these pics at a distance of 80 or 100 miles, they may ascertain the position of their vessel by simple elevations and angles of altitude.

After a vain attempt to discover the straits mentioned in the account of the apocryphal voyage of Maldonado, and after remaining some time at port Mulgrave, in Bering's bay, (lat. 59° 34′ 20″), Alexander Malaspina directed his course *s.* He anchored at the port of Nootka on the 13th August, sounded the channels round the island of Yucuatl, and determined by observations purely celestial the positions of Nootka, Monterey, and the island of Guadaloupe, at which the galleon of the Philippines (La Nao de China), generally stops, and cape San Lucas. The corvette La Atrevida entered Acapulco, and the corvette La Descubierta entered San Blas in the month of October 1791.

A voyage of six months was no doubt by no means sufficient for discovering and surveying an extensive coast with that minute care which we admire in the voyage of Vancouver, which lasted three years. However, the expedition of Malaspina has one particular merit, which consists not only in the number of astronomical observations, but also in the judicious method employed for attaining certain results. The longitude and latitude of four points of the coast, cape San Lucas, Monterey, Nootka, and port Mulgrave, were ascertained in an absolute manner. The intermediate points were connected with these fixed points by means of four sea-watches of Arnold. This method, employed by the officers of Malaspina's expedition, MM. Espinosa, Cevallos, and Vernaci, is much better than the partial corrections usually made in chronometrical longitudes by the results of lunar distances.

The celebrated Malaspina had scarcely returned to the coast of Mexico, when, discontented with not having seen at a sufficient nearness the extent of coast from the island of Nootka to cape Mendocino, he engaged Count de Revillagigedo, the viceroy, to prepare a new expedition of discovery towards the *n. w.* coast of America. The viceroy, who was of an active and enterprising disposition, yielded with so much the greater facility to this desire, as new information, received from the officers stationed at Nootka, seemed to give probability to the existence of a channel, of which the discovery was attributed to the Greek pilot, Juan de Fuca, in the end of the 16th century. Martinez had indeed, in 1774, perceived a very broad opening under the 48° 20′ of latitude. This opening was successively visited by the pilot of the Gertrudis, by Ensign Don Manuel Quimper, who commanded the bilander La Princesa Real, and in 1791 by Captain Elisa. They even discovered secure and spacious ports in it. It was to complete this survey that the galeras Sutil and Mexicana left Acapulco on the 8th March 1792, under the command of Don Dionisiso Galiano and Don Cayetano Valdes.

These able and experienced astronomers, accompanied by MM. Salamanca and Vernaci, sailed round the large island which now bears the] [name of Quadra and Vancouver, and they employed four months in this laborious and dangerous navigation. After passing the straits of Fuca and Haro, they fell in with, in the channel Del Rosario, called by the English the gulf of Georgia, the English navigators Vancouver and Broughton, employed in the same researches with themselves. The two expeditions made a mutual and unreserved communication of their labours; they assisted one another in their operations; and there subsisted among them till the moment of their separation a good intelligence and complete harmony, of which, at another epocha, an example had not been set by the astronomers on the ridge of the *cordilleras*.

Galiano and Valdes, on their return from Nootka to Monterey, again examined the mouth of the Ascencion, which Don Bruno Eceta discovered on the 17th August 1775, and which was called the river of Colombia by the celebrated American navigator Gray, from the name of the sloop under his command. This examination was of so much the greater importance, as Vancouver, who had already kept very close to this coast, was unable to perceive any entrance from the 45° of latitude to the channel of Fuca; and as this learned navigator began then to doubt of the existence of the Rio de Colombia, or the Entrada de Eceta.

In 1797 the Spanish government gave orders that the charts drawn up in the course of the expedition of MM. Galiano and Valdes should be published, "in order that they might be in the hands of the public before those of Vancouver." However, the publication did not take place till 1802; and geographers now possess the advantage of being able to compare together the charts of Vancouver, those of the Spanish navigators published by the Deposito Hydrografico of Madrid, and the Russian chart published at Petersburgh in 1802, in the depôt of the maps of the charts of the emperor. This comparison is so much the more necessary, as the same capes, the same passages, and the same islands, frequently bear three or four different names; and geographical synonomy has by that means become as confused as the synonomy of cryptogameous plants has become from an analogous cause.

At the same epocha at which the vessels Sutil and Mexicana were employed in examining, in the greatest detail, the shore between the parallels of 45° and 51°, the Count de Revillagigedo destined another expedition for higher latitudes. The mouth of the river of Martin de Aquilar had been unsuccessfully sought for in the vicinity of cape Orford and cape Gregory. Alexánder Malaspina, in place of the famous channel De Maldonado, had only found openings without any outlet. Galiano and Valdes had ascertained that the strait of Fuca was merely an arm of the sea, which separates an island of more than 1700 square leagues, that of Quadra and Vancouver, from the mountainous coast of New Georgia. The extent of this island, calculated according to the maps of Vancouver, is 1730 square leagues, of 25 to the sexagesimal degree. It is the largest island to be found on this *w.* coast of America. There still remained doubts as to the existence of the straits, of which the discovery was attributed to admiral Fuentes or Fonte, which was supposed to be under the 53° of latitude. Cook regretted his want of ability to examine this part of the continent of New Hanover; and the assertions of Captain Colnet, an able navigator, rendered it extremely probable that the continuity of the coast was interrupted in these latitudes. To resolve a problem of such importance, the viceroy of New Spain gave orders to Lieutenant Don Jacinto Caamaño, commander of the frigate Aranzazu, to examine with the greatest care the shore from the 51° to the 56° of *n.* latitude. M. Caamaño set sail from the port of San Blas on the 20th March 1792; and he made a voyage of six months. He carefully surveyed the *n.* part of Queen Charlotte's island, the *s.* coast of the Prince of Wales's island, which he called Isla de Ulloa, the islands of Revillagigedo, of Banks (or De la Calamidad), and of Aristizabal, and the great inlet of Moniño, the mouth of which is opposite the archipelago of Pitt. The considerable number of Spanish denominations preserved by Vancouver in his charts proves that the expeditions, of which we have given a summary account, contributed in no small degree to our knowledge of a coast, which, from the 45° of latitude to cape Douglas, to the *e.* of Cook's creek, is now more accurately surveyed than the most part of the coasts of Europe.

We have now given all the information which we could procure with regard to the voyages undertaken by the Spaniards, from 1553 to our own times, towards the *w.* coast of Nueva España to the *n.* of New California. The assemblage of these materials appears to us to be necessary in a work embracing whatever concerns the political and commercial relations of Mexico.

The geographers, who are eager to divide the world for the sake of facilitating the study of their science, distinguish on the *n. w.* coast an English part, a Spanish part, and a Russian part. These divisions have been made without consulting the chiefs of the different tribes who inhabit these] [countries! If the puerile ceremonies which the Europeans call taking possession, and if astronomical observations made on a recently discovered coast, could give rights of property, this portion of the new continent would be singularly pieced out and divided among the Spaniards, English, Russians, French, and Americans. One small island would sometimes be shared by two or three nations at once, because each might have discovered a different cape of it. The great sinuosity of the coast between the parallels of 55° and 60° embraces the successive discoveries of Gali, Bering, and Tschirekow, Quadra, Cook, La Perouse, Malaspina, and Vancouver!

No European nation has yet formed a solid establishment on the immense extent of coast from cape Mendocino to the 59° of latitude. Beyond this limit the Russian factories commence, the most part of which are scattered and distant from one another, like the factories established by European nations for these last 300 years on the coast of Africa. The most part of these small Russian colonies have no communication with one another but by sea; and the new denominations of Russian America, or Russian Possessions in the New Continent, ought not to induce us to believe that the coast of the basin of Bering, the peninsula Alaska, or the country of the Tschugatschi, have become Russian provinces, in the sense which we give to this word, speaking of the Spanish provinces of Sonora or New Biscay.

The *w.* coast of America affords the only example of a shore of 1900 leagues in length, inhabited by one European nation. The Spaniards, as we have already indicated in the commencement of this work, have formed establishments from fort Maullin in Chile to S. Francis in New California. To the *n.* of the parallel of 38° succeed independent Indian tribes. It is probable that these tribes will be gradually subdued by the Russian colonists, who, towards the end of the last century, passed over from the *e.* extremity of Asia to the continent of America. The progress of these Russian Siberians towards the *s.* ought naturally to be more rapid than that of the Spanish Mexicans towards the *n.* A people of hunters, accustomed to live in a foggy and excessively cold climate, find the temperature of the coast of New Cornwall very agreeable; but this coast appears an uninhabitable country, a polar region, to colonists from a temperate climate, from the fertile and delicious plains of Sonora and New California.

The Spanish government since 1788 has begun to testify uneasiness at the appearance of the Russians on the *n. w.* coast of the new continent. Considering every European nation in the light of a dangerous neighbourr, they examined the situation of the Russian factories. The fear ceased on its being known at Madrid that these factories did not extend *e.* beyond Cook's inlet. When the Emperor Paul, in 1799, declared war against Spain, it was some time in agitation at Mexico to prepare a maritime expedition in the ports of San Blas and Monterey against the Russian colonies in America. If this project had been carried into exeention we should have seen at hostilities two nations, who, occupying the opposite extremities of Europe, approach each other in the other hemisphere on the *e.* and *w.* limits of their vast empires.

The interval which separates these limits becomes progressively smaller; and it is for the political interest of Nueva España to know accurately the parallel to which the Russian nation has already advanced towards the *e.* and *s.* A manuscript which exists in the archives of the viceroyalty of Mexico, and which was seen by Humboldt, gave him only vague and incomplete notions. It describes the state of the Russian establishments as they were 20 years ago. M. Malte Brun, in his Universal Geography, gives an interesting article on the *n. w.* coast of America. He was the first who made known the account of the voyage of Billings, (entitled, "Account of the geographical and astronomical expedition undertaken for exploring the coast of the Icy sea, the land of the Tshutski, and the islands between Asia and America, under the command of Captain Billings, between the years 1785 and 1794, by Martin Sauer, secretary to the expedition. Putetchestwie flota-kapitana Sarytschewa po severowostochnoi tschasti sibiri, ledowitawa mom, i wostochnogo okeana, 1804,") published by M. Sarytschew, which is preferable to that of M. Sauer. The following account of the Russian factories is extracted from an official document, being a chart of discoveries successively made by Russian navigators in the Pacific ocean, and in the Icy sea, published in 1802. It shews the same to be merely collections of sheds and huts, that serve, however, as emporiums for the fur trade.

On the coast nearest to Asia, along Bering's straits, between the 67° and 64° 10′ of latitude, under the parallels of Lapland and Iceland, we find a great number of huts frequented by the Siberian hunters. The principal posts, reckoning from *n.* to *s.* are, Kigiltach, Leglelachtok, Tuguten, Netschich, Tchinegriun, Chibalech, Topar, Pintepata, Agulichan, Chavani, and Nugran, near cape Rodney (Cap du Parent). These habitations] [of the natives of Russian America are only from 30 to 40 leagues distant from the huts of the Tchoutskis of Asiatic Russia. The straits of Bering, which separate them, are filled with desert islands, of which the most *n.* is called Imaglin. The *n. e.* extremity of Asia forms a peninsula, which is only connected with the great mass of the continent by a narrow isthmus between the two gulfs Mitschigmen and Kaltschin. The Asiatic coast which borders the straits of Bering is peopled by great numbers of *cetaceous mammiferi.* On this coast the Tchoutskis, who live in perpetual war with the Americans, have collected together their habitations. Their small villages are called Nukan, Tugulan, and Tschigin.

Following the coast of the continent of America from cape Rodney and Norton creek to cape Malowodan, cape Littlewater, we find no Russian establishment; but the natives have a great number of huts collected together on the shore between the 63° 20′ and 60° 5′ of latitude. The most *n.* of their habitations are Agibaniach and Chalmiagmi, and the most *s.* Kuyncgach and Kuymin.

The bay of Bristol, to the *n.* of the peninsula Alaska (or Aliaska) is called by the Russians the gulf Kamischezkaia. They in general preserve none of the English names given by Captain Cook and Captain Vancouver, in their charts, to the *n.* of the 55° of latitude. They choose rather to give no names to the two great islands which contain the Pic Trubizin (the mount Edgecumbe of Vancouver, and Cerro de San Jacinto of Quadra), and cape Tschiricof (cape San Bartholomé), than adopt the denominations of King George's Archipelago, and Prince of Wales's Archipelago.

The coast from the gulf Kamischezkaia to New Cornwall is inhabited by five tribes, who form as many great territorial divisions in the colonies of Russian America. Their names are Koniagi, Kenayzi, Tschugatschi, Ugalachmiuti, and Koliugi.

The most *n.* part of Alaska, and the island of Kodiak, vulgarly called by the Russians Kichtak, though Kightak, in the language of the natives, in general means only an island, belongs to the Koniagi division. A great interior lake of more than 26 leagues in length, and 12 in breadth, communicates by the river Igtschiagick with the bay of Bristol. There are two forts and several factories on the Kodiak island (Kadiak), and the small adjacent islands. The forts established by Schelikoff bear the name of Karluk and the Three Sanctifiers. M. Malte Brun says, that according to the latest information, the Kichtak archipelago was destined to contain the head place of all the Russian settlements. Sarytschew asserts, that there are a bishop and Russian monastery in the island of Umanak (Umnak). We do not know whether there has been any similar establishment elsewhere; for the chart published in 1802 indicates no factory either at Umnak, Unimak, or Unalaschka. It is, however, read at Mexico, in the manuscript journal of Martinez's voyage, that the Spaniards found several Russian houses, and about 100 small barks, at the island of Unalaschka in 1788. The natives of the peninsula Alaska call themselves the men of the east (Kagataya-Koung'ns).

The Kenayzi inhabit the *w.* coast of Cook creek, or the gulf Kenayskia. The Rada factory, visited by Vancouver, is situated there under the 61° 8′. The governor of the island of Kodiak, a Greek named Ivanitsch Delareff, assured M. Sauer, that, notwithstanding the rigour of the climate, grain would thrive well on the banks of Cook river. He introduced the cultivation of cabbages and potatoes into the gardens at Kodiak.

The Tschugatschi occupy the country between the *n.* extremity of Cook inlet and the *e.* of Prince William's bay (Tschugatskaia gulf.) There are several factories and three small forts in this district: fort Alexander, near the mouth of port Chatham, and the forts of the Tuk islands, Green island of Vancouver), and Tchalca (Hinchinbrook island).

The Ugalachmiuti extend from the gulf of Prince William to the bay of Jakutal, called by Vancouver Bering's bay; and here we must not confound the bay of Bering of Vancouver, situated at the foot of mount St. Elie, with the Bering's bay of the Spanish maps, near mount Fairweather (Nevado de Buentiempó.) Indeed, without an accurate acquaintance with geographical synonymy, the Spanish, English, Russian, and French works on the *n. w.* coast of America, are almost unintelligible; and it is only by a minute comparison of the maps that this synonymy can be fixed. The factor of St. Simon is near cape Suckling, (cape Elie of the Russians). It appears that the central chain of the *cordilleras* of New Norfolk is considerably distant from the coast at the Pic of St. Elie; for the natives informed M. Barrow, who ascended the river Madnaja (Copper river) for a length of 500 werst (120 leagues), that it would require two days journey *n.* to reach the high chain of the mountains.

The Kolingi inhabit the mountainous country of New Norfolk, and the *n.* part of New Cornwall. The Russians mark Burrough bay on their charts (latitude 55° 50′) opposite the Revillagigedo island of Vancouver (Isla de Gravina of the Spanish maps), as the most *s.* and *e.* boundaries of the] [extent of country of which they claim the property. It appears that the great island of the King George archipelago has, in fact, been examined with more care and more minutely by the Russian navigators than by Vancouver. Of this we may easily convince ourselves by comparing attentively the *w.* coast of this island, especially the environs of cape Trubizin (cape Edgecumbe), and of the port of the Archangel St. Michael, in Sitka bay (the Norfolk sound of the English, and Tchinkitané bay of Marchand), on the charts published at Petersburgh in the imperial depôt, in 1802, and on the charts of Vancouver. The most *s.* Russian establishment of this district of the Kolingi is a small fortress (*crupost*) in the bay of Jakutal, at the foot of the *cordillera* which connects mount Fairweather with Mont St. Elie, near port Mulgrave, under the 59° 27′ of latitude. The proximity of mountains covered with eternal snow, and the great breadth of the continent from the 58° of latitude, render the climate of this coast of New Norfolk, and the country of the Ugalachmiuti, excessively cold and inimical to the progress of vegetation.

When the sloops of the expedition of Malaspina penetrated into the interior of the bay of Jakutal as far as the port of Desengaño, they found the *n.* extremity of the port under the 59° of latitude covered in the month of July with a solid mass of ice. We might be inclined to believe that this mass belonged to a glacier which terminated in high maritime Alps; but Mackenzie relates, that on examining the banks of the slave lake, 250 leagues to the *e.* under 61° of latitude, he found the lake wholly frozen over in the month of June. The difference of temperature observable in general on the *e.* and *w.* coast of the new continent, appears only to be very sensible to the *s.* of the parallel of 53°, which passes through New Hanover and the great island of Queen Charlotte.

There is nearly the same absolute distance from Petersburgh to the most *e.* Russian factory on the continent of America, as from Madrid to the port of San Francisco in New California. The breadth of the Russian empire embraces under the 60° of latitude an extent of country of nearly 2400 leagues; but the small fort of the bay of Jakutal is still more than 600 leagues distant from the most *n.* limits of the Mexican possessions. The natives of these *n.* regions have, for a long time, been cruelly harassed by the Siberian hunters. Women and children were retained as hostages in the Russian factories. The instructions given by the Empress Catharine to Captain Billings, drawn up by the illustrious Pallas, breathe the spirit of philanthropy, and the most noble sensibility. The present government is seriously occupied in diminishing the abuses, and repressing the vexations; but it is difficult to prevent these evils at the extremities of a vast empire; and the American is doomed to feel every instant his distance from the capital. Moreover, it appears more than probable that before the Russians shall clear the interval which separates them from the Spaniards, some other enterprising power will attempt to establish colonies either on the coast of New Georgia, or on the fertile islands in its vicinity.]

Chronological series of the Indian Emperors of Mexico.

1. Acamapictli, the first king of the Mexicans; elected when they established themselves on the lake; he married Ilanqueitl, daughter of the king Acolmictli of Cohuetitlan, and having no heir, he married a second time with Tezcatlamiahuatl, daughter of the noble Tetepanco; he reigned 20 years with much despotism, refused to be tributary to the king of Azcapuzalco, and being engaged all his life in keeping up a spirit of harmony among his vassals, died not without great fame.

2. Hutzizhuitl, son of the former; who obtained the crown not by hereditary right, but through the election of the elders and chiefs of the republic; he married Ayanhzihuatl, daughter of the king of Azcapuzalco, and following the maxims of his father, took for his second wife Miahuaxochitl, daughter of Texcacahualtzin, king of Quauhnahuac, so that these two princes uniting their force became the most formidable power of all the other nations: this emperor nominated as captain-general of his armies Quatlecohualtzin, his brother; expressed his abhorrence at the inhumanity of Maxtla, in slaying his infant son, Acolnahucatl; and reigned happily for 22 years.

3. Chimalpopoca, brother of the former, who suffered the greatest indignities from his brother-in-law Maxtla, emperor of Azcapuzalco, who, after having deceived Chimalpopoca, violated one of his wives, and then fled; Chimalpopoca, irritated at this, sent back to his brother-in-law a present of some women's garments, instead of the regular tribute, saying that these were more fitting to him than bows and arrows: this conduct, of course, irritated the emperor, and knowing that Chimalpopoca had made a conspiracy against his life, determined to seize him; when Chimalpopoca, not being able to resist the force brought against him, had recourse to solicit the protection of his god Huitzilopochtli, together with his nobles, making a festival on the occasion, which he was on the eve of celebrating when the troops of Maxtla entered the city and took him prisoner.

He was immediately confined in prison, and very scantily supplied with food, and at last, to deprive his enemy of the triumph of killing him, put an end to his own existence.

4. Izcohuatl, son of the former king Acamapictli, elected on acconnt of the valour and credit he had manifested whilst captain-general of the armies; being born of a female slave, he was legitimized by his father, and was 46 years old when he took the sceptre in his hands; he governed with great prudence, and was one of the happiest of the Mexican kings: he conquered many provinces, gained many battles, and he revenged the affronts offered to his predecessor, destroying the empire of the Tapanecas in one battle, through the death of Maxtla, who flying before the victors, took refuge in some baths called Temascal, and here he was killed by means of poles and stones. Izeohuatl, full of triumphal honours, and after having greatly extended the kingdom, built the temple of the idol Chibuacohuatl, which means mother-snake, and in the year following the famous temple of Huitzilopochtli, the first god of the Mexicans: shortly afterwards he was attacked with an infirmity, and died in a few days.

5. Moctecuhzuma, the first of this name, which means angry man; he was also called Ilhuicamina, or the man who shoots arrows to heaven; he was captain-general of the army, when he was elected through his merit and brilliant valour: his first care, after he was elected king, was to build a temple and a house to the lying deity of the demon in the ward called Huitznahuac, and thinking that his dominions were too small, he extended them by the conquests of the provinces of Chalco, Tlatilalco, Cohuixca, Oztlomantlaca, Cuezalteca, Ichatezipanteca, Teoxahualcas, conquering all the natives of these provinces, as well as those of Tlachco and Tlachmalac. Returning from the conquests of the latter, he enlarged the temple and habitation of his chief god Huitzilopochtli, adorning it with the spoils of victory, and returned to a campaign against the Chilapenecas, the Quauhteopan, and Tzumpahuacan, rendering these also subjeet to him. After this he reigned nine years in peace and quietness, when the waters of the lake rose to such an height as to run through the whole city; he having then consulted with the king of Tezcuco concerning a remedy, had just finished surrounding the city with a dry wall when the Spaniards arrived. To this misfortune succeeded another of a distressing famine, also the rebellion of Chalco and some other provinces, which were always very jealous of this powerful prince: at last he died, crowned with victories, in the 29th year of his reign, according to the computation of the Mexicans, giving wise regulations respecting the election of his successor.

6. Axayacatl, who exercised the office of captain-general, and thought worthy from his valour of ascending the throne; he was not less prosperous than his predecessor, and although the Father Acosta, Herrera, and other historians, do not place him in the sixth order of succession as he stands here, and make him the son of the former, the contrary is the case, according to the Mexican annals, written in their own types and figures, and of whose chronology ours is a counterpart. This emperor made tributary to him the Tlatelulcos, and various other kings and chiefs, and was taken in a battle which he was fighting against the Otomies of the kingdom of Xiquipilco; he was always the first in dangers and the last to fly, a stranger to fear, and inclined rather to inhumanity than to clemency; he at last died full of glory.

7. Tizoc, the seventh king of the Mexicans, elder brother of Axayacatl, on the election of whom to the empire, the present became captain-general, and was actually filling this post when chosen to the throne. Although he was not so warlike and courageous as his predecessors, yet he had a war with the Indians of Tlacotepec, and came off victorious; after this he dedicated himself to peace and to religious culture, determining to build another still more sumptuous temple to Huitzilopochtli; and for this end had collected immense quantities of materials, when his death, which happened about three years after he had projected this undertaking, put an end to his views; he died by some wounds which were given him by some women at the instigation of Tichotlela, a noble of Iztapalapan, and who were sent to him for that purpose, and not by his vassals on account of their being disgusted at his effeminate habits, as Acosta pretends; for, were it so, the women who inflicted the murder would not have been put to death for their crime, as was, in fact, the case.

8. Ahuizotl, brother of the former; also graced with the title of captain-general; he began his reign by busying himself in the completion of the temple of the god Huitzilopochtli, and afterwards declared war against the Mazahuas, who had rebelled, and having conquered these he turned his arms against the Tziuhcoacas and Topacnecas, of the province of Xalisco, keeping the prisoners of these campaigns and of that of Tlacapan for sacrifice in the dedication of the temple, the number of them, as it is said, amounting to 72,000. In the 4th year of his reign Mexico experienced a dreadful earthquake, to which followed an inundation

of the city by the overflowing of the waters of the lake; and in order to guard more effectually against a like misfortune, another stone wall was made, which also served to divide the salt-water from the fresh: he endeavoured to bring to Mexico the water of Huitzilopuchco, putting to death Tzutzumatzin for contradicting and telling him that in so doing he would drown Mexico; and when this proved to be the case, he was so angry at his own weakness that he struck himself a blow which he had afterwards cause to repent. In the mean time, however, he extended his dominions throughout nearly the whole of Nueva España as far as Guatemala; discovered the quarry of the tezontli-stone, of which the houses of the city are built, and were then beautified; and after a reign of 18 years, and being reputed the greatest monarch that had reigned in that kingdom, died from the effects of the blow he had struck himself, although three years after, to the universal regret of his vassals, and was succeeded in the throne by,

9. Moctecuhzuma, the second of this name, the 10th in the series of the kings, and not the 11th, as we are wrongly informed by the ebronologer Don Antonio de Solis; he was son of the King Axayacatl, and nephew of Tizoc and Ahuizotl; he was elected, because from his reputation he was thought likely to equal his predecessor; he was very grave and demure; it was looked upon as a miracle if he spoke, and yet when he made a speech in the council of state, of which he was a member, his eloquence caused universal admiration. He was generally shut up in a great *calpul* or saloon, which he had destined for himself in the grand temple of Huitzilopochtli, with whom he was said to have frequent commune, and to whom he was priest; and when he received intelligence of his election to the throne he was in the act of sweeping the temple. His first act was to sally forth to the punishment of the province of Atlixco, which was in a state of rebellion. On his return thence, his real character appeared, and he manifested a great degree of haughtiness and hypocrisy. He declared war against the republic of Tlaxcala, in which he uniformly met with bad success; and, when he had reigned four years, a most distressing famine was suffered in his kingdom: he renovated the aqueduct by which the water was conveyed to the city, fortifying and enlarging the causeway; he afterwards had continual wars, in which he subjected many provinces, and extended the limits of his empire as far as the provinces of Honduras and Nicaragua; he was looked up to and feared on all sides, and had reigned 18 years when he received news of the arrival of the Spaniards under Hernan Cortes upon the coast; he received them with kindness and affection, and died from a wound which he had received by a stone thrown at the Spaniards from the Indians, as he was going to a party of the latter to order them to lay down their arms.

10. Cuitlahuatlan, and not Guatimozin, as the Spanish historians have it: this emperor was the brother of the former, elected as soon as the death of the other was known, and whilst the Mexicans were at war with the Spaniards; he followed up the war with great eagerness; but his reign was of short duration, for, when the city was in the greatest danger, he escaped by the lake with a numerous fleet of canoes; but was soon afterwards made prisoner, lost his kingdom, and then his life; since Hernan Cortes ordered him to be strangled in his journey to Honduras, having found that he had attempted to regain the empire, after that he had acknowledged for its emperor, and sworn homage to, Charles V.

Catalogue of the Archbishops who have presided in Mexico.

1. Don *Fr.* Juan de Zumarraga, of the order of San Francisco, native of the town of Durango in Vizcaya; he was guardian of his convent of Abrogo, and withdrawn from thence by the Emperor Charles V. through the fame of his heroic virtues; presented to be first bishop, and afterwards archbishop of the holy cathedral of Mexico; he at first refused, but afterwards accepted this office in 1527; his holiness granted him the pall in 1545. So great was his piety and his virtue that he had a vision of the Most Holy Virgin, to whom he dedicated the first hermitage; he died in 1548, at the advanced age of 80, having in his life confirmed no less than 14,500 Indians.

2. Don *Fr.* Alonso de Montufar, of the order of preachers, native of the city of Loja in the kingdom of Granada; he was prior of his convent of Santa Cruz, *calificador* of the holy office, elevated to the dignity of this archbishopric in 1551, laboured with indefatigable zeal, celebrated two provincial councils, one in 1555, the other in 1561, perfected the hermitage of Nuestra Señora de Guadalupe, and died, after a long illness, in 1569.

3. Don Pedro de Moya y Contreras, native of Cordoba, doctor in sacred canons in the university of Salamanca, *maestre-escuela* in the holy church of Canaria, inquisitor of Murcia, and founder of the church of Mexico, elected archbishop in 1573; he celebrated a third provincial council in 1585, and owing to the skill and ability manifested by him, the king was induced to charge him with the visitation of the whole kingdom, and with the

office of viceroy and captainship-general through the death of the Count of Coruña; he was called to Spain to give an account of his visitation, and the king being satisfied thereat, conferred on him the presidency of the supreme council of the Indies: a few months after that he died, in 1591.

4. Don Alonso Fernandez Bonilla, also native of Cordoba, inquisitor, fiscal of the holy tribunal of Mexico, dean of its holy church, bishop of Guadalaxara in Nueva Galicia, nominated visitor-general of Peru, which office he filled with great credit; and presented to the bishopric by King Philip II. in 1592; after having been consecrated in Lima, he received orders to pass over to the city of Quito to pacify the disturbances which had been raised by the establishment of the Alcabala; but he died before he had proceeded on his journey.

5. Don *Fr.* Garcia de Santa Maria y Mendoza, of the order of San Gerónimo, native of Alcalá de Henares, of the honse of the Dukes of Infantado; prior of the royal monastery of the Escorial, general of his order, executor to King Don Philip II. and presented to the archbishopric of Mexico by King D. Philip III. in 1600; he accepted this office rather by compulsion, and manifested such zeal in the discharge of his duty, that he underwent excessive labours during the six years of his government, in the reformation of the clergy and in the defence of the ecclesiastical immunity; he was a religions, charitable, and pious man, and died in apostolic fame in 1606.

6. Don *Fr.* Garcia Guerra, of the order of St. Domingo, native of the town of Fromesta in the bishopric of Palencia; he was prior and master of the province in his convent of Valladolid, and presented to this archbishopric in 1607; governed with singular ability as well in ecclesiastical as secular concerns, as he was also nominated viceroy; endowed a monthly charity for persons of decayed fortune in the church of Nuestra Señora de Guadalupe, and died in consequence of a fall which he had in getting into his carriage, in 1611.

7. Don Juan Perez de la Serna, native of the town of Cervera, in the bishopric of Cuenca, collegiate of San Antonio de Siguenza, and in the *colegio mayor* of Santa Cruz de Valladolid, professor in the same, and magisterial canon of the church of Zamora; elected archbishop in 1613, a charge which he managed with such skill that he left behind him the reputation of a most excellent pastor; he was, accordingly, a great favourite with the chief pontiff, who wrote to him many letters of endearment. He had made himself beloved by his subjects, and used to give them alms with his own hands, removed the body of the venerable servant of God, Gregorio Lopez, and published, at his own expence, the third provincial council that was celebrated by his predecessor; blessed the second chapel which was dedicated to the Virgin of Guadalupe in 1622; and such were his merits, that the king wishing to have him nearer to himself, removed him to the bishopric of Zamora, where he died in 1631.

8. Don Francisco Manso y Zuñiga, native of the town of Cañas in the bishopric of Calahorra, collegiate in the *colegio mayor* of Santa Cruz de Valladolid, evening lecturer of sacred canons in that university, *oidor* of the chancery of Granada; of the council of his Majesty in the *real hacienda*, or royal revenues, and in the supreme council of the Indies; abbot of San Adian, chief priest of La Rioja and of Camero Viejo in that church, and presented to this bishopric by Señor Don Felipe IV. in 1629; he manifested his ardent zeal in the succour which he afforded at the inundation of Mexico in 1630, going out himself in a canoe to distribute food, nor less anxious for the common good in the plague which succeeded; he was promoted to the archbishopric of Badajoz and Cartagena, and to the commissariate of the holy crusade, to the archbishopric of Burgos, and to a seat in the council of the Indies, his Majesty having conferred upon him the title of Conde de Hervias and Vizconde de Negueruela.

9. Don Francisco Verdugo, native of the city of Carmona, collegiate of Santa Maria de Jesus in Sevilla, morning lecturer in sacred canons, inquisitor of Lima, and bishop of Guamanga; presented to the archbishopric of Mexico, but he died in that city before he received the bulls.

10. Don Feliciano de la Vega, native of Lima, jubilee morning lecturer in sacred canons, canon of that holy church and chanter, provisor and vicar-general of the archbishopric, judge of the appeals of its suffragans; elected bishop of Popayán and Vera Cruz in 1628, and presented to this metropolitan bishopric of Mexico in 1638; he was a man of consummate learning, as his works testify, and equally great in his apostolic zeal. As to his qualifications, it will be enough to observe, that, of the 4000 opinions that he had given, not one of them has been revoked; he did not take possession of the bishopric, having died before he entered it at Mazatlan, 30 leagues from Acapulco, in 1640.

11. The venerable Señor D. Juan de Palafox y Mendoza, bishop of La Puebla de los Angeles; promoted to this bishopric, which he did not accept.

12. Don Juan de Mañozca, native of Marquina

in Vizcaya, collegiate of the royal and most ancient college of Bartolomé el Viejo of Salamanca, first inquisitor of Cartagena de Indias, afterwards of Lima, and of La Suprema, and president of the chancery of Granada; presented to this bishopric in 1643; he died in 1654.

13. Don Marcelo Lopez de Azcona, abbot of Roncesvalles; presented to the archbishopric the aforesaid year, 1653, and he died a few days after taking possession.

14. Don Mateo de Sagade Burgueiro, native of Pontevedra in Galicia, collegiate in the *colegio mayor* of Santa Cruz de Valladolid, professor of arts in that of Durango, and of sacred writings in the university there; magisterial canon in the holy churches of Astorga and of Toledo, and elected archbishop of Mexico in 1655; he was a most strenuous defender of the ecclesiastical jurisdiction, and was presented by his Majesty to the bishopric of Cadiz in 1662; shortly afterwards to that of Leon, and lastly to the church of Cartagena in 1663, where he died in 1672.

15. Don Diego Osorio de Escobar y Llamas, native of Coruña in Galicia, advocate of the royal councils, canonical doctor of the church of Toledo, inquisitor-general of its archbishopric; of the council of the government of his most excellent the Señor Cardinal of Sandoval; bishop of La Puebla de los Angeles in 1656, and in 1664 viceroy, governor, and captain-general of Nueva España; elected archbishop, which dignity, although he refused, he kept till the arrival of a successor.

16. Don Alonso de Cueva y Davalos, native of Mexico, magisterial canon, treasurer, and archdeacon of the church of La Puebla, dean of this metropolitan church, bishop of Oaxaca, a man of illustrious birth and singular virtues; but who died before he took the pall, in 1665: he was interred in the same cathedral in which he was baptized.

17. Don *Fr.* Marcos Ramirez de Prado, of the order of San Francisco, native of Madrid; was bishop of Chiapa and Mechoacán, visitor of the tribunal of the holy crusade in Nueva España; promoted to this archbishopric in 1666, entered to take possession with universal jubilee, and whilst expectation was alive to see the effects of that wonderful ability which had graced all his other stations, he died in the following year without receiving the pall.

18. Don *Fr.* Payo Enriquez de Rivera, of the order of San Agustin, native of Sevilla, son of the Duke of Alcalá, prior of various convents, and rector of the college of Doña Maria de Aragon in Madrid, *calificador* of the holy office, bishop of Guatemala and of Mechoacán, and presented to this bishopric in 1668: he was a man extremely modest and charitable, zealous for ecclesiastical discipline, and was elevated to the supreme command of viceroy and captain-general in 1673: his disinterested, useful, and pacific government lasted for some time, and having passed over to Spain, where he was called to fill the bishopric of Cuenca, he retired to the convent of Nuestra Señora del Risco of his order, close to the city of Avila, where bringing himself to an austere religious life, he died in 1684.

19. Don Manuel Fernandez de Santa Cruz Sahagun, native of Palencia, collegiate of the college of Santa Cruz de Valladolid, magisterial canon of Segovia, elected bishop of Chiapa, Guadalaxara, and of La Puebla de los Angeles in 1677; promoted to the archbishopric of Mexico in 1680, but renounced it together with the viceroyship with extreme humility.

20. Don Francisco de Aguiar, native of the town of Betanzos in Galicia, collegiate of the *colegio mayor* of Cuenca in the university of Salamanca, magisterial canon of Astorga, penitentiary of Santiago; presented to the bishopric of Mechoacán, and to this bishopric in 1681; he preserved throughout his life an angelic purity, and a modesty foreign to all pride, and so zealous was he in the performance of his duty that he visited the whole diocese with excessive fatigue, bringing many souls to the bosom of the church: he was an example for prelates, being charitable, devout, edifying, vigilant in the reform of customs, kind to all, and severe only to himself. It was he that put on foot the establishing of the college for female children of San Miguel de Belen, and at his solicitude was built the college called the Seminario Tridentino; he also built the house for the reception of mad women, there called De Sayagos, was a great benefactor to the house of compassion for married women, and laid the first stone of the magnificent temple of Guadalupe, and died in 1698. There are some who speak of his beatification, since, even in his life-time, he deserved the eulogy of the apostolic see, and of the Cardinal Aguirre in the catalogue of the bishops of Mexico.

21. Don Juan de Ortega Montañes, native of Llanes in the principality of Asturias: he arrived through his merits to the bishopric of Durango, afterwards to that of Mechoacán, to that of Guatemala, and the viceroyalty and captainship-general of Nueva España; promoted to this archbishopric in 1701. The integrity with which he governed induced his Majesty to entrust to him the command for the second time, and through a great zeal of finishing the temple of the Virgin of

Guadalupe, he went about personally to collect alms for the purpose through the city, and just as it was finished and ready to be dedicated he died, in 1710.

22. Don Joseph Lanciego y Eguilaz, native of the town of Viana in the kingdom of Navarra, of the order of San Benito, preacher to his Majesty, *calificador* of La Suprema, and abbot of his monastery of Naxera; presented to the bishopric in 1711: he governed with great prudence, and with equal zeal visited the whole of the bishopric; erected at his own expence the greater part of the building of the college of Belén, was watchful over the chapels of the sanctuary of Nuestra Señora, and obtained the bull for the erection of the church into a collegiate; he died in 1728.

23. Don Manuel Joseph de Endaya y Haro, native of Luzon in the Philippine isles, master in philosophy, and doctor in sacred theology in the university of Manila, canon of the holy church of Plasencia, archdeacon of Alarcon, dignitary of the church of Cuenca; presented to the bishopric of Oviedo, from whence he was called to the council which was celebrated by the Pontiff Benedict XIII. in 1725, and in which he did the othee of bishop, assistant, and domestic prelate of the apostolic chapel: he was elected bishop of Mexico in 1728, and, having already in his hands the bull and the sacred pall, died at Benaventa, a town of his diocese, in 1729.

24. Don Juan Antonio de Lardizabal y Elorza, native of Segura in Vizcaya, collegiate-major in the old college of San Bartolomé de Salamanca, professor of philosophy at Durango, and of the disciples of Scotus, in that university, magisterial canon of that holy church, and elected bishop of La Puebla de los Angeles in 1722; promoted to this holy metropolitan church in 1727, which dignity he immediately renounced.

25. Don Juan Antonio de Vizarron y Equiarreta, native of the city and port of Santa Maria, titular archdeacon of the holy patriarchal church of Sevilla, a principal attendant to his Majesty; elected archbishop of Mexico in 1730, was viceroy and captain-general of the kingdom, and, in either employ, paid large sums of money for the benefit of his flock in various foundations, perpetual monuments of his good name, as were those in particular relating to the chapel in the Colegio Tridentino, and the pious work of his having endowed 4000 dollars annually for the support of such females as were bringing up for a religious life: he received and solemnized the oath of the patroness Nuestra Señora de Guadalupe, and died in 1747.

26. Don Manuel Joseph Rubio y Salinas, native of Colmenar Viejo in Castilla la Nueva, visitor-general of the bishopric of Oviedo, and of the abbey of Alcalá la Real, chaplain of honour to his Majesty, *fiscal* of his royal chapel, house, and court, perpetual abbot of regular canons of San Isidro de Leon; presented to this bishopric by Señor Don Fernando VI. in 1747; a man of singular parts, charitable, affable, and zealous for the honour of God; he erected the beautiful royal college of Nuestra Señora de Guadalupe, obtained for it, of the holy see, the confirmation of an universal patronage in N. America, and died in 1765.

27. Don Francisco Antonio de Lorenzana y Buitron, collegiate in the *colegio mayor* of San Salvador de Oviedo, native of Leon, canon of the holy church, primate of Toledo, vicar-general of the same place, abbot of St. Vicente, and bishop of Plasencia, and from thence removed, through his merit, in 1766, to the archbishopric of Mexico, where he governed with the greatest skill and benevolence, when his Majesty elected him to the supreme dignity of the archbishopric of Toledo in 1771.

28. Don Alonso Nuñez de Haro y Peralta, native of Huete, *collegiate-mayor* of Bolonia, canon of Toledo, and archbishop in 1771.

[Mexico, Gulf of, is that part of the N. Atlantic ocean which washes the *s.* and *s. w.* coast of Florida, the *e.* coast of New Leon and New Gallicia, in New Mexico, and the *n. e.* coast of Old Mexico, or New Spain, in N. America. It is properly bounded on the *n.* by the Floridas, and on the *s.* by the gulf of Darien, or perhaps still more properly by the *n. w.* point of the isthmus of Darien, supposing a line to be drawn from one to the other. The gulf of Mexico is therefore to be considered as the *w.* part of the great gulf between the *n.* and *s.* continents of America. This spacious gulf contains a great many islands of various extent and size; and it receives several great rivers, particularly the Mississippi, the river of the N. and a multitude of others of comparatively less note. Its coasts are so irregular and indented that its lesser gulfs and bays are almost innumerable; the chief of these are the gulf of Honduras and Guanajos, and the bays of Campechy, Palaxay, and St. Louis. It is conjectured by some, and we think with great reason, to have been formerly land; and that the constant attrition of the waters in the Gulf stream has worn it to its present form. See Gulf Stream, and Gulf of Florida; also for a table of the longitudes and latitudes of the most important places about this bay, see the end of the general preface.]

Mexico, Nuevo, another extensive kingdom, the most *n.* part of the dominions of Nueva España in America; bounded *s.* by the province of Durango, *e.* by Louisiana, *n.* by unknown regions, and *w.* by the great ridge of stony mountains which divide N. America; is included between long. 103° and 107° 20′ *w.* and from lat. 31° to 38° 15′. Its length is 425 miles from *n.* to *s.* and its average width about 90 from *e.* to *w.*

The first notice of this country was given by some Conchos Indians to the *Fr.* Agustin Ruiz, of the order of San Francisco, in 1581: he accordingly went over to discover it with certain people that were sent him from Mexico for that purpose, under the command of Antonio de Espejo. The natives, who had already been instructed in the Catholic religion by Alvar Nuñez de Cabeza de Vaca, Andres Dorantes, Bernardino del Castillo, and the Negro Estebanico, who had been saved from being wrecked in the ship of Pantilo de Narvacz in Florida, and had traversed the country till they arrived at Mexico, received Ruiz and his party in an amicable manner, and Espejo bore the news of this reception back to Mexico. In 1595, during the viceroyship of Don Luis de Velasco, Juan de Oñate entered to reduce these provinces, which were then in a state of rebellion, having put to death the governor, and many missionaries who were amongst them.

This kingdom enjoys various temperatures, cold, hot, and moderate. It is fertile and pleasant, produces abundance of wheat, maize, delicate fruits and vegetables, and peculiarly fine grapes. The mountains are covered with firs, oaks of different species, savines, and many other trees of different qualities, and from which certain portions of wood are cut. Here are found a variety of animals, such as deer, wolves, bears, apes, mountain sheep, and a species of deer so large as to equal a mule in stature, its horn measuring nearly two yards in length. Here are also many singular birds, particularly in the snowy parts, where they are caught alive the whole year round, being held in great estimation. Although some mines have been discovered, yet they have proved all of tin.

The population of this kingdom consists of 30 settlements of Indians, who are *reducciones* from the nations of the Piros, Tiguas, Mansos, Queres, Suñis, Tolonas, Xernes, Xeres, Picurics, Thanos, Pecos, Teguas, Thaos, and Sumas, and are very numerous. They are of better appearance as to colour and proportion than the other Indians, go always clothed, and wear goat-skin shoes, are very fond of employment; and the women, in particular, are dextrous in weaving mantles of wool and cotton for vesture. They are always on horseback, and their dwellings deserve particular attention, as being different from those of any other Indians. These are a sort of barracks, from three to four stories high, well put together, and having no door; the inhabitants getting into an upper floor by means of a small ladder, which at night they draw up after them, to guard themselves from the attacks of the nations which are their enemies. These huts are erected opposite to each other for the purpose of mutual protection.

This kingdom has many rivers and streams, but the principal and largest is that called Del Norte, which passes through the middle of it. The monks of the order of San Francisco have converted the natives to the Catholic faith, and they have established some extensive missions since the year 1660. The capital is Santa Fé.

[Several geographers confound the kingdom, or, as Humboldt designates it, the province of New Mexico with the *provincias internas;* and they speak of it as a country rich in mines, and of vast extent. The celebrated author of the Philosophic History of the European Establishments in the Two Indies has contributed to propagate this error. Its territorial extent has already been accurately given, and is much less than people of no great information in geographical matters are apt to suppose even in that country. The national vanity of the Spaniards loves to magnify the spaces, and to remove, if not in reality, at least in imagination, the limits of the country occupied by them to as great a distance as possible. In the memoirs which Humboldt procured on the position of the Mexican mines, the distance from Arispe to the Rosario is estimated at 300, and from Arispe to Copala at 400 marine leagues, without reflecting that the whole intendancy of Soñora is not 280 marine leagues in length. From the same cause, and especially for the sake of conciliating the favour of the court, the *conquistadores*, the missionary monks, and the first colonists, gave weighty names to small things. We have elsewhere described one kingdom, that of Leon, of which the whole population does not equal the number of Franciscan monks in Spain. Sometimes a few collected huts take the pompous title of Villa. A cross planted in the forests of Guayana figures on the maps of the missions sent to Madrid and Rome, as a village inhabited by Indians. It is only after living long in the Spanish colonies, and after examining more narrowly these fictions of kingdoms, towns, and villages, that the traveller can form a proper scale for the reduction of objects to their just value.

The Spanish conquerors, shortly after the destruction of the Aztec empire, set on foot solid] [establishments in the *n.* of Anahuac. The town of Durango was founded under the administration of the second viceroy of New Spain, Velasco el Primero, in 1559. It was then a military post against the incursions of the Chichimec Indians. Towards the end of the 16th century, the viceroy, Count de Monterey, sent the valorous Juan de Onate to New Mexico. It was this general who, after driving off the wandering Indians, peopled the banks of the great Rio del Norte.

From the town of Chihuahua a carriage can go to Santa Fé of New Mexico. A sort of caleche is generally used, which the Catalonians call *volantes.* The road is beautiful and level; and it passes along the *e.* bank of the Great river (Rio Grande), which is crossed at the Paso del Norte. The banks of the river are extremely picturesque, and are adorned with beautiful poplars, and other trees peculiar to the temperate zone.

It is remarkable enough to see that, after the lapse of two centuries of colonization, the province of New Mexico does not yet join the intendancy of New Biscay. The two provinces are separated by a desert, in which travellers are sometimes attacked by the Cumanches Indians. This desert extends from the Paso del Norte towards the town of Albuquerque. Before 1680, in which year there was a general revolt among the Indians of New Mexico, this extent of uncultivated and uninhabited country was much less considerable than it is now. There were then three villages, San Pascual, Semillete, and Socorro, which were situated between the marsh of the Muerto and the town of Santa Fé. Bishop Tamaron perceived the ruins of them in 1760; and he found apricots growing wild in the fields, an indication of the former cultivation of the country. The two most dangerous points for travellers are the defile of Robledo, *w.* from the Rio del Norte, opposite the Sierra de Doña Ana, and the desert of the Muerto, where many whites have been assassinated by wandering Indians.

The desert of the Muerto is a plain 30 leagues in length, destitute of water. The whole of this country is in general of an alarming state of aridity; for the mountains De los Mansos, situated to the *e.* of the road from Durango to Santa Fé, do not give rise to a single brook. Notwithstanding the mildness of the climate, and the progress of industry, a great part of this country, as well as Old California, and several districts of New Biscay, and the intendancy of Guadalaxara, will never admit of any considerable population.

New Mexico, although under the same latitude with Syria and central Persia, has a remarkably cold climate. It freezes there in the middle of May. Near Santa Fé, and a little farther *n.* (under the parallel of the Morea), the Rio del Norte is sometimes covered, for a succession of several years, with ice thick enough to admit the passage of horses and carriages. We are ignorant of the elevation of the soil of the province of New Mexico; but Humboldt does not believe that, under the 37° of lat. the bed of the river is more than 7 or 800 metres, or 2296 or 2624 feet, of elevation above the level of the ocean. The mountains which bound the valley of the Rio del Norte, and even those at the foot of which the village of Taos is situated, lose their snow towards the beginning of the month of June.

The great river of the N. rises in the Sierra Verde, which is the point of separation between the streams which flow into the gulf of Mexico, and those which flow into the S. sea. It has its periodical rises (*crecientes*) like the Orinoco, the Mississippi, and a great number of rivers of both continents. The waters of the Rio del Norte begin to swell in the month of April; they are at their height in the beginning of May; and they fall towards the end of June. The inhabitants can only ford the river on horses of an extraordinary size during the drought of summer, when the strength of the current is greatly diminished. These horses in Peru are called *cavallos chimbadores.* Several persons mount at once; and if the horse takes footing occasionally in swimming, this mode of passing the river is called *passar el rio a volapie.*

The water of the Rio del Norte, like that of the Orinoco, and all the great rivers of S. America, is extremely muddy. In New Biscay they consider a small river, called Rio Puerco (*nasty river*), the mouth of which lies *s.* from the town of Albuquerque, near Valencia, as the cause of this phenomenon; but M. Tamaron observed that its waters were muddy far above Santa Fé and the town of Taos. The inhabitants of the Paso del Norte have preserved the recollection of a very extraordinary event which took place in 1752. The whole bed of the river became dry all of a sudden for more than 30 leagues above, and 20 leagues below the Paso; and the water of the river precipitated itself into a newly-formed chasm, and only made its re-appearance near the Presidio de San Elcazario. This loss of the Rio del Norte remained for a considerable time; the fine plains which surround the Paso, and which are intersected with small canals of irrigation, remained without water; and the inhabitants dug wells in the sand, with which the bed of the river was filled.]

[At length, after the lapse of several weeks, the water resumed its ancient course, no doubt because the chasm and the subterraneous conductors had filled up. This phenomenon bears some analogy to a fact which Humboldt was told by the Indians of Jaen de Bracamoros during his stay at Tomependa. In the beginning of the 18th century the inhabitants of the village of Puyaya saw, to their great terror and astonishment, the bed of the river Amazonas completely dried up for several hours. A part of the rocks near the cataract (*pongo*) of Rentema had fallen down through an earthquake; and the waters of the Marañon had stopt in their course till they could get over the dike formed by the fall. In the *n.* part of New Mexico, near Taos, and to the *n.* of that city, rivers take their rise which run into the Mississippi. The Rio de Pecos is probably the same with the Red river of the Natchitoches, and the Rio Napestla is, perhaps, the same river which, farther *e.* takes the name of Arkanas.

The colonists of this province, known for their great energy of character, live in a state of perpetual warfare with the neighbouring Indians. It is on account of this insecurity of the country life that we find the towns more populous than we should expect in so desert a country. The situation of the inhabitants of New Mexico bears, in many respects, a great resemblance to that of the people of Europe during the middle ages. So long as insulation exposes men to personal danger, we can hope for the establishment of no equilibrium between the population of towns and that of the country.

However, the Indians who live on an intimate footing with the Spanish colonists are by no means all equally barbarous. Those of the *e.* are warlike, and wander about from place to place. If they carry on any commerce with the whites, it is frequently without personal intercourse, and according to principles of which some traces are to be found among some of the tribes of Africa. The savages, in their excursions to the *n.* of the Bolson de Mapimi, plant along the road between Chihuahua and Santa Fé small crosses, to which they suspend a leathern pocket, with a piece of stag flesh. At the foot of the cross a buffalo's hide is stretched out. The Indian indicates by these signs that he wishes to carry on a commerce of barter with those who adore the cross. He offers the Christian traveller a hide for provisions, of which he does not fix the quantity. The soldiers of the *presidios*, who understand the hieroglyphical language of the Indians, take away the buffalo hide, and leave some salted flesh at the foot of the cross. (*Diario del Illmo. Señor Tamaron*, MS.) This system of commerce indicates at once an extraordinary mixture of good faith and distrust.

The Indians to the *w.* of the Rio del Norte, between the rivers Gila and Colorado, form a contrast with the wandering and distrustful Indians of the *savannas* to the *e.* of New Mexico. Father Garces is one of the latest missionaries who in 1773 visited the country of the Moqui, watered by the Rio de Yaquesila. He was astonished to find there an Indian town with two great squares, houses of several stories, and streets well laid out, and parallel to one another. Every evening the people assembled together on the terraces of which the roofs of the houses are formed. The construction of the edifices of the Moqui is the same with that of the Casas Grandes on the banks of the Rio Gila, of which we have already spoken. The Indians who inhabit the *n.* part of New Mexico give also a considerable elevation to their houses, for the sake of discovering the approach of their enemies. Every thing in these countries appears to announce traces of the cultivation of the ancient Mexicans. We are informed even by the Indian traditions, that 20 leagues *n.* from the Moqui, near the mouth of the Rio Zaguananas, the banks of the Nabajoa were the first abode of the Aztecs after their departure from Aztlan. On considering the civilization which exists on several points of the *n. w.* coast of America, in the Moqui, and on the banks of the Gila, we are tempted to believe that at the period of the migration of the Toltecs, the Acolhues, and the Aztecs, several tribes separated from the great mass of the people to establish themselves in these *n.* regions. However, the language spoken by the Indians of the Moqui, the Yabipais, who wear long beards, and those who inhabit the plains in the vicinity of the Rio Colorado, is essentially different from the Mexican language; in proof of which assertion, see the testimony of several missionary monks well versed in the knowledge of the Aztec language. (*Chronica Serafica del Collegio de Queretaro*, p. 408.)

In the 17th century several missionaries of the order of St. Francis established themselves among the Indians of the Moqui and Nabajoa, who were massacred in the great revolt of the Indians in 1680. Humboldt states that he had seen in manuscript maps drawn up before that period the name of the Provincia del Moqui.

The province of New Mexico contains three *villas* or towns, (Santa Fé, Santa Cruz de la Cañada y Taos, and Albuquerque y Alameda), 26 *pueblos* or settlements, three *parroquias* or parishes, 19 missions, and no solitary farm (*rancho*).]

[The population in 1803 amounted to 40,200, and the extent of surface in square leagues is 5709; the number of inhabitants to the square league being seven.]

[Mexico, a township in Herkemer county, New York, incorporated in 1796, lying on Canada and Wood creeks, and Oneida lake.]

MEXILLONES, a port of the coast of the S. sea, in the province and *corregimiento* of Atacama and kingdom of Peru.

Mexillones, an island of the S. sea, situate near the coast of the kingdom of Chile, in the province and *corregimiento* of Coquimbo. It is in lat. 29° 20′.

MEXISTLAN, a settlement of the head settlement of the district of Chichicatepec, and *alcaldia mayor* of Villalta, in Nueva España. It contains 62 families of Indians, and is eight leagues *s. w.* of its capital.

MEXORADA. See Cotuy.

MEXTITLAN de la Sierra, a jurisdiction and *alcaldía mayor* of the kingdom of Nueva España. Its territory is covered with *sierras*, mountains, and *barrancas*, and such is its roughness and asperity as to deny all cultivation. This *sierra* is so lofty and extensive that it is known throughout the kingdom by the name of Sierra Madre de Mextitlan, and is in fact one of the largest to be found in those vast regions, and serving as a boundary to many neighbouring jurisdictions. The productions of this country, and those in which a traffic is carried on, are seeds and cotton; but it is sometimes much in want of water, which, as supplied only by a river which runs here, is at times so scarce as to put the inhabitants to very serious inconvenience. The population consists of 10 principal settlements or head settlements of districts, and upon these the following are dependent.

Mextitlan,	Santiago Tepehuacan,
Sta. Maria Molango,	San Lorenzo Ixtacayotla
Colotlán,	Tianguistengo,
Chapuluacán,	Tlanchinol,
Meztitlan,	Zizicaxtla.

The capital is of the same name, of a mild temperature, and situate in a fertile and pleasant glen, by which runs a river descending from the *sierra*, its waters being used in irrigating the fields which are cultivated on its banks, whenever there may be a deficiency of rain. This town contains 2000 families of Mexican Indians, and 15 or 20 of Spaniards, *Mustees*, and Mulattoes; with a good convent of monks of the order of San Agustin. It is 95 miles *e. n. e.* of Mexico, in long. 98° 2′ *w.* and lat. 20° 37′ *n.*

Mextitlan, another settlement, of the head settlement and *alcaldía mayor* of Compostela in the kingdom of Nueva Galicia; situate in the island of San Sebastian.

MEYOPONTE, a settlement and *real* of gold mines of the Portuguese, in the kingdom of Brazil, and country of the Guayazas Indians; situate on the shore of the river of its name; although Mr. D'Anville places it on the side of the river Paranaiba.

Meyoponte. The above river of the same name, which rises in the territory of the Guayazas Indians, runs *n.* and enters the sources of the river Tocantines.

MEZTITLAN, a settlement of the head settlement of the district and *alcaldía mayor* of Mextitlan in Nueva España, annexed to the curacy of Molango. It contains 163 families of Indians, and lies seven leagues to the *e. n. e.* of its capital.

MEZTLA, a small settlement or ward of the *alcaldía mayor* of Guauchinango in Nueva España, annexed to the curacy of the settlement of Tlaola.

MIACATLAN, San Salvador de, a settlement of the head settlement of the district of Mazacatepec, and *alcaldía mayor* of Cuernavaca, in Nueva España. It contains 20 families of Indians and 19 of *Mustees* and Mulattoes, who live by sowing maize, this being the only produce of the place. It is eight leagues *n.* of its head settlement.

MIAHUATLAN, a jurisdiction and *alcaldía mayor* of Nueva España in the province and bishopric of Oaxaca; bounded *w.* by the *corregimiento* of this city, and *n.* by the jurisdiction of Cimatlan; by that of Nexapa on the *e.* and that of Theozaqualco on the *s.* Its length between *e.* and *s.* is more than 40 leagues, and its width somewhat less. It consists of 73 settlements without those of the smaller wards, and its territory is very fruitful, especially in cochineal, which is its principal commerce, and which makes it one of the best and most desirable *alcaldías* of the kingdom.

The capital is a settlement of the same name, with the dedicatory title of San Andrés. It is of a mild and pleasant temperature, inhabited by 615 families of Indians, including those of the wards of its district, and some Spaniards, *Mustees*, and Mulattoes, who live in various estates and farms of its district. It abounds in cochineal, in the commerce of which there are many rich merchants. The flesh of the cattle here, and especially that of the sheep, is very delicious, nor are there wanting plenty of seeds and fruit. It is 108 leagues *s. e.* of Mexico, in long. 275° 15′, and lat. 18° 55′.

The principal or head settlements of the district are as follows:

S. Luis de Amatlan,	S. Justo Otzolotepec,
Losicha,	S. Vicente,
S. Mateo de las S. Pinas,	S. Miguel Zuchitepec,
S. Pablo Coatlan,	Santa Catalina,
Santa Maria Otzolotepec,	Seneguia.

Miahuatlan, with the dedicatory title of San Joseph, another settlement of the head settlement of the district of Naulingo, and *alcaldía mayor* of Xalapa, in the same kingdom; situate in a mountainous tract, the temperature of which is cold, owing to its being near the *sierra* which lies to the *n.* It is inhabited by 140 families of Indians, dedicated to the cultivation of the land, and with the productions of this consists the commerce of the place. It is one league to the *n. n. w.* of its head settlement.

Miahuatlan, another settlement, of the same head settlement of the district and *alcaldía mayor* as the former, which in the Mexican tongue signifies a long ear of maize, from the abundance which this soil yields. It contains 72 families of Indians, and is two leagues to the *n.* of its head settlement.

Miahuatlan, another, a small settlement or ward of the head settlement of the district of Acatepec, and *alcaldía mayor* of Thehuacan, in the same kingdom. It is close to its head settlement.

MIALILU, or Maulu, a river of the kingdom of Chile, being a large arm of the river Diamante, formed by that of the Tenuyán and others. It runs *s. s. e.* and enters the Como-Leuvu or Gran Desaguadero de los Sauces.

[MIAMI River, Little, in the N.W. territory, has a *s. w.* course, and empties into the Ohio, on the *e.* side of the town of Columbia, 20 miles *e.* of the Great Miami, in a straight line, but 27 taking in the meanders of the Ohio. It is too small for batteaux navigation. Its banks are good land, and so high as to prevent in common the overflowing of the water. At the distance of 57 miles from the Ohio, the Miamis approximate each other within eight miles and a half. On this river are several salt springs.]

[Miami River, Great, or Great Mineami, called also Assereniet, or Rocky River, in the N.W. territory, has a *s.* by *w.* course, and empties into the Ohio by a mouth 200 yards wide, 32¼ miles from Big Bones, 63 miles from the Rapids, and 233 from the mouth of the Ohio. It is one of the most beautiful streams in the territory, and is so clear and transparent, at its highest state, that a pin may very plainly be seen at its bottom. It has a very stony channel, a swift stream, but no falls. At the Pieque or Pickawee towns, above 75 miles from its mouth, it is not above 30 yards broad, yet loaded batteaux can ascend 50 miles higher. The portage from the navigable waters of its *e.* branch to Sandusky river is nine miles, and from those of its *w.*-branch to the Miami of the lakes, only five miles. It also interlocks with the Scioto.]

[Miami, or Meames of the Lakes, a navigable river of the N. W. territory, which falls into lake Erie, at the *s. w.* corner of the lake. A *s.* branch of this river communicates with the Great Miami by a portage of five miles. This river is called by some writers Mawmee, also Omee, and Manmick. See Territory N. W. of the Ohio.]

[Miami, or Meame, a village on the Miami of the Lake near the Miami fort. Large canoes can come from Quiatanon, a small French settlement on the *w.* side of the Wabash, 30 miles below the Miami carrying-place, which last is nine miles from this village.]

[MIAMIS, or Meames, an Indian nation who inhabit on the Miami river, and the *s.* side of lake Michigan. They can raise about 300 warriors. In consequence of lands ceded to the United States by the treaty of Greenville, August 3d, 1795, government paid them a sum in hand, and engaged to pay them annually for ever, to the value of 1000 dollars in goods.]

[Miamis Bay, at the mouth of the Miami of the Lakes.]

Miamis, a fort in the same country; situate on the shore of the river of its name, near lake Erie, built by the French in 1750.

Miamis, a small river of the same country, which runs *n. e.* and enters the lake Erie.

All the above take their names from a nation of Indian savages, who dwell at the source of the lake Michigan, where in a place called Chicagou they have a village, in which resides their chief or *cazique*, who can bring into the field 4 or 5000 fighting men; never going abroad himself without a guard of 40 men, the same number keeping watch day and night by his cabin; this chief seldom appears in public, and communicates his orders through his officers.

MIAMO, a settlement of the province of Guayana, and government of Cumaná; one of those of the missions which were held there by the Catalonian Capuchin fathers.

[MIATA Island, one of the Society islands in the S. Pacific ocean. Lat. 17° 40′ *s.* Long. 148° 3′ *w.*]

MICANI, San Francisco de, a settlement of the province and *corregimiento* of Chayanta or Charcas in Peru.

[MICHAEL. See S. MIGUEL and S. MICHEL.]

[MICHAEL, ST. or FOND DES NEGRES, a town on the *s.* peninsula of St. Domingo island, 10 leagues *n. e.* of St. Lonis.]

[MICHAEL, ST. or ST. MIGUEL River, is also on the *s.* coast of the isthmus between N. and S. America, and on the N. Pacific ocean, and 18 leagues to the *w.* of port Martin Lopez, and three *e.* of Guibaltigue. It has three fathoms water at flood. Within the river to the *n. e.* is the burning mountain of St. Miguel, in the midst of an open plain.]

. MICHAEL, ST. a small island of the N. sea, near the coast of the province and colony of Georgia, at the entrance of port Royal, and one of those called the Georgian.

[MICHAEL'S Bay, ST. on the *e.* side of the island of Barbadoes, in the W. Indies, a little *n.* of Foul's bay, *n. e.* of which last bay are Cobler's rocks, in the shape of a horn.]

[MICHAEL'S Bay, ST. in Tierra Firme, in the S. sea.]

[MICHAEL'S, ST. a parish in Charlestown district, S. Carolina.]

[MICHAEL'S, ST. a town in Talbot county, Maryland, eight miles *w.* of Easton, and 21 *s. e.* of Annapolis.]

MICHAPARU, a small river of the province and government of Guayana or Nueva Andalucia. It runs from *s.* to *n.* and enters the Orinoco to the *w.* of the rapid stream of Camiseta.

MICHARDS, a small island of the N. sea; situate near the coast of N. Carolina, at the entrance or mouth of the strait of Albemarle.

MICHATOYATL, a large river of the province and *corregimiento* of Chiquimula in the kingdom of Guatemala. It rises from a lake which is about four leagues from the spot where the capital stood. It gives such a great fall, that, according to *Fr.* Juan de Torquemada, a musket-shot fired from the bottom will not reach its top; and forming an immense cavity in the rock in which it falls, so that in it breed bats of an enormous size, and which, if they find a person or any animal sleeping, will suck their blood: in this cave they are as thick as leaves, and are of the size of a hen: moreover, from the great mischief they do to the calves, the breed of cattle has greatly decreased in these parts.

MICHAU, a port of the *s.* coast of the Royal island or Cape Breton, between the port Toulouse and the island of St. Esprit.

MICHAU, a small island, situate near the coast of the same province as the former port, and at the entrance of the same.

MICHEL, S. a small island, situate in lake Superior, of New France or Canada in N. America, and in the point of Chagovamigon.

[MICHEL, S. See S. MIGUEL, and S. MICHAEL.]

MICHICANI, a settlement and *asiento* of the silver mines of the province and government of Chucuito in Peru, annexed to the curacy of its capital; situated on the shore of the lake Umamarca on the *s.*

MICHIGAN, Lake, one of the five of New France or Canada, of N. America, between a point of the continent, close to Michillimakinak; an establishment of the Huron Indians, stretching towards the *s.* and the other point which is opposite and looks to the *n.* the two points forming a strait, by which the lake Huron communicates with this. Its vicinity is very unpleasant for an establishment, from its excessive coldness, the which no doubt arises from the continual agitation of the lakes by the *n.* wind. This lake, of which we treat, is the least, although it is no less than 300 leagues in circumference, without reckoning the bay of Puants, which runs inland for 28 leagues. The inequality of the tides greatly affects the navigation of this lake; their irregularity has been frequently marked, and it has been found that they observe no rule whatever, being in some places extremely high. Near the island of Michillimakinak they rise during the full of the moon in 24 hours, so as to run completely into the interior of the lake, and it is not less certain, however wonderful, that besides these tides there is another current always running from lake Huron to the adjoining lake, and which is common to both, although it does not impede the natural course of the lake Michigan, which, the same as lake Superior, discharges its waters into the lake Huron. The first of these two currents, that is to say, that which runs from lake Huron to Michigan, is more perceptible when the wind blows a contrary way to it, that is from the *s.*; and then may be seen pieces of ice floating from the former lake to the latter with the same velocity as a ship sailing before the wind; and, indeed, the same is the case in the Bahama channel.

In the channel by which lake Superior empties itself into the Huron, are many streams or currents below the surface of the water, and which are at times so strong as to break the fishermen's nets; from which we may judge that this great lake throws a certain part of its waters into the lake Michigan, by means of subterranean courses, by the same means as it is alleged that the Caspian unites with the Euxine sea, and this with the Mediterranean.

In what relates to lake Superior, this presumption

is well founded, in as much that it receives at least 40 large rivers, of the which 10 or 12 are nearly as wide as the strait itself, from whence it would follow, that were there no other egress for the waters than this channel, it would send out much less water than it received. The same may be asserted of lake Michigan, which also receives a great number of rivers, many of them very large; from which it is remarkable to judge that, besides the outlet afforded to its waters into lake Huron, it must have other subterraneous channels, as we have said of lake Superior. This conjecture is corroborated by a discovery that all the rocks that have been found at a certain depth near the strait called St. Mary's Falls, are as porous as a sponge, and that many of them are washed into large hollows, which must arise from the currents before mentioned.

In the navigation from Michillimakinak to the river St. Joseph, it is found that although the wind is against the course of the vessel, she will make no less than eight or 10 leagues a day, proving that the current must contribute to her course, the same also being the case at the entrance of the bay of Puants. It is not to be doubted but that the waters of this bay, which have no other egress than through one part, run into lake Michigan; and that this, being in the same situation, empties itself into Huron, both of them, besides these waters, receiving, the one and the other, different rivers, some of which are not less than the Seine in France. These currents are not perceptible, save in the middle of the channel, by a kind of reflux or reaction of the waters along either shore, which is very advantageous for the coasting of the small canoes, here made of the trunks of trees, and which first run five leagues to the *w.* to enter the lake Michigan, and afterwards to the *s.* the same being the only course which vessels have for 100 leagues, being the extent of this lake from *n.* to *s.* and from whence they continue their passage till they reach fort St. Joseph. The part of the land which divides the two lakes Huron and Michigan, is a country the most fertile and charming that one can possibly imagine.

[Michigan Lake is the largest and most considerable lake which is wholly within the United States, and lies between lat. 42° 10′ and 45° 40′ *n.* and between 84° 30′ and 87° 30′ *w.* long. It is navigable for shipping of any burden, and communicates with lake Huron, at the *n. e.* part, through the straits of Michillimakinak. The strait is six miles broad, and the fort of its name stands on an island at the mouth of the strait. In this lake are several kinds of fish; particularly trout of an excellent quality, weighing from 20 to 60 pounds, and some have been taken in the strait which weigh 90 pounds. On the *n. w.* part of this lake, the waters push through a narrow strait, and branch out into two bays; that to the *n.* is called Noquet's bay, the other to the *s.* Puants, or Green bay, which last, with the lake, forms a long peninsula, called cape Townsend, or Vermillion point. About 30 miles *s.* of bay de Puants, is lake Winnebago, which communicates with it: and a very short portage interrupts the water communication, *s. w.* from Winnebago lake through Fox river, then through Ooisconsin, into the river Missisippi. Chicago river, also at the *s. w.* extremity of lake Michigan, furnishes a communication interrupted by a still shorter passage with Illinois river. See all these places mentioned under their respective names. Lake Michigan receives many small rivers from the *w.* and *e.* some 150 and even 250 yards broad at their mouths. See Grand, Masticon, Maraне, St. Joseph, &c.]

[Michigan, a newly erected territorial government of the United States, bounded *s.* by a line drawn from the *s.* part of lake Michigan to the bay of Miamis which is the most *w.* bay of lake Erie, *w.* by the lake of its name, *n.* by the straits of S. Mary's, and *e.* by lake Huron, the river and lake St. Clair and part of lake Erie. R is well watered on the *w.* side by a number of rivers which empty into lake Michigan, and in the midst of this territory is a rather fine elevated plain.

The population of this government amounted by the census of 1810 to 4762 souls.]

MICHILLIMAKINAK, a small island of lake Huron in New France or Canada, in N. America, situate in 45° 45′ *n.* lat. It has a moderate-sized town, in which used to be carried on a trade in skins, owing to its being resorted to by, and being very convenient for the meeting of, several savage nations of Indians: this traffic is, at the present day transferred to Hudson's bay by the river Borbon. The situation of the island is very advantageous, from lying between the three great lakes, the Michigan, which is 300 leagues in circumference, exclusive of the great bay of Puants, into which it empties itself; the Huron, which is 350, and of a triangular figure; and the last, the lake Superior, which is 500; the whole of them being navigable for large vessels, and the two first being divided only by a small strait, in the which there is sufficient water for the same vessels, without any obstruction through the whole of the lake Erie to Niagara. Between the lakes Huron and Superior is a communication by means of a canal 22 leagues long, but which is interrupted by many cascades or falls, which impede the canoes from arriving to disembark at Michillimakinak whatever they might bring from lake Superior. [This island, within the line of the United States, was delivered to them by the British, by treaty, in 1794, and retaken in the present year, 1812.]

MICHIMALOYA, a settlement of the *alcaldía mayor* of Tula in Nueva España; annexed to the curacy of its capital, from whence it lies a quarter of a league to the *n. w.* It contains 60 families.

[MICHIPICOTON, a river which empties into lake Superior, on the *n. e.* side of the lake. It has its source not far distant from Moose river, a water of James's bay. It forms at its mouth a bay of its own name; and on the *w.* part of the bay, is a large island so called, close to the land; a small strait only separates it from Otter's head on the *n.*]

[MICHIPICOTON House, in Upper Canada, is situated on the *e.* side of the mouth of the above river, in lat 47° 56′ *n.*]

[MICHISCOUI is the Indian and present name of the most *n.* river in Vermont. It rises in Belvidere, and runs nearly *n. e.* until it has crossed into Canada, where it runs some distance; it turns *w.* then *s.* re-enters the state of Vermont, in Richford, and empties into lake Champlain, at Michiscoui bay at Highgate. It is navigable for the largest boats to the falls at Swan town, seven miles from its mouth. Michiscoui, La Moelle, and Onion rivers, are nearly of the same magnitude.]

[MICHISCOUI Tongue or Bay, a long point of land which extends *s.* into lake Champlain from the *n. e.* corner of the state of Vermont, on the *w.* side of the bay of this name, and forms the township of Allburgh.]

MICHIUILCA, a settlement of the province and *corregimiento* of Tarma in Peru, annexed to the curacy of Tapu.

MICIMPUCHU, a settlement of the province and government of Venezuela; situate on the shore of the river Tucuyo, to the *n.* ¼ to the *n. e.* of the city.

MICKLON, a small island, situate near the *s.* coast of Newfoundland, at the month or entrance of the gulf of St. Lawrence.

[MICKMACKS, an Indian nation which inhabit the country between the Shapody mountains, and the gulf of St. Lawrence in Nova Scotia, opposite to St. John's Island. This nation convey their sentiments by hieroglyphics marked on the rind of the birch and on paper, which the Roman missionaries perfectly understand. Many of them reside at the heads of the rivers in King's and Hants counties.]

[MICOYA Bay is situated on the *s. w.* coast of Mexico, or New Spain, on the N. Pacific ocean. In some charts it is laid down in lat. 10° 15′ *n.* and having cape Blanco and Chira island for its *s. e.* limit.

MICTLAN, a settlement of the head settlement of the district of Tequantepec in Nueva España; its name meaning hell. It was thronged with inhabitants in the time of the Indians, and adorned with very superb edifices; amongst the most celebrated of which was a temple dedicated to the devil, with dwellings for its priests, and having an hall adorned with stones and curiosities with great ingenuity. The doors, which were extremely lofty, were composed of only three pieces of stone, one on each side and one above. It had also another saloon, supported with pillars so thick that two men could scarcely make their arms meet round them, and yet of one solid block of stone. The first Spaniards who discovered this temple affirmed that these pillars were 50 feet high, and that they were very like those in the great church of St. Mary's at Rome. The Emperor Mocthecutzuma had centinels from amongst the people of this settlement, who might give him intelligence of what was going on at sea; and these were the people who gave intelligence of the arrival of Cortes.

MICULAPAYA, a settlement of the province and *corregimiento* of Porco in Peru.

MIDDLE, some small islands of the lake Ontario in New France or Canada, of N. America. They are three, and are situated close to the *n.* coast.

[MIDDLE Bank, a fishing ground in the Atlantic ocean, which lies from *n. e.* to *s. w.* between St. Peter's bank and that of Sable island, and opposite to and *s. e.* of Cape Breton island; laid down in some charts between lat. 44° 32′ and 45° 34′ *n.* and between long. 57° 37′ and 59° 32′.]

[MIDDLE Cape is to the *s. w.* of cape Anthony, in Staten Land, on the strait Le Maire, and the most *w.* point of that island, at the extremity of S. America.]

[MIDDLE Islands, or ILHAS DE EN MEDIO, on the *w.* coast of New Mexico, and are between the islands of Chira and St. Luke. They are in the N. Pacific ocean, in lat 9° 30′ *n.* There is only from six to seven fathoms from Chira to these islands, and all vessels should keep nearer to them than to the main.]

[MIDDLE States, one of the grand divisions of the United States, (so denominated in reference to the *n.* and *s.* states) comprehending the states of New York, New Jersey, Pennsylvania, Delaware, and the Territory N.W. of the Ohio; which see.]

[MIDDLEBERG, a new town of New York in Schoharie county, incorporated in 1797.]

[MIDDLEBOROUGH, the Namaskett of the

ancient Indians, a township in Plymouth county, Massachusetts, bounded *w.* by Freetown and Taunton, *e.* by Carver and Warham, and is 35 miles *s.* of Boston; was incorporated in 1669, and contains 4526 inhabitants. This town was formerly thickly inhabited by Indian natives, governed by the noted sachem Tispacan: there are now only 30 or 40 souls remaining, who, to supply their immediate necessities, make and sell brooms and baskets. The town is remarkable for a large range of ponds, which produce several sorts of fish, and large quantities of iron ore. The bottom of Assowamset pond may be said to be an entire mine of iron ore. Men go out with boats, and use instruments like oyster dredges, to get up the ore from the bottom of the pond. It is now so much exhausted, that half a ton is thought a good day's work for one man; but for a number of years one man could take up four times the quantity. In an adjacent pond there is yet great plenty at 20 feet deep, as well as from shoaler water. Great quantities of nails are made here. In winter, the firmers and young men are employed in this manufacture. Here, and at Milton in Norfolk county, the first rolling and slitting mills were erected about 40 years ago, but were imperfect and unproductive, in comparison with those of the present time. The prints of naked hands and feet are to be seen on several rocks in this town, supposed to have been done by the Indians. These are probably similar to those observed in the states of Tennessee and Virginia.]

[MIDDLEBOURG Key, a small islet, separated from St. Martin's, in the W. Indies, on the *n. e.*]

[MIDDLEBURG, or EOOA, the most *s.* of all the Friendly islands, in the S. Paeihc ocean; and is about 10 leagues in circuit.]

MIDDLEBURGH, NUEVO, a city of the province and government of Guayana, in the part possessed by the Dutch. It is situate on the shore of the river Poumaron, and near the coast, in the point or cape of Nassau.

MIDDLEBURGH, a cape or point of land on the *s.* coast of the straits of Magellan, in the extremity which looks to the *n.* of the island of Luis el Grande.

[MIDDLEBURY, a post town of Vermont, and capital of Addison county. It is 30 miles *n.* by *w.* of Rutland, 12 from Vergennes, and 27 *s. e.* of Burlington. Here is a brewery upon a pretty large scale. The township lies on the *e.* side of Otter creek, and contains 395 inhabitants.]

[MIDDLEFIELD, a township in Hampshire county, Massachusetts, 30 miles *n. w.* of Springfield. It was incorporated in 1783, and contains 608 inhabitants.]

[MIDDLEHOOK, a village in New Jersey, eight miles *w.* of Brunswick, on the cross post-road from Brunswick to Flemington, and on the *n.* bank of Rariton river.]

[MIDDLESEX, a county of Massachusetts, bounded *n.* by the state of New Hampshire, *e.* by Essex county, *s.* by Suffolk, and *w.* by Worcester county. Its figure is nearly equal to a square of 40 miles on a side; its greatest length being 52, and its greatest breadth 42 miles. It has 42 townships, which contain 42,737 inhabitants. The religious societies are, 55 of Congregationalists, six of Baptists, and some Presbyterians. It was made a county in 1643. It is watered by five principal rivers, Merrimack, Charles, Concord, Nashua, and Mystick; besides smaller streams. The chief towns are Charlestown, Cambridge, and Concord. Charlestown is the only sea-port in the county; Concord is the most respectable inland town, and is near the centre of the county, being 16 miles *n. w.* of Boston. There are in the county 24 fulling-mills, about 70 tan-yards, four paper-mills, two snuff-mills, six distilleries, and about 20 pot and pearl ash houses. The *s.* and *n.* sides of the county are hilly, but not mountainous; few of the hills exceeding 100 feet in height, and are covered with wood, or cultivated quite to the summits. The air is generally serene, and the temperature mild. The extreme variation of Fahrenheit's thermometer may be considered as 100° in a year; but it is in very few instances, that in the course of a year it reaches either extreme; 92° may be considered as the extreme summer heat, and 5° or 6° below 0°, as that of the winter cold. In the winter of 1796-97, it sunk to 11° below 0°. The soil is varions, in some parts of rich, black loam; and in others it is light and sandy. It produces the timber, grain, and fruit which are common throughout the state, either by natural growth or cultivation.]

[MIDDLESEX, a maritime county of Connecticut, bounded *n.* by Hartford county, *s.* by Long island sound, *e.* by New London county, and *w.* by Newhaven. Its greatest length is about 30 miles, and its greatest breadth 19 miles. It is divided into six townships, containing 18,855 inhabitants, of whom 221 are slaves. Connecticut river runs the whole length of the county, and on the streams which flow into it are a number of mills. Middletown is the chief town.]

[MIDDLESEX, a county of New Jersey, bounded *n.* by Essex, *n. w.* and *w.* by Somerset, *s. w.* by Burlington, *s. e.* by Monmouth, *e.* by Rariton bay and part of Staten island. It contains 15,956 inhabitants, including 1318 slaves. From the mouth of Rariton river up to Brunswick the land on both sides is generally good, both for pasturage and tillage, producing considerable quantities of every kind of grain and hay.]

[MIDDLESEX, a county of Virginia, on the *s.* side of Rappahannock river, on Chesapeak bay. It is about 35 miles in length, and seven in breadth, containing 4140 inhabitants, including 2558 slaves. Urbana is the chief town.]

[MIDDLESEX, a township in Chittendon county, Vermont, on the *n. e.* side of Onion river. It contains 60 inhabitants.]

[MIDDLESEX Canal, Massachusetts, it is expected, will be of great importance to the states of Massachusetts and New Hampshire. It is now opening at a vast expence by an incorporated company. The design is to open a water communication from the waters of Merrimack river at Chelmsford to the harbour of Boston. The route of the canal will be *s.* through the *e.* parts of Chelmsford and Billerica, the *w.* part of Wilmingtou, and the middle of Woburn; where it comes to some ponds, from which the waters run by Mystick river into Boston harbour. The distance from the Merrimack to these ponds will be 17 miles. The canal will, without meeting with any large hills or deep valleys, be straighter than the country road near it. The distance from the Merrimack to Medford, as the canal will be made, is 27, and to Boston 31 miles. The canal is to be 24 feet wide at the bottom, and 32 at the top, and six feet deep. The boats are to be 12 feet wide and 70 feet long. The toll is to be six cents a mile for every ton weight which shall pass, besides pay for their boats and labour.]

[MIDDLETON, an interior township in Essex county, Massachusetts, 28 miles *n.* of Boston. It was incorporated in 1728, and contains 682 inhabitants.]

[MIDDLETON, a city and post-town of Connecticut, and the capital of Middlesex county, pleasantly situated on the *w.* bank of Connecticut river, 25 miles from its month at Saybrook bar, according to the course of the river; 14 miles *s.* of Hartford, 24 *n. e.* of Newhaven, 27 *n. w.* of New London, and 156 *n. e.* of Philadelphia. Its public buildings are, a Congregational church, an Episcopalian church, a court-honse, and naval office. It contains about 300 houses, and carries on a considerable trade. Here the river has 10 feet water at full tides. Lat. 41° 34′ *n.* Long. 72° 34′ *w.* This place was called Mattabesick by the Indians, and was settled in 1650 or 1651. Two miles from the city is a lead mine which was wrought during the war, and was productive; but it is too expensive to be worked in time of peace.]

[MIDDLETOWN, a township in Stratford county, New Hampshire; about 30 miles *n. n. w.* of Portsmouth. It was incorporated in 1778, and contains 617 inhabitants.]

[MIDDLETOWN, a township in Rutland county, Vermont. It contains 699 inhabitants, and is 39 miles *n.* of Bennington.]

[MIDDLETOWN, a village on Long island, New York state; 12 miles from Smithtown, and 13 from Bridgehampton.]

[MIDDLETOWN, a township in Ulster county, New York, erected from Rochester and Woodstock in 1789, and contains 1019 inhabitants, including six slaves. In 1796 there were 135 of the inhabitants entitled to be electors.]

[MIDDLETOWN, a township in Newport county, Rhode Island state, contains 840 inhabitants, including 15 slaves. In this town, which is on the island which gives name to the state, and about two miles from Newport, is the large and curious cavity in the rocks called Purgatory.]

[MIDDLETOWN, a small post-town in Newcastle county, Delaware, lies on Apoquinimy creek, 19 miles *s. s. w.* of Wilmington, and 41 *s. w.* of Philadelphia.]

[MIDDLETOWN, in Monmouth county, New Jersey; a township which contains two places of worship, one for Baptists and one for the Dutch reformed church, and 3226 inhabitants, including 491 slaves. The centre of the township is 50 miles *e.* by *n.* of Trenton, and 30 *s. w.* by *s.* of New York city. The light-house built by the citizens of New York on the point of Sandy Hook, is in this township. The high lands of Navesink are on the sea-coast near Sandy Hook. They are 600 feet above the surface of the water, and are the lands first discovered by mariners on this part of the coast.]

[MIDDLETOWN, a flourishing town in Dauphin county, Pennsylvania; situated on the *n. w.* side of Swatara creek, which empties into the Susquehannah, two miles below. It contains a German church and above 100 houses, and carries on a brisk trade with the farmers in the vicinity. It is estimated that above 200,000 bushels of wheat are brought down these rivers annually to the landing place, two miles from the town. Contiguous to the town is an excellent merchant-mill, supplied with a constant stream, by a canal cut from the Swatara. It is six miles *s.* of Hummelston, and 73 *w.* by *n.* of Philadelphia. Lat. 10° 13′ *n.* Long. 76° 44′ *w.* There are also other townships of this name in the state; the one in

Delaware county, the other in that of Cumberland.]

[MIDDLETOWN, in Frederick county, Maryland, lies nearly eight miles *w. n. w.* of Frederickstown.]

[MIDDLETOWN, in Dorchester county, Maryland, is about five miles *n.* of the Cedar landing-place, on Transquaking creek, seven *w.* of Vienna, and 8½ *n. w.* of Cambridge.]

[MIDDLETOWN Point, in the above township, lies on the *s. w.* side of the bay, within Sandy Hook, seven miles *e.* by *n.* of Spotswood, and 14 *n. w.* of Shrewsbury. A post-office is kept here.]

[MIDWAY, a village in Liberty county, Georgia, 26 miles *s. w.* of Savannah, and nine miles *n. w.* of Sunbury. Its inhabitants are Congregationalists, and are the descendants of emigrants from Dorchester near Boston, in New England, who migrated as early as 1700.]

[MIDWAY, a township in Rutland county, Vermont, *e.* of and adjoining Rutland.]

MIEL, RIO DE LA, a river in the province and government of Mariquita and Nuevo Reyno de Granada. It rises in the valley of Corpus Christi, passes through the city of Los Remedios, and enters the Grande de la Magdalena.

MIEL, a settlement of the province and government of Venezuela; situate in the road which leads down from Bariquisimeto, between this settlement and that of Tucuyo.

MIER, a settlement of the province and government of Sierra Gorda, in the bay of Mexico and kingdom of Nueva España; founded by the count of that title, Don Joseph de Escandon, colonel of the militia of Queretaro, in 1750.

[MIFFLIN, a county of Pennsylvania, surrounded by Lycoming, Franklin, Cumberland, Northumberland, Dauphin, and Huntingdon counties. It contains 1851 square miles, 1,184,960 acres, and is divided into eight townships. The mountains in this county abound with iron ore, for the manufacturing of which several forges have been erected. It is well watered by the Juniatta, and other streams which empty into the Susquehannah. Chief town Lewistown.]

[MIFFLIN, a small town lately laid out in the above county, on the *e.* side of the Juniatta, nine miles *e.* of Lewistown, and 109 from Philadelphia.]

[MIFFLIN, Fort, in Pennsylvania, is situated on a small island at the mouth of Schuylkill river, about six miles *s.* of Philadelphia.]

MIGUEL, S. called El Grande, a town and capital of the jurisdiction and *alcaldía mayor* of its name in the kingdom of Nueva España, and bishopric of Mechoacán; situate on the skirt of a mountain-plain. It is of a mild temperature, of a large population, fertile, and abounding greatly in commerce. It was founded by the Spaniards and the Tlaxcaltecas, who assisted in conquering the country, and was at that time inhabited by Chichimecas Indians. Its population is composed of 3000 families of Spaniards, and the Indians live in the rich and fertile grazing lands for large and small cattle, as well as amongst the cultivated estates which abound in its district, some being employed as labourers, and others living as renters of the lands.

In its parish church is venerated an image of Christ, representing that period of his divine mission of "Ecce Homo," and very great is the reverence in which it is held by the whole jurisdiction, from the advantages procured by it. It has a convent of the religious order of San Francisco, which has been a seminary for studies, a congregation of clergy of San Felipe Neri, of exemplary virtues, employed in the teaching of the first rudiments, as also the profounder studies; the same body having been founded by the Father Juan Antonio Perez de Espinosa, and the temple here being entitled De Nuestra Señora de la Soledad. Contiguous to this is the chapel of La Santa Casa de Loreto, of beautiful architecture and sumptuous ornaments, the same having been given by the zealous and devout Don Manuel de la Canal. There are in this town many salutary waters, and especially those of a fountain which runs from some rocks close to the settlement, and which is called the Chorillo, and from which the public reservoirs are provided. The principal commerce consists in cattle, of the skins of which are made many saddles for riding, beautifully worked; also in white arms, such as stilettos, swords, knives, spurs, stirrups, and other useful and curious articles in steel. The women employ themselves in making quilts. [Humboldt bears testimony to the great industry of the inhabitants, and he mentions cotton cloth as the chief of their manufactures]. This *alcaldía mayor* has only one other head settlement of the district, which is the town of San Felipe. The capital is 51 leagues to the *n.* ¼ to the *n. w.* of Mexico, in lat. 21° 45′. Long. 273° 46′.

MIGUEL, S. a city of the province and *alcaldía mayor* of San Salvador in the kingdom of Guatemala; situate two leagues from the coast of the S. sea and bay of Fonseca, the same serving it as a port. Its population is small, and it has, besides the parish church, which is very decent, some convents of monks of the religious orders of San Francisco and La Merced, as also one of nuns. It is 22 leagues from its capital, and 62 from Guatemala.

Miguel, S. another city, called also Bridgetown, the capital of the island of Barbadoes; situate in the *s.* part of the same, with a good, convenient, and secure port, and which is capable of receiving 500 ships. The population is large, the streets straight, and the buildings handsome, especially the hall of justice and the exchange. It has plenty of storehouses and shops, from the great number of merchants who reside in it, and abounds in every thing imaginable. It is badly situated; for, lying lower than the banks which form the boundaries to the sea, it is generally full of swamps, which render it unhealthy. It has two very good castles at the entrance of the port for its defence, furnished with artillery, the principal of which is called fort Charles, and is on the point Nedham. It has also a magnificent and well endowed college for students, which was founded by the Colonel Christoval Codrington, native of this city. In lat. 13° 24′. Long. 318° 40′.

Miguel, S. another city, of the Nuevo Reyno de Granada; founded in the province and *corregimiento* of Los Pauches by Anton de Ollala. It is of very hot temperature, and has fallen into such decay, that of it there remains nothing but a mean village. It is 50 miles *n. w.* from Santa Fé.

Miguel, S. a town of the province of Ostimuri in Nueva España; situate between the rivers Mayo and Nacari.

Miguel, S. a settlement, with the surname of Del Valle, or De la Miel, in the province and *corregimiento* of Tunja of the Nuevo Reyno de Granada. It is of an hot temperature, abounding in sugar canes, of which a great portion of sugar and honey are made, as also in the other vegetable productions of a warm climate. It contains 80 inhabitants, who are subject to the disorder of *cotos*, or swellings on the neck. It is close to the settlement of Capitanejo, and 94 miles to the *n. e.* of its capital.

Miguel, S. another, of the province and *corregimiento* of Carangas in Peru, and of the archbishopric of Charcas; annexed to the curacy of the settlement of Colquemar.

Miguel, S. with the surname of Molleambato, in the province and *corregmiento* of Latacunga, and kingdom of Quito, in the district of which, towards the *n.* is the great estate called Tasin.

Miguel, S. another, of the province and *corregimiento* of Chimbo, in the same kingdom as the former.

Miguel, S. another of the missions that are held by the religious order of San Francisco, in the territory of the town of San Christoval of the Nuevo Reyno de Granada; situate on the shore of the river Apure. It is of an hot temperature, very scanty population, and produces nothing but wheat, barley, and maize.

Miguel, S. another, of the head settlement of the district of Tholimán, and *alcaldía mayor* of Queretaro, in Nueva España. It contains 75 families of Indians.

Miguel, S. another, of the province and government of Atacames in the kingdom of Quito.

Miguel, S. another, of the head settlement of the district and *alcaldía mayor* of Tochimilco in Nueva España; situate on the top of a mountain covered with fruit-trees. It is of a mild temperature, and contains 68 families of Indians, who maintain themselves by cutting wood and making charcoal. It is four miles to the *w.* of its capital.

Miguel, S. another, of the head settlement of Tamazunchale, and *alcaldía mayor* of Valles, in the same kingdom; annexed to the curacy of Tampasquin, from whence it is two leagues distant.

Miguel, S. another, of the head settlement of the district of Amatepec, and *alcaldía mayor* of Zultepec, in the same kingdom. It contains 22 families of Indians, and is very close to its head settlement.

Miguel, S. another, of the head settlement of Zumpahuacán, and *alcaldía mayor* of Marinalco, in the same kingdom.

Miguel, S. another, of the head settlement of the district and *alcaldía mayor* of Lerma in the same kingdom. It contains 281 families of Indians, and is three leagues *n. e.* of its head settlement.

Miguel, S. another, of the head settlement and *alcaldía mayor* of Toluca in the same kingdom. It contains 100 families of Indians, and lies a little to the *s.* of its capital.

Miguel, S. another, of the head settlement and *alcaldía mayor* of Tepeaca in the same kingdom; five leagues from its capital.

Miguel, S. another, of the jurisdiction and *alcaldía mayor* of Octupán, in the same kingdom as the former.

Miguel, S. another, of the head settlement of Palmar, and *alcaldía mayor* of Tepeaca, in the same kingdom. It contains 59 families of Indians, and is two leagues from its head settlement.

Miguel, S. another, of the head settlement of Ahuatlan, and *alcaldía mayor* of Zacatlan, in the same kingdom; one league from its head settlement.

Miguel, S. another, of the head settlement of Teutalpán, and former *alcaldía mayor;* three leagues from its head settlement.

Miguel, S. another, of the head settlement of Xalazala, and *alcaldía mayor* of Tlapa, in the same kingdom. It contains 38 families of Indians, employed in agriculture, and is three leagues from its head settlement.

Miguel, S. another, of the head settlement of Tecali, and *alcaldía mayor* of this name, with 39 families of Indians.

Miguel, S. another, a small settlement or ward of the head settlement of the district and *alcaldía mayor* of Juxtlahuaca in the same kingdom.

Miguel, S. another, of the province and *corregimiento* of Canta in Peru; annexed to the curacy of San Buenaventura.

Miguel, S. another, a small settlement or ward of the head settlement of the district and *alcaldía mayor* of Leon, in the province and bishopric of Mechóacan and kingdom of Nueva España. It contains 100 families of Indians, employed in the cultivation of maize and fruit-trees, and as labourers in the estates of its district. It is very close to its capital, by the *s.*

Miguel, S. another, of the *alcaldía mayor* of San Luis de Potosi, in the same kingdom and bishopric. It contains 53 families of Indians, who occupy themselves solely in cutting of fuel and making charcoal. It is to the *s.* of its capital.

Miguel, S. another, with the surname of Alto, of the head settlement of the district and *alcaldía mayor* of Maravatio in the same kingdom. It is of a cold temperature, contains 21 families of Indians, and is five leagues from its capital.

Miguel, S. another, of the head settlement of Quiatoni, and *alcaldía mayor* of Teutitlán, in the same kingdom. It contains 52 families of Indians, and is two leagues *n. e.* of its head settlement.

Miguel, S. another, of the *alcaldía mayor* of Huamelula in the same kingdom; situate at the foot of a lofty mountain, and in its vicinity runs a river so large and deep that it is necessary to pass it in canoes. It is of an hot temperature, and its territory barren in vegetable productions, yielding nothing but cochineal, this being the only branch of its commerce. Its population consists of 54 families of Indians. At two leagues distance, and on the shore of the river, are seen the ruins of the settlement of San Bartolomé, which was abandoned, owing to all its inhabitants having perished in an epidemic disorder in 1736. It is 12 leagues from Pochutla.

Miguel, S. another, of the head settlement and *alcaldía mayor* of Juchipila in the same kingdom; five leagues to the *n.* of the same head settlement.

Miguel, S. another, which is the head settlement of the district of the *alcaldía mayor* of Villalta in the same kingdom. It contains 26 families of Indians, and is eight leagues and an half from its capital.

Miguel, S. another, of the head settlement and *alcaldía mayor* of Juchipila, distinct from that of which we have already spoken; six leagues to the *s. w.* of its head settlement.

Miguel, S. another, of the province and government of Quixos y Macas in the kingdom of Quito, the capital of the missions of the Sucumbios Indians, which were reduced and held under the care of the regulars of the company. It lies on the shore of the river Napo.

Miguel, S. another, of the province of Barcelona, and government of Cumaná, in the Nuevo Reyno de Granada; situate *s.* of the settlement of Piritú.

Miguel, S. another, with the surname of Boqueron, in the district of Chiriqui, of the province and government of Veragua, and kingdom of Tierra Firme; situate in the royal road, three leagues from its head settlement.

Miguel, S. another, of the missions that were held by the regulars of the company, in the province and government of Mainas in the kingdom of Quito.

Miguel, S. another, of the missions that were held by the same regulars, in the Orinoco. It is a *reduccion* of the Guajiva nation, and situate on the shore of the river Meta. In 1734, this settlement was burnt and destroyed by the Caribes.

Miguel, S. another, of the province and *corregimiento* of Pasto in the kingdom of Quito; situate on the shore of the river Telembí.

Miguel, S. another, of the missions that were held by the regulars of the company, in the province of Guairá and government of Paraguay; situate on the shore of the river Curituba; where are seen the ruins that were made by the Portuguese of San Pablo at the end of the last century, (1600.)

Miguel, S. another, of the province and *captainship* of Rey in Brazil; situate on the coast opposite the island of Santa Catalina.

Miguel, S. another, of the missions that were held by the regulars of the company, in the province and government of Paraguay; situate between the settlements of San Juan and San Lorenzo.

Miguel, S. another, also of the missions of the same regulars, in the country of the Chiquitos Indians; situate on the *n.* of a lake.

Miguel, S. another, of the province and *corregimiento* of Cuyo in the kingdom of Chile; situate on the shore of one of the lakes of Huanacache.

Miguel, S. another, of the province and *alcaldia mayor* of Zacapula in the kingdom of Guatemala.

Miguel, S. another, of the province and government of Buenos Ayres in Peru; situate on the shore of the river Paraná, at the mouth where it enters the Carcarañal.

Miguel, S. another, which was once in the province and government of Moxos in the kingdom of Quito, distinct from that we have mentioned, but was depopulated by an epidemic disorder.

Miguel, S. another, of the province and government of Tucumán in Peru; situate on the shore and at the source of the river of its name.

Miguel, S. another, of the province and kingdom of Guatemala.

Miguel, S. another, of the province and government of Santa Marta in the Nuevo Reyno de Granada; situate on a *llano* or plain to the *e.* of the Cienega.

Miguel, S. another, of the province and government of Maracaibo; situate on the shore of the river Masparro, between the cities of Barinas Vieja and Nueva.

Miguel, S. another, of the province and government of Cinaloa; situate on the shore of the river Del Fuerte, between the settlements of Ahome and of Michicauchi.

Miguel, S. another, of the missions that were held by the regular company of Jesuits, in the government and country of the Chiquitos Indians, distinct from that already mentioned. It is situate on the bank of a small river, and at a small distance from the source of the river Capivari.

Miguel, S. another, of the province and government of Maracaibo, distinct from one already mentioned; situate on the shore of the river Tucuyo, and to the *w.* of this city.

Miguel, S. another, of the island of Curazao, one of the Antilles; situate on the *n.* coast.

Miguel, S. another, of the kingdom and *corregimiento* of Quito, in the district of Las Cinco Leguas; situate *w.* of the settlement of Canzacoto.

Miguel, S. another, of the same kingdom and *corregimiento* as the former; situate on the shore of the river San Pedro.

Miguel, S. another, of the province and government of Moxos in the kingdom of Quito, distinct from those above mentioned; situate on the shore of the river Baures, and one of those over which the Portuguese have gained the dominion.

[Miguel, S. a settlement of Indians, of the province and government of Buenos Ayres; situate on a small branch of the river Piratiny, in lat. 28° 32′ 26″ *s.* Long. 54° 39′ 27″ *w.*]

[Miguel, S. a fort of the province and government of Buenos Ayres; situate on a small river at the *s.* end of lake Mini, 90 miles *n. e.* of Maldonado. Lat. 33° 44′ 44″ *s.* Long. 53° 35′ 30″ *w.*]

Miguel, S. a river of the same province and kingdom as the former port. It rises in the country of the Tepuñacas Indians, and enters the Itenes. On its shores is a beautiful estate called Del Francés.

Miguel, S. another river, of the province and government of Santa Cruz de la Sierra in Peru. It rises from two lakes, runs *n.* and, turning *w.* with the name of Sara, enters the Guapaig.

Miguel, S. another, of the province and *corregimiento* of Pasto in the kingdom of Quito. It rises in the Sierra Nevada, and enters the Putumayo, after running many leagues *e.*

Miguel, S. another, of the province and government of Tucumán in Peru. It runs *e.* and enters the Rio Dulce.

Miguel, S. another, of the province and *captainship* of Pernanbuco in Brazil. It rises in the mountains of Itaberaba, runs *s. s. e.* and enters the sea between those of Las Lagunas and Yaqueacú.

Miguel, S. another, of the province and government of Atacames or Esmeraldas in the kingdom of Quito. It enters the Santiago to run into the sea in the port of Limones.

Miguel, S. a gulf of the S. sea, in the province of Darien, and kingdom of Tierra Firme. It is very great and beautiful, having its mouth or entrance closed in by a shoal called El Buey, there being only a narrow channel left for the course of vessels. Within it are many small rocks or reefs, and there runs into it a large river which flows down from the mountains of the same province.

Miguel, S. a small port of the *s.* coast of the island of Jamaica.

Miguel, S. an island of the N. sea, one of the Lesser and most *w.* of the Antilles.

Miguel, S. a long strip of land or point of the coast of the gulf of California or Mar Roxo de Cortés, in the centre of the said coast. It runs into the sea, and forms a side of the mouth of the bay of La Concepcion.

MIGUEL, S. a fortress or castle of the province and *captainship* of Rey in Brazil. It is near the coast, and not far from the lake Imeri or Merin.

MIGUEL, S. another fortress and garrison, in the province and government of Paraguay.

MIGUEL, S. See S. MICHAEL, and S. MICHEL.

MIGUELITO, S. a settlement of the province of Tepeguana, and kingdom of Nueva Vizcaya; situate on the bank of the river Guanabál, and not far from the town of Parras.

[MILFIELD, in Grafton county, New Hampshire, settled in 1774.]

[MILFORD, a township in Mifflin county, Pennsylvania.]

[MILFORD, a post-town of the state of Delaware, pleasantly situated on the *n.* side of Muspilion creek, about eight miles *w.* of its mouth in Delaware bay, 15 *s.* by *e.* of Dover, five *s.* of Frederica, and 65 *s.* by *w.* of Philadelphia. It contains nearly 100 houses, all built since the war, except one. The inhabitants are Episcopalians, Quakers, and Methodists.]

[MILFORD, a town of Northampton county, Pennsylvania, lately laid out on the *n. w.* side of the Delaware, on a lofty situation, at Well's ferry, 85 miles above Philadelphia. In front of the town, which contains as yet only a few houses, the river forms a cove well fitted for sheltering boats and lumber in storms or freshes in the river. A saw-mill and paper-mill have been erected here; the latter belongs to Mr. Biddis, who has discovered the method of making paper and pasteboard, by substituting a large proportion of saw-dust in the composition.]

[MILFORD, a post-town of Connecticut, on Long island sound, and in New Haven county, 17 miles *s. w.* of New Haven, and *e.* of Stratford. The mouth of the creek on which it stands has three fathoms water. This town was called Wopowage by the Indians, and was settled in 1638. It contains an Episcopal church, and two Congregational churches.]

[MILFORD Haven, a deep bay on the coast of Nova Scotia, to the *s. w.* round the point of the strait of Canso. It receives several rivers from the *n. w.* and *s. w.*]

[MILITARY Townships, in the state of New York. The legislature of the state granted 1,500,000 acres of land, as a gratuity to the officers and soldiers of the line of this state. This tract, forming the new county of Onondago, is bounded *w.* by the *e.* shore of the Seneca lake, and the Massachusetts lands in the new county of Ontario; *n.* by the part of lake Ontario near fort Oswego; *s.* by a ridge of the Alleghany mountains and the Pennsylvania line; and *e.* by the Tuscarora creek (which falls nearly into the middle of the Oneida lake) and that part of what was formerly Montgomery county, which has been settling by the New England people very rapidly since the peace. This pleasant county is divided into 25 townships, of 60,000 acres each, which are again subdivided into 100 convenient farms, of 600 acres; making in the whole 2500 farms. This tract is well watered by a multitude of small lakes and rivers.]

MILL, a river of the province and colony of Nova Scotia or Acadia, rising from a small lake near lake Rosignol. It runs *n. w.* and enters the sea in the port of Annapolis Real.

MILLALAB, a settlement of Indians, of the island of Laxa in the kingdom of Chile; situate on the shore of the river Burcu.

MILLAPOA, called by others Millapo, a town of the province and *corregimiento* of Maule in the kingdom of Chile; situate on the shore of the river Biobio. In its vicinity is a large estate called Toro.

MILLAQUI, a settlement of Indians, of the island of Laxa, in the same kingdom as the two former; situate on the shore of the river Tolpan, to the *e.* of the town of Colhue.

MILLER, a river of the province and colony of Massachusetts in New England, of N. America. It runs *w.* forming a curve, and enters the Connecticut.

MILLER, a *paramo* or mountain desert of the *cordillera* in the kingdom of Quito; one of those which were chosen by the academicians of the sciences of Paris in 1738 to fix their instruments for astronomical observations.

MILLEYS, a small river of the province and colony of N. Carolina. It runs *n. w.* and enters the Cutawba.

MILLS, a settlement of the island of Barbadoes, in the district of the parish of San Andres.

MILLS, a bay on the *e.* coast of the same island of Barbadoes, on the shore of which the former settlement is situate.

MILLUIHUAI, a settlement of the province and *corregimiento* of Cicasica in Peru; annexed to the curacy of Yanacache.

MILOCAN, a settlement of the head settlement of the district and *alcaldía mayor* of Zicayán in Nueva España. It is of a warm and dry temperature, contains 46 families of Indians, and is one league to the *n.* of its head settlement.

MILPA, a small settlement or ward (at the present day in a state of ruin), of the head settlement of the district and *alcaldía mayor* of Autlan in Nneva España.

MILPANDUENAS, a settlement of the province and kingdom of Guatemala; annexed to the curacy of Almolonga.

MILPAS ALTAS, a settlement of the province and kingdom of Guatemala in N. America. It has also the dedicatory title of Santo Tomas. Its population consists of 678 Indians, in which are included those which are in four settlements annexed to its curacy.

MILPAS, surnamed Baxas, to distinguish it from the former, a settlement in the same province and kingdom.

MILPILLA, a settlement of the head settlement of the district and *alcaldía mayor* of Acaponeta in Nueva España. It is 22 leagues to the *n. e.* of its capital, to the curacy of which it is annexed.

MILTEPEC, a settlement of the *alcaldía mayor* of Teotales in Nueva España. It contains 89 families of Indians.

MILTON, a city of the county of Suffolk, in the province and bay of Massachusetts; situate on the shore of the river of its name, which enters the sea in the bay of Boston. It is seven miles *s.* of this city, two from Dorchester, and six *n. w.* of Braintree.

MIMBRES, an isle or shoal of the *w.* head or foreland called Del Placer, which is in the island of Cuba, to the *w.* of that of Espiritu Santo.

MIMBRES, a cape or point, called also De Barrancas, on the coast which lies between the river La Plata and the straits of Magellan.

MIMIALPA, SAN MIGUEL DE, a settlement and head settlement of the district of the *alcaldía mayor* of Metepec in Nueva España. It contains 49 families of Indians.

MIMINI, a settlement of the province and *corregimiento* of Arica in Peru; annexed to the curacy of Camiña.

MIN, a river of the province and *corregimiento* of Cuenca in the kingdom of Quito. It rises in the mountain of El Altar, runs *n. e.* and enters by the *s.* into the Lluzin or De las Nieves, just before this joins the Chinchon, in lat. 1° 41′ *s.*

MINA, a small river of the province and government of San Juan de los Llanos in the Nuevo Reyno de Granada. It rises between those of Sinaruco and Cantanapalo, runs *e.* and enters the Orinoco between the mouths of the former, and of the Banahatú.

MINABAUJOU, a settlement of Indians, of New France or Canada; situate on the coast of lake Superior.

MINAGE, a river of Nova Scotia or Acadia. It rises from the lake Mipisigonche, runs *e.* for many leagues, and, inclining afterwards to the *s. e.* enters the sea in the bay of Miramichi.

MINANGUA, a small river of the province and government of Paraguay, which runs into the Paraná, between the rivers Acaray-piti and Yaperibuy.

[MINAS, sometimes also called Le Grand Praye, is a gulf on the *s. e.* side of the bay of Fundy, into which its waters pass by a narrow strait, and set up into Nova Scotia in an *e.* and *s.* direction. It is about 30 leagues from the entrance of Annapolis, and 10 from the bottom of Bedford bay. It is 12 leagues in length, and three in breadth. See BASIN OF MINAS.]

[MINAS, or DE LAS MINAS Hill, is the middlemost of the three hills described as marks within land for Bonaventura bay and river, on the *w.* coast of S. America: these are *s.* of Panama bay.]

MINAS, a fort of the English, in the province of Nova Scotia; situate on the shore of the same bay.

MINAS, a settlement of the jurisdiction of the town of Ibague, and *corregimiento* of Mariquita, in the Nuevo Reyno de Granada. It is of an hot temperature, and contains 100 inhabitants, who employ themselves in the labour of the copper mines, from whence they extract a good quantity of metal of excellent quality, and from whence the settlement has its name. It is annexed to the curacy of the city of Ibague.

MINAS, another, with the dedicatory title of Santa Catalina, in the head settlement of the district of Cuilapa, and *alcaldía mayor* of Quatro Villas, in Nueva España. It has this name from its vicinity to the silver mines of Chichicapa, which were formerly worked, but to-day abandoned. It contains only 12 families of Indians, who are employed in the cultivation and commerce of cochineal, seeds, fruit, and in collecting coal and wood. It is a little more than six leagues *s. w.* of its head settlement.

MINAS, a town of the province and government of Buenos Ayres; situate near the source of the river St. Lucia, about 34 miles *n. e.* of Maldonado. Lat. 34° 21′ 30″ *s.* Long. 55° 5′ 34″ *w.*

MINAS, another, with the addition of Nuevas, of the province of Tepeguana, and kingdom of Nueva Vizcaya, in N. America. It is a *real* of silver mines, and four leagues to the *w.* of the settlement of Parrál.

MINAS, another, with the additional title of Generales, a town of the Portuguese, in the province and *captainship* of Puerto Seguro in Brazil; situate at the source of the river Maranlao, to the *w.* of the lake Parapitinga. [Mr. Mawe tells us, that the province of Minas Generales, or Geraes, is from 6 to 700 miles from *n.* to *s.* and about the same extent from *e.* to *w.*; that it contains a population 360,000 persons, 200,000 of which are Negroes, or of Negro origin. The number of native Indians is not at all known; they neither mix with the colonists, nor give them any disturbance. Indeed the road seems to be so well guarded by those military posts called register houses, where all passengers undergo a strict examination, and the country is so completely scoured by a corps of well mounted *caçadores*, that it is more than probable the poor Indians confine themselves to the mountains. Mr. Mawe seldom mentions them under any other designation than that of the Anthropophagi.]

MINAS, another, a settlement of Indians of the island of Laxa in the kingdom of Chile; situate on the shore of the river of Los Sauces.

MINAS, a cape or point of land on the coast of Nova Scotia or Acadia, within the great bay of Fundy.

MINCHA, a settlement and *asiento* of gold mines, of the province and *corregimiento* of Qnillota in the kingdom of Chile; annexed to the curacy of Hillapel, with two other vice-parishes.

MINCHA, a bay in the province and *corregimiento* of Coquimbo in the same kingdom, on the side of the river Choapa.

[MINDAWARCARTON, Indians of N. America, the only band of Sioux who cultivate corn, beans, &c.; though these even cannot properly be termed a stationary people. They live in tents of dressed leather, which they transport by means of horses and dogs, and ramble from place to place during the greater part of the year. They are friendly to their own traders; but the inveterate enemies of such as supply their enemies, the Chippeways, with merchandise. They also claim the country in which they hunt, commencing at the entrance of the river St. Peter's, and extending upwards, on both sides of the Mississippi river, to the month of the Crow river. The land is fertile and well watered, lies level, and sufficiently timbered. Their trade cannot be expected to increase much.]

MINDO, a settlement of the *corregimiento* and kingdom of Quito, in the district of the jurisdiction of Los Cinco Leguas.

MINDO, another, in the province and government of Esmeraldas of the same kingdom; situate in the district of the Yumbos Indians, on the *n.* shore of the river Pirusay, a little before the union of this with the Nambillo. It is of a very hot climate, but its territory abounds in all kinds of vegetable productions. In lat. 2° 30′ *s.*

MINE, a small river of the province and government of Louisiana in N. America. It runs *s. e.* between the rivers Ovisconsin and Paris, and enters the Misipi.

[MINE AU FER, (or IRON MINES), on the *e.* side of Mississippi river, is 67¼ miles *n.* by *e.* of Chickasaw river, and 15 *s.* by *e.* of the Ohio. Here the land is nearly similar in quality to that bordering on the Chicaksaw river, interspersed with gradual risings or small eminences. There was a post at this place, near the former *s.* boundary of Virginia.]

[MINEHEAD, a township in Essex county, Vermont, on Connecticut river.]

MINER, a settlement of the island of Barbadoes.

MINES, Basin of the, a bay lying in the interior of the bay of Fundy in Nova Scotia or Acadia. It is very capacious, secure, and sheltered from the winds.

[MINETARES, are Indians of N. America, who claim no particular country, nor even assign themselves any limits: their tradition relates that they have always resided at their present villages. In their customs, manners, and dispositions, they are similar to the Mandans and Ahwahhaways. The scarcity of fuel induces them to reside, during the cold season, in large bands, in camps, on different parts of the Missouri, as high up that river as the mouth of the river Yellow Stone, and *w.* of their villages, about the Turtle mountain. These people, as well as the Mandans and Ahwahhaways, might be prevailed on to remove to the mouth of Yellow Stone river, provided an establishment is made at that place. They have as yet furnished scarcely any beaver, although the country they hunt abounds with them; the lodges of these animals are to be seen within a mile of their villages. These people have also suffered considerably by the small-pox; but have successfully resisted the attacks of the Sioux. The N.W. company have lately formed an establishment on the Missouri, near these people.]

MINGAN, a small river of the land or country of Labrador. It runs *s.* and enters the river St. Lawrence.

[MINGO Town, an Indian town on the *w.* bank of the Ohio river, 86 miles *n. e.* of Will's town, by the Indian path, and 33 *s. w.* of Pittsburg. It stands a few miles up a small creek, where there are springs that yield the *petral*, a bituminous liquid.]

MINGOS, a settlement of Indians of the province and country of the Iroquees in N. America; situate on the shore of the river Ohio, where the English have a fort and establishment for their commerce.

[MINGUN Islands, on the *n.* side of the mouth of the river St. Lawrence. They have the island Anticosti *s.* distant 10 leagues. Lat. 50° 15′ *n.* Long. 63° 25′ *w.*]

MINI, a settlement of the missions that were held by the regulars of the company, in the province and government of Paraguay. See SAN IGNACIO.

MINI, another settlement, of the province and government of Yucatán in N. America, in the which the Spaniards found a cross of stone, which the Indians used to adore in 1527, when they (the Spaniards) entered this country under Francisco de Montejo. The story of the Indians was, that an Indian chief and priest, called Chilaucalcatl (and whom the Father Charlevoix wrongly denominates Chilau Combal), a person esteemed for a great prophet, once said, that in a few days time there would come from that part where the sun rises a barbarous and white nation, who would carry before them that signal of the cross, and before which their idols would flee away; that this nation would conquer that land, that they would do no injury to its possessors, but that these would live in amity with the new-comers, would desert their idols, and adore one God. He caused a garment to be woven of cotton, and said that in that manner they would pay tribute to those people; he then ordered the lord of that settlement, whose name was Machauxiuch, to offer the same mantle to the idols, that it might be safely preserved, and he cut out of a piece of stone the sign of the cross, placing it also in the court of the temple, and saying, that that was the true tree of the world. Hence it was that the Indians asked the Spaniards, who arrived here under Francisco Hernandez de Cordoba, if they came from where the sun rose: likewise when Montejo arrived here, and saw the reverence that the Indians paid to the cross, he was assured of the truth of what was said to have been told them by their priest Chilaucalcatl. These Indians looked upon this cross as the god of rain, assuring themselves that they would never want moisture when they prayed to it devoutly.

MINI, a lake, which is also called by the Indians Imeri, in the province and *captainship* of Puerto Seguro in Brazil, near the sea-coast: on its shore the Portuguese have various settlements.

MINIPI, a settlement of the jurisdiction of Las Palmas, and *corregimiento* of Tunja, in the Nuevo Reyno de Granada; situate in a wild country, full of mountains and bogs, but of a benign temperature, abounding in tobacco, cotton, sugar-cane, plantains, and *yucas*, also in pigs, these being its principal commerce. It contains 300 inhabitants and a good number of Indians.

[MINISINK, a village in New Jersey, at the *n. w.* corner of the state, and on the *w.* side of Delaware river; about five miles below Montague, and 57 *n. w.* of Brunswick, by the road.]

[MINISINK, a township in Orange county, New York; bounded *e.* by the Wallkill, and *s.* by the state of New Jersey. It contains 2215 inhabitants, of whom 320 are entitled to be electors, and 51 are slaves.]

[MIQUELON, a small desert island, eight miles *w.* of cape May in Newfoundland island. It is the most *w.* of what have been called the Three Islands of St. Pierre or St. Peter, and is not so high as the other two; but its soil is very indifferent, and it is not more than three-fourths of a league in length. There is a passage or channel from the *w.* along by the *n.* end of this island into Fortune bay on the *s.* coast of Newfoundland. Lat. 47° *n.* Long. 56° 4′ *w.* It is sometimes called Maguelon.]

MIRA, SAN NICOLAS DE, a settlement of the province of Venezuela, and government of Maracaibo, in the Nuevo Reyno de Granada; situate in a valley called De los Obispos. It is of a very hot temperature, though pleasant and delightful, and is surrounded by the copions rivers of the St. Domingo, Masparro, and La Yuca. It is very healthy, and abounds in every kind of fruit and vegetable production, particularly *cacao*, and tobacco of the same quality as that of Barinas, which affords a great profit by its commerce, as being every where much esteemed. It has many machines for making sugar, abounds in neat cattle and horses of good quality, and contains 600 housekeepers. It is in the boundary which divides the archbishopric of Santa Fé from the bishopric of Caracas. Six leagues from the city of Barinas.

MIRA, another settlement, of the province and *corregimiento* of Ibarra in the kingdom of Quito, celebrated for the multitude of asses bred in its territory, and from whence the other settlements of the jurisdiction are supplied for the purposes of commerce. They traverse the country in troops, and the masters of the estates in that district allow, for a small acknowledgment, the natives to go and catch them; and this they effect with great facility, surrounding them by numbers on foot and horseback, although it costs them great labour afterwards to tame them, owing to their natural courage. These animals never permit a horse to be amongst them, and should one join them they bite and kick him to death. In the vicinity of this settlement is a mountain called Pachon, from whence not many years since great riches were extracted. It lies in a valley so hot that the Indians are dispensed from the *mita* or personal service. In lat. 32° 30′ *n*.

MIRA, another settlement, of the province and government of Mainas in the kingdom of Quito, called also Mamos; situate on the shore of the river Guayabeno.

MIRA, a river of the same province and kingdom as the former settlement, from whence it takes its name. It rises in the *n*. skirt of the mountain of Mojanda, of the mountains of Los Cofanes, from two large streams which enter the lake of San Pablo, out of which it issues; and, collecting the waters of the rivers Pisco, Angel, Taguando, Escudillas, Caguasqui, and Chiles or Mayasquer, which flows down from the heights of Pellizo, turns its course to the *n. e.* and receives the rivers Camunixi, Gualpi, Nulpe, and Puespi, taking the name of the province, and afterwards changing it to Mira, till it reaches the spot where it enters the Pacific or S. sea by nine months, between the point of Manglares and the island and port of Tumaco. Its shores in the territory of the town of Ibarra are very delightful and pleasant, from being full of gardens and sugar-cane plantations. This river, after it incorporates itself with the Chiles, enters the province of Esmeraldas, and divides it from that of Barbacoas.

MIRA POR VOS, some reefs or rocky shoals in the N. sea, by the *s*. part of the *w*. head of the island of Cuba, between this island and that of Yuma.

MIRACA, a lake of the province and government of Venezuela in the Nuevo Reyno de Granada. It is near the coast, in the cape of San Roman of the peninsula of Paraguana.

MIRAFLORES, SANTIAGO DE SANA DE, a town of the province and *corregimiento* of Saña in Peru, and of the bishopric of Truxillo; founded by order of the viceroy the Count de Nieva, in 1546, in a pleasant valley on the *n*. side of the river of its name, seven leagues from the sea: the valley was separated from the bishopric of Truxillo at the time that it was added to the district of Chiclayo. It was the capital of its province and very opulent, but at the present day reduced to a miserable settlement, the greater part of its inhabitants having established themselves in the settlement of Lambayeque. The English pirate Edward David sacked it in 1680, and in 1720 it was completely ruined by a deluge of rain which lasted several days, and which caused the river to swell to such a degree as to inundate it. In this town died Santo Toribio, archbishop of Lima, whilst upon his visitation, and before the establishment of the bishopric of Truxillo. It has four convents of the following religious orders, San Francisco, San Agustin, La Merced, and San Juan de Dios, all of which are almost in a state of ruin and extreme poverty. It is 357 miles from Lima.

MIRAFLORES, with the dedicatory title of San Esteban, a settlement of the province and government of Tucumán in Peru, and of the district and jurisdiction of the city of Salta. It is a *reduccion* of the Lules and Toconotes Indians, of the district of Gran Chaco, and of the missions that were held by the regulars of the company; but at present under the charge of the monks of San Francisco. It is of a benign and agreeable temperature, has very large breeds of cattle, formerly contained upwards of 600 Indians, and is situate on the shore of the river Salta. It lies in the direct and necessary way from Buenos Ayres to Lima. In its vicinity is the fort of San Joseph to restrain on that side the incursions of the infidel Indians.

MIRAFLORES, another, of the province and *corregimiento* of Tunja in the Nuevo Reyno de Granada. It is of an hot temperature, abundant and fertile in sugar-cane and cotton, contains 200 inhabitants, and a multitude of Negroes in the various estates of its district. It is half a league from its capital.

MIRAFLORES, another, of the province and government of Popayán, in the same kingdom as the former. It is on a fertile, pleasant spot, near the river Timbo, and the climate is very healthy. It had in former times a considerable population, but is at present reduced to a miserable state. It lies to the *n. e.* of the Palo Bobo, in lat. 2° 13′ *n*.

MIRAFLORES, another, of the province and *corregimiento* of Guamalies in Peru; annexed to the curacy of Llacta.

MIRAFLORES, another, with the dedicatory title of San Marcos, in the province and *corregimiento* of Chayauta in the same kingdom,

MIRAFLORES, another, of the province and *corregimiento* of Cercado in the same kingdom; annexed to the curacy of La Magdalena.

MIRAFLORES, a river of the province and *corregimiento* of Conchucos. It rises near the settlement of this name, as will be found mentioned above, runs *n*. and enters the Marañon.

MIRAFLORES, a fort, with the dedicatory title

of San Esteban, in the province and government of Tucumán.

[MIRAGOANE, a town on the *n.* side of the *s.* peninsula of the island of St. Domingo, and *s.* side of the bight of Leogane, at the head of a bay of its name. It is on the road from Jeremie to Port au Prince, about 31 leagues *e.* by *s.* of the former, and 23 *w.* by *s.* of the latter. Lat. 18° 27′ *n.*]

MIRAGUANA, an isle situate near the *n.* coast of the island of St. Domingo, in the part possessed by the French, opposite the island Goanava, between the point of Petez and the Tron Forban.

[MIRAMICHI, or MIRACHI, a port, bay, and river, on the *n. e.* coast of New Brunswick. The port is at the mouth of the river. The entrance into the bay is very wide; it has point Portage for its *n.* entrance, and its *s.* side is formed by Escuminax point, which is 53 miles *n. e.* of Shediac harbour, and 34 *s. e.* of the mouth of Nippisighit river, which empties into Chaleur bay. There is a salmon fishery in Miramichi river.]

MIRAQUANE, a settlement of the province and government of Louisiana in N. America; situate on the coast at the entrance of the bay of La Mobila.

[MIRAY Bay, on the coast of the island of Cape Breton, is to the *s.* from Morienne bay. Large vessels may go up six leagues, and have good anchorage, and lie secure from all winds. Lat. 46° 5′ *n.* Long. 59° 49′ *w.*]

[MIREBALAIS, an interior town in the French part of the island of St. Domingo; situated nearly 12 leagues *n.* of Port au Prince, on the road from that city to Varettes; from which last it is 14 leagues *s. e.*]

MIRGAS, a settlement of the province and *corregimiento* of Conchucos in Peru; annexed to the curacy of Llamellin.

MIRIBIRA, a large island of the river Marañon or Amazonas; situate near its entrance into the sea, and almost opposite the city of Pará.

MIRINAI, a river of the province and government of Paraguay. It rises from the lake Iberia, runs *s.* and then turning *e.* enters the Uruguay.

MIRLIGUECHE, Bay of, on the *s.* coast of Nova Scotia or Acadia, between cape Rage and the island of La Croix.

MISCHAUALLI, a small and reduced settlement of the province and government of Quixos y Macas in the kingdom of Quito; situate on the shore of the river of its name.

MISCHAUALLI. This river runs *e.* and opposite the city of Archidona unites itself with another small torrent, and takes this name, afterwards collecting the waters of the rivers Hollin, Tena, and Pano, until it enters by the *n.* side into the Napo, in lat. 1° 1′ 18″ *s.*

[MISCOTHINS, a small tribe of Indians who inhabit between lake Michigan and the Mississippi.]

[MISCOU, or MISCO, an island which forms the *s.* side of the entrance of Chaleur bay, and is now called Muscow island. The gut of Chepayan, about two or three leagues in length, and in some parts near a league wide, separates it from the *n. e.* coast of New Brunswick. It abounds with salt marsh hay.]

MISERICORDIA, a port on the coast of the straits of Magellan; discovered by Admiral Pedro Sarmiento in 1579, and then taken possession of for the seventh time, for the crown of Spain, after that it had been abandoned by Villalobos, with whom the aforesaid admiral had established other six colonies in the gulf of La Santisima Trinidad.

[MISERY, an isle between Salem and cape Ann in Massachusetts.]

MISHUM, a river of the province and colony of New England in N. America.

MISINA, a small river of the country or land of Labrador. It runs *s.* between the Ovatessaou and the Esquimaux, and enters the sea in the gulf of St. Lawrence.

MISION GRANDE, a settlement of the missions that were held by the French regulars of the company, in New France or Canada; situate on the shore of the river St. Lawrence, between the city of Tadoussac and the point of Ocramane.

MISION, another settlement, of the province and *corregimiento* of Maule in the kingdom of Chile; situate on the shore of the river Biobio, to the *w.* of the town of Millapoa.

MISION, another, with the surname of Nueva, in the province and country of Las Amazonas, and in the territory which is occupied by the Portuguese, being a *reduccion* of the missions which were held by the Carmelite fathers of this nation; situate on the shore of the river Guatuma.

MISIPINAC, a river of New France or Canada in N. America. It rises from a lake in the country of the Papinachois Indians, runs *s. e.* and enters the grand river St. Lawrence at its mouth or entrance.

MISISAGAN, or BUADE, a lake of the province and government of Luisiana, formed of various rivers. It runs into a river which is called the river of the Lake, to enter the Mississippi.

MISISAGUES, formerly a settlement of Indians of New France or Canada in N. America; situate on the shore of the strait of its name, but which is now called the river St. Clair. In it the French built a fort and establishment for their commerce.

Misisagues. The aforesaid strait or river is a large canal of water which runs from lake Huron, on the *s.* side, and communicates with lake Erie, forming in the middle the lake of St. Clair.

[MISKO, an island on the *s. w.* side of Chaleur bay, at its mouth.]

MISKOUAKIMINA, a settlement of the same province and country as the former lake; situate on the shore of the lake Michigan, and at the mouth of the river Mellaki.

MISKOUASKANE, a lake of New France or Canada in N. America; situate between the lakes Beauharnois and Begon in the territory of the Chemonchovanistes Indians.

MISOA, a settlement of the government of Maracaibo, and province of Venezuela, in the Nuevo Reyno de Granada; situate on the *e.* coast of the great lake of that name.

MISPILION, a river of the province and colony of Pennsylvania in N. America.

MISQUITIC, a settlement of the head settlement of the district and *alcaldía mayor* of Tecpatitlan in Nueva Galicia; nine leagues to the *n. e.* of its capital.

[MISSINABE Lake is situated in the *n.* part of N. America, in lat. 48° 29′ 42″ *n.* and long. 84° 2′ 42″ *w.*]

[Missinabe House is situated on the *e.* side of Moose river, eight miles from Missinabe lake, and 80 *w.* by *s.* of Frederic house; and is a station belonging to the Hudson bay company.]

MISSISSIPPI, a large and abundant river in N. America. It rises at the high land which separates the waters running into the Hudson's bay from those running into the bay of Mexico: its origin is Turtle lake, from whence it traverses this spacious country till it disembogues itself in the sea at the gulf of Mexico. It is navigable, and its course is very winding and irregular for the space of more than 2000 miles, including its windings. The French took possession of it in 1712; and at the peace of Versailles, in 1763, it was stipulated that its navigation should be free to the English and French, and that a line being drawn down the middle of its course, a frontier of division should be marked between the possessions of the one and the other; expressing further that all the continent to the *w.* of this river should remain to the French, and the country to the *e.* to the English. In the soundings which were made at its entrance there were found 16 feet water, and immediately a French ship, the Neptune, which had just arrived from France, entered the river, and sailed up as far as New Orleans. Mr. Decan and the Father Hennepin of the company of the Jesuits also entered this river from the fort of Crevecœur, and navigated up as far as lat. 44° 10′ *n.* where their course was impeded by a very lofty cascade of water, occupying the whole width of the river called Antonio de Padua.

This river traverses nearly the whole of N. America. The Baron Touti, who navigated it in 1680, dwells much on the pleasantness of the countries that it irrigates, and says that it is 300 leagues from its origin to where it is entered by the Ilines or Ilinois; and, until it enters the sea in the bay of Mexico, 800. The French, under the regency of the duke of Orleans, flattered themselves they should accumulate great riches through some establishments they thought to make in 1719, under the name of the Company of Mississippi, similar to the S. sea company of the English; and to this intent they were eager to put their money into a fund, and to purchase shares; but they were all ruined, and the projected establishments fell to the ground.

This river receives in its course infinite others, which augment its stream; such as the Ohio, nearly equal to the Danube, the Ovacache little inferior, the grand river of Alibama, the Mobila, and others, some of which are so impetuous and abundant as to bring down such a quantity of mud as to obscure the water for a distance of 20 leagues. In this river breed a multitude of alligators and other amphibious animals, and also aquatic birds: the country on either side is very fertile, inhabited by an infinite number of different nations of Indians, the best known of whom are the Hadovesaves, the Hanetons, Ovas, and Thuntolas. It disembogues itself into the sea by a great many mouths, which form a number of islands, some of which are of a considerable size. The aforesaid Baron Tonti places its three principal months between lat. 28° and 29°, and the best geographers between 28° 50′ and 29° 15′ and between long. 89° 5′ and 89° 38° *w.* The country on either side of these mouths is full of woods, and uncultivated through the frequent inundation and barrenness of the territory, this producing nothing but shrubs, and a species of trees, all of which are stripped of their bark, &c. by the force of the waters. Some leagues further up, beyond the lake, the country is represented by travellers as being very agreeable, covered with vines and all kinds of fruit-trees, and producing maize in abundance, with pulse and other grain, of which two crops are gathered yearly.

[Mr. Ashe represents this river as exhibiting, in

its scenery and current, an almost continued succession of beauty, richness, and grandeur. The navigation, like that of the Ohio, is interrupted with islands; of which the number is increasing. During its floods, which are periodical, the same author asserts, a "first-rate man of war may descend with safety." The country on both sides of the Mississippi, and on its tributary streams, is equal in goodness to any in N. America. This river is navigable to St. Anthony's falls without any obstruction, and some travellers describe it as navigable above them. On both sides this river are salt springs or licks, which produce excellent salt; and on its branches are innumerable such springs. Besides the coal mines in the upper parts of the Ohio country, there are great quantities of coal on the upper branches of this river. Some account of the valuable productions on the banks of this majestic river, and the lands which its branches water, will be seen under the description of Louisiana, W. Florida, Tennessee, Georgia, &c. &c.

Much, indeed, has been written, and much still remains to be said, of the course and soundings of this river, and of the properties of the soil through which it runs. It is a subject highly interesting and important; and we shall not therefore be afraid of extending this article to an undue length by inserting

Some Observations made in a Voyage, commencing at St. Catherine's landing, on the e. bank of the Mississippi, proceeding downwards to the mouth of Red river, and from thence ascending that river, the Black river, and the Washita river, as high as the Hot springs, in the proximity of the last-mentioned river, extracted from the Journals of William Dunbar, Esq. and Dr. Hunter.

'Mr. Dunbar, Dr. Hunter, and the party employed by the United States to make a survey of, and explore the country traversed by, the Washita river, left St. Catherine's landing, on the Mississippi, in lat. 31° 26′ 30″ *n.* and long. 6h. 5′ 56″ *w.* from the meridian of Greenwich, on Tuesday the 16th of October 1804. A little distance below St. Catherine's creek, and five leagues from Natches, they passed the White cliffs, composed chiefly of sand, surmounted by pine, and from 100 to 200 feet high. When the waters of the Mississippi are low, the base of the cliff is uncovered, which consists of different coloured clays, and some beds of ochre, over which there lies, in some places, a thin lamina of iron ore. Small springs, possessing a petrifying quality, flow over the clay and ochre, and numerons logs and pieces of timber, converted into stone, are strewed about the beach. Fine pure argil of various colours, chiefly white and red, is found here.

'On the 17th they arrived at the mouth of Red river, the confluence of which with the Mississippi, agreeably to the observations of Mr. de Ferrer, lies in lat. 31° 1′ 15″ *n.* and long. 6h. 7′ 11″ *w.* of Greenwich. Red river is here about 500 yards wide, and without any sensible current. The banks of the river are clothed with willow, the land low and subject to inundation, to the height of 30 feet or more above the level of the water at this time. The mouth of the Red river is accounted to be 75 leagues from New Orleans, and three miles higher up than the Chafalaya or Opelousa river, which was probably a continuation of the Red river when its waters did not unite with those of the Mississippi but during the inundation.

'On the 18th the survey of the Red river was commenced, and on the evening of the 19th the party arrived at the mouth of the Black river, in lat. 31° 15′ 48″ *n.* and about 26 miles from the Mississippi. The Red river derives its name from the rich fat earth or marl of that colour, borne down by the floods; the last of which appeared to have deposited on the high bank a stratum of upwards of half an inch in thickness. The vegetation on its banks is surprisingly luxuriant; no doubt owing to the deposition of marl during its annual floods. The willows grow to a good size; but other forest-trees are much smaller than those seen on the banks of the Mississippi. As you advance up the river it gradually narrows; in lat. 31° 8′ *n.* it is about 200 yards wide, which width is continued to the mouth of Black river, where each of them appears 150 yards across. The banks of the river are covered with pea-vine and several sorts of grass bearing seed, which geese and ducks eat very greedily; and there are generally seen willows growing on one side, and on the other a small growth of black oak, paccawn, hiccory, elm, &c. The current in the Red river is so moderate as scarcely to afford an impediment to its ascent.

'On sounding the Black river, a little above its mouth, there was found 20 feet of water, with a bottom of black sand. The water of Black river is rather clearer than that of the Ohio, and of a warm temperature, which it may receive from the water flowing into it from the valley of the Mississippi, particularly by the Catahoola. At noon on the 23d, by a good meridian observation, they ascertained their latitude to be 30° 36′ 29″ *n.* and [were then a little below the mouths of the Catahoola, Washita, and Bayan Tenza, the united waters of which form the Black river. The current is very gentle the whole length of the Black river, which in many places does not exceed 80 yards in width. The banks on the lower part of the river present a great luxuriance of vegetation and rank grass, with red and black oak, ash, paccawn, hiccory, and some elms. The soil is black marl, mixed with a moderate proportion of sand, resembling much the soil on the Mississippi banks; yet the forest-trees are not lofty, like those on the margin of the Great river, but resembling the growth on the Red river. In lat. 31° 22′ 46″ *n.* they observed that canes grew on several parts of the right bank, a proof that the land is not deeply overflowed; perhaps from one to three feet: the banks have the appearance of stability; very little willow, or other productions of a newly formed soil on either side. On advancing up the river, the timber becomes larger, in some places rising to the height of 40 feet; yet the land is liable to be inundated, not from the waters of this small river, but from the intrusion of its more powerful neighbour the Mississippi. The lands decline rapidly, as in all alluvial countries, from the margin to the cypress swamps, where more or less water stagnates all the year round. On the 21st they passed a small but elevated island, said to be the only one in this river for more than 100 leagues ascending. On the left bank, near this island, a small settlement of a couple of acres has been begun by a man and his wife. The banks are not less than 40 feet above the present level of the water in the river, and are but rarely overflowed: on both sides they are clothed with rich cane brake, pierced by creeks fit to carry boats during the inundation.

'They saw many cormorants, and the hooping crane; geese and ducks were not yet abundant, but are said to arrive in myriads, with the rains and winter's cold. They shot a fowl of the duck kind, whose foot was partially divided, and the body covered with a bluish or lead-coloured plumage. On the morning of the 22d they observed green matter floating on the river; supposed to come from the Catahoola and other lakes and bayaus of stagnant water, which, when raised a little by rain, flow into the Black river; and also many patches of an aquatic plant, resembling small islands, some floating on the surface of the river, and others adhering to, or resting on the shore and logs. On examining this plant it was found a hollow-jointed stem, with roots of the same form, extremely light, with very narrow willow-shaped leaves projecting from the joint, embracing, however, the whole of the tube, and extending to the next inferior joint or knot. The extremity of each branch is terminated by a spike of very slender, narrow, seminal leaves, from one to two inches in length, and one-tenth, or less, in breadth, producing its seed on the underside of the leaf, in a double row almost in contact; the grains alternately placed in perfect regularity: not being able to find the flower, its class and order could could not be determined, although it is not probably new. Towards the upper part of the Black river the shore abounded with muscles and perriwinkles. The muscles were of the kind called pearl muscles. The men dressed a quantity of them, considering them as an agreeable food; but Mr. D. found them tough and unpalatable.

'On arriving at the mouth of the Catahoola, they landed to procure information from a Frenchman settled there. Having a grant from the Spanish government, he has made a small settlement, and keeps a ferry-boat for carrying over men and horses travelling to and from Natchez, and the settlements on Red river, and on the Washita river. The country here is all alluvial. In process of time, the rivers, shutting up ancient passages, and elevating the banks over which their waters pass, no longer communicate with the same facility as formerly: the consequence is, that many very large tracts, formerly subject to inundation, are now entirely exempt from that inconvenience. Such is the situation of a most valuable tract upon which this Frenchman is settled. His honse stands on an Indian mount, with several others in view. There is also a species of rampart surrounding this place, and one very elevated mount, a view and description of which was postponed till the travellers return; their situation not allowing of the requisite delay. The soil is equal to the best Mississippi bottoms.

'From this place they proceeded to the mouth of Washita, in lat. 35° 37′ 7″ *n.* and encamped on the evening of the 23d.

'This river derives its appellation from the name of an Indian tribe formerly resident on its banks; the remnant of which, it is said, went into the great plains to the *w.* and either compose a small tribe themselves, or are incorporated into another nation. The Black river loses its name at the junction of the Washita, Catahoola, and Tenza, although our maps represent it as taking place of the Washita. The Tenza and Catahoola are also named from Indian tribes now extinct. The latter is a creek 12 leagues long, which is the issue of a lake of the same name, eight leagues in length, and about two leagues in breadth. It lies] [*w.* from the mouth of the Cataboola, and communicates with the Red river during the great annual inundation. At the *w.* or *n. w.* angle of the lake, a creek called Little river enters, which preserves a channel with running water at all seasons, meandering along the bed of the lake; but in all other parts its superficies, during the dry season from July to November, and often later, is completely drained, and becomes covered with the most luxuriant herbage; the bed of the lake then becomes the residence of immense herds of deer, of turkeys, geese, cranes, &c. which feed on the grass and grain. Bayau Tenza serves only to drain off a part of the waters of the inundation from the low lands of the Mississippi, which here communicate with the Black river during the season of high water.

Between the mouth of the Washita and Villemont's prairie on the right, the current of the river is gentle, and the banks favourable for towing. The lands on both sides have the appearance of being above the inundation; the timber, generally such as high lands produce, being chiefly red, white, and black oaks, interspersed with a variety of other trees. The magnolia grandiflora, that infallible sign of the land not being subject to inundation, is not, however, among them. Along the banks a stratum of solid clay or marl is observable, apparently of an ancient deposition. It lies in oblique positions, making an angle of nearly 30 degrees with the horizon, and generally inclined with the descent of the river, although in a few cases the position was contrary. Timber is seen projecting from under the solid bank, which seems indurated, and unquestionably very ancient, presenting a very different appearance from recently formed soil. The river is about 80 yards wide. A league above the mouth of the Washita, the Bayau Ilaha comes in unexpectedly from the right, and is one of the many passages through which the waters of the great inundation penetrate and pervade all the low countries, annihilating, for a time, the currents of the lesser rivers in the neighbourhood of the Mississippi. The vegetation is remarkably vigorous along the alluvial banks, which are covered with a thick shrubbery, and innumerable plants in full blossom at this late season.

'Villemont's prairie is so named in consequence of its being included within a grant under the French government to a gentleman of that name. Many other parts on the Washita are named after their early proprietors. The French people projected and began extensive settlements on this river, but the general massacre planned, and in part executed, by the Indians against them, and the consequent destruction of the Natchez tribe by the French, broke up all these undertakings, and they were not recommenced under that government. Those prairies are plains, or savannas, without timber; generally very fertile, and producing an exuberance of strong, thick, and coarse herbage. When a piece of ground has once got into this state in an Indian country, it can have no opportunity of re-producing timber, it being an invariable practice to set fire to the dry grass in the fall or winter, to obtain the advantage of attracting game when the young tender grass begins to spring: this destroys the young timber, and the prairie annually gains upon the wood-land. It is probable that the immense plains known to exist in America, may owe their origin to this custom. The plains of the Washita lie chiefly on the *e.* side, and being generally formed like the Mississippi land, sloping from the bank of the river to the Great river, they are more or less subject to inundation in the rear; and in certain great floods the water has advanced so far as to be ready to pour over the margin into the Washita. This has now become a very rare thing, and it may be estimated, that from a quarter of a mile to a mile in depth, will remain free from inundation during high floods. This is pretty much the ease with those lands nearly as high as the post of the Washita, with the exception of certain ridges of primitive high land; the rest being evidently alluvial, although not now subject to be inundated by the Washita river, in consequence of the great depth which the bed of the river has acquired by abrasion. On approaching towards the Bayau Louis, which empties its waters into the Washita on the right, a little below the rapids, there is a great deal of high land on both sides, which produces pine and other timber, not the growth of inundated lands. At the foot of the rapids the navigation of the river is impeded by beds of gravel formed in it. The first rapids lie in lat. 31° 48′ 57.5″ *n.* a little above which there is a high ridge of primitive earth, studded with abundance of fragments of rocks, or stone, which appear to have been thrown up to the surface in a very irregular manner. The stone is of a friable nature, some of it having the appearance of indurated clay; the outside is blackish from exposure to the air; within, it is a greyish white. It is said that in the hill the strata are regular, and that good grindstones may be here obtained. The last of the rapids, which is formed by a ledge of rocks crossing the entire bed of the river, was passed in the evening of the 27th; above it the water became again like a mill-pond, and about 100 yards wide.]

[The whole of these first shoals or rapids embraced an extent of about a mile and a half; the obstruction was not continued, but felt at short intervals in this distance. On the right, about four leagues from the rapids, they passed the Bayau aux Bœufs, a little above a rocky hill: high lands and savanna are seen on the right. On sounding the river they found three fathoms water on a bottom of mud and sand. The banks of the river, above the bayau, seem to retain very little alluvial soil; the high land earth, which is a sandy loam of a light grey colour, with streaks of red sand and clay, is seen on the left bank; the soil not rich, bearing pines, interspersed with red oak, hiccory, and dog wood. The river is from 60 to 100 yards wide here, but decreases as you advance. The next rapid is made by a ledge of rocks traversing the river, and narrowing the water channel to about 30 yards. The width between the high banks cannot be less than 100 yards, and the banks from 30 to 40 feet high. In lat. 32° 10′ 13″, rapids and shoals again occurred, and the channel was very narrow; the sand bars, at every point, extended so far into the bend as to leave little more than the breadth of the boat of water sufficiently deep for her passage, although it spreads over the width of 70 or 80 yards upon the shoal.

'In the afternoon of the 31st they passed a little plantation or settlement on the right, and at night arrived at three others adjoining each other. These settlements are on a plain or prairie, the soil of which we may be assured is alluvial from the regular slope which the land has from the river. The bed of the river is now sufficiently deep to free them from the inconvenience of its inundation; yet in the rear, the waters of the Mississippi approach, and sometimes leave dry but a narrow strip along the bank of the river. It is, however, now more common, that the extent of the fields cultivated (from ¼ to ½ mile) remains dry during the season of inundation; the soil here is very good, but not equal to the Mississippi bottoms; it may be esteemed second rate. At a small distance to the *e.* are extensive cypress swamps, over which the waters of the inundation always stand to the depth of from 15 to 25 feet. On the *w.* side, after passing over the valley of the river, whose breadth varies from a quarter of a mile to two miles, or more, the land assumes a considerable elevation, from 100 to 300 feet, and extends all along to the settlements of the Red river. These high lands are reported to be poor, and badly watered, being chiefly what is termed a pine barren. There is here a ferry and road of communication between the post of the Washita and the Natchez, and a fork of this road passes to the settlement called the Rapids, on Red river, distant from this place, by computation, 150 miles.

'On this part of the river, lies a considerable tract of land, granted by the Spanish government to the Marquis of Maison Rouge, a French emigrant, who bequeathed it with all his property to M. Bouligny, son of the late colonel of the Louisiana regiment, and by him sold to Daniel Clarke. It is said to extend from the post of Washita with a breadth of two leagues, including the river, down to the Bayau Calumet; the computed distance of which along the river is called 30 leagues, but supposed not more than 12 in a direct line.

'On the 6th of November, in the afternoon, the party arrived at the post of the Washita, in lat. 32° 29′ 37″ *n.* where they were politely received by Lieutenant Bowmar, who immediately offered the hospitality of his dwelling, with all the services in his power.

'From the ferry to this place, the navigation of the river is, at this season, interrupted by many shoals and rapids. The general width is from 80 to 100 yards. The water is extremely agreeable to drink, and much clearer than that of the Ohio. In this respect it is very unlike its two neighbours, the Arkansa and Red rivers, whose waters are loaded with earthy matters of a reddish brown colour, giving to them a chocolate-like appearance; and, when those waters are low, are not potable, being brackish from the great number of salt springs which flow into them, and probably from the beds of rock salt over which they may pass. The banks of the river presented very little appearance of alluvial land, but furnished an infinitude of beautiful landscapes, heightened by the vivid colouring they derive from the autumnal changes of the leaf. Mr. Dunbar observes, that the change of colour in the leaves of vegetables, which is probably occasioned by the oxygen of the atmosphere acting on the vegetable matter, deprived of the protecting power of vital principle, may serve as an excellent guide to the naturalist who directs his attention to the discovery of new objects for the use of the dyer. For he has always remarked, that the leaves of those trees whose bark or wood is known to produce a dye, are changed in autumn to the same colour which is extracted in the dyer's vat from the woods; more especially by the use of mordants, as alum, &c. which yields oxygen: thus the foliage of the hiccory and oak, which produces the quercitron bark, is changed] [before its fall into a beautiful yellow. Other oaks assume a fawn colour, a liver colour, or a blood colour, and are known to yield dyes of the same complexion.

'In lat. 32° 18' *n*. Dr. Hunter discovered along the river side a substance nearly resembling mineral coal; its appearance was that of the carbonated wood described by Kirwan. It does not easily burn; but on being applied to the flame of a candle, it sensibly increased it, and yielded a faint smell, resembling in a slight degree that of the gum lac of common sealing-wax.

'Soft friable stone is common, and great quantities of gravel and sand, upon the beaches in this part of the river. A reddish clay appears in the strata, much indurated and blackened by exposure to the light and air.

'The position called fort Miro being the property of a private person, who was formerly civil commandant here, the lieutenant has taken post about 400 yards lower; has built himself some log houses, and inclosed them with a slight stockade. Upon viewing the country *e.* of the river, it is evidently alluvial; the surface has a gentle slope from the river to the rear of the plantations. The land is of excellent quality, being a rich black mould to the depth of a foot, under which there is a friable loam of a brownish liver colour.

'At the post of the Washita, they procured a boat of less draught of water than the one in which they ascended the river thus far; at noon, on the 11th of November, they proceeded on the voyage, and in the evening encamped at the plantation of Baron Bastrop.

'This small settlement on the Washita, and some of the creeks falling into it, contains not more than 500 persons, of all ages and sexes. It is reported, however, that there is a great quantity of excellent laud upon these creeks, and that the settlement is capable of great extension, and may be expected, with an accession of population, to become very flourishing. There are three merchants settled at the post, who supply, at very exorbitant prices, the inhabitants with their necessaries. These, with the garrison, two small planters, and a tradesman or two, constitute the present village. A great proportion of the inhabitants continue the old practice of hunting during the winter season, and they exchange their peltry for necessaries, with the merchants, at a low rate. During the summer, these people content themselves with raising corn, barely sufficient for bread during the year. In this manner they always remain extremely poor. Some few who have conquered that habit of indolence, which is always the consequence of the Indian mode of life, and attend to agriculture, live more comfortably, and taste a little of the sweets of civilized life,

'The lands along the river above the post are not very inviting, being a thin poor soil, and covered with pine wood. To the right, the settlements on the Bayan Barthelemi and Siard, are said to be rich laud.

'On the morning of the 13th they passed an island and a strong rapid, and arrived at a little settlement below a chain of rocks, which cross the channel between an island and the mainland, called Roque Raw. The Spaniard and his family settled here, appear, from their indolence, to live miserably. The river acquires here a more spacions appearance, being about 150 yards wide. In the afternoon they passed the Bayau Barthelemi on the right, above the last settlements, and about 12 computed leagues from the post. Here commences Baron Bastrop's great grant of land from the Spanish government, being a square of 12 leagues on each side, a little exceeding a million of French acres. The banks of the river continue about 30 feet high, of which 18 feet from the water are a clayey loam of a pale ash colour, upon which the water has deposited 12 feet of light sandy soil, apparently fertile, and of a dark brown colour. This description of land is of small breadth, not exceeding half a mile on each side of the river, and may be called the valley of the Washita, beyond which there is high land covered with pines.

'The soil of the Bayau des Buttes continues thin with a growth of small timber. This creek is named from a number of Indian mounts discovered by the hunters along its course. The margin of the river begins to be covered with such timber as usually grows on inundated land, particularly a species of white oak, vulgarly called the over-cup oak; its timber is remarkably hard, solid, ponderous, and durable, and it produces a large acorn in great abundance, upon which the bear feeds, and which is very fattening for hogs.

'In lat. 32° 50' 8" *n*. they passed a long and narrow island. The face of the country begins to change; the banks are low and steep; the river deep and more contracted, from 30 to 50 yards in width. The soil in the neighbourhood of the river is a very sandy loam, and covered with such vegetables as are found on the inundated lands of the Mississippi. The tract presents the appearance of a new soil, very different from what they passed] [below. This alluvial tract may be supposed the site of a great lake, drained by a natural channel, from the abrasion of the waters: since which period the annual inundations have deposited the superior soil: 18 or 20 feet are wanting to render it habitable for man. It appears, nevertheless, well stocked with the beasts of the forest, several of which were seen.

'Quantities of water-fowl were beginning to make their appearance, which are not very numerous here until the cold rains and frost compel them to leave a more *n.* climate. Fish is not so abundant as might be expected, owing, it is said, to the inundation of the Mississippi, in the year 1799, which dammed up the Washita, some distance above the post, and produced a stagnation and consequent corruption of the waters that destroyed all the fish within its influence.

'At noon, on the 15th of November, they passed the island of Mallet, and at 90 yards *n. e.* from the upper point of the island, by a good observation, ascertained their latitude to be 32° 59′ 27″ *n.* or two seconds and a half of latitude *s.* of the dividing line between the territories of Orleans and Louisiana. The bed of the river along this alluvial country is generally covered with water, and the navigation uninterrupted; but in the afternoon of this day, they passed three contiguous sand bars, or beaches, called Les Trois Battures, and before evening the Bayan de Grand Marias, or Great Marsh creek, on the right, and La Cypreri Chattelrau, a point of high land on the other side, which reaches within half a mile of the river. As they advanced towards the Marias de Saline, on the right, a stratum of dirty white clay under the alluvial tract shewed them to be leaving the sunken, and approaching the high land country. The Salt Lick marsh does not derive its name from any brackishness in the water of the lake or marsh, but from its contiguity to some of the licks sometimes called saline, and sometimes glaise, generally found in a clay compact enough for potters' ware. The Bayau de la Tulipe forms a communication between the lake and the river. Opposite to this place, there is a point of high land forming a promontory, advancing within a mile of the river, and to which boats resort when the low grounds are under water. A short league above is the mouth of the grand Bayau de la Saline (Salt Lick creek). This creek is of a considerable length, and navigable for small boats. The hunters ascend it, to 100 of their leagues, in pursuit of game, and all agree that none of the springs which feed this creek are salt. It has obtained its name from the many buffalo salt licks which have been discovered in its vicinity. Although most of these licks, by digging, furnish water which holds marine salt in solution, there exists no reason for believing, that many of them would produce nitre. Notwithstanding this low and alluvial tract appears in all respects well adapted to the growth of the long moss (tilandsia), none was observed since entering it in lat. 32° 52′; and as the pilot informed them, none would be seen in their progress up the river, it is probable that the latitude of 33° is here the *n.* limit of vegetation. The long-leaf pine, frequently the growth of rich and even inundated land, was here observed in great abundance: the short-leaved or pitch pine, on the contrary, is always found upon arid lands, and generally in sandy and lofty situations.

'This is the season when the poor settlers on the Washita turn out to make their annual hunt. The deer is now fat and the skins in perfection; the bear is now also in his best state, with regard to the quality of his fur, and the quantity of fat or oil he yields, as he has been feasting luxuriantly on the autumnal fruits of the forest. It is here well known, that he does not confine himself, as some writers have supposed, to vegetable food; he is particularly fond of hog's flesh: sheep and calves are frequently his prey, and no animal escapes him which comes within his power, and which he is able to conquer. He often destroys the fawn, when chance throws it in his way; he cannot, however, discover it by smelling, notwithstanding the excellence of his scent, for nature has, as if for its protection, denied the fawn the property of leaving any effluvium upon its track, a property so powerful in the old deer. The bear, unlike most other beasts of prey, does not kill the animal he has seized upon before he eats it; but, regardless of its struggles, cries, and lamentations, fastens upon, and, if the expression is allowable, devours it alive. The hunters count much on their profits from the oil drawn from the bear's fat, which, at New Orleans, is always of ready sale, and much esteemed for its wholesomeness in cooking, being preferred to butter or hog's lard. It is found to keep longer than any other animal oil, without becoming rancid; and boiling it, from time to time, upon sweet bay leaves, restores its sweetness, or facilitates its conservation.

'In the afternoon of the 17th they passed some sand beaches, and over a few rapids. They had cane brakes on both sides of the river; the canes were small, but demonstrated that the water does not surmount the bank more than a few feet. The river begins to widen as they advance; the banks of the river shew the high land soil, with a stratum] [of three or four feet of alluvion deposited by the river upon it. This superstratum is greyish, and very sandy, with a small admixture of loam, indicative of the poverty of the mountains and uplands where the river rises. Near this they passed through a new and very narrow channel, in which all the water of the river passes, except in time of freshes, when the interval forms an island. A little above this pass is a small clearing, called Cache la Tulipe (Tulip's hiding place); this is the name of a French hunter who here concealed his property. It continues the practice of both the white and red hunters to leave their skins, &c. often suspended to poles, or laid over a pole placed upon two forked posts, in sight of the river, until their return from hunting. These deposits are considered as sacred, and few examples exist of their being plundered. After passing the entrance of a bay, which within must form a great lake during the inundation, great numbers of the long-leaf pine were observed; and the increased size of the canes along the river's bank, denoted a better and more elevated soil; on the lett was a high hill (300 feet) covered with lofty pine trees.

'The banks of the river present more the appearance of upland soil, the under-stratum being a pale yellowish clay, and the alluvial soil of a dirty white, surmounted by a thin covering of a brown vegetable earth. The trees improve in appearance, growing to a considerable size and height, though yet inferior to those on the alluvial banks of the Mississippi. After passing the Bayau de Hachis, on the left, points of high land, not subject to be overflowed, frequently touch the river, and the valley is said to be more than a league in breadth on both sides. On the left are pine hills called Code de Champignole. The river is not more than 50 or 60 yards wide. On the morning of the 20th they passed a number of sand beaches, and some rapids, but found good depth of water between them. A creek called Chemin Convert, which forms a deep ravine in the high lands, here enters the river; almost immediately above this is a rapid where the water in the river is confined to a channel of about 40 yards in width: above it they had to quit the main channel, on account of the shallowness and rapidity of the water, and pass along a narrow channel of only 60 feet wide: without a guide, a stranger might take this passage for a creek.

'Notwithstanding the lateness of the season, and the *n.* latitude they were in, they this day met with an alligator. The banks of the river are covered with cane or thick under-brush, frequently so interwoven with thorns and briars as to be impenetrable. Birch, maple, holly, and two kinds of wood to which names have not yet been given, except water side wood, are here met with; as also persimons and small black grapes. The margin of the river is fringed with a variety of plants and vines, among which are several species of convolvulus.

'On the left they passed a hill and cliff 100 feet perpendicular, crowned with pines, and called Cote de Finn's (Fin's hill), from which a chain of high land continues some distance. The cliff presents the appearance of an ash-coloured clay. A little farther to the right is the Bayau d'Acasia (Locust creek). The river varies here from 80 to 100 yards in width, presenting frequent indications of iron along its banks, and some thin strata of iron ore. The ore is from half an inch to three inches in thickness.

'On the morning of the 22d of November, they arrived at the road of the Chadadoquis Indian nation, leading to the Arkansa nation; a little beyond this is the Ecor à Fabri (Fabri's cliffs) from 80 to 100 feet high; and a little distance above a smaller cliff called Le Petit Eeor à Fabri (the Little cliff of Fabri): these cliffs appear chiefly to be composed of ash-coloured sand, with a stratum of clay at the base, such as runs all along under the banks of this river. Above these cliffs are several rapids; the current is swifter, and denotes their ascent into a higher country; the water becomes clear, and equal to any in its very agreeable taste, and as drinking water. In the river are immense beds of gravel and sand, over which the water passes with great velocity in the season of its floods, carrying with it vast quantities of drift wood, which it piles up, in many places, to the height of 20 feet above the present surface, pointing out the difficulty and danger of navigation in certain times of the flood; accidents, however, are rare with the canoes of the country.

'As the party ascended they found the banks of the river less elevated, being only from nine to 12 feet: they are probably surmounted by the freshes some feet. The river becomes more obstructed by rapids, and sand and gravel beaches, among which are found fragments of stone of all forms, and a variety of colours, some highly polished and rounded by friction. The banks of the river in this upper country suffer greatly by abrasion, one side and sometimes both being broken down by every flood.

'At a place called Auges d' Arclon, (Arclon's troughs) is laminated iron ore, and a stratum of black sand, very tenacious, shining with minute crystals. The breadth of the river is here about]

[80 yards: in some places, however, it is enlarged by islands, in others contracted to 80 or 100 feet. Rocks of a greyish colour, and rather friable, are here found in many places on the river. On the banks grow willows of a different form from those found below, and on the margin of the Mississippi; the last are very brittle; these, on the contrary, are extremely pliant, resembling the osier, of which they are probably a species.

'At noon on the 24th, they arrived at the confluence of the Lesser Missouri with the Washita; the former is a considerable branch, perhaps the fourth of the Washita, and comes in from the left hand. The hunters often ascend the Little Missouri, but are not inclined to penetrate far up, because it reaches near the great plains or prairies upon the Red river, visited by the Lesser Osage tribes of Indians, settled on Arkansa; these last frequently carry war into the Chadadoquis tribe settled on the Red river, about *w. s. w.* from this place, and indeed they are reported not to spare any nation or people. They are prevented from visiting the head waters of the Washita by the steep hills in which they rise. These mountains are so difficult to travel over, that the savages not having an object sufficiently desirable, never attempt to penetrate to this river, and it is supposed to be unknown to the nation. The Chadadoquis (or Cadaux as the French pronounce the word) may be considered as Spanish Indians: they boast, and it is said with truth, that they never have imbrued their hands in the blood of a white man. It is said that the stream of the Little Missouri, some distance from its mouth, flows over a bright splendid bed of mineral of a yellowish white colour (most probably martial pyrites); that 30 years ago, several of the inhabitants, hunters, worked upon this mine, and sent a quantity of the ore to the government at New Orleans, and they were prohibited from working any more.

'There is a great sameness in the appearance of the river banks; the islands are skirted with osier, and immediately within, on the bank, grows a range of birch trees and some willows; the more elevated banks are covered with cane, among which grow the oak, maple, elm, sycamore, ash, hiccory, dog-wood, holly, ironwood &c. From the pilot they learned that there is a body of excellent land on the Little Missouri, particularly on the creek called the Bayau à Terre Noire, which falls into it. This land extends to Red river, and is connected with the great prairies which form the hunting grounds of the Cadaux nation, consisting of about 200 warriors. They are warlike, but frequently unable to defend themselves against the tribe of Osages, settled on the Arkansa river, who, passing round the mountains at the head of the Washita, and along the prairies which separated them from the main chain on the *w.* where the waters of the Red and Arkansa rivers have their rise, pass into the Cadaux country, and rob and plunder them.

'The water in the river Washita rising, the party are enabled to pass the numerous rapids and shoals which they meet with in the upper country; some of which are difficult of ascent. The general height of the main banks of the river is from six to 12 feet above the level of the water: the land is better in quality, the canes, &c. shewing a more luxuriant vegetation. It is subject to inundation, and shews a brown soil mixed with sand. Near Cache Macon (Maison's hiding place) on the right, they stopped to examine a supposed coal mine: Dr. Hunter, and the pilot, set out for this purpose, and about a mile and a half *n. w.* from the boat, in the bed of a creek, they found a substance similar to what they had before met with under that name, though more advanced towards a state of perfect coal. At the bottom of the creek, in a place then dry, were found detached pieces of from 50 to 100 pounds weight, adjoining to which lay wood, changing into the same substance. A stratum of this coal, six inches thick, lay on both sides of this little creek, over another of yellow clay, and covered by one foot of gravel; on the gravel are eight inches of loam, which bear a few inebes of vegetable mould. This stratum of coal is about three feet higher than the water in the creek, and appears manifestly to have been, at some period, the surface of the ground. The gravel and loam have been deposited there since, by the waters. Some pieces of this coal were very black and solid, of an homogeneous appearance, much resembling pit coal, but of less specific gravity. It does not appear sufficiently impregnated with bitumen, but may be considered as vegetable matter in the progress of transmutation to coal.

'Below the Bayau de l'Eau Froide, which runs into the Washita from the right, the river is 100 and 70 yards, flowing through tolerably good land. They passed a beautiful forest of pines, and on the 28th fell in with an old Dutch hunter and his party, consisting in all of five persons.

'This man has resided 40 years on the Washita, and before that period had been up the Arkansa river, the White river, and the St. Francis: the two last, he informs, are of difficult navigation, similar to the Washita, but the Arkansa river is] [of great magnitude, having a large and broad channel, and when the water is low, has great sand banks, like those in the Mississippi. So far as he has been up it, the navigation is safe and commodious, without impediments from rocks, shoals, or rapids; its bed being formed of mud and sand. The soil on it is of the first rate quality. The country is easy of access, being lofty open forests, unembarrrassed by canes or undergrowth. The water is disagreeable to drink, being of a red colour, and brackish when the river is low. A multitude of creeks which flow into the Arkansa furnish sweet water, which the voyager is obliged to carry with him for the supply of his immediate wants. This man confirms the accounts of silver being abundant up that river: he has not been so high as to see it himself, but says he received a silver pin from a hunter, who assured him that he himself collected the virgin silver from the rock, out of which he made the epinglete by hammering it out. The tribe of the Osage live higher up than this position, but the hunters rarely go so high, being afraid of these savages, who are at war with all the world, and destroy all strangers they meet with. It is reported that the Arkansa nation, with a part of the Choctaws, Chickasaws, Shawneese, &c. have formed a league, and are actually gone, or going, 800 strong, against these depredators, with a view to destroy or drive them entirely off, and possess themselves of their fine prairies, which are most abundant hunting grounds, being plentifully stocked with buffalo, elk, deer, bear, and every other beast of the chase common to those latitudes in America. This hunter having given information of a small spring in their vicinity, from which he frequently supplied himself with salt by evaporating the water, Dr. Hunter, with a party, accompanied him, on the morning of the 29th of November, to the place. They found a saline, about a mile and a half *n.* of the camp from whence they set out, and near a creek which enters the Washita a little above. It is situated in the bottom of the bed of a dry gully. The surrounding land is rich, and well timbered, but subject to inundation, except an Indian mount on the creek side, having a base of 80 or 100 feet diameter, and 20 feet high. After digging about three feet, through blue clay, they came to a quicksand, from which the water flowed in abundance: its taste was salt and bitter, resembling that of water in the ocean. In a second hole it required them to dig six feet before they reached the quicksand, in doing which they threw up several broken pieces of Indian pottery. The specific gravity, compared with the river, was, from the first pit, or that three feet deep, 1.02720, from the second pit, or that six feet deep, 1.02104, yielding a saline mass, from the evaporation of 10 quarts, which, when dry, weighed eight ounces; this brine is, therefore, about the same strength as that of the ocean on our coast, and twice the strength of the famous licks in Kentucky, called Bullet's Lick and Mann's Lick, from which so much salt is made.

'The Fourche de Cadaux (Cadadoquis fork), which they passed on the morning of the 30th, is about 100 yards wide at its entrance into the Washita, from the left; immediately beyond which, on the same side, the land is high, probably elevated 300 feet above the water. The shoals and rapids here impede their progress. At noon they deduced their latitude, by observation, to be 30° 11′ 37″ *n.* Receiving information of another salt lick, or saline, Dr. Hunter landed, with a party, to view it. The pit was found in a low flat place, subject to be overflowed from the river; it was wet and muddy, the earth on the surface yellow, but on digging through about four feet of blue clay, the salt water oozed from a quicksand. Ten quarts of this water produced, by evaporation, six ounces of saline mass, which, from taste, was principally marine salt; to the taste, however, it shewed an admixture of soda, and muriated magnesia, but the marine salt greatly preponderated. The specific gravity was about 1.076, probably weakened from the rain which had fallen the day before. The ascent of the river becomes troublesome, from the rapids and currents, particularly at the Isle du Bayau des Roches (Rocky Creek island), where it required great exertions, and was attended with some hazard, to pass them. This island is three-fourths of a mile in length. The river presents a series of shoals, rapids, and small cataracts; and they passed several points of high land, full of rocks and stones, much harder and more solid than they had yet met with.

'The rocks were all silicious, with their fissures penetrated by sparry matter. Indications of iron were frequent, and fragments of poor ore were common, but no rich ore of that, or any other metal, was found. Some of the hills appear well adapted to the cultivation of the vine; the soil being a sandy loam, with a considerable proportion of gravel, and a superficial covering of good vegetable black earth. The natural productions are, several varieties of oak, pine, dog-wood, holly, &c. with a scattering undergrowth of whortleberry, hawthorn, China briar, and a variety of small vines.

'Above the Isle de Mallon, the country wears]

[another prospect; high lands and rocks frequently approach the river. The rocks, in grain, resemble free stone, and are hard enough to be used as hand-mill stones, to which purpose they are frequently applied. The quality of the lands improves, the stratum of vegetable earth being from six to 12 inches, of a dark brown colour, with an admixture of loam and sand. Below Deer island they passed a stratum of free stone, 50 feet thick, under which is a quarry of imperfect slate in perpendicular layers. About a league from the river, and a little above the slate quarry, is a considerable plain, called Prairie de Champignole, often frequented by buffaloes. Some salt licks are found near it, and in many situations on both sides of this river, there are said to be salines, which may hereafter be rendered very productive, and from which the future settlements may be abundantly supplied.

'About four miles below the Chuttes (falls) they, from a good observation, found the latitude 34° 21′ 25″. The land on either hand continues to improve in quality, with a sufficient stratum of dark earth of a brownish colour. Hills frequently rise out of the level country, full of rocks and stones, hard and flinty, and often resembling Turkey oil stones. Of this kind was a promontory which came in from the right hand a little below the Chuttes; at a distance it presented the appearance of ruined buildings and fortifications, and several insulated masses of rock conveyed the idea of redoubts and out-works. This effect was heightened by the rising of a flock of swans, which had taken their station in the water at the foot of these walls. As the voyagers approached, the birds floated about majestically on the glassy surface of the water, and in tremulous accents seemed to consult upon means of safety. The whole was a sublime picture. In the afternoon of the third of December, the party reached the Chuttes, and found the falls to be occasioned by a chain of rocks of the same hard substance seen below, extending in the direction of *n. e.* and *s. w.* quite across the river. The water passes through a number of branches worn by the impetuosity of the torrent, where it forms so many cascades. The chain of rock or hill on the left appears to have been cut down to its present level by the abrasion of the waters. By great exertion, and lightening the boat, they passed the Chuttes that evening, and encamped just above the cataracts, and within the hearing of their incessant roar.

'Immediately above the Chuttes, the current of the water is slow, to another ledge of hard free stone; the reach between is spacious, not less than 200 yards wide, and terminated by a hill 300 feet high, covered with beautiful pines: this is a fine situation for building. In lat. 34° 25′ 48″ they passed a very dangerous rapid, from the number of rocks which obstruct the passage of the water, and break it into foam. On the right of the rapid is a high rocky hill covered with very handsome pine woods. The stratum of the rock has an inclination of 30° to the horizon, in the direction of the river descending. This hill may be 300 or 350 feet high: a border, or list, of green cane skirts the margin of the river, beyond which generally rises a high, and sometimes a barren hill. Near another rapid they passed a hill on the left, containing a large body of blue slate. A small distance above the Bayau de Saline they had to pass a rapid of 150 yards in length, and four feet and a half fall, which, from its velocity, the French have denominated La Cascade. Below the cascade there are rocky hills on both sides, composed of very hard free stone. The stone in the bed of the river, and which has been rolled from the upper country, was of the hardest flint, or of a quality resembling the Turkey oil stone. Fourche au Tigre, (Tyger's creek), which comes in from the right, a little above the cascade, is said to have many extensive tracts of rich level land upon it. The rocky hills here frequently approach the Washita on both sides; rich bottoms are nevertheless unfrequent, and the upland is sometimes of moderate elevation and tolerably level. The stones and rocks here met with have their fissures filled by sparry and crystalline matter.

'Wild turkeys become more abundant and less difficult of approach than below; and the howl of the wolves is heard during the night.

'To the Fourche of Calfat, (Caulker's creek) where the voyage terminates, they found level and good land on the right, and high hills on the left hand. After passing over a very precipitous rapid, seemingly divided into four steps or falls, one of which was at least 15 inches in perpendicular height, and which together could not be less than five and a half feet, they arrived at Ellis's camp, a small distance below the Fourche au Calfat, where they stopped on the sixth of December, as the pilot considered it the most convenient landing from whence to carry their necessary baggage to the hot springs, the distance being about three leagues. There is a creek about two leagues higher up, called Bayau des Sources Chauds, (Hot Spring creek) upon the banks of which the hot springs are situated at about two leagues from its mouth. The banks of it are hilly, and the road less eligible than from Ellis's camp.]

[' On ascending the hill, to encamp, they found the land very level and good, some plants in flower, and a great many evergreen vines; the forest oak with an admixture of other woods. The latitude of this place is 34° 27′ 31″. The ground on which they encamped was about 50 feet above the water in the river, and supposed to be 30 feet higher than the inundations. Hills of considerable height, and clothed with pine, were in view, but the land around, and extending beyond their view, lies handsomely for cultivation. The super-stratum is of a blackish-brown colour, upon a yellow basis, the whole intermixed with gravel and blue schistus, frequently so far decomposed as to have a strong aluminous taste. From their camp, on the Washita, to the hot springs, a distance of about nine miles, the first six miles of the road is in a *w.* direction without many sinuosities, and the remainder *n.* which courses are necessary to avoid some very steep hills. In this distance, they found three principal salt licks, and some inferior ones, which are all frequented by buffalo, deer, &c. The soil around them is a white tenacious clay, probably fit for potters' ware; hence the name of *glaise*, which the French hunters have bestowed upon most of these licks, frequented by the beasts of the forest, many of which exhibit no saline impregnation. The first two miles from the river Camp is over level land of the second rate quality; the timber chiefly oak, intermixed with other trees common to the climate, and a few scattered pines. Further on, the lands, on either hand, rise into gently swelling hills, covered with handsome pine woods. The road passes along a valley frequently wet by the numerous rills and springs of excellent water which issue from the foot of the hills. Near the hot springs the hills become more elevated, steeper of ascent, and rocky. They are here called mountains, although none of them in view exceed 4 or 500 feet in altitude. It is said that mountains of more than five times the elevation of these hills are to be seen in the *n. w.* towards the sources of the Washita. One of them is called the Glass, Crystal, or Shining mountain, from the vast number of hexagonal prisms of very transparent and colourless crystal which are found on its surface; they are generally surmounted by pyramids at one end, rarely on both. These crystals do not produce a double refraction of the rays of light. Many searches have been made over these mountains for the precious metals, but it is believed without success.

' At the hot springs they found an open log cabin, and a few huts of split boards, all calculated for summer encampment, and which had been erected by persons resorting to the springs for the recovery of their health.

' They slightly repaired these huts, or cabins, for their accommodation during the time of their detention at the springs, for the purpose of examining them and the surrounding country, and making such astronomical observations as were necessary for ascertaining their geographical positiou.

' It is understood that the hot springs are included within a grant of some hundred acres, granted by the late Spanish commandant of the Washita to some of his friends, but it is not believed that a regular patent was ever issued for the place; and it cannot be asserted that residence, with improvement of the land here, forms a plea upon which any claim to it can be founded.

' On their arrival they immediately tasted the waters of the hot springs, that is, after a few minutes cooling, for it was impossible to approach it with the lips when first taken up, without scalding: the taste does not differ from that of good water rendered hot by culinary fire.

' On the 10th they visited all the hot springs. They issue on the *e.* side of the valley, where the huts are, except one spring, which rises on the *w.* bank of the creek, from the sides and foot of a hill. From the small quantity of calcarious matter yet deposited, the *w.* spring does not appear to be of long standing; a natural conduit probably passes under the bed of the creek, and supplies it. There are four principal springs rising immediately on the *e.* bank of the creek, one of which may be rather said to spring out of the gravel-bed of the run; a fifth, a smaller one than that above-mentioned, as rising on the *w.* side of the creek; and a sixth, of the same magnitude, the most *n.* and rising near the bank of the creek; these are all the sources that merit the name of springs, near the huts; but there is a considerable one below; and all along, at intervals, the warm water oozes out, or drops, from the bank into the creek, as appears from the condensed vapour floating along the margin of the creek where the drippings occur.

' The hill from which the hot springs issue is of a conical form, terminating at the top with a few loose fragments of rock, covering a flat space 25 feet in diameter. Although the figure of the hill is conical, it is not entirely insulated, but connected with the neighbouring hills by a very narrow ridge. The primitive rock of this hill, above the base, is principally silicious, some part of it being of the hardest flint, others a free stone, extremely compact and solid, and of various colours. The base of the hill, and for a considerable extent,] [is composed of a blackish blue schistus, which divides into perpendicular lamina like blue slate. The water of the hot springs is, therefore, delivered from the silicious rock, generally invisible at the surface, from the mass of calcarious matter with which it is incrusted, or rather buried, and which is perpetually precipitating from the water of the springs: a small proportion of iron, in the form of a red calx, is also deposited; the colour of which is frequently distinguishable in the lime.

'In ascending the hill several patches of rich black earth are found, which appear to be formed by the decomposition of the calcarious matter: in other situations the superficial earth is penetrated, or incrusted, by limestone, with fine lamina or minute fragments of iron ore.

'The water of the hot springs must formerly have issued at a greater elevation in the hill, and run over the surface, having formed a mass of calcarious rock 100 feet perpendicular, by its deposition. In this high situation they found a spring, whose temperature was 140° of Fahrenheit's thermometer. After passing the calcarious region, they found the primitive hill covered by a forest of not very large trees, consisting chiefly of oak, pine, cedar, holly, hawthorn, and others common to the climate, with a great variety of vines, some said to produce black, and others yellow grapes, both excellent in their kinds. The soil is rocky, interspersed with gravel, sand, and fine vegetable mould. On reaching the height of 200 feet perpendicular, a considerable change in the soil was observable; it was stony and gravelly, with a superficial coat of black earth, but immediately under it lies a stratum of fat, tenacious, soapy, red clay, inclining to the colour of bright Spanish snuff, homogeneous, with scarcely any admixture of sand, no saline, but rather a soft agreeable taste: the timber diminishes, and the rocks increase in size to the summit. The whole height is estimated at 300 feet above the level of the valley.

'On examining the four principal springs, or those which yield the greatest quantity of water, or of the highest temperature, No. 1 was found to raise the mercury to 150°, No. 2 to 154°, No. 3 to 136°, and No. 4 to 132° of Fahrenheit's thermometer; the last is on the *w.* side of the creek: No. 3 is a small basin, in which there is a considerable quantity of green matter, having much the appearance of a vegetable body, but detached from the bottom, yet connected with it by something like a stem, which rests in calcarious matter. The body of one of these pseudo plants was from four to five inches in diameter; the bottom a smooth film of some tenacity, and the upper surface divided into ascending fibres of half, or three fourths of an inch long, resembling the gills of a fish, in transverse rows. A little further on was another small muddy basin, in which the water was warm to the finger: in it was a vermes about half an inch long, moving with a serpentine or vermicular motion. It was invariably observed, that the green matter forming on the stones and leaves covered a stratum of calcarious earth, sometimes a little hard or brittle, at others soft and imperfect. From the bottom of one of the hot springs a frequent ebullition of gas was observed, which not having the means of collecting, they could not ascertain its nature: it was not inflammable, and there is little doubt of its being carbonic acid, from the quantity of lime, and the iron, held in solution by the water.

'They made the following rough estimate of the quantity of water delivered by the springs. There are four principal springs, two of inferior note; one rising out of the gravel, and a number of drippings and drainings, all issuing from the margin, or from under the rock which overhangs the creek. Of the four first mentioned, three deliver nearly equal quantities, but No. 1, the most considerable, delivers about five times as much as one of the other three; the two of inferior note may, together, be equal to one; and all the drippings, and small springs, are probably under-rated at double the quantity of one of the three; that is, all together, they will deliver a quantity equal to eleven times the water issuing from the one most commodiously situated for measurement. This spring filled a vessel of 11 quarts in 11 seconds; hence the whole quantity of hot water delivered from the springs at the base of the hill is 165 gallons in a minute, or 3774 hogsheads in 24 hours, which is equal to a handsome brook, and might work an overshot mill. In cool weather condensed vapour is seen rising out of the gravel-bed of the creek, from springs which cannot be taken into account. During the summer and fall, the creek receives little or no water but what is supplied by the hot springs; at that season itself is a hot bath, too hot, indeed, near the springs; so that a person may choose the temperature most agreeable to himself, by selecting a natural basin near to, or farther from, the principal springs. At three or four miles below the springs the water is tepid and unpleasant to drink.

'From the *w.* mountain, estimated to be of equal height with that from which the hot springs flow, there are several fine prospects. The valley of the Washita, comprehended between the hills on] [either side, seemed a perfect flat, and about 12 miles wide. On all hands were seen the bills, or mountains, as they are here called, rising behind each other. In the direction of *n.* the most distant were estimated to be 50 miles off, and are supposed to be those of the Arkansa river, or the rugged mountains which divide the waters of the Arkansa from those of the Washita, and prevent the Osage Indians from visiting the latter, of whom they are supposed ignorant; otherwise their excursions here would prevent this place from being visited by white persons, or other Indians. In a *s. w.* direction, at about 40 miles distance, is seen a perfectly level ridge, supposed to be the high prairies of the Red river.

'Notwithstanding the severity of the weather, a considerable number, and some variety of plants were in flower, and others retained their verdure: indeed the ridge was more temperate than the valley below; there it was cold, damp, and penetrating; here dry, and the atmosphere mild. Of the plants growing here was a species of cabbage; the plants grow with expanded leaves, spreading on the ground, of a deep green, with a shade of purple; the taste of the cabbage was plainly predominant, with an agreeable warmth inclining to that of the radish; several tap-roots penetrated into the soil of a white colour, having the taste of horse-radish, but much milder. A quantity of them taken to the camp and dressed, proved palatable and mild. It is not probable that cabbage seed has been scattered on this ridge; the hunters ascending this river have always had different objects. Until further elucidation, this cabbage must be considered as indigenous to this sequestered quarter, and may be denominated the cabbage radish of the Washita. They found a plant, then green, called by the French racine rouge, (red root), which is said to be a specific in female obstructions; it has also been used, combined with the China root, to dye red; the last probably acting as a mordant. The top of this ridge is covered with rocks of a flinty kind, and so very hard as to be improper for gun-flints, for when applied to that use it soon digs cavities in the hammer of the lock. This hard stone is generally white, but frequently clouded with red, brown, black, and other colours. Here and there fragments of iron stone were met with, and where a tree had been overturned, its roots brought to view fragments of schistus, which were suffering decomposition from exposure to the atmosphere. On digging where the slope of the bill was precipitous, they found the second stratum to be a reddish clay, resembling that found on the conical hill *e.* of the camp. At two-thirds down the hill, the rock was a hard free-stone, intermixed with fragments of flint which had probably rolled from above. Still lower was found a blue schistus, in a state tending to decomposition where exposed to the atmosphere, but hard and resembling coarse slate in the interior. Many stones had the appearance of Turkey oil stones: at the foot of the hill the country expands into good farming lands.

'Dr. Hunter, upon examining the waters of the hot springs, obtained the following results:

'It differed nothing from the hot water in smell or taste, but caused a slight eructation shortly after drinking it.

'Its specific gravity is equal to rain or distilled water.

'It gave to litmus paper a slight degree of redness, evincing the presence of the carbonic acid, or fixed air sulphuric, and threw down a few detached particles. Oxylat of ammoniac caused a deposition and white cloud, shewing the presence of a small portion of lime. Prusiat of potash produced a slight and scarcely perceptible tinge of blue, designating the presence of a small quantity of iron.

'Sixteen pounds of water evaporated to dryness left ten grains of a grey powder, which proved to be lime.

'The myrtle wax tree grows in the vicinity of the springs. At the season in which the voyagers were there, the wax was no longer green, but had changed its colour to a greyish white, from its long exposure to the weather. The berry, when examined by a microscope, is less than the smallest garden pea, approaching to an oval in form. The nucleus, or real seed, is the size of the seed of a radish, and is covered with a number of kidney-shaped glands, of a brown colour and sweet taste; these glands secrete the wax, which completely envelops them, and at this season gives to the whole the appearance of an imperfectly white berry. This is a valuable plant and merits attention: its favourite position is a dry soil, rather poor, and looking down upon the water. It is well adapted to ornament the margins of canals, lakes, or rivulets. The cassina yapon is equally beautiful, and proper for the same purpose; it grows here along the banks of this stony creek, intermingled with the myrtle, and bears a beautiful little red berry, very much resembling the red currant.

'The rock through which the hot springs either pass or trickle over, appears undetermined by the waters of the creek. The hot water is continually depositing calcarious, and perhaps some silicious matter, forming new rocks, always augmenting] [and projecting their promontories over the running water of the creek, which prevents its formation below the surface. Wherever this calcarious crust is seen spreading over the bank and margin of the creek, there, most certainly, the hot water will be found, either running over the surface, or through some channel, perhaps below the new rock, or dropping from the edges of the overhanging precipice. The progress of nature in the formation of this new rock is curious, and worthy the attention of the mineralogist. When the hot water issues from the fountain, it frequently spreads over a superficies of some extent: so far as it reaches on either hand, there is a deposition or growth of green matter. Several lamina of this green matter will be found lying over each other; and immediately under, and in contact with the inferior lamina, which is not thicker than paper, is found a whitish substance resembling a coagulum; when viewed with a microscope, this last is also found to consist of several, sometimes a good number of lamina, of which that next the green is the finest and thinnest, being the last formed; those below increasing in thickness and tenacity until the last terminates in a soft earthy matter, which reposes in the more solid rock. Each lamina of the coagulum is penetrated in all its parts by calcarious grains, extremely minute, and divided in the more recent web, but much larger and occupying the whole of the inferior lamina. The understratum is continually consolidating, and adding bulk and height to the rock. When this acquires such an elevation as to stop the passage of the water, it finds another course over the rock, hill, or margin of the creek, forming, in turn, accumulations of matter over the whole of the adjacent space. When the water has found itself a new channel, the green matter, which sometimes acquires a thickness of half an inch, is speedily converted into a rich vegetable earth, and becomes the food of plants. The surface of the calcarious rock also decomposes and forms the richest black mould, intimately mixed with a considerable portion of soil; plants and trees vegetate luxuriantly upon it.

'On examining a piece of ground upon which the snow dissolved as it fell, and which was covered with herbage, they found, in some places, a calcarious crust on the surface; but in general a depth of from five inches to a foot of the richest black mould. The surface was sensibly warm to the touch. In the air the mercury in the thermometer stood at 44°; when placed four inches under the surface, and covered with earth, it rose rapidly to 68°; and upon the calcarious rock, eight inches beneath the surface, it rose to 80°. This result was uniform over the whole surface, which was about a quarter of an acre.

'On searching they found a spring, about 15 inches under the surface, in the water of which the thermometer shewed a temperature of 150°. Beneath the black mould was found a brown mixture of lime and silex, very loose and divisible, apparently in a state of decomposition, and progressing towards the formation of black mould; under this brownish mass it became gradually whiter and harder, on the depth of from six to 12 inches, where it was a calcarious sparkling stone. It was evident that the water had passed over this place, and formed a flat superficies of silicious lime-stone: and that its position, nearly level, had facilitated the accumulation of earth, in proportion as the decomposition advanced. Similar spots of ground were found higher up the hill, resembling little savannas, near which hot springs were always discovered, which had once flowed over them. It appears probable that the hot water of the springs, at an early period, had all issued from its grand reservoir in the hill, at a much greater elevation than at present. The calcarious crust may be traced up, in most situations on the *w.* side of the hill looking down the creek and valley, to a certain height, and perhaps 100 feet perpendicular; in this region the hill rises precipitously, and is studded with hard silicious stones; below, the descent is more gradual, and the soil a calcarious black earth. It is easy to discriminate the primitive hill from that which has accumulated, by precipitation, from the water of the springs; this last is entirely confined to the *w.* side of the hill, and washed at its base by the waters of the creek, no hot spring being visible in any other part of its circumference. By actual measurement along the base of the hill, the influence of the springs is found to extend 70 perches, in a direction a little to the *e.* of *n.*: along the whole of this space the springs have deposited stony matter, calcarious, with an addition of silex, or crystallized lime. The accumulation of calcarious matter is more considerable at the *n.* end of the hill than the *s.*; the first may be above 100 feet perpendicular, but sloping much more gradually than the primitive hill above, until it approaches the creek, where not unfrequently it terminates in a precipice of from six to 20 feet. The difference between the primitive and secondary hill is so striking that a superficial observer must notice it; the first is regnlarly very steep, and studded with rock and stone of the hardest flint, and other silicious compounds, and a superficies] [of two or three inches of good mould covers a red clay; below, on the secondary bill, which carries evident marks of recent formation, no flint or silicious stone is found; the calcarious rock conceals all from view, and is itself frequently covered by much fine rich earth. It would seem that this compound, precipated from the hot waters, yields easily to the influence of the atmosphere; for where the waters cease to flow over any portion of the rock, it speedily decomposes; probably more rapidly from the heat communicated from the interior part of the bill, as insulated masses of the rock are observed to remain without change.

'The cedar, the wax myrtle, and cassina yapon, all evergreens, attach themselves particularly to the calcarious region, and seem to grow and thrive even in the clefts of the solid rock.

'A spring, enjoying a freedom of position, proceeds with great regularity in depositing the matter it holds in solution; the border or rim of its basin forms an elevated ridge, from whence proceeds a glacis all around, where the waters have flowed for some time over one part of the brim; this becomes more elevated, and the water has to seek a passage where there is less resistance; thus forming, in miniature, a crater, resembling in shape the conical summit of a volcano. The hill being steep above, the progress of petrifaction is stopped on that side, and the waters continue to flow and spread abroad, incrusting the whole face of the hill below. The last formed calcarious border of the circular basin is soft, and easily divided: at a small depth it is more compact; and at the depth of six inches it is generally hard white stone. If the bottom of the basin is stirred up, a quantity of the red calx of iron rises, and escapes over the summit of the crater.

'Visitants to the hot springs, having observed shrubs and trees with the roots in the hot water, have been induced to try experiments, by sticking branches of trees in the run of hot water. Some branches of the wax myrtle were found thrust into the bottom of a spring run, the water of which was 130° by Fahrenheit's thermometer; the foliage and fruit of the branch were not only sound and healthy, but, at the surface of the water, roots were actually sprouting from it: on pulling it up the part which had penetrated the hot mud was found decayed.

'The green substance discoverable at the bottom of the hot springs, and which at first sight has the appearance of plush, on examination by the microscope, was found to be a vegetable production. A film of green matter spreads itself on the calcarious base, from which rise fibres more than half an inch in length, forming a beautiful vegetation. Before the microscope it sparkled with innumerable nodules of lime, some part of which was beautifully crystallized. This circumstance might cause a doubt of its being a true vegetable, but its great resemblance to some of the mosses, particularly the byssi, and the discovery which Mr. Dunbar made of its being the residence of animal life, confirmed his belief of its being a true moss. After a diligent search he discovered a very minute shell-fish, of the bivalve kind, inhabiting this moss; its shape nearly that of the fresh water muscle; the colour of the shell a greyish brown, with spots of a purplish colour. When the animal is undisturbed it opens the shell, and thrusts out four legs, very transparent, and articulated like those of a quadruped; the extremities of the fore legs are very slender and sharp, but those of the hind legs somewhat broader, apparently armed with minute toes: from the extremity of each shell issues three or four forked hairs, which the animal seems to possess the power of moving; the fore legs are probably formed for making incisions into the moss for the purpose of procuring access to the juices of the living plant, upon which, no doubt, it feeds; it may be provided with a proboscis, although it did not appear while the animal was under examination: the hind legs are well adapted for propelling in its progress over the moss, or through the water.

'It would be desirable to ascertain the cause of that perpetual fire which keeps up the high temperature of so many springs as flow from this hill, at a considerable distance from each other: upon looking around, however, sufficient data for the solution of the difficulty are not discoverable. Nothing of a volcanic nature is to be seen in this country; neither could they learn that any evidence in favour of such a supposition was to be found in the mountains connected with this river. An immense bed of dark blue schistus appears to form the base of the hot spring hill, and of all those in its neighbourhood: the bottom of the creek is formed of it; and pieces are frequently met with rendered soft by decomposition, and possessing a strong aluminous taste, requiring nothing but lixiviation and crystallization to complete the manufacture of alum. As bodies undergoing chemical changes generally produce an alteration of temperature, the heat of these springs may be owing to the disengagement of caloric, or the decomposition of the schistus. Another, and perhaps a more satisfactory cause may be assigned: it is well known, that within the circle of] [the waters of this river, vast beds of martial pyrites exist: they have not yet, however, been discovered in the vicinage of the hot springs, but may, nevertheless, form immense beds under the bases of these hills; and as in one place at least, there is evidence of the presence of bitumen, the union of these agents will, in the progress of decomposition, by the admission of air and moisture, produce degrees of heat capable of supporting the phenomena of the hot springs. No sulphuric acid is present in this water; the springs may be supplied by the vapour of heated water, ascending from caverns where the heat is generated, or the heat may be immediately applied to the bottom of an immense natural caldron of rock, contained in the bowels of the hill, from which, as a reservoir, the springs may be supplied.

'A series of accurate observations determined the latitude of the hot springs to be 34° 31′ 4.16″ *n.* and long. 6 h. 11′ 25″, or 92° 50′ 45″ *w.* from the meridian of Greenwich.

'While Mr. Dunbar was making arrangements for transporting the baggage back to the river camp, Dr. Hunter, with a small party, went on an excursion into the country. He left the hot springs on the morning of the 27th, and after travelling sometimes over hills and deep craggy mountains, with narrow valleys between them, then up the valleys, and generally by the side of a branch emptying into the Washita, they reached the main branch of the Calfat in the evening, about 12 miles from the springs. The stones they met with during the first part of the day were silicions, of a whitish grey, with flints, white, cream-coloured, red, &c. The beds of the rivulets, and often a considerable way up the hills, shewed immense bodies of schistus, both blue and grey, some of it efflorescing and tasting strongly of alum. The latter part of the day, they travelled over and between hills of black, hard, and compact flint in shapeless masses, with schist as before. On ascending these high grounds, you distinctly perceive the commencement of the piney region, beginning at the height of 60 or 70 feet, and extending to the top. The soil in these narrow valleys is thin and full of stones. The next day, which was stormy, they reached a branch of the Bayan de Saline, which stretches towards the Arkansa, and empties into the Washita many leagues below, having gone above 12 miles. The mountains they had passed being of the primitive kind, which seldom produce metals, and having hitherto seen nothing of a mineral kind, a little poor iron ore excepted, and the face of the country, as far as they could see, presenting the same aspect, they returned to the camp and the hot springs, on the evening of the 30th, by another route, in which they met with nothing worthy notice.

'In consequence of the rains which had fallen, Mr. Dunbar, and those who were transporting the baggage to the river camp, found the road watery. The soil on the flat lands, under the stratum of vegetable mould, is yellowish, and consists of decomposed schistus, of which there are immense beds in every stage of dissolution, from the hard stone recently uncovered and partially decomposed, to the yellow and apparently homogeneous earth. The covering of vegetable earth between the hills and the river is, in most places, sufficiently thick to constitute a good soil, being from four to six inches; and it is the opinion of the people upon the Washita, that wheat will grow here to great perfection. Although the higher hills, 300 to 600 feet in height, are very rocky, yet the inferior hills, and the sloping bases of the first, are generally covered with a soil of a middling quality. The natural productions are sufficiently luxuriant, consisting chiefly of black and red oak, intermixed with a variety of other woods, and a considerable undergrowth. Even on these rocky hills are three or four species of vines, said to produce annually an abundance of excellent grapes. A great variety of plants which grow here, some of which in their season are said to produce flowers highly ornamental, would probably reward the researches of the botanist.

'On the morning of the 8th of January 1805, the party left Ellis's on the river camp; where they had been detained for several days, waiting for such a rise in the waters of the river, as would carry their boat in safety over the numerous rapids below. A rise of about six feet, which had taken place the evening before, determined them to move this morning; and they passed the Chuttes about one o'clock. They stopped to examine the rocky promontory below these falls, and took some specimens of the stone which so much resembles the Turkey oil stone. It appears too hard. The strata of this chain were observed to run perpendicularly nearly *e.* and *w.* crossed by fissures at right angles from five to eight feet apart; the lamina from one-fourth of an inch to five inches in thickness. About a league below, they landed at Whetstone hill, and took several specimens. This projecting hill is a mass of greyish blue schistus of considerable hardness, and about 20 feet perpendicular, not regularly so, and from a quarter to two inches in thickness, but does not split with an even surface.

'They landed again on the morning of the 9th,]

[in sight of the Bayau de la Prairie de Champignole, to examine and take specimens of some freestone and blue slate. The slate is a blue schistus, hard, brittle, and unfit for the covering of a house: none proper for that purpose have been discovered, except on the Calfat, which Dr. Hunter met with in one of his excursions.

'On the evening of the 10th they encamped near Arclon's troughs, having been only three days in descending the distance which took them 13 to ascend. They stopped some time at the camp of a Mr. Le Fevre. He is an intelligent man, a native of the Illinois, but now residing at the Arkansas. He came here with some Delaware and other Indians, whom he had fitted out with goods, and receives their peltry, fur, &c. at a stipulated price, as it is brought in by the hunters. Mr. Le Fevre possesses considerable knowledge of the interior of the country; he confirms the accounts before obtained, that the bills or mountains which give rise to this little river, are, in a manner, insulated; that is, they are entirely shut in and inclosed by the immense plains or prairies which extend beyond the Red river, to the *s.* and beyond the Missouri, or at least some of its branches, to the *n.* and range along the *e.* base of the great chain, or dividing ridge, commonly known by the name of the Sand hills, which separate the waters of the Mississippi from those which fall into the Pacific ocean. The breadth of this great plain is not well ascertained. It is said by some to be at certain parts, or in certain directions, not less than 200 leagues; but it is agreed by all who have a knowledge of the *w.* country, that the mean breadth is at least two-thirds of that distance. A branch of the Missouri, called the river Plate or Shallow river, is said to take its rise so far *s.* as to derive its first waters from the neighbourhood of the sources of the Red and Arkansa rivers. By the expression plains or prairie, in this place, is not to be understood a dead flat, resembling certain savannas, whose soil is stiff and impenetrable, often under water, and bearing only a coarse grass resembling reeds; very different are the *w.* prairies, which expression signifies only a country without timber. These prairies are neither flat nor hilly, but undulating into gentle swelling lawns, and expanding into spacious valleys, in the centre of which is always found a little timber growing on the banks of the brooks and rivulets of the finest waters.

'The whole of these prairies are represented to be composed of the richest and most fertile soil; the most luxuriant and succulent herbage covers the surface of the earth, interspersed with millions of flowers and flowering shrubs, of the most ornamental kinds. Those who have viewed only a skirt of these prairies, speak of them with enthusiasm, as if it was only there that nature was to be found truly perfect: they declare, that the fertility and beauty of the rising grounds, the extreme richness of the vales, the coolness and excellent quality of the water found in every valley, the salubrity of the atmosphere, and above all the grandeur of the enchanting landscape which this country presents, inspire the soul with sensations not to be felt in any other region of the globe. This paradise is now very thinly inhabited by a few tribes of savages, and by the immense herds of wild cattle (bison) which people these countries. The cattle perform regular migrations, according to the seasons, from *s.* to *n.* and from the plains to the mountains; and in due time, taught by their instincts, take a retrograde direction.

'The Indian tribes move in the rear of the herds, and pick up stragglers, and such as lag behind, which they kill with the bow and arrow for their subsistence. This country is not subjected to those very sudden deluges of rain which in most hot countries, and even in the Mississippi territory, tear up and sweep away, with irresistible fury, the crop and soil together: on the contrary, rain is said to become more rare in proportion as the great chain of mountains is approached; and it would seem that within the sphere of the attraction of those elevated ridges, little or no rain falls on the adjoining plains. This relation is the more credible, as in that respect the new country of the United States resembles other flat or low countries similarly situated; such as the country lying between the Andes and the *w.* Pacific. The plains are supplied with nightly dews so extremely abundant, as to have the effect of refreshing showers of rain; and the spacious valleys, which are extremely level, may, with facility, be watered by the rills and brooks, which are never absent from these situations. Such is the description of the better known country lying to the *s.* of Red river, from Nacogdoches towards St. Antonio, in the province of Taxus; the richest crops are said to be procured there without rain; but agriculture in that country is at a low ebb; the small quantity of maize furnished by the country, is said to be raised without cultivation. A rude opening is made in the earth, sufficient to deposit the grain, at the distance of four or five feet, in irregular squares, and the rest is left to nature. The soil is tender, spongy, and rich, and seems always to retain humidity sufficient, with the bounteous dews of heaven, to bring the crops to maturity.]

['The Red and Arkansa rivers, whose courses are very long, pass through portions of this fine country. They are both navigable to an unknown distance by boats of proper construction; the Arkansa river is, however, understood to have greatly the advantage with respect to the facility of navigation. Some difficult places are met with in the Red river below the Nakitosh, after which it is good for 150 leagues (probably computed leagues of the country, about two miles each); there the voyager meets with a very serious obstacle, the commencement of the "raft," as it is called; that is, a natural covering which conceals the whole river for an extent of 17 leagues, continually augmenting by the driftwood brought down by every considerable fresh. This covering, which, for a considerable time was only driftwood, now supports a vegetation of every thing abounding in the neighbouring forest, not excepting trees of a considerable size; and the river may be frequently passed without any knowledge of its existence. It is said that the annual inundation is opening for itself a new passage through the low grounds near the hills; but it must be long before nature, unaided, will excavate a passage sufficient for the waters of Red river. About 50 leagues above this natural bridge, is the residence of the Cadaux or Cadadoquis nation, whose good qualities are already mentioned. The inhabitants estimate the post of Nakitosh to be half way between New Orleans and the Cadaux nation. Above this point the navigation of Red river is said to be embarrassed by many rapids, falls, and shallows. The Arkansa river is said to present a safe, agreeable, and uninterrupted navigation, as high as it is known. The lands on each side are of the best quality, and well watered with springs, brooks, and rivulets, affording many situations for mill-seats. From description it would seem that along this river there is a regular gradation of hill and dale, presenting their extremities to the river; the hills are gently swelling eminences, and the dales spacious valleys with living water meandering through them; the forests consist of handsome trees, chiefly what is called open woods. The quality of the land is supposed superior to that on Red river, until it ascends to the prairie country, where the lands on both rivers are probably similar.

'About 200 leagues up the Arkansa is an interesting place called the Salt prairie: there is a considerable fork of the river there, and a kind of savanna where the salt-water is continually oozing out and spreading over the surface of a plain. During the dry summer season the salt may be raked up in large heaps; a natural crust, of a hand breadth in thickness, is formed at this season. This place is not often frequented, on account of the danger from the Osage Indians: much less dare the white hunters venture to ascend higher, where it is generally believed that silver is to be found. It is further said, that high up the Arkansa river salt is found in form of a solid, and may be dug out with the crow-bar. The waters of the Arkansa, like those of Red river, are not potable during the dry season, being both charged highly with a reddish earth or mould, and extremely brackish.

'This inconvenience is not greatly felt upon the Arkansa, where springs and brooks of fresh water are frequent; the Red river is understood not to be so highly favoured. Every account seems to prove that immense natural magazines of salt must exist in the great chain of mountains to the *w.*; as all the rivers, in the summer season, which flow from them, are strongly impregnated with that mineral, and are only rendered palatable after receiving the numerous streams of fresh water which join them in their course. The great *w.* prairies, besides the herds of wild cattle, (bison, commonly called buffalo), are also stocked with vast numbers of wild goat (not resembling the domestic goat), extremely swift-footed. As the description given of this goat is not perfect, it may from its swiftness prove to be the antelope, or it possibly may be a goat which has escaped from the Spanish settlements of New Mexico. A Canadian, who had been much with the Indians to the *w.* speaks of a wool-bearing animal larger than a sheep, the wool much mixed with hair, which he had seen in large flocks. He pretends also to have seen a unicorn, the single horn of which, he says, rises out of the forehead and curls back, conveying the idea of the fossil cornn ammonis. This man says he has travelled beyond the great dividing ridge so far as to have seen a large river flowing to the *w.* The great dividing mountain is so lofty that it requires two days to ascend from the base to its top: other ranges of inferior mountains lie before and behind it; they are all rocky and sandy. Large lakes and valleys lie between the mountains. Some of the lakes are so large as to contain considerable islands; and rivers flow from some of them. Great numbers of fossil bones, of very large dimensions, are seen among the mountains, which the Canadian supposes to be the elephant.

'He does not pretend to have seen any of the precious metals, but has seen a mineral which he supposes might yield copper. From the top of the high mountain the view is bounded by a curve, as] [upon the ocean, and extends over the most beautiful prairies, which seem to be unbounded, particularly towards the *e.* The finest of the lands he has seen are on the Missouri; no other can compare in richness and fertility with them. This Canadian, as well as Le Fevre, speaks of the Osages of the tribe of Whitehairs, as lawless and unprincipled; and the other Indian tribes hold them in abhorrence as a barbarous and uncultivated race, and the different nations who hunt in their neighbourhood, have their concerting plans for their destruction. On the morning of the 11th, the party passed the Petit Ecor à Fabri. The osier which grows on the beaches above, is not seen below upon the river; and here they began to meet with the small tree called Charnier, which grows only on the water side, and is met with all the way down the Washita. The latitude of 33° 40″ seems the *n.* boundary of the one, and the *s.* boundary of the other of these vegetables. Having noticed the limit set to the long moss, (telandsia) on the ascent of the river, in lat. 33°, Mr. Dunbar made inquiry of Mr. Le. Fevre, as to its existence on the Arkansa settlement, which is known to lie in about the same parallel: he said, that its growth is limited about 10 miles *s.* of the settlement, and that as remarkably, as if a line had been drawn *e.* and *w.* for the purpose; as it ceases all at once, and not by degrees. Hence it appears, that nature has marked with a distinguishing feature, the line established by congress between the Orleans and Louisiana territories. The cypress is not found on the Washita higher than lat. 34° *n.*

'In descending the river, they found their rate of going to exceed that of the current about six miles and a half in 24 hours; and that on the 12th, they had passed the apex of the tide or wave occasioned by the fresh, and were descending along an inclined plain; as they encamped at night, they found themselves in deeper water the next morning, and on a more elevated part of the inclined plain, than they had been in the preceding evening, from the progress of the apex of the tide during their repose.

'At noon, on the 16th, they reached the post of the Washita.

'Mr. Dunbar being anxious to reach the Natchez as early as possible, and being unable to procure horses at the post, took a canoe with one soldier and his own domestic, to push down to the Cataboola, from whence to Concord there is a road of 30 miles across the low grounds. He set off early on the morning of the 20th, and at night reached the settlement of an old hunter, with whom he had conversed on his way up the river. This man informed him, that at the place called the Mine, on the Little Missouri, there is a smoke which ascends perpetually from a particular place, and that the vapour is sometimes insupportable. The river, or a branch of it, passes over a bed of mineral, which from the description given is no doubt martial pyrites. In a creek or branch of the Fourche à Luke, there is found on the beaches and in the cliffs, a great number of globular bodies, some as large, or larger, than a man's head, which, when broken, exhibit the appearance of gold, silver, and precious stones; most probably pyrites and crystallized spar. And at the Fourche des Glaises à Paul, (higher up the river than Fourche à Luke,) near the river there is a cliff full of hexagonal prisms, terminated by pyramids which appear to grow out of the rock: they are from six to eight inches in length, and some of them are an inch in diameter. There are beds of pyrites found in several small creeks communicating with the Washita, but it appears that the mineral indications are greatest on the Little Missouri; because, as before noted, some of the hunters actually worked on them, and sent a parcel of the ore to New Orleans. It is the belief here, that the mineral contains precious metal, but that the Spanish government did not choose a mine should be opened so near to the British settlements. An express prohibition was issued against working these mines.

'At this place, Mr. Dunbar obtained one or two slips of the bois de arc, (bow wood or yellow wood), from the Missouri. The fruit, it seems, had fallen before the maturity, and lay upon the ground. Some were the size of a small orange, with a rind full of tubercles; the colour, though it appeared faded, still retained a resemblance to pale gold.

'The tree in its native soil, when laden with its golden fruit, (nearly as large as the egg of an ostrich,) presents the most splendid appearance; its foliage is of a deep green, resembling the varnished leaf of the orange tree; upon the whole, no forest tree can compare with it in ornamental grandeur. The bark of the young tree resembles, in texture, the dog wood bark; the appearance of the wood recommends it for trial as an article which may yield a yellow dye. It is deciduous; the branches are numerous, and full of short thorns or prickles, which seem to point it out as proper for hedges or live fences. This trece is known to exist near the Nakitosh (perhaps in lat. 32°) and upon the river Arkansa, high up (perhaps in lat. 36°); it is therefore probable that it may thrive from] [latitude 38° to 40°, and will be a great acquisition to the United States if it possesses no other merit than that of being ornamental.

'In descending the river, both Mr. Dunbar and Dr. Hunter searched for the place said to yield gypsum, or plaster of Paris, but failed. The former gentleman states, that he has no doubt of its existence, having noted two places where it has been found; one of which is the first hill or high land which touches the river on the *w.* above the Bayau Calumet, and the other is the second high land on the same side. As these are two points of the same continued ridge, it is probable that an immense body of gypsum will be found in the bowels of the hills where they meet, and perhaps extending far beyond them.

'On the evening of the 22d, Mr. Dunbar arrived at the Catahoola, where a Frenchman of the name of Hebrard, who keeps the ferry across Black river, is settled. Here the road from the Washita forks, one branch of it leading to the settlement on Red river, and the other up to the post on the Washita. The proprietor of this place has been a hunter and a great traveller up the Washita into the *w.* country; he confirms generally the accounts received from others. It appears, from what they say, that in the neighbourhood of the hot springs, but higher up, among the mountains, and upon the Little Missouri, during the summer season, explosions are very frequently heard, proceeding from under the ground, and not rarely a curious phenomenon is seen, which is termed the blowing of the mountains; it is confined elastic gas forcing a passage through the side or top of a hill, driving before it a great quantity of earth and mineral matter. During the winter season the explosions and blowing of the mountains entirely cease, from whence we may conclude, that the cause is comparatively superficial, brought into action by the increased heat of the more direct rays of the summer sun.

'The confluence of the Washita, Catahoola, and Tenza is an interesting place. The last of these communicates with the Mississippi low lands, by the intervention of other creeks and lakes, and by one in particular, called Bayau d'Argent, which empties into the Mississippi, about 14 miles above Natchez. During high water there is a navigation for batteaux of any burthen along the bayau. A large lake, called St. John's lake, occupies a considerable part of the passage between the Mississippi and the Tenza: it is in a horse-shoe form, and has, at some former period, been the bed of the Mississippi: the nearest part of it is about one mile removed from the river at the present time. This lake, possessing elevated banks similar to those of the river, has been lately occupied and improved. The Catahoola Bayau is the third navigable stream: during the time of the inundation there is an excellent communication by the lake of that name, and from thence, by large creeks, to the Red river. The country around the point of union of these three rivers is altogether alluvial, but the place of Mr. Hebrard's residence is no longer subject to inundation. There is no doubt, that as the country augments in population and riches, this place will become the site of a commercial inland town, which will keep pace with the progress and prosperity of the country. One of the Indian mounts here is of a considerable elevation, with a species of rampart, surrounding a large space, which was, no doubt, the position of a fortified town.

'While here, Mr. Dunbar met with an American who pretended to have been up the Arkansa river 300 leagues. The navigation of this river he says is good to that distance, for boats drawing three or four feet water. Implicit faith, perhaps, ought not to be given to his relation, respecting the quantity of silver he pretends to have collected there. He says he has found silver on the Washita, 30 leagues above the hot springs, so rich, that three pounds of it yielded one pound of silver, and this was found in a cave. He asserts, also, that the ore of the mine upon the Little Missouri was carried to Kentucky, by a person of the name of Bon, where it was found to yield largely in silver. This man says he has been up the Red river likewise, and that there is a great rapid just below the raft, or natural bridge, and several others above it; that the Caddo nation is about 50 leagues above the raft, and near to their village commences the country of the great prairies, which extend 4 or 500 miles to the *w.* of the sand mountains, as they are termed. These great plains reach far beyond the Red river to the *s.* and *n.* over the Arkansa river, and among the numerous branches of the Missouri. He confirms the account of the beauty and fertility of the *w.* country.

'On the morning of the 25th, Mr. Dunbar set out, on horseback, from the Catahoola to Natchez. The rain which had fallen on the preceding days rendered the roads wet and muddy, and it was two in the afternoon before he reached the Bayau Crocodile, which is considered half way between the Black river and the Mississippi. It is one of the numerous creeks in the low grounds, which assist in venting the waters of the inundation. On the margins of the water courses the lands are highest, and produces canes; they fall off, in the] [rear, into cypress swamps and lakes. The waters of the Mississippi were rising, and it was with some difficulty that they reached a house near Concord that evening. This settlement was begun since the cession of Louisiana to the United States, by citizens of the Mississippi territory, who have established their residence altogether upon newly acquired lands taken up under the authority of the Spanish commandant, and have gone to the expence of improvement either in the names of themselves or others, before the 20th of December 1803, hoping thereby to hold their new possession under the sanction of the law.

' Exclusive of the few actual residents on the banks of the Mississippi, there are two very handsome lakes in the interior, on the banks of which similar settlements have been made. He crossed at the ferry, and at mid-day of the 26th reached his own house.

' Dr. Hunter, and the remainder of the party, followed Mr. Dunbar, down the Washita, with the boat in which they ascended the river, and ascending the Mississippi, reached St. Catharine's landing on the morning of the 31st January 1805.]

MISSOURI, a settlement of Indians of the province and government of Louisiana; situate on the shore of the river of its name, and where the French have built a fort for the defence of their establishment there. The Spaniards, in 1721, attempted to take this fort, and attacked two settlements of the Octotatas Indians; but the missionaries came to their succour, and finding the Spaniards asleep cut off all their heads, with the exception of one religious person, whom they suffered to accompany them. He afterwards escaped by a stratagem, in pretending to shew the Indians his way of managing a horse; it was by taking to flight.

[MISSOURI River, in Louisiana, falls into the Mississippi from the *w.* 18 miles below the mouth of the Illinois, 130 above the mouth of the Ohio, and above 1160 miles from the Balize, or months of the Mississippi, in the gulf of Mexico. In Captain Hutchins's map, it is said to be navigable 1300 miles. Late travellers up this river, (among whom is a French gentleman, a general officer, who has made a map of his expedition) represent that the progress of settlement by the Spaniards on the *s.* and *w.* and by the English on the *n.* and *e.* of the Missouri, is astonishing. People of both these nations have trading-houses 6 or 700 miles up this river. Mr. M'Kenzie performed a tour from Montreal to the S. sea; and it appears by his map that by short portages, and these not very numerous, there is a water communication, without great interruption, from the Upper lakes to Nootka sound, or its neighbourhood; but the most correct notion respecting the navigation of a river is always to be derived from the account as given verbatim by those who have visited it. The following copious information is therefore extracted from the *Travels of Captains Lewis and Clarke, from St. Louis, by way of the Missouri and Columbia Rivers to the Pacific Ocean, performed in the years* 1804, 1805, *and* 1806, *containing some Delineations of the Manners, Customs, Religion, &c. of the Indians.* N. B. At the end will be found a statement of the commerce of the Missouri.

' On the 14th of May 1804, (as these travellers observe), we embarked from St. Louis on the expedition, having, previous to our setting out, provided ourselves with every thing requisite for the prosecution of the voyage, particularly with large quantities of ammunition and fire-arms, for the purpose of protecting us from the hostile attacks of the natives, and for procuring food. We likewise took a large quantity of ornaments, consisting of medals, trinkets, &c. for the purpose of gaining a favourable reception among the Indians, and obtaining such articles of use as our situation might require.

' Our party, consisting of 43, was generally divided into two companies, the one for hunting, who travelled by land; the other to remain in our water conveyance, which consisted only of two small perogues and a batteau. Larger vessels would have obstructed us in ascending the Missouri near its source. Both companies joined at night, when we were compelled to encamp by the banks of the river; our vessel being too light to sail except by day.

' The country bordering on the Missouri produces immense quantities of fur, which can be purchased of the Indians for a mere trifle, and may be easily transported from the head of this river to the Columbia river at a small expence, on account of the low rate at which horses might be purchased for the purpose from the Snake Indians, who inhabit this mountainous district: from the Columbia river they may be conveyed to China by a very short route.

' This trade would give employment to an immense number of inhabitants; and the country is sufficiently luxuriant for the population of an immense colony.

' The Missouri is already ranked among the greatest rivers. It is an object of astonishment to the whole world. The uninformed man admires its rapidity, its lengthy course, and the salubrity] [of its waters, and is amazed at its colour; while the reflecting mind admires the innumerable riches scattered on its banks, and, foreseeing the future, beholds already this rival of the Nile flowing through countries as fertile, as populous, and as extensive as those of Egypt.

'The Missouri joins the Mississippi five leagues above the town of St. Louis, about lat. 40° *n.* It is necessary to observe, that after uniting with the Mississippi, it flows through a space of 1200 miles, before it empties itself into the gulf of Mexico. As this part of its course is well known, I shall speak (writes Captain Lewis) of the Missouri only.

'I ascended about 600 leagues, without perceiving a diminution either in its width or rapidity. —The principal rivers which empty into the Missouri, are, as you ascend, the Gasconade, the river of the Osages, the two Charaturns, the Great river, the river Des Canips, Nichinen, Batoney, the Great and Little Nimaha, the river Plate, the river De Sioux, the L'Eau-qui-court.

'As far as 25 leagues above its junction with the Mississippi, are to be found different settlements of American families, viz. at Bonhomme, and Femme Osage, &c.; beyond this, its banks are inhabited only by savage nations—the Great and Little Osages, settled 120 leagues on the river of that name; the Canips, the Ottos, the Panis, the Loupes or Panis Mahas, the Mahas, the Poukas, the Ricaras, the Mandanes, the Sioux: the last nation is not fixed on the banks of the Missouri, but habitually goes there to hunt.

'The banks of the Missouri are alternately woods and prairies: it is remarked, that the higher you ascend this river, the more common are these prairies; and they seem to increase every year by the fires which are kindled every autumn by the savages, or white hunters, either by chance, or with the design of facilitating their hunting.

'The waters of the Missouri are muddy, and contain throughout its course a sediment of very fine sand, which soon precipitates; but this circumstance, which renders them disagreeable to the sight, takes nothing from their salubrity.

'Experience has proved, that the waters of the Missouri are more wholesome than those of the Ohio and the Upper Mississippi. The rivers and streams which empty into the Missouri below the river Plate, are clear and limpid; above this river they are as muddy as the Missouri itself. This is occasioned by beds of sand, or bills of a very fine white earth, through which they take their course.

'The bed of the Missouri is obstructed with banks, sometimes of sand, and sometimes of gravel, which frequently change their place, and consequently render the navigation always uncertain. Its course is generally *w.* by *n. w.*

'To give a precise idea of the incalculable riches scattered on the banks of the Missouri, would require unbounded knowledge.

'The flats are covered with huge trees; the liard or poplar; the sycamore, out of one piece of which are made canoes, which carry nearly 18 cwt; the maple, which affords the inhabitants an agreeable and wholesome sugar; and the wild cherry-tree, and the red and black walnut, so useful in joiners' work; the red and white elm, necessary to cartwrights; the triacanthos, which, when well trimmed, forms impenetrable hedges; the water-willow, the white and red mulberry-tree, &c. &c.

'On the shores are found in abundance the white and black oak, proper for every kind of ship-wrights' and carpenters' work; the pine, so easily worked; and, on the stony mountains, the durable cedar.

'It would be impossible to detail all the species of trees, even those unknown in other countries, and the use that can be made of them, of which we are still ignorant.

'The plants are still more numerous; we will pass lightly over this article, for the want of sufficient botanical knowledge. The Indians are well acquainted with the virtues of many of them; they make use of them to heal their wounds, and to poison their arrows; they also use various kinds of savoyannes to dye different colours; they have one which is a certain and prompt cure for the venereal disease.

'The lands on the borders of the Missouri are excellent, and when cultivated are capable of yielding abundantly all the productions of the temperate, and even some of the warm climates; wheat, maize, and every species of grain, Irish potatoes, and excellent sweet potatoes. Hemp seems here to be an indigenous plant: even cotton succeeds, though not so well as in more *s.* countries. Its culture, however, yields a real advantage to the inhabitants settled on the banks of the Missouri, who find in the crop of a field of about two acres sufficient for the wants of their families.

'The natural prairies are a great resource, being of themselves excellent pasturages, and facilitating the labours of the man who is just settled, who can thus enjoy, with little labour, from the first year a considerable crop. Clay fit for making bricks is very common. There is also Fayance clay, and another species of clay, which in the opinion of intelligent persons is the real koaolin to] [which the porcelain of China owes the whole of its reputation.

'There are found on the borders of the Missouri many springs of salt-water of every kind, which will yield more than sufficient salt for the consumption of the country, when it shall become inhabited.

'Salt-petre is found here in great abundance, in numberless caves, which are met with along the banks of the river.

'The stones are generally calcarious and gates. There is found one also, which is believed to be peculiar to the banks of the Missouri. It is of a blood red colour, compact, solt under the chisel, and hardens in the air, and is succeptible of a most beautiful polish. The Indians use it for their calumets; but from the extent of its layers, it might be easily employed in more important works. They have also quarries of marble, of which we only know the colour; they are streaked with red. One quarry is well known, and easily worked, consisting of a species of plaster, which we are assured is of the same nature as that of Paris, and of which the United States make a great use: we also found volcanic stones, which demonstrate the ancient existence of unknown volcanoes.

'We were confirmed in the belief that there were volcanoes in some of their mountains, by the intelligence that we received from the Indians, who informed us, "that the Evil Spirit was mad at the red people, and caused the mountains to vomit fire, sand, gravel, and large stones, to terrify and destroy them; but the Good Spirit had compassion on them, and put out the fire, chased the Evil Spirit out of the mountains, and left them unhurt; but when they returned to their wickedness, the Great Spirit had permitted the Evil Spirit to return to the mountains again, and vomit up fire; but on their becoming good, and making sacrifices, the Great Spirit chased away the Evil Spirit from disturbing them, and for 40 snows (40 years) he had not permitted him to return."

'The short stay we have generally made among the savage nations has prevented us from making those researches which would have supplied us with more extensive information respecting the various mines found on the borders of the Missouri. We know with certainty only of those of iron, lead, and coal; there is, however, no doubt, but that there are some of tin, of copper, of silver, and even of gold, according to the account of the Indians, who have found some particles or dust of these metals either on the surface of the earth, or on the banks of small torrents.

'I consider it a duty at the same time to give an idea of the salt mines, and the salines, which are found in the same latitude on the branches of the river Arkansas. At about 300 miles from the village of the Great Osages, in a *w.* direction, after having passed several branches of the river Arkansas, we find a flat of about 15 leagues in diameter, surrounded by hills of an immense extent: the soil is a black sand, very fine, and so hard that the horses hardly leave a trace. During a warm and dry season there exhales from this flat, vapours, which, after being condensed, fall on this black sand, and cover it with an incrustation of salt, very white and fine, and about half an inch thick: the rains destroy this phenomenon.

'At about 18 miles from this flat are found mines of genuine salt near the surface of the earth. The Indians, who are well acquainted with them, are obliged to use levers to break and raise it.

'At a distance of about 15 leagues from the flat of which we have just spoken, and in a *s.* direction, there is a second mine of genuine salt of the same nature as the other. These two mines differ only in colour: the first borders on a blue, the second approaches a red. Much further *s.* and still on the branches of the Arkansas, is a saline, which may be considered as one of the most interesting phenomena in nature.

'On the declivity of a small hill there are five holes, about a foot and a half in diameter, and two in depth, always full of salt water, without ever overflowing. If a person were to draw any of this water the hole would immediately fill itself; and about ten feet lower, there flows from this same hill a large stream of pure and sweet water.

'If this country were peopled, the working of these genuine salt mines would be very easy by means of the river Arkansas. This species of salt is found by experience to be far preferable to any other for salting provisions.

'Should these notes, imperfect and without order as they are, but in every respect founded on truth, and observations made by myself, excite the curiosity of men of intelligence, capable of investigating the objects which they have barely suggested, I do not doubt but that incalculable advantages would result to the United States, and especially to the district of Louisiana.

'It is impossible to give an exact account of the peltries which are brought down the Mississippi, as they are all immediately transported to Canada, without passing any port of this country; we can obtain a true statement only from the settlements on the lakes. It is but a short time since the Red river has been explored.

'After leaving the river Des Moens the fur]

[trade from the Upper Missouri is carried on entirely by British houses, and almost the whole of the fur which is obtained from the other Indian traders is also sent to Canada, where it commands much higher prices than at New Orleans; where, in fact, there is no demand. It is also necessary to observe, that the further *n.* we go, the greater is the value of the peltries. It is but a few years since peltries have been exported from America by way of the Ohio. It is to be thought that the *e.* part of America will encourage this exportation, by raising the prices of peltries to nearly those of Canada.

'The countries at the head of the Missouri and of the Columbia rivers bear a great similarity; being cold and very sterile, except in pasturage only. At the foot of the mountain, at the head of the Missouri, lives a tribe of Indians called Serpentine or Snake Indians; who are the most abject and miserable of the human race, having little besides the features of human beings.

'They live in a most wretched state of poverty, subsisting on berries and fish; the former they manufacture into a kind of bread, which is very palatable, but possesses little nutritious quality. Horses form the only article of value which they possess,—in these the country abounds; and in very severe winters they are compelled to subsist on them for the want of a better substitute for food. They are a very harmless inoffensive people; when we first made our appearance among them they were filled with terror, many of them fled, while the others who remained were in tears, but were soon pacified by tokens of friendship, and by presents of beads, &c. which soon convinced them of our friendly disposition.

'The Snake Indians are in their stature crooked, which is a peculiarity, as it does not characterise any other tribe of Indians that came within the compass of our observation. To add to this deformity, they have high cheek bones, large light coloured eyes, and are very meagre, which gives them a frightful aspect.

'For an axe we could purchase of them a good horse. We purchased 27 from them, that did not cost more than 100 dollars; which will be a favourable circumstance for transporting fur over to the Columbia river.

'At the head of the Columbia river, resides a tribe by the name of Pallotepallors, or Flatheads; the latter name they derive from an operation that renders the top of the head flat; which is performed while they are infants, when the bones of the cranium are soft and elastic, and are easily brought to the desired deformity. The operation is performed by tying boards, hewn to a proper shape for the purpose, which they compress on the head. In performing this eccentric operation, many infants, it is thought, without doubt lose their lives. The more they get the head misshapen, the greater do they consider its beauty.

'They are a very kind and hospitable people. We left in charge with them, when we descended the Columbia river, our horses, which they kept safe. They likewise found where we had concealed our ammunition in the earth; and had they not been an honest people, and preserved it safe, our lives must have been inevitably lost; they delivered up the whole, without wishing to reserve any, or to receive for it a compensation.

'They, like the Snake Indians, abound in horses, which subsist in the winter season on a shrub they call evergreen, which bears a large leaf, that is tolerably nutritious; they likewise feed upon the side of hills out of which gush small springs of water that melt the snow and afford pasture. In this manner our horses subsisted while going over the rocky mountains.

'The country inhabited by the Snake and Flatheaded Indians produces but very little game.

'Captain Clark kept an account of the distances of places from one to another; which were not kept by myself, for which reason I hope it will be a sufficient apology for subjoining two of his statements.

Letter from Captain Clark to his Excellency Governor Harrison.

'"Dear Sir, Fort Mandan, April 2d.

'"By the return of a party which we sent from this place with dispatches, I do myself the pleasure of giving you a summary view of the Missouri, &c.

'"In ascending as high as the Kauzas river, which is 334 miles up the Missouri: on the *s. w.* side, we met a strong current, which was from five to seven miles an hour, the bottom is extensive, and covered with timber, the high country is interspersed with rich handsome prairies, well watered, and abounding in deer and bears; in ascending as high as the river Plate, we met a current less rapid, not exceeding six miles an hour; in this distance we passed several small rivers on each side, which water some finely diversified country, principally prairie, as between Vincennes and Illinois, the bottoms continuing wide, and covered with timber: this river is about 6000 yards wide at the month, not navigable; it heads in the rocky mountains, with the N. river, and Yellow Stone river, and] [passes through an open country; 15 leagues up this river the Ottocs and 30 Missouries live in one village, and can raise 200 men; 15 leagues higher up, the Paneas and Panea republicans live in one village, and can raise 700 men; up the Wolf fork of this river, Papia Louisis live in one village, and can raise 280 men; these Indians have partial ruptures frequently; the river Plate is 630 miles up the Missouri on the *s. w.* side. Here we find the antelope or goat; the next river of size ascending, is the Stone river, commonly called by the Ingaseix, Little river Desious; it takes its rise in lake Dispice, 15 miles from the river Demoir, and is 64 yards wide; here commences the Sioux country. The next by note is the Big Sioux river, which heads with the St. Peter's, and waters of lake Winnepie, in some high wooded country; about 90 miles, still higher, the river Jacqua falls on the same side, and about 100 yards wide; this river heads with the waters of lake Winnepie, at no great distance *e.* from the place, the head of the river Demon in Pelican lake, between the Sioux rivers and St. Peter's; the country on both sides of the Missouri from the river Plate to that place has very much the same appearance; extensive fertile plains, containing but little timber, and that little, principally confined to the river bottoms and streams; the country *e.* of this place, and off from the Missouri as low as Stone river, contains a number of small trees, many of which are said to be so much impregnated with Glauber's salt as to produce all its effects; certain it is, that the water in the small streams from the hill below on the *s. w.* side possesses this quality. About the river Jacqua Bruff, the country contains a great quantity of mineral, cobalt, cinnabar, alum, copperas, and several other things; the stone coal which is on the Missouri is very indifferent. Ascending 52 miles above the Jacqua, the river Quicum falls in on the *s. w.* side of this river, is 1026 miles up, 150 yards wide, not navigable; it heads in the black mountains which run nearly parallel to the Missouri from about the head of the Kanzas river, and ends *s. w.* of this place. Quicum waters a broken country 122 miles by water higher. White river falls in on the *s. w.* side, and is 300 yards wide, and navigable, as all the other streams are which are not particularly mentioned; this river heads in some small lakes, short of the black mountains. The Mahan and Poncan nations rove on the heads of this river and the Quicum, and can raise 250 men; they were very numerous a few years ago, but the small-pox and the Sioux have reduced them to their present state; the Sioux possess the *s. w.* of the Missouri, above White river, 132 miles higher, and on the *w.* side. Teton river falls into it, it is small, and heads in the open plains; here we met a large band of Sioux, and the second which we had seen, called Tetons; these are rascals, and may be justly termed the pirates of the Missouri; they made two attempts to stop us; they are subdivided, and stretch on the river near to this place, having reduced the Racres and Mandans, and driven them from the country they now occupy.

'" The Sioux bands rove in the country to the Mississippi. About 47 miles above the Teton river, the Chyanne river falls in from the *s. w.* 4000 yards wide, is navigable to the black mountains, in which it takes its rise, in the third range; several bands of Indians but little known, rove on the head of this and the river Plate, and are stated to be as follows: Chaoenne 800 men; Stactons 100; Canenaviech 400; Cayanwa and Wetahato 200; Cataha 70; Detame 30; Memesoon 50; Castahana 1300 men; it is probable that some of those bands are the remains of the Padoucar nation. At 1440 miles up the Missouri, (and a short distance above two handsome rivers which take their rise in the black mountains), the Kicaras live in three villages, and are the remains of 10 different tribes of Paneas, who have been reduced and driven from their country lower down by the Sioux; their number is about 500 men, they raise corn, beans, &c. and appear friendly and well-disposed; they were at war with the nations of this neighbourhood, and we have brought about peace. Between the Recars and this place, two rivers fall in on the *s. w.* and one on the *n. e.* not very long, and take their rise in the open country; this country abounds in a great variety of wild animals, but a few of which the Indians take; many of those animals are uncommon in the United States, such as white, red, and grey bears; long-eared mules, or black-tail deer, (black at the end of the tail only) large hares, antelope or goat; the red fox; the ground prairie dogs, (who burrow in the ground) the braroca, which has a head like a dog, and the size of a small dog; the white brant, magpie, calumet eagle, &c. and many others are said to inhabit the rocky mountains.

'" I have collected the following account of the rivers and country in advance of this, to wit: two days march, in advance of this, the Little Missouri falls on the *s.* side, and heads at the *n. w.* extremity of the black mountains; six days march further, a large river joins the Missouri, affording] [as much water as the main river. This river is rapid, without a fall, and navigable to the rocky mountains, its branches head with the waters of the river Plate; the country in advance is said to be broken.

'" The trade of the nations at this place is from the *n. w.* and Hudson's bay establishments, on the Assinneboin river, distant about 150 miles: those traders are nearly at open war with each other, and better calculated to destroy than promote the happiness of those nations to which they have latterly extended their trade, and intend to form an establishment near this place in the course of this year.

" Your most obedient servant,

" WM. CLARK."

Letter from Captain Clark to his Brother.

'" St. Louis, 23d Sept. 1806.

'" Dear Brother,

'" We arrived at this place at 12o'clock to-day from the Pacific ocean, where we remained during the last winter, near the entrance of the Columbia river. This station we left on the 27th of March last, and should have reached St. Louis early in August, had we not been detained by the snow, which barred our passage across the rocky mountains until the 24th of June. In returning through these mountains, we divided ourselves into several parties, digressing from the route by which we went out, in order the more effectually to explore the country, and discover the most practicable route which does exist across the continent, by the way of the Missouri and Columbia rivers: in this we were completely successful, and have therefore no hesitation in declaring, that, such as nature has permitted, we have discovered the best route which does exist across the continent of N. America in that direction. Such is that by way of the Missouri to the foot of the rapids, below the great falls of that river, a distance of 2575 miles, thence by land, passing by the rocky mountains to a navigable part of the Kooskooske, 340; and with the Kooskooske 73 miles, Lewis's river 154 miles, and the Columbia 413 miles, to the Pacific ocean, making the total distance, from the confluence of the Missouri and Mississippi, to the discharge of the Columbia into the Pacific ocean, 3555 miles. The navigation of the Missouri may be deemed good; its difficulties arise from its falling banks, timber imbedded in the mml of its channel, its sand bars, and the steady rapidity of its current, all which may be overcome with a great degree of certainty, by using the necessary precautions. The passage by land of 340 miles, from the falls of the Missouri to the Kooskooske, is the most formidable part of the track proposed across the continent. Of this distance, 200 miles is along a good road, and 140 miles over tremendous mountains, which for 60 miles are covered with eternal snows. A passage over these mountains is, however, practicable from the latter part of June to the last of September; and the cheap rate at which horses are to be obtained from the Indians of the rocky mountains, and the *w.* of them, reduces the expences of transportation over this portage to a mere trifle. The navigation of the Kooskooske, Lewis's river, and the Columbia, is safe and good, from the 1st of April to the middle of August, by making three portages on the latter river; the first of which, in descending, is 1200 paces at the falls of Columbia, 261 miles up that river; the second of two miles at the long narrows, six miles below the falls; and a third, also of two miles, at the great rapids, 65 miles still lower down. The tide flows up the Columbia 183 miles, and within seven miles of the great rapids. Large sloops may with safety ascend as high as the tide water; and vessels of 300 tons burthen reach the entrance of the Multnomah river, a large *s.* branch of the Columbia, which takes its rise on the confines of New Mexico, with the Colorado and Apostle's rivers, discharging itself into the Columbia, 125 miles from its entrance into the Pacific ocean. I consider this track across the continent of immense advantage to the fur trade, as all the furs collected in nine-tenths of the most valuable fur country in America, may be conveyed to the mouth of the Columbia, and shipped from thence to the East Indies, by the 1st of August in each year; and will of course reach Canton earlier than the furs which are annually exported from Montreal arrive in Great Britain.

'" In our outward-bound passage we ascended to the foot of the rapids below the great falls of the Missouri, where we arrived on the 14th of June 1805. Not having met with any of the natives of the rocky mountains, we were of course ignorant of the passes by land, which existed through those mountains to the Columbia river: and had we even known the route, we were destitute of horses, which would have been indispensably necessary to enable us to transport the requisite quantity of ammunition and other stores to ensure the remaining part of our voyage down the Columbia; we therefore determined to navigate the Missouri as far as it was practicable, or until we met with some of the natives, from whom we could obtain horses and information of the country. Accordingly, we undertook a most laborious] [portage at the falls of the Missouri, of 18 miles, which we effected with our canoes and baggage by the 3d of July. From hence, ascending the Missouri, we penetrated the rocky mountains at the distance of 71 miles above the upper part of the portage, and penetrated as far as the three forks of that river, a distance of 180 miles further. Here the Missouri divides into three nearly equal branches at the same point. The two largest branches are so nearly of the same dignity, that we did not conceive that either of them could with propriety retain the name of the Missouri; and therefore called these streams Jefferson's, Madison's, and Gallatin's rivers. The confluence of those rivers is 2858 miles from the mouth of the Missouri, by the meanders of that river. We arrived at the three forks of the Missouri on the 27th of July. Not having yet been so fortunate as to meet with the natives, although I had previously made several excursions for that purpose, we were compelled still to continue our route by water.

' " The most *n.* of the three forks, that to which we had given the name of Jefferson's river, was deemed the most proper for our purpose, and we accordingly ascended it 248 miles, to the upper forks, and its extreme navigable point; making the total distance to which we had navigated the waters of the Missouri 3096 miles, of which 429 lay within the rocky mountains. On the morning of the 17th of August 1805, I arrived at the forks of Jefferson's river, where I met Captain Lewis, who had previously penetrated, with a party of three men, to the waters of the Columbia, discovered a band of the Shoshone nation, and had found means to induce 35 of their chiefs and warriors to accompany him to that place. From these people we learned that the river on which they resided was not navigable, and that a passage through the mountains in that direction was impracticable. Being unwilling to confide in this unfavourable account of the natives, it was concerted between Captain Lewis and myself, that one of us should go forward immediately with a small party, and explore the river; while the other in the interim should lay up the canoes at that place, and engage the natives with their horses to assist in transporting our stores and baggage to their camp. Accordingly I set out the next day, passed the dividing mountains between the waters of the Missouri and Columbia, and descended the river which I call the East fork of Lewis's river, about 70 miles. Finding that the Indian's account of the country, in the direction of this river, was correct, I returned and joined Captain Lewis on the 29th of August, at the Shoshone camp, excessively fatigued, as you may suppose; having passed mountains almost inaccessible, and compelled to subsist on berries during the greater part of my route. We now purchased 27 horses of these Indians, and hired a guide, who assured us that he could in 15 days take us to a large river in an open country, *w.* of these mountains, by a route some distance to the *n.* of the river on which they lived, and that by which the natives *w.* of the mountains visit the plains of the Missouri, for the purpose of hunting the buffalo. Every preparation being made, we set forward with our guide on the 31st of August, through those tremendous mountains, in which we continued until the 22d of September, before we reached the lower country beyond them; on our way we met with the Olelachshoot, a band of the Tuchapaks, from whom we obtained an accession of seven horses, and exchanged eight or ten others. This proved of infinite service to us, as we were compelled to subsist on horse beef about eight days before we reached the Kooskooske.

' " During our passage over those mountains, we suffered every thing which hunger, cold, and fatigue could impose; nor did our difficulties, with respect to provision, cease on our arrival at the Kooskooske, for although the Pallotepallors, a numerous nation inhabiting that country, were extremely hospitable, and for a few trifling articles furnished us with an abundance of roots and dried salmon, the food to which they were accustomed, we found that we could not subsist on these articles, and almost all of us grew sick on eating them; we were obliged, therefore, to have recourse to the flesh of horses and dogs, as food, to supply the deficiency of our guns, which produced but little meat, as game was scarce in the vicinity of our camp on the Kooskooske, where we were compelled to remain, in order to construct our perogues, to descend the river. At this season the salmon are meagre, and form but indifferent food. While we remained here, I was myself sick for several days, and my friend Captain Lewis suffered a severe indisposition.

' " Having completed four perogues and a small canoe, we gave our horses in charge to the Pallotepallors until we returned, and on the 7th of October re-embarked for the Pacific ocean. We descended by the route I have already mentioned. The water of the river being low at this season, we experienced much difficulty in descending: we found it obstructed by a great number of difficult and dangerous rapids, in passing which our perogues several times filled, and the men escaped] [narrowly with their lives. However, this difficulty does not exist in high water, which happens within the period I have previously mentioned. We found the natives extremely numerous, and generally friendly, though we have on several occasions owed our lives and the fate of the expedition to our number, which consisted of 31 men. On the 17th of November we reached the ocean, where various considerations induced us to spend the winter; we therefore searched for an eligible situation for that purpose, and selected a spot on the *s.* side of a little river, called by the natives Netul, which discharges itself at a small bar on the *s.* side of the Columbia, and 14 miles within point Adams. Here we constructed some log-houses, and defended them with a common stockade work. This place we called fort Clatsop, after a nation of that name who were our nearest neighbours. In this country we found an abundance of elk, on which we subsisted principally during the last winter. We left fort Clatsop on the 27th of March. On our homeward-bound voyage, being much better acquainted with the country, we were enabled to take such precautions as in a great measure to secure us from the want of provisions at any time, and greatly to lessen our fatigues, when compared with those to which we were compelled to submit in our outward-bound journey. We have not lost a man since we left the Mandans, a circumstance which I assure you is a pleasing consideration to me. As I shall shortly be with you, and the post is now waiting, I deem it unnecessary here to attempt minutely to detail the occurrences of the last 18 months.

"I am, &c. your affectionate brother,

"Wm. Clark."

'The treatment we received from the Indians, during nearly three years that we were with them, was very kind and hospitable; except the ill treatment we received from the Sioux tribe, who several times made attempts to stop us; and we should have been massacred, had we not terrified them from their murderous intention, by threatening them with the small-pox, in such a manner as would kill the whole tribe. Nothing could be more horrible to them, than the bare mention of this fatal disease. It was first communicated to them by the Americans, and it spread from tribe to tribe with an unabated pace until it extended itself across the continent.

'"This fatal infection spread around with a baneful rapidity, which no flight could escape, and with a fatal effect that nothing could resist. It destroyed with its pestilential breath whole families and tribes; and the horrid scene presented to those who had the melancholy and affecting opportunity of beholding it, a combination of the dead and dying, and such as to avoid the horrid fate of their friends around them, prepared to disappoint the plague of its prey, by terminating their own existence. The habits and lives of those devoted people, who provide not to-day for the wants of to-morrow, must have heightened the pains of such an affliction, by leaving them not only without remedy, but even without alleviation. Nothing was left them, but to submit in agony and despair. To aggravate the picture, if aggravation were possible, may be added the sight of the helpless child, beholding the putrid carcase of its beloved parents dragged from their huts by the wolves, who were invited hither by the stench, and satiated their hunger on the mangled corpse; or, in the same manner, the dog serving himself with food from the body of his once beloved master. Nor was it uncommon for the father of a family, whom the infection had just reached, to call his family around him, to represent their sufferings and cruel fate from the influence of some evil spirit, who was preparing to extirpate their race; and to invite them to baffle death with all its horrors, with their own weapons; and at the same time, if their hearts failed in this necessary act, he was himself ready to perform the deed of mercy with his own hand, as the last act of his affection, and instantly follow them to the chambers of death." (*A Western Traveller.*)—The Indians being destitute of physicians, living on animal food, and plunging themselves into cold water, on the first discovery of the disease, rendered it generally mortal.

'While we were at fort Mondan, the Sioux robbed several of our party when they were returning to the fort, with the fruits of an excursion after game; and murdered several of the Mandan tribe in cold blood, without provocation, while reposing on the bosom of friendship. On hearing of this massacre, Captain Clark and the greater part of us volunteered to avange the murder; but were deterred by not receiving succour from the Mandan warriors, who declined to avenge the outrage committed on them. The probable reason of their not enlisting was, that they were too much afraid of the superior number of the Sioux to venture an engagement.

'Soon after this massacre, we received authentic intelligence, that the Sioux had it in contemplation (if their threats were trne) to murder us in the spring; but were prevented from making the attack, by our threatening to spread the small pox, with all its horrors, among them. Knowing that it first originated among the white people, and] [having heard of inoculation, and the mode of keeping the infection in phials, which they had but an imperfect idea of, a bare threat filled them with horror, and was sufficient to deter them from their resolute and bloody purpose. This stratagem may appear insignificant to the reader, but was of the greatest consequence to us; for to it alone we owed not only the fate of the expedition, but our lives.

'Most of the tribes of Indians, that we became acquainted with (except the Sioux), after being introduced by our interpreter, and having found that our intentions were friendly towards them, never failed of greeting us with many tokens of their friendly disposition. Soon after our interview, we were invited to smoke the calumet of peace, and to partake freely of their venison. The women and children, in particular, were not wanting in shewing tokens of friendship, by endeavouring to make our stay agreeable. On our first meeting, they generally held a council, as they term it, when their chief delivers a talk, in which they give their sentiments respecting their new visitors, which were filled with professions of friendship, and often were very eloquent, and abounded with sublime and figurative language.

'When we departed, after taking leave, they would often put up a prayer; of which the following is a sample, which was put up for us by a Mandan :—That the Great Spirit would favour us with smooth water, with a clear sky by day, and a bright star-light by night; that we might not be presented with the red hatchet of war; but that the great pipe of peace might ever shine upon us, as the sun shines in an unclouded day, and that we might be overshadowed by the smoke thereof; that we might have sound sleep, and that the bird of peace might whisper in our ears pleasant dreams; that the deer might be taken by us in plenty; and that the Great Spirit would take us home in safety to our women and children. These prayers were generally made with great fervency, often smiting with great vehemence their hands upon their breast; their eyes fixed in adoration towards heaven. In this manner they would continue their prayers until we were out of sight.

'In the fore part of autumn we experienced slight typhus indispositions, caused by great vicissitudes of weather, which at times was very damp.

'Our affectionate companion Serjeant Floyd was seized with a severe astenic disease, to which he fell a victim. He was seized with an acute pain in his intestines, accompanied with great suppression of the pulmonary function. Every effort that our situation allowed, was in vain used for his recovery; we buried him in the most decent manner that our circumstances would admit: he was universally lamented by us.

'Several times, many of our party were in imminent danger of being devoured by the wild beasts of prey; but happily we escaped. Frequently we were annoyed by a kind of light-coloured bear, of which the country, near the head of the Missouri, abounds. After being attacked, they give no quarter, but rush with great fury toward their enemy. One of our party shot at one of them, and wounded him; the bear, instead of being intimidated by the smart of the wound, was stimulated into rage, and rushed with great fury to devour the assailant; who saved his life by running headlong down a steep precipice, that formed the bank of the river; but was severely bruised by this precipitate retreat.

'The following narrative of an encounter with a snake, is told by a companion, whose veracity can be relied on; I will give it in his own words, as he related it in a letter to his friend.

'"Some time," says he, "before we reached fort Mandan, while I was out on an excursion of hunting, one of the greatest monsters that ever shocked the mind with horror was presented to my sight. When passing deliberately in a forest that bordered on a prairie, I heard a rustling in the bushes; I leaped towards the object, delighted with the prospect of acquiring game. But on proceeding a few paces further, my blood was chilled by the appearance of a serpent of an enormous size; on discovering me, he immediately erected his head to a great height; his colour was of a yellower hue than the spots of a rattlesnake, and on the top of his back were spots of a reddish colour; his eyes emitted fire, his tongue darted, as though he menaced my destruction. He was evidently in the attitude of springing at me, when I levelled my rifle at him; but probably owing to my consternation, I only wounded him; but the explosion of the gun and the wound turned to flight the awful enemy. Perhaps you may think, that my fright has magnified the description. I can candidly aver, that he was in bulk half as large as a middle-sized man."

'In the Indian tribes there is so great a similarity in their stature, colour, government, and religious tenets, that it will be requisite for perspicuity, to rank them under one general head; and when there is a contrast in the course of the description it will be mentioned.

'They are all (except the Snake Indians) tall in]

stature, straight, and robust; it is very seldom they are deformed, which has given rise to the supposition, that they put to death their deformed children, which is not the case. Their skin is of a copper colour, their eyes large, black, and of a bright and sparkling colour, indicative of a subtle and discerning mind. Their hair is of the same colour, and prone to grow long, straight, and seldom or never curled; their teeth are large and white. I never observed any decayed among them, which makes their breath as sweet as the air they inhale. The women are about the stature of the English women, and much inclined to corpulency, which is seldom the case with the other sex.

'I shall not enter into a discussion about the cause of their hue. I shall barely mention the suppositions that are made respecting it. Some have asserted, that it is derived principally from their anointing themselves with fat in the summer season, to prevent profuse perspiration, and this, combined with the influence of the sun, has given the tincture of their complexion. To support the hypothesis, they assert that the abovementioned causes repeated give colour to the parent, who procreates his own likeness, until at length it is entailed on posterity. But notwithstanding this curious reasoning, others are of opinion, that the hand of the Creator gave the reddish hue to the Indians, the sable colour to the African, and that of white to the civilized nations.

'They esteem a beard exceedingly unbecoming, and take great pains to get rid of it, nor is there ever any to be perceived on their faces, except when they grow old and become inattentive to their appearance. Every crinose excrescence on other parts of their body is held in as great abhorrence by them, and both sexes are equally careful to extirpate it, in which they often employ much time.

'The Pallotepallors, Serpentine, Mandan, and other interior tribes of Indians, pluck them out with bent pieces of hard wood, formed into a kind of nippers made for that purpose; while those that have a communication with Americans or Europeans procure from them wire, which they ingeniously make into an instrument resembling a screw, which will take so firm a hold of the beard, that with a sudden twitch they extirpate it by the roots, when considerable blood never fails to flow.

'The dress of the Indians varies according to the tribe they belong to; but in general it is made very commodious, not to encumber them in pursuing the chase, or their enemy; those that inhabit the Missouri, I have often seen, in cold weather, without any apparel to screen themselves from the inclemency of the weather. The lower rank of the Pallotepallors and Clatsops, wear nothing in the summer season, but a small garment about their hips, which is either manufactured out of bark or skins, and which would vie with, if not excel, any European manufature, being diversified with different colours, which give it a gay appearance. Their kings are generally dressed in robes made out of small skins (which takes several hundred for a garment) of different colours, neatly tanned; these they hang loosely over their shoulders.

'In deep snows they wear skins that entirely cover their legs and feet, and almost answer for breeches, being held up by strings tied to the lower part of their waist. Their bodies, in the winter season, are covered with different kinds of skins, that are tanned with the fur on, which they wear next to the skin. Those of the men, who wish to appear more gay than others, pluck out the greatest part of their hair, leaving only small locks, as fancy dictates, on which are hung different kinds of quills, and feathers of elegant plumage superbly painted. The Sioux and Osages, who traffic with the Americans, wear some of our apparel, such as shirts and blankets; the former they cannot bear tied at the wristbands and collars, and the latter they throw loosely over their shoulders. Their chiefs dress very gay: about their heads they wear all kinds of ornaments that can well be bestowed upon them, which are curiously wrought, and in the winter long robes of the richest fur, that trail on the ground.

'In the summer there is no great peculiarity, only that what the higher rank wear is excessively ornamented.

'The Indians paint their heads and faces yellow, green, red, and black; which they esteem very ornamental. They also paint themselves when they go to war; but the method they make use of on this occasion differs from that which they employ merely for decoration.

'The Chipaway young men, who are emulous of excelling their companions in finery, slit the outward rim of both ears; at the same time they take care not to separate them entirely, but leave the flesh thus cut, still untouched at both extremities: around this spungy substance, from the upper to the lower part, they twist brass wire, till the weight draws the amputated rim in a bow of five or six inches diameter, and drags it down almost to the shoulder. This decoration is esteemed gay and becoming.

'It is also a custom among them to bore their noses, and wear in them pendants of different sorts.]

[Shells are often worn, which when painted are reckoned very ornamental.

'The Indians who inhabit the borders of Louisiana, make for their legs a kind of stocking, either of skins or cloth; these are sewed up as much as possible in the shape of their leg, so as to admit of being drawn on and off; the edges of the stuff of which they are composed, are left annexed to the seams, and hang loose about the breadth of a hand; and this part, which is placed on the outside of the leg, is generally ornamented with lace and ribbons, and often with embroidery and porcupine quills variously coloured. The hunters from Louisiana find these stockings much more convenient than any others. Their shoes are made of the skins of deer or elk; these, after being dressed with the hair on, are cut into shoes, and fashioned so as to be easy to their feet, and convenient for walking. The edges round the ancle are decorated with pieces of brass or tin, fixed round a leather string about an inch long, which being placed very thick make a very pleasing noise when they walk or dance.

'The dress of the women in the summer season, consists only of a petticoat that does not reach down to their knees. In the winter they wear a shift made of skins, which answers a very good purpose when they stand erect, as it is sufficiently low; but when they stoop they often put modesty to the blush. Their feet and legs are covered in a manner similar to the other sex.

'Most of the female Indians who dwell on the *w.* side of the Mississippi, near its confluence with the Missouri, decorate their heads by enclosing their hair in plates of silver; these are costly ornaments, and used by the highest rank only. Those of the lower rank make use of bones, which they manufacture to resemble those of silver. The silver made use of is formed into thin plates of about four or five inches broad, in several of which they confine their hair. That plate which is nearest to the head is of considerable width; the next narrower, and made so as to pass a little way under the other, and so gradually tapering until they get to a very considerable magnitude.

'This decoration proves to be of great expence, for they often wear it on the back part of the head, extending to the full length of their hair, which is commonly very long.

'The women of every nation generally paint a spot against each ear, about the size of a crown piece; some of them paint their hair, and sometimes a spot on the middle of their forehead.

'The Indians have no fixed habitations when they are hunting; but build where conveniency directs; their houses are made so low as not to admit one to stand erect, and are without windows. Those that are built for a permanent residence are much more substantial; they are made of logs and bark, large enough to contain several apartments. Those built for their chiefs are often very elegant. That of the chief warrior of the Mahas is at least 60 feet in circumference, and lined with furs and painting. The furs are of various colours, many of which I had never seen before, and were extremely beautiful; the variety in colour formed a contrast that much added to its elegance. The paintings were elegant, and would adorn the dwellings of an opulent European prince. But the houses of the common people are but very indifferent.

'They have also moveable houses, which they use for fishing, and sometimes for hunting, made of deer skins or birch bark sewed together, which they cover over poles made for the purpose; they are bent over to form a semicircle, resembling those bent by the Americans for beans or hops to grow on, and are covered over as before mentioned; they are very light, and easily transported where necessity requires.

'The best of their cabins have no chimneys, but a small hole to let the smoke through, which they are compelled to stop up in stormy weather; and when it is too cold to put out their fire, their huts are filled with clouds of smoke, which render them insupportable to any but an Indian.

'The common people lie on bear skins, which are spread on the floor. Their chiefs sleep on beaver skins, which are sometimes elevated.

'Their utensils are few, and in point of usefulness very defective; those to hold water in are made of the skins of animals, and the knotty excrescences of hard wood; their spoons are manufactured out of wood, or the bones of a buffalo, and are tolerably commodious, and I have often seen them elegant, and sometimes painted.

'The Flatheads and Clatsops make baskets out of rushes that will hold water, if they are not very dry. These two nations appear to have more of a mechanical genius, than any other people that I have ever been acquainted with; and I think they are not outrivalled by any nation on earth, when taking into consideration their very limited mechanical instruments.

'Many of the Indian nations make no use of bread, salt, and spices, and many live to be old without seeing or tasting of either. Those that live near the snowy mountains, live in a great measure on berries, which clothe the fields in great abundance.]

['The Taukies, and other *e.* tribes, where Indian corn grows, take green corn and beans, boil them together with bear's flesh, the fat of which gives a flavour, and renders it beyond comparison delicious: they call this dish Succatosh.

'In general they have no idea of the use of milk, although great quantities might be collected from the buffalo and elk. They only consider it proper for the nourishment of the young of these animals in their tender state. It cannot be perceived, that any inconvenience arises from the disuse of articles so much esteemed by civilized nations, which they employ to give a relish and flavour to their food. But on the contrary, the great healthiness of the Indians, and the unhealthiness of the sons of Epicurus, prove that the diet of the former is the most salutary.

'They preserve their meat by exposing it to the sun in the summer, and in the winter by putting it between cakes of ice, which keep it sweet, and free from any putrefactive quality.

'Their food consists, in a great measure, of the flesh of the bear, buffalo, and deer. They who reside near the head of the Missouri and Columbia rivers, chiefly make use of the buffalo and elk, which are often seen from 50 to 100 in a drove. Where there are plenty of the two last mentioned animals there are but few of the former, and where there are many of the former, but few of the latter.

'The mode of roasting their meat is by burning it under ground, on the side of a hill, placing stones next to the meat: the mode of building to heat it somewhat resembles the fire-place made under a limekiln. In this manner they roast the largest of their animals.

'The mode of cooking smaller pieces is to roast them in stones, that are hewn out for the purpose.

'The Flatheads and Clatsops procure a root about the size of a potato, spontaneously and in great abundance, which is tolerably palatable, and perfectly agrees with the natives; but made us all sick, while we were among them. Before we descended the Columbia river, we were unable to procure game, and had recourse to the flesh of dogs and horses to preserve life, as those roots would, without doubt, have destroyed us, and we were unable to procure any other kind of food.

'Many of the tribes of Indians are extremely dirty. I have seen the Maha Indians bring water in the paunches of animals that were very dirty, and in other things equally so. But the Maha chiefs are very neat and cleanly in their tents, apparel, and food.

'The Indians commonly eat in large parties, so that their meals may with propriety be termed feasts; they have not set hours for their meals, but obey the dictates of nature.

'Many of the tribes dance before or after their meals, in devotion to the Great Spirit for the blessings they receive. Being informed of the mode of our saying grace, they answered, that they thought we were stupid and ungrateful, not to exercise our bodies for the great benefits we received; but muttering with our lips, they thought was an unacceptable sacrifice to the Great Spirit, and the stupid mode of the ceremony ridiculous in the extreme. In their feasts, the men and women eat apart; but in their domestic way of living, they promiscuously eat together.

'Instead of getting together and drinking, as the Americans do, they make use of feasting as a substitute.

'When their chiefs are assembled together on any occasion, they always conclude with a feast, at which their hilarity and cheerfulness know no bounds.

'No people on earth are more hospitable, kind, and free, than the Indians. They will readily share with any of their own tribe, and even with those of a different nation, the last part of their provisions. Though they do not keep one common stock, yet that community of goods, which is so prevalent among them, and their generous dispositions, render it nearly of the same effect.

'They strike fire by rubbing together sticks of wood, of a particular kind, which will yield it with ease; from other kinds it is impossible to procure it.

'They are extremely circumspect and deliberate in every word and action; nothing hurries them into any intemperate wrath, but that inveteracy to their enemies, which is rooted in every Indian's breast, and never can be eradicated. In all other instances they are cool and deliberate, taking care to suppress the emotions of the heart. If an Indian has discovered that a friend of his is in danger of being cut off by a lurking enemy, he does not tell him of his danger in direct terms, as though he were in fear, but he first coolly asks him which way he is going that day, and having his answer, with the same indifference tells him, that he has been informed, that a noxious beast lies on the route he is going, which might probably do him mischief. This hint proves sufficient, and his friend avoids the danger with as much caution, as though every design and motion of his enemy had been pointed out to him.

'This apathy often shews itself on occasions]

[that would call forth the favour of a susceptible heart. If an Indian has been absent from his family for several months, either on a war or hunting party, and his wife and children meet him at some distance from his habitation, instead of the affectionate sensations that naturally arise in the breast of more refined beings, and give rise to mutual congratulations, he continues his course without looking to the right or left; without paying the least attention to those around him, till he arrives at his house: he there sits down, and with as much unconcern as if he had not been absent a day smokes his pipe; those of his friends who followed him do the same; perhaps it is several hours before he relates to them the incidents that have befallen him during his absence, though perhaps he has left a father, a brother, or a son dead on the field, (whose loss he ought to have lamented) or has been successful in the undertaking that called him from his home.

'If an Indian has been engaged for several days in the chase, or any other laborious expedition, and by accident continued long without food, when he arrives at the hut of a friend, where he knows that his wants will be immediately supplied, he takes care not to shew the least symptoms of impatience, or betray the extreme hunger that he is tortured with; but, on being invited in, sits contentedly down, and smokes his pipe with as much composure as if his appetite was cloyed, and he was perfectly at ease; he does the same if among strangers. This custom is strictly adhered to by every tribe, as they esteem it a proof of fortitude, and think the reverse would entitle them to the appellation of old women.

'If you tell an Indian, that his children have greatly signalized themselves against an enemy, have taken many scalps, and brought home many prisoners, he does not appear to feel any strong emotions of pleasure on the occasion; his answer generally is, "they have done well," and makes but very little inquiry about the matter; on the contrary, if you inform him that his children are slain or taken prisoners, he makes no complaints, he only replies, "it is unfortunate," and, for some time, asks no questions about how it happened.

'This seeming indifference, however, does not proceed from a suppression of the natural affections, for, notwithstanding they are esteemed savages, I never saw among any other people greater proofs of filial tenderness; and although they meet their wives after a long absenee with the stoical indifference just mentioned, they are not, in general, void of conjugal affection.

'Anothr peculiarity is observable in the manner of payng their visits. If an Indian goes to visit a parcular person in a family, he mentions to whom is visit is intended, and the rest of the family imediately retire to the other end of the hut or tet, and are careful not to come near enough tc interrupt them during the whole conversation. The same method is pursued when a young ma goes to pay his addresses to a young woman; ut then he must be careful not to let love be th subject of his discourse whilst the daylight remins.

'Theydiscover an amazing sagacity, and acquire wit the greatest readiness any thing that depends upn the attention of the mind. By expericuce, ad an acute observation, they attain many perfectios, to which Americans are strangers. For instnce, they will cross a forest or a plain, which is'00 miles in breadth, so as to reach with great exctness the point at which they intend to arrive, keping during the whole of that space in a direct ie, without any material deviations; and this theyvill do with the same ease, let the weather be fir or cloudy.

'Witl equal acuteness they will point to that part of te heavens the sun is in, though it be intercepter by clouds or fogs. Besides this, they are abldo pursue, with incredible facility, the traces ofman or beast, either on leaves or grass; and on tis account it is with great difficulty a flying chew escapes discovery.

'The are indebted for these talents, not only to nature, out to an extraordinary command of the intellectal faculties, which can only be acquired by an urcmitted attention, and by long experience.

'The are in general very happy in a retentive memory They can recapitulate every particular that hasoeen treated of in council, and remember the exat time when they were held. Their belts of wampm preserve the substance of the treaties they hae concluded with the neighbouring tribes, for age back, to which they will appeal and refer, wit as much perspicuity and readiness, as Europens can to their written records.

'Evry nation pays great respect to old age. The adice of a father will never receive any extraordinry attention from the young Indians; probabl they receive it with only a bare assent; but the will tremble before a grandfather, and submit to iis injunctions with the utmost alacrity. The wcds of the ancient part of their community are estemed by the young as oracles. If they take, uring hunting parties, any game that is reckond by them uncommonly delicious, it is]

[immediately presented to the eldest of their relations.

'They never suffer themselves to be overburdened with care; but live in a state of perfect tranquillity and contentment, being naturally indolent. If provisions, just sufficient for their subsistence, can be procured with little trouble, and near at hand, they will not go far, or take any extraordinary pains for it, though by so doing they might acquire greater plenty, and of a more estimable kind.

'Having much leisure time, they indulge this indolence to which they are so prone, by sleeping and rambling about among their tents. But when necessity obliges them to take the field, either to oppose an enemy, or to procure food, they are alert and indefatigable. Many instances of their activity on these occasions, will be given when we treat of their wars.

'The greatest blemish in their character, is that savage disposition which impels them to treat their enemies with a severity that every other nation shudders at. But if they are thus barbarous to those with whom they are at war, they are friendly, hospitable, and humane in peace. It may with truth be said of them, that they are the worst enemies, and the best friends, of any people in the world.

'They are, in general, strangers to the passion of jealousy, and brand a man with folly that is distrustful of his wife. Among some tribes the very idea is not known; as the most abandoned of their young men very rarely attempt the virtue of married women, nor do these put themselves in the way of solicitations: yet the Indian women in general are of an amorous disposition; and before they are married, are not the less esteemed for the indulgence of their passions.

'The Indians, in their common state, are strangers to all distinction of property, except in the articles of domestic use, which every one considers as his own, and increases as circumstances admit. They are extremely liberal to each other; and supply the deficiency of their friends with any superfluity of their own.

'In dangers they readily give assistance to those of their band who stand in need of it, without any expectation of return, except those just rewards that are always conferred by the Indians on merit. Governed by the plain and equitable laws of nature, every one is rewarded according to his deserts; and their equality of condition, manners, and privileges, with that constant and sociable familiarity which prevails throughout every Indian nation, animates them with a pure and patriotic spirit, that tends to the general good of the society to which they belong.

'If any of their neighbours are bereaved, by death, or by an enemy, of their children, those who are possessed of the greatest number of prisoners, who are made slaves, supply the deficiency; and these are adopted by them, and treated in every respect as if they really were the children of the person to whom they are presented.

'The Indians can form to themselves no idea of the value of money; they consider it, when they are made acquainted with the uses to which it is applied by other nations, as the source of innumerable evils. To it they attribute all the mischiefs that are prevalent among Europeans, such as treachery, plundering, devastation, and murder.

'They esteem it irrational, that one man should be possessed of a greater quantity than another, and are amazed that any honour should be annexed to the possession of it.

'But that the want of this useless metal should be the cause of depriving persons of their liberty, and that on the account of this particular distribution of it, great numbers should be shut up within the dreary walls of a prison, cut off from society, of which they constitute a part, exceeds their belief; nor do they fail, on hearing this part of the United States' system of government related, to charge the institutors of it with a total want of humanity, and to brand them with the names of savages and brutes.

'They show almost an equal degree of indifference for the productions of art. When any of these are shewn them, they say, "It is pretty, I like to look at it," and often are not inquisitive about the construction of it, neither can they form proper conceptions of its use. But if you tell them a person runs with great agility, that he is skilled in hunting, can direct with unerring aim a gun, or bend with ease a bow; that he can dexterously work a canoe, understands the art of war, is acquainted with the situations of a country, and can make his way without a guide through an immense forest, subsisting during the time on a small quantity of provisions, they are in raptures; they listen with great attention to the pleasing tale, and bestow the highest commendation on the hero of it.

'They make but little use of physicians and medicine, and consequently have few diseases amongst them. There is seldom an Indian but that blooms with the appearance of health. They have no midwives among them; and among several tribes the mother is without the assistance of any person being with her at the time of her delivery, not having even a female attendant.]

[‘ Soon after the birth of a child, it is placed on a board, which is covered with a skin stuffed with soft moss: the child is laid on its back, and tied to it. To these machines are fastened strings, by which they hang them to branches of trees: or, if they do not find trees handy, they place them against a stump or stone while they dress the deer or fish, or do any domestic business. In this position they are kept until they are several months old. When taken out they are suffered to go naked, and are daily bathed in cold water, which renders them vigorous and active.

‘ An Indian child is generally kept at the breast until it is two years old, and sometimes, though rarely, a year longer.

‘ The Indians often occasion inflammatory diseases by excessive eating, after a fast of three or four days, when retreating from, or pursuing an enemy.

‘ The inequality of riches, the disappointments of ambition, and merciless oppression, are not with them exciting causes of insanity. I made great inquiry, but was not able to learn, that a single case of melancholy or madness was ever known among them.

‘ The dreadful havoc that the small-pox has made has necessarily been mentioned.

‘ The mode of curing a fever is by profuse perspiration, which is effected by the patient being confined in a close tent or wigwam, over a hole in the earth, in which red-hot stones are placed; a quantity of hot water is then thrown upon the stones, which involves the patient in a cloud of vapours and sweat; in this situation he rushes out, and plunges into a river, and from thence retires into a warm bed.

‘ They never think of giving medicine, until they have first made an attempt to remove the disease by sacrifices and prayer, and if the patient recovers soon, it is attributed to the holy management of the priest; and if medicine is to be used as the last alternative, they never administer it without its being accompanied with prayer, and a large quantity of meat, which they consume on the fire for a sacrifice.

‘ They have a plant among them, which has the power of producing abortion. It is related by Mr. Jefferson, in his Notes on Virginia, that the Indians inhabiting the frontiers possess a plant that produces the same effect.

‘ Considering their ignorance of astronomy, time is very rationally divided by the Indians. Those in the interior parts (and of those I would generally be understood to speak) count their years by winters; or, as they express themselves, by snows.

‘ Some nations among them reckon their years by moons, and make them consist of 12 synodical or lunar months, taking care, when 30 moons have waned, to add a supernumerary one, which they term the lost moon; and then begin to count as before. They pay a great regard to the first appearance of every moon, and on the occasion always repeat some joyful sounds, stretching at the same time their hands towards it.

‘ Every month has with them a name expressive of its season; for instance, they call the month of March (in which their year generally begins, at the first new moon after the vernal equinox) the worm month or moon; because at this time the worms quit their retreats in the bark of the trees, &c. where they have sheltered themselves during the winter.

‘ The month of April is termed by them the month of plants. May, the month of flowers. June, the hot moon. July, the buck moon. Their reason for thus denominating these is obvious.

‘ August, the sturgeon moon; because in this month they catch great numbers of that fish.

‘ September, the corn moon; because in that month they gather in their Indian corn.

‘ October, the travelling moon; as they leave at this time their villages, and travel towards the place where they intend to hunt during the winter.

‘ November, the beaver moon; for in this month the beavers begin to take shelter in their houses, having laid up a sufficient store of provisions for the winter season.

‘ December, the hunting moon, because they employ this month in pursuit of their game.

‘ January, the cold moon, as it generally freezes harder, and the cold is more intense in this than in any other month.

‘ February, they call the snow moon, because more snow commonly falls during this month than any other in the winter.

‘ When the moon does not shine, they say the moon is dead; and some call the three last days of it the naked days. The moon's first appearance they term its coming to life again.

‘ They make no division of weeks; but days they count by sleeps, half days by pointing to the sun at noon, and quarters by the rising and the setting of the sun; to express which in their traditions they make use of very significant hieroglyphics.

‘ The Indians are totally unskilled in geography as well as all the other sciences, and yet they draw on their birch bark very exact charts or maps of] [the countries they are acquainted with. The latitude and longitude only are wanting to make them tolerably complete.

'Their sole knowledge in astronomy consists in being able to point out the polar star; by which they regulate their course when they travel in the night.

'They reckon the distance of places, not by miles or leagues, but by a day's journey, which, according to the best calculations I could make, appears to be about 20 English miles. These they also divide into halves and quarters, and will demonstrate them in their maps with great exactness, by the hieroglyphics just mentioned, when they regulate in council their war parties, or their most distant hunting excursions.

'They have no idea of arithmetic; and though they are able to count to any number, figures as well as letters appear mysterious to them, and above their comprehension.

'Every separate body of Indians is divided into bands or tribes; which band or tribe forms a little community with the nation to which it belongs. As the nation has some particular symbol by which it is distinguished from others, so each tribe has a badge from which it is denominated; as that of the eagle, the panther, the tiger, the buffalo, &c. One band is represented by a snake, another a tortoise, a third a squirrel, a fourth a wolf, and a fifth a buffalo. Throughout every nation they particularize themselves in the same manner, and the meanest person among them will remember his lineal descent, and distinguish himself by his respective family.

'Did not many circumstances tend to confute the supposition, I should be almost induced to conclude, from this distinction of tribes, and the particular attachment of the Indians to them, that they derive their origin, as some have asserted, from the Israelites.

'Besides this, every nation distinguishes itself by the manner of constructing its tents or huts. And so well versed are all the Indians in this distinction, that though there appears to be no difference on the nicest observations made by an American, yet they will immediately discover, from the position of a pole left in the ground, what nation has encamped on the spot many months before.

'Every band has a chief, who is termed the great chief, or the chief warrior; and who is chosen in consideration of his experience in war, and of his approved valour, to direct their military operations, and to regulate all concerns belonging to that department. But this chief is not considered as the head of the state. Besides the great warrior, who is elected for his warlike qualifications, there is another who enjoys a pre-eminence as his hereditary right, and has the more immediate management of their civil affairs. This chief might, with great propriety, be denominated the Sachem; whose assent is necessary in all conveyances and treaties, to which he affixes the mark of the tribe or nation.

'Though these two are considered as the heads of the band, and the latter is usually denominated their king, yet the Indians are sensible of neither civil nor military subordination. As every one of them entertains a high opinion of his consequence, and is extremely tenacious of his liberty, all injunctions that carry with them the appearance of a positive command, are instantly rejected with scorn.

'On this account it is seldom that their leaders are so indiscreet as to give out any of their orders in a peremptory style; a bare hint from a chief that he thinks such a thing necessary to be done, instantly arouses an emulation among the inferior ranks, and it is immediately executed with great alacrity. By this method the disgustful part of the command is evaded, and an authority that falls little short of absolute sway instituted in its room.

'Among the Indians no visible form of government is established; they allow of no such distinction as magistrate and subject, every one appearing to enjoy an independence that cannot be controuled. The object of government among them is rather foreign than domestic, for their attention seems more to be employed in preserving such an union among members of their tribes as will enable them to watch the motions of their enemies, and act against them with concert and vigour, than to maintain interior order by any public regulations. If a scheme that appears to be of service to the community is proposed by the chief, every one is at liberty to choose whether or not he will assist in carrying it on; for they have no compulsory laws that lay them under any restrictions. If violence is committed, or blood is shed, the right of revenge is left to the family of the injured: the chiefs assume neither the power of inflicting nor of moderating the punishment.

'Some nations, where the dignity is hereditary, limit the succession to the female line. On the death of a chief, his sister's son sometimes succeeds him in preference to his own son; and if he happens to have no sister, the nearest female relation assumes the dignity. This accounts for a woman being at the head of the Winnebago nation,] [which, before I was acquainted with their laws, appeared strange to me.

'Each family has a right to appoint one of its chiefs to be an assistant to the principal chief, who watches over the interest of his family, and without whose consent nothing of a public nature can be carried into execution. These are generally chosen for their ability in speaking; and such only are permitted to make orations in their councils and general assemblies.

'In this body, with the hereditary chief at its head, the supreme authority appears to be lodged; as by its determination every transaction relative to their hunting, to their making war or peace, and to all their public concerns, is regulated. Next to these the body of warriors, which comprehends all that are able to bear arms, hold their rank. This division has sometimes at its head the chief of the nation, if he has signalized himself by any renowned action, if not, some chief that has rendered himself famous.

'In their councils, which are held by the foregoing members, every affair of consequence is debated, and no enterprise of the least moment undertaken, unless it there meets with the general approbation of the chiefs. They commonly assemble in a hut or tent appropriated to this purpose, and being seated in a circle on the ground, the eldest chief rises and makes a speech; when he has concluded another gets up, and thus they all speak, if necessary, by turns.

'On this occasion their language is nervous, and their manner of expression emphatical. Their style is adorned with images, comparisons, and strong metaphors, and is equal in allegorics to that of any of the *e.* nations. In all their set speeches they express themselves with much vehemence, but in common discourse according to our usual method of speech.

'The young men are suffered to be present at the councils, though they are not allowed to make a speech till they are regularly admitted; they however listen with great attention, and to show that they both understand and approve of the resolutions taken by the assembled chiefs, they frequently exclaim, "That is right;" "That is good."

'The customary mode among all ranks of expressing their assent, and which they repeat at the end of almost every period, is by uttering a kind of forcible aspiration, which sounds like an union of the letters OAH.

'Dancing is a favourite exercise among the Indians; they never meet on any public occasion, but this makes a part of the entertainment: and when they are not engaged in war or hunting, the youth of both sexes amuse themselves in this manner every evening.

They always dance, as I have just observed, at their feasts. In these as well as all other dances, every man rises in his turn, and moves about with great freedom and boldness; singing, as he does so, the exploits of his ancestors. During this, the company, who are seated on the ground in a circle round the dancer, join with him in making the cadence, by an odd tone, which they utter all together, and which sounds, "Heh, heh, heh." These notes, if they might be so termed, are articulated with a harsh accent, and strained out with the utmost force of their lungs; so that one would imagine their strength must be soon exhausted by it; instead of which they repeat it with the same violence during the whole of their entertainment.

'The women, particularly those of the *w.* nations, dance very gracefully. They carry themselves erect, and with their arms hanging down close to their sides, move first a few yards to the right, and then back again to the left. This movement they perform without taking any steps as an American would do, but with their feet conjoined, moving by turns their toes and heels. In this manner they glide with great agility to a certain distance and then return; and let those who join in the dance be ever so numerous, they keep time so exactly with each other that no interruption ensues. During this, at stated periods, they mingle their shrill voices with the hoarser ones of the men, who sit around (for it is to be observed that the sexes never intermix in the same dance), which, with the music of the drums and chichicoes, make an agreeable harmony.

'The Indians have several kinds of dances, which they use on different occasions, as the pipe or calumet dance, the war dance, the marriage dance, and the dance of the sacrifice. The movements in every one of these are dissimilar; but it is almost impossible to convey any idea of the points in which they are unlike.

'Different nations likewise vary in their manner of dancing. The Chipaway throw themselves into a greater variety of attitudes than any other people; sometimes they hold their heads erect, at others they bend them almost to the ground; then recline on one side, and immediately after on the other. Others carry themselves more upright, step firmer, and move more gracefully; but they all accompany their dances with the disagreeable noise just mentioned.

'The pipe dance is the principal, and the most]

[pleasing to a spectator of any of them, being the least frantic, and the movements of it most graceful. It is but on particular occasions that it is used; as when ambassadors from an enemy arrive to treat of peace, or when strangers of eminence pass through their territories.

'The war dance, which they use both before they set out on their war parties and on their return from them, strikes terror into strangers. It is performed, like the others, amidst a circle of the warriors; a chief generally begins it, who moves from the right to the left, singing at the same time both his own exploits and those of his ancestors. When he has concluded his account of any memorable action, he gives a violent blow with his war club against a post that is fixed in the ground, near the centre of the assembly, for this purpose.

'Every one dances in his turn, and recapitulates the wondrous deeds of his family, till they all at last join in the dance. Then it becomes truly alarming to any stranger that happens to be among them, as they throw themselves into every horrible and terrifying posture that can be imagined, rehearsing at the same time the parts they expect to act against their enemies in the field. During this they hold their sharp knives in their hands, with which, as they whirl about, they are every moment in danger of cutting each other's throats; and did they not shun the threatened mischief with inconceivable dexterity, it could not be avoided. By these motions they intend to represent the manner in which they kill, scalp, and take their prisoners. To heighten the scene, they set up the same hideous yells, cries, and war whoops they use in time of action; so that it is impossible to consider them in any other light than as an assembly of demons.

'After some hours spent in dancing, the feast begins; the dishes being brought near me, I perceived that they consisted of dog's flesh; and I was informed that at all their public grand feasts they never use any other kind of food.

'In this custom of eating dog's flesh on particular occasions, they resemble the inhabitants of some of the countries that lie on the *n. e.* borders of Asia. The author of the account of Kamschatka, published by order of the Empress of Russia, informs us, that the people inhabiting Koreka, a country *n.* of Kamschatka, who wander about in hordes like the Arabs, when they pay their worship to the evil being, kill a rein-deer or a dog, the flesh of which they eat, and leave the head and tongue sticking on a pole with the front towards the *e.*: also, that when they are afraid of any infectious distemper, they kill a dog, and winding the guts about two poles pass between them. These customs, in which they are nearly imitated by the Indians, seem to add strength to my supposition, that America was first peopled from this quarter.

'"I know not," says a traveller among them, "under what class of dances to rank that performed by the Indians who came to my tent when I landed near lake Pepin, on the banks of the Mississippi. When I looked out, as I there mentioned, I saw about 20 naked young Indians, the most perfect in their shape, and by far the handsomest of any I had ever seen, coming towards me, and dancing as they approached to the music of their drums. At every ten or twelve yards they halted, and set up their yells and cries.

'"When they reached my tent I asked them to come in; which, without deigning to make me any answer, they did. As I observed that they were painted red and black, as they usually are when they go against an enemy, and perceived that some parts of the war dance were intermixed with their other movements, I doubted not but they were set on by the inimical chief who had refused my salutation; I therefore determined to sell my life as dear as possible. To this purpose, I received them sitting on my chest with my gun and pistols beside me, and ordered my men to keep a watchful eye on them, and to be also upon their guard.

'"The Indians being entered they continued their dance alternately, singing at the same time of their heroic exploits, and the superiority of their race over every other people. To enforce their language, though it was uncommonly nervous and expressive, and such as would of itself have carried terror to the firmest heart; at the end of every period they struck their war clubs against the poles of my tent with such violence that I expected every moment it would have tumbled upon us. As each of them, in dancing round, passed by me, they placed their right hand above their eyes, and coming close to me looked steadily in my face, which I could not construe into a token of friendship. My men gave themselves up for lost, and I acknowledge, for my own part, that I never found my apprehensions more tumultuous on any occasion.

'"When they had nearly ended their dance, I presented to them the pipe of peace, but they would not receive it. I then, as my last resource, thought I would try what presents would do; accordingly I took from my chest some ribbons and trinkets, which I laid before them. These seemed to stagger their resolutions, and to avert, in some] degree their anger; for after holding a consultation together they sat down on the ground, which I considered as a favourable omen.

' " Thus it proved, for in a short time they received the pipe of peace, and lighting it, first presented it to me, and then smoked with it themselves. Soon after they took up the presents, which had hitherto lain neglected, and appearing to be greatly pleased with them departed in a friendly manner; and never did I receive greater pleasure than at getting rid of such formidable guests.

' " It never was in my power to gain a thorough knowledge of the designs of my visitors. I had sufficient reason to conclude that they were hostile, and that their visit, at so late an hour, was made through the instigation of the Grand Sautor; but I was afterwards informed that it might be intended as a compliment which they usually pay to the chiefs of every other nation who happen to fall in with them, and that the circumstances in their conduct which had appeared so suspicious to me, were merely the effects of their vanity, and designed to impress on the minds of those whom they thus visited, an elevated opinion of their valour and prowess. In the morning before I continued my route, several of their wives brought me a present of some sugar, for whom I found a few more ribbons.

' " The dance of the sacrifice is not so denominated from their offering up at the same time a sacrifice to any good or evil spirit, but is a dance to which the Naudowessies give that title, from being used when any public fortunate circumstance befals them. Whilst I resided among them, a fine large deer accidentally strayed into the middle of their encampment, which they soon destroyed. As this happened just at the new moon, they esteemed it a lucky omen; and having roasted it whole, every one in the camp partook of it. After their feast, they all joined in a dance, which they termed, from its being somewhat of a religious nature, a dance of the sacrifice." (See *Dr. Hubbard's Compilation of Indian History*.)

' Hunting (continues Lewis) is the chief employ of the Indians; they are trained to it from their youth, and it is an exercise which is esteemed no less honourable than necessary towards their subsistence. A dexterous and resolute hunter is held in nearly as great estimation by them as a distinguished warrior. Scarcely any device, which the ingenuity of man has discovered, for ensnaring or destroying those animals that supply them with food, or whose skins are valuable, is unknown to them.

' Whilst they are engaged in this exercise, they shake off the indolence peculiar to their nature, and become active, persevering, and indefatigable. They are equally sagacious in finding their prey, and in the means they use to destroy it. They discern the footsteps of the beasts they are in pursuit of, although they are imperceptible to every other eye, and can follow them with certainty through the pathless forest.

' The beasts that the Indians hunt, both for their flesh, on which they subsist, and for their skins, which serve them either for apparel, or to barter with Europeans for necessaries, are the buffalo, elk, deer, moose, carriboo, bear, beaver, otter, martin, &c. I defer giving a description of these animals here, and shall only, at present, treat of the manner of hunting them.

' The route they shall take for this purpose, and the parties that shall go on the different expeditions, are fixed in their general councils, which are held some time in the summer, when all the operations for the ensuing winter are settled. The chief warrior, whose province it is to regulate their proceedings on this occasion, with great solemnity issues out an invitation to those who choose to attend him; for the Indians, as before observed, acknowledge no superiority, nor have they any idea of compulsion; and every one that accepts the invitation, prepares himself by fasting during several days.

' The Indians do not fast, as some other nations do, on the richest and most luxurious food, but totally abstain from every kind, either of victuals or drink; and such is their patience and resolution, that the most extreme thirst could not induce them to taste a drop of water; yet amidst this severe abstinence they appear cheerful and happy.

' The reasons they give for thus fasting are, that it enables them freely to dream, in which dreams they are informed where they shall find the greatest plenty of game; also that it averts the displeasure of the evil spirits, and induces them to be propitious. They also on these occasions blacken those parts of their bodies that are uncovered.

' The fast being ended, and the place of hunting made known, the chief, who is to conduct them, gives a grand feast to those who are to form the different parties; of which none of them dare to partake till they have bathed. At this feast, notwithstanding they have fasted so long, they eat with great moderation; and the chief who presides employs himself in rehearsing the feats of those who have been most successful in the business they are about to enter upon. They soon after set out on the march towards the place appointed, painted or rather bedaubed with black, amidst the acclamations of all the people.

'It is impossible to describe their agility or perseverance, whilst they are in pursuit of their prey; neither thickets, ditches, torrents, pools, nor rivers stop them; they always go straight forward in the most direct line they possibly can, and there are few of the savage inhabitants of the woods that they cannot overtake.

'When they hunt for bears, they endeavour to find out their retreats; for during the winter, these animals conceal themselves in the hollow trunks of trees, or make themselves holes in the ground, where they continue without food whilst the severe weather lasts.

When the Indians think they have arrived at a place where these animals usually haunt, they form themselves into a circle according to their number, and moving onward, endeavour, as they advance towards the centre, to discover the retreats of their prey. By this means, if any lie in the intermediate space, they are sure of arousing and bringing them down, either with their bows or their guns. The bears will take to flight at sight of a man or a dog, and will only make resistance when they are extremely hungry, or after they are wounded.

The Indian method of hunting the buffalo is, by forming a circle or a square, nearly in the same manner as when they search for the bear. Having taken their different stations, they set the grass, which at this time is rank and dry, on fire, and these animals, who are extremely fearful of that element, flying with precipitation before it, great numbers are hemmed in a small compass, and scarcely a single one escapes.

They have different ways of hunting the elk, the deer, and the carriboo. Sometimes they seek them out in the woods, to which they retire during the severity of the cold, where they are easily shot from behind the trees. In the more *n.* climates they take the advantage of the weather to destroy the elk; when the sun has just strength enough to melt the snow, and the frost in the night forms a kind of crust on the surface, this animal being heavy, breaks it with his forked hoofs, and with difficulty extricates himself from it: at this time, therefore, he is soon overtaken and destroyed.

'Some nations have a method of hunting these animals which is more easily executed, and free from danger. The hunting party divides into two bands, and choosing a spot near the borders of some river, one party embarks on board their canoes, whilst the other, forming themselves into a semicircle on the land, the flanks of which reach the shore, let loose their dogs, and by this means rouse all the game that lies within these bounds; they then drive them towards the river, into which they no sooner enter, than the greatest part of them are immediately dispatched by those who remain in the canoes.

'Both the elk and buffalo are very furious when they are wounded, and will return fiercely on their pursuers, and trample them under their feet, if the hunter finds no means to complete their destruction, or does not seek for security in flight to some adjacent tree; by this method they are frequently avoided, and so tired with the pursuit that they voluntarily give it over.

'But the hunting in which the Indians, particularly those who inhabit the *n.* parts, chiefly employ themselves, and from which they reap the greatest advantage, is that of the beaver. The season for this lasts the whole of the winter, from November to April; during which time the fur of these animals is in the greatest perfection. A description of this extraordinary animal, the construction of their huts, and the regulations of their almost rational community, I shall give in another place.

'The hunters make use of several methods to destroy them. Those generally practised, are either taking them in snares, cutting through the ice, or opening their causeways.

'As the eyes of these animals are very quick, and their hearing exceedingly acute, great precaution is necessary in approaching their bodies; for as they seldom go far from the water, and their houses are always built close to the side of some large river or lake, or dams of their own construction, upon the least alarm they hasten to the deepest part of the water, and dive immediately to the bottom; as they do this they make a great noise by beating the water with their tails, on purpose to put the whole fraternity on their guard.

'They are taken with snares, in the following manner:—though the beavers usually lay up a sufficient store of provision to serve for their subsistence during the winter, from time to time they make excursions to the neighbouring woods, to procure further supplies.

'The hunters having found out their haunts, place a trap in their way, baited with small pieces of bark, or young shoots of trees, which the beaver has no sooner laid hold of, than a large log of

[wood falls upon him and breaks his back; his enemies, who are upon the watch, soon appear, and instantly dispatch the helpless animal.

'At other times, when the ice on the rivers and lakes is about half a foot thick, they make an opening through it with their hatchets, to which the beavers will soon hasten, on being disturbed at their houses, for a supply of fresh air. As their breath occasions a considerable motion in the water, the hunter has sufficient notice of their approach, and methods are easily taken for knocking them on the head the moment they appear above the surface.

'When the houses of the beavers happen to be near a rivulet, they are more easily destroyed: the hunters then cut the ice, and spreading a net under it, break down the cabins of the beavers, who never fail to make towards the deepest part, where they are entangled and taken. But they must not be suffered to remain there long, as they would soon extricate themselves with their teeth, which are well known to be excessively sharp and strong.

'The Indians take great care to hinder their dogs from touching the bones of the beavers. The reasons they give for these precautions are, first, that the bones are so excessively hard, that they spoil the teeth of the dogs; and secondly, that they are apprehensive they shall so exasperate the spirits of the beavers by this permission, as to render the next hunting season nnsuccessful.

'When the Indians destroy buffaloes, elks, deer, &c. they generally divide the flesh of such as they have taken among the tribe to which they belong. But in hunting the beaver a few families usually unite, and divide the spoil between them. Indeed, in the first instance they generally pay some attention in the division to their own familes; but no jealousies or murmurings are ever known to arise on account of any apparent partiality.

'Among the Naudowessies, if a person shoots a deer, buffalo, &c. and it runs a considerable distance before it drops, where a person belonging to another tribe, being nearer, first sticks a knife into it, the game is considered as the property of the latter, notwithstanding it had been mortally wounded by the former. Though this custom appears to be arbitrary and unjust, yet that people cheerfully submit to it. This decision is, however, very different from that practised by the Indians on the back of the colonies, where the first person that hits is entitled to the best share.

'The Indians begin to bear arms at the age of 15, and lay them aside when they arrive at the age of 60. Some nations to the *s.* I have been informed, do not continue their military exertions after they are 50.

'In every band or nation there is a select number who are styled the warriors, and who are always ready to act either offensively or defensively, as occasion requires. These are well armed, bearing the weapons commonly used among them, which vary according to the situation of their countries. Some make use of tomahawks, knives, and fire-arms; but those who have not an opportnity of purchasing these kinds of weapons, use bows and arrows, and also the casse-tete, or war club.

'The Indians that inhabit still further to the *w.* a country which extends to the S. sea, use in fight a warlike instrument that is very uncommon. Having great plenty of horses, they always attack their enemies on horseback, and encumber themselves with no other weapon than a stone of a middling size, curiously wrought, which they fasten by a string, about a yard and a half long, to their right arms, a little above the elbow. These stones they conveniently carry in their hands, till they reach their enemies, and then swinging them with great dexterity, as they ride full speed, never fail of doing execution. The country which these tribes possess abounding with large extensive plains, those who attack them seldom return, as the swiftness of the horses on which they are mounted, enables them to overtake even the fleetest of their invaders.

'I was informed, that unless they found morasses or thickets, to which they could retire, they were sure of being cut off; to prevent this they always took care, whenever they made an onset, to do it near such retreats as are impassable for cavalry, having then a great advantage over their enemies, whose weapons would not there reach them.

'Some nations make use of a javelin, pointed with bone, worked into different forms; but the Indian weapons in general are bows and arrows, and the short club already mentioned. The latter is made of a very hard wood, and the head of it fashioned round like a ball, about 3½ inches diameter; in this rotund part is fixed an edge resembling that of a tomahawk, either of steel or flint, whichsoever they can procure.

'The dagger is peculiar to some nations, and of ancient construction; but they can give no account how long it has been in use among them. It was originally made of flint or bone, but since] [they have had communication with the European traders, they have formed it of steel. The length is about 10 inches, and that part close to the handle nearly three inches broad. Its edges are keen, and it gradully tapers towards a point. They wear it in a sheath made of deer's leather, neatly ornamented with porcupine quills; and it is usually hung by a string, decorated in the same manner, which reaches as low as the breast. This curious weapon is worn by a few of the principal chiefs alone, and considered both as a useful instrument, and an ornamental badge of superiority.

'I observed among them a few targets, or shields, made of raw buffalo hides, and in the form of those used by the ancients: but as the number of these was small, and I coald gain no intelligence of the era in which they first were introduced among them, I suppose those I saw had descended from father to son for many generations.

'The reasons the Indians give for making war against one another, are much the same as those urged by more civilized nations for disturbing the tranquillity of their neighbours. The pleas of the former are, however, in general more rational and just, than such as are brought by Europeans in vindication of their proceedings.

'The extension of empire is seldom a motive with these people to invade, and to commit depredations on the territories of those who happen to dwell near them. To secure the rights of hunting within particular limits, to maintain the liberty of passing through their accustomed tracks, and to guard those lands which they consider from a long tenure as their own, against any infringement, are the general causes of those dissensions that so often break out between the Indian nations, and which are carried on with so much animosity.

'Though strangers to the idea of separate property, yet the most uncultivated among them are well acquainted with the rights of their community to the domains they possess, and oppose with vigour every encroachment on them.

'Notwithstanding it is generally supposed that from their territories being so extensive, the boundaries of them cannot be ascertained, yet I am well assured that the limits of each nation in the interior parts are laid down in their rude plans with great precision. By theirs, as I have before observed, was I enabled to regulate my own; and after the most exact observations and inquiries I found but a very few instances in which they erred.

'But interest is not either the most frequent or most powerful incentive to their making war on each other. The passion of revenge, which is the distinguishing characteristic of these people, is the most general motive. Injuries are felt by them with exquisite sensibility, and vengeance pursued with unremitted ardour. To this may be added that natural excitation which every Indian is sensible of, as soon as he approaches the age of manhood, to give proof of his valour and prowess.

'As they are early possessed with a notion that war ought to be the chief business of their lives, that there is nothing more desirable than the reputation of being a great warrior, and that the scalps of their enemies, or a number of prisoners, are alone to be esteemed valuable, it is not to be wondered at, that the young Indians are continually restless and uneasy if their ardour is repressed, and they are kept in a state of inactivity. Either of these propensities, the desire of revenge or the gratification of an impulse, that by degrees become habitual to them, is sufficient, frequently, to induce them to commit hostilities on some of the neighbouring nations.

'When the chiefs find any occasion for making war, they endeavour to arouse their prejudices, and by that means soon excite their warriors to take arms. For this purpose they make use of their martial eloquence, nearly in the following words, which never fail of proving effectual: "The bones of our deceased countrymen lie uncovered, they call out to us to revenge their wrongs, and we must satisfy their request. Their spirits cry out against us. They must be appeased. The genii, who are the guardians of our honour, inspire us with a resolution to seek the enemies of our murdered brothers. Let us go and devour those by whom they were slain. Sit here no longer inactive, give way to the impulse of your natural valour, anoint your hair, paint your faces, fill your quivers, cause the forest to resound with your songs, console the spirits of the dead, and tell them they shall be revenged."

'Animated by these exhortations, the warriors snatch their arms in a transport of fury, sing the songs of war, and burn with impatience to imbrue their hands in the blood of their enemies.

'Sometimes private chiefs assemble small parties and make excursions against those with whom they are at war, or such as have injured them. A single warrior, prompted by revenge or a desire to show his prowess, will march unattended for several hundred miles, to surprise and cut off a straggling party.

'These irregular sallies, however, are not al-]

[ways approved of by the elder chiefs, though they are often obliged to connive at them.

'But when a war is national, and undertaken by the community, their deliberations are formal and slow. The elders assemble in council, to which all the head warriors and young men are admitted, where they deliver their opinions in solemn speeches, weighing with maturity the nature of the enterprise they are about to engage in, and balancing with great sagacity the advantages or inconveniences that will arise from it.

'Their priests are also consulted on the subject, and even, sometimes, the advice of the most intelligent of their women is asked.

'If the determination be for war, they prepare for it with much ceremony.

'The chief warrior of a nation does not on all occasions head the war party himself, he frequently deputes a warrior of whose valour and prudence he has a good opinion. The person thus fixed on, being first bedaubed with black, observes a fast of several days, during which he invokes the Great Spirit, or deprecates the anger of the evil ones, holding, whilst it lasts, no converse with any of his tribe.

'He is particularly careful at the same time to observe his dreams, for on these do they suppose their success will in a great measure depend; and from the firm persuasion every Indian, actuated by his own presumptuous thoughts, is impressed with, that he shall march forth to certain victory, these are generally favourable to his wishes.

'After he has fasted as long as custom prescribes, he assembles the warriors, and holding a belt of wampum in his hand, thus addresses them:—

'"Brothers! by the inspiration of the Great Spirit I now speak unto you, and by him am I prompted to carry into execution the intentions which I am about to disclose to you. The blood of our deceased brothers is not yet wiped away; their bodies are not yet covered, and I am going to perform this duty to them."

'Having then made known all the motives that induced him to take up arms against the nation with whom they are to engage, he thus proceeds: "I have therefore resolved to march through the war path to surprise them. We will eat their flesh, and drink their blood; we will take scalps, and make prisoners; and should we perish in this glorious enterprise we shall not be for ever hidden in the dust, for this belt shall be a recompense to him who buries the dead." Having said this, he lays the belt on the ground, and he who takes it up declares himself his lieutenant, and is considered as the second in command; this, however, is only done by some distinguished warrior who has a right, by the number of his scalps, to the post.

'Though the Indians thus assert that they will eat the flesh and drink the blood of their enemies, the threat is only to be considered as a figurative expression; notwithstanding they sometimes devour the hearts of those they slay, and drink their blood, by way of bravado, or to gratify in a more complete manner their revenge.

'The chief is now washed from his sable covering, anointed with bear's fat, and painted with their red paint, in such figures as will make him appear most terrible to his enemies. He then sings the war song, and enumerates his warlike actions. Having done this he fixes his eyes on the sun, and pays his adoration to the Great Spirit, in which he is accompanied by all the warriors.

'This ceremony is followed with dances, such as I have before described; and the whole concludes with a feast, which usually consists of dog's flesh.

'This feast is held in the hut or tent of the chief warrior, to which all those who intend to accompany him in his expedition send their dishes to be filled; and during the feast, notwithstanding he has fasted so long, he sits composedly with his pipe in his mouth, and recounts the valorous deeds of his family.

'As the hopes of having their wounds, should they receive any, properly treated and expeditiously cured, must be some additional inducement to the warriors to expose themselves more freely to danger, the priests, who are also their doctors, prepare such medicines as will prove efficacious. With great ceremony they carry various roots and plants, and pretend that they impart to them the power of healing.

'Notwithstanding this superstitions method of proceeding, it is very certain that they have acquired a knowledge of many plants and herbs that are of a medicinal quality, and which they know how to use with great skill.

'From the time the resolution of engaging in war is taken, to the departure of the warriors, the nights are spent in festivity, and their days in making the needful preparations.

'If it is thought necessary by the nation going to war, to solicit the alliance of any neighbouring tribe, they fix upon one of their chiefs, who speaks the language of that people well, and who is a good orator, and send to them by him a belt of wampum, on which is specified the purport of the embassy, in figures that every nation is well ac-]

[quainted with. At the same time he carries with him a hatchet painted red.

'As soon as he reaches the camp or village to which he is destined, he acquaints the chief of the tribe with the general tenor of his commission, who immediately assembles a council, to which the ambassador is invited. There having laid the hatchet on the ground he holds the belt in his hand, and enters more minutely into the occasion of his embassy. In his speech he invites them to take up the hatchet, and as soon as he has finished speaking, delivers the belt.

'If his hearers are inclined to become auxiliaries to his nation, a chief steps forward and takes up the hatchet, and they immediately espouse with spirit the cause they have thus engaged to support. But if on this application neither the belt nor hatchet are accepted, the emissary concludes that the people whose assistance he solicits have already entered into an alliance with the foes of his nation, and returns with speed to inform his countrymen of his ill success.

'The manner in which the Indians declare war against each other, is by sending a slave with a hatchet, the handle of which is painted red, to the nation which they intend to break with; and the messenger, notwithstanding the danger to which he is exposed from the sudden fury of those whom he thus sets at defiance, executes his commission with great fidelity.

'Sometimes this token of defiance has such an instantaneous effect on those to whom it is presented, that in the first transports of their fury a small party will issue forth, without waiting for the permission of the elder chiefs, and slaying the first of the offending nation they meet, cut open the body, and stick a hatchet of the same kind as that they just received, into the heart of their slaughtered foe. Among the more remote tribes this is done with an arrow or spear, the end of which is painted red. And the more to exasperate, they dismember the body, to show that they esteem them not as men, but as old women.

'The Indians seldom take the field in large bodies, as such numbers would require a greater degree of industry to provide for their subsistence, during their tedious marches through dreary forests, or long voyages over lakes and rivers, than they would care to bestow.

'Their armies are never encumbered with baggage or military stores. Each warrior, besides his weapons, carries with him only a mat, and, whilst at a distance from the frontiers of the enemy, supports himself with the game he kills or the fish he catches.

'When they pass through a country where they have no apprehensions of meeting with an enemy, they use very little precaution; sometimes there are scarcely a dozen warriors left together, the rest being in pursuit of their game; but though they should have roved to a very considerable distance from the war path, they are sure to arrive at the place of rendezvous by the hour appointed.

'They always pitch their tents long before sunset; and being naturally presumptuous, take very little care to guard against a surprise. They place great confidence in their Manitous, or household gods, which they always carry with them; and being persuaded that they take upon them the office of centinels, they sleep very securely under their protection.

'These Manitous, as they are called by some nations, but which are termed Wakons, that is, spirits, by the Naudowessies, are nothing more than the otter and martin skins I have already described, for which, however, they have a great veneration.

'After they have entered the enemy's country, no people can be more cautious and circumspect; fires are no longer lighted, no more shouting is heard, nor the game any longer pursued. They are not even permitted to speak; but must convey whatever they have to impart to each other, by signs and motions.

'They now proceed wholly by stratagem and ambuscade. Having discovered their enemies, they send to reconnoitre them; and a council is immediately held, during which they speak only in whispers, to consider of the intelligence imparted by those who were sent out.

'The attack is generally made just before day-break, at which period they suppose their foes to be in the soundest sleep. Throughout the whole of the preceding night they will lie flat upon their faces, without stirring; and make their approaches in the same posture, creeping upon their hands and feet till they are got within bow-shot of those they have destined to destruction. On a signal given by the chief warrior, to which the whole body makes answer by the most hideous yells, they all start up, and discharging their arrows in the same instant, without giving their adversaries time to recover from the confusion into which they are thrown, pour in upon them with their war clubs or tomahawks.

'The Indians think there is little glory to be acquired from attacking their enemies openly in the field; their greatest pride is to surprise and destroy. They seldom engage with a manifest appearance of disadvantage. If they find the enemy on their] [guard, too strongly entrenched, or superior in numbers, they retire, provided there is an opportunity of doing so. And they esteem it the greatest qualification of a chief warrior, to be able to manage an attack, so as to destroy as many of the enemy as possible, at the expence of a few men.

'When the Indians succeed in their silent approaches and are able to force the camp which they attack, a scene of horror that exceeds description ensues. The savage fierceness of the conquerors, and the desperation of the conquered, who well know what they have to expect should they fall alive into the hands of their assailants, occasion the most extraordinary exertions on both sides. The figure of the combatants, all besmeared with black and red paint, and covered with the blood of the slain, their horrid yells and ungovernable fury, are not to be conceived by those who have never seen them. Though the Indians are negligent in guarding against surprise, they are alert and dexterous in surprising their enemies. To their caution and perseverance in stealing on the party they design to attack, they add that admirable talent, or rather instinctive qualification I have already described, of tracing out those they are in pursuit of. On the smoothest grass, on the hardest earth, and even on the very stones, will they discover the traces of an enemy, and by the shape of the footsteps, and the distance between the prints, distinguish not only whether it is a man or woman who has passed that way, but even the nation to which they belong. However incredible this might appear, yet, from the many proofs I received, whilst among them, of their amazing sagacity in this point, I see no reason to discredit even these extraordinary exertions of it.

'When they have overcome an enemy, and victory is no longer doubtful, the conquerors first dispatch all such as they think they shall not be able to carry off without great trouble, and then endeavour to take as many prisoners as possible; after this they return to scalp those who are either dead, or too much wounded to be taken with them.

'At this business they are exceedingly expert. They seize the head of the disabled or dead enemy, and placing one of their feet on the neck, twist their left hand in the hair; by this means, having extended the skin that covers the top of the head, they draw out their scalping knives, which are always kept in good order for this cruel purpose, and with a few dexterous strokes take off the part that is termed the scalp. They are so expeditions in doing this, that the whole time required scarcely exceeds a minute. These they preserve as monuments of their prowess, and at the same time as proofs of the vengeance they have inflicted on their enemies.

'If two Indians seize in the same instant a prisoner, and seem to have an equal claim, the contest between them is soon decided; for to put a speedy end to any dispute that might arise, the person that is apprehensive he shall lose his expected reward, immediately has recourse to his tomahawk or war club, and knocks on the head the unhappy cause of their contention.

'Having completed their purposes, and made as much havoc as possible, they immediately retire towards their own country, with the spoil they have acquired, for fear of being pursued.

'Should this be the case, they make use of many stratagems to elude the searches of their pursuers. They sometimes scatter leaves, sand, or dust over the prints of their feet; sometimes tread in each other's footsteps; and sometimes lift their feet so high, and tread so lightly, as not to make any impression on the ground. But if they find all these precautions unavailing, and that they are near being overtaken, they first dispatch and scalp their prisoners, and then dividing, each endeavours to regain his native country by a different route. This prevents all farther pursuit; for their pursuers now despairing, either of gratifying their revenge, or of releasing those of their friends who were made captives, return home.

'If the successful party is so lucky as to make good their retreat unmolested, they hasten with the greatest expedition to reach a country where they may be perfectly secure; and that their wounded companions may not retard their flight, they carry them by turns in litters, or, if it is in the winter season, draw them on sledges.

'The prisoners, during their march, are guarded with the greatest care. During the day, if the journey is over land, they are always held by some of the victorious party; if by water, they are fastened to the canoe. In the night time they are stretched along the ground quite naked, with their legs, arms, and neck fastened to hooks fixed in the ground. Besides this, cords are tied to their arms or legs, which are held by an Indian, who instantly awakes at the least motion of them.

'During their march they oblige their prisoners to sing their death song, which generally consists of these or similar sentences: "I am going to die, I am about to suffer: but I will bear the severest tortures my enemies can inflict, with becoming fortitude. I will die like a brave man, and I shall then go to join the chiefs that have suffered on the] [same account." These songs are continued with necessary intervals, until they reach the village or camp to which they are going.

'When the warriors are arrived within hearing, they set up different cries, which communicate to their friends a general history of the success of the expedition. The number of the death cries they give, declare how many of their own party are lost; the number of war whoops, the number of prisoners they have taken.

'It is difficult to describe these cries, but the best idea I can convey of them is, that the former consists of the sounds whoo, whoo, whoop, continued in a long shrill tone, nearly till the breath is exhausted, and then broken off with a sudden elevation of the voice. The latter is a loud cry, of much the same kind, which is modulated into notes by the hand being placed before the mouth. Both of them might be heard to a very considerable distance.

'Whilst these are uttering, the persons to whom they are designed to convey the intelligence, continue motionless and all attention. When this ceremony is performed, the whole village issue out to learn the particulars of the relation they have just heard in general terms; and accordingly, as the news proves mournful, or the contrary, they answer by so many acclamations or cries of lamentation.

'Being by this time arrived at the village or camp, the women and children arm themselves with sticks and bludgeons, and form themselves into two ranks, through which the prisoners are obliged to pass. The treatment they undergo before they reach the extremity of the line, is very severe. Sometimes they are so beaten over the head and face, as to have scarcely any remains of life; and happy would it be for them if by this usage an end was put to their wretched beings. But their tormentors take care that none of the blows they give prove mortal, as they wish to reserve the miserable sufferers for more severe inflictions.

'After having undergone this introductory discipline, they are bound hand and foot, whilst the chiefs hold a council in which their fate is determined. Those who are decreed to be put to death by the usual torments, are delivered to the chief of the warriors; such as are to be spared, are given into the hands of the chief of the nation: so that in a short time all the prisoners may be assured of their fate, as the sentence now pronounced is irrevocable. The former they term being consigned to the house of death, the latter to the house of grace.

'Such captives as are pretty far advanced in life, and have acquired great honour by their warlike deeds, always atone for the blood they have spilt by the tortures of fire. Their success in war is readily known by the blue marks upon their breasts and arms, which are as legible to the Indians as letters are to Americans.

'The manner in which these hieroglyphics are made, is by breaking the skin with the teeth of fish, or sharpened flints, dipped in a kind of ink, made of the soot of pitch pine. Like those of the ancient Picts of Britain, these are esteemed ornamental; and at the same time they serve as registers of the heroic actions of the warrior, who thus bears about him indelible marks of his valour.

'The prisoners destined to death are soon led to the place of execution, which is generally in the centre of the camp or village; where, being stript, and every part of their bodies blackened, the skin of a crow or raven is fixed on their heads. They are then bound to a stake, with faggots heaped around them, and obliged, for the last time, to sing their death song.

'The warriors, for such only commonly suffer this punishment, now perform in a more prolix manner this sad solemnity. They recount with an audible voice all the brave actions they have performed, and pride themselves in the number of enemies they have killed. In this rehearsal they spare not even their tormentors, but strive by every provoking tale they can invent, to irritate and insult them. Sometimes this has the desired effect, and the sufferers are dispatched sooner than they otherwise would have been.

'There are many other methods which the Indians employ to put their prisoners to death, but these are only occasional; that of burning is most generally used.

'This method of tormenting their enemies is considered by the Indians as productive of more than one beneficial consequence. It satiates, in a greater degree, that diabolical lust of revenge, which is the predominant passion in the breast of every individual of every tribe, and it gives the growing warriors an early propensity to that cruelty and thirst for blood, which is so necessary a qualification for such as would be thoroughly skilled in their savage art of war.

'Notwithstanding these acts of severity exercised by the Indians towards those of their own species who fall into their hands, some tribes have been very remarkable for their moderation to such female prisoners, belonging to the English colonies, as have happened to be taken by them. Women of great beauty have frequently been carried off] [by them, and during a march of 3 or 400 miles, through their retired forests, have lain by their sides without receiving any insult, and their chastity has remained inviolate. Instances have happened, where female captives, who have been pregnant at the time of their being taken, have found the pangs of child-birth come upon them in the midst of solitary woods, and savages their only companions; yet from these, savages as they were, have they received every assistance their situations would admit of, and been treated with a degree of delicacy and humanity they little expected.

'Those prisoners that are consigned to the house of grace, and these are commonly the young men, women, and children, await the disposal of the chiefs, who, after the execution of such as are condemned to die, hold a council for this purpose.

'A herald is sent round the village or camp, to give notice that such as have lost any relations in the late expedition, are desired to attend the distribution which is about to take place. Those women who have lost their sons or husbands, are generally satisfied in the first place; after these, such as have been deprived of friends of a more remote degree of consanguinity, or who choose to adopt some of the youth.

'The division being made, which is done, as in other cases, without the least dispute, those who have received any share lead them to their tents or huts; and having unbound them, wash and dress their wounds, if they happen to have received any; they then clothe them, and give the most comfortable and refreshing food their store will afford.

'Whilst their new domestics are feeding, they endeavour to administer consolation to them, they tell them that as they are redeemed from death, they must now be cheerful and happy; and if they serve them well, without murmuring or repining, nothing shall be wanting to make them such atonement for the loss of their country and friends as circumstances will allow.

'If any men are spared, they are commonly given to the widows that have lo-t their husbands by the hand of the enemy, should there be any such, to whom, if they happen to prove agreeable, they are soon married. But should the dame be otherwise engaged, the life of him who falls to her lot is in great danger; especially if she fancies that her late husband wants a slave in the country of spirits to which he is gone.

'When this is the case, a number of young men take the devoted captive to some distance, and dispatch him without any ceremony. After he has been spared by the council, they consider him of too little consequence to be entitled to the torments allotted to those who have been judged worthy of them.

'The women are usually distributed to the men, from whom they do not fail of meeting with a favourable reception. The boys and girls are taken into the families of such as have need of them, and are considered as slaves; and it is not uncommon that they are sold in the same capacity to the American traders who come among them.

'The Indians have no idea of moderating the ravages of war, by sparing their prisoners and entering into a negociation with the band from whom they have been taken, for an exchange. All that are taken captive by both parties, are either put to death, adopted, or made slaves of. And so particular is every nation in this respect, that if any of their tribe, even a warrior, should be taken prisoner, and by chance be received into the house of grace, either as an adopted person or a slave, and should afterwards make his escape, they will by no means receive him, or acknowledge him as one of their band.

'The condition of such as are adopted, differs not in any one instance from the children of the nation to which they now belong. They assume all the rights of those whose places they supply, and frequently make no difficulty of going in the war parties against their own countrymen. Should, however, any of those by chance make their escape, and be afterwards retaken, they are esteemed as unnatural children, and ungrateful persons, who have deserted and made war upon their parents and benefactors, and are treated with uncommon severity.

'That part of the prisoners which are considered as slaves, are generally distributed among the chiefs; who frequently make presents of some of them to the American governors of the out-posts, or to the superintendants of Indian affairs. I have been informed that the Jesuits and French missionaries first occasioned the introduction of these unhappy captives into the settlements, and who, by so doing, taught the Indians that they were valuable.

'Their views indeed were laudable, as they imagined that by this method they should not only prevent much barbarity and bloodshed, but find the opportunities of spreading their religion among them increased. To this purpose they have encouraged the traders to purchase such slaves as they met with.

'The good effects of this mode of proceeding, were not however equal to the expectations of these] [pious fathers. Instead of being the means of preventing cruelty and bloodshed, it only caused dissensions between the Indian nations to be carried on with a greater degree of violence, and with unremitted ardour. The prize they fought for being no longer revenge or fame, but the acquirement of spirituous liquors, for which their captives were to be exchanged, and of which almost every nation is immoderately fond, they sought for their enemies with unwonted alacrity, and were constantly on the watch to surprise and carry them off.

'It might still be said, that fewer of the captives are tormented and put to death, since these expectations of receiving so valuable a consideration for them have been excited, than there usually had been; but it does not appear that their accustomed cruelty to the warriors they take, is in the least abated; their natural desire of vengeance must be gratified: they now only become more assiduous in securing a greater number of young prisoners, whilst those who are made captive in their defence, are tormented and put to death as before. And this, even in despite of the disgraceful estimation; for the Indians consider every conquered people as in a state of vassalage to their conquerors. After one nation has finally subdued another, and a conditional submission is agreed on, it is customary for the chiefs of the conquered, when they sit in council with their subduers, to wear petticoats, as an acknowledgment that they are in a state of subjection, and ought to be ranked among the women. Their partiality to the French has, however, taken too deep root for time itself to eradicate it.

'The wars that are carried on between the Indian nations are in general hereditary, and continue from age to age with a few interruptions. If a peace becomes necessary, the principal care of both parties is to avoid the appearance of making the first advances.

'When they treat with an enemy, relative to a suspension of hostilities, the chief who is commissioned to undertake the negociation, if it is not brought about by the mediation of some neighbouring band, abates nothing of his natural haughtiness; even when the affairs of his country are in the worst situation, he makes no concessions, but endeavours to persuade his adversaries that it is their interest to put an end to the war.

'Accidents sometimes contribute to bring about a peace between nations that otherwise could not be prevailed on to listen to terms of accommodation.

'Sometimes the Indians grow tired of a war, which they have carried on against some neighbouring nation for many years without much success, and in this case they seek for mediators to begin a negociation. These being obtained, the treaty is thus conducted:

'A number of their own chiefs, joined by those who have accepted the friendly office, set out together for the enemy's country; such as are chosen for this purpose, are chiefs of the most extensive abilities, and of the greatest integrity. They bear before them the pipe of peace, which I need not inform my readers is of the same nature as a flag of truce among the Americans, and is treated with the greatest respect and veneration, even by the most barbarous nations. I never heard of an instance wherein the bearers of this sacred badge of friendship were ever treated disrespectfully, or its rights violated. The Indians believe that the Great Spirit never suffers an infraction of this kind to go unpunished.

'The pipe of peace, which is termed by them the *calumet*, for what reason I could never learn, is about four feet long. The bowl of it is made of red marble, and the stem of a light wood, curiously painted with hieroglyphics in various colours, and adorned with feathers of the most beautiful birds; but it is not in my power to convey an idea of the various tints and pleasing ornaments of this much esteemed Indian implement.

'Every nation has a different method of decorating these pipes, and they can tell at first sight to what band it belongs. It is used as an introduction to all treaties, and great ceremony attends the use of it on these occasions.

'The assistant or aid-de-camp of the great warrior, when the chiefs are assembled and seated, fills it with tobacco mixed with herbs, taking care at the same time that no part of it touches the ground. When it is filled, he takes a coal that is thoroughly kindled, from a fire which is generally kept burning in the midst of the assembly, and places it on the tobacco.

'As soon as it is sufficiently lighted, he throws off the coal. He then turns the stem of the pipe towards the heavens, after this towards the earth, and now holding it horizontally, moves himself round till he has completed a circle. By the first action he is supposed to present it to the Great Spirit, whose aid is thereby supplicated; by the second, to avert any malicious interposition of the evil spirits; and by the third to gain the protection of the spirits inhabiting the air, the earth, and the waters. Having thus secured the favour of these invisible agents, in whose power they] [suppose it is either to forward or obstruct the issue of their present deliberations, he presents it to the hereditary chief, who having taken two or three whiffs, blows the smoke from his mouth, first towards heaven, and then around him upon the ground.

'It is afterwards put in the same manner into the mouths of the ambassadors or strangers, who observe the same ceremony, then to the chief of the warriors, and to all the other chiefs in turn, according to their gradation. During this time the person who executes this honourable office holds the pipe slightly in his hand, as if he feared to press the sacred instrument; nor does any one presume to touch it but with his lips.

'When the chiefs who are entrusted with the commission for making peace, approach the town or camp to which they are going, they begin to sing and dance the songs and dances appropriated to this occasion. By this time the adverse party are apprised of their arrival, and, at the sight of the pipe of peace, divesting themselves of their wonted enmity, invite them to the habitation of the great chief, and furnish them with every conveniency during the negociation.

'A council is then held; and when the speeches and debates are ended, if no obstructions arise to put a stop to the treaty, the painted hatchet is buried in the ground, as a memorial that all animosities between the contending nations have ceased, and a peace taken place. Among the ruder bands, such as have no communication with the Americans, a war club painted red is buried, instead of the hatchet.

'A belt of wampum is also given on this occasion, which serves as a ratification of the peace, and records to the latest posterity, by the hieroglyphics into which the beads are formed, every stipulated article in the treaty.

'These belts are made of shells found on the coasts of New England and Virginia, which are sawed out into beads of an oblong form, about a quarter of an inch long, and round like other beads. Being strung on leathern strings, and several of them sewed neatly together with fine sinewy threads, they compose what is termed a belt of wampum.

'The shells are generally of two colours, some white and others violet; but the latter are more highly esteemed than the former. They are held in as much estimation by the Indians, as gold, silver, or precious stones are by the Americans.

'The belts are composed of 10, 12, or a greater number of strings, according to the importance of the affair in agitation, or the dignity of the person to whom it is presented. On more trifling occasions, strings of these beads are presented by the chiefs to each other, and frequently worn by them about their necks as a valuable ornament.

'The Indians allow of polygamy, and persons of every rank indulge themselves in this point. The chiefs in particular have a seraglio, which consists of an uncertain number, usually from six to 12 or 14. The lower rank are permitted to take as many as there is a probability of their being able, with the children they may bear, to maintain. It is not uncommon for an Indian to marry two sisters; sometimes, if there happen to be more, the whole number; and notwithstanding this (as it appears to civilized nations) unnatural union, they all live in the greatest harmony.

'The younger wives are submissive to the elder; and those who have no children, do such menial offices for those who are fertile, as causes their situation to differ but little from a state of servitude. However, they perform every injunction with the greatest cheerfulness, in hopes of gaining thereby the affections of their husbands, that they in their turn may have the happiness of becoming mothers, and be entitled to the respect attendant on that state.

'It is not uncommon for an Indian, although he takes to himself so many wives, to live in a state of continence with some of them for several years. Such as are not so fortunate as to gain the favour of their husband, by their submissive and prudent behaviour, and by that means to share in his embraces, continue in their virgin state during the whole of their lives, except they may happen to be presented by him to some stranger chief, whose abode among them will not admit of his entering into a more lasting connection. In this case they submit to the injunction of their husband without murmuring, and are not displeased with the temporary union. But if at any time it is known that they take this liberty without first receiving his consent, they are punished in the same manner as if they had been guilty of adultery.

'This custom is more prevalent among the nations which lie in the interior parts, than among those that are nearer the settlements, as the manners of the latter are rendered more conformable in some points to those of the Americans, by the intercourse they hold with them.

'The Indian nations differ but little from each other in their marriage ceremonies, and less in the manner of their divorces. The tribes that inhabit the borders of Canada, make use of the following custom.

'When a young Indian has fixed his inclinations]

[on one of the other sex, he endeavours to gain her consent, and if he succeeds, it is never known that her parents obstruct their union. When every preliminary is agreed on, and the day appointed, the friends and acquaintance of both parties assemble at the house or tent of the oldest relation of the bridegroom, where a feast is prepared on the occasion.

'The company who meet to assist at the festival are sometimes very numerous: they dance, they sing, and enter into every other diversion usually made use of on any of their public rejoicings.

'When these are finished, all those who attended merely out of ceremony depart, and the bridegroom and bride are left alone with three or four of the nearest and oldest relations of either side; those of the bridegroom being men, and those of the bride, women.

'Presently the bride, attended by these few friends, having withdrawn herself for the purpose, appears at one of the doors of the house, and is led to the bridegroom, who stands ready to receive her. Having now taken their station, on a mat placed in the centre of the room, they lay hold of the extremities of a wand, about four feet long, by which they continue separated, whilst the old men pronounce some short harangues suitable to the occasion.

'The married couple then make a public declaration of the love and regard they entertain for each other, and still holding the rod between them, dance and sing. When they have finished this part of the ceremony, they break the rod into as many pieces as there are witnesses present, who each take a piece and preserve it with great care.

'The bride is then reconducted out of the door at which she entered, where her young companions wait to attend her to her father's house; there the bridegroom is obliged to seek her, and the marriage is consummated. Very often the wife remains at her father's house till she has a child, when she packs up her apparel, which is all the fortune she is generally possessed of, and accompanies her husband to his habitation.

'When from any dislike a separation takes place, for they are seldom known to quarrel, they generally give their friends a few days notice of their intentions, and sometimes offer reasons to justify their conduct. The witnesses who were present at the marriage, meet on the day requested, at the house of the couple that are about to separate, and bringing with them the pieces of rod which they had received at their nuptials, throw them into the fire in the presence of all the parties.

'This is the whole of the ceremony required, and the separation is carried on without any murmurings or ill will between the couple or their relations; and after a few months they are at liberty to marry again.

'When a marriage is thus dissolved, the children which have been produced from it, are equally divided between them; and as children are esteemed a treasure by the Indians, if the number happens to be odd, the woman is allowed to take the better half.

'Though this custom seems to encourage fickleness and frequent separations, yet there are many of the Indians who have but one wife, and enjoy with her a state of connubial happiness not to be exceeded in more refined societies. There are also not a few instances of women preserving an inviolable attachment to their husbands, except in the cases before mentioned, which are not considered as either a violation of their chastity or fidelity.

'Although I have said that the Indian nations differ very little from each other in their marriage ceremonies, there are some exceptions. The Naudowessies have a singular method of celebrating their marriages, which seems to bear no resemblance to those made use of by any other nation I passed through. When one of their young men has fixed on a young woman he approves of, he discovers his passion to her parents, who give him an invitation to come and live with them in their tent.

'He accordingly accepts the offer, and by so doing engages to reside in it for a whole year, in the character of a menial servant. During this time he hunts, and brings all the game he kills to the family; by which means the father has an opportunity of seeing whether he is able to provide for the support of his daughter and the children that might be the consequence of their union. This however is only done whilst they are young men, and for their first wife, and not repeated like Jacob's servitude.

'When this period is expired, the marriage is solemnized after the custom of the country, in the following manner: three or four of the oldest male relations of the bridegroom, and as many of the bride's, accompany the young couple from their respective tents, to an open part in the centre of the camp.

'The chiefs and warriors being here assembled to receive them, a party of the latter are drawn up in two ranks on each side of the bride and bridegroom immediately on their arrival. Their principal chief then acquaints the whole assembly with the design of their meeting, and tells them that the couple before them, mentioning at the same time] [their names, are come to avow publicly their intentions of living together as man and wife. He then asks the two young people alternately, whether they desire that the union might take place. Having declared with an audible voice that they do so, the warriors fix their arrows, and discharge them over the heads of the married pair; this done, the chief pronounces them man and wife.

'The bridegroom then turns round, and bending his body, takes his wife on his back, in which manner he carries her amidst the acclamations of the spectators to his tent. The ceremony is succeeded by the most plentiful feast the new-married man can afford, and songs and dances, according to the usual custom, conclude the festival.

'Among the Indiau as well as European nations, there are many that devote themselves to pleasure, and notwithstanding the accounts given by some modern writers of the frigidity of an Indian's constitution, become the zealous votaries of Venus. The young warriors that are thus disposed, seldom want opportunities for gratifying their passion; and as the mode usually followed on these occasions is rather singular, I shall describe it.

'"When one of these young debauchees imagines, from the behaviour of the person he has chosen for his mistress, that he shall not meet with any great obstruction to his suit from her, he pursues the following plan.

'"It has been already observed, that the Indians acknowledge no superiority, nor have they any ideas of subordination, except in the necessary regulations of their war or hunting parties; they consequently live nearly in a state of equality, pursuant to the first principles of nature. The lover therefore is not apprehensive of any check or controul in the accomplishment of his purposes, if he can find a convenient opportunity for completing them.

'"As the Indians are also under no apprehension of robbers, or secret enemies, they leave the doors of their tents or huts unfastened during the night, as well as in the day. Two or three hours after sunset, the old people cover over their fire, that is generally burning in the midst of their apartment with ashes, and retire to their repose.

'"Whilst darkness thus prevails, and all is quiet, one of these sons of pleasure, wrapt up closely in his blanket, to prevent his being known, will sometimes enter the apartment of his intended mistress. Having first lighted at the smothered fire a small splinter of wood, which answers the purpose of a match, he approaches the place where she reposes, and gently pulling away the covering from the head, jogs her till she awakes. If she then rises up, and blows out the light, he needs no further confirmation that his company is not disagreeable; but if, after he has discovered himself, she hides her head, and takes no notice of him, he might rest assured that any further solicitations will prove vain, and that it is necessary immediately for him to retire. During his stay he conceals the light as much as possible in the hollow of his hands, and as the tents or rooms of the Indians are usually large and capacious, he escapes without detection. It is said, that the young women who admit their lovers on these occasions, take great care, by an immediate application to herbs, with the potent efficacy of which they are well acquainted, to prevent the effects of these illicit amours from becoming visible; for should the natural consequences ensue, they must for ever remain unmarried."

'The children of the Indians are always distinguished by the name of the mother; and if a woman marries several husbands, and has issue by each of them, they are called after her. The reason they give for this is, that as their offspring are indebted to the father for their souls, the invisible part of their essence, and to the mother for their corporeal and apparent part, it is more rational that they should be distinguished by the name of the latter, from whom they indubitably derive their being, than by that of the father, to which a doubt might sometimes arise whether they are justly entitled.

'There are some ceremonies made use of by the Indians at the imposition of the name, and it is considered by them as a matter of great importance; but what these are, I could never learn, through the secrecy observed on the occasion. I only know that it is usually given when the children have passed the state of infancy.

'Nothing can exceed the tenderness shown by them to their offspring; and a person cannot recommend himself to their favour by any method more certain, than by paying some attention to the younger branches of their families.

'Some difficulty attends an explanation of the manner in which the Indians distinguish themselves from each other. Besides the name of the animal by which every nation and tribe is denominated, there are others that are personal, and which the children receive from their mother.

'The chiefs are also distinguished by a name that has either some reference to their abilities, or to the hieroglyphic of their families; and these are acquired after they arrive at the age of manhood. Such as have signalized themselves either in their war or hunting parties, or are possessed of some] [eminent qualifications, receive a name that serves to perpetuate the fame of those actions, or to make their abilities conspicuous.

'It is certain the Indians acknowledge one Supreme Being, or Giver of Life, who presides over all things; that is, the Great Spirit; and they look up to him as the source of good, from whom no evil can proceed. They also believe in a bad spirit, to whom they ascribe great power, and suppose that through his means all the evils which befal mankind are inflicted. To him therefore do they pray in their distresses, begging that he would either avert their troubles, or moderate them when they are no longer avoidable.

'They say that the Great Spirit, who is infinitely good, neither wishes nor is able to do any mischief to mankind; but on the contrary, that he showers down on them all the blessings they deserve; whereas the evil spirit is continually employed in contriving how he may punish the human race; and to do which, he is not only possessed of the will, but of the power.

'They hold also that there are good spirits of a lower degree, who have their particular departments, in which they are constantly contributing to the happiness of mortals. These they suppose to preside over all the extraordinary productions of nature, such as those lakes, rivers, or mountains that are of an uncommon magnitude; and likewise the beasts, birds, fishes, and even vegetables or stones, that exceed the rest of their species in size or singularity. To all of these they pay some kind of adoration.

'But at the same time I fancy that the ideas they annex to the word spirit, are very different from the conceptions of more enlightened nations. They appear to fashion to themselves corporeal representations of their gods, and believe them to be of a human form, though of a nature more excellent than man.

'Of the same kind are their sentiments relative to a futurity. They doubt not but they shall exist in some future state; they however fancy that their employments there will be similar to those they are engaged in here, without the labour and difficulties annexed to them in this period of their existence.

'They consequently expect to be translated to a delightful country, where they shall always have a clear unclouded sky, and enjoy a perpetual spring; where the forests will abound with game, and the lakes with fish, which might be taken without a painful exertion of skill, or laborious pursuit; in short, that they shall live for ever in regions of plenty, and enjoy every gratification they delight in here, in a greater degree.

'To intellectual pleasures they are strangers; nor are these included in their scheme of happiness. But they expect that even these animal pleasures will be proportioned and distributed according to their merit; the skilful hunter, the bold and successful warrior, will be entitled to a greater share than those who, through indolence or want of skill, cannot boast of any superiority over the common herd.

'The priests of the Indians are at the same time their physicians, and their conjurers; whilst they heal their wounds, or cure their diseases, they interpret their dreams, give them protective charms, and satisfy that desire which is so prevalent among them, of searching into futurity.

'How well they execute the latter part of their professional engagements, and the methods they make use of on some of these occasions, I have already shewn in the exertions of the priest of the Killistiones, who was fortunate enough to succeed in his extraordinary attempt near lake Superior. They frequently are successful likewise in administering the salubrious herbs they have acquired a knowledge of; but that the ceremonies they make use of during the administration of them, contributes to their success, I shall not take upon me to assert.

'When any of the people are ill, the person who is invested with this triple character of doctor, priest, and magician, sits by the patient day and night, rattling in his ears goad shell, filled with dry beans, called a chichicoue, and making a disagreeable noise that cannot well be described.

'This uncouth harmony one would imagine would disturb the sick person and prevent the good effects of the doctor's prescription; but on the contrary they believe that the method made use of, contributes to his recovery, by diverting from his malignant purposes the evil spirit who has inflicted the disorder; or at least that it will take off his attention, so that he shall not increase the malady. This they are credulous enough to imagine he is constantly on the watch to do, and would carry his inveteracy to a fatal length if they did not thus charm him.

'I could not discover that they make use of any other religious ceremonies than those I have described; indeed on the appearance of the new moon they dance and sing; but it is not evident that they pay that planet any adoration; they only seem to rejoice at the return of a luminary that] [makes the night cherful, and which serves to light them on their way when they travel during the absence of the sun.

'Notwithstanding Mr. Adair has asserted that the nations among whom he resided, observe with very little variation all the rites appointed by the Mosaic law, I own I could never discover among those tribes that lie but a few degrees to the *n. w.* the least traces of the Jewish religion, except it be admitted that one particular female custom, and their divisions into tribes, carry with them proof sufficient to establish this assertion.

'The Jesuits and French missionaries have also pretended, that the Indians had, when they first travelled into America, some notions, though these were dark and confused, of the Christian institution; that they have been greatly agitated at the sight of a cross, and given proofs by the impressions made on them, that they were not entirely unacquainted with the sacred mysteries of Christianity. I need not say that these are too glaring absurdities to be credited, and could only receive their existence from the zeal of those fathers, who endeavoured at once to give the public a better opinion of the success of their missions, and to add support to the cause they were engaged in.

'The Indians appear to be in their religious principles rude and uninstructed. The doctrines they hold are few and simple, and such as have been generally impressed on the human mind, by some means or other, in the most ignorant ages. They, however, have not deviated, as many other uncivilized nations, and too many civilized ones, have done, into idolatrous modes of worship: they venerate indeed and make offerings to the wonderful parts of the creation, as I have before observed; but whether those rights are performed on account of the impression such extraordinary appearances make on them, or whether they consider them as the peculiar charge, or the usual places of residence of the invisible spirits they acknowledge, I cannot positively determine.

'The human mind in its uncultivated state is apt to ascribe the extraordinary occurrences of nature, such as earthquakes, thunder, or hurricanes, to the interposition of unseen beings; the troubles and disasters also that are annexed to a savage life, the apprehensions attendant on a precarious subsistence, and those numberless inconveniences which man in his improved state has found means to remedy, are supposed to proceed from the interposition of evil spirits; the savage, consequently, lives in continual apprehensions of their unkind attacks, and to avert them has recourse to charms, to the fantastic ceremonies of his priest, or the powerful influence of his Manitous. Fear has of course a greater share in his devotions than gratitude, and he pays more attention to deprecating the wrath of the evil, than to securing the favour of the good beings.

'The Indians, however, entertain these absurdities in common with those of every part of the globe who have not been illuminated with that religion, which only can disperse the clouds of superstition and ignorance, and they are as free from error as people can be, who have not been favoured with its instructive doctrines.

'In Penobscot, a settlement in the province of Maine, in the *n. e.* part of New England, the wife of a soldier was taken in labour, and notwithstanding every necessary assistance was given her, could not be delivered. In this situation she remained for two or three days, the persons around her expecting that the next pang would put an end to her existence.

'An Indian woman, who accidentally passed by, heard the groans of the unhappy sufferer, and inquired from whence they proceeded. Being made acquainted with the desperate circumstance attending the case, she told the informant, that if she might be permitted to see the person, she did not doubt but that she should be of great service to her.

'The surgeon that had attended, and the midwife who was then present, having given up every hope of preserving their patient, the Indian woman was allowed to make use of any methods she thought proper. She accordingly took a handkerchief, and bound it tight over the nose and mouth of the woman: this immediately brought on a suffocation; and from the struggles that consequently ensued, she was in a few seconds delivered. The moment this was achieved, and time enough to prevent any fatal effect, the handkerchief was taken off. The long-suffering patient, thus happily relieved from her pains, soon after perfectly recovered, to the astonishment of all those who had been witnesses to her desperate situation.

'The reason given by the Indian for this hazardous method of proceeding, was, that desperate disorders require desperate remedies; that as she observed the exertions of nature were not sufficiently forcible to effect the desired consequence, she thought it necessary to augment their force, which could only be done by some mode that was violent in the extreme.]

[' An Indian meets death when it approaches him in his hut, with the same resolution as he evinces when called to face him in the field. His indifference under this important trial, which is the source of so many apprehensions to almost every other nation, is truly admirable. When his fate is pronounced by the physician, and it remains no longer uncertain, he harangues those about him with the greatest composure.

' If he is a chief and has a family, he makes a kind of funeral oration, which he concludes by giving to his children such advice for the regulation of their conduct as he thinks necessary. He then takes leave of his friends, and issues out orders for the preparation of a feast, which is designed to regale those of his tribe that can come to pronounce his eulogium.

' The character of the Indians, like that of other uncivilized nations, is composed of a mixture of ferocity and gentleness. They are at once guided by passions and appetites, which they hold in common with the fiercest beasts that inhabit their woods, and are possessed of virtues which do honour to human nature.

' In the following estimate I shall endeavour to forget on the one hand the prejudices of Americans, who usually annex to the word Indian, ideas that are disgraceful to human nature, and who view them in no other light than as savages and cannibals, whilst with equal care I avoid my partiality towards them, as some must naturally arise from the favourable reception I met with during my stay among them.

' That the Indians are of a cruel, revengeful, inexorable disposition; that they will watch whole days unmindful of the calls of nature, and make their way through pathless, and almost unbounded woods, subsisting only on the scanty produce of them, to pursue and revenge themselves of an enemy; that they hear unmoved the piercing cries of such as unhappily fall into their hands, and receive a diabolical pleasure from the tortures they inflict on their prisoners, I readily grant: but let us look on the reverse of this terrifying picture, and we shall find them temperate both in their diet and potations, (it must be remembered that I speak of those tribes who have little communication with Americans); that they withstand, with unexampled patience, the attacks of hunger, or the inclemency of the seasons, and esteem the gratification of their appetites but as a secondary consideration.

' We shall likewise see them social and humane to those whom they consider as their friends, and even to their adopted enemies; aud ready to share with them the last morsel, or to risk their lives in their defence.

' In contradiction to the report of many other travellers, all of which have been tinctured with prejudice, I can assert, that notwithstanding the apparent indifference with which an Indian meets his wife and children after a long absence, an indifference proceeding rather from custom than insensibility, he is not unmindful of the claims either of connubial or parental tenderness.

' Accustomed from their youth to innumerable hardships, they soon become superior to a sense of danger, or the dread of death; and their fortitude, implanted by nature, and nurtured by example, by precept and accident, never experiences a moment's allay.

' Though slothful and inactive whilst their stores of provisions remain unexhausted, and their foes are at a distance, they are indefatigable and persevering in pursuit of their game, or in circumventing their enemies.

' If they are artful and designing, and ready to take every advantage, if they are cool and deliberate in their councils, and cautious in the extreme, either of discovering their sentiments, or of revealing a secret, they might at the same time boast of possessing qualifications of a more animated nature, of the sagacity of a hound, the penetrating sight of a lynx, the cunning of a fox, the agility of a bounding roe, and the unconquerable fierceness of the tiger.

' In their public characters, as forming part of a community, they possess an attachment for that band to which they belong, unknown to the inhabitants of any other country. They combine, as if they were actuated only by one soul, against the enemies of their nation, and banish from their minds every consideration opposed to this.

' They consult without unnecessary opposition, or without giving way to the excitements of envy or ambition, on the measures necessary to be pursued for the destruction of those who have drawn on themselves their displeasure. No selfish views ever influence their advice, or obstruct their consultations. Nor is it in the power of bribes or threats to diminish the love they bear their country.

' The honour of their tribe, and the welfare of their nation, are the first and most predominant emotions of their hearts; and from hence proceed in a great measure all their virtues and their vices. Actuated by these, they brave every danger, endure the most refined torments, and expire triumphing in their fortitude, not as a personal quality, but as a national characteristic.]

[' From hence also flows that insatiable revenge towards those with whom they are at war, and all the consequent horrors that disgrace their name. Their uncultivated mind being incapable of judging of the propriety of an action, in opposition to their passions, which are totally insensible of the controul of reason or humanity, they know not how to keep their fury within any bounds, and consequently that courage and resolution, which would otherwise do them honour, degenerate into a savage ferocity.

' But this short dissertation,' continues Captain Lewis, ' must suffice: the limits of my work will not permit me to treat the subject more copiously, or to pursue it with a logical regularity. The observations already made by my readers on the preceding pages, will, I trust, render it unnecessary; as by them they will be enabled to form a tolerably just idea of the people I have been describing. Experience teaches that anecdotes, and relations of particular events, however trifling they might appear, enable us to form a truer judgment of the manners and customs of a people, and are much more declaratory of their real state, than the most studied and elaborate disquisition, without these aids.'

Statement of the Commerce of the Missouri.

The following statement of the commerce of the Missouri is extracted from the work of the author whom we have already so liberally quoted.

' The products which are drawn from the Missouri are obtained from the Indians and hunters in exchange for merchandize. They may be classed according to the subjoined table.

				d.	c.	dolls.	cts.
Castor - -	12281	lbs.	at	1	20	14737	20
Otters - -	1267	skins		4	—	5068	—
Foxes, Pouha Foxes, Tiger cats	802	skins		0	50	401	—
Raccoons -	4248	skins		0	25	1062	—
Bears, black, grey, and yellow	2541	skins		2	—	5082	—
Puces - -	2541	skins		2	—	5082	—
Buffaloes -	1714	skins		3	—	5142	—
Dressed cow hides	189	skins		1	50	283	50
Shorn deer skins	96926	lbs.		0	50	38770	40
Deer skins, with hair	6381	skins		0	50	3190	50
					Carry forward	78818	60

			d.	c.	dolls.	cts.
Brought forward					78818	60
Tallow and fat	8313	lbs.	0	20	1662	60
Bears oil -	2310	galls.	1	28	2472	—
Muskrats -	—		—		—	—
Martens -	—		—		—	—
				Total	82953	20

' The calculation in this table, drawn from the most correct accounts of the products of the Missouri, during fifteen years, makes the average of a common year about 77,971 dollars.

' On calculating, in the same proportion, the amount of merchandize entering the Missouri, and given in exchange for peltries, it is found that it amounts to 61,250 dollars, including expences, equal to one-fourth of the value of the merchandize.

' The result is, that this commerce gives an annual produce of 16,721 dollars, or about 27 per cent.

' If the commerce of the Missouri, without encouragement, and badly regulated, gives annually so great a profit, may we not rest assured that it will be greatly augmented, should government direct its attention to it. It is also necessary to observe, that the price of peltry fixed by this table is the current price in the Illinois: if it were regulated by the prices of London, deducting the expences of transportation, the profit, according to our calculation, would be much more considerable.

' If the Missouri, abandoned to savages, and presenting but one branch of commerce, yields such great advantages, in proportion to the capital employed in it, what might we not hope, if some merchants or companies with large capital, and aided by a population extended along the borders of the river, should turn their attention to other branches of the trade, which they might undertake (I dare say) with a certainty of success, when we consider the riches buried in its banks, and of which I have endeavoured in these notes to give an idea?

' AN ESTIMATE OF THE PRODUCE OF THE SEVERAL MINES.

Mine at Burton, 550,000lbs. mineral, estimated to produce $66\frac{2}{3}$, is $336,666\frac{2}{3}$lbs. lead, at 5 dollars, is -	18,333 33
To which add 30 dollars, (on 120,000lbs, manufac-	
Carry forward	18,333 33]

[Brought forward	18,333 33	
tured) to each thousand, is - - - - -	3,600 —	
		21,933 33
Old Mines, 200,000lbs. mineral, estimated to produce $66\frac{2}{3}$, is $133,333\frac{1}{2}$lbs. lead, at 5 dollars per cwt. is - - - - -	6,666 67	
Mine à la Mott, 200,000lbs. lead, at 5 dollars per cwt. is - - - - -	10,000 —	
Suppose at all the other mines 30,000lbs. lead, at 5 dollars, is - -	1,500 —	
		18,166 67
Total amount is	Dollars	40,100 —

'When the manufacture of white and red lead is put into operation, the export valuation will be considerably augmented on the quality of lead.']

[MISSOURI Indians, properly so called, are the remnant of the most numerous nation inhabiting the Missouri, when first known to the French. Their ancient and principal village was situated in an extensive and fertile plain, on the *n.* bank of the Missouri, just below the entrance of the grand river. Repeated attacks of the small-pox, together with their war with the Saukees and Renars, have reduced them to their present state of dependence on the Ottoes, with whom they reside, as well in their village as on their hunting excursions. The Ottoes view them as their inferiors, and sometimes treat them amiss. These people are the real proprietors of an extensive and fertile country lying on the Missouri, above their ancient village, for a considerable distance, and as low as the mouth of the Osage river, and thence to the Mississippi. For an account of other nations inhabiting the borders of the Missouri, see that river.]

[MISSQUASH River. Nova Scotia and New Brunswick provinces are separated by the several windings of this river, from its confluence with Beau Basin (at the head of Chignecto channel) to its rise or main source; and from thence by a due *e.* line to the bay of Verte, in the straits of Northumberland. See NEW BRUNSWICK.]

[MISTAKE Bay, a large bay on the *w.* side of the entrance of Davis's straits, and to the *n.* of Hudson's straits; from which it is separated by a peninsula of the *n.* main on the *w.* and Resolution island on the *s.* It is to the *n. e.* of Nieva island, and *n. w.* of cape Elizabeth.]

[MISTAKEN Cape, the *s.* point of the easternmost of the Hermit's islands, is about 23 miles *n.* from cape Horn, at the extremity of S. America.]

[MISTAKEN Point, to the *w.* of cape Race, at the *s. e.* point of the island of Newfoundland, and to the *e.* of cape Pine, is so called because it has been frequently mistaken by seamen for cape Race when they first make the island from the *s.* though it is two leagues *w. n. w.* from it.]

MISTAN, a settlement of the head settlement of the district and *alcaldía mayor* of Huauchinango in Nueva España; situate on the *s.* side of the said head settlement.

MISTASSINS, a great lake of New France or Canada in N. America; formed from the river Rupert, in the country of the Indians of its name, and is divided into three parts, which afterwards communicate.

MISTECAPA, a small settlement or ward of the head settlement of the district of San Luis, of the coast and *alcaldía mayor* of Tlapa in Nueva España. It contains 16 families of Indians, and is a little more than a league from the settlement of Quauzoquitengo.

MISTEPEC, a settlement and head settlement of the district of the *alcaldía mayor* of Juxtlahuaca in Nueva España. It contains 232 families of Indians, including those of five wards of its district.

MISTEPEC, another settlement, of the *alcaldía mayor* of Guajuapa in the same kingdom; containing 12 families of Indians.

MISTEPEQUE, SAN AGUSTIN DE, a settlement and head settlement of the district of the *alcaldía mayor* of Nexapa in Nueva España; composed of six other settlements.

MISTEPEQUE, another, with the dedicatory title of San Agustin, of this head settlement and *alcaldía mayor*. It is situate on an extensive lofty plain, having on either side two very deep and craggy glens, in the which the Indians cultivate cochineal and other seeds. Every eight days there is celebrated here a *tianguis* or fair, visited by traders as well of other jurisdictions as this, for the buying and selling of seeds, household utensils, fruit, flesh, mantles, cloths, and cotton stuffs. It is 34 leagues *s.* with a small inclination *w.* of its capital.

MISTEPEQUE, another, with the dedicatory title of San Andres, of the same head settlement and *alcaldía mayor*; containing 56 Indian families.

MISTERIOSA, a small island of the N. sea; between the coast of Honduras, or rather between the shoal of Santillana and El Placer.

[MISTIC, or Mystic, a short river which falls into the *n.* side of Boston harbour by a broad mouth on the *e.* side of the peninsula of Charlestown. It is navigable for sloops four miles to the industrious town of Medford; and is crossed a mile above its mouth by a bridge 130 rods in length, through which vessels pass by means of a draw.]

[MISTINSINS, an Indian nation who inhabit on the *s.* side of the lake of the same name in Lower Canada.]

[MISTISSINNY Lake, in Canada; on the *s. e.* side of which is a Canadian house, or station for trade.]

MISTLAN, San Juan de, a settlement of the province and *alcaldía mayor* of Guazacapan in the kingdom of Guatemala; annexed to the curacy of Nuestra Señora de la Concepcion of Escuintla; formerly of the monks of St. Domingo.

MITARE, a settlement of the province and government of Venezuela; situate on the shore of the river of its name, in the part where this unites with the Seco, to the *w.* of the city of Coro.

Mitare. The aforesaid river is large, and enters the N. sea near the mouth of the gulf of Maracaibo, in lat. 11° 27′ *n.*

[MITCHELL'S Eddy, the first falls of Merrimack river, 15 miles from its mouth, and six above the new bridge which connects Haverhill with Bradford. Thus far it is navigable for ships of burden.]

MITCHIGAMES, a barbarous nation of Indians of the province and government of Louisiana in N. America. They retired and fled from the Tchicachas to the territory of the Arkansas, and these finding them valorous and useful to them in their wars allowed them to domesticate, and thus the two tribes became confounded, to the extinction of this of which we treat.

MITIC, a settlement of the head settlement of the district and *alcaldía mayor* of Lagos in Nueva España; four leagues *n. e.* of its capital.

MITIMARES, certain Indians, who removed themselves from different provinces to others, a practice which was much encouraged and enforced by the Incas, when, after taking any new provinces, they doubted of the loyalty of some of the subjects.

MITLA, a settlement and head settlement of the district of the *alcaldía mayor* of Teutitlan in Nueva España. It contains 150 families of Indians, and is four leagues *w.* of its capital.

MITLANTONGO, Santa Cruz de, a settlement and head settlement of the district of the *alcaldía mayor* of Nochiztlan in Nueva España; containing 58 families of Indians, and being eight leagues *e.* and *s.* of its capital.

Mitlantongo, with the dedicatory title of Santiago, another settlement of the same head settlement and *alcaldía mayor* as the former; containing 48 Indian families, and being nine leagues *e.* with an inclination to *s.* of its capital.

MITLAZINCO, a settlement of the head settlement of the district of Otengo, and *alcaldía mayor* of Chilapa, in Nueva España; two leagues *n.* of its head settlement.

MITMAS, a settlement of the province and *corregimiento* of Chachapoyas in Peru; annexed to the curacy of Chisquilla.

MITO, a settlement of the province and *corregimiento* of Tarma in Peru; annexed to the curacy of Tapu.

MITOTO, a settlement of the province and *corregimiento* of Xauja in Peru.

MITQUITLAN, a settlement of the province of Cuextlan in Nueva España, in the time of the Indian gentilism; conquered by the King Ahuitzotl, although, from the valour of its natives, it cost him a great many lives of his best troops.

MIXAPA, a settlement of the province and *alcaldía mayor* of Los Zoques in the kingdom of Guatemala.

MIXATA, a settlement of the province and government of Sonsonate in the kingdom of Guatemala; annexed to the curacy of San Pedro Chipilapa, and containing 60 Indians.

MIXCO, a settlement of the province and kingdom of Guatemala; founded in an extensive valley, from which it takes its name, and on the shore of the river Las Vacas. It contains 300 families, and is very rich, being the decided pass to Mexico. The opulence of its inhabitants is acquired chiefly from the keeping of mule-droves for the purposes of forwarding merchandize, and the Father Tomas Gage, who was curate here for five years, relates, that one person, named Juan Palomequel, himself kept no less than 300 of these animals. What contributes, however, no less to its opulence is, that the Indians here are very expert in making earthen-ware articles of great beauty, and such as are eagerly bought by a greater part of the jurisdictions of the kingdom. It abounds in wheat, fruit, seeds, and all sorts of fowl. Eleven leagues from Guatemala.

Mixco. The aforesaid valley is five leagues long and three quarters of a league wide, watered by the river Las Vacas, and thereby rendered very fertile and delightful. It has some breeds of cattle, and produces the best wheat in the kingdom, and from it the capital is supplied. In it are 36 or 40 scattered houses which belong to so many masters, and all being of the curacy of a church which is at a small distance, called Nuestra Señora del Carmen.

MIXCONTIQUE, a settlement of the province and *alcaldía mayor* of Chiapa in the kingdom of Guatemala.

MIXO, a small river of St. Domingo, rising in the mountains of Ciboo, and running *s. s. w.* with the river San Juan to enter the Neiba.

MIXQUIAHUACAN, a settlement of the province of Cempoala, and of the nation of Totonacas Indians, in Nueva España. After the conquest of the kingdom by the Spaniards, it changed its name to that of San Francisco, which it preserves.

MIXQUIC, a province of Nueva España, conquered by the Emperor Thechotlatzin, the fifth of the Chichimecas and Aculhuas. These united themselves with Monquihuix, king of the Tlatelucas, to defend themselves against Axayacatl, king of Mexico; who, however, reduced them to obedience, and subjected them to the empire. They afterwards entered into an alliance with Cortes, and were greatly instrumental to the carrying his purposes, by assisting him with canoes and whatever else they had it in their power to afford.

MIXTAN, a settlement of the province and *alcaldía mayor* of Guazacapan in the kingdom of Guatemala; annexed to the curacy of San Pedro Chipilapa, and containing 60 Indian inhabitants.

MIXTECA, a province of Nueva España; situate on the coast of the S. sea, divided into *Alta* and *Baxa;* the first being in the *serranía*, and its settlements belonging to the jurisdiction of the bishopric of La Puebla de los Angeles; the second being of the bishopric of Oaxaca, and lying in the *llanuras* contiguous to the coast; bounded by the province and *alcaldía mayor* of Xicayán, and by Oaxaca, with the province of Huizo. Its district comprehends the settlements of Teposcolula, Nochitlan, and others, to the which are subject various principal settlements, such as Nanguitlán and Tlaxiaco. The temperature is for the most part cold throughout. It abounds in wheat, maize, fruit, and particularly in fine cochineal and silk-worms; and these, with some manufactures and some goat cattle which they kill, constitute the chief articles of commerce. In the capital settlements of the *alcaldías mayores* above mentioned live some Spaniards; but in all the others there are none but Indians, speaking the Mixtecan tongue, the language of this country. These are, generally speaking, docile, civil, and industrious, and less averse to labour than other Indians. In one of the aforesaid settlements of this province was born the illustrious Don Nicolas del Puerto, an Indian and celebrated lawyer, and such as merited to become the bishop of Oaxaca, a man of great virtue and science, and who destroyed the prejudice that no Indian was competent in ability to the offices of Europeans.

MIZANTLA, a jurisdiction and *alcaldía mayor* of Nueva España, called also Vera Cruz Vieja, from there being in it the city thus named and founded by Hernan Cortes, but which was since removed. It consists of seven settlements, which are,

Santa Maria Talixcoya,	S. Martin Tlacotepeque,
Cempoala,	
S. Francisco Tenampa,	Chicontepec y Colipa.
Santa Maria Tetela,	

All these are small, though heads of districts, and to them are annexed different wards; their commerce consisting in the several breeds of cattle, as also in maize and other seed. The natives equally apply themselves to the cultivation of cotton and to fishing, since they are girt by the sea, and have a small port which serves only for little vessels. In this port Hernan Cortes landed at the conquest of this place; and here he effected the stratagem of sending back his vessels as soon as his men were ashore, for fear they might be induced to fly to them for succour.

The capital is the settlement of the same name; situate on the spot where the city of Vera Cruz stood. It is of an hot and somewhat moist temperature, inhabited by 230 families of Spaniards, *Mustees*, and Mulattoes, and 260 of Mexican Indians. One hundred and forty-six miles *e.* of Mexico, and 53 *n.* by *w.* of Vera Cruz. Lat. 19° 54′ *n.* Long. 96° 36′ *w.*

MIZAPA, Punta de, a very lofty mountain of the coast of Nueva España, between the river Guazacoalco and the *sierras* of San Martin, and serving as a land-mark to vessels.

MIZQUE, a province of the government of Santa Cruz de la Sierra in Peru; bounded *s.* by the province of Yamparaes, the river Grande intersecting the two, *s. w.* by that of Charcas, *w.* by that of Cochabambas, and *n.* by the *serranias* of the *cordillera*. Its temperature is for the most part hot, although it has some places mild. It produces wheat, maize, pulse, and garden herbs, some sugar-cane, and vines of which wine is made; and has various estates of large and small cattle brought from Santa Cruz. This province is very poor, and all its commerce consists in the cultivation of the fields for the mere maintenance of the families which inhabit it. It has no mine whatever, and has no person of any consequence residing in it; even those who lived there once being either dead or removed to Potosi and other parts. It has, however, in its woods several sorts of trees good for building; such as cedars, *carobe* trees, *quinaquinas*, and others; also many tigers, leopards, foxes, ounces, turkeys, doves, parrots, ducks, herons, and other animals and birds. Near the settlement of Pocóna is a lake two leagues in circumference, and is watered by rivers abounding in fish sufficient to supply the jurisdiction. The inhabitants amount to 12,000, and its *corregidor* used to enjoy a *repartimiento* of 11,512 dollars. The settlements of its district are Pocóna, Tintin, Aiguile, Totora, Omereque; also those of its ecclesiastical jurisdiction, Punata and Tarata, which in their civil government belong to the province of Cochabamba.

The capital is of the same name; a small city, founded in a beautiful and extensive valley of eight leagues long, abounding in grain, wax, and honey. It was formerly large and opulent, as no few vestiges of its ancient grandeur testify. It has besides the parish churches convents of monks of St. Domingo, San Francisco, San Juan de Dios, and of barefooted Augustins, the which, at the present day, can scarcely maintain a single monk. The principal cause of this decay has arisen from the continued affliction of tertian fevers, to which these hot valleys are subject, and which are here called *chuchu*. Near the city pass two rivers, which come from the province of Cochabamba and enter Marañon, in which fish are caught. [It is situate on the shore of the river Grande, in lat. 18° 40′ *s.* Long. 56° 42′ *w.*]

MIZQUIAHUALA, a settlement and head settlement of the district and *alcaldía mayor* of Tepetango in Nueva España; comprehending settlements, and containing 50 families of Indians, 20 of Spaniards, *Mustees*, and Mulattoes. Eighteen leagues *n.* of Mexico.

MIZQUITIC, a settlement of the province of Zacatecas in Nueva España; founded by order of the viceroy Don Luis de Velasco, for which purpose he sent some Indians from the province of Tlaxcala. It has in it a convent of the religious order of San Francisco, under whom it was long dependent for religious instruction. Three leagues from the settlement of Tlaxcalilla, and 75 *n.* of Mexico.

MIZQUIYAHUALAN, a settlement near the city of Mexico in the time of the Indians, celebrated for having been one of the places founded by the Chichemacas during their peregrinations under Xolotl.

[M'KESSENSBURG, a town of Pennsylvania, York county, on Tom's creek, 32 miles *w.* *s. w.* of York.]

MOA, Cayo de, an isle of the N. sea, near the coast of the island of Cuba and the point of Las Mulas.

MOABAS, a settlement of the missions that were held by the regulars of the company, in the province of Ostimuri of N. America; four leagues from the river Chico.

[MOAGES Islands, on the *n.* coast of S. America, in the entrance of the gulf of Venezuela. They extend from *n.* to *s.* and lie *w.* of the island of Aruba; are eight or nine in number, and all, except one, low, flat, and full of trees. The southernmost is the largest.]

MOALCACHI, a settlement of the missions which were held by the Jesuits, in the province of Taraumara and kingdom of Nueva Vizcaya, *s. e.* of the town and *real* of mines of Chiguagua. In its vicinity are three large cultivated estates called Cosaguinoa, lying four leagues to the *s.* Calichiqui four and an half to the *w.* and Cochunigui eight to the *s. w.*

[MOBILE, a large navigable river, formed by two main branches of the Alabama and Tombeckbee, in the *s. w.* part of Georgia. It pursues a *s.* course into W. Florida; the confluent stream enters the gulf of Mexico at Mobile point, in lat. 30° 17′ *n.* 11 leagues below the town of Mobile. Large vessels cannot go within seven miles of the town. The breadth of the bay is in general about three or four leagues. Vast numbers of large alligators bask on the shores, as well as swim in the rivers and lagoons. See Georgia, Alabama, Tombeckbee, &c. From the *n. e.* source of the waters of the Alabama, to Mobile point at the mouth of Mobile bay, is, according to the best maps, about 460 miles: large boats can navigate 350 miles, and canoes much farther.]

[Mobile, a city of W. Florida, formerly of considerable splendour and importance, but now in a state of decline. It is pretty regular, of an oblong figure, and situated on the *w.* bank of the river. The bay of Mobile terminates a little to the *n. e.* of the town, in a number of marshes and lagoons; which subject the people to fevers and agues in the hot season. It is 33 miles *n.* of Mobile point, about 32 below the junction of the two principal branches of Mobile river, and 46 *w. n. w.* of Pensacola. There are many very elegant houses here, inhabited by French, English, Scotch, and Irish. Fort Conde, which stands very near the bay, towards the lower end of the town, is a regular fortress of brick; and there is a neat square of barracks for the officers and soldiers. Mobile, when in possession of the British, sent yearly to London skins and furs to the value of from 12 to 15,000*l*. sterling. It surrendered to the Spanish forces in 1780.]

MOBJACK, a bay on the coast, province, and colony of N. Carolina, at the entrance of the bay of Chesapeak.

MOCA, a province of Peru in the time of the Indian gentilism, on the confines of the kingdom of Quito; at present confounded in the division made by the Spaniards. It was conquered and united to the empire by the Inca Tapac Yupanqui, eleventh emperor.

MOCALASA, a settlement of Indians of the province of S. Carolina; situate on the shore of the river Albama.

MOCANACO, a point of the coast of the N. sea, and kingdom of Nueva España in N. America, two leagues and an half from Vera Cruz. At the distance of one league and an half is the mouth of the river Medellin, on the shore of which is a small settlement inhabited by 30 families of Indians, who maintain themselves by the trade of fishing in the said river, and carrying their stock to the market of the city. It is of a warm and dry temperature, in lat. 19° 4′ *n*. Long. 96° 3′ *w*.

MOCHA, a settlement of the province and *corregimiento* of Arica in Peru; annexed to the curacy of Cibaya.

MOCHA, another settlement, of the province and *corregimiento* of Ambato in the kingdom of Quito, on the *s*. It is near the river Pachanlica, which runs by the *s*. and is of a cold temperature from its vicinity to the *paramo* or mountain desert of Chimborazo, which is always covered with snow. The inhabitants are almost all *Mustees*, and have the credit throughout the province of being notorious robbers; on which account it is said, that in Mocha they sow grain and gather mules, meaning that they do not there maintain themselves by what they sow, but rather by what they plunder, inasmuch as these depredators constantly take the mules from passengers proceeding to Guaranda or Ambato, which lie in the road from Guayaquil and from Quito. In lat. 1° 27′ *s*.

MOCHA, another, with the dedicatory title of Santa Lucia, in the province and *corregimiento* of Truxillo in Peru. It is very small; situate near the sea, and two leagues from the capital to the *s. e.* in the road leading to Lima. The natives are all people of colour, (excepting here and there a Spaniard), and living by agriculture; so that its limited district is nothing but a pleasant, cultivated garden, of a mild and salutary climate. It lies in the direct road to Lima, and is the place where the passports must be shewn to the lieutenant of the *corregidor;* in default of which no person whatever is permitted to pass.

MOCHA, another, of the kingdom of Chile; situate on the coast, at the mouth of the river Villagran.

MOCHA, an island of the S. sea, belonging to the kingdom of Chile, six leagues from the continent. It is small, but well peopled with Indians, who cultivate it with success, the soil being extremely fertile. The climate is benign and salutary, and its water is supplied by a most delicate fountain. In lat. 38° 21′ *s*.

MOCHA, a river of the province and *corregimiento* of Truxillo in Peru, which rises from the lakes Guaigaicocha and San Lorenzo, and, running 22 leagues, passes through the fertile valley of Chimo, where the capital is founded, a league's distance from the river. It collects the waters of many other rivers and streams, and being united with the Trapiche, takes the name of Minocucho. It so overflows the valley in the winter time that it must be passed in rafts, as by here runs the royal road to Lima. It empties itself into the S. sea.

MOCHARA, a settlement of the province and *corregimiento* of Chichas and Tarija in Peru.

MOCHICAUCHI, or MOCHICAHUI, a settlement of the province and *alcaldía mayor* of Cinaloa in N. America; situate on the shore of the river Fuerte, between the settlements of Charay and San Miguel.

MOCHICOS, a very numerous nation of Indians of Peru, who used to dwell in the valleys on the coast of the S. sea.

MOCHIMA, a port of the coast of the N. sea, in the province and government of Cumaná. It is large, convenient, and frequented by vessels which carry on an illicit trade on those coasts; situate between cape Cordera and point Araya.

MOCO, a river which flows down from the mountains of Bogota, in the Nuevo Reyno de Granada, runs *e*. and, after having collected the waters of several others, enters by the *n*. shore into the Orinoco.

MOCOA, a large and extensive province of the kingdom of Quito, in the jurisdiction and *corregimiento* of Pasto, discovered by Hernan Perez de Quesada in 1541. In this province the rivers Caquetá to the *n*. and the Putumayo or Iza to the *s*. take their rise. It has also a large lake of the same name as itself, in which are found pearls, which, although small, are extremely fine. The territory here is but little known and less peopled;

bounded *n.* by the province of Popayán, *w.* by that of Pasto, *s. w.* by the town of Ibarra, *s.* by the province of Sucumbios, and *e.* by the mountains of the infidel Indians. It is irrigated by the two rivers aforesaid, as also by those of Sucia, Tango, Pato, Labaquero, Piedras, Guinchoa, Vides, Quino, Pischilin, Yaca, and others of less note. The natives are expert at making beautiful wooden utensils, giving them a varnish which they fix in by the fire, and having the appearance of japan, and which work they call *de Mocoa.* In the mountains is found a small animal of the figure of a beetle, which becomes converted into a plant, and in the city of Pasto they have frequently been seen half in an animal and half in a vegetable state, previous to the perfect metamorphosis of the animal to the plant with roots and tendrils. The capital of this province was the city of the same name, the which is at present destroyed, though once situate on the *s.* shore of the river Caqueta, in lat. 1° 32′ *n.* At the present day the capital is the settlement of Sibundoy.

MOCOBI, a river of the province and government of Moxos in the kingdom of Quito. It runs *n. w.* near the settlement of La Santisima Trinidad, and empties itself into the Marmore, in lat. 14° 53′ *s.*

MOCOBIS, a barbarous nation of Indians, of the province and government of Tucumán in Peru, to the *n.* of the city of Cordoba. These Indians are ferocious, and in their incursions infest the whole province. They extend *e.* as far as the river Ocloyas, and *n.* as far the district of the city of Salta, to the *w.* as far as the river Salado, and to the *s.* as the fort of San Joseph. They go in troops through the woods, and burst suddenly upon a settlement and destroy it. In 1668, the governor Don Alonso Mercado attempted their reduction by means of the Jesuits the Fathers Agustin Fernandez and Pedro Patricio: these succeeded in forming with them a treaty of peace, but which was soon broken, and when they immediately returned to their hostilities. The Jesuits of the college of the town of Santa Fé catechised in 1744 one of their principal caciques named Anacaiqui, and he returned a short time after, requesting the Spaniards would send back with him a missionary of the Jesuits, and found a settlement to which he might induce those of his nation to come. The governor of Buenos, Don Miguel de Salcedo, acceeded to his wishes, and giving his commands to the provincial of the abolished order of the Jesuits, the Father Francisco Burgher was nominated to the mission, and he established a numerous *reduction,* with the name of San Francisco Xavier, when he was afterwards joined by a brother of his order, Miguel de Zea. What operated against a great number of conversions, was the circumstance of the contiguity of this new settlement to the city of Santa Fé; for the Indians, who had frequent occasion to go to this city, and observing the customs there, would tell their instructors that the Spaniards acted very differently from what they were told to do. This induced the missionaries to remove the settlement to a greater distance; and then the Mocobis and many Abiporis and other nations of Indians flocked to it, and embraced the Christianity, in which they have since persevered.

MOCODONE, a port of the *s.* coast of Nova Scotia or Acadia in N. America, between the islands Poland and Liscomb.

MOCOMO, a settlement of the province and *corregimiento* of Larecaja in Peru.

[MOCOMOKO, or Little Orinoco, a river to the *s. e.* of the great river Orinoco, on the *e.* coast of S. America. Four leagues *w.* of Amacum.]

MOCONDINO, called by some Mojondino, a settlement of the province and *corregimiento* of Pasto in the kingdom of Quito.

MOCORCA, a lake of the province and *corregimiento* of Collabuas in Peru, of the district of Arequipa, belonging to the settlement of Cabacondo. It is three leagues in circumference, and in it breeds a certain fish serving as a common food for the natives, and which in their language is called *ilpi.*

MOCORETA-GUAZU, a small river of the province and government of Buenos Ayres, which runs *e.* and enters the Uruguay between the Timboy and the following river.

Mocoreta-Rimi, or Mini, a small river of the same province and government as the former, also entering the Uruguay, between the former river and the Maudozobó.

MOCORIPE, a port on the coast of the province and *captainship* of Seara in Brazil, between the cape of Las Sierras and the river Koko.

MOCORITO, a settlement of the missions which were held by the Jesuits of the province and government of Cinaloa in N. America.

MOCOYAGUI, a settlement of the missions that were held by the Jesuits in the same province and government as the former.

MOUTUN, a settlement of the head settlement of the district and *alcaldía mayor* of Villalta in Nueva España, of a cold temperature, and containing 15 Indian families. Seven leagues *e.* of its capital.

MOCUL, a settlement of the province and *cor-*

regimiento of Maule in the kingdom of Chile; situate on the shore of the river Colorado.

MOCUPE, a settlement of the province and *corregimiento* of Saña in Peru.

[MODER AND DAUGHTERS Islands, a long island two leagues *e.* by *s.* of the Father, or Vaader island, with two small ones, so called, near Cayenne, on the *e.* coast of S. America, not far from the Constables, and in about lat. 5° *n.* Long. 52° *w.*]

[MOGHULBUGHKITUM, or MUHULBUCKTITUM, a creek which runs *w.* to Alleghany river in Pennsylvania. It is passable in flat-bottomed boats to the settlements in Northumberland county. Wheeling is its *n.* branch.]

MOGI, a small river of the province and *captainship* of San Vicente in Brazil. It rises in the mountains near the coast, and runs *n.* to enter the Sapocay.

MOGIROIRI, a settlement of the same province and kingdom as the former; situate on the *n.* of the bay of San Vicente.

MOGOTES, SANTA BARBA DE, a settlement of the province and *corregimiento* of Tunja in the Nuevo Reyno de Granada. It is of an hot temperature, but subject to wet, with frequent tempests of thunder and lightning. It produces maize, *yucas*, plantains, and sugar canes, of which are made good sugars and conserves, called here *panelas;* this being its principal article of commerce, though not without some woven cotton stuffs. It is a very healthy country, and where people generally live to the age of 80 years, and some to 100. The inhabitants of this settlement amount to 400, who are very poor, and it is 93 miles *n. e.* of Tunja, and three *e.* of the town of San Gil.

MOGOTES, a river, called also De las Fortelezas, in the province and government of Chocó, and Nuevo Reyno de Granada. It rises in the grand *cordillera*, and traverses the whole kingdom of Tierra Firme; running *n. w.* it follows its course to Peru, and enters the S. sea near the cape Corrientes.

MOHANET, a settlement of Indians of the province and colony of Pennsylvania in N. America; situate on the bank of the *e.* arm of the river Susquehannah.

[MOHAWK River, in New York, rises about 10 miles to the *e.* of lake Ontario, about eight miles from Black or Sable river, a water of lake Ontario, and runs *s.* 25 miles to fort Schuyler, then *e.* 80 miles, and after receiving many tributary streams, falls into Hudson river, by three mouths opposite to the cities of Lansinburgh and Troy, from 7 to 10 miles *n.* of Albany. The produce that is conveyed down this river is landed at Schenectady, on its *s.* bank, and is thence conveyed by land 16 miles over a barren, sandy, shrub plain to Albany. It is in contemplation either to cut a canal from Schenectady to the navigable waters of Hudson river, or to establish a turnpike road between Schenectady and Albany. This fine river is now navigable for boats, from Schenectady, nearly or quite to its source, the locks and canals round the Little falls, 56 miles above Albany, having been completed in the autumn of 1795; so that boats full loaded now pass them. The canal round them is nearly three quarters of a mile, cut almost the whole distance through an uncommonly hard rock. The opening of this navigation is of great advantage to the commerce of the state. A shore of at least 1000 miles in length is, in consequence of it, washed by boatable waters, exclusive of all the great lakes; and many millions of acres of excellent tillage land, rapidly settling, are accommodated with water communication for conveying their produce to market. The intervales on both sides of this river are of various widths; and now and then, interrupted by the projection of the hills quite to the banks of the river, are some of the richest and best lands in the world. The fine farms which embrace these intervales, are owned and cultivated principally by Dutch people, whose mode of managing them would admit of great improvement. The manure of their barns they consider as a nuisance, and instead of spreading it on their upland, which they think of little value, (their meadow lands do not require it) they either let it remain for years in heaps, and remove their barns, when access to them becomes difficult, or else throw it into the river, or the gullies and streams which communicate with it. The banks of this river were formerly thickly settled with Indians. At the period when Albany was first settled, it has been said by respectable authority, that there were 800 warriors in Schenectady; and that 300 warriors lived within a space which is now occupied as one farm. The Cohoez in this river are a great curiosity. They are three miles from its entrance into the Hudson. The river is about 1000 feet wide; the rock over which it pours, as over a mill-dam, extends from *s. w.* to *n. e.* almost in a line from one side of the river to the other, and is about 40 feet perpendicular height; and including the descent above, the fall is as much as 60 or 70 feet. About a mile below the falls, is a handsome bridge, finished in July 1795. It is 1100 feet in length, 24 in breadth, and 15 feet above the bed of the river, which for the most part is rock, and is supported by 13 solid stone pillars. It is a free bridge, and including the expence of cutting through a ledge on the *n.e.* side of the river, cost 12,000 dollars. The river, immediately below the bridge, divides into three branches, which form several large islands. The branches are fordable at low water, but are dangerous. From the bridge you have a fine view of the Cohoez on the *n.w.*]

[Mohawk, a branch of the Delaware river. Its course from its source in lake Uttayantha is *s.w.* 45 miles, thence *s.e.* 12 miles, when it mingles with the Popachton branch; thence the confluent stream is called Delaware.]

[Mohawk, a town on the *s.* side of the river of its name in Montgomery county, New York; situated in one of the most fertile countries in the world. It was abandoned by the Mohawk Indians in the spring of 1780. See Hunter Fort. The township is bounded *n.* by Mohawk river, *e.* and *s.* by Albany county. In 1790, it contained 4440 inhabitants, including 111 slaves.]

[Mohawks, an Indian nation, acknowledged by the other tribes of the Six Nations to be "the true old heads of the confederacy." They were formerly very powerful, and inhabited on Mohawk river. As they were strongly attached to the Johnson family, on account of Sir William Johnson, a part of them emigrated to Canada with Sir John Johnson, as early as the year 1776. About 300 of this nation now reside in Upper Canada. See Hunter Fort and Six Nations.]

[MOHEGAN, situated between Norwich and New London in Connecticut. This is the residence of the remains of the Mohegan tribe of Indians. A considerable part of the remains of this tribe lately removed to Oneida with the late Mr. Occom. See Brothertown.]

[MOHICCONS, a tribe of Indians who inhabit on a branch of the Susquehannah, between Chagnet and Owegy. They were reckoned by Hutchins, about 50 years ago, at 100; but by Imlay, in 1773, at only 70 fighting men. They were formerly a confederate tribe of the Delawares. Also an Indian tribe, in the N.W. territory, who inhabit near Sandusky, and between the Sciota and Muskingum; warriors, 60.]

MOHICKANS, a settlement of Indians, of the same province and kingdom as the former; situate on the shore of the river Bever.

MOHOCAMAC, a settlement of the province and colony of New York in N. America; situate on the shore of the *e.* arm of the river Delaware.

MOHOSA, a settlement of the province and *corregimiento* of Cochabamba in Peru.

MOINA, a small lake of the province and *corregimiento* of Quispicanchi in Peru, where a fort has been built to restrain the incursions of the infidel Indians. See Oropesa.

MOINES, a small island within the bay of the Cul de Sac Royal in Martinique, very close to the coast.

MOINGONA, an abundant river of the province and government of Louisiana, its origin not being known for certain. It runs *s.e.* for many leagues, and enters the Mississippi, traversing some immense *llanuras*, which abound with buffaloes.

[MOINS, a river of Louisiana, which empties from the *n.w.* into the Mississippi, in lat. 39° 38′ *n.* The Sioux Indians descend by this river.]

[MOISIE River, on the *n.* shore of the St. Lawrence, is about three leagues *w.s.w.* of Little Saguena river, from which to the *w.n.w.* within the Seven Islands, is a bay so called from these islands.]

MOITACO, a settlement of the missions that were held by the religious observers of San Francisco, in the province and government of Guayana. It is the first of the establishments which were formed there on the shores of the Orinoco, and near the mouth of the Caura.

MOJIBIO, a settlement of the province and government of Popayán in the Nuevo Reyno de Granada.

MOJICA, a settlement of the province and government of Venezuela; situate on the shore of the river Guarico.

MOJOIN, a settlement of the province and country of Las Amazonas; situate on the coast, near the entrance of this river into the sea, at the cape of Miguari and territory of the Portuguese.

MOLANGO, Santa Maria de, a settlement of the jurisdiction and *alcaldía mayor* of Meztitlan in Nueva España. It contains a very good convent of the religious order of San Agustin, and 480 families of Indians. Fifteen leagues *n.n.e.* of its capital.

MOLCOCHINECON, a small river of the province and colony of Virginia, running *w.* and entering the Ohio.

[MOLE, The, is the *n.w.* cape of the island of St. Domingo, to the *n.* of cape St. Nicholas, and is often called by that name. The Mole, though inferior, by a great deal, to cape François and Port au Prince, is the first port in the island for safety in time of war, being strongly fortified both by nature and art. Count D'Estaing, under whose direction these works were constructed, intended to have established here the seat of the French government; but the productions of its dependencies were of too little value to engage his successors to carry his plan into effect; so that it is now no more than a garrison. It has a beautiful and safe port, and is considered as the healthiest situation in St. Domingo, by reason of the purity of its springs. The exports from January 1, 1789, to December 31, of the same year, were only 265,615 lb. coffee, 26,861 lb. cotton, 2,823 lb. indigo, and other small articles to the value of 129 livres. The value of duties on exportation 1,250 dollars, 21 cents. It is 14 miles *s.* by *w.* of Jean Rabel, 69 *w.* of cape François, and 48 *w.* by *s.* of Port de Paz. Lat. 19° 51′ *n.* Long. 73° 26′ 30″ *w.*]

MOLEQUES, Rio de los, a small river of the district of Matogroso in Brazil, which rises in the mountains, and runs *s. w.* to enter the Itenes.

MOLINA, a river of the province and government of Esmeraldas in the kingdom of Quito. It runs between the rivers Santiago and Vainillas to the *n. w.* and enters the Pacific or S. sea, to the *n.* of the equator.

Molina, a settlement of the province and government of Costarica in the kingdom of Guatemala; situate on the shore of the river of Los Anzuelos, on the coast.

[MOLINE'S Gut, on the *s. w.* side of the island of St. Christopher's in the W. Indies, is the first rivulet to the *s. e.* of Brimstone hill, near the mouth of which is anchorage in five and 10 fathoms, and a clear shore; but to the *e.* of it are some sunken rocks.]

MOLINO, a settlement of the province and *corregimiento* of Ica in Peru, annexed to the curacy of San Juan de Ica.

Molino, another settlement, of the kingdom of Chile; situate on the shore of the river Cauten.

Molino, a small island, situate near the *n.* coast of the island St. Domingo, and the promontory of Monte Christi, between the islands Arenosa and Crisin.

MOLI-TATUBA, a small river of the province and *captainship* of Seara in Brazil. It runs *n.* and enters the sea on the coast of Los Humos, between the island of Corubán and the settlement of Manapirange.

MOLLEAMBATO, San Miguel de, a settlement of the province and *corregimiento* of Latacunga in the kingdom of Quito.

MOLLEBAMBA, a settlement of the province and *corregimiento* of Aimaraez in Peru.

MOLLEBAYA, a settlement of the province and *corregimiento* of Moquehua in Peru; annexed to the curacy of Pocsi.

MOLLEPATA, a settlement of the province and *corregimiento* of Guamachuco in Peru; one of the four districts into which the curacy of Estancias is divided.

Mollepata, another settlement, in the province and *corregimiento* of Abancai of the same kingdom.

Mollepata, another, of the province and *corregimiento* of Conchucos in the same kingdom; situate on the shore of the river Tablachica.

MOLLEPONGO, a settlement of the province and *corregimiento* of Chimbo in the kingdom of Quito, of the district of Alausi.

MOLLETURO, a settlement of the province and *corregimiento* of Cuenca in the kingdom of Quito.

MOLLOBAMBA, a settlement of the province and *corregimiento* of Chancay in Peru; annexed to the curacy of Canchas.

MOLOACAN, Santiago de, a settlement and head settlement of the district of the *alcaldía mayor* of Acayuca in Nueva España. It contains 109 families of Ahualulcos Indians, including those of its adjoining ward. It is 18 leagues *e.* ¼ *s. e.* of its capital.

MOLUEDEC, a settlement of the province and *corregimiento* of Chillan in the kingdom of Chile; opposite the lakes of the Desaguadero on the *w.*

MOMAS, a settlement of the head settlement of the district of Tlaltenango, and *alcaldía mayor* of Colotlan, in Nueva España. Three leagues *n. n. w.* of its head settlement.

MOMBACHA, a settlement of the province and government of Nicaragua in the kingdom of Guatemala.

MOMIL, a settlement of the province and government of Cartagena in the Nuevo Reyno de Granada, of the district of the town of Sinú; situate on the shore of the river of this name, between the settlements of San Juan and San Andres.

MOMOSTENANGO, a settlement of the province and *alcaldía mayor* of Gueguetenango in the same kingdom as the former.

MOMPON, Senal de, a mountain of the coast of Peru, in the province and *corregimiento* of Santa; serving as a land-mark for vessels off that coast.

MOMPOX, Santa Cruz de, a great and rich town of the province and government of Cartagena in the Nuevo Reyno de Granada; founded on the *w.* shore of the river of La Magdalena, on the *n.* of Honda and *s. e.* of Cartagena, by Gerónimo de Santa Cruz, who gave it his name, in 1540. It is of an healthy temperature, although warm and very moist, as being surrounded with swamps. It is the embarking place of the river which leads to the provinces of the Nuevo Reyno; by a commerce with which it is rendered very rich and flourishing. It is fertile and abounds in vegetable productions, and especially in *cacao*, cotton, sugar-cane, and tobacco. The natives make mats of straw which they call *petates*, fans, and other articles, much esteemed for their beauty throughout the kingdom. In this town is a royal custom-house, where goods and merchandize going to the Nuevo Reyno pay a certain duty. It has a very good parish church, some convents of the religious orders of San Francisco, San Agustin, San Juan de Dios, and had a college of the Jesuits. It is inhabited by various noble and rich families, but the natives have the character of being haughty and litigious. It is greatly infested with musquitoes and by alligators, which come up the river to feed upon what is thrown from the city and the *albarrada*, which is a strong stone wall to keep the water from running into the streets. Indeed, this city has been frequently inundated by the swelling of the waters, and particularly in 1762, when the inhabitants were obliged to desert their houses and save themseves in canoes.

[MONA, or La Guenon, or The Monk, a small island, 38 miles *s. e.* of cape Engano, which is the most *e.* point of the island of St. Domingo, and 38 miles *w.* by *n.* of cape Morrilos in the island of Porto Rico. It is five miles from *e.* to *w.* and a little less from *n.* to *s.* It has several ports for small vessels, plenty of good water, and all that would be necessary for settlements of culture, and the breeding of cattle. Its fruit-trees, and particularly the orange, are much extolled. Two miles and a half *n. w.* of Mona is a very small island, called Monito, or the Little Monkey. The king Don Fernando the Catholic gave this island as a property to the admiral Don Christopher Columbus, with a *repartimiento* of 200 Indians. The English disembarked upon it in 1521. It is at an equal distance from St. Domingo and Puerto Rico.]

MONACACI, a small river of the province and colony of Maryland in N. America, of the district of Frederick county. It runs *s.* and enters the Patowmac.

[MONADNOCK, Great, a mountain, situated in Cheshire county, New Hampshire, between the towns of Jaffray and Dublin, 10 miles *n.* of the Massachusetts line, and 22 miles *e.* of Connecticut river. The foot of the hill is 1395 feet, and its summit 3254 feet above the level of the sea. Its base is five miles in diameter from *n.* to *s.* and three from *e.* to *w.* On the sides are some appearances of subterraneous fires. Its summit is a bald rock.]

[Monadnock, Upper Great, a high mountain in Canaan, in the *n. e.* corner of the state of Vermont.]

[MONAHAN, a township in York county, Pennsylvania.]

MONATOMY, a town of the county of Middlesex and bay of Massachusetts, three miles *n.* of Waterton, and four *n. w.* of Cambridge.

MONBATECEI, a river of the province and government of Paraguay, rising in the *serranías* between the rivers Paraguay and Parana. It runs *w.* and enters the former between the Monbemboi and the Taquari.

MONBEMBOI, a small river of the same province and government as the former. It has the same rise and course, and also enters the Paraguay.

MONCLOUA, a town and capital of the province and *alcaldía mayor* of Coaguila or Cohahuila, founded by order of the viceroy of Nueva España, with this title. It has in it a garrison of troops for the defence of the frontiers, and to restrain the infidel Indians. It contains 150 families of Spaniards, and is 258 leagues to the *n.* of Mexico, in lat. 27° 50′. Long. 270° 5′. [This military post is now, according to Humboldt, under the intendancy of San Luis Potosi.]

MONDAHU, a small river of the province and *captainship* of Seara in Brazil, which rises near the coast, runs *n.* and enters the Curú in its mid-course.

MONDAI, a river of the province and government of Paraguay, running *s. e.* and entering the Paraná.

MONDAQUE, a settlement of Indians of the province and government of Louisiana; situate in the road which leads to Nuevo Mexico, not far from the river of La Trinidad.

[MONDAY Bay, on the *s.* shore of the straits of Magellan, in that part of the straits called the Long Reach, and four leagues *w.* of Pisspot bay. It is nearly *s.* of Buckley point, on the *n.* side of the strait, and affords good anchorage in 20 fathoms.]

[Monday, a cape in the above straits, seven leagues *w. n. w.* of cape North. Lat. 53° 12′ *s.* Long. 73° 30′ *w.*]

MONFERRATO, Nuestra Senora de, a settlement of the province and *captainship* of Pernambuco in Brazil; situate on the coast, on the shore of the Bahia Grande or Puerto Calvo.

MONGA-AGUA, a river of the province and *captainship* of San Vicente in Brazil. It enters the sea opposite the island of Nuestra Señora.

MONGANGAPE, a small river of the province and *captainship* of Paraiba in Brazil. It rises near the coast, runs *e.* and enters the sea at the cape of Leda and settlement of Jorge Pinto.

MONGAGUABA, a settlement of the province and *captainship* of Pernambuco in Brazil; situate on the coast and at the point of Las Piedras.

MONGAGUEIRA, Sierra de, a *cordillera* of mountains of the province and *captainship* of Todos Santos in Brazil. It runs *n. n. e.* following the course of the river Del Real.

MONGAUEIRAS, a settlement of the island of Joanes or Marajo in Brazil; situate on an arm of the river of Las Amazonas, opposite the Gran Parà.

MONGE, a river of the province and government of Buenos Ayres. It runs *e.* and enters the Paraná close to the settlement of Calchaqui.

MONGES, some *farallones* or isles of the N. sea; situate near the coast of the province and government of Santa Marta, 10 leagues from the point of Mazola.

Monges, a river of the kingdom of Brazil, which rises at the foot of the Sierra Grande, runs *n. n. e.* and enters the Tocantines between that of Santa Lucia and that of Corijas.

MONGON, a cape or point and extremity of the island of Cuba, close to the Caico Grande or del N. looking *s.* and near to that of Caico Pequeño.

[Mongon, on the coast of Peru and province of Santa, on the S. Pacific ocean, is 20 miles *n.* of the harbour of Guarmey, and four leagues from Bermejo island, which lies between the former places. Casma is four leagues *n.* of it. Mongon is known at sea by a great mountain just over it, which is seen farther than any others on this part of the coast. Lat. 9° 49′ *n.*]

[Mongon, Cape, on the *s.* side of the island of St. Domingo, is 3000 fathoms *n.* of point Bahoruco and the river Nayauco, and nearly *s.* of the little port of Petit Trou.]

MONGRAUE, a small island of the N. sea, one of the Lucayas, and the last at the mouth or entrance of the canal of Bahama.

MONGUA, a settlement of the province and *corregimiento* of Tunja in the Nuevo Reyno de Granada. It is of a very cold temperature, abounding in wheat, barley, beans, maize, and pig-nuts, the which when put into a hole with water, and this being often changed, make a kind of fetid oil, which they call *futes*, and which is taken as an excellent stomachic. Its population may amount to 80 persons, and about as many Indians. Ten leagues *n. e.* of Tunja.

MONGUI, Nuestra Senora de, a settlement of the province and *corregimiento* of Tunja in the Nuevo Reyno de Granada; of a cold temperature, producing some fruits of this climate. It has a good convent of monks of San Francisco, in which is venerated an image of the Virgen del Rosario, with the child Jesus in her arms, and St. Joseph on the side, the same having been painted by the Emperor Charles V. who sent it to this settlement with some rich ornaments, as an acknowledgment to it as having been the first settlement of that kingdom which had voluntarily offered obedience to the crown of Castille. At its entrance runs a large river called La Quebrada, over which there is a stone bridge of beautiful architecture. Eight leagues *n. e.* of Tunja.

Mongui, another settlement, with the additional title of Charala, to distinguish it from the former; in the jurisdiction and district of the town of San Gil and of the same kingdom. It is a large population, of a moderately hot temperature, and healthy, though subject to rains; produces great quantity of cotton, maize, and tartuffles, with which the neighbouring towns are provided, especially that of Socorro. It abounds equally in plantains, *uyamas*, and a variety of fruits, as also in exquisite kinds of woods. Some cotton-stuffs are made here, though little esteemed. Its population consists of 2000 souls, and it is situate between two fine rivers abounding with delicious water. It is seven leagues *s. e.* of San Gil, and three *e.* of Socorro.

Mongui, another, of the province and *corregimiento* of Parinacochas in Peru; annexed to the curacy of Pampamarca.

[MONHEGAN, or Menhegan, a small island in the Atlantic ocean, 12 miles *s. e.* of Pemaquid point, in Lincoln county, district of Maine, and in lat. 43° 42′. North of it are a number of small isles at the mouth of St. George's river. Captain Smith landed his party here in 1614. The chimneys and remains of the houses are yet to be seen.]

[MONIIETOU Islands, in the N. W. territory, lie towards the *e.* side of the Michigan lake, towards its *n.* end, and *s.* of Beaver islands.]

MONI, a large and copious river of the province and government of Marañan in Brazil. Its shores are delightfully pleasant, and the immediate soil yields the best sugar-cane in the whole kingdom. It empties itself in the gulf of San Luis de Marañan. On the *w.* near its source and amidst some very thick woods, dwells a nation of barbarian Indians, the Tapayos.

MONICA, Santa, a settlement of the head settlement of the district and *alcaldia mayor* of

Marinalco in Nueva España, from whence it is a little more than half a league's distance.

Monica, Santa, another, a small settlement or ward of the head settlement of the district of Ocuila, and of the same *alcaldia mayor* and kingdom as the former.

Monica, Santa, a port on the *s.* coast of the strait of Magellan, close to cape Pilares, and at its entrance by the S. sea.

MONIGOTE, a settlement of the province and government of Buenos Ayres; situate on the shore of the river Saladillo, between this river and the lake Brava.

MONIQUIRI, a settlement of the *corregimiento* of the jurisdiction of Velez in the Nuevo Reyno de Granada, of an hot temperature, but healthy, fertile, and abounding in all the fruits of a warm climate, especially sugar-canes, of which are made great quantities of sugar by the engines or mills for the purpose. It also abounds in excellent water, *yucas*, plantains, and maize, and they make here delicate conserves, sweet-meats, and honey, which are much esteemed in the other provinces where they are sold. It contains 500 house-keepers, and is eight leagues to the *e.* of Velez, and four from Leiba.

MONITO, El, a small island of the N. sea; situate close to that of La Mona, to the *n.* between those of Saona and Puerto Rico.

MONK'S-NECK, a small river of the province and colony of Virginia, in the county of Brunswick. It runs *s. e.* and enters the Nottaway.

[MONKTON, a township in Addison county, Vermont, *e.* of Ferrisburgh, and contains 450 inhabitants.]

[Monkton, a township in Annapolis county, Nova Scotia, inhabited by Acadians, and a few families from New England. It lies partly on the basin of Annapolis, and partly on S. Mary's bay, and consists chiefly of wood-land and salt-marsh. It contains about 60 families.]

MONLEO, a settlement of the province and *alcaldia mayor* of San Miguel in the kingdom of Guatemala; annexed to that of Yayantique.

[MONMOUTH, a large maritime county of New Jersey, of a triangular shape, 80 miles in length, and from 25 to 40 in breadth; bounded *n.* by part of Rariton bay, *n. w.* by Middlesex county, *s. w.* by Burlington, and *e.* by the ocean. It is divided into six townships, and contains 16,918 inhabitants, including 1596 slaves. The face of the country is generally level, having but few hills. The most noted of these are the highlands of Navesink and Centre hill. See Middletown. A great part of the county is of a sandy soil, but other parts are fertile. There is a very curious cave, now in ruins, at the mouth of Navesink river, 30 feet long and 15 wide, and contains three arched apartments.]

[Monmouth, or Freehold, a post-town of New Jersey, and capital of the above county; situated 18 miles *n. e.* by *e.* of Allentown, 25 *e.* of Trenton, 12 *s. w.* by *s.* of Shrewsbury, and 46 *n. e.* by *e.* of Philadelphia. It contains a court-house and gaol, and a few compact dwelling-houses. This town is remarkable for the battle fought within its limits on the 27th of June 1778, between the armies of General Washington and Sir Henry Clinton. The latter having evacuated Philadelphia, was on his march to New York. The loss of the Americans, in killed and wounded, was about 250; that of the British, inclusive of prisoners, was about 350. The British pursued their march the night after, without the loss of their covering party or baggage. See Freehold.]

[Monmouth, a small post-town in Lincoln county, district of Maine; situated to the *e.* of Androscoggin river, 10 miles *w.* of Hallowell court-house, five *w.* of Winthrop, seven *n. e.* by *n.* of Greene, 39 *n.* of Portland, and 125 *n.* by *e.* of Boston.]

[Monmouth Cape, on the *s. e.* side of the straits of Magellan, about half way from the *s.* entrance of the second narrows, to the *s. e.* angle of the straits opposite cape Forward.]

[Monmouth Island, one of the four islands of Royal reach, in the straits of Magellan, and the second from the *w.*]

MONO and Mona. Some small barren and desert isles of the N. sea, amongst the Antilles; 12 leagues to the *w.* of the point of La Aguada of Puerto Rico, in long. 308° 4′. Lat. 18° 4′.

[MONOCACY, a river which after a *s. s. w.* course, empties into the Patowmac, about 50 miles above Georgetown.]

MONONEPIOUI, a small river of New France or Canada, which runs *s. w.* and enters lake Superior.

[MONONGAHELA River, a branch of the Ohio, is 400 yards wide at its junction with the Alleghany at Pittsburg. It is deep, gentle, and navigable with batteaux and barges beyond Red Stone creek, and still further with lighter craft. It rises at the foot of the Laurel mountain in Virginia, thence meandering in a *n.* by *e.* direction, passes into Pennsylvania, and receives Cheat river from the *s. s. e.* thence winding in a *n.* by *w.* course, separates Fayette and Westmoreland from Washington county, and passing into Alleghany county, joins the Alleghany river at Pittsburg, and forms the Ohio. It is 300 yards wide 12 or 15 miles from its mouth, where it receives the Youghiogany from the *s. e.* which, is navigable with batteaux and barges to the foot of Laurel hill. Thence to Red Stone, at fort Byrd, by water, is 50 miles, by land 30, and 18 in a straight line. Thence to the mouth of Cheat river, by water 40 miles, by land 28; the width continuing at 300 yards, and the navigation good for boats. Thence the width is about 200 yards to the *w.* fork, 50 miles higher, by water, and the navigation frequently interrupted by rapids; which, however, with a swell of two or three feet, become very passable for boats. It then admits light boats, except in dry seasons, 65 miles farther, by water, to the head of Tygart's valley, presenting only some small rapids and falls of one or two feet perpendicular, and lessening in its width to 20 yards. The *w.* fork is navigable in the winter, towards the *n.* branch of the Little Kanhaway, and will admit a good waggon road to it. From the navigable waters of the *s.* easternmost branch of the Monongahela, there is a portage of 10 miles to the *s.* branch of Patowmac river. The hills opposite Pittsburg on the banks of this river, which are at least 300 feet high, appear to be one solid body of coal. On the Pike run of this river, a coal-hill has been on fire 10 years, yet it has burnt away only 20 yards.]

[MONONGALIA, a county in the *n. w.* part of Virginia, about 40 miles long and 30 broad, and contains 4768 inhabitants, including 154 slaves.]

MONOS, Islas de. Some islands in the N. sea, near the coast of the kingdom of Tierra Firme, in the bay of Garrote, in the jurisdiction and government of Portovelo. They are many, all small, barren and desert, and peopled only by monks; from whence they are so called.

Monos, one of the months of Los Dragos to enter the gulf Trieste, between the point of Paria on the coast of Tierra Firme, and the *w.* point of the island Trinidad; situate between this island and a very small isle lying at the same rhumb and mouth.

MONPATAR, a settlement of the island and government of La Margarita; situate at the point of Ballenas, opposite the coast of Tierra Firme.

[MONPOX, a city of Tierra Firme, about 75 miles *s. e.* by *e.* of Tolu.]

MONQUIRA, a settlement of the *corregimiento* of Sachica, in the jurisdiction of the town of Leiba and Nuevo Reyno de Granada, only a quarter of a league distant from the latter; of a benign temperature, and producing wheat and seeds.

MONSAUILACHA, a small river of the province and government of Buenos Ayres, which runs *n. n. w.* and enters the Cota.

[MONSEAG Bay, in Lincoln county, district of Maine, is separated from Sheepscut river by the island of Jeremysquam.]

MONSEFU, a small but pleasant and pretty settlement of the province and *corregimiento* of Saña in Peru, and of the district of Lambayeque, from whence it is five leagues distant.

MONSERRAT. [See Montserrat.]

[MONSIES, the third tribe in rank of the Delaware nation of Indians.]

MONSIEUR, a small isle of the N. sea, situate near the *s. e.* coast of Martinique. It forms, with two other islands and the point of Rose, a great port called Cul de Sac Robert.

[MONSON, a township in Hampshire county, Massachusetts, *e.* of Brimfield, and 55 miles *s. w.* by *w.* of Boston. It was incorporated in 1760, and contains 1331 inhabitants.]

MONT Louis, a road of the river St. Lawrence in New France or Canada. It is the mouth of a river which enters into the aforesaid river, and offers a good sounding, though exposed to the *n.* wind, but this seldom blows in the spring. It is capable of admitting vessels of 100 tons, where they may lie secure from tempests and from enemies, but they cannot enter but at high tide, since at the ebb there is no more than two feet water. It is equally easy to be defended, having on one side inaccessible mountains, and on the other a peninsula, on which might be built a fort. From these advantages, Mr. Riverin was induced in 1697 to establish a cod-fishery, for which purpose he established a company; but just as he was about to put his design into execution, and the ships with all their necessaries were setting sail for the new establishment, the Count of Frontenac received advice of a probable rupture with the English, and he broke up the expedition. In 1700, the same Mr. Riverin attempted it again, but having arrived at Mont Louis at a late season for the fish, and those engaged with him not deriving the advantages they were led to expect, he found himself, through want of assistance, to abandon, for the second time, his project.

[MONTAGUE, a township in Hampshire county, Massachusetts, on the *e.* bank of Connecticut river, between Sunderland and Wendel, about 18 miles *n.* of Northampton, and 70 miles *w.* by *n.* of Boston. It was incorporated in 1753, and contains 906 inhabitants. A company was incorporated in 1792, to build a bridge over here. The work has not yet been completed.]

[MONTAGUE, the northernmost township in New Jersey, is situated in Sussex county, on the *e.* side of Delaware river, about five miles *n. e.* of Minisink, and 17 *n.* of Newtown. It contains 543 inhabitants, including 25 slaves.]

[MONTAGUE, the largest of the small islands in Prince William's sound, on the *n. w.* coast of N. America.]

MONTALUAN, S. BAPTISTA DEL PAO DE, a town of the province and government of Venezuela in the Nuevo Reyno de Granada, founded in this century (1700), after the establishment of the company of Guipuzcoa.

MONTANA, S. FRANCISCO DE LA, a settlement of the province and government of Santiago de Veragua in the kingdom of Tierra Firme. It is of an hot temperature, fertile, abounding in vegetable productions, and in gold mines, which are named De la Libertad of Aguacatál and of San Francisco, and in the vicinity of which there is a mill for grinding metals, erected in 1749 by Don Geronimo Sancho. The settlement is situate on the top of a mountain towards the *n.* coast, three leagues from its capital.

MONTANA, another settlement, in the province and government of Popayán, of the district of the city of Pasto.

MONTANAS, SANTIAGO DE LAS, a city of the province and government of Jaen de Bracamoros in the kingdom of Quito, founded by the Captain Juan de Salinas; of a very unequal temperature, and of a rough and rocky territory, as its name shews, and so reduced and poor that it does not deserve the name of a city. The soil is, however, fertile; nor is it without mines of gold of excellent quality, but which are not worked through scarcity of hands and means. One hundred and fifty-eight miles *e.* of Loyola, on the *n.* bank of the river S. Yago, and six miles *n. w.* of S. Borja.

[MONTAUK Point, the *e.* extremity of Long island, New York. A tract here, called Turtle Hill, has been ceded to the United States for the purpose of building a light-house thereon.]

MONTE, [the Spanish word for mountain: for the chief mountains in America, see MOUNTAINS.]

[MONTE, a fort of the province and government of Buenos Ayres; situate near a small lake, about 60 miles *s. w.* of Buenos Ayres. Lat. 35° 25′ 40″. Long. 59° 50′ 54″.]

[MONTE CHRIST, a cape, bay, town, and river, on the *n.* side of the island of St. Domingo. The cape is a very high hill, in the form of a tent, called by the French Cape la Grange, or Barn. It is situated in lat. 19° 54′ 30″ *n.* and in long. 71° 44′ *w.* A strip of level land joins it to the territory of Monte Christ, and it is owing to this that the cape has been taken for an island. It is 29 miles *n. e.* by *e.* of cape François, where it may be seen in a clear day with the naked eye. After doubling this cape, we find the bay of Monte Christ running nearly *s. w.* It is formed by cape La Grange on one side, and point Des Dunes (Down point) on the other, about 6500 fathoms asunder. The bay is about 1400 fathoms deep, and its winding is nearly four leagues. About 900 fathoms from the cape, descending the bay, we find the little island of Monte Christ, 350 fathoms from the shore. One may sail between the two, with two, four, and five fathoms water; and about 250 fathoms further on, is anchorage in from six to ten fathoms. A league and a quarter from cape La Grange, is a battery intended to protect a landing place of 100 fathoms wide, which is below and opposite the town of Monte Christ. The town of Monte Christ, standing at 800 fathoms from the sea-side, rises in form of an amphitheatre on the side of the coast, which is very high all round this bay. The town is 200 fathoms square, which space is divided into nine parts, cut by two streets running from *e.* to *w.* and two others from *n.* to *s.* It was founded in 1533, abandoned in 1606, and now but a poor place, destitute of every resource but that of cattle raised in its territory, and sold to the French. The town and territory contain about 3000 souls. There is a trifling garrison at Monte Christ. About a league from the battery, following the winding of the bay, is the river of Monte Christ, or more properly, the river Yaqui. The land round the town is barren and sandy, and the river contains great numbers of crocodiles. Monte Christ is a port well known to American smugglers, and carries on a great commerce from its vicinity to the French plantations. In the time of peace, all the produce of the plain of Mariboux, situated between port Dauphin and Mancenille bay, is shipped here, and in a war between France and Britain, it used to be a grand market, to which all the French in the *n.* part of the island sent their produce, and where purchasers were always ready.]

[MONTE CHRIST, a chain of mountains which extend parallel to the *n.* coast of the island of St. Domingo, from the bay of Monte Christ to the bay of Samana on the *e.* Two large rivers run in opposite directions along the *s.* side of this chain:

the river Monte Christ or Yaqui in a *w.* by *s.* direction, and Yuna river in an *e.* by *s.* course to the bay of Samana. They both rise near La Vega, and have numerous branches.]

MONTE CHRISTI, a settlement of the district of the Puerto Viejo, in the province and government of Guayaquil, and kingdom of Quito; situate on the sea-shore. It was formerly in Manta, and large and populous, from traffic done here by the vessels going from Panamá to the ports of Peru; but having been destroyed and sacked by pirates, with whom those seas at that time were covered, the inhabitants retired to the spot where the settlement now stands, at the foot of the mountain from whence it takes its name, and which serves as a land-mark to vessels for making the port of Manta, it being one of the loftiest mountains on that coast.

MONTE CHRISTI, a small island close to the *n.* coast of St. Domingo, between the island Madera and the port La Granja.

MONTE DE PLATA, a port on the coast of California of the S. sea, discovered by General Sebastian, a Biscayan, in 1602, when he was sent by the viceroy of Nueva España, the Count de Monterrey, to reconnoitre that coast, and in honour of whom he thus named it. It is large, convenient, and sheltered from all the winds, abounds in wood, and has some fine straight firs fit for ship-masts, oaks, beech, and various other kinds of fine timber; also many fountains of rich water, lakes, fertile glens and meadows, and excellent land for agriculture. In its woods are found a variety of animals, and particularly great bears, and some animals as large as a young ox, in shape like a stag, with hair like a pelican of a quarter of a yard in length, long neck, and on the head some very large horns, with a tail of a yard long and half a yard wide, and the hoof cleft like that of an ox. Here are also deer, stags, hares, rabbits, wild cats, bustards, ducks, swallows, geese, doves, thrushes, sparrows, goldfinches, cardenals, quails, partridges, feldfares, and various other birds, also Indian fowl of a large sort, measuring from the extremity of one wing to the other seven palms. On the sea-shore are various sorts of shell-fish, and some with shells of most beautiful mother of pearl: here are also taken oysters, lobsters, crabs, marine wolves, and whales. The whole of the port is surrounded by Indians; who live in huts, are of a docile disposition, kind and liberal, and using the bow and arrow for weapons. They seemed to express great sorrow when the Spaniards left them.

MONTE DE PLATA, a settlement of the island of St. Domingo, taking its name from a lofty mountain of white stone discovered by Admiral Don Christopher Columbus in his first voyage in 1493, and who thought it was silver. It has a good port, which the French call Portoplate, from the name that had been given it by the Spaniards of Puerto de Plata, and which was also the name of the settlement. This, on account of its being exposed to the incursions of pirates, was removed in 1606 to the place where it now stands, and nearer to the mountain.

MONTE ESTANCIA, a lake of the province and government of Buenos Ayres, between the rivers Hueque Leuvu and Saladillo.

MONTE LEON, a settlement and head settlement of the district of the *alcaldía mayor* of Tepozcolula in Nueva España; of a cold temperature, inhabited by 52 families of Indians, who maintain themselves by cultivating and selling seeds. Five leagues *s. e.* of its capital.

MONTE, REAL DEL, a settlement of the jurisdiction and *alcaldía mayor* of Pachuca in Nueva España. It contains 80 families of Spaniards, *Mustees*, and Mulattoes, employed in mining and preparing the silver when extracted. This is their principal commerce. The Indians are also, some of them, employed in this labour, although they are rather dedicated to agriculture, the sowing of wheat, maize, and other seeds. This settlement is much frequented by traders who carry to Mexico cloths and other articles, taking silver in exchange. Two leagues *e. n. e.* of its capital.

MONTE, REAL DEL, another settlement, with the dedicatory title of San Rafael, in the head settlement of the district and *alcaldía mayor* of Guejozingo in the same kingdom; contains 54 families of Indians, and is to the *w.* of its capital.

MONTE, REAL DEL, another, with the dedicatory title of San Miguel, of the head settlement of the district of Etequaro, and *alcaldía mayor* of Valladolid, of the province and bishopric of Mechoacán. It is six leagues to the *n.* of its head settlement, and in its district are some cultivated estates, by which and the cutting of wood the natives maintain themselves.

MONTE, REAL DEL, another, of the province and government of Popayán in the Nuevo Reyno de Granada.

MONTE, REAL DEL, another, called Boca del Monte, in the province and government of Maracaibo, and Nuevo Reyno de Granada; situate *s.* of the city of Pedraza.

MONTE, REAL DEL, another, with the addition of San Juan, in the province and *captainship* of Rey in Brazil; situate on the shore and at the source of the river Tajay.

MONTE, REAL DEL, a town of the province and

corregimiento of Melipilla in the kingdom of Chile; situate to the *w.* of its capital.

Monte Redondo, a settlement of the province and government of Tucumán, in the jurisdiction of the city of Cordoba; situate on the bank of a small river.

Monte Trigo, a small island of the N. sea; situate near the coast of the province and *captainship* of San Vicente in Brazil; between the island San Felipe and Santiago, and the great island of San Sebastian.

Monte Video, a city of the province and government of Buenos Ayres in Peru; founded on the *n.* shore of the river La Plata, at its mouth or entrance, by order of the field-marshal Don Bruno de Zavala, in 1726, by Don Domingo de Vasavilvaso. It is a small place, having only one parish and a convent of the religious order of San Francisco, although it had once an house of entertainment of the regulars of the company of Jesuits. It is situate on a lofty spot, upon a great and convenient bay, which is frequented by vessels going to Buenos Ayres. It has a citadel or castle, which is badly constructed, with four bulwarks and some batteries for its defence: the same is the residence of the governor. The town, which is well fortified with a strong wall and sufficient artillery, is inhabited by more than 1000 souls, amongst whom are some rich and noble families. The climate is excellent, cheerful, and healthy, the soil fertile and abounding in vegetable productions, and flesh and fish are so plentiful as to cost almost nothing. Its principal commerce consists in the hides of cattle, and these are killed merely for the above perquisites. It is 111 miles *e. s. e.* from Buenos Ayres, in lat. 34° 50′ 30″ *s.* and long. 56° 16′ *w.*

[Few places in Spanish America have experienced a greater change in their political consequence and physical energies, since the time Alcedo wrote, than that of which we are now treating. Independently of its present litigations with Buenos Ayres, it has been rendered famous in history by the English expedition which visited the river La Plata in 1806. It was for some little time in possession of the British troops, and finally evacuated at the beginning of September 1807. The following description of Monte Video and the adjoining country is extracted from the Travels of Mr. Mawe; we divide the information under the following heads. *Site of the town.---Population.---Character of the inhabitants.--- Trade. --- Geological remarks.--- Ditto on the country n. e. of Monte Video.--- Limestone, and mode of burning it.---Peons.--- Horses.---Defective state of agriculture.---Manners and dress of the natives.---Wild animals.*

Monte Video is a tolerably well-built town, standing on a gentle elevation at the extremity of a small peninsula, and is walled entirely round. Its population amounts to between 15,000 and 20,000 souls. The harbour, although shoal, and quite open to the *pamperos*, is the best in the Rio de la Plata; it has a very soft bottom of deep mud. When the wind continues for some time at *n. e.* ships drawing 12 feet water are frequently aground for several days, so that the harbour cannot be called a good one for vessels above 300 or 400 tons.

There are but few capital buildings; the town in general consists of houses of one story, paved with brick, and provided with very poor conveniences. In the square is a cathedral, very handsome, but awkwardly situated; opposite to it, is an edifice divided into a town-house or cabildo, and a prison. The streets, having no pavement, are always either clouded with dust or loaded with mud, as the weather happens to be dry or wet. In seasons of drought the want of conduits for water is a serious inconvenience, the well, which principally supplies the town, being two miles distant.

Provisions here are cheap and in great abundance. Beef in particular is very plentiful, and, though rarely fat or fine, makes excellent soup. The best parts of the meat may, indeed, be called tolerable, but they are by no means tender. The pork is not eatable. Such is the profusion of flesh-meat, that the vicinity for two miles round, and even the purlieus of the town itself, present filthy spectacles of bones and raw flesh at every step, which feed immense flocks of sea-gulls, and in summer breed myriads of flies, to the great annoyance of the inhabitants, who are obliged at table to have a servant or two continually employed in fanning the dishes with feathers, to drive away those troublesome intruders.

The inhabitants of Monte Video, particularly the Creolians, are humane and well disposed, when not actuated by political or religious prejudices. Their habits of life are much the same with those of their brethren in Old Spain, and seem to proceed from the same remarkable union of two opposite, but not incompatible qualities, indolence and temperance. The ladies are generally affable and polite, extremely fond of dress, and very neat and cleanly in their persons. They adopt the English costume at home, but go abroad usually in black, and always covered with a large veil or mantle. At mass they invariably appear in black silk, bordered with deep fringes. They delight in conversation, for which their vivacity eminently qualifies them, and they are very courteous to strangers.]

[The chief trade of Monte Video consists in hides, tallow, and dried beef; the two former of these articles are exported to Europe, and the latter is sent to the W. Indies, especially to the Havannah. Coarse copper from Chile in square cakes is sometimes shipped here, as well as a herb called *matté* from Paraguay, the infusion of which is as common a beverage in these parts as tea is in England.

The inhabitants were by no means opulent before the English took the garrison, but through the misfortunes of the latter at Buenos Ayres, and the losses of our commercial adventurers by ill-judged and imprudent speculations, they were considerably enriched. The great prospects indulged in England, before the expedition to the Plata, of immense profits by trade to that river, have generally ended in ruin; very few, indeed, of the speculators have escaped without considerable loss. Property, once litigated, might be considered in a fair way for confiscation; and in case of its having been deposited until certain questions were decided, restitution was generally obtained at the loss of one half. It frequently happened that goods detained in the custom-houses or lodged in private stores in the river were opened, and large quantities stolen. The party on whom suspicion seemed most reasonably to fall was the consignee, who, even with a few cargoes, was generally observed to get rich very rapidly. Not contented with the profits accruing from his commission, he seldom scrupled to take every advantage which possession of the property afforded him, of furthering his own interests at the expence of his correspondent. The dread of a legal process could be but a slight check upon him; for in the Spanish courts of justice, as well as in others, a native and a stranger are seldom upon equal terms. Other circumstances have concurred to enrich the inhabitants of Monte Video. It is a fact that the English exported thither goods to the amount of a million and a half sterling, a small portion of which, on the restoration of the place to the Spaniards, was re-shipped for the ap of Good Hope and the W. Indies; the remainder was for the most part sacrificed at whatever price the Spaniards chose to give. As their own produce advanced in proportion as ours lowered in price, those among them who speculated gained considerably. The holders of English goods sold their stock at upwards of fifty per cent. profit immediately after the evacuation of the place.

The climate of Monte Video is humid. The weather, in the winter months (June, July, and August), is at times boisterous, and the air in that season is generally keen and piercing. In summer the serenity of the atmosphere is frequently interrupted by tremendous thunder-storms, preceded by dreadful lightning, which frequently damages the shipping, and followed by heavy rain, which sometimes destroys the harvest. The heat is troublesome, and is rendered more so to strangers by the swarms of mosquitoes, which it engenders in such numbers that they infest every apartment.

The town stands on a basis of granite, the feldspar of which is for the most part of an opaque milk-white colour, in a decomposing state; in some places it is found of a flesh-red colour and crystallized. The mica is generally large and foliated, in many places imperfectly crystallized. It is obvious that the excessive quantity of mud in the harbour and throughout the banks of the river cannot have been formed from this stratum. The high mount on the opposite side of the bay, which is crowned with a light-house, and gives name to the town, is principally composed of clay-slate in laminæ perpendicular to the horizon. This substance appears much like basalt in texture, but its fracture is less conchoidal; it decomposes into an imperfect species of wakke, and ultimately into ferruginous argil. Beds of clay, from which flows much water, are observable in various parts of the mountain.

The vicinity of Monte Video is agreeably diversified with low gently-sloping hills, and long valleys watered by beautiful rivulets; but the prospects they afford are rarely enlivened by traces of cultivation; few enclosures are seen except the gardens of the principal merchants. The same defect appears in a *n. e.* direction from the town, where similar varieties of hill, valley, and water prevail, and seem to want only the embellishment of silvan scenery to complete the landscape. Some wood, indeed, grows on the margin of the Riachuelo, which is used for the building of hovels and fot fuel. There is a pleasant stream about 10 leagues from Monte Video called the Louza, the banks of which seem to invite the labour of the planter, and would certainly produce abundance of timber. It is to be remarked that the almost entire want of this article here, occasions great inconvenience and expence: wood for mechanical purposes is extremely scarce, and planks are so dear that hardly one house with a boarded floor is to be found.

About 25 leagues *n. e.* from Monte Video, is an irregular ridge of granite mountains in a direction nearly *n.* and *s.* and the country from this distance gradually assumes a rugged appearance. Mica is very common upon the road, and in some places] [quartz; on one hill are found several detached crystals of the latter substance. The ravines of these stony wilds and the wooded margins of the rivers afford shelter to many ferocious animals, such as jaguars, here called tigers, lions, and ounces. Here are also great numbers of wild dogs which breed in the rocks, and at times make great havoc among the young cattle. The farms in this district, for the most part, include tracts of land from 20 to 30 miles in length by half that extent in breadth, watered by pleasing streams. Vast herds of cattle are bred upon them; it is calculated that each square league sustains 1500 or 2000 head.

At the distance of about 40 leagues from Monte Video, in the direction above mentioned, the range of hills gradually lessens and disappears; the country opens finely on the left, and is intersected by numerous rivulets. After crossing several of these you arrive at the head of a little brook called Polancos, which a few miles below assumes the name of Barriga Negra. It there receives several small streams, and in the course of 10 leagues is augmented by the confluence of some others: becoming thus a considerable river, about as large as the Trent at Gainsborough, it is denominated Godoy, but on passing into the Portuguese territories it changes its name to that of Zebolyati and flows into the Lagun Meni.

The country here in general may be termed stony and mountainous, though its inequalities do not exceed those of Derbyshire. No traces of either volcanic or alluvial matter are to be found; the solid rock frequently appears on the surface, and in many places projects in masses of various sizes. The mountains and rocks are of granite; no veins of metallic substance have hitherto been discovered, but fine red and yellow jasper, chalcedony, and quartz, are not unfrequently found loose on the surface. Some fossils of the asbestos kind, and some very poor oxides of iron, are likewise to be met with occasionally. The bases of many of the conical granite mountains are overlaid with (apparently) primitive lime-stone of an obscure blue colour, in laminæ; Mawe found in this substance many capillary veins of calx-spar, and sometimes crystals of pyrites. In one part of the vicinity there is a plain about half a mile square, on the surface of which are found large quantities of white lime-stone in nodules; it is of a very close texture, but being considered inferior in quality to the other species it is never converted into lime. The summits of these mountains are no where calcareous, excepting those of one ridge, the singular appearance of which induced the above traveller to trace it as far as was practicable. The lime-stone on these summits is of a close compact kind, united to transparent quartz in a tabular form, standing, as it were, in laminæ perpendicular to the horizon, and thus presenting to the view a number of upright slabs somewhat similar to the grave-stones in a country church-yard. This singular ridge apparently commences at a mountain of very unusual form, and, extending about two miles, in which it crosses two or three valleys, terminates in a ravine of considerable depth. No vestige of calcareous crystallization appeared in this lime-stone. It is singular to remark, that the cavities formed by the laminæ afford refuge for reptiles, particularly rattle-snakes: it is said that a person employed here in getting the stone, destroyed upwards of 27 serpents of that species in the course of a few weeks.

The lime-stone is loosened by the wedge and lever, and brought away in large slabs to the kilns, where it is broken into fragments of a convenient size, and burnt with wood. The kilns are capacious, but so badly constructed that the process of calcination is very slow and tedious. The lime, when slaked, is measured, put into sacks made of green hides, and sent in large carts, drawn by oxen, principally to Colonia, Monte Video, and Buenos Ayres.

Barriga Negra is distant about 160 miles *n. e.* from Monte Video, about 120 from Maldonado, and 90 from the town of Minas. The country around it is mountainous, well watered, and not destitute of wood. The banks of the streams are thickly covered with trees, rarely, however, of large size; for the creeping plants, interweaving with the shoots, check their growth and form an impenetrable thicket. Here are numbers of great breeding estates, many of which are stocked with from 60,000 to 200,000 head of cattle. These are guarded principally by men from Paraguay called Peons, who live in hovels built for the purpose at convenient distances. Ten thousand head are allotted to four or five Peons, whose business it is to collect them every morning and evening, and once or twice a month to drive them into pens where they are kept for a night. The cattle by this mode of management are soon tamed; a ferocious or vicious beast is never seen among them. Breeding is alone attended to; neither butter nor cheese is made, and milk is scarcely known as an article of food. The constant diet of the people, morning, noon, and night, is beef, eaten almost always without bread, and frequently without salt. This habitual subsistence on strong food would probably engender diseases, were it not corrected by copious] draughts of an infusion of their favourite herb matté, which are frequently taken.

The dwellings of the Peons are in general very wretched, the walls being formed by a few upright posts interwoven with small branches of trees, plastered with mud inside and out, and the roof thatched with long grass and rushes. The door is also of wicker-work, or, in its stead, a green hide stretched on sticks and removable at pleasure. The furniture of these poor hovels consists of a few scalps of horses, which are made to serve for seats; and of a stretched hide to lie upon. The principal, if not the sole, cooking utensil is a rod or spit of iron stuck in the ground in an oblique position, so as to incline over the fire. The beef when spitted on this instrument is left to roast until the part next the fire is supposed to be done enough, then a twist is given to the rod, which is occasionally repeated, until the whole is cooked. The juices of the meat, by this mode of roasting, help to mend the fire, and indeed the people seem to think that they are fit for nothing else. The meat, which is naturally poor and coarse, being thus dried to a cake, bears little affinity to the boasted roast beef of England. Fuel, in some parts, is so extremely scarce that the following strange expedient is resorted to for a supply. As the mares in this country are kept solely for breeding, and are never trained to labour, they generally exceed the due proportion; a flock of them is frequently killed, and their carcases, with the exception of the hides and tails, are used as firing.

The Peons are chiefly emigrants from Paraguay, and it is a singular fact that, among the numbers that are here settled, very few women are to be found. A person may travel in these parts for days together without seeing or hearing of a single female in the course of his journey. To this circumstance may be attributed the total absence of domestic comfort in the dwellings of these wretched men, and the gloomy apathy observable in their dispositions and habits. It is true that the mistress of an estate may occasionally visit it for a few months, but she is obliged, during her stay, to live in great seclusion, on account of the dreadful consequences to be apprehended from being so exposed.

The dexterous mode in which the Peons catch their cattle, by throwing a noose over them, has been frequently detailed, but certainly no description can do full justice to their agility. They throw with equal precision and effect, whether at full gallop or at rest. Their method of catching horses by means of balls attached to leather thongs, is as unerring as it is surprising; and scarcely an instance has been known of its failure, except in those frequent trials which are requisite to acquire perfect skill in the practice.

They have a very singular and simple way of training mules and horses to draw light carts, coaches, &c. No harness is made use of; a saddle or pad is girted on, and a leather thong is fastened to the girth on one side, so that the animal moving forward with his body in a rather oblique direction, keeps his legs clear of the apparatus which is attached to him, and draws with a freedom and an agility that in a stranger excite great surprise. A similar contrivance is used in the catching of cattle. The Peon fastens one end of his *lazo* (or noosed thong) to the girth of his horse, who soon learns to place himself in such an attitude as to draw the ox which his rider has caught, and even should the latter dismount, he keeps the thong on the stretch.

The horses in this country are very spirited, and perform almost incredible labour. They seldom work longer than a week at a time, being then turned out to pasture for months together. Their sole food is grass, and the treatment they meet with from their masters is most harsh and unfeeling. They are frequently galloped until their generous fire is spent, and they drop through exhaustion and fatigue. The make of the bridle is alone sufficient to torture the animal, being of the heavy Spanish fashion. They are never shod. The girths of the saddles are of a curious construction; they are generally formed of shreds of green hide, or of the sinew of the neck; the middle part is 20 inches broad, terminated at each end by an iron ring. One of these ends is made fast to the saddle by its ring; to the other side of the saddle is attached a third ring and a pliable strap, which, being passed through it and the girth-ring three or four times, affords the rider great purchase, and enables him to gird the saddle very tight, which is thus kept so firm in its place that a crupper is unnecessary, and indeed is never used.

Trained horses are here from five to seven dollars each; horned cattle, in good condition, by the herd of 1000, at two dollars a head; mares at three rials (1*s*. 6*d*. sterling) each. Sheep are very scarce, and never eaten; they are kept by some families merely for the sake of their wool, which is made into flocks for bedding. It is worthy of remark, that in the remote parts of the interior, where no settlements have been made, the cattle are found of a dark dirty brown colour, except on a small part of the belly, which is white, but when they become domesticated they produce breeds of [a lighter colour, with hides beautifully spotted and variegated. The fine herds bred in many parts of this district have often tempted the Portuguese to make predatory incursions, and the country being accessible by fine open passes to the frontier, as well as to the *n.* side of the Plata, these violations of territory have been carried on to a very serious extent. So frequent were they at one period that it became necessary to appoint a military force to parade the boundaries, and to defend the Spanish settlements against these inroads.

In taking a general view of the country, a stranger cannot but observe, with regret, that while Nature has been profuse in her blessings, the inhabitants have been neglectful in the improvement of them. Here is, for instance, abundance of excellent clay and plenty of wood on the margin of the rivers, yet it is rare to meet with an enclosure, even for a kitchen-garden, much more so for a corn-field. They generally choose their grounds for tillage by the bank of a rivulet, so as to have one side or sometimes two sides bounded by it; the remainder is fenced in the most clumsy and bungling manner imaginable. Ploughing is performed by the help of two oxen yoked to a crooked piece of wood, about four inches in diameter, and pointed at the end. After the ground has been rooted up, the wheat is sown, without any previous attempt to clear it from noxious seeds. While it grows up, it is never weeded; so that wild oats, poppies, and other pernicious herbs, thriving among it in thick luxuriance, obstruct the sun's rays and hinder it from ripening kindly. Indian corn, beans, melons, &c. are all treated in a similar way. The wheat, when ripe, is cut down with sickles and gathered into heads or sheaves. A circular pen of from 40 to 60 yards in diameter is then formed with rails and hides; in the centre of this enclosure is placed a quantity of about 100 or 200 quarters of wheat in the straw. The pile is so formed as to have the ears on the outside as much as possible. A small quantity is pulled down towards the circumference of the circle, and a herd of about 20 mares are driven in, which, being untamed, are easily frightened and made to gallop round. At this pace they are kept by means of whips for four or five hours, until the corn is trod out of the ears, and the straw is completely reduced. Another parcel of the sheaves is then pulled down, and a fresh herd of mares is let in, and this operation is repeated until the whole heap is reduced, and the straw is broken as small as chaff. In this state it is left until a brisk wind happens to rise; and then the winnowing is performed by emptying baskets of the mixed grain and chaff at an elevation of eight feet from the ground. While the chaff is borne away by the current of air, the grain falls, and at the close of the operation is sewed up in green hides. In this state it is sent to the sea-ports, where a considerable quantity of biscuit is prepared for shipping. It is obvious, that by the above mode of separating the grain, a considerable quantity must be lost by abrasion, and by mixture with a large portion of earth which cannot be blown away by the wind.

The climate and soil are equally favourable for the growth of grapes, apples, peaches, and in short every species of fruit belonging to the temperate zone, but these are known here only as rarities. That inestimable root, the potato, would thrive abundantly, if once introduced; but, though much has been said in recommendation of it, the people remain totally averse to this or any other proposal for improving their means of subsistence, and seem to wish for nothing beyond the bare necessaries of life. Indeed the state of society among them weakens those ties which naturally attach men to the soil on which they are accustomed to subsist. The Peons, brought from Paraguay in their infancy, grow up to the age of manhood in a state of servitude, uncheered by domestic comfort; at that period they generally wander in search of employment toward the coast, where money is in greater plenty. They are for the most part an honest and harmless race, though equally as liable, from the circumstances of their condition, to acquire habits of gambling and intoxication, as the higher classes of the people, numbers of whom fall victims to those seductive vices. Such is their excessive propensity to gambling, that they frequently carry cards in their pocket, and, when an opportunity occurs, form parties, and retire to a convenient place, where one of them spreads his *pancho* or mantle on the ground, in lieu of a table. When the loser has parted with his money, he will stake his clothes, so that the game generally continues until one of them goes away almost naked. This bad practice often leads to serious consequences. On one occasion, a party playing in the neighbourhood of a chapel after mass had been said, the clergyman came and kicked away the cards in order to put an end to the game. On this one of the Peons rose up, and retiring a few paces, thus accosted the intruder: "Father, I will obey you as a priest; but (drawing his knife) you must beware how you molest our diversion." The clergyman knew the desperate character of these men too well to remonstrate, and retired very hastily, not a little chagrined. It] [is usual for a Peon who has been fortunate at play, to go to Monte Video and clothe himself anew in the shop of a slop-seller. While the man is looking out the articles he calls for, he deliberately places his dollars on the counter, in separate piles, assigning each to its destined purpose. He then retires to a corner, and attires himself; an unfortunate comrade invariably attends him, who examines his cast clothes, and, if better than his own, puts them on. After passing a few days in idleness, he sets out on his return home, where he appears in his new dress. The various evils resulting from the above vices are multiplied by the lax administration of the laws; even in case of murder the criminal has little to fear if he can escape to a distance of 20 or 30 leagues; he there lives in obscurity, probably for the remainder of his life, without ever being brought to justice. It is to be feared that this want of vigilance in the magistracy is a temptation for the numerous refugees who seek shelter here, such as European Spaniards, who have deserted from the service, or have becu ba-nished for their crimes. These wretches, loaded with guilt, flee into the interior, where they seldom fail to find some one or other of their countrymen who is willing to give them employment, though frequently at the peril of his life. By the corrupt example of these refugees, the innocent Creolian is soon initiated in vice, and becomes a prey to all those violent passions which are engendered and fostered by habitual intoxication.

The common dress of the people is such as might be expected from their indolence and poverty. They generally go without shoes and stockings; indeed, as they rarely go on foot they have seldom occasion for shoes. Some of them, particularly the Peons, make a kind of boots from the raw skins of young horses, which they frequently kill for this sole purpose. When the animal is dead, they cut the skin round the thigh, about 18 inches above the gambrel; having stripped it, they stretch and dress it until it loses the hair and becomes quite white. The lower part, which covered the joint, forms the heel, and the extremity is tied up in a bunch to cover the toes. These boots, when newly finished, are of a delicate colour, and very generally admired. The rest of their apparel consists of a jacket, which is universally worn by all ranks, and a shirt and drawers made of a coarse cotton cloth brought from the Brazils. Children run about with no dress but their shirts until their fifth or sixth year. Their education is very little attended to, and is confined to mere rudiments; a man who is able to read and write is considered to have all the learning he can desire.

Among the many natural advantages which this district possesses, are the frequent falls in the rivulets and larger streams, which might be converted to various mechanical purposes, if the population were more numerous and better instructed. Some of these streams, as was before stated, join the various branches of the Godoy, and flow into the lake Meni; those on the other side the mountains in a *n.* direction empty themselves principally through the Riachuelo and the St. Lucia, into the Plata.

The want of cultivation in this vast territory may be inferred from the numbers and varieties of wild animals which breed upon it. Tigers, ounces, and lions are common. The former are heavy sluggish animals; their chief prey is the young cattle, which they find in such abundance that they rarely attack a man. Hence little danger is to be apprehended from them by any person travelling on horseback, unless when inadvertently approaching the haunt of a female with young. The ounce has the same character, and the lion is considered less vicious than either. Among the many daring and active feats performed by the Peons, we cannot forbear commemorating one of the most extraordinary of late years, being the capture of a tiger by a female of that tribe. She was a mulatto woman, brought up in the vicinity of Barriga Negra. She was accustomed at an early age to ride horses, and prided herself in doing offices which belonged to the stronger sex, such as catching cattle with the noose, killing them, &c. Her form was masculine, and she became so inured to men's work that she was hired as a Peon, and fulfilled that office much to the satisfaction of her employers. She was noted for selecting spirited horses, and for riding them at full speed. One day on her return from labour, as she was passing a rivulet, she observed a large tiger at no great distance. Surprised that the animal did not steal away, as is generally the case when he sees a person mounted, she drew nearer, still keeping her horse's head from him, so as to be ready to gallop off if he should make a spring. He was still inattentive and motionless; the woman observing this, and thinking he ailed something, after some minutes' pause, backed her horse until she came within 20 yards of him, loosening at the same time her noose from the saddle, which she threw most dexterously over his neck, and immediately galloped away with him to a considerable distance. Whether ill or not before, she knew he must now.]

[be dead; she therefore alighted, flayed him, and carried home the skin as a trophy. The animal was above the ordinary size, and not smaller than a calf of six weeks old. This exploit was long the talk of the neighbourhood, and Mawe, the traveller, asserts that he heard the woman herself relate it. Besides the animals above mentioned there is one of the pig kind, called the pig of the woods, which has an aperture on its back, whence it emits a most intolerable stench when closely pursued. If on killing the animal the part be instantaneously cut out, the flesh affords good eating, but should that operation be neglected, even for a short period, the taint contaminates the whole carcase. The domestic pigs are by no means good, for they feed so much upon beef that their flesh is very hard and coarse. There is an animal of the opossum kind, about the size of a rabbit, called a zurilla, the skin of which is streaked black and white, and is considered of some value. When attacked, it ejects a fetid liquor, which is of so pungent a nature that if it falls on any part of the dress of its pursuers, there is no possibility of getting rid of the stench but by continual exposure to the weather for some months. The zurilla is very fond of eggs and poultry, and sometimes enters a house in quest of its prey; the inhabitants immediately hasten out and leave their unwelcome visitant in quiet possession as long as she chooses to stay; well aware that the slightest attempt to drive her out would expose them to an ejectment from the premises for ever. Eagles, both of the grey and blue species, as well as other birds of prey, are found here in great numbers. Here are also parroquets in immense flocks, pigeons, great red-legged partridges, small partridges, wild ducks, and wild turkeys. Ostriches of a large species are very numerous; they are so fleet and active that even, when well mounted, it is impossible to get near them but by surprise; the stroke of their wing is said to be inconceivably strong.

Here are considerable herds of small deer, which in this fine country would afford the sportsman excellent diversion, but unfortunately the dogs are good for nothing, as there is no attention paid to the preservation of the breed. The rivers produce tortoises and other amphibious animals, but they are chiefly noted for a variety of singularly ugly fish, which afford tolerable, but by no means good eating.]

Monte Video, a mountain of the same province and government, on the coast of the river La Plata, from whence the former city takes its name.

MONTEGA, a bay on the *n.* coast of the island of Jamaica.

[MONTEGO Bay is on the *n.* side of the island of Jamaica, 12 miles *e.* of Lucca harbour, and 19 *w.* of Falmouth harbour. This was formerly a flourishing and opulent town; it consisted of 225 houses, 33 of which were capital stores, and contained about 600 white inhabitants. The number of topsail vessels which cleared annually at this port were about 150, of which 70 were capital ships; but in this account are included part of those which entered at Kingston. This fine town was almost totally destroyed by an accidental fire, in July 1795; the damage was estimated at 200,000*l.* sterling.]

MONTEREI, a city and capital of the Nuevo Reyno de Leon in N. America; founded in 1599 by order of the Count de Monterei, viceroy of Nueva España, who gave it his name. It has two parishes, one for Spaniards, and another, which is a convent of San Francisco, for the Indians. In its ecclesiastical concerns it belongs to the bishopric of Guadalaxara, and in its civil to the audience of Mexico. [Three hundred and ninety-seven miles *n.* of the latter, and 198 *n. n. e.* of Zacatecas, in lat. 25° 59′ *n.* Long. 100° 7′ *w.*]

Monterei, a settlement of the province and *corregimiento* of Coquimbo in the kingdom of Chile, where there is a fort and garrison to restrain the incursions of the Araucanos Indians, who border on that part. It is situate on the shore and at the source of the river Limari. In 1612, a flourishing mission was established here by the Father Luis de Valdivia, of the extinguished company, he having been sent by the King Philip III. to make a treaty of peace with the Indians, the which he effected. He was accompanied by a brother of the same order, Orazio Vecchi; and he was the first who moistened these territories with the blood of a martyr, having suffered under the hands of the Cazique Ungunamon, in the place of Elicura, that same year.

[MONTEREY, a town of the intendancy of San Luis Potosi, the seat of a bishop, in the small kingdom of Leon.]

[MONTERO, a diamond-work of the province and *captainship* of Rio de Janeiro in Brazil; a few miles up the river Jigitonhonha, and very near to Tejuco, the capital of Minas Novas. It was visited by Mawe, the traveller, in 1809; for an account of whose journey thither, see Villa Rica.]

MONTES, a river of the province and government of Paraguay in Peru.

MONTES-CLAROS, Jesus de, a city, also

called Valle Grande, of the province and government of Santa Cruz de La Sierra in Peru; bounded by the province of Tomina. It contains 3500 souls, the greater part people of colour. It was founded by the viceroy of Peru, the Marquis Montes Claros, who gave it his name. It is small and poor, and is 12 leagues *s.* of the settlement of Samapaita.

Montes-Claros, a town of the province and government of Cinaloa in N. America.

[Montes-Claros, a town of the intendancy of Sonora; the same as Villa del Fuerte.]

MONTESILLO, S. Christoval de, a small settlement or ward of the *alcaldía mayor* of San Luis del Potosi. It contains 30 families of Indians, and eight of *Mustees*, all of whom are weavers, shoe-makers, and hat-makers. It is very close to its capital between *n.* and *s.*

[MONTGOMERY, a new county in the upper district of Georgia.]

[Montgomery, a county of New York, at first called Tryon, but its name was changed to Montgomery in 1784, by act of the legislature. It consisted of 11 townships, which contained 28,848 inhabitants, according to the census of 1791. Since that period the counties of Herkemer and Otsego have been erected out of it. It is now bounded *n.* and *w.* by Herkemer, *e.* by Saratoga, *s.* by Schoharie, and *s. w.* by Otsego county. By the state census of 1796, it is divided into eight townships; and of the inhabitants of these 3379 are qualified electors. Chief town, Johnston.]

[Montgomery, a township in Ulster county, New York, bounded *e.* by New Windsor and Newburgh, and contains 3563 inhabitants, including 236 slaves. By the state census of 1796, 497 of the inhabitants were qualified electors.]

[Montgomery, a fort in New York state; situated in the high lands, on the *w.* bank of Hudson's river, on the *n.* side of Popelop's creek, on which are some iron works, opposite to St. Anthony's Nose, six miles *s.* of W. point, and 52 from New York city. The fort is now in ruins. It was reduced by the British in October 1777. See Anthony's Nose.]

[Montgomery, a township in Franklin county, Vermont.]

[Montgomery, a township in Hampshire county, Massachusetts, 100 miles from Boston. It was incorporated in 1780, and contains 449 inhabitants.]

[Montgomery, a county in Pennsylvania, 33 miles in length and 17 in breadth, *n. w.* of Philadelphia county. It is divided into 26 townships, and contains 22,929 inhabitants, including 114 slaves. In this county are 96 grist mills, 61 saw mills, four forges, six fulling mills, and 10 paper mills. Chief town, Norritown.]

[Montgomery, a township in the above county. There is also a township of this name in Franklin county.]

[Montgomery, a county in Salisbury district, N. Carolina, containing 4725 inhabitants, including 854 slaves.]

[Montgomery, a county of Virginia, *s.* of Botetourt county. It is about 100 miles in length, 44 in breadth, and contains some lead mines. Chief town, Christiansburg.]

[Montgomery Court-house, in Virginia, is 28 miles from Wythe court-house, and 81 from Salisbury, in N. Carolina. It is on the post-road from Richmond to Kentucky. A post-office is kept here.]

[Montgomery, a county of Maryland, on Patowmac river. It contained 18,003 inhabitants, including 6030 slaves.]

[Montgomery Court-house, in the above county, is 22 miles *s. e.* by *s.* of Frederickstown, 12 *n.* by *w.* of George-town on the Patowmac, and 23 *s. w.* of Baltimore.]

[Montgomery, a new county in Tennessee state, Mero district. This and Robertson county are the territory formerly called Tennessee county, the name of which ceases since the state has taken that name.]

[Montgomery, a court-house in N. Carolina, 28 miles from Salisbury and 18 from Anson court-house. It is now known by the name of Stokes Court-house.]

MONTIJO, a settlement of the province and government of Veragua in the kingdom of Tierra Firme.

[MONTMORENCY, Fall of. The fall of Montmorency, which is situated about eight miles to the *n. e.* of Quebec, derives its elegant and majestic appearance more from its height than from the body of water that flows over the precipice. According to the most accurate computation, it is 250 feet high and 80 feet wide. Its breadth is, however, increased or diminished according to the quantity of water supplied by the river, which is a narrow stream, and in many parts extremely shallow. In spring and autumn, when the melting of the snow, or much rain, swells the current, the fall is increased, and seen at those periods to great advantage. In winter but a small portion of the fall is visible, in consequence of the cones of ice which are formed by the rising spray, and intercept the view nearly half way up.

The river Montmorency falls between a large

cleft in the mountain, which appears to have been formed by the shock of an earthquake. The waters thus precipitate themselves into a kind of basin, upwards of 300 yards wide, many parts of which are fordable towards the entrance at low water; but under the fall there is an immense chasm. The mountain consists of the black lime slate, which as it becomes exposed to the air continually moulders away. Near the summit of the falls, the banks of the cleft are ornamented with a variety of shrubs, fir-trees, and other evergreens, whose dark foliage form an agreeable contrast to the snowy whiteness of the fall, and give to the *tout ensemble* a pleasing and romantic appearance. The fall of Montmorency has, however, more of the elegant and beautiful in it than of the "awfully grand, or wonderfully sublime!"

If, turning your attention altogether from the fall of Montmorency, you direct it up the river, the scenery is not to be surpassed any where. After viewing the fall, if you turn your attention towards the St. Lawrence and the island of Orleans, and, following the course of the river, direct your view towards the lower end of the island, by Chateau Riché, till you reach the mountain called Cap Tourment, it must be allowed that it is difficult to imagine an assemblage of objects more interesting, or better calculated to inflame the fancy of the poet, or give life to the canvas of the painter.

Both the Montmorency and the Chaudiere may be viewed either from the top or bottom of the fall. The latter, it is generally thought, is seen to greatest advantage from below. The Montmorency, too, viewed from below, is truly sublime, as it is thought to be so much the more famous than the Chaudiere, in as much as it is seen at a distance by all who sail up the St. Lawrence.]

[MONTMORIN, a new town on the *n.* bank of Ohio river, 18 miles below Pittsburgh; situated on a beautiful plain, very fertile, and abounding with coal.]

MONTOOK, PUNTA DE, an extremity or *e.* head of Long island, in the province and colony of New York, running many leagues into the sea.

[MONTPELIER, a township in Caledonia county, Vermont, on the *n. e.* side of Onion river. It has 118 inhabitants, and is 27 miles from lake Champlain.]

MONTREAL, a large island of the river St. Lawrence in New France or Canada; situate at the conflux of the two rivers Utawas and Cataxakui. It is 27 miles long and 12 wide, takes the name of a very lofty mountain, situate in the middle of the island, appearing to command the island, and so called by the French, Mont-Real. The river St. Lawrence is here a league wide, and its shores are covered with trees and settlements, and many small islands, some cultivated, others uncultivated, and altogether affording a very pleasing prospect. From the city of Quebec to this island the French have many establishments upon very level roads, as also several towns and settlements of different lordships; but the river is navigable only as far as Montreal from the number of cascades and rocks which there impede its course. The territory produces much maize, and all the European fruits thrive here; although the French have fixed on none of these as a principal branch of commerce, being employed in a traffic with the Indians for skins of castors, foxes, racoons, deer, and other articles of the same class, by which they make a good profit, and live very agreeably in this delightful country. The Indians barter the skins for brandy, tobacco, muskets, powder and ball, and the French have certain traders whom they call Runners of the Woods, who, traversing immense lakes and rivers in their canoes made of the barks of trees with an incredible patience and ingenuity, carry their effects to the most remote parts of America, amongst nations to us entirely unknown; and these, on the other hand, establish a fair at Montreal, at which people assemble from 100 miles distance. This fair is celebrated in the month of June, and sometimes is kept up for three months with great solemnity: guards are stationed at different parts, and the governor himself assists to restrain any incursion that so numerous a nation of savages might make: but these precautions are more particularly to guard against the violent behaviour, which generally partakes of something like madness, common to the Indians when they are inebriated. Notwithstanding this, the fair has been carried on at Montreal for many years in a very flourishing manner; and although many of the tribes of Indians who assemble here must necessarily pass the English establishments of New York, Albany, &c. where they might provide themselves with what they want much more readily and at half the price for which they can obtain them at Montreal, owing to the great expences of a long land-carriage, yet do they prefer buying them there at the increased rate, and at the second hand. This is curious, and the French have lately found it more to their advantage to buy the goods and merchandize for sale at New York than in their own country, which only proves that the French have a greater skill in conciliating and pleasing these barbarians than have the English.

The capital of this island is of the same name.

It is of an oblong form, having long and straight streets, and buildings of a good construction. It is surrounded by a strong wall and flanked by 11 redoubts serving as bastions. The ditch which surrounds the whole wall, except the part which is opposite the river, is about eight feet deep and of a proportionate width: besides this it has a citadel or fortress, the batteries of which command the streets of the city on either side; and upon the river called San Pedro there is a bridge. The shore of the river St. Lawrence, upon the which the city stands, has a gradual slope from the water's edge to the further end of the city. This is divided into two parts, the superior and inferior, although the pass from the one to the other is scarcely perceptible. The merchants commonly dwell in the inferior part, and here is the place of arms, the hospital, and the king's storehouses; but the chief buildings are in the superior part, together with the convent of the reformed Franciscans, the parish church, the public school, the college which belonged to the Jesuits, the governor's palace, and the greater part of the houses of the officers of the garrison. The convent, of which we have spoken, of the religious order of St. Francis, is very large, and has a numerous community. The parish church is large and of beautiful architecture, entirely of cut stone, and the public school, which is contiguous to it, is very convenient. The church, which belonged to the Jesuits, is small, but richly adorned. The governor's palace is a magnificent building, as are also many other edifices with which the city is adorned; but, amongst the rest, we must not forget to mention the hospital, assisted by the religions lay-sisters from the city of Fleche, in the county of Anjou. Outside of the city and on the other side of the river St. Pedro, are some pretty country houses, especially one belonging to Mr. de Calliere, and the public hospital called Charron Bretheren, from having been founded by a person of this name, who, in conjunction with other pious and devout persons, undertook this work of charity, as also the providing with masters the school for the instruction of Indian children, having had the satisfaction before his death, which occurred in 1719, to see the hospital established, although his companions had abandoned him in the undertaking. This city belongs to the seminary of the St. Sulpice, of Paris. The English took possession of it in 1760, after they took Quebec, and remained in the enjoyment of it, together with the greater part of the country. [It is 116 miles *s. w.* of Quebec, in lat. 46° 33′ *n.* Long. 73° 18′ 30″ *w.*]

[INDEX TO ADDITIONAL INFORMATION RESPECTING MONTREAL, AND FURTHER DESCRIPTION OF THE ISLAND AND TOWN.

Inhabitants.--Mechanics.---Markets.---Water carriage.---Government stores.---Indians of Cachenonaga.---Boundary line between Canada and the United States.---Commerce.

MONTREAL is justly considered at the present day the second city in rank in Lower Canada. While the French had possession of Canada, both the city and island of Montreal belonged to private proprietors, who had improved them so well that the whole island had become a delightful spot, and produced every thing that could administer to the convenience of life. The city, around which is a very good wall, built by Louis XIV. of France, forms an oblong square, divided by regular and well formed streets; and when taken by the British, the houses were built in a very handsome manner; and every house might be seen at one view from the harbour, or from the southernmost side of the river, as the hill on the side on which the town stands falls gradually to the water. Montreal contains at present about 1200 houses, few of them elegant; but since it fell into the hands of the British in 1760, it has suffered much from fire. A regiment of men are stationed here, and the government of the place borders on the military. It is about half a league from the *s.* shore of the river, 120 miles *s. w.* of Quebec, Trois Rivieres being about half way, 98 *n.* of Crown point, 220 *n.* by *w.* of Boston, and 286 *n. e.* of Niagara. The lat. and long. are mentioned above. See ST. LAWRENCE. The river St. Lawrence is about three miles wide at Montreal.

Near Bout de l'Isle, or the end of the island of Montreal, the river is intersected by a number of small isles and islets. One, named Eagle island, is the property of Captain Cartwright of the Canadian fencibles, and was celebrated for some excellent horses which he reared upon it. It contains only his own house, in which he resided for several years with his family. The surrounding scenery is beautiful, and must afford a delightful retreat to those who are fond of rural felicity. Within view of this island is the ferry which crosses from the post-road at Repentigny to the end of the island of Montreal. A bridge was formerly built over the river in the vicinity of this place by Mr. Porteous, of Terrebonne, but was carried away two or three years ago by the ice. The provincial parliament have recently passed an act permitting him to build another from Repentigny to isle Bourdon.

The shores of the island of Montreal are elevated]

several feet above the level of the river. The soil is uncommonly rich and fertile, and yields more abundant harvests than any other part of Lower Canada. The price of land averages from 20 to 30 dollars per acre. The island is 30 miles in length and about seven in breadth.

The opposite island of St. Helen belongs to the Baroness de Longueil: this lady married a gentleman of the name of Grant, and brought him very extensive and valuable landed property. Since his death it has been divided between her and the children. The eldest son goes by the familiar appellation of Baron Grant.

The town of Montreal has a singular appearance when viewed from the water, in consequence of the light-grey stone of the new buildings, and the tin-covered roofs of the houses, which emit a strong glare when the sun shines. The shipping lie close to the shore, which is very steep, and forms a kind of natural wharf, upon which the vessels discharge their cargoes. About 20 yards back the land rises to the height of 15 or 20 feet; and an artificial wharf has been constructed and faced with plank; the goods are, however, all shipped from, and landed upon, the beach below. A great many English vessels arrive annually at Montreal, but it is a voyage that few captains are willing to make a second time, if they can possibly avoid it, the navigation up the river above Quebec being very hazardous, and the pilots unskilful and inattentive.

The interior of Montreal is extremely heavy and gloomy. The buildings are ponderous masses of stone, erected with very little taste and less judgment. They are seldom more than two stories above the ground floor, including garrets. The doors and window-shutters are covered with large sheets of tin, painted of a red or lead colour, corresponding with the gloomy darkness of the stone of which most of the old houses are built. There is a heavy sameness of appearance which pervades all the streets, whether new or old, nor are they remarkable for width, though they are for the most part laid out in a regular manner. The only open place or square in the town, except the two markets, is the Place d'Armes, and which, under the French government, was the place where the garrison troops paraded. The French Catholic church occupies the whole of the *e.* side of the square, and on the *s.* side, adjoining some private houses, is a very good tavern, called the Montreal Hotel, kept by Mr. Dillon.

The town walls and fortifications which were erected to protect the inhabitants against the irruptions of the Iroquois and other hostile Indians, are now falling to decay. A great part have been levelled with the ground, and an act has lately passed the provincial parliament to remove the remainder.

At the back of the town, just behind the new court-house, is the parade, where the troops are exercised. The ground is considerably elevated along this part, and forms a steep bank for several hundred yards in length. Here the inhabitants walk of an evening, and enjoy a beautiful view of the suburbs of St. Lawrence and St. Antoine, and the numerous gardens, orchards, and plantations of the gentry, adorned with neat and handsome dwelling-houses. Large green fields are interspersed amidst this rich variety of objects, which are concentrated in an extensive valley, gradually rising towards a lofty mountain, that stands about two miles and a half distant, at the back of the town: from this mountain the island has taken its name of Montreal, or Royal Mount. It is said by some recent authors, but most erroneously, to be elevated 700 miles above the level of the river; it is upwards of two miles in length from *n.* to *s.* It is covered with trees and shrubs, except towards its base, where some parts have been cleared and cultivated. A large handsome stone building, belonging to the widow of the late Mr. M'Tavish, of the N.W. company stands at the foot of the mountain, in a very conspicuous situation. Gardens and orchards have been laid out, and considerable improvements made, which add much to the beauty of the spot. Mr. M'Tavish is buried in a tomb a short distance from his house on the side of the mountain, in the midst of a thick shrubbery. A monumental pillar is erected over the vault, and may be seen a long way off.

The town, including all its suburbs, occupies a considerable extent of ground, and the number of inhabitants is computed at 12,000. The principal public buildings are, the general hospital; the Hotel Dieu; the convent of Notre Dame; the French cathedral; the English church, an unfinished building; the old monastery of Franciscan friars, converted into barracks; the seminary; the court-house; government-house, &c.

The general hospital was founded by Madame Yonville, a widow lady, in 1753, and contains a superior and 19 nuns; it is situated on the banks of the river, near a small rivulet, which divides it from the town. There is also a college for the education of young men, founded in 1719 by the Sieur Charron.

The Hotel Dieu was established in 1644, by Madame de Bouillon, for the purpose of administering relief to the sick poor; it contains a superior [and 39 nuns, who attend and nurse the patients. An apartment in the upper part of the house is appropriated to the females, and a large room below for the men. The establishment is now chiefly supported by a slender income, arising from landed property: the funds, upon which it formerly relied, being vested in Paris, were lost during the revolution.

The convent of Notre Dame contains a superior and upwards of 40 nuns. It was founded about the year 1650, by Mademoiselle Marguerite Bourgeois, for the instruction of female children. The sisters of this situation are not confined in so strict a manner as at the other convents, but have the liberty of going out. They attend mass at the French church on Sunday morning and afternoon. They are dressed in black gowns and hoods, and are chiefly elderly women.

There are two of the old Franciscan friars still living in one corner of their monastery, the remainder of which has been converted into barracks for the troops quartered in the city. Upon the arrival of several additional regiments at Quebec, the 49th and 100th were sent up to Montreal to do duty in that town, and to garrison the out-posts near the American line.

The French cathedral in the Place D'Armes is a large substantial stone building, built with little taste. The interior is, however, plentifully decorated in the Catholic style, with all the appropriate decorations of that religion; and the size of the building renders it a very commodious place of worship, and well adapted for the accommodation of its numerous congregation. In summer a great many people kneel outside the church in preference to being within. The service of the English church is performed at present in a small chapel, which is also used by the Presbyterians. A handsome new church is partly built, but for want of funds remains in an unfinished state.

The court-house is a neat and spacious building, and an ornament to the town; a gaol is building on one side of it, upon the site of the old college of the Jesuits. The city (as before observed) is divided into Upper and Lower Towns, though there is very little difference in their elevation. The principal street of the latter extends from *n.* to *s.* the whole length of the city, nearest the water-side, and is called Paul-street. Here are situated the wholesale and retail stores of the merchants and traders; the lower market-place; the post-office; the Hotel Dieu; and a large tavern, formerly kept by Hamilton, but now in the possession of Mr. Holmes. There are several smaller taverns in this street and in the market-place, but they are frequented principally by the American traders who visit Montreal. Paul-street, though narrow, presents a scene of greater bustle than any other part of the town, and is the chief mart of the trade and commerce carried on in Montreal.

Several short streets proceed *w.* from Paul-street, and communicate with that of Notre Dame, which runs in a parallel line, extending the whole length of the city. This street forms what is called the Upper Town, and contains the Recollet monastery, the French seminary, the Catholic church, the Place d'Armes; the new English church, the convent of Notre Dame, the court-house and gaol, and the old building called the government-house, which latter has no claim to particular notice. The dwelling-houses of the principal merchants are mostly situated in Notre Dame street, and other parts of the Upper Town, their stores being stationed near the water-side. These two parallel streets are considerably lengthened to the *n.* by the suburb of Quebec; and to the *s.* by the suburbs of St. Antoine and Recollet. In the centre of Notre Dame street, a long street branches off to the *w.* and forms the suburb of St. Lawrence. It is also the high road to the interior of the island, and crossing the intermediate valley, passes over the foot of the mountain. In one of the short streets leading to the Upper Town, and situated opposite the court-house, a new market-place, and rows of convenient stalls, have been recently constructed: it will be a great accommodation to the town, as the old market in Paul-street is too much confined for the increased population of the place. The streets of Montreal are, for the most part, well paved, and the improvements which are going on throughout the town, will render it more commodious and agreeable than it is at present. The town itself will always be gloomy, but the environs are beautiful.

All the principal N. W. merchants reside at Montreal, which is the emporium of their trade, and the grand mart of the commerce carried on between Canada and the United States. They, and other respectable merchants, have country-houses a few miles from the city, which, with their numerous orchards and gardens, well stocked with every variety of fruit-trees, shrubs, and flowers, render the surrounding country extremely beautiful and picturesque. The succession of rich and variegated objects that are presented to the eye of the spectator, from the base of the neighbouring mountain, cannot be surpassed in any part of Canada, with the exception, perhaps, of the view from cape Diamond, at Quebec. They are, however, both of a different nature, and may be des-]

[cribed like Homer and Virgil; the one grand, bold, and romantic, the other serene, beautiful, and elegant. Quebec has more of the majesty of nature, Montreal more of the softness of art.

A large store has been converted into a theatre, in which Mr. Prigmore's company occasionally perform. Society is reckoned more friendly and agreeable in Montreal than in any other town in Lower Canada. The N.W. merchants live in a superior style to the rest of the inhabitants, and keep very expensive tables. They are friendly and hospitable to strangers who are introduced to them, and whom they entertain in a sumptuous manner. The envious, however, consider their apparent generosity as flowing more from pride and ostentation than from real hospitality, and they have often been the subjects of newspaper criticism.

A public assembly is held at Holmes's tavern during the winter; and these, with private dances, tea, and card parties, and cariole excursions out of town form the whole amusements of that season. In summer pleasure gives way to business, which at that period of bustle affords full employment to all. A few excursions, and dinner parties in the country, occur sometimes to relieve the weight of mercantile affairs. Concerts are very rare, and never take place unless the regimental bands are in town. The inhabitants, like those of Quebec and Three Rivers, possess very little knowledge of the polite and liberal accomplishments necessary to form the complete lady or gentleman. They however labour under the disadvantage of the want of proper masters, and institutions to instruct and complete them in the higher branches of education; yet it is, perhaps, their fault that they have them not, for without proper reward and encouragement they never can have them.

Ship-building is successfully carried on by Mr. Munn, who generally launches two or three vessels from 200 to 500 tons every year. They have of late taken French Canadians as apprentices, who are highly praised for their capacity. This is a very good plan, for European ship-builders have very high wages, and are besides a very drunken dissolute set. The Canadian workmen, on the contrary, are sober, steady men, and attend regularly to their work from break of day to sun-set. There is an island near the middle of the river opposite the city, at the lower end of which is a mill with eight pair of stones, all kept in motion, at the same time, by one wheel. The works are said to have cost 11,000*l.* sterling. A large mound of stone, &c. built out into the river, stops a sufficiency of water to keep the mill in continual motion. And what is very curious, at the end of this mound or dam, vessels pass against the stream, while the mill is in motion. Perhaps there is not another mill of the kind in the world.

One of the greatest errors committed by persons who go to Canada to settle, is the taking of European servants with them; for experience has fully proved, in innumerable instances, that no obligations whatever are sufficient to ensure a master the labour of his European servants, more especially if he is in advance to them for any part of their wages. The inducements to leave him, in such cases, become so great, that the servant must be more than commonly virtuous, or have strong motives for staying, if he does not break his engagement. This complaint is so general at Quebec, that little or nothing is done to remedy the grievance, which seems to set the laws at defiance: yet the magistrates have sufficient power to punish both masters and servants; but they seldom or never give a satisfactory decision in cases where the latter are to blame.

One very great mischief in this town is occasioned by the low price of spirits, particularly rum, which may be obtained for less than 5*s.* a gallon. Hence few of the lower order of Europeans who arrive at Quebec, but become drunkards in a very short time, and drunkenness never fails to precipitate them into worse vices. If they have a little money, it is soon squandered, either in liquor with their dissolute companions, or in going to law with their masters, in which case it seldom fails to find its way into the pocket of the before mentioned advocate, and the account is generally wound up by some crimp for the shipping, or recruiting serjeant for the army.

The scarcity of hands for labour is certainly considerable, yet by no means so great as is generally represented; it is therefore more to the interest of gentlemen settling in Canada, to engage the native artisans, than to take out men who will never remain in their service. The French mechanics and farmers may be, and indeed are, greatly inferior in abilities to Europeans; but they are superior to them in sobriety, industry, and civility. The French Canadians, however, have great ingenuity, and it only requires cultivation to render them excellent artists. Some clever American mechanics are also frequently to be met with in Canada, particularly mill-wrights; these people are sometimes steady workmen, but they will often give their employers the slip in the middle of their work, if they happen to meet with a more lucrative offer from another person. The practice of enticing away each other's servants, is but too] [much the custom in Canada, and it is owing as much to this want of good faith, that strangers on their arrival find it so difficult to retain their servants, as to any other cause.

The markets of Montreal are plentifully supplied with all kinds of provisions, which are sold much cheaper than at Quebec or Three Rivers: large supplies are brought in every winter from the States, particularly cod-fish, which is packed in ice and conveyed in sleighs from Boston. Hay and wood are sold in the Place d'Armes. Two newspapers are printed weekly at Montreal, the Gazette, and Canadian Courant, both on Monday afternoon.

From Montreal to La Chine is a turnpike road, about seven or eight miles in length. This is the only turnpike in Lower Canada, and the road is not very well kept up for the toll that is demanded; fourpence is charged for a horse, and eightpence for a horse and chaise; but for a subscription of one or two dollars per annum, an inhabitant of the island may be exempted from the daily toll. A great traffic is maintained on this road by the carters who carry all the goods for the upper country, from Montreal to La Chine, where they are put on board batteaux.

For the first mile or two out of town, the road passes partly over a common, which is beginning to be inclosed and cultivated. After passing through the turnpike, the road proceeds up a steep ascent, and continues along a lofty height for nearly four miles, when it descends rather abruptly, and passes again over a low, flat country, until it reaches La Chine, which is situated along the shore of the river St. Lawrence. The road is lined with the houses and farms of the habitans, and along the height, the eye wanders with pleasure over an extensive cultivated valley, bordered by the St. Lawrence, which disappears amidst the thick foliage of the trees, while a small serpentine stream meanders prettily through the fields. This low country was, ages ago, probably a part of the river, and the high land, along which the turnpike road now runs, was most likely the boundary within which it was confined. Its flat and marshy soil affords some foundation for this conjecture. There is another road to La Chine which winds along the shore of the St. Lawrence, and passes the rapids of St. Louis; situated about half way. It is about a league longer than the turnpike road. A few years ago, before the road was made, it was nearly a day's journey for carts to go from Montreal to La Chine. The road is certainly now in a better condition, but there is still room for improvement.

La Chine is delightfully situated upon the banks of the river. It is of considerable extent, in consequence of the houses being built in the same straggling manner as the other small settlements in Canada, where the dwellings are regulated by the situations of the farms, and are seldom formed into an assemblage of houses laid out in streets. All the goods and merchandise sent to Upper Canada are embarked at this village, to which they are carted from Montreal, as the rapids of St. Louis prevent vessels from passing up the river from that city. The goods are put on board large batteaux, or flat-bottomed boats, each of which is worked by four men and a guide, who make use of paddles and long poles, as the depth or rapidity of the current requires. A gentleman of the name of Grant, who resides at La Chine, is the owner of the batteaux, and shipper of the goods for the merchants, who pay him freight for the transportation of their merchandise. Upwards of 50 batteaux are employed in the voyage to and from Kingston, on lake Ontario, in the course of the year. Mr. Grant also ships off the goods for the N.W. merchants in large bark canoes belonging to the company; these goods, which consist of provisions, cloth, blankets, fowling-pieces, powder and shot, and other articles for the Indian trade, are exchanged for furs.

Between 40 and 50 canoes, deeply laden with the above articles, and navigated by Canadian and Indian voyagers, are dispatched in the course of the spring from La Chine, and proceed up the Outaouais, or Grand river, through rapids, and over portages or carrying places, into lake Nipissing. From thence they pass through Riviere des François into lake Huron, and arrive at the company's post in lake Superior, from whence the goods are afterwards transported to the Lake of the Woods, and distributed to the several trading posts, far in the interior of the continent.

The government stores belonging to the Indian department are kept at La Chine, under the care of Mr. Hawdon the store-keeper general. About 30 batteaux, laden with Indian presents, are dispatched every spring to Kingston, York, Niagara, and other posts belonging to the king in Upper Canada, as far as lake St. Joseph's, near Michillimakinak; where store-keepers and clerks reside, for the delivery of the presents in their respective districts. The presents are delivered out of the stores at La Chine, by an order from Sir John Johnson, who is the superintendant-general of the Indian department. They consist] [chiefly of the following articles:—Scarlet and blue cloth, strouds, molton, blankets of various sizes, Irish linen, flannel, Russia and English sheeting, hats, laced coats, rifles and fowling-pieces, powder, shot and flints, swords, spears, harpoons, hooks and fishing-lines, copper and tin kettles, vermilion, looking-glasses, pins, needles, tapes, thread, scissars, knives, nests of trunks, boxes, &c.

In the stores are sometimes also included many pieces of fine French cambric, a quantity of tea, Jew's harps, razors, &c.; but it is thought that articles of this description seldom or never reach the Indians, being much oftener used by the store-keepers and agents of the Indian department for their own families. The great abuses which formerly existed in that branch of the public service were shameful, but are now greatly abolished. The former enormous requisitions are also reduced to little more than 10,000*l*. for Upper and Lower Canada; and together with the salaries of the officers and agents of the Indian department, the expences do not amount to half the sum stated by Mr. Weld in 1796, which he computed at 100,000*l*.

Opposite to La Chine stands the Indian village of Cachenonaga. Its inhabitants, who amount in all to 1200, are descended from the Agniers, one of the Iroquois nations, who, though bitter enemies to the French, were, by the indefatigable zeal and abilities of the Jesuits, partly civilised, and converted to the Christian faith. They were originally settled at La Prairie, but the land producing very indifferent maize, they removed to Sault St. Louis, and from thence to the situation they now occupy. Idleness reigns in every part of their village, nor is there man, woman, or child to be found ever employed at any sort of work. Their habitations are dirty, miserable, and destitute of furniture; and the whole village, which is divided into two or three streets, presents a most forlorn and wretched appearance.

These Indians are under the care of Mr. Vanfelson the curé of the village. He lives in a tolerable house adjoining a small chapel, in which service is regularly performed by him on Sundays and festivals. The Indians, who happen to be at home, attend with their wives and children, and behave in a very respectful and becoming manner. The women, particularly, are solemn and devout in their deportment, and are strongly attached to the Holy Virgin; for whom they seem to have a remarkable veneration. They have good voices, and sing their Indian hymns in an agreeable manner.

Mr. Vanfelson is a most respectable young priest, and attends, with much diligence, to the improvement of the Indians. His brother at Quebec is an advocate of some eminence.

The Indians of Cachenonaga cultivate a little corn, and breed hogs and poultry; but the principal part of them subsist upon hunting and fishing. A chief resides among them, called Captain Thomas: his house is but little better furnished than the rest; and he is a very drunken character.

The boundary line between Canada and the United States is about 18 miles from St. John's, and passes across the Richlieu river, within a few miles of lake Champlain. Hence the Canadians are completely shut out from the lake in case of war, and even from the water communication with their own territory in Missisqui bay. The greatest part of this bay lies in Canada, and is thus cut off by the line of demarcation allowed by the English negotiators in the treaty of peace with the American states in 1783. In case of war, the Americans have every advantage over the Canadians, by confining them to the narrow channel of Richlieu river; and the ill effects of it have been already experienced since the embargo, as the rafts of timber were not permitted to come out of Missisqui bay, for the purpose of passing down the Richlieu river. The laws however were broken in several instances; but the parties were liable to fine and imprisonment. If the line had been drawn across the wide part of lake Champlain, the Americans could never have stationed their gun-boats with such effect, as they have of late years, in the Richlieu river, by which means they interrupted the communication between the two countries by water, and seized great quantities of goods.

From St. John's to the entrance of the lake, there are scarcely any settlements. Both shores are lined with woods, consisting chiefly of pines, which grow to a great height. A few straggling log-huts are seen at intervals, but otherwise it is completely in a state of nature. The Isle au Noix is situated near the line. Upon it are the remains of a small fortification, which had been successively occupied by the French, English, and American armies, during the several wars which have occurred in that country. The name of the island used sometimes to be given out for the parole upon those occasions; and it is related of an English officer during the American war, that upon being challenged by the sentinel, he gave the word, "Isle au Noix," in the true pronunciation, but the sentinel refused to let him pass. The officer persisted he was right, and the soldier maintained he was wrong; till at length the for-]

[mer recollecting himself, cried out, "Isle of Nox." —" Pass," said the soldier; "you have hit it at last!"

For account of the commerce of Montreal, see the section of this title under article CANADA.]

[MONTREAL, a river which runs *n. e.* into lake Superior, on the *s.* side of the lake.]

[MONTREAL Bay lies towards the *e.* end of lake Superior, having an island at the *n. w.* side of its entrance, and *n. e.* of Caribou island.]

MONTRONIS, a river of the island of St. Domingo, in the part possessed by the French. It rises in the *w.* head, near the mountains of Tapion, runs to that rhumb, and enters the sea in the port of Trou Forban.

MONTROUIS, a bay of the *w.* coast of the island of St. Domingo.

[MONTROUIS, a town in the *w.* part of the island of St. Domingo, at the head of the Bight of Leogane, 14 miles *s. e.* of St. Marcos, and 33 *n. w.* of Port au Prince.]

MONTSERRAT, an island of the N. sea, one of the Caribes, of the Atlantic ocean, discovered by Christopher Columbus in 1493. It is of an oval figure, and is nine miles long, and five and an half wide, and from 18 to 20 in circumference. The mountains are covered with cedar, *caoba*, and other trees, and the valleys are fertile and pleasant, and similar to those of the other islands; the climate is also the same. The principal productions are cotton, although of an interior quality, indigo, tobacco, and a great quantity of sugar and spirit, made of the sugar-cane, which is very general. It is surrounded by shoals and rocks, which render its navigation in tempestuous weather very dangerons; for, indeed, it cannot be said to have any port whatever. Its population consists of 5000 Europeans, and of about 10 to 20,000 Negro slaves. In 1733 it experienced an hurricane, the damage done by which, without counting that which affected the vessels, amounted to 50,000*l.* sterling.

The first who established themselves on this island were Irish, [in the year 1632,] whose descendants, and some persons from other countries, are its present inhabitants; but the common language is Irish, even amongst the Negroes. In the war of 1700 it was sacked ten days successively by the French, but in the 11th article of the treaty of Utrecht, it was stipulated, that they should render satisfaction to the English, although it does not appear that such was the case. [It was again invaded, and with most of the other islands captured by the French in the late war, and restored with the rest.] Its government is composed of a lieutenant-governor, and a council or assembly of eight representatives, two for each of the four districts into which it is divided.

It has only three open roads for vessels; and these are called Plymouth, Old-harbour, and Kers, where, both in the embarking and disembarking, the same precaution is necessary to be observed as in the road of San Christoval, and as we have noticed under that article. In 1770 the productions which were sent to England and Ireland amounted to 90,000*l.* sterling, and those to N. America to 12,000*l.*

[In the report of the privy council on the slave trade, in 1788, the British property vested here is estimated at 38,400 taxed acres of patented estates, and the Negroes are computed at 9500, to the value of 50*l.* each Negro. The cultivation of sugar occupies 6000 acres; cotton, provision, and pasturage have 2000 acres allotted for each. No other tropical staples are raised. The productions were, on an average, from 1784 to 1788, 2737 hhds. of sugar, of 16 cwt. each, 1107 puncheons of rum, and 275 bales of cotton.]

The following is an account of the number of vessels, &c. that have cleared outwards from the islands of Montserrat and Nevis, between the 5th January 1787, and the 5th January 1788; together with an account of their cargoes, and the value thereof.

Whither bound.	SHIPPING.			Sugar.			Rum	Molas.	Indigo.	Cotton.	Dying Woods, in Value.			Miscellaneous Articles, in Value.			TOTAL.		
	No.	*Ton.*	*Men.*	*Cwt.*	*qrs.*	*lbs.*	*Galls.*	*Galls.*	*lbs.*	*lbs.*	£.	*s.*	*d.*	£.	*s.*	*d.*	£.	*s.*	*d.*
Great Britain —	25	5371	341	108,325	—	21	4,406	1313	140	91,972	352	7	6	1162	3	2	185,709	10	11
American States	20	1850	138	1,895	—	—	122,710	—	—	—	—	—	—	70	10	—	13,981	12	6
Brit. Col. in Amer.	7	379	40	64	—	—	21,300	—	—	500	—	—	—	41	6	3	2,055	14	5
Foreign W. Indies	71	3085	377	—	—	—	140,660	—	—	—	—	—	—	89	4	—	12,896	19	—
Africa — —	1	102	8																
Tot. from Montserrat and Nevis.	122	10,787	904	110,284	—	21	289,076	1313	140	92,172	352	7	6	1363	3	5	214,141	16	6

By return to house of commons 1806, the hogsheads of sugar of 13 cwt. exported, were as follows,

In 1789,	3150
1799,	2595
1805,	2000

The official value of the imports and exports of Montserrat were, in

	Imports.	*Exports.*
1809,	£35,407	£10,460
1810,	£62,462	£16,816.

And the quantities of the principal articles imported into Great Britain were, in

	Coffee.		Sugar.		Rum.	Cotton Wool.
	Brit. Plant.	For. Plant.	Brit. Plant.	For. Plant.		
	Cwt.	Cwt.	Cwt.	Cwt.	Galls.	Lbs.
1809,	—	—	21,915	—	51,132	29,455
1810,	—	—	41,112	—	48,880	46,313

By report of the privy council in 1788, and by a subsequent estimate, the population of Montserrat amounted to

Years.	Whites.	People of Colour.	Slaves.
In 1787	1300	260	10,000
1805	1000	250	9,500

Montserrat is 26 miles *s. w.* of Antigua, about the same distance *s. e.* of Nieves, and lies in lat. 16° 45′ *n.* and long. 62° 17′ *w.*

MONTSERRAT, another island, of the gulf of California or Mar Roxo de Cortés; situate near the coast, between those of Carmen and La Catalina.

MONTSERRAT, a settlement of the island and government of Trinidad; situate near the *w.* coast.

MONTSERRAT, another, of the province and *captainship* of San Vicente in Brazil; situate on the shore at the source of the river Tiete or Añembi.

MONTSINERI, a river of the province of Guayana, in the part possessed by the French.

[MONTSIOUGE, a river or bay in Lincoln county, district of Maine, which communicates with the rivers Sheepscut and Kennebeck.]

[MONTVILLE, a township in New London county, Connecticut, about nine miles *n.* of New London city. It has 2053 inhabitants.]

MONTUOSA, a *real* of gold mines of the district of the city of Pamplona, in the jurisdiction of the *alcaldía mayor* of mines of the Nuevo Reyno de Granada, established at Bocaneme. It is of a very cold temperature, and produces some vegetable productions, but with scarcity, since the principal labour is confined to the gold mines; and from these much riches have been extracted; some silver mines have also been discovered in its territory.

MONTUOSA, a small island of the S. sea, near the coast of the province and government of Veragua, in the kingdom of Tierra Firme.

[MONUMENT Bay, on the *e.* coast of Massachusetts, is formed by the bending of cape Cod. It is spacious and convenient for the protection of shipping.]

MONZON, a settlement of the province and *corregimiento* of Guamalies in Peru; annexed to the curacy of Chavin de Pariarca.

[MOORE, a county of N. Carolina, in Fayette district. It contains 3770 inhabitants, including 371 slaves. Chief town, Alfordston.]

[MOORE Court-house, in the above county, where a post-office is kept, is 38 miles from Randolph court-house, and 40 from Fayetteville.]

[MOORE Fort, a place so called in S. Carolina, is a stupendous bluff, or high perpendicular bank of earth on the Carolina shore of Savannah river, perhaps 90 or 100 feet above the common surface of the water, exhibiting the singular and pleasing spectacle to a stranger, of prodigious walls of party-coloured earths, chiefly clays and marl, as red, brown, yellow, blue, purple, white, &c. in horizontal strata, one over the other. A fort formerly stood here, before the erection of one at Augusta, from which it stood a little to the *n. e.* The water now occupies the spot on which the fort stood.]

[MOORE's Creek is 16 miles from Wilmington, in N. Carolina. Here General M'Donald with about 2000 royalists were defeated (after a retreat of 80 miles, and a desperate engagement) by General Moore, at the head of 800 continentals. General M'Donald and the flower of his men were killed.]

[MOORFIELD, in New Jersey, 11 miles *e.* of Philadelphia.]

[MOORFIELDS, a post-town, and the capital of Hardy county, Virginia; situated on the *e.* side of the *s.* branch of Patowmac river. It contains a court-house, a goal, and between 60 and 70 houses. It is 16 miles from Romney, 28 from Winchester, and 116 from Richmond.]

MOOSE, Factory of the River, an establishment of the English, in the province of New S. Wales in N. America, founded in 1740 near the mouth or entrance of the river of its name, in lat. 51° 15′, on the shore of another navigable river, which, at 12 miles from the fort which has been built for the defence of the founders, divides itself into two arms, the one running from the *s.* the other from the *s. w.* On the shore of this *s.* arm are produced all kinds of vegetable productions, which are carried to the factory, such as barley, beans, and common pease, notwithstanding the cold winds blowing over the ice in the bay. On the same shore, and above the cascade, grows wild a certain corn resembling rice, and in the woods of the interior of Moose and Albany bay, as well as upon the shore of the river Rupert, are large trees of every kind, such as oaks, cedars, firs, &c. also an excellent grass for making hay; and throughout the whole territory may be raised the different European grains and fruits. The ice breaks at the factory about the beginning of March, but higher up, in the middle of the month. The river is navigable for canoes as far as the cascades; and 50 leagues up there is one of a fall of 50 feet, but after this the river runs deep and navigable through a fine healthy country.

[Moose River, a short stream in Grafton county, New Hampshire, which runs *n. e.* from the White mountains into Amariscoggin river.]

[Moose Island, on the coast of the district of Maine, at the mouth of Schoodick river, contains about 30 families. On the *s.* end of this island is an excellent harbour suitable for the construction of dry docks. Common tides rise here 25 feet.]

[MOOSEHEAD Lake, or Moose Pond, in Lincoln county, district of Maine, is an irregular shaped body of water, which gives rise to the *e.* branch of Kennebec river, which unites with the other above Norridgewock, about 17 miles *s.* of the lake. There are very high mountains to the *n.* and *w.* of the lake; and from these the waters run by many channels into the St. Lawrence.]

[MOOSEHILLOCK, the highest of the chain of mountains in New Hampshire, the White mountains excepted. It takes its name from its having been formerly a remarkable range for moose, and lies *w.* of the White mountains. From its *n. w.* side proceeds Baker's river, a branch of Pemigewasset, which is the principal branch of Merrimack. On this mountain, snow has been seen from the town of Newbury, Vermont, on the 30th of June and 31st of August; and on the mountains intervening, snow, it is said, lies the whole year.]

MOPORA, a settlement of the jurisdiction of Muzo, and *corregimiento* of Tunja, in the Nuevo Reyno de Granada. It is much reduced; of a warm temperature, and produces sugar-canes, cotton, *yucas*, maize, and plantains, on which the natives, who are very poor, subsist.

MOPORO, a settlement of the province and government of Maracaibo, in the same kingdom as the former; situate within the great lake of Maracaibo. It is small, and there is this singular circumstance attached to it, namely, that all the posts and rafters on which it is built, and which are of a kind of wood called *vera*, become petrified after having been a certain time in the water.

MOPOSPAN, Santiago de, a settlement of the head settlement of the district and *alcaldia mayor* of Cholula in Nueva España. It contains 39 families of Indians, and is a quarter of a league *n.* of its capital.

MOQUEHUA, a province and *corregimiento* of the kingdom of Peru, bounded *n.* by the province of Lampa, *n. e.* by that of Paucarcolla or Puno, *e.* by that of Chucuico, *s.* and *s. e.* by that of Arica, and *w.* by that of Arequipa. It is 42 leagues long, and its temperature is for the most part cold, from its being situate on the heights and sides of the *cordillera*, the tops of which are always covered with snow. Towards the lower part, where it is bounded by the province of Arica, and in some degree by that of Arequipa, the temperature is moderate, as also in some of the valleys formed by the windings of the *cordillera*, and in one of which the capital stands. In the aforesaid *cordillera* are many volcanoes, which are almost continually vomiting fire, and, in 1600, one called Omate exploded with such violence as to scatter its ashes over nearly the whole province, rendering useless many estates which were before fertile and productive, and carrying its destruction as far as the city of Arequipa and some of its settlements, which suffered dreadful damages. There are in this province some silver mines, which are worked, but to little profit. Its vegetable productions are quantities of maize, which is carried to the neighbouring provinces, and wines, which are for the most part converted into brandy, to send to the provinces of the sierra; and in the valley of its name alone they usually make about 60,000 *arrobas*. It also produces some sugar, wheat, and other seeds, a good number of large and small cattle, and other fruits peculiar to the *serranía*. It is watered by several streams which flow down from the *cordillera*, from the greater part of which are formed two rivers; the larger running into the sea by the valley of Tambo, of the jurisdiction of Arequipa; but its waters being bad, since near its source it is impregnated by some hot streams of a fetid quality; the other, which is less, being principally formed of three streams which pass by the capital and its vicinity, and which, after watering the greater and better territories, fall into the port of Ilo, belonging to the province of Arica. It was conquered by famine by the Emperor Maita Capac. Its *corregidor* used to have a *repartimiento* of 110,650 dollars, and its population consists of 10 settlements.

Torata,	Quinistacas,
Carumas,	Ubinas,
Puquina,	Pocsi,
Coalaque,	Mollebaya,
Omate,	Socay.

The capital is the town of the same name, at least so called from the time of its foundation by the aforesaid Emperor Maita Capac; but the Spaniards call it Santa Catarina de Guadalcazar, from its having been rebuilt by the Marquis of this title, the viceroy of Peru in 1626. It is situate at the foot of the *cordillera*, in a pleasant and fertile valley, abounding in vegetable productions, and especially in vines, upwards of 60,000 *arrobas* of wine being made yearly. It has a very good parish church, three convents, namely, of the religious orders of San Francisco, San Domingo, Betletmitas, and an hospital, and a college formerly of the Jesuits. It suffered severely by an earthquake in 1715. Its climate is mild and healthy, and it contains more than 6000 souls, amongst whom are some rich and noble families. In lat. 17° 13′ *s.* and long. 70° 48′ *w.*

MOQUI ARAYVE, a province and country of barbarian Indians in N. America; bounded *s.* by the river Gila as far as La Primeria, *e.* by Nuevo Mexico, *n.* and *w.* by the extensive regions as yet unknown, save by the confused advices of certain Indians, who said that they journeyed *w.* for six moons, and from whose further informations it might be conjectured that they reach as far as the confines of Tartary by the strait of Uriz. This province is inhabited by various barbarous nations of infidel Indians, from whom, in 1743, a deputation to the number of 44 came to Nuevo Mexico to entreat the governor, who was then the Lieutenant-colonel Don Gaspar Domingo, that he would send amongst them some missionaries who might reduce them to the catholic faith. This he acceded to, defraying out of his private purse the expences of cattle, seeds, and instruments, which he gave them to cultivate their land, at the same time establishing the settlements of

Hualpi,	Quianna,
Tanos,	Aguatubi,
Moxonavi,	And the Rio Grande
Xongopavi,	de Espelcta.

In 1748 the commissary of the missions of San Francisco entered by Nuevo Mexico to continue these reductions, as also those of the province of Navajoos, to the *n.* of that of Moqui and *n. w.* of Santa Fé.

MORAGA, a small settlement of the district and jurisdiction of Anserma in the Nuevo Reyno de Granada; situate on an eminence on the shore of the river Cauca; and in its vicinity are some gold mines celebrated for the abundance of this metal. Seven leagues from its capital.

MORAL, a point on the *s.* coast, and in the French possessions of the island of St. Domingo, between cape Jaquemel and the river Benet.

MORALES, a settlement of the province and government of Santa Marta in the Nuevo Reyno de Granada; situate on the *e.* shore of the Rio Grande de la Magdalena. On the *e.* a small river runs near to it, but immediately enters the aforesaid river. Its climate is warm and moist, and consequently unhealthy. In lat. 8° 15′ *n.*

[MORANT Keys, off the island of Jamaica, in the W. Indies. Lat. 17° 26′ *n.* Long. 75° 57′ *w.*]

[MORANT Point, the most *e.* promontory of the island of Jamaica. On the *s. w.* side of the point is a harbour of the same name. From point Morant it is usual for ships to take their departure that are bound through the Windward passage, or to any part of the *w.* end of the island of St. Domingo. Lat. 17° 57′ *n.* Long. 76° 7′ *w.*]

[MORANT HARBOUR, Port, is about 10 miles *w.* of point Morant, on the *s.* coast of the island of Jamaica. Before the mouth of it is a small island, called Good island, and a fort on each point of the entrance.]

[MORANT River is about six miles *w.* of the *w.* point of point Morant. The land here forms a bay, with an anchorage along the shore.]

MORAVIAYS, a settlement of the province and colony of New York in N. America; situate on the shore and at the source of the river Delaware.

MORAVO, a settlement of the province and *corregimiento* of Chichas and Tarija in Peru, of the district of the former; annexed to the curacy of Talina.

MORCHIQUEJO, a settlement of the province and government of Cartagena, in the division and district of Mompox; situate on the shore of the Rio Grande de la Magdalena.

MORCOT, a settlement of the island of Barbadoes, in the district of the parish of San Jorge.

MORCOTE, a settlement of the province and government of San Juan de los Llanos in the Nuevo Reyno de Granada. It is of a moderately hot temperature; situate at the foot of the mountains of Bogotá, very fertile, pleasant, and salutary, abounding in vegetable productions, and particularly in cotton, which the natives spin with much nicety, making excellent linen, white and striped mantles, delicate napkins, pavilions, and other articles of curious manufacture, vying with those of Tarma, which are esteemed the best in the kingdom. It also produces *aguecates*, and a species of small plantains, which may be eaten at one mouthful, and are called *cambures*, the like not being found elsewhere, and highly esteemed. The dates also are very fine, being as good as those of Africa and Palestine. It contains 100 housekeepers, and more than 400 Indians, who are the most docile, laborious, and well inclined of any in the province. This settlement lies in the road leading to Tunja, upon an extensive and beautiful lofty plain.

[MORE, a township in Northumberland county, Pennsylvania.]

[MORELAND, the name of two townships of Pennsylvania; the one in Philadelphia county, the other in that of Montgomery.]

[MORENA, a cape on the coast of Chile, S. America, is in lat. 23° 20′ *s.* and long. 70° 32′ *w.* It is 15 miles *n.* by *e.* of cape George. The bay between these capes seems very desirable to strangers to go in; but in a *n. w.* wind is very dangerous, because the wind blows right on the shore, and makes a very heavy sea in the road. Here is a very convenient harbour, but exceedingly narrow, where a good ship might be careened.]

MORENO, a port of the Morro, on the coast of the S. sea, of the province and *corregimiento* of Atacama in Peru.

MORETI, a river of the province and government of Darien, and kingdom of Tierra Firme. It rises in the mountains of the interior, runs *w.* and enters the Grande de Chucunaqui.

MORETOWN, a settlement of the province and colony of Georgia in N. America; situate on a small island formed by an arm of the river Pompon.

MORGAN, a settlement of the island of Barbadoes, in the district of the parish of St. Philip, distinct from two others which are there; one of the district of the parish of St. Thomas, the other on the *w.* coast.

MORGAN, a river of the province and government of Cumaná, running *w.* and entering the San Jacome.

[MORGAN District, in N. Carolina, is bounded *w.* by the state of Tennessee, and *s.* by the state of S. Carolina. It is divided into the counties of Burke, Wilkes, Rutherford, Lincoln, and Buncomb; and contains 33,292 inhabitants, including 2693 slaves.]

[MORGANS, a settlement in Kentucky, 38 miles *e.* of Lexington, and 18 *n. e.* of Boonsborough.]

[MORGANTOWN, a post-town and the chief town of the above district, is situated in Burke county near Catabaw river. Here are about 30 houses, a court-house and gaol. It is 30 miles from Wilkes, 31 from Lincolntown, 74 from Salem, and 408 from Philadelphia. Lat. 35° 47′ *n.*]

[MORGANTOWN, a post-town of Virginia, and shire-town of Monongalia county, is pleasantly situated on the *e.* side of Monongahela river, about six miles *s.* of the mouth of Cheat river; and contains a court-house, a stone-gaol, and about 40 houses. It is 57 miles from Romney, 17 from Union-town in Pennsylvania, 55 from Cumberland fort in Maryland, and 219 from Philadelphia.]

[MORGANZA, a town now laying out in Washington county, Pennsylvania; situated in, and almost surrounded by the *e.* and *w.* branches of Charter's river, including the point of their confluence; 13 miles *s. w.* of Pittsburg, and on the post-road from thence to Washington, the county town distant 10 miles. Boats carrying from 200 to 300 barrels of flour, have been built at Morganza, laden at the mill tail there, and sent down the Chartiers into the Ohio, and so to New Orleans. By an act of the legislature of Pennsylvania, the Chartiers, from the Ohio upwards as far as Morganza, is declared to be a high-way. This town is surrounded by a rich country, where numbers of grist and saw mills are already built, and the lands in its environs well adapted to agriculture and grazing; and is spoken of as a country that is, or will be, the richest in Pennsylvania. Morganza, from its situation and other natural advantages, must become the centre of a great manufacturing country; especially as considerable bodies of iron ore, of a superior quality, have been already discovered in the neighbourhood, and have been assayed. The high waving hills in this country are, from the quality of the soil, convertible into the most luxuriant grazing lands, and are already much improved in this way. These hills will be peculiarly adapted to raise live stock, and more particularly the fine long-woolled breed of sheep; such as that of the Cotswold hills in England, whose fleeces sell for 2*s.* sterling per pound; when others fetch only 1*s.* or 1*s.* 3*d.* The wheat of this country is said to weigh, generally, from 62 to 66 lb. the bushel of eight gallons. From hence considerable exports are already made to New Orleans, of flour, bacon, butter, cheese, cider, rye, and apple spirits. The black cattle raised here are sold to the new settlers, and to cattle merchants, for the Philadelphia and Baltimore markets; many have also been driven to Niagara and Detroit, where there are frequent demands for live stock, which suffer much in those *n.* countries, from hard winters, failures in crops, and other causes.]

MORGNE, or TUERTO, a settlement of the parish of the French, in the part they possess in the island of St. Domingo; situate on the *n.* coast, on the shore of the river of its name.

[MORGUE Fort, or FORTABEZA DE MORGUE, on the *s.* shore of the entrance to Valdivia bay, on the coast of Chile, on the S. Pacific ocean. The channel has from nine to six fathoms.]

[MORIENNE, a bay on the *e.* coast of the island of Cape Breton, near Miray bay, from which it is separated only by cape Brule. It is a tolerably deep bay.]

MORIN, a settlement of the province and *corregimiento* of Truxillo in Peru, to the *w.* of the mountian of Pelagatos.

MORIN, another settlement, in the island of St. Domingo, and part possessed by the French; situate on the *n.* coast, between the settlement of Limonade, and that of La Petite Ance. It is in the plain of cape Frances, and one of the 12 parishes of the name of Santa Rosa.

MORINECA, a settlement of the province and government of Darien, and kingdom of Tierra Firme; situate by the *s.* and on the shore of the Rio Grande de Tuira, near the *real* of Santa Maria.

MORINI, or MAROWINE, a river of the province and government of La Guayana, which serves as a limit of division between the territories of the Dutch and French. It runs *n. e.* then turns *n.* and empties itself into the Atlantic, in lat. 5° 55′ 4″ *n.*

MORIS, a settlement and *reduccion* of Indians, of the missions that were held by the regulars of the Jesuits in the province and government of Cinaloa in N. America.

MORLAND, WEST, a county of the province and colony of Virginia in N. America.

MORNE, GROS, a very lolly mountain of the island of Martinique, covered with points, similar to those of the mountains of Montserrat in Cataluna. It is near the coast which looks *s. e.* opposite the bay or Ance du Gallion.

MORNE, GROS, another mountain, on a point of land of the island of San Christoval, one of the Lesser Antilles; situate on the *n. e.* coast, between the river Cabrito and the bay of Caret.

MORNE, GROS, some other mountains, of the island of Guadalupe, with a very lofty mountain at their extremity, and on the coast which looks to *n. w.* between the port of Mouillage and La Grand Ance, or Great bay.

MORNE, GROS, a settlement and parish of the French, in the part which they possess in the island of St. Domingo; situate on the *n.* coast.

MORO, a settlement of the province and *corregimiento* of Santa in Peru.

[MORO, Castle, is on the point or head-land on the *e.* side of the channel of the Havannah, in the *n. w.* part of the island of Cuba, and is the first of two strong castles for the defence of the channel against the approach of an enemy's ships. It is a kind of triangle, fortified with bastions, on which are mounted about 60 pieces of cannon, 24 pounders. From the castle there also runs a wall or line, mounted with 12 long brass cannon, 36 pounders; called, by way of eminence, "The Twelve Apostles:" and at the point, between the castle and the sea, there is a tower where a man stands and gives signals of what vessels approach. See HAVANNAH.]

[MORO QUEMADO, FAZENDA DO, a farm about 60 miles to the *n. e.* of Rio de Janeiro, visited by Mawe in 1809. He passed through it in his way to Canta Gallo; and the following is a description of his route, which we extract nearly verbatim, in order to give our readers, as far as we are able, an accurate idea of the qualities of the soil and nature of the territory in these parts.

'Some time (observes this traveller) after my return from Santa Cruz (to Rio de Janeiro), a circumstance of a singular nature took place, which occasioned me to undertake a journey to a district called Canta Gallo, distant about 40 leagues from [the capital, and one of the latest discovered in this part of Brazil. Two men reported that they had there found a mine of silver, and brought to the mint a quantity of earthy matter reduced to powder, from which was smelted a small ingot of that metal. This report being officially laid before his Excellency Don Rodrigo, I was solicited to go to Canta Gallo and investigate the business on the spot, the two men being ordered to meet me there. Before I proceed to relate the result of my inquiry, I shall briefly describe whatever I observed worthy of note in the course of the journey.

'Being provided with a passport, and also a sketch of the route, taken from a MS. map in the archives, I departed from Rio on the 10th of April 1809, accompanied by Dr. Gardner, lecturer on chemistry at the college of St. Joaquim. Having to pass to the bottom of the harbour, towards the *n.* we embarked in a small vessel, and being favoured with a strong sea-breeze, ran down to the entrance of the fine river Maccacu, which we reached after a five hours' sail. The wind then dying, our boatmen took to their oars, and proceeding up the river we reached a house called Villa Nova, where numbers of market-boats for Rio were waiting for the land-wind and the turn of the tide. After taking some refreshment here, we rowed onward until the river became so narrow that the vessel frequently touched the bank on each side, and the men were obliged to push her along with poles. At day-break we reached Porto dos Caxhes, a place of great resort from the interior, being the station where the mules discharge their loads of produce from the many plantations in the neighbourhood. The town consists of several poor houses, and of stores where goods are deposited for embarkation. The stratum hereabouts is primitive granite, covered with fine strong clay. Leaving this place, we proceeded for some distance and came to a large swamp, which we navigated in a canoe, with very little difficulty, and shortly afterwards arrived at the village of Maccacu. It stands on a small eminence in the midst of a fine plain, watered by a considerable stream, over which there are two good bridges. Though almost at the base of the chain of mountains that forms a barrier along the coast, the neighbourhood affords some fine situations; the land in general consists of a strong clay, but appears much worn out. The commander, Colonel José, to whom I introduced myself, gave me a very polite reception, as did also the brethren of the convent, to whom I paid a visit.

'On the following day, being accommodated by the colonel with a horse and guide, I proceeded along the winding banks of the river, which in many places present most beautiful views. Here was more cultivated land than I expected to see; but the sugar plantations, and, in general, the low pasture grounds, are quite neglected. We passed several farms belonging to convents, which, from their apparent condition, and the accounts we received, do little more than maintain the Negroes and incumbents upon them. There was rarely a milch cow to be met with: pigs and poultry were equally scarce. The population of these fine valleys is deplorably thin and poor; there was a general sickliness in the looks of the women and children we met with, which may be imputed to their miserable diet and inactive life. I ought to state that the manners of the people here are mild and gentle; we were every where treated with civility, and all our inquiries were answered with the most friendly marks of respect and attention.

The air, as we drew nearer the mountains, was fresh, and indeed cold. Towards evening we arrived at a farm belonging to a convent of nuns in Rio de Janeiro, where we were kindly accommodated for the night. This place is most agreeably situated, and might, under skilful and industrious management, be rendered a paradise. It has excellent clay, fine timber, a good fall of water, which forms a beautiful rivulet, and runs into a navigable river within 100 yards of the house; a fine extent of arable land, and a still finer of pasture, which peculiarly qualifies it for dairy farming. It is distant only one day's journey from port Caxhes, whence there is a navigable communication with the metropolis. What a scene for an enterprising agriculturist! At present all is neglected: the house, the out-buildings, and other conveniences, are in a state of decay, and the people who manage the land appear, in common with the animals that feed upon it, to be half famished.

'The next morning we proceeded *e.* and crossing the stream, which was at least 60 yards broad and full three feet deep, rode along the farther margin, which is rather more elevated, and presents a view of some fine plains, stretching from thence to the base of the mountains. Journeying in that direction we reached the fine plantation of Captain Ferreta, who received us very politely, and shewed us every attention. This place, bounded by the alpine ridge behind it, is the extreme point to which the river Maccacu is navigable. It is six or seven leagues from the village of that name. The estate maintains about 100 Negroes, who are chiefly employed in raising sugar, cotton, and coffee; but] [to me the situation appeared much better calculated for growing grain and feeding cattle, as the weather is at times cold, the evenings are frequently attended with heavy dews, and owing to the proximity of the mountains, here are frequent rains, accompanied by thunder and lightning. Numbers of fine springs burst forth from various parts of the hills, and form rivulets with falls, which, as here is plenty of fine timber, afford every means for working machinery. The owner lives in opulence, and is so humane and liberal to his people, that they seem to revere him as a father. We were much pleased with the air of domestic comfort and contented industry, which we observed among them on visiting their dwellings in the evening. Some of the Negro children were at play; others of more advanced age were assisting the women to pick cotton; and the men were scraping and preparing mandioca. Their cheerfulness was not at all interrupted by our approach, nor did they betray any uneasy feeling of constraint in the presence of their superiors. In lieu of candles, which are seldom to be met with but in the capital, they burn oil, extracted from the bean of the palm, or from a small species of ground-nut here called meni.

'About noon, on the following day, horses being provided, and a soldier appointed for our guide, we left the fazenda, accompanied by its hospitable owner, Captain Ferrera, who conducted us half a league on our way. The river, along which we passed in an *e.* direction, bursts through vast masses of rock with great force, and in some parts forms considerable falls. The captain, ere we parted, led me to a water-course, in which were found pieces of granite covered with manganese in a botryoidal form. After crossing the river twice, we arrived at what is called the first register, or searching-honse, distant about two miles from the fazenda. This station is guarded by a corporal and a private soldier, who are charged with the receipt of various tolls, and are empowered to search passengers, in order to prevent the smuggling of gold dust. After shewing my passport, I took leave of Captain Ferrera, who made me promise to pay him a longer visit on my return.

'We had been warned of the badness of the roads, and were by no means agreeably deceived in them, for we were nearly four hours in going the next six miles. At the close of day, after a laborious and dangerous passage through abrupt ravines, and along the sides of steep hills, our guide announced that we were in sight of the second register, where it was proposed that we should pass the night. On arriving we found it a most miserable place, inhabited by five or six soldiers under the command of a serjeant. This good man gave us a hearty welcome, and with the assistance of his comrades, cooked us a supper of fowls, and regaled us with whatever else their scanty store afforded. We were not without music to our repast, for the house is built on the edge of a roaring torrent, which, bursting through a ravine, has washed away every thing except some huge masses of rock. A bit of ground, about 10 yards square, is all the garden these poor people have, and even this is much neglected, for the guards here are so often changed, that no one thinks of adding to the comforts and conveniences of an abode, which others are to enjoy.

'At day-break, we found that our mules had strayed into a wood adjoining, but as the road was stopped, we were under no apprehension of losing them, for the thickets on each side were impervious. This occurrence gave me an opportunity of seeing more of these remote regions, and certainly the imagination of Salvator Rosa himself never pictured so rude a solitude. On one side rose the great barrier of mountains, which we had yet to cross, covered to their summits with trees and underwood, without the smallest trace of cultivation; on the other lay the broken country, between this ridge and the plain, presenting the same wild features of silvan scenery. The miserable hut, at which we lodged, partook of the savage character of the neigbourhood, and seemed formed for the abode of men cut off from all intercourse with their fellows. On our return we were provided with a breakfast of coffee and eggs; as to milk, there was no possibility of procuring any; a cow would have been considered here as an incumbrance, nor would any one of the six idle soldiers have given himself the trouble of milking her, though they all had been dying of hunger,

'On resuming our journey, we enteredon a road still more sleep and rugged than that which we had passed. We were often obliged to dismount and lead our mules up almost perpendicular passes, and along fearful declivities. In some places, the thick foliage of the trees and that of the underwood, which grew higher than our heads, sheltered us from the sun, and indeed scarcely admitted the light. Not a bird did we see, nor the trace of any living thing, except some wild hogs. We passed several bare granite rocks of a gneiss-like formation.

'In journeying to the next station, we observed nothing worthy of note excepting a small saw mill, worked by an overshot wheel, of very clumsy construction. The frame, which contains a single]

[saw of very thick iron, moves in a perpendicular direction; at every stroke a boy brings the timber up, by pulling a cord attached to a crank that moves the cylinder on which it rests. How readily, thought I, would the meanest Russian peasant improve this machine!

'We proceeded on our way up an ascent so precipitous that we were obliged to walk more than ride; after two hours toiling along the side of a granite mountain, in which we observed some beds of fine clay, we reached the summit, from whence we saw the bay of Rio de Janeiro, the sugar-loaf mountain, and the city itself, to all appearance not more than four or five leagues distant from us, though, in reality, more than 20. At this elevation, which we may state to be at 4 or 5000 feet above the level of the sea, the air was sharp and keen; the thermometer stood at 58°. Continuing in a *n. e.* direction, we passed two poor farms, and entered upon a range of grand scenery, composed of bare abrupt conical mountains, with immense water-falls in every direction.

At the close of day we arrived at a farm-house, called Fazenda do Moro Quemado, the manager of which received us hospitably, and accommodated us for the night. The weather was so cold, that a double supply of bed-clothes scarcely produced sufficient warmth; in the morning the thermometer was at 48° Fahrenheit. After the heavy dew cleared away we took a view of the grounds, in company with the manager; they appeared well-suited for a grazing farm, but the temperature of the atmosphere is too severe for growing the common produce of the country; particularly cotton, coffee, and bananas, which are frequently blasted. I was informed that some wheat has been grown here, though the people are quite unacquainted with the European method of farming. Indian corn, for the feed of hogs, is the staple article. This plantation is infested by ounces, which at times prey upon young cattle; the manager, who is a great hunter, keeps dogs, though of a poor race, for the express purpose of destroying them, which is thus practised:—When the carcase of a worried animal has been found, or when an ounce has been seen prowling about, the news is soon proclaimed among the neighbours, two or three of whom take fire arms loaded with heavy slugs, and go out with the dogs in quest of the animal, who generally lurks in some thicket near the carcase he has killed, and leaves so strong a scent that the dogs soon find. When disturbed he retreats to his den, if he has one, the dogs never attempting to fasten on him, or even to face him, but, on the contrary, endeavouring to get out of his way, which is not difficult, as the ounce is heavy and slow of motion. If he caves, the sport is at an end, and the hunters make up the entrance; but he more commonly has recourse to a large tree, which he climbs with great facility; here his fate is generally decided, for the hunters get near enough to take a steady aim, and seldom fail to bring him down, one of them reserving his fire to dispatch him, if required, after he has fallen. It generally happens that one or two of the dogs are killed in coming too near, for even in his dying struggles, a single stroke of his paw proves mortal. The skin is carried home as a trophy, and the neighbours meet and congratulate each other on the occasion.

'This farm, in the hands of an experienced and skilful agriculturist, might be managed so as to produce amazing returns. Its soil is wet, adapted to the growth, not only of Indian corn, but of wheat, barley, potatoes, &c. and it is so well irrigated, by numerous mountain streams, that the pastures are always luxuriant. Here are fine falls of water, and abundance of excellent timber, so that corn-mills might be erected at little more expence than what would arise from the purchase of mill-stones. Connected with the nun's farm below, this establishment might be rendered one of the most complete and advantageous in Brazil.

'Leaving Moro Quemado at noon, and descending on the other side of the ridge of mountains, we passed through an unequal tract, formed of hills and ravines. Onward the land appeared finer, and the timber of a superior growth, but there were few cultivated spots, and not many houses. The first extensive fazenda we reached was that of Manuel José Pereira, a native of the Azores, who managed his agricultural concerns much better than the other farmers whom we visited. We were shewn a large field of Indian corn ready for cutting; the quantity that had been sown was about 11 *fanegas*, or bushels, and the produce was estimated at 1500 bushels, about 150 for one. This was an ordinary crop; in good years the harvest yields 200 for one. The corn, as before stated, is chiefly consumed in the fattening of pigs; the quantity requisite for this purpose is six or seven bushels each, and the time 10 or 12 weeks. The curing of bacon is performed by cutting all the lean from the flitches, and sprinkling them with a very little salt. This food has the peculiar effect of giving great solidity to the fat, which of itself is not liable to putrefaction.' See *Mawe's Travels.*]

MOROCA, a river of the province and government of Guayana. It rises in the *serranía* of Imataca, and enters the sea on the *e.* coast.

MOROCOLLA, a settlement of the province and *corregimiento* of Lucanas in Peru; annexed to the curacy of Huacaña.

MOROI, a settlement of the province and *corregimiento* of Chichas and Tarija in Peru.

[MOROKINNEE, or MOROTINNEE, in the island of Mowee, one of the Sandwich islands. It is in the N. Pacific ocean, and lies in lat. 20° 29′ *n.*

MOROMORO, a settlement of the province and *corregimiento* of Cayanta or Charcas in the kingdom of Peru.

MORON, a river of the province and *corregimiento* of Cuenca in the kingdom of Quito. It takes its origin in some mountains to the *e.* of that city, and runs *s. e.* to enter the Marañon in the province of Mainas.

[MORON, a parish of the province and government of Buenos Ayres; situate on a small river emptying itself into the La Plata about 20 miles *s. w.* of Buenos Ayres. Lat 34° 40′ 10″. Long. 59° 54′ 45″.]

MORONA, a large river of the kingdom of Quito to the *s.* It rises in the province of Alausi to the *e.* and, after collecting the waters of the Zuña, Jubal, Puentehonda, Bolcan, and Avenico, and taking itself the name of Upango, runs *s.* then receives the Apatenoma, Guachiyuca, and Amaga on its *s. w.* side, and on its *n. e.* the Arrabima, Atassari, Yanassa, Hechizero, Chipanga, Apiaga, and Puschaga, and then, with the name of Morona, becomes navigable and laves the lands of the mountains of the country of Los Xibaros, where many barbarians of this nation dwell: it passes very near the city of Macas, the capital of this province, and enters with a large body into the Marañon or Amazon, 41 miles *s.* by *e.* of San Borja, between the rivers Pastaza to the *e.* and Santiago to the *w.* in lat. 4° 38′ 30″ *s.*

[MOROSQUILLO Bay is to the *s.* of Carthagena, on the coast of the Spanish main, and in the bight of the coast coming out of Darien gulf, on the *e.* shore. It is large, but very open.]

MOROTOCOS, a barbarous nation of Indians, of the province and government of Paraguay, discovered by the Father Juan Baptista Zea, of the abolished order of Jesuits, in 1711. They are very different in their customs from the other Indians, are taller and of a redder complexion. They make their darts and lances of a wood extremely hard, and they manage these with the greatest dexterity, as also their bows and arrows. Amongst these Indians the women had the entire authority, and the husbands were not only obedient to them, but managed all the household affairs. The women never kept more than two children, one of each sex, and the rest they put to death as soon as born, avoiding thereby the trouble of rearing them.

Although this nation, in common with the others, had its caciques or captains, they preserved no form of government whatever, and their authority was only limited to affairs of war. The country, which is in lat. 20° 30′ *s.* is dry, barren, and surrounded by mountains, on the which are thick woods of palms, in the trunks of which is a kind of spongy marrow, from which they squeeze out a juice which serves them for drink. Notwithstanding that it freezes much in the winter, they all, men and women, go naked; from whence it has been said of them that they had a very hard skin and two fingers thick. The Boxos Indians, who were reduced to the faith, took two children from the Morotocos, and presented them to the Father Suarez, a missionary in the *reduccion* of the Chiquitos, and making use of these as interpreters, entered the country of the Morotocos to preach; and such were the fruits of his labours, that by the end of the year 1711, he had converted the whole of them, when they established themselves in the aforesaid settlement of San Joseph.

[MOROTOI, or MOROKOI, one of the Sandwich islands in the Pacific ocean, is about 2¼ leagues *w. n. w.* of Mowee island, and has several bays on its *s.* and *w.* sides. Its *w.* point is in lat. 21° 20′ *n.* and long. 157° 14′ *w.* and is computed to contain 36,000 inhabitants. It is seven leagues *s. e.* of Woahoo island.]

MORRILLO, PUNTA DEL, a cape or extremity of the island of Inagua; thus called from a small mountain in its vicinity of this name.

[MORRIS, a county on the *n.* line of New Jersey, *w.* of Bergen county. It is about 25 miles long, and 20 broad, is divided into five townships, and contains about 156,809 acres of improved, and 30,429 acres of unimproved land. The *e.* part of the county is level, and affords fine meadows, and good land for Indian corn. The *w.* part is more mountainous, and produces crops of wheat. Here are seven rich iron mines, and two springs famous for curing rheumatic and chronic disorders. There are also two furnaces, two slitting and rolling mills, 35 forges and fire works, 37 saw mills, and 43 grist mills. There are in the county 16,216 inhabitants, of whom 636 are slaves.]

Morris, a settlement of the island of Barbadoes, in the parish and district of St. Joseph; situated to the *s.* on the *e.* coast.

Morris, another, in the same island, in the district of the parish of Todos Santos.

[Morris Bay, on the *w.* coast of the island of Antigua, in the W. Indies. It cannot be recommended to ships to pass this way, as there is in one place *s.* from the Five islands only two fathoms water. Vessels drawing more than nine feet water must not attempt it.]

[MORRISSINA, a village in W. Chester county, New York, contiguous to Hell-gate, in the sound. In 1790 it contained 133 inhabitants, of whom 30 were slaves. In 1791 it was annexed to the township of W. Chester.]

[MORRISTOWN, a post-town and capital of the above county, is a handsome town, and contains a Presbyterian and Baptist church, a court-house, an academy, and about 50 compact houses; 18 miles *w.* by *n.* of Newark, and about 56 *n. e.* of Philadelphia. The head-quarter of the American army, during the revolution war, was frequently in and about this town.]

[MORRISVILLE, a village in Pennsylvania; situated in Berk's county, on the *w.* bank of Delaware river, one mile from Trenton, nine from Bristol, and 24 from Philadelphia. A post-office is kept here.]

MORRITOS, some mountains of the coast of the Nuevo Reyno de Granada, in the province and government of the Rio del Hacha; they may be discovered at a great distance, and are close to cape Chichibacoa on the *w.*

MORRO, a settlement of the government and jurisdiction of Merida, in the Nuevo Reyno de Granada. It is of a cold but healthy temperature, produces much wheat on its hilly sides, maize, and other vegetable productions of a cold climate; a tolerable number of neat cattle, goats, and sheep; and contains 40 housekeepers and 80 Indians.

Morro, another settlement, of the district of La Punta of Santa Elena, in the province and government of Guayaquil and kingdom of Quito.

Morro, another, of the province and *corregimiento* of Chachapoyas in Peru.

Morro, a river of the kingdom of Chile, which rises in the mountains of the *cordillera*, and enters the sea in the bay of Concepcion.

Morro, another, of the province and *corregimiento* of Chachapoyas in Peru. It rises in the *sierra*, and incorporates itself with another river to enter the Mocobamba.

Morro, a mountain, with the surname of Hermoso, on the coast of the province and government of Cartagena, on a point of land which runs into the sea between the point of Zamba and the island Verde.

Morro, another mountain, with the same additional title as the former, on the coast of California in N. America.

Morro, another, with the surname of Quemado, on the coast of the province and *corregimiento* of Nasca in Peru.

Morro, another, on the coast of the province and *corregimiento* of Arica in Peru.

Morro, another, with the surname of Hermoso as well as the former, on the coast of the province and government of Costa Rica and kingdom of Guatemala, by the S. sea, between the port of Las Velas and cape Guiones.

Morro Chico, a mountain of the coast of the province and government of Honduras and kingdom of Guatemala, between the river Seco and that of Callera.

[Morro Viejo. See St. Gallan.]

MORROA, a settlement of the province and government of Cartagena in the Nuevo Reyno de Granada; situate on the shore of the stream Pichelin, near the settlement of San Christoval.

MORRON, a settlement of the province and government of Cartagena in the Nuevo Reyno de Granada; situate on the *n.* of the town of San Benito Abad.

Morron, another settlement, in the province and government of Venezuela; situate on the sea-shore and to the *w.* of the port Cabello. This settlement has also a port, which, although small, is sheltered from the winds.

MORROPE, a large settlement of the province and *corregimiento* of Saña and bishopric of Truxillo in Peru. It is of the best climate of any settlement on that coast; near it runs the river Pozuelos, which fertilizes the territory and renders it extremely delightful. The natives employ themselves in digging lime out of a quarry in the desert of Sechura and Lito, the which is used for making soap; this privilege having been granted by the government, and extended to the settlement of Pacora, which is annexed to this settlement.

MORROPON, a river of the province and *corregimiento* of Piura in Peru, to the *e.* It runs *s. w.* and enters the Piura between the rivers Frias and Sauri, by its *w.* shore, in lat. 5° 24′ *s.*

Morropon, a settlement of this province and kingdom, so called from the former river.

MORROS, some *farallones* or isles of the N. sea, lying opposite the coast of Santa Marta, and about the distance of a cannon-shot from it.

MORROSQUILLO, a gulf on the coast of the

province and government of Cartagena and Nuevo Reyno de Granada, between the river Sinú and the islands of San Bernardo.

MORT, a port or bay on the *s.* coast of the island of Newfoundland, within the great bay of Plaisance.

MORT, another bay, on the *s.* coast of the straits of Magellan, between that of San Martin and the creek of Sweet Water.

MORTAGUA, a river of the province and government of Honduras and kingdom of Guatemala. It runs *n.* and enters the sea between the cape of Las Puntas and the bay of Omoa.

MORTALLA, an island of the bay of Nassau in Florida. See NASSAU.

[MORTIER'S Rocks, on the *s.* coast of Newfoundland island. Lat. 47° 2′ *n.* Long. 54° 52′ *w.*]

MORTIGURA, a settlement of the province and *captainship* of Pará in Brazil; situate on the island Samauna.

[MORTO Island, on the coast of Peru, so called by the Spaniards, from its striking resemblance to a dead corpse extended at full length. It is also called St. Clara. It is about 13 miles *n. w.* from the mouth of the river Tumbez; and is two miles in length, and 72 miles from Guayaquil.]

[MORTON Bay, on the *n. w.* coast of the island of Nevis in the W. Indies, is near the narrows or channel between that island and St. Christopher's, to the *n. w.* of which there is from three to eight fathoms, according to the distance from shore.]

MORUAS, a barbarous nation of Indians but little known, who inhabit the woods near the river Yetan to the *w.* of Paraguay. They go naked, without fixed abode, and maintain themselves by the chase.

[MORUES Bay, on the *s.* shore of the river St. Lawrence, *s.* of Gaspee bay, and *w.* of Bonaventura and Miscan islands.]

MORUGA, a river of the province and government of Cumaná. It rises in the *sierra* of Imataca, and enters the sea near the river Poumaron, in the district possessed by the Dutch.

[MORUGO, a small river to the *w.* and *n. w.* of the gulf of Essequibo, on the coast of Surinam in S. America.]

MORUI, a settlement of the province and government of Venezuela in the Nuevo Reyno de Granada, situate in the peninsula of Paraguana, nearly in the centre of the same.

MORUNGABA, a settlement of the province and *captainship* of San Vicente in Brazil; situate between those of Samambay and Rio Verde.

MOSCARI, SANTIAGO DE, a settlement of the province and *corregimiento* of Chayanta or Charcas in the same kingdom as the former.

MOSCAS, MOZCAS, or MUISCAS, an ancient nation of Indians, and very numerous, of the Nuevo Reyno de Granada, who dwell to *e.* of the mountains and *llanuras* of Bogotá. They were the most civilized of all the nations of this kingdom; were clothed with a sort of cotton shirt, and over that a square mantle of the same fabric, and upon their heads, with the skins of animals they had killed, adorned with beautiful plumes, and in the front of the same an half moon of gold or silver, with the points upwards; also on their arms they wore bracelets of stone or bone, in their nostrils rings of gold, which they called *chaqualas;* and the height of their gala or luxury was to paint their faces and body with *vija*, a kind of paint, and with the juice of *jaqua*, a fruit which produces a black tint. The women made use of the square mantle, which they called *chircarte*, and which was fastened round their waist by a clasp, which in their language was called *chumbe* or *maure*, and upon their shoulders another mantle which was smaller, named *liquira*, and which was fastened to their breast by a large gold buckle. The men wore their hair long upon their shoulders and parted in the Nazarene form, and the women carried it loose, availing themselves of the use of certain herbs, which, by the help of the fire, might render it of a deeper black; and the greatest affront that could be offered them was to cut it.

These Indians, as well men as women, are of an ingenuous countenance, of a good disposition. Their arms were slings, swords of *macana*, a wood as hard and as shining as steel; also a certain kind of darts of light wood. They believed that there was a general Creator of all, but they nevertheless adored the sun and moon, calling the former Zupé and the latter Chia; neither did they doubt of the immortality of the soul, but they imagined that it passed to other countries, and thus they buried their dead with certain portions of victuals, gold, emeralds, &c. They were in continual warfare with the Muzos and Colimas nations, and were feared and respected by all.

The Moscan tongue, formerly called Chibcha, was the general language of the whole kingdom; and this was governed by a king or *zipa*, who was elective. This language is now almost entirely lost. Nearly all the settlements of the Nuevo Reyno de Granada are of Mozcan Indians, reduced to the Catholic faith. They are of a generous nature, bold, faithful, and robust, but much inclined to drunkenness. Some authors believed that the name of Mozcas had been given them by the Spaniards to signify their numbers; but it is certain that this was their own name, as may be seen in the history of the Nuevo Reyno de Granada, which has been written with infinite ability by the most illustrious Señor Don Lucas Fernandez de Piedrahita, bishop of Santa Marta and of Panama, and where may also be seen further particulars concerning these Indians.

Moscas, a settlement of the province and *corregimiento* of Tarma in Peru; annexed to the curacy of Parianchacra.

[MOSCHKOS. See Kikapus.]

[MOSE, or Vlla del Mose, a town on the bank of the river Tabasco, in the bottom of Campechy gulf, to which small barges may go up. Great quantities of cocoa are shipped here for Spain; which brings a great many sloops and small vessels to the coast.]

[MOSES Point, a head or cape of land, on the *e.* side of the entrance into Bonavista bay, on the *e.* coast of Newfoundland island. It is to the *s.* of the rocks called Sweers, and five miles *s. w.* of cape Bonavista.]

MOSINA, a settlement of the province and government of Cumaná; situate on the *w.* coast of the capital.

MOSLEVIN, a settlement of the province and government of Tucumán in Peru, at present ruined by the infidel Indians; to the *n. n. e.* of San Joseph de Vilelas.

[MOSLEY'S, a place on Roanoke river, nine miles below St. Tammany's, and three above Eaton's. The produce of the upper country is brought to these places, and sent from thence by waggons to Petersburg in Virginia.]

MOSNACHO, a settlement of the government and jurisdiction of Maracaibo. It is very reduced and poor; annexed to the curacy of Chachopo. It produces some seeds and fruits peculiar to its climate, which is temperate.

MOSQUITO, a bay on the *n.* coast, and in the part possessed by the French, of the island of St. Domingo, between the port of Paz and Agua.

MOSQUITOS, a country of N. America, between Truxillo and Honduras, of the kingdom of Guatemala, in lat. 13° and 15° *n.* and between long. 85° and 88° *w.* bounded *n.* and *e.* by the N. sea, *s.* by the province of Nicaragua, and *w.* by that of Honduras. The Spaniards consider it as part of the latter province, but they have no establishment or settlement whatever in it; since the enmity which the natives possess towards the Spaniards, inclines them easily to enter into alliance with any other nation, and particularly the English, who most frequently are upon their coasts, in order to make extortions upon them.

These Indians are excellent fishermen, and are much given to the fishery of the marine cow, and they moreover frequently make a voyage in English vessels to Jamaica. The Duke of Albemarle, being governor of this province, admitted the Mosquitos Indians under the protection of England, and their prince received a special commission. After his death his successor proceeded to Jamaica to restore the treaty, but his vassals were not willing to acknowledge it. The English have at various times projected the establishment of a colony here. The pirate William Dampiere, speaking of these Indians, says that they have so acute a sight that they can see vessels at a much greater distance than can Europeans. Their dexterity also is such, that with a little bar of iron like the ramrod of a gun they can stop every hit that is made at them; so that they are quite secure except that in case the said bar should break.

Mosquitos, some islands near the coast and government of Honduras, inhabited by the Mosquitos Indians. They are many and small, and close to those of Los Manglares.

Mosquitos, another island, which is one of the Little Virgin isles; situate near the *n.* coast of La Virgen Gorda, on which it is dependent.

Mosquitos, another island, near the *e.* coast of Florida, just without the Bahama channel.

Mosquitos, a bay on the *n.* coast of the island of Cuba, between the port of the Havana and the bay of Mariel.

Mosquitos, a point of the coast in the province and government of Darien, and kingdom of Tierra Firme, between the island of Pinos and the *rancho* of Harpones.

Mosquitos, a bay, called Rincon de Mosquitos, on the coast of the province and government of Nicaragua, and kingdom of Guatemala, close to the cape of Gracias a Dios, and opposite the shoal of Tiburones.

Mosquitos, a river of the province and government of Venezuela in the Nuevo Reyno de Granatla. It rises from two lakes at the foot of the *sierra* of Carrizal, by the *s.* side, and enters the Orinoco.

MOSTARDAS, a settlement of the province and *captainship* of Rey in Brazil; situate opposite the great lake of Los Patos.

MOSTAZAL, a river of the province and *corregimiento* of Coquimbo in the kingdom of Chile; which runs *w.* and enters the Limari.

MOSTAZAS, a settlement of the province and

government of Venezuela in the Nuevo Reyno de Granada; founded, in 1740, in the *serranía*, after the Real Compañia de Guipuzcoa.

MOTA, a settlement of the jurisdiction and *alcaldía mayor* of Pilon in the Nuevo Reyno de Leon.

MOTA, another settlement, on the *s.* coast of the island of Cuba, with a good port.

MOTATAN, a large river of the province and government of Venezuela in the Nuevo Reyno de Granada. It rises from the mountains of Merida, runs towards the settlement of Bocono, and receiving in its course the waters of the Nequitao, fertilizes the fields of Truxillo, which place it laves on the *e.* side, and then with a stream increased by several other rivers from the mountains of Merida, empties itself into the lake of Maracaibo by the *e.* side, in lat. 9° 45′ *n.*

MOTAUITA, a settlement of the province and *corregimiento* of Tunja in the Nuevo Reyno de Granada. It is of a very cold temperature, and produces wheat, maize, *papas*, and barley, contains 50 housekeepers and as many Indians, and is a little less than a league's distance from its capital.

MOTE, SAN JOSEPH DE, a settlement of the province and government of Quixos and Macas in the kingdom of Quito. It is much reduced, of a cold temperature, and produces only maize, *papas*, and potatoes, which are there called *camotes*. It is situate at the foot of a very lofty mountain called Sumaco.

MOTE, a river of the province of Guayana, in the Dutch possessions.

MOTE, a small island of the lake of the Iroquees Indians in N. America, near the *n.* coast.

MOTEPORE, a settlement and *real* of silver mines of the province and government of Sonora in N. America.

[MOTHER Creek, in Kent county, Delaware. See FREDERICA.]

MOTILONES, a barbarous and ferocious nation of Indians, of the province and government of Venezuela in the Nuevo Reyno de Granada. It is equally formidable from its numbers as from its intrepidity. These Indians wander over a vast tract of country, which is bounded *n.* by the province of Maracaibo, *e.* by the city of Merida, *s.* by those of Cucuta and Salazar, of Las Palmas, and *w.* by those of Ocaña and Tamalameque, of the province of Santa Marta.

These barbarians continually infest the public roads leading to the above-named settlements by their incursions, stopping all the traders, and more especially on the mountains which lie between Pamplona and Merida, and also in the navigation of the celebrated river Sullia.

In 1737 Machen Barrena proposed to the viceroy of Santa Fé to make an expedition against these Indians, so as to facilitate the commerce and security of those provinces, and although he went upon the expedition with three bodies of troops from three different places, namely, San Faustino, Salazar de las Palmas, and Merida, yet he failed on account of a fourth body not coming from Ocaña, which was kept back for want of ammunition, since the governor of Santa Marta refused to allow them any; and thus the Motilones, although surrounded by the three other bodies, effected their escape.

MOTINES, a jurisdiction and *alcaldía mayor* of the kingdom of Nueva España, in the province and bishopric of Mechoacán. It produces a great quantity of large and small cattle, *copale*, wax, *cocos*, and other seeds. It consists of five principal settlements or head settlements of districts, on the S. sea, and has upon the coast a signal-house to give intelligence of vessels, and particularly of the arrival of the bark from California. On its coast is a port before you come to that of Acapulco, called Santelmo. The aforesaid settlements of its district are,

Zixamitlan,	Guacoman,
Xolotlan,	Maquili.
Chiamila,	

The capital is the settlement of the same name, and which the Indians call Pomaro. It is of an hot temperature, inhabited by 10 families of Indians, and about 15 of Spaniards, *Mustees*, and Mulattoes. It lies 10 leagues from the coast; although its proximity to the same is conducive to the fishery of *robalos*, in which nearly all the natives are engaged, selling their stock in the neighbouring provinces and jurisdictions of the bishopric. The territory of this *alcaldía mayor* is very rough and uneven, and the climate is hot. The natives used abominable sacrifices; but these were exterminated by the labour and exertions of *Fr.* Pedro de las Garrovillas of the order of San Francisco, native of the town of this name in Estremadura, and who was the person who entered to preach the doctrine to these infidels, with such zeal and effect, that in one day he burnt upwards of 100 of their idols. This jurisdiction is about 240 miles to the *w.* one quarter to the *s.* of Mexico, and between the settlements of Zacatula and Purificacion.

MOTOBAR, a small river of the province and government of Venezuela in the Nuevo Reyno de Granada. It runs to *n. n. w.* and enters the lake of Maracaibo, between the settlement of Las Barbacoas and the city of Gibraltar.

MOTOZINTA, a settlement of the province and *alcaldia mayor* of Gueguetenango in the kingdom of Guatemala; annexed to the curacy of the settlement of Santa Ana Cuilco.

[MOTTE Isle, a small island in lake Champlain, about eight miles in length and two in breadth, distant two miles *w.* of N. Hero island. It constitutes a township of its own name in Franklin county, Vermont, and contains 47 inhabitants.]

MOTUPE, a settlement of the province and *corregimiento* of Piura in Peru. It has this name from a province in which it was in the time of the Indians, between the provinces of Plura and Truxillo, and in the valleys of which Pizarro refreshed his troops when going to the conquest of Peru. At present its territory is incorporated with the two provinces aforesaid.

MOTUPE, a settlement of the province and *corregimiento* of Piura in Peru.

[MOUCHA, LA, a bay on the coast of Chile, on the *w.* coast of S. America.]

MOUILLAGE, a settlement of the island Martinique, one of the Antilles; situate on the *w.* coast, with a good port. It is a curacy of the religious order of St. Domingo, between the bay of Touche and the river of the fort of S. Pierre.

MOUILLAGE, a small river of the island of Guadalupe, on the *n. w.* coast, between the point of Gros Morne and the river of Lancesan des Hayes.

MOULE, a large bay or port of the island Guadalupe, on the coast which looks to the *n. e.* between the rock of La Corona and the port of Las Chalupas.

MOULINET, a great fall of the river Catarakui, between lake St. François and fort Augusta in N. America.

[MOULTONBOROUGH, a post-town in Strafford county, New Hampshire; situated at the *n. w.* corner of lake Winnipiseogee, 15 miles *e.* of Plymouth, and 50 *n. w.* by *n.* of Portsmouth. This township was incorporated in 1777, and contains 565 inhabitants.]

[MOULTRIE Fort. See SULLIVAN's Island.]

[MOUNT Island, on the above coast. Lat. 50° 5′ *n.* Long. 61° 35′ *w.*]

[MOUNT BETHEL, UPPER and LOWER, two townships in the county of Northampton, Pennsylvania.]

[MOUNT DESERT, an island on the coast of Hancock county, district of Maine, about 13 miles long and 10 broad. It is a valuable tract of land, intersected in the middle by the waters flowing into the *s.* side from the sea. There are two considerable islands on the *s. e.* side of Mount Desert island, called Cranberry islands, which assist in forming a harbour in the gulf which sets up on the *s.* side of the island. In 1790, it contained 744 inhabitants. The *n.* part of the island was formed into a township, called Eden, in 1796. The *s.* easternmost part of the island lies in about lat. 44° 18′ *n.* On the mainland, opposite the *n.* part of the island, are the towns of Trenton and Sullivan. It is 178 miles *n. e.* of Boston.]

MOUNT DESERT Rock, a rock in the N. Atlantic ocean, near the coast of Maire, about 27 miles *s.* of Mount Desert island, in lat. 43° 48′ *n.* and long. 68° 3′ 30″ *w.*

[MOUNT HOLLY, a village in Burlington county, New Jersey; situated on the *n.* bank of Anocus creek, about seven or eight miles *s. e.* of Burlington.]

[MOUNT HOPE Bay, in the *n. e.* part of Naraganset bay.]

[MOUNT HOPE, a small river of Connecticut, a head branch of the Shetucket, rising in Union.]

[MOUNT JOLI, on the *n.* coast of the gulf of St. Lawrence, in Labrador.]

MOUNT JOY, a dependence of the county of Newcastle, in the province and colony of Pennsylvania, from whence the first calcareous stone was brought from America to Europe. This country is notorious for its excellent sand.

[MOUNT JOY, the name of two townships in Pennsylvania, the one in Lancaster, the other in York county.]

[MOUNT JOY, a Moravian settlement in Pennsylvania, 16 miles from Litiz.]

[MOUNT MISERY, a barren mountain of the island of St. Christopher, evidently a decayed volcano. Its perpendicular height is 3711 feet, and it has an immense crater on the top, the bottom of which is nearly level, and supposed to contain 50 acres, of which seven are covered with water; the rest are clothed with high grass and trees, among which the mountain cabbage is very conspicuous. From the crannies or fissures of this crater still flow streams of hot water, which are strongly impregnated with sulphur, alum, and vitriolic acid.]

[MOUNT PLEASANT, a township in W. Chester county, New York; situated on the *e.* side of Hudson river: bounded *s.* by Greensburg, and *n.* and *e.* by Philipsburg. It contains 1924 inhabitants, of whom 275 are qualified electors, and 84 slaves. Also the name of a township in York county, Pennsylvania.]

[MOUNT PLEASANT, a village of Maryland;

situated partly in each of the counties of Queen Ann and Caroline, about 11 miles *e.* of the town of Church-hill.]

[Mount Tom, a noted mountain on the *w.* bank of Connecticut river, near Northampton. Also the name of a mountain between Litchfield and Washington in Connecticut.]

[Mount Vernon, the seat of George Washington, late president of the United States. It is pleasantly situated on the Virginia bank of Patowmac river, in Fairfax county, Virginia, where the river is nearly two miles wide; eight miles below Alexandria; four above the beautiful seat of the late Colonel Fairfax, called Bellevoir; 52 from point Look-out, at the mouth of the river. The area of the mount is 200 feet above the surface of the river; and after furnishing a lawn of five acres in front, and about the same in rear of the buildings, falls off rather abruptly in those two quarters. On the *n.* end it subsides gradually into extensive pasture grounds; while on the *s.* it slopes more steeply, in a short distance, and terminates with the coach-house, stables, vineyard, and nurseries. On either wing is a thick grove of different flowering forest trees. Parallel with them, on the land side, are two spacious gardens, into which one is led by two serpentine gravel walks, planted with weeping willows and shady shrubs. The mansion house itself (though much embellished by, yet not perfectly satisfactory to the chaste taste of the present possessor) appears venerable and convenient. The superb banqueting room was finished just after he returned home from the army. A lofty portico 96 feet in length, supported by eight pillars, has a pleasing effect when viewed from the water; the whole assemblage of the green-house, school-house, offices, and servants' halls, when seen from the land side, bears a resemblance to a rural village; especially as the lands on that side are laid out somewhat in the form of English gardens, in meadows and grass grounds, ornamented with little copses, circular clumps, and single trees. A small park on the margin of the river, where the English fallow deer and the American wild deer are seen through the thickets, alternately with the vessels as they are sailing along, add a romantic and picturesque appearance to the whole scenery. On the opposite side of a small creek to the *n.* an extensive plain, exhibiting corn-fields and cattle grazing, affords in summer a luxuriant landscape; while the blended verdure of wood-lands and cultivated declivities, on the Maryland shore, variegates the prospect in a charming manner. Such are the philosophic shades to which the commander in chief of the American army retired in 1783, at the close of a victorious war; which he again left in 1789, to dignify with his unequalled talents the highest office in the gift of his fellow-citizens; and to which he again retreated (1797) loaded with honours, and the benedictions of his country, to spend the remainder of his days as a private citizen, in peace and tranquillity.]

[Mount Vernon, a plantation in Lincoln county, district of Maine, in the neighbourhood of Sidney and Winslow.]

[Mount Washington, in the upper part of the island of New York.]

[Mount Washington, one of the highest peaks of the White mountains, in New Hampshire.]

[Mount Washington, the *s.* westernmost township of Massachusetts, in Berkshire county, about 104 miles *w.* by *s.* of Boston. It was incorporated in 1779, and contains 67 inhabitants.]

MOUNTAINS. The number of mountains in Spanish America are infinite, which in different *cordilleras* traverse the whole country through various parts. The principal of these are,

Abides,	Guanta,
Abipi,	Guanacas,
Abitanis,	Huanacauri,
Acacuña,	Huantajaya,
Acochala,	Huatzapa,
Altar,	Lampangui,
Añapuras,	Ligua,
Andes,	Llanganate,
Antisana,	Llaon,
Antojo,	Mohanda,
Asuay,	Notuco,
Avitahua,	Omate,
Buritaca,	Opon,
Caruairasu,	Osorno,
Caxamima,	Paragoana,
Cayambe,	Peteroa,
Cequin,	Pichinche,
Chima,	Picurú,
Chimborasu,	Pintac,
Chocayas,	Porco,
Chuapa,	Potosi,
Chumbilla,	Purasé,
Collanes,	Quechucavi,
Corazon,	Quelendana,
Cotacache,	Quindio,
Cotopacsi,	Sahuancuca,
Cucunuco,	Sanguay,
Cumbal,	San Pedro,
Elenisa,	Santa Juana,
Fosca,	San Antonio,
Gachaneque,	Saporovis,
Guanas,	Sierra Nevada,

Sincholagua, Ucuntaya,
Sinu, Uritusinga,
Sunchuli, Vacarima,
Tampaya, Villagran,
Tiscan, Imbabura,
Tioloma, Itoco.
Tunguragua,

For the other mountains of America, See North America.

MOURE, a fort, of the English in the province and colony of Georgia; situate on the shore of the river Savannah, opposite the city and fort of Augusta.

MOURISCA, a settlement of the province and *captainship* of Paraiba in Brazil; situate on the shore of the river Paraiba.

MOUSA, a lake of the province and government of Mòxos in Peru, on the shore of the river San Xavier, where this unites with the Travesia.

[MOUSE Harbour, at the *e.* side of the island of St. John's, and at the *s. w.* angle of the gulf of St. Lawrence, is between E. point and Three Rivers, and goes in with a small creek that is moderately spacious within.]

[MOUSOM, a small river of York county, district of Maine, which falls into the ocean between Wells and Arundel.]

MOUSTIQUE, a small river of the island of Guadalupe, which rises in the mountains of the *e.* coast, runs *e.* and enters the sea in the bay and port of Cul de Sac Grand.

MOUTON, Le, a shoal of rocks of the N. sea, near the island S. Christoval, one of the Antilles, and off its *n. e.* coast, opposite that of Morne.

Mouton, Le, a port of Nova Scotia or Canada in N. America. It is little and only fit for small vessels, and this only in case of distress. On the *e.* coast, near the port of the Ileve, in lat. 44°. Long. 64° 30′ *w.*

[MOWEE, one of the Sandwich isles, next in size to, and *n. w.* of Owhyhee. It has a large bay of a semicircular form, opposite to which are the islands Tahoorowa and Morokinnee. It is about 162 miles in circumference, and is thought to contain nearly 70,000 inhabitants.]

MOXANDA, a very lofty mountain, always covered with snow, in the province and *corregimiento* of Otavalo and kingdom of Quito. Its summit is divided into two tops, the one of which looks to the *e.* the other to the *w.* and from each of them runs a *cordillera*. In this mountain the rivers Batan and Emacyacu have their source. In lat. 12′ *n.*

MOXI, a river of the province and *captainship* of Puerto Seguro in Brazil, which runs *n. n. w.* and enters the Supacay-guazu.

MOXICONES, a bay of the coast of the kingdom of Chile, in the district of the province and *corregimiento* of Atacama in Peru.

MOXIMO, a river of the province of Cinaloa in the kingdom of Quito. It runs to *s. s. e.* and enters the river Belleno, in lat. 1° 32′ *s.*

MOXO, a settlement of the province and *corregimiento* of Paucarcolla in Peru.

Moxo, another settlement, of the province and *corregimiento* of Chichas and Tarija in the same kingdom, of the district of the jurisdiction of the former; annexed to the curacy of Talina.

MOXOCAYA, a settlement of the province and *corregimiento* of Tomina in Peru.

MOXON, a settlement of the province and *corregimiento* of Xauja in Peru; annexed to the curacy of its capital.

MOXONAUI, a settlement of the province of Moqui, in the kingdom of Nuevo Mexico in N. America.

MOXOS, an extensive province and country of the kingdom of Peru; bounded *s.* by the province of Santa Cruz de la Sierra, *n. e.* by the river Itenes or Huapore, *s. e.* by the intervention of many woods, by the Chiquitos Indians, *s. w.* by the *cordillera*, at the back of which is the province of Cochabamba, *w.* by the missions of Apolabamba, the river Beni running between, and *n.* by the rivers Iruiame and Exaltatio. It is about 420 miles long from *e.* to *w.* and about 300 wide from *n.* to *s.*; although to travel them it would add considerably to the above distances, owing to the difficulties of the roads. It is watered by four large rivers, besides innumerable rivers of less note: the first is the Marmoré, which rises in the *sierras* of Altissimas; the second the Itenes, which is also called Huapore; the third the Beni, towards the *w.* part; and the fourth Branco, or S. Miguel.

The temperature of this province is hot and moist, owing to the number of woods and rivers, and these form innumerable lakes and swamps, especially in the rainy season, which begins in October and lasts till May, when the inundations are so great, that in many parts nothing but the tops of trees are to be seen, and all communication between one settlement and another must be made by rafts, when you may swim about for two or three days without finding a dry place to tread on. At this season the cattle become sick and languid from want of pasture, and many of them die, whilst the great moisture combined with the parching heat which through want of a generous air is experienced, excites such a degree of putrefaction in the stagnant waters, that there is never a year that passes but which generally brings with it some fatal epidemic fever or disorder, which at times destroys whole settlements at once, as was the case with those of San Luis Gonzaga, San Pablo, and San Miguel, which no longer exist; and, indeed, were it not for the natural fecundity of the women and the exertions of the missionaries in drawing together these barbarian Indians to dwell in societies, there would scarcely be any population whatever.

The territory is as unkind in the production of bread and wine, as it is favourable to those plants which require great heat and moisture, such as maize, sugar canes, *yucas* or *mandioca*, rice, *camotes*, plantains, green *ajies*, *mani*, &c. In some of the settlements they gather very good crops of *cacao*, and every where of cotton; the grain of the former is so large, tender and rich, that the chocolate made of it is of the most delicate flavour and strong nourishment; but it has the defect of becoming rancid if kept long, which is the case with every kind that is very oily. In the woods are found many trees, the wood and fruits of which are much esteemed, such as *guayucanes*, cinnamon, *marias*, from whence is extracted the oil of this name, the *quinaquina*, the seed of which is very fragrant when burnt, cedars, palms, *tajibos*, almonds different from those of Europe, *copaibos*, *bainillas*, dragon plants, and others. In the trunks of the trees various kinds of bees lay their wax: some of these insects are white, others yellow, and others, which live under ground and are less esteemed, of a grey colour. Here are many wild animals, tigers, *antas*, deer, rabbits, wild boars, and ant-eaters, thus called because they have a very long snout with which they devour thousands of the above little insects. These animals have no other defence than their claws, which are like daggers, and when they fight with the tiger, the conflict generally proves mortal to both; for the bear grapples with the tiger, and darts its claws into his heart and bowels, whilst the latter tears to pieces with his jaws the head and face of its adversary. Here are also very large snakes, called *bobas*, rattlesnakes, vipers, small and large spiders, scorpions, mosquitoes, large and fierce ants, *gegenes*, bats of an extraordinary size, and various other venomous insects, many rare birds of fine song and beautiful plumage, and others well known in Europe. In the rivers and lakes are abundance of fish, alligators, thornbacks, *palometas*, and *toñinas* or dolphins, the which the Indians kill with arrows.

This province is divided into three districts, which are, Moxos, Baures, and Pampas. The former consists of six settlements on either side of the river Marmore, with the names of

Loreto,	S. Pedro,
Trinidad,	La Exâltacion,
S. Xavier,	Santa Ana.

The second, of six other settlements on the *w.* shore of the same river, called

Magdalena,	S. Martin,
Concepcion,	S. Simon,
S. Joaquin,	S. Nicolas.

And the third of three, which are,

S. Ignacio,	Los Santos Reyes.
S. Francisco de Borja,	

In all of which there are 22,000 Indians of the following nations,

Moxos,	Sapis,
Tapacuras,	Cayubabas,
Bolepas,	Canacures,
Coriciaras,	Ocoronos,
Baures,	Chumanos,
Itonamas,	Mayacamas,
Heriboconos,	Tibois,
Meques,	Nairas,
Boyomas,	Norris,
Huarayos,	Pacarabas,
Rotoroños,	Pacanabos,
Mures,	Sinabus,
Erirumas,	Cuizaras,
Canicianas,	Cabinas.
Pechucos,	

These Indians rather resembled wild beasts than human creatures, lived without any appearance of religion or worship, and adored nothing but the devil and tigers. Some of them called themselves priests and sorcerers, also physicians, without more knowledge of disorders or remedies than to suck the sore part. They made others believe a thousand stories of visions that they had had with the devil, in order to induce them to multiply their offerings: their altars were nothing but some miserable huts, adorned with *tutumas*, spears, feathers, bows, arrows, and darts, and the chief act of adoration consisted in making themselves drunk with *chicha*, a drink of maize and *yuca*, for many days together, when some fatal and melancholy results would close the solemnities. Whenever they were angry they took up their arms and inflicted instant death; and as they had no ideas of civil life, there was nothing thought of amongst them concerning a common good; but each man was master of his own family, and here he lorded it as his whim directed.

Thus in their political affairs they had no head whatever, and though in time of war they would go into the field under a commander, who was selected as being the most savage and furious amongst them, yet, as soon as the fight was begun, would they every one of them, individually, take the command upon themselves. Their advance was as rapid as their retreat, and if by the violence of the former they gained the victory, they made so cruel and barbarous a use of it, that they not only eat their prisoners, but took pains to put them to the severest tortures in killing them.

Even the matrimonial tie was broke upon the slightest pretence; polygamy was carried to a great extent, but no part of their conduct was more disorderly than that of the education of the children. The father, in respect to these, was merely a slave, and when the infirmities of age required the attentions of the son, and had a claim upon his veneration and respect, the latter would with the greatest insolence put his hoary sire to death, alleging that he would now be no longer useful: in the same way would he kill his younger brothers and sisters, saying, that he thus liberated them from the disgrace and the misery of orphans.

In short, it is scarcely possible to imagine any abomination of drunkenness, lasciviousness, superstition, and cruelty of the most barbarous nature, which was not practised by these savages previous to the time that the light of the gospel began to shed its influence amongst them. There were certain distinctions of manners in the aforesaid nations, which will be found explained under their proper articles.

About the middle of the 16th century the missionaries of the Jesuits began to attempt by bribes, persuasions, and promises, to reduce to the Catholic faith this savage multitude, and after great labours and fatigues, and not without the loss of several lives, did they at last succeed in domesticating them and diffusing amongst them a rational and christian-like spirit, forming large and regular settlements, and selecting for these purposes those parts of the country which were least exposed to inundations. They also built magnificent temples, which were richly ornamented, and where on festival days they would cause to be performed fine concerts of music, vocal and instrumental, with organs, harps, violins, flutes, trumpets, &c.

The spiritual government is the same as that of the Chiquitos Indians. They hear mass every day early in the morning, and are afterwards instructed in their religion: again they all meet at night-fall to say the rosary and hear sermons. The political government consists of a governor, nominated by the curate on the first day of the year, with two *alcaldes*, *aguazils*, and *capitulars*, whose care it is to guard against public disorders. They visit the *chacras* or huts of the Indians, to see that they are industrious and take care of their families; and these are supplied with abundance of flesh meat from the herds of neat cattle which are kept in the neighbourhood. The delinquent, after he has been made to know the measure of his crime, is punished by a flogging, or other way that may be thought necessary, and the obstinate or incorrigible Indian is banished from the settlement. Every Indian, after his marriage, is obliged to form a *chacra*, or small estate, where, amongst other things, he must cultivate cotton sufficient for the clothing of his family. The curate's *chacra* is large, and is kept in order by the community. Its productions are devoted to his use, and to that of the mechanics and other servants living immediately under him. Here they manufacture sugar, refine the wax collected in the woods, make chocolate, very fine cotton stuffs for table-cloths, handkerchiefs, towels, and napkins; and many of these articles are carried for sale, by order of the curate, to the neighbouring provinces, and with their product other necessaries are purchased, especially salt, of which there is none here.

In this province the Indians go better clothed than in any other: many are seen with waistcoats and breeches of leather and even of silk, especially such as are of higher rank, being masters of liberal or mechanic arts, which are very celebrated. In some of the settlements there are not only musicians but compositors; and some are so dexterous that they imitate whatever they see, though they are rare who know how to write; notwithstanding a breviary has been seen which was done by them so nicely that it is impossible to discover it from print.

The common arms of these Indians are the bow and arrow, and at the entrance of the Spaniards in 1762 and 1766, to dislodge the Portuguese from the station they had taken upon the other side of the river Itenes, near the settlement of Santa Rosa, they accustomed themselves to the use of fire-arms, and were extremely useful to the Spaniards against the Portuguese, of whom numbers were taken prisouers, and doomed to work in the mines of Cuyaba and Matogroso, they being now known by the name of Certanistas. This province was conquered and united to the empire of Peru by the Inca Yupanqui, eleventh emperor.

Moxos, a settlement of the province and *corregimiento* of Chichas and Tarija in Peru; situate on the lofty part of a mountain much exposed to the winds. It has at its entrance a river which passes through the settlement by an aqueduct erected at great cost. Twenty-eight leagues from the city of Santiago de Cotagaita.

Moxos, another, with the dedicatory title of S. Juan de Sahagun, in the missions that were held by the religious order of San Francisco in the province of Apolabamba.

MOXO-TORO, a settlement of the province and *corregimiento* of Yamparaes, and archbishopric of Charcas, in Peru.

MOYA, a settlement of the province and *corregimiento* of Angaraes in Peru; annexed to the curacy of Conaica.

Moya, another, a small settlement or ward of the head settlement of the district and *alcaldía mayor* of Lagos in the kingdom and bishopric of Nueva Galicia; situate to the *e.* of its capital.

Moya, a small river of the province and government of Jaen de Bracamoros in the kingdom of Quito, which enters the Marañon.

MOYAGUA, a settlement of the head settlement of the district and *alcaldía mayor* of Juchipila in Nueva España. Six leagues to the *s.* of the said head settlement.

MOYALEC, Leuvu, or Colorado, also called Desaguadero de Mendoza, a river of the province and government of Tucumán in Peru. It rises in the territory of the Aucaes Indians, runs in a large stream to *s. s. e.* for many leagues, and then turns its course to *s.*

[MOYAMENSING, a township in Philadelphia county, Pennsylvania.]

MOYEN, a large sand-bank on the coast of the island of Newfoundland, one of those which serve for the cod-fishery. It is to the *w.* of Green bank.

MOYOBAMBA, or Santiago de los Valles, a city, the capital of the district of this name in the province and *corregimiento* of Chachapoyas and kingdom of Peru. It is of an hot temperature, moist, and unhealthy, but abounding in vegetable productions, in cattle, cotton, sugar, tobacco, of which alone there were gathered 200 load annually, before it was monopolized by the crown, and of such excellent quality is it as to be preferred to all of the other provinces; it likewise produces many kinds of fruit. It has, besides the parish church, a chapel of Nuestra Señora de Belen. [It is 192 miles *e.* by *n.* of Truxillo, on the shore of the river of its name, and 310 miles *n. n. e.* of Lima. In lat. 7° *s.* and long. 75° 51′ *w.*]

Moyobamba. The aforesaid river, in the same province and *corregimiento*, rises *s.* of the capital, close to the settlement of Naranjos, runs *e.* and enters with a large stream into the Guallaga.

Moyobamba, a valley of the same province, of a triangular figure, shut in by the *cordillera* of the Andes and the rivers Moyobamba and Negro.

MOYOC-MARCA, a name given by the Indians of Peru, in the time of their gentilism, to one of the great towers of the fortress of Cuzco.

MOYOTEPEC, a settlement of the head settlement of the district of San Luis de la Costa; containing 16 families of Indians, and a little more than a league's distance from Quauzoquitengo.

MOYUTA, San Juan Baptista de, a settlement of the *alcaldía mayor* of Jutiapa, and kingdom of Guatemala; annexed to the curacy of Conguaco.

MUBERRY, a small river of the province and colony of S. Carolina, which runs *e.* and enters that of Thirty Miles.

MUCABUSA, a settlement of the province and country of Las Amazonas, in the part possessed by the Portuguese. It is situate on the shore of the river Madera, opposite the river Uvirabasú.

MUCARAS, some isles or rocky shoals, lying between the Lucayas islands, and *n.* of that of Cuba. They are many, and are between cape Lobos and the island San Andres.

MUCARI, a bay on the *s.* coast of the island Jamaica.

MUCCIA. See Moche.

MUCHIMILCO, a settlement of the province of Huejotzinco in Nueva España, in the time of the Indian gentilism; situate near the Sierra Nevada.

MUCHIPAI, a small and poor settlement of the jurisdiction of the city of La Palma, and *corregimiento* of Tunja, in the Nuevo Reyno de Granada. It produces some vegetable productions, such as maize, cotton, *yucas*, and plantains, all of a warm climate; this being its temperature.

MUCHUCHIS, or Mucuchíes, a settlement of the government and jurisdiction of Merida in the Nuevo Reyno de Granada, near the source of the river Cama. It is of a fine temperature, rather warm than cold, of a very fertile soil, and abounding in excellent *cacao*, wheat, maize, and other vegetable productions; contains 50 housekeepers and 200 Indians. In lat. 8° *n.*

MUCHUMI, a settlement of the province and *corregimiento* of Saña in Peru.

MUCUNO, a settlement of the government and jurisdiction of Merida in the Nuevo Reyno de Granada; situate in the valley of Azequias: of a cold temperature, producing much wheat, maize, *turmas*, beans, lentils, &c. It has also abundance of cattle, and contains 40 housekeepers and 100 Indians.

MUCURES, a settlement of the province of Barcelona and government of Cumaná, one of those which are under the charge of the religious observers of San Francisco, the missionaries of Piritu. It is situate on the shore of the river Pao, in the bend it makes before its entrance into the Orinoco.

MUCURUBA, a settlement of the government of Merida in the Nuevo Reyno de Granada. It is of a mild but healthy temperature, producing much maize and other vegetable productions peculiar to its climate. It contains 50 Indians, and as many other inhabitants; and is annexed to the curacy of the settlement of Muchuchís. The regulars of the company had in the district of this settlement some rich cattle farms.

MUCURURI, a settlement of the province and government of Guayana or Nueva Andalucia; situate on the *s.* shore of the river Caroni, near its mouth or entrance into the Orinoco.

[MUD Island, in Delaware river, is six or seven miles below the city of Philadelphia; whereon is a citadel, and a fort not yet completed. On a sandbar, a large pier has been erected, as the foundation for a battery, to make a cross fire.]

[MUD Lake, in the state of New York, is small, and lies between Seneca and Crooked lakes. It gives rise to a *n.* branch of Tioga river.]

MUDURA, a small river of the province and government of Guayana, one of those which enter by the *s.* side into the Usupania.

MUELLAMUES, a settlement of the province and government of Popayan in the Nuevo Reyno de Granada.

MUERTES, RIO DE LAS, a river in the province and *captainship* of the Rio Janeiro in Brazil. It rises to the *w.* of the town of Jubaraba, runs *s. s. w.* and enters the Paraná.

MUERTOS, CAYOS DE LOS, a small island of the N. sea, close to the *s.* coast of the island Puerto Rico.

MUGERES, a small island of the N. sea, near the coast of Yucatán; situate about 18 miles *s. e.* of cape Cotoche; discovered by Francisco Hernandez Giron in 1517, who gave it this name, from having found in it several Indian idols well clothed, and which appeared to resemble *mugeres*, or women. This island has always been the common refuge of the Zambos and Mosquitos pirates for careening their vessels. It is in lat. 21° 18′ *n.* and long. 86° 40′ *w.*

MUISNE, a river of the province and government of Esmeraldas in the kingdom of Quito. It runs *n.* and, just before it enters the sea, turns its course *s.* On its shores are a great number of very lofty palms; and its entrance into the sea is between the river San Francisco to the *n.* and the Potete to the *s.* In lat. 37° 30′ *n.*

MUITACON, a settlement of the province and government of Guayana or Nueva Andalucia; situate on the shore of the river Orinoco, and to the *n.* of the city of Real Corona.

MUITO, a small river of the province and government of Paraguay, which enters the Piratini.

MUJA, a settlement of the province and government of Antioquia in the Nuevo Reyno de Granada.

MUJU, a river of the province and government of Pará in Brazil. It runs *n.* and enters the Marañon by the *s.* side in the bay of Pará, near the fort Capi. In lat. 1° 33′ *s.*

MULAHALO, a settlement of the province and *corregimiento* of Latacunga in the kingdom of Quito, in the district of which is the mountain or volcano of Cotopaxi, notorious for the mischief it has done in that province. From it rises the river San Felipe, which traverses the province, as also another river called Guapante, which with the Ambato forms the large stream of the Patate. In the vicinity of this settlement many veins of silver ore have been discovered, though none have been worked. On the *w.* at no great distance, is a very large estate, called El Callo.

MULAS, a point on the *n.* coast of the island of Cuba, between port Sama and the river of Los Plátanos.

MULATAS, some islands of the N. sea, and of the province and government of Darien in the kingdom of Tierra Firme; situate close to the point of San Blas and to the *e.* They are many, small, and one of them larger than the rest, are very dangerous in the sailing from Portovelo to Cartagena, and on them several vessels have been wrecked.

[MULATRE Point, in the island of Dominica in the W. Indies. Lat. 15° 16′ *n.* Long. 61° 21′ *w.*]

MULATTO, a cast of people of America, produced by a black mother and white father, or by a black father and white mother, but the latter very rarely, although the former very commonly, so that America abounds with Mulattoes: they are thus the offspring of a libidinous intercourse between Europeans and the female slaves, which the authority of the one and the sensuality of the other tend to make very general. The colour of the children thus produced participate of both white and black, or are rather of a dingy brown colour. Their hair is less crisp than that of the Negro and of a clear chesnut tint. The Mulatto is regularly well made, of fine stature, vigorous, strong, industrious, intrepid, ferocious, but given to pleasure, deceitful, and capable of committing the greatest crimes without compunction.

It is a certain fact, that throughout the vast dominions of the king of Spain in America there are no better soldiers than the Mulattoes, nor more infamous men. When the mother is a slave the offspring is also, by the principle of the law that *partus sequitur ventrem;* but inasmuch as that they are in general the offspring of the master of the mother they are made free, and from their earliest infancy are brought up in all kinds of vice. As the Mulatto, as well as the Negro, is at the time of its birth nearly white, not taking its real colour till nearly 10 days after; the difference is distinguished by the private parts, for these in the Negro child, together with the extremities of its toes and fingers, are already of a dark colour, which is not the case with the Mulatto. The French, in order to keep down the numbers of this cast in their colonies, established a law that the father of a Mulatto should pay a fine of 2000 lbs. of sugar, and further, that if he were master of the slave, that he should forfeit her as well as the child, the money arising from the fine to be paid into the funds of the hospital of La Charité.

There have been many Europeans, Spaniards, French, English, and other nations of America, who have married Negro women; and the sons of these alone are admitted by law to the offices of the state, and although there is a general prohibition against all Mulattoes whatever, yet has this been in several cases dispensed with. Notwithstanding the bad qualities of the Mulatto, some of them have been found, who from their extraordinary virtues and qualifications have deserved great marks of approbation and distinction from the viceroys, bishops, and other persons of eminence. Such were Miguel Angel de Goenaga, captain of militia in the city of Portovelo, whose merits had gained him a universal title to respect at home and in the English, French, and Dutch colonies; also in Puerto Rico another person, named Miguel Enrriques, who, although in the humble employment of a shoemaker, had done such services to the king, that he was honoured with a royal medal, and allowed to put to his name the title of Don. These examples we conceive to be sufficient to shew how little influence the colour of a man has over the endowments of his soul.

[MULATTO Point, on the *w.* coast of S. America, is the *s.* cape of the port of Ancon, 16 or 18 miles *n.* of Cadavayllo river.]

MULDEN, a city of the province and colony of New England in N. America.

MULEGE, a river of the province of California in N. America. It rises in the centre of the province, and enters the sea in the bay of Concepcion.

MULEQUES, ISLAS DE LOS, three small islands, situate in the river La Plata, near the *n.* coast, close to the islands of Los Ingleses and those of Anton Lopez.

[MULGRAVE Port. See ADMIRALTY Bay. Lat. 67° 45′ *n.* Long. 165° 9′ *w.*]

[MULHEGAN River, in Vermont, rises in Lewis, and empties into Connecticut river at Brunswick.]

[MULLICUS River, in New Jersey, is small, and has many mills and iron-works upon it, and empties into Little Egg harbour bay, four miles *e.* of the town of Leeds. It is navigable 20 miles for vessels of 60 tons.]

MULLONES, an ancient settlement of the nation of Indians of this name now extinguished, or at least of whom nothing remains but this settlement, in the province of Pasto, and kingdom of Quito. It is close to the mountain of Cumbal, which it has to the *s.* and to the *n.* the settlement of Mullama. Its territory is laved by the river Telembi by the *w.* and it is in lat. 57° 22′ *n.*

MULMUL, a *paramo* or mountain covered with snow, of the province and *corregimiento* of Riobamba in the kingdom of Quito, and one of those which were used by the academicians of the sciences at Paris, to fix their instruments for their mathematical observations. On its skirt are some cow-herds huts or Indian cottages, where they watch the cattle which graze in those parts.

MULOT, a shoal of rock always covered by the water near the coast of Nova Scotia or Acadia, three quarters of a league to the *s.* of the point of Fourché.

MUMU, a name which the Indians of the province of Veragua in the kingdom of Tierra Firme gave to the village or small settlement.

MUNAMESA, a small river of the province and government of Mainas in the kingdom of Quito, rising between the Chambire and the Tigre. It runs *s. s. e.* and enters the Marañon.

MUNANI, a settlement of the province and

corregimiento of Asangaro in Peru; annexed to the curacy of its capital.

[MUNCY, a creek which empties into the Susquehannah from the *n. e.* about 20 miles *n.* of the town of Northumberland.]

MUNICHES, a settlement of the missions which were held by the Jesuits in the province and government of Maiuas, of the kingdom of Quito; annexed to the curacy and settlement of Nuestra Señora de Loreto de Paranapuras; situate on the shore of the river of this name.

MUNIGITURA, a settlement of the province and *captainship* of Pará in Brazil; situate on the sea-coast to the *e.* of the city of Caete.

[MUNSIES, DELAWARES, and SAPOONES, three Indian tribes, who inhabit at Diagho and other villages up the *n.* branch of Susquehannah river. About 20 years ago the two first could furnish 150 warriors each, and the Sapoones 30 warriors.]

MUQUIYAUIO, a settlement of the province and *corregimiento* of Jauxa in Peru; annexed to the curacy of Huaripampa.

MURA, CANO DE, an arm of the river Barima, which communicates with the Guarini, in the province and government of Cumaná. It runs *e.*

MURA, RANDAL DE, a very dangerous whirlpool of the river Caura.

MURAPARAXIA, an island of the river Madera in the province and country of Las Amazonas, very near its shore of the *w.* side.

MURATAS, a barbarous nation of Indians who dwell in the woods of the river Pastaza to the *s. w.* and lying to the *n. n. e.* of the river Morona, near the source of the Guassaga. The abolished order of the Jesuits, the missionaries of Mainas, discovered these Indians in 1757, and formed of them a settlement of 250 persons, to which they gave the name of Nuestra Señora de los Dolores de Muratas. These Indians are of a docile and quiet disposition, notwithstanding that they are at continual war with the barbarous and ferocious nation of the Xibaros Indians, their neighbours.

MURCO, a settlement of the province and *corregimiento* of Collahuas in Peru; annexed to the curacy of Llauta.

[MURDERERS' Creek, in the state of New York.]

MURES, a barbarous nation of Indians of Peru, bounded by that of Los Moxos. They are ferocious and treacherous, and it is said that some of them have been reduced to the Catholic faith by the Jesuits.

[MURFRESBOROUGH, a post-town of N. Carolina, and capital of Gates county. It is situated on Meherrin river, near the Virginia line, contains a few houses, a court-house, gaol, and tobacco warehouse. It carries on a small trade with Edenton, and the other sea-port towns. It is three miles from Princeton, seven from Winton, 29 *n.* by *w.* of Edenton, and 234 *s. s. w.* of Philadelphia.]

[MURGA-MORGA River, on the coast of Chile, in S. America, is *s.* of the *s.* point of Quintero bay, and not far from the entrance into Chile river. It is not navigable, but is very good to water in.]

MURIBIRA, a settlement of the province and *captainship* of Pará in Brazil; situate on the shore of the arm of the river of Las Amazonas, which forms the island of Marajo and the bay Del Sol.

MURICHAL, a river of the province and government of Guayana. It rises in the table-land of Guanipa on the *e.* runs *n.* and enters the Guarapiche.

MURITATI, a settlement of the province of Tepeguana and kingdom of Nueva Vizcaya in N. America.

MURRI, SAN JOSEPH DE, a settlement of the province and government of Darien in the kingdom of Tierra Firme; situate on the shore of the river of its name.

MURRI. This river rises in the mountains of Chocó, runs *w.* and enters the Atrato.

MURUACI, a small river of the province and colony of Surinam, or part of Guayana possessed by the Dutch. It joins various others and enters the Cuyuri by the *s.* side.

MURUCURI, a settlement of the province of Guayana and government of Cumana, one of the missions held there by the Catalanian Capuchin fathers. It is situated on the shore of the river Caroni, near the mouth where this runs into the Orinoco.

MURUCUTACHI, a dry part of the *serranía* in the province and government of Sonora in N. America.

MURUMURU, an ancient province of Peru, in the time of the Indians, in the district of Collasuyu; conquered and united to the empire by the Inca Capac Yupanqui.

MUSCADOBOIT, a bay on the *s.* coast of Nova Scotia or Acadia, between that of Chebouclo and cape Charles.

MUSCLE, a small island; situate near the coast of the province of Sagadahock, between the river George and the bay of Pénobscot.

[MUSCLE Bank, at the entrance into Trinity,

bay or harbour, in the direction of *s. w.* on the *e.* coast of Newfoundland island.]

[MUSCLE Bay, in the straits of Magellan, in S. America, is half way between Elizabeth's bay and York road; in which there is good anchorage with a *w.* wind.]

[MUSCLE Bay, or MESSILONES, on the coast of Chile or Peru, in S. America, five leagues *s.* by *w.* of Atacama.]

[MUSCLE Shoals, in Tennessee river, about 250 miles from its mouth, by the course of the river, but only 145 in a direct line, about 20 miles in length, and derive their name from the number of shell-fish found there. At this place the river spreads to the breadth of three miles, and forms a number of islands; and the passage is difficult, except when there is a swell in the river. From this place up to the whirl, or suck, where the river breaks through the Great ridge, or Cumberland mountain, is 250 miles, including the turnings, the navigation all the way excellent.]

[MUSCONECUNK, a small river of New Jersey, which empties into the Delaware six miles below Easton.]

MUSCONGUS, a small river of the same province as that of the former island. It runs *s.* between rivers George and Sheepscut, and enters the sea.

MUSINAM, a settlement of the province and *corregimiento* of Copiapo in the kingdom of Chile.

[MUSKINGUM, that is, Elk's Eye, a navigable river of the N. W. Territory. It is 250 yards wide at its confluence with the Ohio, 172 miles below Pittsburgh, including the windings of the Ohio, though in a direct line it is but 103 miles. At its mouth stands fort Harmar and Marietta. Its banks are so high as to prevent its overflowing, and it is navigable by large batteaux and barges to the Three Legs, 120 miles from its mouth, and by small boats to the lake at its head, 45 miles farther, including windings. From thence, by a portage of about one mile, a communication is opened to lake Erie, through Cayahoga, a stream of great utility, navigable the whole length, without any obstruction from falls. From lake Erie the avenue is well known to Hudson's river in the state of New York. The land on this river and its branches is of a superior quality, and the country abounds in springs and conveniences fitted to settlements remote from sea navigation, viz. salt-springs, coal, free-stone, and clay. A valuable salt-spring has been very lately discovered, eight miles from this river, and 50 from Marietta, called the Big Spring. Such a quantity of water flows as to keep 1000 gallons constantly boiling. Ten gallons of this water will, as experiment has proved, afford a quart of salt of superior quality to any made on the sea-coast.]

MUSKINGUN, a town of the Owendoos Indians in N. America, where the English have a fort and establishment at Virginia, near the river of this name.

[MUSKOGULGE, MUSKOGEE, or, as they are more commonly called, CREEK Indians, inhabit the middle parts of Georgia. The Creek or Muskogulge language, which is soft and musical, is spoken throughout the confederacy, (although consisting of many nations, who have a speech peculiar to themselves) as also by their friends and allies the Natchez. The Chickasaw and Chactaw language the Muskogulges say is a dialect of theirs. The Muskogulges eminently deserve the encomium of all nations for their wisdom and virtue, in expelling the greatest, and even the common enemy of mankind, viz. spirituous liquors. The first and most cogent article in all their treaties with the white people is, that " there shall not be any kind of spirituous liquors sold or brought into their towns." Instances have frequently occurred, on the discovery of attempts to run kegs of spirits into their country, of the Indians striking them with their tomahawks, and giving the liquor to the thirsty sand, not tasting a drop of it themselves. It is difficult to account for their excellent policy in civil government; it cannot derive its efficacy from coercive laws, for they have no such artificial system. Some of their most favourite songs and dances they have from their enemies, the Chactaws; for it seems that nation is very eminent for poetry and music.

The Muskogulges allow of polygamy in the utmost latitude; every man takes as many wives as he pleases, but the first is queen, and the others her handmaids and associates. The Creek or Muskogulge confederacy have 55 towns, besides many villages. The powerful empire of the Muskogulges established itself upon the ruin of that of the ancient Natchez. The Oakmulge fields was the first settlement they sat down upon after their emigration from the *w.* beyond the Mississippi, their original native country. They gradually subdued their surrounding enemies, strengthening themselves by taking into confederacy the vanquished tribes. Their whole number, some years since, was 17,280, of which 5860 were fighting men. They consist of the Appalachies, Alibamas, Abecas, Cawittaws, Coosas, Conshacks, Coosactees, Chacsihoomas, Natchez, Ocones,

Oakmulges, Okohoys, Pakanas, Taensas, Talepoosas, Weetumkas, and some others. Their union has rendered them victorious over the Chactaws, and formidable to all the nations around them. They are a well-made, expert, hardy, sagacious, politic people, extremely jealous of their rights, and averse to parting with their lands. They have abundance of tame cattle and swine, turkeys, ducks, and other poultry; they cultivate tobacco, rice, Indian corn, potatoes, beans, peas, cabbage, melons, and have plenty of peaches, plums, grapes, strawberries, and other fruits.

They are faithful friends, but inveterate enemies; hospitable to strangers, and honest and fair in their dealings. No nation has a more contemptible opinion of the white mens faith in general than these people, yet they place great confidence in the United States, and wish to agree with them upon a permanent boundary, over which the *s.* states shall not trespass.

The country which they claim is bounded *n.* by about the 34th degree of latitude; and extends from the Tombeckbee or Mobile river to the Atlantic ocean, though they have ceded a part of this tract on the sea-coast, by different treaties, to the state of Georgia. Their principal towns lie about lat. 32° and long. 86° 20′. They are settled in a hilly but not mountainous country. The soil is fruitful in a high degree, and well watered, abounding in creeks and rivulets, from whence they are called the Creek Indians.]

[MUSQUAKIES Indians inhabit the *s.* waters of lake Michigan, having 200 warriors.]

[MUSQUATONS, an Indian tribe inhabiting near lake Michigan.]

[MUSQUITO Cove, in N. America, lies in lat. 65° 2′. Long. 53° 3′ 45″ *w.*]

[MUSQUITO River and Bay lie at a small distance *n.* of cape Canaverel, on the coast of E. Florida. The banks of Musquito river towards the continent abound in trees and plants common to Florida, with pleasant orange groves; whilst the narrow strips of land towards the sea are mostly sand hills.]

[MUSQUITONS, an Indian nation in the neighbourhood of the Piankeshaws and Outtagomies; which see.]

MUTANAMBO, a settlement of the province and government of Cumana; situate near the settlement of San Joseph de Leonisa to the *e.* the river Curuma running between.

MUTARNATI, a river of the province and government of Darien, and kingdom of Tierra Firme. It rises in the mountains of the interior of this province, runs *w.* and enters the grand river Chucunaqui.

MUTCA, a settlement of the province and *corregimiento* of Aimaraez in Peru; annexed to the curacy of Chuquinga.

MUTON, a port of the *s.* coast of Nova Scotia or Acadia in N. America, between the port of Rosignol and the bay of Santa Catalina.

MUTQUIN, a settlement of the province and government of Tucumán in Peru; of the district of its capital, to the *n. n. e.* of the city of S. Francisco de Catamarca.

MUTUANIS, a barbarous nation of Indians of the province and country of Las Amazonas, bounded by that of Los Moxos. We have little sound intelligence concerning them, but there are plenty of fabulous accounts, stating that they are giants, and are possessed of extremely rich gold mines, which lie two months journey from the mouth of the river Omopalcas.

MUTUPI, a large valley of the kingdom of Peru; between Pascamayu and Tumbez. Its natives were conquered and reduced to the empire by the Inca Huaina Capac.

MUXIA, a river of the province and government of Antioquia in the Nuevo Reyno de Granada. It enters the Cauca just before the city of Caramanta on the opposite shore.

MUYSCAS. See MOSCAS.

MUYUMUYU, an ancient province in the time of the Indians, and of little extent, in the kingdom of Peru; comprehended in the present day in the province of Charcas to the *s.* of Cuzco: conquered by the Inca Roca, sixth emperor of the Incas.

MUYUPAMPA, an ancient province of the Indians, comprehended at the present day under the name of Moyobamba, in the province of Chachapoyas; conquered and united to the empire by the Inca Tupac Yupanqui.

MUZA, a settlement of the province and *corregimiento* of Abancay in Peru; annexed to the curacy of Paccho.

MUZOS, a barbarous nation of Indians of the Nuevo Reyno de Granada, who gave name to a province much celebrated for its rich emerald mines, which have produced and still produce the finest of these stones in the world. It is 24 leagues *n. w.* of Santa Fé, and is 25 leagues long, and 11 wide; is entirely of a mountainous country, and hot and moist: very barren in the productions, animal and vegetable, of a cold climate, but abounding in all those peculiar to its own. From all its *sierras* may be seen the *n.* and *s.* polar stars, and at the end of August and in the middle of March the sun throws no shade throughout the whole day in any part. Its inhabitants are very numerous, extremely barbarous, and of peculiar customs. They say that at the beginning of the world, there was on the other side of the river Magdalena the shadow of a man, whom they called in their language *Are*, who was always recumbent, and who cut out in wood the images of some men and women, which being thrown into the river became animated bodies; that they married one with another, and that he taught them how to cultivate the land; after which he disappeared, leaving them as the first peoplers of all the Indies.

They had no gods, neither did they adore the sun and moon like other nations, affirming that these bodies were created since themselves; but they nevertheless called the sun father, and the moon mother. When the husband died a natural death, the brother became heir, taking the wife of the defunct, save when she might be the cause of the death. One of their most singular customs was the following, relating to their marriages. When the girl had reached her 16th year, an agreement of marriage was concerted between the parents without consulting her in any degree, and all being settled between them, the bridegroom paid a visit to the bride, where he made his court assiduously for three days, offering presents and ornaments, for which she would as cordially return cudgelling and blows; but this amusement being over, she would become more pacific, and set about dressing the dinner, to which were invited the friends and relations who lived nearest. To this it is added, that for a whole moon the new-wedded pair would sleep together without consummating the marriage rites, the bride thinking that in that case she would be looked upon as a bad woman. The husband, in the mean time, would devote himself to the manual labour of agriculture, assisted with his new mother-in-law, for the benefit of his bride, and he would offer her fresh presents of petticoats embroidered with a kind of beads called by them *suches*, and which, when the person walked, made a jingling noise.

If the woman committed adultery, the husband in his wrath would destroy himself, or else would be satisfied with breaking all the pots and pans of earthen ware and of wood, and would retire to the mountain, where he remained for the space of about a month, till the wife might have new furnished the house, and when she would go forth to look after him; and when she found him, she would drag him by the hair of his head, and would give him a good kicking, after which ceremonies they returned home mutually satisfied and content. When the husband died, the parents would put the wife upon her knees, where she was obliged to cry for three days successively without eating or drinking any thing more than a little *chicha*; when this was accomplished, they took the body, burnt it over a fire, and then laid it on a scaffold, which served as a tomb, and around it hung the bows and arrows and other weapons and ornaments of the deceased, and, after a year was passed, buried it. But it was not then followed by the bride, who all this while had fled, no one speaking to her, nor giving her ought to eat; so that she would starve, did she not contrive to cultivate the land for her support; but when the body was interred, her parents would seek her out, bring her home, and prepare for her a second nuptials.

The Indians of this province were subject to the Nauras and to the Moscas, but such was their valour that they drove each of these nations from their territory. The first Spaniard who found his way hither was Captain Luis Lanchero in 1539, (and not Bernardo de Fuentes, in 1547, as the ex-jesuit Coleti asserts; our information being taken from the most illustrious Piedrahita); but such was the resistance that Lanchero met with, that his men were routed with great slaughter, himself being severely wounded. A better fortune did not await Melchor Valdes, who by the order of Gonzalo Ximinez de Quesada undertook the reduction of these Indians in 1544, he being obliged to retreat in a similar way to his predecessor. In 1551, Pedro de Ursua entered with better fortune, and founded the city of Tudela, in memory of his country, but it was abanonded shortly alter by its inhabitants, who were shocked at the barbarities of the Muzos; and thus the final conquest of this people was left to the aforesaid Captain Luis Lanchero, who manifested feats of valour on the occasion in 1559. This country abounds in rice, maize, cotton, tobacco, and some *cacao*, and it is provided with flesh-meat from the immediate province of Ubate. It is watered by the abundant river Zarbe, besides others of less note.

The capital is the city of the same name, with the dedicatory title of Santisima Trinidad, belonging to the *corregimiento* of Tunja, and founded by Captain Luis Lanchero. It was the seat of the government, which was afterwards removed to Tunja; is of a mild tempcrature, contains a tolerable church and three convents of the religious orders of San Francisco, St. Domingo, and San Agustin, which, with the rest of the population, are very poor. This is composed of 200 families, and all of them being devoted to the working of the mines of its emeralds, so highly esteemed in Europe, and which have rendered this city notorious since their first discovery by Captain Juan de Penagos, they neglected its agriculture, to which the extreme fertility of the soil offers every advantage, until at last that finding themselves checked in their darling pursuit by some fallacious appearances of certain mines, they had recourse to the cultivation of the land for their sustenance. The soil produces rice, *cacao*, sugar-cane, maize, *yucas*, plantains, and many vegetable productions, and excellent fruits; and in its woods are found ebony, walnut, and cedar trees, and sweet-scented gums, although in cattle it is scarce. In 1764 the viceroy of Peru, Don Manuel Amat sent to Don Joseph Antonio de Villegas y Avendaño to reconnoitre these emerald mines, and having re-discovered the lost vein, resumed the working them at the expence of the crown. This mine is nine miles *n. w.* from the city, and 60 miles *n. n. w.* of Santa Fé, and 43 nearly *w.* of Tunja. In lat. 5° 31′ *n.* and long. 74° 28′ *w.*

MUZUPIES, or MONZUPIES, a barbarous nation of Indians of the province of Guanuco in Peru; who dwell to the *n. n. e.* bounded by the provinces of Los Panataguas and Cocmonomas, with whom they are at continual war. It is but little known.

[MYERSTOWN, a village of Dauphin county, Pennsylvania; situated on the *n.* side of Tulpehockon creek, a few miles below the canal. It contains about 25 houses, and is 28 miles *e.* by *n.* of Harrisburg, and 57 from Philadelphia.]

[MYNOMANIES, or MINOMANIES, an Indian tribe, who with the tribes of the Chipewas and Sankeys live near bay Puan, and could together furnish about 20 years ago 550 warriors. The Mynomanies have about 300 fighting men.]

[MYRTLE Island, one of the Chandeleurs or Myrtle islands, in Nassau bay, on the coast of Florida, on the *w.* side of the peninsula.]

N

[NAAMAN'S Creek, a small stream which runs *s. e.* into Delaware river, at Marcus' hook.]

[NAB'S Bay, near the *w.* limit of Hudson's bay, known by the name of the Welcome sea. Cape Eskimaux is its *s.* point or entrance.]

NABA, a settlement of the province and *corregimiento* of Caxatambo in Peru; annexed to the curacy of Churin.

NABAN, a settlement of the same province and kingdom as the former; annexed to the curacy of Andajaes.

NABON, a settlement of the province and *corregimiento* of Cuenca in the kingdom of Quito; situate in the road which leads to the province of Jaen.

NABUAPO, a river of the province and country of the Iquitos Indians in Peru. It has its origin to the *n.* of the settlement of San Xavier, runs *s.* and enters the Marañon a little above the river Tigre by the *n.* side, in lat. 3° 17′ *s.*

NABUSO, a *paramo* or mountain always covered with snow, of the province and *corregimiento* of Riobamba in the kingdom of Quito, on which the academicians of the sciences at Paris fixed their mathematical instruments.

NACARI, a small river of the province of Ostimuri in Nueva España. It rises near the town of San Miguel, and after running a little way, enters the Hyaqui.

NACARNERI, a settlement of the province and government of Sonora in N. America; situate near the river of this name.

NACATCHES, a settlement of Indians of the province and government of Texas in N. America; situate on the shore of the river Rouge, and to the *n.* of the fort Natchitoches.

NACATLAN, a settlement of the head settlement of the district of Zapotitlan, and *alcaldia mayor* of Zacatlan, in Nueva España, half a league from its capital.

NACAUNE, a settlement of Indians of the province and government of Louisiana in N. America; situate on the shore of the river Trinidad, in the way which leads to Nuevo Mexico.

NACAUTEPEC, a settlement of the head settlement of the district and *alcaldia mayor* of Cuicatlan in Nueva España. It is of a moist temperature, and contains 33 families of Indians; 11 leagues to the *e.* of its capital.

NACHAPALAN, a settlement of the province

and *alcaldia mayor* of Panuco in Nueva España: It was large and populous in the time of the Indians. Here it was that the soldiers of Hernan Cortes took the 40 men of the nation of Francisco Garay, who wished to effect the conquest of these Indians.

NACHEGO, a large lake of the province and government of Mainas in the kingdom of Quito, to the *s.* of the river Marañon. Into this lake run the two rivers Sungoto and Manguy, and it empties itself by a narrow channel into the river Cahuapanas by the *w.* side, in lat. 5° 23′ *s.*

NACIMIENTO, a settlement and fortress of the kingdom of Chile; situate on the further side of the river Biobio as a frontier against the Araucanos Indians, but who burnt and destroyed it in 1601.

NACO, a settlement of the province and government of Honduras, founded by Christoval Olid, captain of Hernan Cortes in 1524; situate in a valley of the same name. When this general went from Mexico to chastise the aforesaid founder, he having rebelled against his master, the Cacique Canek observed to Cortes, that he would lead him to a settlement of people with white beards, meaning the Spaniards, and those of this settlement of Naco. Cortes arrived under his conductor, but found Olid already dead under the hands of Francisco de las Casas.

NACODOCHES, a settlement and *reduccion* of Indians, of the missions that were held there by the religious order of San Francisco, in the province of Texas in N. America.

NACORI, a settlement of the province of Ostimuri in N. America.

NACOSARI, a settlement of the province and government of Sonora in N. America; situate on the *s.* of the garrison of Coro de Guachi.

NACOSARI, another settlement and *real* of silver mines, of the province of Ostimuri, nine leagues *n. e.* of the river Chico.

NADACO, a settlement of Indians of the province and government of Texas in N. America; situate between the sources of the rivers Adayes and La Trinidad.

NADAIMA, a settlement of the province and government of Nicaragua in the time of the gentilism of the Indians; situate near where the capital stands.

NADIO, a settlement of the head settlement of the district of Zitaquaro, and *alcaldía mayor* of Maravatio, in the kingdom and bishopric of Mechoacán. It is of an extremely hot temperature, and abounding in sugar-canes; contains 80 families of Indians, and is four leagues to the *s.* of its head settlement.

NAFOLI, a settlement of Indians of the province and colony of S. Carolina; situate on the shore of the river Albama.

NAGARANDO, a name given by the Indians of the province of Nicaragua to the spot where the Spaniards founded the city of Leon, the capital of the same province.

NAGUALAPA, a settlement of the head settlement of the district of Almoloyan, and *alcaldia mayor* of Colima, in Nueva España. It is of an hot temperature, contains 22 Indian families, who trade in wood and maize. In its vicinity are many *cocales* estates, the productions of which are sold in the other jurisdictions. Five leagues *w.* of its head settlement.

NAGUAPO, SAN SIMON DE, a settlement of the province and government of Mainas in the kingdom of Quito; situate on the shore of the river Trocamana.

NAGUATZEN, S. LUIS DE, a settlement of the head settlement of the district of Siguinam, and *alcaldía mayor* of Valladolid, in the province and bishopric of Mechoacan. It contains five families of Spaniards and 139 of Indians, who are curriers and make beautiful saddles. It is half a league from its head settlement.

NAGUERACHI, a settlement of the missions which were held by the Jesuits in the province of Taraumara, and kingdom of Nueva Vizcaya. Forty-five leagues *n.* of the town and *real* of the mines of Chiguagua.

[NAHANT Point forms the *n. e.* point of Boston harbour, in Massachusetts; nine miles *e. n. e.* of Boston. Lat. 42° 27′ *n.* Long. 70° 57 *w.* See LYNN Beach.]

NAHUAS, a nation of Indians of Nueva España, one of those which spoke the Mexican language. They believed in the immortality of the soul, and said that this had different places to visit according to the death the body underwent: thus, that those who were killed by a flash of lightning, went to a place called *toocan*, where resided the deities presidiog over water, called *taloques;* that those who died in war, went to the house of the sun; and that those who died of infirmities, wandered over the earth for a certain time, so that their relations took care to provide them well with clothes, victuals, and other necessaries in their sepulchres; and after this they said that they descended into the infernal regions, these being divided into nine parts, and having a very wide river running through it. Moreover, that from thence they never escaped, being constantly guarded by a red-coloured dog: a fable which bears much resemblance to the celebrated river Styx and the dog Cerberus of the ancients.

NAHUATLACAS, a nation of Indians of Nueva España, in former times: one of the primites nations, and from whom it is thought the Mexicans are descended.

NAHUELHUAFI, a settlement of the province and *corregimiento* of Chiloe in the kingdom of Chile, to the *e.* and 90 miles from the sea; a *reduccion* of the Pulches and Poyas Indians, amongst whom the missions of the Jesuits met with very great success. It is situate on the *n.* shore of the lake of its name, in lat. 41° 22′ 30″ *s.* and long. 70° 40′ *w.*

[NAHUNKEAG, a small island in Kennebeck river, 38 miles from the sea, signifies, in the Indian language, the land where eels are taken.]

NAICUCU, a small river of the province and government of Guayana or Nueva Andalucia. It rises near that of the Tocome, runs parallel with it from *s.* to *n. e.* and then turning *e.* enters the Oaroni on the *w.* side, about 33 miles before this river enters the Orinoco on the *s.* side.

NAIGUADA, a settlement of the province and government of Venezuela, of the Nuevo Reyno de Granada; situate on the *e.* side of the city of Caracas.

[NAIN, a Moravian settlement, which was established in 1763, on Lehigh river, in Pennsylvania.]

[Nain, a settlement of the Moravians on the coast of Labrador, near the entrance of Davis' straits, being *s. s. w.* of cape Farewell. It was begun under the protection of the British government, but is now deserted.]

[NAMASKET, a small river which empties into Narraganset bay.]

NAMBALLE, a settlement of the province and government of Jaen de Bracamoros in the kingdom of Quito.

NAMBE, a settlement of Nuevo Mexico in N. America; situate on the bank of a small river which enters the Grande del Norte, between the settlements of Pasuque and Tesuque.

NAMIQUIPA, a settlement of the missions which are under the charge of the religious order of San Francisco, in the province of Taraumara, and kingdom of Nueva Vizcaya. Twenty-five leagues *n. w.* of the town and *real* of mines of San Felipe de Chiguagua.

NANAHUATIPAC, a settlement of the head settlement of the district of Teutitlan, and *alcaldia mayor* of Cuicatlan, in Nueva España. It contains 49 families of Indians, and is one league from its head settlement.

NANASCA. See Nasca.

NANAY, a large and navigable river of the province and government of Mainas in the kingdom of Quito. It rises from the lake Pachina, and from another small lake near to the same, and runs more than 85 leagues to the *e. s. e.* augmenting its stream by the rivers Necanumú, Blanco, and various others of less note. In the woods of its vicinity, towards the *n.* and *s.* dwell some barbarian Indians of the nation of the Iquitos, and on the *n. n. e.* are some Paranos Indians. This river takes its name from the many firs on its shores, called by the Indians *nanay*. It enters the Marañon by the *n.* part, to the *w.* of the settlement of Napeanos, in lat. 3° 27′ *s.*

NANCAGUA, a settlement of the province and *corregimiento* of Colchaqua in the kingdom of Chile. It has two vice-parishes, and in one of them is the celebrated gold mine of Apaltas. It is situate on the shore of the river Tinguiririca.

NANCOKE, a small river of the province and colony of Maryland in N. America.

NANCOOK, a settlement of the island of Barbadoes; situate on the *w.* coast.

[NANDAKOES are Indians of N. America, who live on the Sabine river, 60 or 70 miles to the *w.* of Yattassees, near where the French formerly had a station and factory. Their language is Caddo: about 40 men only of them remain. A few years ago they suffered very much by the small-pox. They consider themselves the same as Caddos, with whom they intermarry, and are occasionally visiting one another in the greatest harmony: have the same manners, customs, and attachments.]

NANDUIQUAZU, a river of the province and government of Paraguay in Peru, which rises near the ruins of the settlement of La Cruz de Bolaños, runs *e.* and incorporates itself with the following.

NANDUI-MINI, a river of the same province and kingdom as is the former, with which it unites, entering together into the Pardo or Colorado.

NANEGAL, a settlement of the province and government of Pastos in the kingdom of Quito, and of the district and jurisdiction of its audience.

Nanegal, another settlement, of the province and government of Esmeraldas in the same kingdom.

NANIS, a settlement of the province and *cor-*

regimiento of Caxatambo in Peru; annexed to the curacy of Mangas.

[NANJEMY River, a short creek which empties into the Patowmac in Charles county, Maryland, *s. w.* of Port Tobacco river.]

NANOUCHI, a settlement of Indians of the Cherokees nation, in the province and colony of Carolina; situate at the source of the river Apalachicola, where the English have a fort and establishment for their commerce.

[NANSEMOND, a county of Virginia, on the *s.* side of James's river, and *w.* of Norfolk county, on the N. Carolina line. It is about 44 miles in length, and 24 in breadth, and contains 9010 inhabitants, including 3817 slaves.]

[NANSEMOND, a short river of Virginia, which rises in Great Dismal swamp, and pursuing a *n.* then a *n. e.* direction, empties into James's river, a few miles *w.* of Elizabeth river. It is navigable to Sleepy hole, for vessels of 250 tons; to Suffolk, for those of 100 tons; and to Milner's, for those of 25 tons.]

[NANTASKET Road may be considered as the entrance into the channels of Boston harbour; lies *s.* of the light-house near Kainsford or Hospital island. A vessel may anchor here in from seven to five fathoms in safety. Two huts are erected here with accommodations for shipwrecked seamen.]

[NANTIKOKE, a navigable river of the *e.* shore of Maryland, empties into the Chesapeak bay.]

[NANTIKOKES, an Indian nation who formerly lived in Maryland, upon the above river. They first retired to the Susquehannah, and then farther *n.* They were skilled in the art of poisoning; by which shocking art nearly their whole tribe was extirpated, as well as some of their neighbours. These, with the Mobickons and Conoys, 20 years ago inhabited Utsanango, Chagnet, and Owegy, on the *e.* branch of the Susquehannah. The two first could at that period furnish 100 warriors each, and the Conoys 30 warriors.]

[NANTMILL, EAST and WEST, two townships in Chester county, Pennsylvania.]

NANTOUNAGAN. See TONNAGANE.

[NANTUCKET Island, belonging to the state of Massachusetts, is situated between lat. 41° 13′ and 41° 22′ 30″ *n.* and between long. 69° 56′ and 70° 13′ 30″ *w.* and is about 43 miles *s.* of cape Cod, and lies *e.* of the island of Martha's Vineyard. It is 14 miles in length, and nine in breadth, including Sandy point; but its general breadth is 3½ miles. This is thought to be the island called Nauticon by ancient voyagers. There is but one bay of any note, and that is formed by a long sandy point, extending from the *e.* end of the island to the *n.* and *w.* (on which stands a light-house, which was erected by the state in 1784), and on the *n.* side of the island as far as Eel point. This makes a fine road for ships, except with the wind at *n. w.* when there is a heavy swell. The harbour has a bar of sand, on which are only 7½ feet of water at ebb tide, but within it has 12 and 14 feet. The island constitutes a county of its own name, and contains 4620 inhabitants, and sends one representative to the general court. There is a duck manufactory here, and 10 spermaceti works. The inhabitants are, for the most part, a robust and enterprising set of people, mostly seamen and mechanics. The seamen are the most expert whale-men in the world. The whale fishery originated among the white inhabitants in the year 1690, in boats from the shore. In 1715, they had six sloops, 38 tons burden, and the fishery produced 1100*l.* sterling. From 1772 to 1775, the fishery employed 150 sail from 90 to 180 tons, upon the coast of Guinea, Brazil, and the W. Indies; the produce of which amounted to 167,000*l.* sterling. The late war almost ruined this business. They have since, however, revived it again, and pursue the whales even into the great Pacific ocean. There is not here a single tree of natural growth; they have a place called the Woods, but it has been destitute of trees for these 60 years past. The island had formerly plenty of wood. The people, especially the females, are fondly attached to the island, and few wish to migrate to a more desirable situation. The people are mostly Friends or Quakers. There is one society of Congregationalists. Some part of the *e.* end of the island, known by the name of Squam, and some few other places, are held as private farms. At present there are near 300 proprietors of the island. The proportional number of cattle, sheep, &c. put out to pasture, and the quantity of ground to raise crops, are minutely regulated; and proper officers are appointed, who in their books debit and credit the proprietors accordingly. In the month of June, each proprietor gives in to the clerks the number of his sheep, cattle, and horses, that he may be charged with them in the books; and if the number be more than he is entitled to by his rights, he hires ground of his neighbours who have less. But, if the proprietors all together have more than their number, the overplus are either killed or transported from the island.

In the year 1659, when Thomas Macy removed

with his family from Salisbury in Essex county to the *w.* end of the island, with several other families, there were nearly 3000 Indians on the island, who were kind to strangers, and benevolent to each other, and lived happily until contaminated by the bad example of the whites, who introduced rum; and their number soon began to decrease. The whites had no material quarrel or difficulty with them. The natives sold their lands, and the whites went on purchasing, till, in fine, they have obtained the whole, except some small rights, which are still retained by the natives. A mortal sickness carried off 222 of them in 1764; and they are now reduced to four males, and 16 females.]

[NANTUCKET, (formerly Sherburne), a post-town, capital, and port of entry in the above island. The exports in the year ending September 30, 1794, amounted to 20,517 dollars. It is 56 miles *e. s. e.* of Newport, 75 *s. e.* of Boston, and 255 *e. n. e.* of Philadelphia.]

[NANTUCKET Shoal, a bank which stretches out above 15 leagues in length, and six in breadth, to the *s. e.* from the island of its name.]

NANTUE, a port of the coast of the province and colony of Maryland, within the bay of Chesapeak.

NANTUXET Bay, New Jersey, is on the *e.* side of Delaware bay, opposite Bombay hook.]

NANZUITA, a settlement of the head settlement of the district of Santa Isabel de Sinacatan, in the province and *alcaldia mayor* of Guazapan, and kingdom of Guatemala; annexed to the curacy of its head settlement.

NAOS, a port on the coast of the province and kingdom of Tierra Firme, very convenient and capacious, frequented by strange vessels which carry on an illicit commerce. It is to the *e.* of the mouth of the river Chagre.

NAOS, a small island of the S. sea, in the bay of Panama, of the province and kingdom of Tierra Firme; one of those which form the port of Perico.

NAOUADICHES, a settlement of Indians of the province and government of Texas in N. America; situate between the rivers Adaes and Trinidad, in the road which leads to Mexico.

NAPAUECHI, a settlement of the missions which were held at the expence of the Jesuits, in the province of Taraumara and kingdom of Nueva Vizcaya. Twenty-two leagues *s. w.* of the *real* of mines and town of San Felipe de Chiguagna.

NAPEANOS, SAN PABLO DE, a settlement of the province and government of Mainas in the kingdom of Quito; situate at the source of the river Nanay.

[NAPESTLE, a river spoken of by Humboldt, who asserts that it is not known at New Mexico by what name it is denominated in Louisiana. It is, however, thought to be the Arkansas.]

NAPO, a large and abundant river of the province and government of Quixos y Macas in the kingdom of Quito, and one of the most considerable in that kingdom. It rises from the mountain and volcano of Cotopacsi, and flows down to the valley Vicioso, running constantly *e.* through some very large rocks; and therefore not navigable, save only from the settlement and port of its name, facilitating the communication between this province and the capital. In its course it collects on the *s.* the waters of the Ansupi, Puni, Araoma, Umuyacu, Ayrunni, Canoa-yacu, Anangú, Serenú, Yutury-yacu, Tiputini, Curaray, and others of less note; and on the *n.* the Hollin, Fusuuñ, Sinú, Payamino, Coca, Itaya, Aguarico, and many which are smaller. On its shores are the settlements of Napo, Napotoas, Santa Rosa de Oas, San Juan Nepomuceno, and El Dulce Nombre de Jesus, all *reduccions* made by the regulars of the Jesuits; but the climate is there very warm and moist, and causing great sickness. Both on one and the other shores dwell various barbarian nations of savage Indians, all having distinct idioms difficult to be learnt. This river, thus enlarged by those aforesaid, enters with so large a body into the Marañon or Amazonas by the *n.* shore, as to have been frequently mistaken for the same. Where it is entered by the river Cacao, is the spot where Francisco de Orellana separated himself from his chief Gonzalo Pizarro and went to sea. This river is most abundantly stocked with delicate fish, and in 1774, at the bursting of the volcano of Cotopaxi, it was so swelled by the melting of the snows and ice, that it burst its boundaries and inundated an immense tract of country, doing infinite damage in the settlements. Its mouth is in lat. 3° 26′ *s.*

NAPO. The settlement aforesaid, one of the missions established by the Jesuits; situate on the shore of the above river, and where the inhabitants catch much fish. It is very fertile, and abounding in *yucas*, maize, rice, and plantains. In 1744 it suffered much in the inundation beforementioned, when the river carried away the greater part of the houses.

NAPOTOAS, a settlement of the same province and kingdom as the former, belonging to the district of Quijos; situate also on the shore of the river Napo: one of the missions founded there by the Jesuits.

NAQUASEE, a settlement of Indians of the province and colony of N. Carolina, on the confines of that province and that of S. Carolina.

NARAGUASET, an ancient name of a territory or district of New England, *e.* of the river Connecticut, now the county of New London.

NARANJA, a settlement of the head settlement of the district of Tirindaro, and *alcaldía mayor* of Valladolid, in the province and bishopric of Mechoacán. It contains 76 families of Indians, and is a quarter of a league *n.* of its head settlement.

NARANJAL, a settlement of the province and government of Guayaquil and kingdom of Quito, in the district of the island of La Puña, abounding in woods of excellent quality for ship-building, and in which its commerce consists. Seven leagues from Guayaquil.

Naranjal, another settlement, of the head settlement of the district and *alcaldía mayor* of Orizava in Nueva España, in which are 108 Indian families and only two Spanish. In its district is the celebrated sugar-mill, called De Tuzpanco, at which there assist no less than eight families of Negro slaves; this great population and extensive boundary belonging to the inheritance of the Marquis de Sierra Nevada. Four leagues *s. w.* of its capital.

Naranjal, another, of the province and government of Popayán in the Nuevo Reyno de Granada; situate near the coast of Timana, and at the source of the Rio Grande de la Magdalena.

Naranjal, another, of the province and government of Antioquia, in the same kingdom as the former; situate on the shore of the river Nechi, near the pass of La Angostura.

Naranjal, a river of the province and government of Guayaquil, which rises *w.* of the settlement of Inca, in the *corregimiento* of Cuenca, and enters the sea near the mouth of the river Guayaquil, in the gulf of its name.

Naranjal, an island of the S. sea, in the gulf of Panama, and province and kingdom of Tierra Firme, one of those called Del Rey, or De las Perlas, and the larger of these. Five leagues in length from *n.* to *s.* desert, and inhabited only by a few Negro slaves of the families of Panama, employed in the search for pearls, and for their maintenance they grow a little maize, this being the only vegetable production. It has a good port on the *e.* side, opposite the coast of the continent, from whence it is distant five leagues.

NARANJO, a settlement of the province and government of Popayán and Nuevo Reyno de Granada; situate on the shore of a small river to the *n.* of the city of Buga.

Naranjo, a river of the island of St. Domingo. It is small, and rises near the coast of the great bay of Samaná, and enters the sea between the river De Estero and the port of S. Lawrence.

NARANJOS, a settlement of the province and *corregimiento* of Chachapoyas in Peru; situate at the source and on the bank of the river Moyobamba.

Naranjos, a river of the island St. Domingo, in the French part. It rises near the coast of the *w.* and, running to this rhumb, enters the sea in the bay of Pozo.

Naranjos, another river, of the island of Cuba, which enters the sea on the *n.* coast, between the ports Sama and Timones.

NARE, a river of the province and government of Antioquia in the Nuevo Reyno de Granada. It is navigable for small vessels, and abounds in good fish: also in its vicinity is gathered good *cacao.* It enters by the *w.* into the Rio Grande de la Magdalena, between the town of Honda and the settlement of Carari.

NAREO, a settlement of the head settlement of the district of Tlapacoya, and *alcaldía mayor* of Quatro Villas, in Nueva España. It contains 28 families of Indians, who cultivate some cochineal, seeds, and fruits, and cut some woods. Three leagues *n. w.* of its head settlement.

NARIGUERA, San Penno de Alcantara de la, a settlement of the province and government of Quixos y Macas in the kingdom of Quito; a *reduccion* of the Sucumbios Indians, and one of the missions which were held there by the Jesuits.

NARIS, an isle of the N. sea, close to the island of Christoval, one of the Antilles.

NARITO, a river of the province and *alcaldía mayor* of Acaponeta or Chiametla in Nueva España. It runs from the province of Cinaloa and Culiacan, and enters the gulf of California or Mar Roxo de Cortes. Although the Indians give it this name the Spaniards call it Toluca. It is very large and abundant.

NARINA, a river of the island and government of Trinidad. It rises from a lake in the *e.* part, not far from the coast, and enters the sea close to the point of Cocos.

NARRAGANSET, a city of the county of Hampshire, in the bay of Massachusetts, of N. America. Five miles *e.* of Sunderland and 10 *w.* of Petersham.

[Narraganset Bay, Rhode island, makes up

from *s.* to *n.* between the mainland on the *e.* and *w.* It embosoms many fruitful and beautiful islands, the principal of which are Rhode island, Canonicut, Prudence, Patience, Hope, Dyers, and Hog islands. The chief harbours are Newport, Wickford, Warren, Bristol, and Greenwich, besides Providence and Patuxet; the latter is near the month of Patuxet river, which falls into Providence river. Taunton river and many smaller streams fall into this capacious bay. It affords fine fish, oysters, and lobsters in great plenty.]

[NARRAGUAGUS Bay. A part of the bay between Goldsborough and Machias, in Washington county, district of Maine, goes by this name. From thence for the space of 30 or 40 miles, the navigator finds, within a great number of fine islands, a secure and pleasant ship-way. Many of these islands are inhabited and make a fine appearance. A river of the same name falls into the bay.]

[NARRAGUAGUS, a post-town; situate on the above bay, 15 miles *n. e.* of Goldsborough, 39 *e.* by *n.* of Penobscot, and five from Pleasant river.]

[NARROWS, The. The narrow passage from sea, between Long and Staten islands, into the bay which spreads before New York city, formed by the junction of Hudson and East rivers, is thus called. This strait is nine miles *s.* of the city of New York.]

[NARROWS, The, a strait about three miles broad, between the islands of Nevis and St. Christopher's, in the W. Indies.]

NARUAEZ, SAN MIGUEL DE, a settlement of the province and government of Quixos and Macas in the kingdom of Quito, belonging to the district of the second.

NASAS, a large and abundant river of the kingdom of Nueva Vizcaya in N. America. It rises near the Real de Minas of Guanavi, 15 leagues *w.* of the city of Guadiana, the capital of the kingdom, and runs from *n. w.* to *s. e.* until it enters the great lake of San Pedro. On its shore are many settlements of Spaniards, *Mustees*, and Mulattoes, and others of Indians, reduced by the missions that were held there by the Jesuits; and its waters are made, by means of aqueducts, serviceable for the irrigation of many gardens and lands, where there are some vineyards which yield abundantly. There was formerly in this part a strong garrison, but which was abolished through the offer of the Count S. Pedro del Alamo, to undertake the defence of the country against the infidel Indians.

NASCA, a celebrated town and port of the province and district thus called in Peru, which is formed from the territories of Ica and Pisco, and extends for more than 50 leagues along the strands of the Pacific sea. This port is called one of the Puertos Intermedios, or intermediate ports, as lying between the kingdom of Peru and Chile. The soil is very fertile, and abounds in vines and olives, of which the crops are excellent, and of which its commerce consists, and which tend to make this port much frequented by vessels which come to lade with these cargoes. The valley in which the vines grow, consists of a pebbly sand, having some streams of water, which never swell to an inordinate height nor diminish, without their origin ever having been discovered, although it is found that they sprout out of some subterranean channels, which were formed by the Indians in the time of their gentilism. The town is well peopled, and in it are many noble and rich families. It has, besides the parish church, a convent of the religions order of San Agustin. It suffered much in an earthquake in 1765: is of a mild and healthy temperature; and its territory was conquered and united to the empire of Peru by Capac Yupanqui, fifth emperor of the Incas. In lat. 15° 7′ 30″ *s.* Long. 75° 24′ *w.*

NASCA, a river of this province, which runs *w.* and enters the sea opposite the promontory also of this name.

NASCA, a mountain on the coast of the said province, at the entrance of the port above mentioned.

NASCATICH, a small lake of New France or Canada, in N. America; formed by a waste-water of the lake St. Peter, and others in the country and territory of the Nekoubanistes Indians.

[NASH, a county of Halifax district, containing 7393 inhabitants, of whom 2009 are slaves. There is a large and valuable body of iron ore in this county; but only one bloomery has yet been erected.]

[NASH Court-house, in N. Carolina, where a post-office is kept, 21 miles *w.* by *n.* from Tathorough, and 22 *s. e.* from Lewisburg.]

[NASHAUN, or NAWSHAWN, one of the Elizabeth isles, the property of the Hon. James Bowdoin, Esq. of Boston; situated at the month of Buzzard's bay, and three miles from the extremity of the peninsula of Barnstaple county. Considerable numbers of sheep and cattle are supported upon this island; and it has become famous for its excellent wool and cheese. Here Capt. Bartholomew Gosnold landed in 1602, and took up his abode for some time.]

[NASHUA, River, is a considerable stream in

Worcester county, Massachusetts, and has rich intervale lands on its banks. It enters Merrimack river at Dunstable. Its course is *n. n. e.*]

[NASHVILLE, the chief town of Mero district in the state of Tennessee, is pleasantly situated in Davidson's county, on the *s.* bank of Cumberland river, where it is 200 yards broad. It was named after Brig. Gen. Francis Nash, who fell on the 4th of October 1777, in the battle of Germantown. It is regularly laid out, and contains 75 houses, a court-house, an academy, and a church for Presbyterians, and one for Methodists. It is the seat of the courts held semi-annually for the district of Mero, and of the courts of pleas and quarter sessions for Davidson county. It is 160 miles *w.* of Knoxville, 66 from Big Salt lick garrison, and 166 *s.* by *w.* of Lexington in Kentucky. Lat. 36° 3′ *n.* Long. 86° 58′ *w.*]

[NASKEAG Point, in Lincoln county, district of Maine, is the *e.* point of Penobscot bay.]

[NASPATUCKET River. See WANASPATUCKET.]

[NASQUIROU River, on the Labrador coast, is to the *w.* of Esquimaux river.]

NASQUIROU, a small river of the country or land of Labrador. It runs *s.* and enters the sea in the gulf of St. Lawrence.

NASSAU, a cape or point of land on the coast of the province and government of Guayana or Nueva Andalucia, one of those which form the mouth or entrance of the river Paumaron, near Esquivo and Demerary. According to some maps, it is the same as that which others call cape of Orange, but which is very erroneous, for it lays 490 miles *w. n. w.* of cape Orange. It is in lat. 7° 36′ *n.* Long. 48° 45′ *w.*

NASSAU, a city, the capital of the island of Providence, one of the Lucayas; situate on the *n.* part, on the sea-coast, defended by a castle well furnished with artillery, with a good port, which has in its neighbourhood various small isles, and where ships may be well sheltered and lie secure, although its entrance is difficult and fit only for small vessels, or such as draw not more than from 10 to 12 feet water. This city was taken by the Spaniards in 1782, but it was restored to the English in the peace of the following year.

[NASSAU Cape, on the *n.* shore of Tierra Firme, S. America.]

[NASSAU, a small town in Dauphin county, Pennsylvania. It contains a German church, and about 35 houses. It is also called Kemp's town.]

[NASSAU Island, at the mouth of Byram river, in Long Island sound.]

[NASSAU, the chief town of Providence island, one of the Bahamas, and the seat of government. It is the only port of entry except at Turk's island. See BAHAMAS and NEW PROVIDENCE.]

[NASTLA, a town of Mexico. See ANGELOS.]

NATA, or SANTIAGO DE LOS CABALLEROS, a city, and capital of the *alcaldía mayor* and jurisdiction of its name in the province and kingdom of Tierra Firme; situate upon the coast of the gulf of Parita, in a beautiful and agreeable spot; the territory being fertile and abounding in cattle, seeds, and fruit, and of an hot temperature. It is called Nata from one of the caciques of that territory; which was discovered by Olonso de Ojeda, in 1515, and settled, in 1517, by Gaspar de Espinosa. The infidel Indians destroyed the town in 1529, but it was rebuilt with the title of city in 1531.

Here they make some sorts of crockery of an earth of a beautiful red colour, forming them of different shapes and figures, and of such beauty as to be in great estimation in Peru and even in Europe; this consequently forms a considerable branch of commerce. In 1748, the president Don Dionisio de Alçedo inflicted an exemplary chastisement on three very numerous companies of smugglers, who had maintained an open commerce with the English, these having furnished them with artillery, arms, and ammunition, so that they built for themselves a fort, and actually opposed and defeated a detachment of the regiment of Granada, putting to death the officer Don Alonzo de Murga, the commander. It is 73 miles *s. w.* from Panamá, in lat. 8° 21′ 50″ *n.* Long. 80° 17′ *w.*

[NATA Point, or CHAMA, or CHAUMU Cape, is at the *w.* point of the gulf of Panamá, from whence the coast tends *w.* to Haguera point seven leagues. All ships bound to the *n. w.* and to Acapulco make this point. It is also called the *s.* point of the bay, which lies within on the *w.* side of this great gulf of Panamá.]

[NATACHQUOIN River, a large river of the coast of Labrador, in N. America, to the *w.* of Nasquirou river, under mount Joli, where it forms a *s.* cape. The little Natachquoin is to the *w. s. w.* of this.]

NATAGA, a settlement of the government of Neiba in the Nuevo Reyno de Granada; situate on an eminence, of a mild temperature, and abounding in vegetable productions and gold mines, and in this metal the Indians here pay their tribute. The natives, who may amount to little more than 50, have some of them established themselves in a neighbouring place, called Los Organos; since they assert that the gold is there more abundant and most easily procured. Sixteen leagues from its capital and near the city of La Plata.

NATAGAIMAS, an ancient nation of Indians of the Nuevo Reyno de Granada, who used to dwell in the *llanuras* of Neiba, and were at continual warfare with the Pijaes. Some of them were reduced to the faith. They are strong, warlike, and of a fierce aspect, but faithful. Very few of them now remain in a settlement of the *corregimiento* of Coyaima, which is of an hot temperature, and produces *cacao*, maize, *yucas*, and plantains, and has good breeds of neat cattle. The Indians, when they have to pay their tribute, sally forth in large companies to Santa Fé, and on their way spend four or five days in fishing in the great river of Saldaña, and in this time they collect all the gold which is necessary for their purposes. Indeed such is the ease with which they collect this metal, that they must infallibly become soon rich, were they not so much given to the vice of drunkenness. This settlement is close to the town of La Purificacion.

NATAGAME, a settlement of the kingdom of Nueva Vizcaya in N. America.

[NATAL, a cape and town on the *s.* shore of the Rio Grande, on the *n. e.* coast of Brazil in S. America, is to the *s. w.* of the four-square shoal, at the month of the entrance of that river, which contains some dangerous rocks. On this point is the castle of the Three Kings, or Fortaleza des Tres Magos. The town of Natal is three leagues from the castle, before which is good anchorage for ships in from four to five fathoms, and well secured from winds.]

NATCHES, a nation of barbarian Indians of Louisiana in N. America ; who occupied the most fertile and best peopled canton. At a short distance from the coast rise two hills, one behind the other, and beyond these are valleys of fertile meadows, interspered with beautiful groups of woods, forming a very enchanting prospect. The most common of the trees are the walnut and the oak.

Mr. de Iberbille, a Frenchman, was the first, who, navigating the Mississippi from its mouth, discovered this nation and country of the Natches in 1701, and who, finding it to have so many advantages, determined to found a colony and town which might be the capital of the establishment that might be formed by the French. Accordingly, having formed his plan, he determined to give to this new settlement the name of Rosalia, which was that of Madame de Pontchartrain ; but it was never founded, although some geographers of the French nation wrongly give it a place in their charts.

The character of these Natches Indians differs much from that of all the other nations, since they are very pacific and really hate war, and never make it unless obliged, deeming it no glory to destroy their fellow-creatures. The form of their government is despotic, and such is the subordination of the vassals as to border upon slavery. They say that their chiefs are descended from the sun, and, indeed, these take the name of this luminary ; and both chief and his wife have the power of inflicting death on all on the slightest pretence. All treat him with the most excessive veneration, and, when he dies, all those of his family think it the greatest honour to die with him, whilst those not related, and who cannot pretend to this happiness, sometimes make themselves a cord by which they may hang themselves.

They have a temple in which a sacred fire is continually burning ; and should it perchance happen to go out, the priest entrusted with the care of it is immediately put to death. There is no nation in the world in which the women are so luxurious as in this; and the sun or chief can oblige them to prostitute themselves to any stranger without the least breach of propriety or decency. Although polygamy is allowed without limitation as to the number of women, they seldom have more than one, but the chief alone can repudiate and cast her off at his fancy. The women are prettily made and dress well, and the noble amongst them may not marry save with plebeians, but they may throw off the alliance whenever their husband displeases them, and take another, should she not be a mother. The wife may break her husband's head if unfaithful to his marriage bed, but the husband has not the same power over his wife, for he generally looks up to her as a slave to his mistress, and may not eat in her presence.

In their wars they have two chiefs. They have two masters of the ceremonies for the temple, and two officers to regulate the treaties of peace and war, one to inspect the works and another to manage the public festivities. The great sun or chief gives these employments, and the persons fulfilling them are respected much by the commonalty. The harvests are made for the general good ; the chief appoints the day of the gathering, and calls together all the people, and at the end of July he fixes another time for the celebration of a feast which lasts three days ; and at which each individually assists, bringing with him some game, fish, and other provisions, consisting of maize, beans, and melons. The sun and his principal wife preside, sitting under a lofty covering of leaves; the former having in his hand a sceptre adorned with feathers of various colours, and all the nobility arranged around them both in the most respectful order. On the last day the chief makes an oration, exhorting all to fulfil their several duties, and most particularly to testify their veneration to the spirits of the temple, and to labour in the education of their children: then, if any one has been instrumental to the public good, the chief proceeds to make his eulogium.

In 1700 the temple was set on fire by a flash of lightning, and eight women threw their children into the flames, thereby thinking to appease the deities. This was one of those actions which was particularly extolled, and the women were looked upon as complete heroines, nor did the chief, in this instance, forget to recommend strongly that all mothers should adopt the same conduct in a similar emergency.

Garcilaso Inca speaks of the nation of the Natches as of a powerful and numerous people; but the fact is, that they are now much reduced as well by the epidemic disorders that have prevailed amongst them as by their wars. At present they have no other population than that where the French have built a fort for their establishment; and Mr. de Iberbille destined the Father Paul de Rude, a Jesuit, to undertake the conversion of these Indians; but he finding that he obtained little fruit, passed over to preach to the Bayagoulas. Some years after this, the same object was had in view by Mr. de S. Cosme, a priest, but he was killed by the Indians; and, indeed, such has been the uniform resistance on their part to any plans adopted for their reduction, as to preclude all possibility of attaining that end.

[NATCHES, a town so called, on the banks of the Mississippi, which, according to Mr. Ashe, contains 2500 inhabitants, much given to luxurions and dissolute propensities, for which they have become proverbial.]

NATCHITOCHES, or NACTCHITOCHES, as some pronounce it, a barbarous nation of Indians of the province and government of Louisiana, in N. America, who dwell 50 leagues up the Red river, which is also known by their name. This tribe of Indians, who have always been the friends of the French and enemies to the Spaniards, is very numerous and composed of more than 200 cabins. The French military, who had fulfilled their time of service, established themselves on an island of the Red river, where they built a fort which they called Natchitoches; but having sowed some tobacco, and found that the sand that was blown upon it made it of a bad quality, they removed their establishment to Tierra Firme, where they have so succeeded in the cultivation of this plant that it is of peculiar estimation. This nation is 60 leagues from New Orleans.

NATICK, an ancient township of the county of Middlesex in the colony and bay of Massachusetts; situate on the shore of Charles river. [It is 18 miles *s. w.* of Boston, and 10 *n. w.* of Dedham. Its name in the Indian language signifies "the place of hills." The famous Mr. Eliot formed a religious society here; and in 1670, there were 50 Indian communicants. At his motion, the general court granted the land in this town, containing about 6000 acres, to the Indians. Very few of their descendants, however, now remain. It was incorporated into an English district in 1761, and into a township in 1781; and now contains 615 inhabitants.]

NATIGAN, a small river of the province and country of Labrador in N. America, which runs *s.* and enters the sea in the gulf of St. Lawrence.

NATIGUANAGUA, a river of the province and government of Darien in the kingdom of Tierra Firme. It rises in the mountains of the *n.* part, and enters the sea opposite the Mulatto isles.

NATISCOTEC, a bay in the island of Anticosti of N. America, on the *e.* coast.

NATIVIDAD, a settlement of the province and government of Sonora in N. America; situate on the shore of the river Bezany.

NATIVIDAD, a small island of the S. sea, discovered by Admiral Sebastian, a Vizcayan, in 1602, when he went by order of the viceroy, Count of Monterrey, to reconnoitre the coast of Nueva España by that sea. This island is small, desert, and abounding only in a sort of wild fennel.

NATIVIDAD, an island of the straits of Magellan.

NATIVITAS, SANTA MARTA DE, a settlement of the head settlement of the district of Tlapacoya and *alcaldía mayor* of Quatro Villas in Nueva España. It contains 64 Indian families, who cultivate some cochineal, seeds and fruit, and cut wood, and in which they trade. Two leagues *n. w.* of its head settlement.

NATIVITAS, an hermitage of Nueva España, at less than a league's distance from the city of Xuchimilco, and four to the *s.* of Mexico; in the which are two or three fountains of excellent water, and in the largest and deepest a stone cross, fixed there by the first of the monks of S. Francisco who passed through that kingdom. This fountain swarms with fish, and the country around being delightfully woody and pleasant, with many orchards and cultivated grounds, is such as to induce the inhabitants of Mexico frequently to visit this spot, and indeed all persons of distinction, going to that city, alight here to examine the cross; which has the following peculiarity attending it, namely, that being fixed upright in a canoe, and this being agitated by the motion caused in the water by the number of the fish, the cross is also seen to move about, whereas its fixture in the canoe not being visible from the shore, it should seem that it ought to be stable.

NATOUAGAMIOU, a lake of New France or Canada, formed from various other small lakes to the *s.* of the great lake of S. Juan.

[NATTENAT, an Indian village on Nootka sound, on the *n. w.* coast of N. America. It has a remarkable cataract, or water-fall, a few miles to the *n.* of it.]

[NATURAL Bridge. See ROCKBRIDGE County, Virginia.]

NAU, a settlement of the province and country of Las Amazonas, in the Portuguese possessions; a *reduccion* of the Indians, and made by the Carmelite missionaries of that nation. It is on the shore of the river Negro, very near the settlement of Baracoa.

NAUAGANTI, a river of the province and government of Darien, and kingdom of Tierra Firme. It rises in the mountains on the *n.* runs nearly to this *rhumb*, and enters the sea opposite the island of Pinos.

NAUAJOA, or NAVAJOOS, a province and territory of Indians of this nation, in N. America; bounded *n.* by that of Moqui, *n. w.* by the town of Santa Fé, the capital of the kingdom of Nuevo Mexico. It is peopled by *rancherias* or farms of barbarian and gentile Indians; but who were easily reduced to the Catholic faith, as was proved by the attempts made in 1748 by the friar Juan Menchero of the order of San Francisco, who with an apostolic zeal went to preach amongst them.

NAUAJOA, a settlement of the province and government of Cinaloa; a *reduccion* of Indians of the aforesaid nation, and of the missions which were held by the Jesuits.

NAUCALPAN, SAN BARTOLOME DE, a settlement of the *alcaldía mayor* of Tacuba in Nueva España; annexed to the curacy of San Antonio de Huixquilucan. It contains 273 families of Indians, and is nine leagues and an half to the *w. s. w.* of its capital.

[NAUDOWESIES, an Indian nation inhabiting lands between lakes Michigan and Superior. Warriors, 500].

[NAUGATUCK River, a *n. e.* branch of Housatonic river in Connecticut. A great number of mills and iron-works are upon this stream and its branches.]

NAUHTECAS, a nation of Indians of Nueva España, who inhabited the coast of the N. sea; conquered and subjected to the empire of Mexico by Moctheuczuma II.; to impede by that direction the entrance of the Tlaxcaltecas, when the conquest of this republic was in agitation. The Nauhtecas were bounded by the Mixcaltzincas.

NAUHTLAN, a settlement of the province and *alcaldía mayor* of Panuco in Nueva España; situate near the sea-coast by the Indians, before the arrival of the Spaniards, who afterwards changed its name to Almeria. The emperor Motheuczuma used to have posted watches or centinels to give notice of what was happening at sea.

NAULINGO, a settlement and head settlement of the district of the *alcaldía mayor* of Xalapa in Nueva España; situate on the top of a stony mountain of a league and an half high; of a cold and moist temperature from its lofty situation, but as fertile as any other settlements of this jurisdiction. Its population is composed of 142 families of Spaniards, 19 of *Mustees* and Mulattoes, and 90 of Indians, devoted for the most part to the cultivation of the soil. Its name, which signifies "four eyes," arises from so many springs of water which rise in a hill contiguous to the settlement. Five leagues *n. e.* of its capital.

NAULINGO, another settlement, with the dedicatory title of Santiago, in the head settlement of the district of Caluco, of the *alcaldía mayor* of Sonsonate, in the kingdom of Guatemala. It is annexed to the curacy of its head settlement, and its natives are Mexican Indians.

NAUMBI, a river of the province and government of Paraguay, which enters the Uruguay between those of Itay and Mbutuay.

NAUNAS, a barbarous nation of Indians, but little known, who dwell in the province and country of Las Amazonas, in the forests and woods close to the river Itau, where they live dispersed and wandering about like wild beasts.

NAUOGAME, a settlement of the missions which were held by the regulars of the company, in the province and government of Sonora in N. America.

NAUPAN, a settlement of the head settlement of the district and *alcaldía mayor* of Guauchinango in uNeva España, of a mild temperature. It has a convent of the religious order of S. Agustin, and contains 334 families of Indians, including those of eight wards annexed to its curacy, who live by cultivating seed and cotton, as also by making loaf-sugar. Three leagues *n.* of its head settlement.

NAUPAN, a very lofty mountain of the *cordillera*, in the *corregimiento* and district of Alause, of the kingdom of Quito.

NAURAS, a barbarous nation of Indians, of the Nuevo Reyno de Granada, who live near the river Carari. They are cannibals and warlike, and sometimes wander as far as the shores of the grand river Magdalena. These barbarians at the present day are far from numerous, and their customs are but little known.

NAUSA, a settlement of the district of Yaguache, in the province and government of Guayaquil and kingdom of Quito.

NAUTA, a river of Nueva España, in the jurisdiction and *alcaldía mayor* of Tampico. It enters the sea between the month of this river and the point Deglada.

NAUUSHAUUN, an island of the N. sea, one of those called Isabella, at the mouth or entrance of the bay of Plymouth and New England; three miles *s. w.* of the peninsula of the county of Barnstable, which forms the cape of Cod bay.

NAUZA, a settlement of the province and *corregimiento* of Guanuco in Peru; annexed to the curacy of Santa Maria del Valle.

NAUZALCO, SAN JUAN DE, a settlement and head settlement of the district of the *alcaldía mayor* of Sonsonate in the kingdom of Guatemala. It contains 2650 Indians, with those contained in three other settlements annexed to its curacy, which belonged to the religious order of S. Domingo, before the clergy had been appointed to it by order of the king.

[NAVARRE, a province of New Mexico, on the *n. e.* side of the gulf of California, which separates it from the peninsula of California, on the *s. w.*]

NAVATIA, a settlement of the province and government of Nicaragua, and kingdom of Guatemala, in the time of the Indian gentilism.

NAVAZA, a small island of the N. sea, to windward of the strait formed by the islands of Cuba and St. Domingo. It is desert, and the English come to it from Jamaica in boats to catch *iguanas*, an amphibious animal resembling a lizard, and which is found here in great abundance, breeding in the roots of old trees: their flesh is white, but hard to masticate, and the sailors say they make good broth. Some of these animals are found three feet long. [It is 67 miles *e. n. e.* of the *e.* end of the island of Jamaica, and 30 miles from Tuburon in the island of St. Domingo. Lat. 18° 33′ *n.* Long. 75° 3′ *w.*]

[NAVESINK Harbour, on the sea-coast of Monmouth county, New Jersey, lies in lat. 40° 24′ *n.* having Jumping point on the *n.* and is 2½ miles *s.* of the *n.* end of Sandy Hook island; and its mouth is five miles from the town of Shrewsbury. The small river of its name falls into it from the *w.* and rises in the same county. Navesink hills extend *n. w.* from the harbour on the Atlantic ocean, to Rariton bay; and are the first land discovered by mariners when they arrive on the coast. They are 600 feet above the level of the sea, and may be seen 20 leagues off.]

NAVIDAD, a settlement of the province and bishopric of Mechoacán in Nueva España, with a good port on the coast of the S. sea. It belongs to the *alcaldía mayor* of La Purificacion, and is 156 miles *w.* of Mexico. In lat. 18° 51′ *n.* Long. 111° 10′ *w.*

NAVIDAD, another port, in the province and *corregimiento* of Itala of the kingdom of Chile.

NAVIO QUEBRADO, a point of land of the coast of the province and government of the Rio del Hacha, and Nuevo Reyno de Granada, between the aforesaid river and the settlement of La Ramada.

NAVIOS, ISLA DE, an island near the coast of the province and government of Louisiana in N. America, close to the falls of St. Diego.

NAVIOS, a bay of the *n. w.* coast of the island of Martinique, between port Case Pilote, and the point De Negres.

[NAVIRES, or CAS DE NAVIRES Bay, in the island of Martinico, in the W. Indies.]

NAVISCALCO, a settlement of the province and *alcaldía mayor* of Zedales in the kingdom of Guatemala.

NAVITO, a port of the coast of Nueva España; opposite the province of California.

[NAVY, a township in Orleans county, in Vermont.]

[NAVY Hall, in Lower Canada, stands on the *s.* side of lake Ontario, at the head and *w.* side of Niagara river, which last separates it from fort Niagara, on the *e.* side, in the state of New York. It is 20 miles *n.* by *w.* of fort Erie, and 50 *s. e.* by *s.* of York.]

[NAVY Island lies in the middle of Niagara river, whose waters separate it from fort Slusher, on the *e.* bank of the river, and the same waters divide it from Grand island, on the *s.* and *s. e.* It is about one mile long, and one broad, and is about three miles *n.* by *e.* of Navy Hall.]

NAYARITH, a large and extensive province of N. America; bounded *e.* by the borders of Nueva Vizcaya, and part by Nueva Galicia; *w.*

by the provinces of Copala and Culiacan; *s.* by the jurisdictions and *alcaldias* of the audience of Guadalaxara; and *n.* by the *sierra* Madre, in which it is situate, and the settlements of Taraumara. The territory is rough and mountainous, but fertile and abounding in rich mines, which are however not worked, and are useless, owing to the want of population.

In this province the Jesuits held a large mission dispersed through several settlements, having for their defence a garrison with two captains, two lieutenants, two serjeants, and 38 soldiers. It was discovered in 1718 by the circumstance of an Indian having come from it to the Spaniards, dressed in all the insignia peculiar to the Chichimecas kings, and asserting that he came from the Nayaritas: he presented himself with a large retinue before the Marquis de Valero, then viceroy of Mexico, to render voluntary obedience to the king of Spain, with all his vassals in those unknown countries, and such as were never guessed at by any Spaniard, owing to the thick and almost inaccessible *serranía* which blocked up the road to the interior provinces. This chief then entreated that his nation might be instructed in the Catholic religion, and asked a supply of troops to aid them in the defence against their enemies. All this was immediately granted, but as they were proceeding on their journey, this king with all his vassals, all of a sudden, took to flight, carrying with them a great part of our equipage: they were of course pursued, but the Spaniards soon lost sight of them in those intricate *serranias*, and found themselves at last on a mountain called the Mesa del Tonati; where, in a very capacious cave, they discovered the place of their sacrifices, and amongst other things a skeleton to which they used to pay adoration, and which was the remains of one of their kings, the fifth grandfather of him we have above mentioned; this figure was covered with a mantle set with precious stones, according to their custom, which reached from the shoulders to the feet, and was seated upon what they call a throne, with a shoulder-belt, bracelets, necklace, and girdles of silver; on its head a crown of beautiful and vari-coloured plumage, with the left hand on the arm of the throne and the right holding a scimitar studded with silver: at his feet were some precious vessels of stone, marble, and alabaster, in the which were offered the human flesh and blood at the sacrifices. This idol was taken to Mexico, where it was publicly burnt in the court of the inquisition by the decree of the judge provisor of the Indians, D. Ignacio de Castorena, dignitary of the holy metropolitan church and afterwards bishop of Yucatán. He celebrated an *auto de fe* on the occasion in the convent of San Francisco, commanding several Indians, who were afterwards taken when Mexico was over-run in 1723, to assist at the same.

The settlements which have been founded in this province by the aforesaid missions of the Jesuits, are

Mesa del Tonati,	Los Dolores,
Santa Teresa,	San Francisco de Paula,
Jesus Maria,	San Joaquin,
Huaynamota,	Santa Ana,
San Pedro,	Peyotán,
San Juan,	San Lucas.
Tecualmes,	

NAYAUCO, a river of S. Domingo; which rises in the *sierra* of Baruco on the *s.* coast, runs to this *rhumb*, and enters the sea in the point of Beata.

NAZARENO, a settlement of the province and *captainship* of Rio Janeiro in Brazil; situate on the coast at cape Frio.

NAZARENO, a very lofty mountain on the coast of the province and government of Sonora in N. America.

NAZARET, NUESTRA SENORA DE, a settlement of the province and *captainship* of Pernambuco in Brazil, on the coast, near the cape San Agustin.

[NAZARETH, a beautiful town in Northampton county, Pennsylvania, inhabited by Moravians or United Brethren. It is situated eight miles *n.* of Bethlehem, and 49 *n.* by *w.* of Philadelphia. It is a tract of good land, containing about 5000 acres, purchased by the Rev. G. Whitfield, in 1740, and sold two years after to the brethren. They were however obliged to leave this place the same year, where it seems they had made some settlements before. Bishop Nitchman arrived from Europe this year (1740) with a company of brethren and sisters, and purchased and settled upon the spot which is now called Bethlehem.

The town of Nazareth stands about the centre of the manor, on a small creek, which loses itself in the earth about a mile and a half *e.* of the town. It was regularly laid out in 1772, and consists of two principal streets which cross each other at right angles, and form a square in the middle, of 340 by 200 feet. The largest building is a stone house, erected in 1755, named Nazareth hall, 98 feet by 46 in length, and 54 in height. The lower floor is formed into a spacious hall for public worship, the upper part of the house is fitted up for a boarding school, where youth from different parts are under the inspection of the minister of the place and several tutors, and are instructed in the English, German, French, and Latin languages; in history, geography, bookkeeping, mathematics, music, drawing, and other sciences. The front of the house faces a large square open to the *s.* adjoining a fine piece of meadow ground, and commands a most delightful prospect. Another elegant building on the *e.* of Nazareth hall is inhabited by the single sisters, who have the same regulations and way of living as those at Bethlehem. Besides their principal manufactory for spinning and twisting cotton, they have lately begun to draw wax tapers. At the *s.w.* corner of the aforesaid square, in the middle of the town, is the single brethren's house, and on the *e. s. e.* corner a store. On the southernmost end of the street is a good tavern. The dwelling houses are, a few excepted, built of lime-stone, one or two stories high, inhabited by tradesmen and mechancis, mostly of German extraction. The inhabitants are supplied with water conveyed to them by pipes from a fine spring near the town. The situation of the town, and the salubrious air of the adjacent country, render this a very agreeable place.

The number of inhabitants in the town and the farms belonging to it, (Shoeneck included) constituting one congregation, and meeting for divine service on Lord's days and holidays, at Nazareth hall, was, in the year 1788, about 450.]

NAZINTLA, a settlement of the head settlement of the district of Xocatla, and *alcaldia mayor* of Chilapa, in Nueva España. One league to the *s.* of its head settlement.

NEALE, a settlement of the island of Barbadoes, in the parish of St. George.

NEBACH, Santa Maria de, a settlement and head settlement of the district of the *alcaldia mayor* and province of Quiche in the kingdom of Guatemala. It contains 1210 Indians, including those of two other settlements annexed to its curacy, and which were formerly of the religions order of S. Domingo.

NEBOME, a nation of Indians of N. America, dwelling in the *sierras* and mountains, 80 leagues from the town of Cinaloa: 360 of whom, men, women, and children, entered in 1615 to establish themselves in the settlement of Aborozas, of the missions which were held by the Jesuits in that province, voluntarily applying to be taken into the lap of the church, and being excited to this by the instructions they had received from certain Indians who attended Alvar Nuñez Cabeza de Vaca, Miguel Dorantes, and the Negro, Estebanico, in their perigrinations through Florida to Mexico. These Indians, after that they were converted to the faith, returned to their country to see their relations, and a few years after this their example was followed by the whole nation, who embraced the faith and were instructed under the Father Diego Vanderspie, a German, but who met with a violent death at their hands.

Previous to their adopting catholicity, these Indians were far less barbarous than any of those regions. They had houses with clay walls, they cultivated the ground, with the fruits of which and by the chase they maintained themselves. They clothed themselves with the skins of stags and other animals, which they adorned with great nicety. They wore a sort of petticoat which trailed on the ground, and from their waist upwards a cotton mantle. The women were equally modest in their dress as in their appearance and deportment.

NECENDELAN, a settlement of the province and *alcaldia mayor* of Ixcuintepeque in the kingdom of Guatemala, conquered by Pedro de Alvarado in 1523. The natives had the custom, according to Francisco Lopez de Gomara, of playing on some bells which they carried in their hands at the same time that they fought.

[NECESSITY, Fort, in Virginia, is situated in the great meadow, within four miles of the *w.* bounds of Maryland, and on the *n.* side of the head water of Red Stone creek, which empties from the *e.* into the Monongahela, in lat. 39° 43′ *n.* about 26 miles from the spot where this fort was erected. It is 238 miles *e.* by *n.* of Alexandria, and 258 *n. w.* of Fredericksburgh by road distances. This spot will be for ever famous in the history of America, as one of the first scenes of General Washington's abilities as a commander. In 1753, it was only a small unfinished entrenchment, when Mr. Washington, then a colonel, in the 22d year of his age, was sent with 300 men towards the Ohio. An engagement with the enemy ensued, and the French were defeated. M. de Villier, the French commander, sent down 900 men besides Indians, to attack the Virginians. Their brave leader, however, made such an able defence with his handful of men in this unfinished fort, as to constrain the French officer to grant him honourable terms of capitulation.]

NECHAS, San Francisco de, a settlement of the missions which are held by the religious order of S. Francisco, in the province and government of Texas in N. America; situate on the shore of the river of its name. Six leagues from the garrison of S. Antonio de Bejar.

NECHI, a settlement of the province and government of Antioquia in the Nuevo Reyno de

Granada; situate in a long strip or point of land formed by the rivers of its name and that of San Jorge, in the *sierras* of Guamoco.

NECOXTLA, San Francisco de, a settlement of the head settlement of the district of Tequilan, and *alcaldía mayor* of Orizava, in Nueva España, in the middle of a *sierra;* of a very cold temperature, and containing 261 families of Indians, whose trade consists in providing the whole jurisdiction with coals, wood, and torches made of the pine-tree. Three leagues *s. w.* of its capital.

NECOYA, San Bartolome de, a settlement of Indians of the province and government of Mainas and kingdom of Quito; a *reduccion* made by the missions held there by the Jesuits, on the shore of the river Napo.

NECTA, San Pedro de, a settlement of the head settlement of the district and *alcaldía mayor* of Gueguetenango in the kingdom of Guatemala. It is of the Indians of the division of Uzumacintla; annexed to the curacy of its head settlement.

[NEDDICK Cape, or Neddock, lies between York river and Well's bay, on the coast of York county, district of Maine.]

[Neddick River, Cape, in the above county, is navigable about a mile from the sea, and at full tide only for vessels of any considerable burden, it having a bar of sand at its mouth, and at an hour before and after low water, this rivulet is generally so shallow as to be fordable within a few rods of the sea.]

[NEEDHAM'S Point, on the *s. w.* angle of the island of Barbadoes in the W. Indies, is to the *s. e.* from Bridgetown, having a fort upon it called Charles fort.]

[Needham, a township in Norfolk county, Massachusetts, 11 miles from Boston. It is about nine miles in length and five in breadth, and is almost encompassed by Charles river. The lower fall of the river, at the bridge between Newton and Needham, is about 20 feet in its direct descent. Here the river divides Middlesex from Norfolk county. It was incorporated in 1711, and contains 1130 inhabitants. A slitting and rolling mill has lately been erected here.]

[NEEHEEHEOU, one of the Sandwich islands, about five leagues to the *w.* of Atooi, and has about 10,000 inhabitants. Its place of anchorage is in lat. 21° 50′ *n.* and long. 160° 15′ *w.* Sometimes it is called Neheeow or Oneeheow.]

[NEEMBUCU, a town of the province and government of Paraguay; situate on the *e.* bank of the Paraguay, and 28 miles from its junction with the Parana. In lat. 26° 52′ 54″ *s.* and long. 58° 11′ 28″ *w.*]

[NEGADA, or Anegada, one of the Caribbee islands in the W. Indies. It is low and desert, encompassed with shoals and sand banks. It is called Negada, from its being mostly overflown by high tides. It is 69 miles *n. w.* of Anguilla, and abounds with crabs. Lat. 18° 46′ *n.* Long. 64° 22′ *w.*]

NEGELOL, a river of the district of Maquegua in the kingdom of Chile. It runs *w.* and unites itself with the Pivinco to enter the Rapamilahue, changing its name for the Reñaico.

NEGRA, a point of the coast of the province and *captainship* of the Rio Janeiro in Brazil, between the capital and cape Frio.

Negra Muerta, a settlement of the province and government of Tucumán, in the jurisdiction of Xuxuy; situate on the shore of the river Laquiaca.

NEGRETE, a town of the island of Laxa in the kingdom of Chile; situate between the rivers Culavi and Duqueco. On the *s.* it has a fort on the shore of the river Biobio to restrain the Araucanos Indians.

NEGRILLO, Puntas del. The *w.* head of the island Jamaica; consisting of two remarkable points, with the names of North and South, three leagues apart, and forming in the intermediate space a semicircular bay, called Long bay, in the which is a small island close upon the shore. Vessels do not enter this bay but under absolute necessity, as it is much exposed to the *w.*, *n.*, and *s.* winds. In lat. 18° 27′ *n.* Long. 78° 17′ *w.*

Negrillo, another point, on the coast of the province and *corregimiento* of Paita in Peru.

Negrillo, a shoal of rock near the coast of the province and government of Cartagena and Nuevo Reyno de Granada; between this city and the point of Canoa.

Negrillo, another shoal of rock in the sound of Campeche.

NEGRILLOS, a settlement of the province and *corregimiento* of Carangas in Peru, of the archbishopric of Charcas, annexed to the curacy of Huachacalla; situate near the source of the river Camorones.

Negrillos, some isles or shoals of rocks of the gulf or bay of Mexico, to the *w.* of the Alacranes isles.

NEGRO, a large and navigable river to the *n.* of the Marañon or Amazon. It runs from *w.* to *e.* laves many and extensive countries inhabited by barbarian Indians, and communicates with the Orinoco by a channel discovered by the Father Manuel Roman, of the Jesuits, native of Olmedo in Castilla la Vieja, missionary of the province of

Santa Fé in the Orinoco, where he was for more than 30 years. Once, navigating the aforesaid channel, he found himself in the river Negro, where he met with some Portuguese who had penetrated as far as this spot from Para on discoveries. This river, Negro, collects in its course the waters of the Ijie, Iquiari, Yurubesch, Nuissi, Casiari, Catabulú, Aravidá, Blanco, and Yaguapiri; and, being much enlarged by these, it becomes at its mouth a league and an half wide. Although geographers vary in describing the course of this river, we have followed Don Carlos de la Condamine, of the royal academy of the sciences at Paris, who reconnoitred it on his return to Europe from Peru by the river Amazonas. It is at its greatest width 1203 toises, as measured by this geometrician; this being the place where the Portuguese have built a fort, in lat. 3° 9′, maintaining in it a detachment of the garrison of Pará for the purpose of catching Indians for the working of the mines. On the shores of this river are different settlements of the missions established by the religious Carmelites of Portugal.

According to the investigations of the Father *Fr.* Antonio Caulin, in his Modern History of Nueva Andalucia, this river rises in the *serranías* of Yaquesa near Popayán, and in which he agrees with the aforesaid academician Don Carlos de la Condamine. It receives on the *n.* shore the Pativita, which runs in the same direction as the Inirícha, so close to it as to be separated only by a very narrow isthmus, and having on its shores the nations of the Civitencs, Guarinimanases, and Maipures. It is then entered by the Aqui and the Itivini, bringing along with it those of the Jehani, Equegani, and Mee, on the borders of which dwell the Borepaquinavis Indians. Before the Mee falls into the Itivini, it throws out by its *w.* shore a river of its name into the Casiquiare, and in the island thus formed, a stream called the Itiriquiri falls into the Negro, on the shores of which dwell the nation of the Avinavis Indians: also, at a short distance, is the union of the Casiquiare and the Negro, beyond which lies the mouth of the Cavapono, and then the mouth of the Guivaro, inhabited by Cogenas Indians. Three days journey down the river is found a torrent, caused by a reef of rocks, which is a continuation from the skirts of the mountain Nuea, and lower still are the mouths of the river Blanco or of Agnas Blancas, called by the natives Aguapiri, which enters the river Negro 35 leagues before this enters the Marañon. By the *s.* it receives the Mapicoro, then the Matrichi and the Danigua, amongst the which dwell the Mauisipitana nation.

The Portuguese, as we before observed, come hither to catch Indians to make them slaves in the mines. They enter by the mouth of the Casiquiare, pass the channel of Mee, and, leaving their vessels, pass by land to the port Manuteso of the river Cimite, an arm of the Atabapo; also others going by the river Negro, enter the mouth of the Itivini, and pass from thence to the river Temi. The Negro enters the Marañon or Amazonas by the *n.* part, in about lat. 3° 16′ *s.*

[For a table of longitudes and latitudes of the most important places in these parts, see the end of the general preface.]

Negro, another, a large and abundant river of the province and government of Buenos Ayres, in the mountains of Brazil. It runs *w.* and then turning *s. w.* and after collecting the waters of various others, so as greatly to increase its stream, unites itself with the Uruguay to enter in a very much increased body the river La Plata.

Negro, another, of the province and *corregimiento* of Tunja in the Nuevo Reyno de Granada. It rises close to the settlement of Las Guadas, runs *n.* and enters the Grande of the Magdalena, to the *w.* of the city of Velez.

Negro, another, in the province of Ubaque, in the same kingdom as the former, which rises near Santa Fé, in the mountains to the *e.* and enters the Meta about 75 miles from its source. This river is called also Caquesa, as it passes near the settlement of this name, and again, because it soon after that, receives a stream of black waters; and the small difference between the words Caquesa and Caqueta having caused foreigners to confound this river Negro (or Black) with the former of which we have treated, so that a great confusion has arisen amongst geographers, as also a doubt whether there was any communication by that river with the Orinoco and the Marañon.

Negro, another, of the province and government of Veragua in the kingdom of Tierra Firme, which rises in the interior of the mountains and runs into the sea, between the Cocle and the Escudo de Veragua.

Negro, another, of the province and government of Texas in N. America, which rises in the mountains of Caligoa, runs *s.* for many leagues, and bending its course with many windings to *n. n. w.* enters the Colorado very near its mouth.

Negro, another, of the province and government of Tucumán in Peru, of the district of the city of Xuxuy, which runs *e.* and enters the Vermejo.

Negro, another, a small river of the province and government of Neiva in the Nuevo Reyno de

Granada. It rises near the settlement of Otaz, and enters the Grande de la Magdalena a little from its source.

NEGRO, another, also a small river of the province and government of Maracaibo in the same kingdom as the former, which rises in the valley of Perija, runs *e.* and enters the great lake of Maracaibo by the *s.* part.

NEGRO, a small river, of the province and *captainship* of San Pablo in Brazil, distinct from that of which we have spoken, in that kingdom. It rises in the mountains of the coast, runs *n. w.* and unites itself with the Itapeba to enter the grand river Curitiva or Iguazú.

NEGRO, another, a small river of the province and *corregimiento* of Chachapoyas in Peru, which runs *n. n. e.* and enters the Moyobamba.

NEGRO, another, a small river of the province and government of Paraguay, which runs *e.* and enters the Grande de Paraná.

NEGRO, another, a small river, called Arroyo Negro, of the province and government of Buenos Ayres in Peru. It runs *w.* and enters the Uruguay, between those of S. Francisco and Bellaco.

NEGRO, another, of the Nuevo Reyno de Granada, distinct from the above, in the district and jurisdiction of the city of San Juan Jiron. It is small, and enters the Lebrija a little above this city.

NEGRO, another, of the kingdom of Brazil, which rises in the country of the Barbados Indians, runs *n. n. w.* and enters the Topayos a little before it does the Yaguaricara.

NEGRO, a settlement, called also Rio Negro, of the district and government of San Juan Jiron, in the Nuevo Reyno de Granada; of an hot temperature, abounding in vegetable productions, particularly *cacao*, the best crops being here of any in the province. Its population is reduced; it is situate near the river Negro, of which we have before treated, and which gives it its name. Six leagues from its capital.

NEGRO, a fort of the province and government of Tucumán in Peru.

NEGRO, a cape or point of land on the exterior coast of the straits of Magellan, on an island formed by the entrance of the channel of S. Barbara.

NEGRO, another cape, of the *s.* coast of Nova Scotia or Acadia, opposite Brown bank.

NEGRO, another cape, of the coast of Brazil, in the province and *captainship* of Rio Grande, between this and the settlement of Natal, where the Portuguese have a fort, called De Los Reyes.

NEGRO, a very lofty mountain, called Cerro Negro, in the province and *corregimiento* of Itata, and kingdom of Chile, between the rivers Itata and Claro.

NEGROES, different nations of various kingdoms and provinces of Africa, who, although not aborigines of America, have a place in this history, as forming a principal part of the inhabitants of these regions, and who, at the present day, if they do not exceed, at least equal in numbers the natives. For these are the people who labour in the mines, who cultivate the land, who are employed in all the servile offices in America, in the dominions of Spain, Portugal, France, England, Holland, &c. They are bought by these nations on the coasts of Africa, and are carried to America, where they are treated and considered as slaves with the greatest rigour and inhumanity, and as though they were not rational creatures. The celebrated *Fr.* Bartolome de las Casas, bishop of Chiapa, was the person who, with a discreet zeal, proposed to free the Indians from servitude, and to procure Negroes for the laborious employments; as though, forsooth, this part of the human species should, on account of their difference of colour, want the privileges of humanity. The shades of complexion amongst themselves vary much, according to their different provinces; and they are distinguished by casts, called the Congos, Mandingas, Chalaes, Ararares, and many others.

They are, in general, well made, muscular, strong, and capable of bearing much labour. They have a flat nose, pouting lips, black and woolly hair, and white teeth. These casts have features peculiar to themselves; thus, for instance, the Chalaes have certain marks or scarifications on their cheeks, made whilst they were yet children; the Ararares file the points of their teeth, &c.

The English, Dutch, and Portuguese, carry on this infamous commerce on the coasts of Guinea, and sell the Negroes in America and in the islands, where, after certain years of slavery and servitude, they may ransom themselves of their master, paying for their freedom the same sum at which they were bought; but, notwithstanding this alleviation, and which was propagated by the Spanish government, little redress is procured to their sufferings, through the interestedness and cruelty of the masters.

It is certain that the propensities of the Negro are most vicious, that they are fraudulent, superstitious, vindictive, cruel, and thievish, and that without the rigour manifested towards them, it would be impossible to manage them; but the love of liberty and the injuries of servitude plead loudly in their exculpation; nor, indeed, have there been wanting examples of some who for their moral virtues might vie even with the beings of civilized nations.

The Spaniards, who, amongst all the rest, are those who treat them the least cruelly, have a short time since the conquest of their provinces supplied themselves with Negroes under different contracts, entered into first with the Genoese, afterwards with Don Domingo del Grillo, the council of Sevilla, Don Nicolas Porcio, Don Bernardo Marin y Guzman, the company of Portugal, the French Guinea company, as far down as the year 1713; when by the peace of Utrecht the trade was granted to the English company for 30 years, namely to 1743: after this the person employed in this business was Don Joseph Ruiz de Noriega, and after him the company of merchants of Cadiz. The first Negroes brought to America by the Spaniards, was through the grant of Charles V. made in 1525 to Lorenzo Garrebood his *mayor domo;* by which he was empowered to introduce 4000, and although, owing to the inconvenience found to arise from the practice, it was ordered to be discontinued for eight years, a certain recompence being paid to the aforesaid person as an indemnification, yet necessity obliged its readoption, as the Indians were not equal to the fatigues required of them, and as, now, their numbers were sensibly diminishing.

In nearly all the settlements, the Negroes are divided into two classes, which are slaves and freemen, and both of these into Criollos and Bozales: a part of the former (the slaves) are employed in tilling the ground, and all the rest in different hard labours, by which to procure their livelihood, giving to their masters so much daily, and keeping the remainder for their own sustenance. The violence of the heat and their own natural warmth of temperature will not permit them to wear any clothing whatever; they, consequently, go quite naked, with the exception of a small cloth round their middle. The same is also the ease with the women slaves, some of whom married, live in the huts with their husbands, and others being employed in the cities, where they gain their livelihood by labour, or by selling in the market-places and through the streets all kinds of eatables, sweetmeats, fruit, and different kinds of broths and drinks, maize-broth, and *cazave*, which serves as bread. Those women who have infant children (and there are hardly any without them) sling them behind their backs, so that they may not interfere in their daily labours or use of their arms; they also give the child the breast by offering the dug under the arm or throwing it over the shoulder. Thus they, without trouble, rear their offspring, nor is this practice to be wondered at, inasmuch as some of their breasts are pendulent below their waist, arising, no doubt, from their never using any stay whatever.

In order to avoid a contraband trade of Negroes, or that they might be imported without paying the regular duties, it was established that a mark should be put upon them, namely the letter R, with a crown above, branded on the left breast; but this practice, so detestable in a civilized and Catholic nation, was abolished by order of Charles III. that generous hearted protector of humanity. Animated by this example, the English endeavoured to abolish this infamous commerce, but the whole of that nation not agreeing on the subject, it was at last recommended that provisions should be made by government for their better treatment and condition, a reward being stated for such persons as should bring the most Negroes alive from the coast of Guinea out of a certain number. On this occasion a porcelain medal was made in England, representing one of these unfortunate creatures, with the motto of, "Am I not man as thou art; am I not thy brother?" Many English, French, and Spaniards, enlightened by the reason of the present age, have given liberty to their slaves; and we may hope for the day when this miserable race shall no longer be shut from the privileges to which they are by nature entitled.

[That the English traders are at last checked in this inhuman commerce, we believe cannot be doubted. They will not risk a conviction of felony, and sentence of transportation to Botany bay. The American government too, having abolished the traffic, and the decision in the noted case of the Amedie, having shewn British crusiers in what manner they may enforce the American prohibition; few vessels bearing that flag are engaged in it, compared with the former amount. But, on the other hand, a prodigious slave trade is still carried on by the Portuguese and Spaniards, and, in the sixth report of the African institution, the directors have no hesitation in stating, from their own information, that between 70,000 and 80,000 Negroes were carried over to America by the above nations in the year 1810.]

NEGROS, an extremity of the *n. w.* coast of the island of Martinique, between the bay of Navires and fort Real.

NEIQUITOS, a settlement of the province and government of Maracaibo in the Nuevo Reyno de Granada; situate *s.* of the city of Truxillo, and near the settlement of Esemxaque.

NEIVA, a province and government of the

Nuevo Reyno de Granada, called De los Pantagoros in the time of the Indians. It is entirely of a level territory, extending 80 leagues from *n.* to *s.* on either side of the river Grande de la Magdalena; this dividing it into High and Low. It is irrigated by many streams, which descend from the *cordilleras*, surrounding it as it were with a wall. One of these *cordilleras* is by the extensive *llanos* of San Juan, and the other by the equinoctial provinces, at 20 leagues distance, though in some parts less, according to the uncertain manner in which the mountains run more or less far into the *llanuras*. Its jurisdiction is bounded by that of the cities of Tocaima, Mariquita, and La Plata. It is very abundant in gold mines, and fertile in vegetable productions, such as maize, *yucas*, potatoes, *cacao*, tobacco, and a variety of fruits and sugar canes, of the which are made delicious sweetmeats and conserves. In the woods are found fine timber, such as cedar, walnut, and *guayacanes*, which has a tendency to become petrified. The neat cattle bred in the *llanos* or plains, is in such abundance as to furnish with supplies the whole kingdom, and particularly the capital of Santa Fé, and notwithstanding the prohibition against carrying any of this food to Popayán, yet is it constantly done.

The temperature of this province is very hot and unhealthy, and the disease of the *carate* is very common here, being a scrophula of various colours breaking out over the whole body, causing great heat and irritability, infecting the blood to such a degree that the malady becomes hereditary. It is also inflicted with the plague of mosquitos, spiders, gnats, centipeds, flies, hornets, ants, various kinds of snakes, and particularly with an insect similar to that known in Spain by the name of *cochinilla de San Anton*, of a red colour and black head, and called here *coya*, which, although it does not bite, yet should it burst and its blood touch any part of the body, save the soles of the feet and the palms of the hands, it is so active a poison as to produce instant death, causing the whole of the blood of the human body to coagulate. It is remarkable the instinct, by which the neat cattle, the horses and mules, shun this venomous insect. As its poison acts as a coagulator, a method has been discovered by some muleteers, of passing the body of the person who has been bitten gradually through the flames of a small tiro made of straw, and this with some success. [According to Mr. Bouker's voyage, this account of the *coya* is merely fibulous.]

Neiva, the capital of the above province, is called La Concepcion del Valle de Neiva; founded in 1550 by Captain Juan Alonso, in the part where at present stands the settlement of Villa Vieja, and where it remained until 1569, when it was destroyed by the Pijaos Indians. In 1612 the governor Don Diego de Hospina began to resettle it in the place where it now is, eight leagues from the former, on the shore of the Rio Grande de la Magdalena. It is of an hot temperature, abounding in vegetable productions, gold and cattle, as does all the province. It has besides the parish church an hospital of the religious order of San Francisco. The population consists of 2000 housekeepers, the greater part being people of colour, although there are not wanting some noble families. It is 107 miles *s. w.* from Santa Fé, 63 *s. s. w.* from Tocaima, in lat. 3° 14′ *n.*

Neiva, with the addition of Vieja, a settlement of the same province and kingdom; situate on the margin of the river Magdalena, where stood the city previous to its removal to its present spot. This settlement is much reduced and very poor.

Neiva, a river of the island S. Domingo; which rises in the mountains of the centre, and near to those of Ciboo, runs *w.* many leagues, and passes to the *s.* with an abundant stream through the valley of its name, and enters the sea in the bay which is also so called.

Neiva. The aforesaid valley is large and beautiful, and running from *n.* to *s.* towards the coast of the latter rhumb, its sides being hemmed in by the rivers of its name, and of Las Damas, as also by the lake Enriquillo or Henriquille. [This valley contains about 80 square leagues, abounds with game, and is a chosen spot for flamingoes, pheasants, and royal or crowned peacocks. These last have a more delicate flavnur and more brilliant plumage than the peacocks of Europe. Nine leagues from the *w.* bank of the Neiva is the town, containing about 200 houses, and can turn out 300 men fit to bear arms. This town is 15 leagues *w.* by *n.* of Azu, and 16 from the point where the line of demarcation ents Brackish pond. This territory produces a sort of plaster, talc, and fossil salt. The natural re-production of the salt is so rapid that a pretty large hollow is absolutely filled up again in the course of a year. The river might be rendered navigable for small craft, and the plain is able to afford eligible situations for 150 sugar plantations.]

Neiva, a bay on the *s.* coast of the same island of S. Domingo, between that of Ocoa and that of Petit Trou. [It is also situated at *n. n. e.* from cape Beata. Lat. 18° 16′ *n.* Long. 70° 56′ *w.*]

Neiva, a river of the province and government of its name in the Nuevo Reyno de Granada, which rises *s.* of the capital, passes opposite to it, and shortly after enters the Magdalena.

[NELSON, a county of Kentucky. Chief town, Bairdstown.]

[NELSON'S Fort, a settlement on the *w.* shore of Hudson's bay; situate at the mouth of a river of the same name, 250 miles *s. e.* of Churchill fort, and 600 *n. w.* of Rupert's fort, in the possession of the Hudson's bay company. It is in lat. 57° 12′ *n.* and long. 92° 42′ *w.* The shoals so called are said to be in lat. 57° 35′ *n.* and long. 92° 12′ *w.* and to have high water at full and change days at 20 minutes past eight o'clock.]

[NELSON'S River is the *n. w.* branch of Hayes river, on the *w.* shore of Hudson's bay, which is separated into two channels by Hayes island, at the mouth of which Nelson's fort is situated.]

NEMBUCHU, a settlement of the province and government of Paraguay; situate on the shore of this river before it reaches the city of Corrientes.

NEMEOUGAMIOU, a small lake of the country of Hudson's bay, between the great lake Mistasins and that of Nemiscau, and formed by the river Rupert.

NEMISCAU, a small lake of the same county as the former; also formed by the river Rupert at its mid course, to enter the Mistasin.

NEMOCON, a settlement of the *corregimiento* of Zipaquira in the Nuevo Reyno de Granada. It is of a cold and moist temperature, celebrated for the capital merchandise which it had, as well as for its very white salt found in some large saline earths, and which are formed by certain fountains abounding in its territory. From hence all the other provinces are supplied with this article, it being esteemed superior to any other; so that it produces upwards of 20,000 dollars annually. This settlement was conquered by Gonzalo Ximenez de Quesada in 1537. Its population is small, since, amongst the rest, we find only 80 Indians. It has, besides the parish church, a chapel, with the dedicatory title of Nuestra Señora de Checua, which is a vice-parish. Two leagues *s.* of Guatavita, and nine *n.* of Santa Fé, in the road which leads to Tunja.

[NENAWEWHCK Indians inhabit near Severn river, *s.* of Severn lake.]

NENINCO, a settlement of Indians of the island of Laxa in the kingdom of Chile; situate on the shore of the river Pecoiquen.

NEOCOYAES, a barbarous nation of Indians of the province and country of Las Amazonas, who dwell amidst the woods to the *n.* of the river Napo. Of some of these has been formed the settlement of San Miguel de los Neocoyaes, by the missionaries of the Jesuits, dependent or annexed to the settlement of El Nombre de Jesus, in lat. 1° 33′ *s.*

[NEOMINAS River, on the coast of Peru, is 12 or 14 leagues to the *n. w.* of Bonaventura river. It is a large river, and empties into the ocean by two mouths. The shore is low, but there is no landing upon it, as it is inhabited only by savages whom it would not be very safe to trust, as their peaceable or hostile disposition towards Europeans cannot be easily known. The coast, though in the vicinity of the most flourishing Spanish colonies, remains unfrequented and wild. Palmas island is opposite to this river, being low land, and having several shoals about it; and from hence to cape Corrientes is 20 leagues to the *n. w.* The river and island are in lat. about 4° 30′ *n.*]

NEOUISACOAUT, a river of Canada in N. America. It runs *n. e.* and enters lake Superior.

[NEPEAN Island, a small island of the S. Pacific ocean, opposite to port Hunter, on the *s.* coast of Norfolk island.]

[NEPEAN Sound, an extensive water on the *n. w.* coast of N. America, having a number of islands in it, in some charts called Princess Royal islands. It opens *e.* from cape St. James, the southernmost point of Washington's or Queen Charlotte's islands. Fitzhugh's sound lies between it and Queen Charlotte's sound to the *s.*]

NEPENA, a settlement of the province and *corregimiento* of Santa in Peru.

NEPIGON, a lake of Canada, to the *n.* of lake Superior, with which it communicates by a large arm.

NEPOHUALCO, a settlement of the province of Cempoala in the time of the gentilism of the Indians, and where the Chichimecos established themselves when they left Chicomoztoc or Siete Cuevas. They gave a name in their language signifying Counter, since there they counted the numbers of those who had arrived. They lived on friendly terms with the Totonaques, a noble of whom, named Xatontan, gave them clothes to cover themselves with, also flesh of different animals, which they used to eat raw.

[NEPONSET, a river of Massachusetts, originates chiefly from Muddy and Punkapog ponds in Stoughton, and Mashapog pond in Sharon, and after passing over falls sufficient to carry mills, unites with other small streams, and forms a very constant supply of water for the many mills situated on the river below, until it meets the tide in Milton, from whence it is navigable for vessels of 150 tons burden to Boston bay, distant about four miles.

There are six paper mills, besides many others of different kinds, on this small river.]

NEPOS, a settlement of the province and *corregimiento* of Caxamarca in Peru.

NEQUE, an island of the N. sea; one of the Lucayas; to the *e.* of Bahama.

NEQUEHUAYOCONDOR, a settlement of the province and *corregimiento* of Guanta in Peru; annexed to the curacy of Tambillo.

NERAGANSAT, a river of the province and colony of New England in N. America.

[NERUKA, a port in the island of Cape Breton, where the French had a settlement.]

[NESBIT'S Harbour, on the coast of New Britain, in N. America, where the Moravians formed a settlement in 1752; of the first party, some were killed and others were driven away. In 1764, they made another attempt under the protection of the British government, and were well received by the Esquimaux, and by the last account the mission succeeded.]

[NESCOPEC River falls into the *n. e.* branch of Susquehannah river, near the mouth of the creek of that name, in Northumberland county, Pennsylvania, and opposite to the town of Berwick, 83 miles *n. w.* of Philadelphia, and in lat. 41° 3'. An Indian town, called Nescopec, formerly stood near the site of Berwick.]

NESHIMENECK, a river of the colony and province of Pennsylvania in N. America.

NESKY, a point on the coast of the province of Sagadahook, one of those which form the great bay of Penobscot.

[NETHERLANDS, New, is the tract now included in the states of Now York, New Jersey, and part of Delaware and Pennsylvania, and was thus named by the Dutch. It passed first by conquest and afterwards by treaty into the hands of the English.]

NEUERI, a river of the province of Barcelona, and government of Cumaná, which rises in the mountains of Bergantin, runs *n.* collecting the waters of various others, and empties, much increased, into the sea, between the cities of Barcelona and Cumaná, but nearest to the former, forming a port which was discovered by Gerónimo de Ortal.

[The Neueri lies 16 leagues *e.* of the Unare. Its source is about 20 leagues *s.* of its mouth. The narrowness of its channel, and the waters it receives from other rivers, give it a rapidity and force which defies all the efforts of navigation, until a little above Barcelona.]

NEUF Port, a new settlement of New France or Canada in N. America, on the shore of the river S. Lawrence, with a good port. W. of Quebec.

Neuf, another port, of Long Island, in the province of New York, to the *s.* and in the strait which this island forms with the continent.

Neuf, a cape or point of land on the coast of Newfoundland, between Cataline bay and cape Lorian.

NEULTRA, or Neuter, a narrow strait of New N. Wales, in the reign of the Arctic pole, between lat. 62° and 63° *n.* discovered by Thomas Roe, an Englishman.

[NEUS, a river of N. Carolina, which empties into Pamlico sound below the town of Newbern. It is navigable for sea vessels 12 miles above Newbern, for scows 50 miles, and for small boats 200 miles.]

[NEUSTRA Sennora, Baia de, or Our Lady's Bay, on the coast of Chile, on the S. Pacific ocean, in S. America, is 30 leagues from Copiapa, and 20 *s. s. w.* of cape George. It is indifferent riding in this bay, as the *n. w.* winds blow right in, and the gusts from the mountains are very dangerous.]

NEVADA, a very lofty and extensive *sierra* of the province and government of Santa Marta, in the Nuevo Reyno de Granada. It is one of the three arms or branches of the *cordillera*, of the highest mountains traversing the whole of America for more than 2000 leagues. The third of these branches begins in the province of Santa Marta, runs through the kingdom of Tierra Firme, narrowing between Panamá and Portovelo, and forming the isthmus which divides the two seas, the N. and S. and then extends itself along into N. America and through the provinces of Nueva España. These mountains of the *sierra* Nevada abound in mines of gold and silver, exquisite and bulky timber, strange birds and animals, the same as will be found enumerated under the article Annes, the same being a part of the *sierra*; but it must be observed, that the climate of these mountains varies considerably in the different kingdoms and provinces through which they run.

Nevada, a mountain perpetually covered with snow, in the province and government of Tucumán in Peru, of the jurisdiction of the city of Córdoba, to the *s. s. w.* of this capital.

NEVADAS, some islands situate near the *s.* coast of the strait of Magellan. They are various, and form the said coast from the month of the channel of San Juan to Monday cape.

NEVAS, a barbarous nation of Indians of the

province and country of Las Amazonas; being descendants of the Semigayes and inhabiting the woods between the rivers Tigre and Curaray.

[NEVERSINK Creek, a stream in the Hardenberg patent, in Ulster county, New York. On an island in this creek Mr. Baker having cut down a hollow beech tree, in March 1790, found near two barrels full of chimney swallows in the cavity of the tree. They were in a torpid state, but some of them being placed near a fire, were pleasantly reanimated by the warmth, and took wing with their usual agility.]

[NEVIL Bay, on the *w.* shore of Hudson's bay, is nearly due *w.* a little *n.* from cape Digges and Mansel island at the entrance into the bay.]

NEVIS, an island of the N. sea; one of the Lesser Antilles; situate a league *s.* of the island S. Christopher or Christoval. It is two leagues long and one wide; is nothing but one lofty mountain, the skirts of which are very fertile for the space of half a league and upwards, the soil losing its property as it approaches the top.

At its first establishment it was very flourishing, and contained 30,000 inhabitants, but owing to what it suffered by the invasion of the French in 1706, and some epidemic distempers, as well as some revolutions, its population has been so much diminished as to consist at the present day of no more than 3000 whites, and 6 or 7000 Negro slaves. It produces much cotton, sugar, and tobacco; these being the only articles of its commerce. The natives have great credit in America, as being active and industrious, and they are particularly distinguishable for the cleanliness of their persons and houses. It has some very good roads, which lead to various small ports, at which are the towns of Newcastle, Littleborough or Moreton, and Charlestown, the capital. The island is divided into three parishes or districts, and employs annually in its traffic 20 vessels. The money arising from the exports of cotton and sugar to England in 1770 amounted to 44,000*l.*; and the value of the rum, lemons, and molasses was 14,000*l.* more. The French restored this island to its former possessors at the peace of Utrecht.

[It is generally believed that Columbus bestowed on it the appellation of Nieves or the Snows, from its resemblance to a mountain of the same name in Spain, the top of which is covered with snow; but it is not an improbable conjecture, that in those days a white smoke was seen to issue from the summit, which at a distance had a snow-like appearance, and that it rather derived its name from thence. It is generally thought that the island was produced from some volcanic explosion, as there is a hollow or crater near the summit still visible, which contains a hot spring strongly impregnated with sulphur; and sulphur is frequently found in substance in the neighbouring gullies and cavities of the earth.

The country is well watered, and the land in general fertile, a small proportion towards the summit of the island excepted, which answers however for the growth of ground provisions, such as yams and other esculent vegetables. The soil is stony; the best is a loose black mould, on a clay. In some places, the upper stratum is a stiff clay, which requires labour, but properly divided and pulverised, repays the labour bestowed upon it. The general produce of sugar (its only staple production) is one hogshead of sixteen cwt. per acre from all the canes that are annually cut, which being about 4000 acres, the return of the whole is an equal number of hogsheads, and this was the average fixed on by the French government in 1782, as a rule for regulating the taxes. As at St. Christopher's, the planters seldom cut *ratoon* canes.

This island, small as it is, is now divided into five parishes, though perhaps only three at the time Alçedo wrote. It contains, as he observes, a town called Charlestown, the seat of government and a port of entry, and there are two other shipping places, called Indian-castle, and Newcastle. The principal fortification is at Charlestown, and is called Charles fort. The commandant is appointed by the crown, but receives a salary from the island.

The government, which is included in that of the Leeward Charaibean islands, in the absence of the governor-general, is administered by the president of the council. This board is composed of the president, and six other members. The house of assembly consists of 15 representatives; three for each parish.

The administration of common law is under the guidance of a chief justice, and two assistant judges, and there is an office for the registry of deeds.

The number of white inhabitants in 1798 did not exceed 600, while the Negroes amounted to about 10,000; a disproportion which necessarily converts all such white men as are not exempted by age or decrepitude, into a well-regulated militia, among which there is a troop consisting of 50 horse, well mounted and accoutred. English forces, on the British establishment, they have none.]

[The English first established themselves in this island in the year 1628, under the protection and encouragement of Sir Thomas Warner; but it was under the administration of his immediate successor, Mr. Lake, that Nevis rose to opulence and importance. He made this island the place of his residence, and it flourished beyond example. It is said, that about the year 1640, it possessed 4000 whites: so powerfully are mankind invited by the advantages of a mild and equitable system of government. The inhabitants of this little island, observes Mr. Bryan Edwards, live amidst the beauties of an eternal spring, beneath a sky serene and unclouded, and in a spot inexpressibly beautiful.

In the report of the privy council on the slave trade in 1788, the British property vested in this island is estimated at 30,000 taxed acres of patented estates, and the Negroes are computed at 8000, at 50*l.* each Negro.

By return to house of commons 1806, the hogsheads of sugar of 13 cwt. exported, were as follows,

In 1789,	4000
1799,	3850
1805,	2400

The official value of the imports and exports of Nevis were, in

	Imports.	*Exports.*
1809,	£ 89,062	£20,500
1810,	£126,443	£11,764.

And the quantities of the principal articles imported into Great Britain were, in

	Coffee.		Sugar.		Rum.	Cotton Wool.
	Brit. Plant.	For. Plant.	Brit. Plant.	For. Plant.		
	Cwt.	Cwt.	Cwt.	Cwt.	Galls.	Lbs.
1809,	—	31	60,872	—	52,478	17,463
1810,	18	—	87,393	—	67,010	11,160

A detailed account of the vessels, &c. that have cleared outwards from this island between January 1807, and January 1808, with their cargoes, is included in the island Montserrat.

According to what has been above stated, the population of this island amounted, in 1640, to 4000 whites besides Negroes; in 1780, to 3000 whites and 6 or 7000 Negroes, and in 1798, to 600 whites and 10,000 Negroes. The first and last of these accounts are derived from Bryan Edwards, the former from our author Alçedo. The following statements are official.

By report of the privy council in 1788, and by a subsequent estimate, the population of Nevis amounted to

Years.	Whites.	People of Colour.	Slaves.
In 1787	1514	140	8420
1805	1300	150	8000

The import of slaves into Nevis by report of privy council 1788, at a medium of four years, and by a return to house of commons in 1805, at a medium of two years to 1803, was

Average of	Imports.	Re-exports.	Retained.
Four years to 1787	544	—	544
Two years to 1803	228	—	228

The middle of this island is in lat. 17° 8′ *n.* and long. 62° 38′ *w.*]

Nevis, a small lake of Canada in N. America, between that of Natovagamiou and the river St. Lawrence.

[NEW, a river of N. Carolina, which empties, after a short course, into the ocean, through New River inlet. Its mouth is wide and shoal. It abounds with mullet during the winter season.]

[New Albion, a name given to a country of indefinite limits, on the *w.* coast of N. America, lying *n.* of California. See Albion.]

[New Andalusia, a province of Tierra Firme, S. America, lying on the coast of the N. sea, opposite to the Leeward islands; bounded by the river Orinoco on the *w.* This country is called Paria by some writers. Its chief town is St. Thomas. Some gold mines were discovered here in 1785.]

[New Andover, a settlement in York county, district of Maine, which contains, including Hiram and Potterfield, 214 inhabitants.]

[NEW ANTICARIA, a town of New Spain, 34 leagues *n.* of Acapulco.]

[NEW ANTIGUERA, an episcopal city of New Spain, in the province of Guaxaca, erected into a bishopric by Paul III. 1547. It has a noble cathedral, supported by marble pillars.]

[NEW ATHENS, or TIOGA Point, stands on the post-road from Cooperstown to Williamsburg, in Luzerne county, Pennsylvania, on the point of land formed by the confluence of Tioga river with the *e.* branch of Susquehannah river, in lat. 41° 54′ and long. 76° 32′ *w.* and about three miles *s.* of the New York line, 16 miles *s. e.* by *e.* of Newtown in New York, 14 *s. w.* of Owego, and 82 *s. w.* of Cooperstown.]

[NEW BARBADOES, a township in Bergen county, New Jersey.]

[NEW BEDFORD, a post-town and port of entry in Bristol county, Massachusetts, situated on a small bay which sets up *n.* from Buzzard's bay, 45 miles *s.* of Boston. The township was incorporated in 1787, and is 13 miles in length and four in breadth; bounded *e.* by Rochester, *w.* by Dartmouth, of which it was originally a part, and *s.* by Buzzard's bay. Acchusnutt was the Indian name of New Bedford, and the small river of that name, discovered by Gosnold in 1602, runs from *n.* to *s.* through the township, and divides the villages of Oxford and Fairhaven from Bedford village. A company was incorporated in 1796, for building a bridge across this river. From the head to the mouth of the river is seven or eight miles. Fairhaven and Bedford villages are a mile apart, and a ferry constantly attended is established between them. The harbour is very safe, in some places 17 or 18 feet of water; and vessels of 3 or 400 tons lie at the wharfs. Its mouth is formed by Clark's neck on the *w.* side, and Sconticutt point on the other. An island between these points renders the entrance narrow, in five fathoms water. High water at full and change of the moon 37 minutes after seven o'clock. Dartmouth is the safest place to lie at with an *e.* wind; but at New Bedford you will lie safe at the wharfs. The river has plenty of small fish, and a short way from its mouth they catch cod, bass, black fish, sheep's head, &c. The damage done by the British to this town in 1778 amounted to the value of 97,000*l.* It is now in a flourishing state. In the township are a post-office, a printing-office, three meetings for Friends, and three for Congregationalists, and 3313 inhabitants. The exports to the different States and to the W. Indies for one year, ending September 30, 1794, amounted to 82,085 dollars. It is 218 miles *n. e.* by *e.* of Philadelphia.]

[NEW BISCAY, a province in the audience of Galicia in Old Mexico or New Spain. It is said to be 100 leagues from *e.* to *w.* and 120 from *n.* to *s.* It is a well watered and fertile country. Many of the inhabitants are rich, not only in corn, cattle, &c. but also in silver mines. See VISCAY.]

[NEW BOSTON, a township in Hillsborough county, New Hampshire, about 70 miles *w.* of Portsmouth. It was incorporated in 1763, and contains 1202 inhabitants.]

[NEW BRAINTREE, a township in Worcester county, Massachusetts, consisting of about 13,000 acres of land, taken from Braintree, Brookfield, and Hardwick, and was incorporated in 1751. It contains 490 inhabitants, mostly farmers, and lies 17 miles *n. w.* of Worcester, and 50 *n. w.* of Boston.]

[NEW BRITAIN. See AMERICA, LABRADOR, and BRITAIN, NEW.]

[NEW BRITAIN, a township in Buck's county, Pennsylvania.]

[NEW BRUNSWICK, in the state of New York, is situated on Paltz kill, about eight miles *s. w.* of New Paltz, and 69 *n.* of New York city.]

[NEW BRUNSWICK, in Middlesex county, New Jersey. See BRUNSWICK.]

[NEW BRUNSWICK, a British province in N. America, the *n. w.* part of Nova Scotia; bounded *w.* by the district of Maine, from which it is separated by the river St. Croix, and a line drawn due *n.* from its source to the Canada line, *n.* by the *s.* boundary of the province of Lower Canada, until it touches the sea-shore at the *w.* extremity of Chaleur bay; then following the various windings of the sea-shore to the bay of Verte, in the straits of Northumberland: on the *s. e.* it is divided from Nova Scotia by the several windings of the Missiquash river, from its confluence with Beau basin (at the head of Chegnecto channel) to its main source; and from thence by a due *e.* line to the bay of Verte. The *n.* shores of the bay of Fundy constitute the remainder of the *s.* boundary. All islands included in the above limits belong to this province. According to Arrowsmith's map, it is about 200 miles long and 170 broad. The chief towns are St. John's, at the mouth of the river of the same name; St. Anne's, the present seat of government, 62 miles up the river; and Fredericks-town, a few miles above St. Anne's. The chief rivers are St. John's, Merrimichi, Petitcodiac, Memramcook, Ristigouche, and Nipisiguit. The coast of this province is indented with numerous bays and commodious harbours, the chief are Chaleur, Merrimichi, Verti, which last is separated from the bay of Fundy by a narrow isthmus of [about 18 miles wide; bay of Fundy, which extends 50 leagues into the country; Chegnecto bay, at the head of the bay of Fundy; Passamaquoddy bay, bordering upon the district of Maine. At the entrance of this bay is an island granted to several gentlemen in Liverpool, in Lancashire, who named it Campo Bello. At a very considerble expence they attempted to form a settlement here, but failed. On several other islands in this bay there are settlements made by people from Massachusetts. Here are numerous lakes, as yet without names. Grand lake, near St. John's river, is 30 miles long and eight or ten broad; and in some places 40 fathoms deep.

The general assembly of this province have granted to the crown the sum of 10,000*l.* in aid of the defence of the province, in the present hostilities with the United States. The ordinary revenues of the colony do not exceed 6000*l.* a year. But we cannot give a better view of the trade and resources of this colony and its interests compared with, and opposed to, those of the United States, than by the publication of the following authentic document, transmitted to this government in 1804, viz.

'*The Memorial and Petition of the Merchants and other Inhabitants of New Brunswick, to Lord Hobart*,

'Humbly sheweth,

'That after the settlement of this province by the American loyalists in the year 1783, its inhabitants eagerly engaged in endeavouring to supply with fish and lumber the British possessions in the West Indies, and by their exertions they had, within the first 10 years, built 93 square-rigged vessels, and 71 sloops and schooners, which were principally employed in that trade. There was the most flattering prospect that this trade would have rapidly increased, when the late war breaking out, the governors of the West India islands admitted, by proclamation, the vessels of the United States of America to supply them with every thing they wanted; by which means the rising trade of this province has been materially injured, and the enterprising spirit of its inhabitants severely checked. For the citizens of the United States, having none of the evils of war to encounter, are not subject to the high rates of insurance on their vessels and cargoes, nor to the great advance in the wages of seamen, to which, by the imperious circumstances of the times, British subjects are unavoidably liable. And being admitted by proclamation, they are thereby exempt from a transient and parochial duty of two and a half to five per cent. exacted in the West India islands from British subjects.

'Admission into the British ports in the West Indies having been once obtained by the Americans, their government has spared neither pains nor expence to increase their fisheries, so essential to that trade. By granting a bounty of nearly 20*s.* per ton on all vessels employed in the cod fishery, they have induced numbers to turn their attention to that business, and now the principal part of the cod fishery in the bay of Fundy is engrossed by them.

'The county of Charlotte being separated from the United States only by a navigable river, the Americans have, under the foregoing advantages, been enabled to carry off annually (to be reshipped for the West India market) nearly three millions of feet of boards cut in that part of this province, and also a large proportion of the fish caught and cured by British subjects in the bay of Passamaquoddy.

'These discouraging circumstances have prevented the trade in fish and lumber from this province to the West Indies from increasing since the year 1793, and would have totally annihilated it, had not the province possessed advantages in point of situation so favourable for that trade, as to enable its inhabitants to continue the establishments already made for that purpose. What those advantages are, your memorialists now beg leave to state to your Lordship.

'The sea-coast of this province abounds with cod and scale fish, and its rivers are annually visited by immense shoals of herrings, shad, and salmon. The numerous harbours along the coast are most conveniently situated for carrying on the cod fishery, which may be prosecuted to any extent imaginable. The herrings which frequent the rivers of this province are a species peculiarly adapted for the West India market; being equally nutritious with the common herrings, and possessed of a greater degree of firmness, they are capable of being kept longer in a warm climate. In such abundance are they annually to be found, that the quantity cured can only be limited by the insufficient number of hands employed in the business.

'The interior of this province, as well as the parts bordering on the sea-coast, is every where intersected by rivers, creeks, and lakes, on the margin of which, or at no great distance from them, the country for the most part is covered with inexhaustible forests of pine, spruce, birch, beech, maple, elm, fir, and other timber, proper for masts of any size, lumber, and ship-building. The smaller rivers afford excellent situations for saw-mills, and every stream, by the melting of the snow in the spring, is rendered deep enough to] [float down the masts and lumber of every description, which the inhabitants have cut and brought to its banks, during the long and severe winters of this climate, when their agricultural pursuits are necessarily suspended. The lands in the interior of the province are generally excellent, and where cleared, have proved very productive.

' Great advances have not hitherto been made in agriculture for want of a sufficient number of inhabitants, yet within a few years there has remained, beyond our domestic supply, a considerable surplus in horses, salted provisions, and butter, for exportation. And your memorialists look forward with confidence to a rapid increase in the exports of those articles, for which the soil and climate of this country are well adapted.

Possessing so many local advantages, your memorialists feel themselves warranted in stating to your Lordship, that, were not the Americans admitted into the British ports in the West Indies, the fisheries of this and the neighbouring colonies, if duly encouraged, would, with the regular supply from the united kingdoms, furnish the British West India islands with all the fish they would require; and that in a few years the supply of lumber from this province, which already exceeds 10,000,000 of feet annually, would with the exception of staves only be equal to the demand in the said islands. And your memorialists farther confidently state, that these provinces would furnish shipping sufficient to carry from the United States all the flour, corn, and staves, which the British West Indies would stand in need of beyond what the Canadian provinces could furnish.

' During the peace from 1783 to 1793, American vessels were not admitted into the British West India islands, (the whole trade of those islands being carried on during that period in British bottoms), and at no time have the supplies been more abundant or more reasonable. Were the Americans excluded from those islands, this and the neighbouring provinces could now furnish a much larger proportion than formerly of the supplies required, and a rapid and progressive increase might annually be expected. But should the Americans obtain by treaty a right to participate in that trade, not only will the farther progress of improvement in this province be interrupted, but many of its most industrious inhabitants, unable to procure a subsistence here, will be urged to forego the blessings of the British constitutien, to which they are most sincerely and zealously attached, and to seek for an establishment in the United States of America. That great advantages would result to the British nation from providing a sure and permanent supply of those essential articles for its West India islands, independent of foreign assistance, must be obvious. The inhabitants of those islands, forming commercial connections only with their fellow-subjects, would continue the more unalterably attached in their dutiful affection and loyalty to the parent state; and there would be the less reason to dread the consequences of any misunderstanding that might hereafter arise between Great Britain and the United States of America. The introduction into the West Indies of contraband articles, particularly teas, and all kinds of East India manufactures, (a traffic which the Americans now carry on to an enormous extent), would thereby be checked, and the whole benefit of the trade of those islands secured to British subjects. If thus aided and supported against the views of the Americans, the trade of these *n.* provinces would speedily acquire new and increasing vigour, and (which may be an important consideration) soon render them valuable nurseries of seamen for the British navy, that grand security to the commerce and prosperity of his Majesty's kingdoms and colonies.

' Your memorialists therefore must humbly pray, &c.

' *Saint John, New Brunswick*, 11*th May*, 1804.'

See Canada, St. John's River, &c.]

[New Caledonia, the name given by the Scotch to the ill-fated settlement which that nation formed on the isthmus of Darien, and on the *s. w.* side of the gulf of that name. It is situated *e.* of of the narrowest part of the isthmus which is between Panama and Porto Bello, and lies *s. e.* of the latter city. The settlement was formed in 1698. See Darien.]

[New Canton, a small town lately established in Buckingham county, Virginia, on the *s.* side of James's river, 41 miles above Richmond. It contains a few houses, and a ware-house for inspecting tobacco.]

[New Carlisle. See Bonaventure.]

[New-Castle, the most *n.* county of Delaware state. It is about 40 miles in length and 20 in breadth, and contains 19,686 inhabitants, including 2562 slaves. Here are two snuff-mills, a slitting-mill, four paper-mills, 60 for grinding different kinds of grain, and several fulling-mills. The chief towns of this county are Wilmington and New-Castle. The land is more broken than any other part of the state. The heights of Christiana are lofty and commanding.]

[New-Castle, a post-town and the seat of justice of the above county. It is situated on the *w.* side of Delaware river, five miles *s.* of Wilmingtou, and 26 *s. w.* of Philadelphia. It contains about 70 houses, a court-house, and goal; a church for Episcopalians and another for Presbyterians. This is the oldest town on Delaware river, having been settled by the Swedes, about the year 1627, who called it Stockholm, after the metropolis of Sweden. When it fell into the hands of the Dutch, it received the name of New Amsterdam; and the English, when they took possession of the country, gave it the name of New-Castle. It was lately on the decline; but now it begins to flourish. Piers are to be built, which will afford a safe retreat to vessels during the winter season. These, when completed, will add considerably to its advantages. It was incorporated in 1672, by the governor of New York, and was for many years under the management of a bailiff and six assistants. Lat. 39° 40′ *n.*]

[NEW-CASTLE, a township in West Chester County, New York, taken from North Castle in 1791, and incorporated. In 1796, there were 151 of the inhabitants qualified electors.]

[NEW-CASTLE, a small town in the county of Rockingham, New Hampshire, eight miles distant from Portsmouth, was incorporated in 1693, and contains 534 inhabitants.]

[NEW-CASTLE, a small post-town in Lincoln county, district of Maine, situated between Damariscotta and Skungut rivers. It is 10 miles *e.* of Wiscasset, 38 *n. e.* of Portland, and eight *n.* by *e.* of Boston. The township contains 896 inhabitants.]

[NEW-CASTLE, a post-town of Hanover county, Virginia; situated at the mouth of Assequin creek, on the *s. w.* side of Pamunky river, and contains about 36 houses. It is 41 miles *n. w.* of Wiliamsburgh, 19 *n. e.* of Richmond, and 170 from Philadelphia.]

[NEW CHESTER, a township in Grafton county, New Hampshire; situated on the *w.* side of Pemigewasset river. It was incorporated in 1778, and contains 312 inhabitants. It is about 11 miles below the town of Plymouth.]

[NEW CONCORD, formerly called Gunthwaite, a township in Grafton county, New Hampshire, on Amonoosuck river, and was incorporated in 1768, and contains 147 inhabitants.]

[NEW CORBUDA, a town of the province of Tucaman in S. America.]

[NEW CORNWALL, a township in Orange county, New York; bounded *n.* by Ulster county, and *e.* by Hudson's river and Haverstraw. It contains 4225 inhabitants, inclusive of 167 slaves.]

[NEW DUBLIN, a township in Lunenburg county, Nova Scotia; situated on Mahone bay; first settled by Irish, and afterwards by Germans.]

[NEW DURHAM, in Strafford county, New Hampshire, lies on the *e.* coast of Winnepisseoga lake, *w.* of Merry-Meeting bay, nearly 40 miles *n. w.* of Portsmouth. Incorporated in 1762, having 554 inhabitants.]

[NEW EDINBURGH, a new settlement in Nova Scotia.]

NEW ENGLAND, a province and colony which belonged to the English in N. America, and at present one of those composing the republic of the United States, being one of the most flourishing of all the establishments belonging to the English in that part of the world; bounded *n. e.* by Nova Scotia, *e.* and *s.* by the Atlantic, *w.* by New York, and *n.* and *n. w.* by Canada or New France. It is 450 miles long and nearly 200 wide at its broadest part; but the cultivated part, and that which deserves most to be mentioned, is somewhat more than 60 miles in extent from the coast. The first discoverer of this country was Sebastian Gabot in 1497, and in 1587 it was taken possession of in the name of Queen Elizabeth, of England, by Philip Amadas and Arthur Barlow. In the following year a colony was brought hither by Richard Grenville, who gave it the name of New Plymouth.

In 1621 many Puritans flocked hither, who, flying from the religious persecutions in England under King James I. went over to Holland, but not finding there the reception they looked for, fixed on this part of America, where the greater part of them perished the first winter through the rigour of the season, being without food or clothing; the rest, however, surmounted these difficulties, and the colony began to increase and flourish about the year 1629.

Next followed a ferment between the Quakers, Anabaptists, and the other religious sects, which caused a kind of civil war. It was then that a disorder arose from a charge against certain Puritans of witchcraft, when in order to take cognisanee of these offences, a tribunal was erected, at which, by the mere impeachment, an infinite number were ordered to be put to death, the governor being William Phipps, a man of low extraction, and who, in his religious zeal, spared neither age, sex, or condition.

This province had the privilege of electing for itself a governor, magistrates, &c. but having abused the same, it was taken from it by Charles

II. king of England, in 1684; but some time after the revolution, which had led to this precaution, it was again granted, though with less licence.

The climate, compared with that of Virginia, is like that of the *s.* of England, compared with the *e.* part, and notwithstanding that it is in the torrid zone, it is very irregular, neither very hot nor very cold, and the air is healthy. When the English first entered it, it was an immense wood, of which only some small parts had been cleared by the Indians for sowing maize, but it did not want, for fertile and well irrigated valleys. The land immediately on the coast is generally low, and in some parts swampy, but about half way it begins to rise into hills, and in the *n. e.* part it is even mountainous.

Few countries are so fertile in rivers, lakes, and springs as this; the former abound in excellent fish, and there are seven of them navigable for many leagues, and would be for more, were it not for the innumerable cascades and cataracts. The names of these rivers are, Connecticut, Thames, Patuxet, Merrimack, Piscataway, Saco, and Cask; and, besides these, in the *e.* part, Sagadohock, Kenebec, Penobscot, and many others, to the advantages of which may be attributed the great number of populous cities found here: besides, in the spaces between the above rivers the ground is so irrigated with streams and fountains, that it is almost impossible to stir 12 feet without finding good sweet water.

New England produces cod-fish in great abundance, fish-oil, whales, cedar-wood, tallow, salt meat, maize, neat cattle, and swine; pulse and fruit of all kinds, masts and yard-arms for vessels, woods of infinite sorts, many fine skins of castors, hares, rabbits, and other animals, of which they make fine hats and various woven articles. All these things provided New England with a plentiful means of carrying on a great trade with all the nations of Europe and others in America, and there used, previously to its independence, to enter every year regularly into its ports more than a thousand vessels. New England is divided into four provinces, which are, Connecticut, Massachusetts, Rhode Island, and New Hampshire; and these contain more than 350,000 souls.

[New England, or as it is now generally known under the title of Northern and Eastern States, lies between lat. 41° and about 48° *n.* and between long. 66° 53′ and 74° 8′ *w.* It lies in the form of a quarter of a circle. Its *w.* line beginning at the mouth of Byram river, which empties into Long Island sound, at the *s. w.* corner of Connecticut, lat. 41°, runs a little *e.* of *n.* until it strikes the 45° of latitude; and then curves to the *e.* almost to the gulf of St. Lawrence. This grand division of the United States comprehends the states of Vermont, New Hampshire, Massachusetts, (including the district of Maine,) Rhode Island and Providence plantations, and Connecticut.

The climate of New England is so healthful that it is estimated that about one in seven of the inhabitants live to the age of 70 years; and about one in 13 or 14 to 80 and upwards. North-west, *w.* and *s. w.* winds are the most prevalent. East and *n. e.* winds, which are unelastic and disagreeable, are frequent at certain seasons of the year, particularly in April and May, on the sea-coasts. The weather is less variable than in the middle, and especially in the *s.* states, and more so than in Canada. The extremes of heat and cold, according to Fahrenheit's thermometer, are from 20° below to 100° above 0. The medium is from 48° to 50°. The diseases most prevalent in New England are alvine fluxes, St. Antony's fire, asthma, atrophy, catarrh, cholic, inflammatory, slow, nervous, and mixed fevers, pulmonary consumption, quinsy, and rheumatism. A late writer has observed, that "in other countries, men are divided according to their wealth or indigence, into three classes; the opulent, the middling, and the poor; the idleness, luxuries, and debaucheries of the first, and the misery and too frequent intemperance of the last, destroy the greater proportion of these two. The intermediate class is below those indulgencies which prove fatal to the rich, and above those sufferings to which the unfortunate poor fall victims: this is therefore the happiest division of the three. Of the rich and poor, the American republic furnishes a much smaller proportion than any other district of the known world. In Connecticut, particularly, the distribution of wealth and its concomitants, is more equal than elsewhere, and therefore, as far as excess, or want of wealth, may prove destructive or salutary to life, the inhabitants of this state may plead exemption from diseases." What this writer, Dr. Foulke, says of Connecticut in particular, will, with very few exceptions, apply to New England at large.

New England is a high, hilly, and in some parts a mountainous country, formed by nature to be inhabited by a hardy race of free, independent republicans. The mountains are comparatively small, running nearly *n.* and *s.* in ridges parallel to each other. Between these ridges, flow the great rivers in majestic meanders, receiving the innumerable rivulets and larger streams which pro-]

[ceed from the mountains on each side. To a spectator on the top of a neighbouring mountain, the vales between the ridges, while in a state of nature, exhibit a romantic appearance. They seem an ocean of woods, swelled and depressed in its surface like that of the great ocean itself. A richer, though less romantic, view is presented when the valleys have been cleared of their natural growth by the industrious husbandmen, and the fruit of their labour appears in loaded orchards, extensive meadows covered with large herds of sheep and neat cattle, and rich fields of flax, corn, and the various kinds of grain. These valleys are of various breadths, from two to 20 miles; and by the annual inundations of the rivers and smaller streams which flow through them, there is frequently an accumulation of rich, fat soil left upon the surface when the waters retire. The principal rivers have been already mentioned. New England, generally speaking, is better adapted for grazing than for grain, though a sufficient quantity of the latter is raised for home consumption, if we except wheat, which is imported in considerable quantities from the middle and *s.* states. Indian corn, rye, oats, barley, buck-wheat, flax, and hemp, generally succeed very well. Apples are common, and in general plenty in New England; and cider constitutes the principal drink of the inhabitants. Peaches do not thrive so well as formerly. The other common fruits are more or less cultivated in different parts. The high and rocky ground is in many parts covered with clover, and generaly affords the best of pasture; and here are raised some of the finest cattle in the world. The quantity of butter and cheese made for exportation is very great. Considerable attention has lately been paid to the raising of sheep. This is the most populous division of the United States. It contained, according to the census of 1790, 1,009,522 souls; and the number, according to the census of 1810, was as follows, viz.

	Souls.
In Vermont, - - -	217,913
New Hampshire, - -	214,414
Massachusetts, - -	472,040
Maine, - - -	228,705
Rhode Island and Providence plantations, - - -	76,931
Connecticut, - - -	261,942
Total,	1,471,945

The great body of these are landholders and cultivators of the soil. As they possess, in fee simple, the farms which they cultivate, they are naturally attached to their country: the cultivation of the soil makes them robust and healthy, and enables them to defend it. New England may, with propriety, be called a nursery of men, whence are annually transplanted, into other parts of the United States, thousands of its natives. Vast numbers of them, since the war, have emigrated into the *n.* parts of New York, into Kentucky and the W. Territory, and into Georgia, and some are scattered into every state and every town of note in the union.

The inhabitants of New England are, almost universally, of English descent; and it is owing to this circumstance, and to the great and general attention that has been paid to education, that the English language has been preserved among them so free from corruption. Learning is diffused more universally, among all ranks of people here, than in any other part of the globe; arising from the excellent establishment of schools in almost every township, and the extensive circulation of newspapers. The first attempt to form a regular settlement in this country was at Sagadahock, in 1607, but the year after, the whole number who survived the winter returned to England. The first company that laid the foundation of the New England states, planted themselves at Plymouth, November 1620. The founders of the colony consisted of but 101 souls. In 1640, the importation of settlers ceased. The persecution of which Alcedo speaks (the motive which had led to transportation to America) was over, by the change of affairs in England. At this time the number of passengers who had come over, in 298 vessels, from the beginning of the colony, amounted to 21,200 men, women, and children; perhaps about 4000 families. In 1760, the number of inhabitants in Massachusetts bay, New Hampshire, Connecticut, and Rhode Island, amounted, probably, to 500,000. For a copious history of the states included in New England, see Index to additional history concerning Massachusetts.]

List of the capes, points, bays, and ports, on the coast of New England.

Point of Pemaquid,	Cape Cod,
Little point,	Gooseberry point,
Cape Elizabeth,	Point Watch,
Cape Porpus,	Cape Sachem,
Cape Nidduck,	Lion's Tongue,
Cape York,	Cape Anne,
Lock's point,	Cape Alderton,
Great Boar point,	Cape Monument,
Mount Pigeon,	Point Billingsgate,
Cape Pullin,	Cape Pamet,
Cape Gurnet,	Point Ninigret,
Point Murray,	Point Black,

South point,
Cape Poge,
Cape Nathan,
Cape Marshfield,
Sandy point,
Race point,
Cape Malabar,
Point Quakhoragok,
Pipe point,
Point Hemunaseth,
Long Neck point.

Bays.

Penobscot,
Sawko,
Cod and Plymouth,
Connecticut,
Mussequoif,
Exeter,
Nahunt,
Naskintucket,
Guilford,
Homes,
Kennebeck,
Wells,
Narraganset,
Winipisoketpond,
Harrasecket,
Little bay,
Oyster River,
Clerks,
Fairfield,
Calko,
Massachusetts,
Long Island,
Merry-meeting,
Broad-cove,
Sandy,
Falmouth,
Nathantick,
Tarpaulin.

Ports.

Winter,
Konochaset,
New Haven,
Piscataqua,
Scituate,
Ship,
Cape Anne,
Yarmouth,
Old Town,
Boston,
Slokom.

[New Fairfield, the north-westernmost township in Fairfield county, Connecticut.]

[New Fane, the chief town of Windham county, Vermont, is situated on West river, a little to the *n. w.* of Brattleborough. It has 660 inhabitants.]

[New Garden, a township in Chester county, Pennsylvania.]

[New Garden, a settlement of the Friends in Guildford county, N. Carolina.]

[New Geneva, a settlement in Fayette county, Pennsylvania.]

[New Germantown, a post-town of New Jersey; situated in Hunterdon county. It is 19 miles *n. w.* of Brunswick, 30 *n.* of Trenton, and 46 *n. e.* by *n.* of Philadelphia.]

[New Gloucester, a small post-town in Cumberland county, district of Maine, 25 miles *n.* of Portland, and 110 *n.* by *e.* of Boston. It was incorporated in 1774, and contains 1355 inhabitants.]

[New Gottingen, a town of Georgia; situated in Burke county, on the *w.* bank of Savannah river, about 18 miles *e.* of Waynesborough, and 35 *n. e.* of Ebenezer.]

[New Granada, a province in the *s.* division of Tierra Firme, S. America, whose chief town is Santa Fe de Bagota. See Cibola.]

[New Grantham, a township in Cheshire county, New Hampshire, was incorporated in 1761, and contains 333 inhabitants, and is about 15 miles *s. e.* of Dartmouth college.]

[New Hampshire, one of the United States of America, is situated between lat. 42° 38′ and 45° 18′ *n.* and between long. 70° 42′ and 72° 32′ *w.* from Greenwich; bounded *n.* by Lower Canada, *e.* by the district of Maine, *s.* by Massachusetts, and *w.* by Connecticut river, which separates it from Vermont. Its shape is nearly that of a right angled triangle; the district of Maine and the sea its leg, the line of Massachusetts its perpendicular, and Connecticut river its hypothenuse. It contains 9491 square miles, or 6,074,240 acres; of which at least 100,000 acres are water. Its length is 162 miles; its greatest breadth 78, and its least breadth 15 miles.

This state is divided into five counties, viz. Rockingham, Strafford, Cheshire, Hillsborough, and Grafton. The chief towns are Portsmouth, Exeter, Concord, Dover, Amherst, Keen, Charlestown, Plymouth, and Haverhill. Most of the townships are six miles square, and the whole number of townships and locations is 214; containing, in 1796, 141,885 persons, including 158 slaves. In 1767, the number of inhabitants was estimated at 52,700, and by the census of 1810, the population amounted to 214,414 souls. This state has but about 14 miles of sea-coast, at its *s. e.* corner. In this distance there are several coves for fishing vessels, but the only harbour for ships is the entrance of Piscataqua river, the shores of which are rocky. The shore is mostly a sandy beach, adjoining to which are salt marshes, intersected by creeks, which produce good pasture for cattle and sheep. The intervale lands on the margin of the great rivers are the most valuable, because they are overflowed and enriched by the water from the uplands, which brings a fat slime or sediment. On Connecticut river these lands are from a quarter of a mile to a mile and an half on each side, and produce corn, grain, and grass, especially wheat, in greater abundance and perfection than the same kind of soil does in the higher lands. The wide spreading hills are esteemed as warm and rich; rocky moist land is accounted good for pasture; drained swamps have a deep mellow soil; and the valleys between the hills are generally very productive. Agriculture is the chief occupation of the inhabitants; beef, pork, mutton, poultry, wheat, rye, Indian corn, barley, pulse, butter, cheese, hops, esculent roots and plants, flax, hemp, &c. are articles which will always find a market, and are raised in immense quantities in New Hampshire, both for home consumption and ex-]

[portation. Apples and pears are the most common fruits cultivated in this state, and no husbandman thinks his farm complete without an orchard. Tree fruit of the first quality cannot be raised in such a *n.* climate as this, without particular attention. New York, New Jersey, and Pennsylvania have it in perfection. As you depart from that tract, either *s.* or *n.* it degenerates. The uncultivated lands are covered with extensive forests of pine, fir, cedar, oak, walnut, &c. For climate, diseases, &c. see NEW ENGLAND.

Several kinds of earths and clays are found in this state, chiefly in Exeter, Newmarket, Durham, and Dover. Marl abounds in several places, but is little used. Red and yellow ochres are found in Somersworth, Chesterfield, Rindge, and Jaffray. Steatites or soap rock is found in Orford. The best lapis specularis, a kind of talc, commonly called ising-glass, is found in Grafton and other parts. Crystals have been discovered at Northwood, Rindge, and Conway; alum, at Barrington, Orford, and Jaffray; vitriol, at Jaffray, Brentwood, and Rindge, generally found combined in the same stone with sulphur. Freestone fit for building is found in Orford; also a grey stone fit for mill-stones. Iron ore is found in many places; black lead in Jaffray, and some lead and copper ore has been seen; but iron is the only metal which has been wrought to any advantage.

New Hampshire is intersected by several ranges of mountains. The first ridge, by the name of the Blue hills, passes through Rochester, Barrington, and Nottingham, and the several summits are distinguished by different names. Behind these are several higher detached mountains. Farther back the mountains rise still higher, and among the third range, Chocorua, Ossapy, and Kyarsarge, are the principal. Beyond these is the lofty ridge which divides the branches of Connecticut and Merrimack rivers, denominated the "Height of Land." In this ridge is the celebrated Monadnock mountain, 30 miles *n.* of which is Sunapee, and 48 miles further is Moosehillock, called also Mooshelock mountain. The ridge is then continued *n.* dividing the waters of the river Connecticut from those of Saco and Amariscoggin. Here the mountains rise much higher, and the most elevated summits in this range are the White mountains, which are 9000 feet above the sea. The lands *w.* of this last mentioned range of mountains, bordering on Connecticut river, are interspersed with extensive meadows, rich and well watered. Ossapy mountain lies adjoining the town of Moultonborough on the *n. e.* In this town it is observed, that in a *n. e.* storm the wind falls over the mountain, like water over a dam; and with such force, as frequently to unroof houses. People who live near these mountains, by noticing the various movements of attracted vapours, can form a pretty accurate judgment of the weather; and they hence style these mountains their almanack. If a cloud is attracted by a monutain, and hovers on its top, they predict rain; and if, after rain, the mountain continues capped, they expect a repetition of showers. A storm is preceded for several hours by a roaring of the mountain, which may be heard 10 or 12 miles. But the White mountains are undoubtedly the highest land in New England, and in clear weather, are discovered before any other land, by vessels coming into the *e.* coast; but by reason of their white appearance, are frequently mistaken for clouds. They are visible on the land at the distance of 80 miles, on the *s.* and *s. e.* sides; they appear higher when viewed from the *n. e.* and it is said, they are seen from the neighbourhood of Chamblee and Quebec. The Indians gave them the name of Agiocohook. The number of summits in this cluster of mountains cannot at present be ascertained, the country around them being a thick wilderness. The greatest number which can be seen at once is at Dartmouth, on the *n. w.* side, where seven summits appear at one view, of which four are bald. Of these the three highest are the most distant, being on the *e.* side of the cluster: one of these is the mountain which makes so majestic an appearance all along the shore of the *e.* counties of Massachusetts. It has lately been distinguished by the name of mount Washington. During the period of nine or 10 months, these mountains exhibit more or less of that bright appearance, from which they are denominated white. In the spring, when the snow is partly dissolved, they appear of a pale blue, streaked with white; and after it is wholly gone, at the distance of 60 miles, they are altogether of the same pale blue, nearly approaching a sky colour; while at the same time, viewed at the distance of eight miles or less, they appear of the proper colour of the rock. These changes are observed by people who live within constant view of them; and from these facts and observations, it may with certainty be concluded, that the whiteness of them is wholly caused by the snow, and not by any other white substance, for in fact there is none.

The reader will find an elegant description of these mountains in the 3d vol. of Dr. Belknap's History of New Hampshire, from which the above is extracted.

The most considerable rivers of this state are]

[Connecticut, Merrimack, Piscataqua, Saco, Androscoggin, Upper and Lower Amonoosuck, besides many other smaller streams. The chief lakes are Winnipiseogee, Umbagog, Sunapee, Squam, and Great Ossipee. Before the war, ship-building was a source of considerable wealth to this state; about 200 vessels were then annually built, and sold in Europe and in the W. Indies, but that trade is much declined. Although this is not to be ranked among the great commercial states, yet its trade is considerable. Its exports consist of lumber, ship timber, whale oil, flax seed, live stock, beef, pork, Indian corn, pot and pearl ashes, &c. &c. In 1790, there belonged to Piscataqua 33 vessels above 100 tons, and 50 under that burden. The tonnage of foreign and American vessels cleared out from the 1st of October 1789, to 1st of October 1791, was 31,097 tons, of which 26,560 tons were American vessels. The fisheries at Piscataqua, including the isle of Shoals, employ annually 27 schooners and 20 boats. In 1791, the produce was 25,850 quintals of cod and scale fish. The exports from the port of Piscataqua in two years, viz. from 1st of October 1789, to 1st of October 1791, amounted to the value of 296,839 dollars, 51 cents; in the year ending September 30th, 1792, 181,407 dollars; in 1793, 198,197 dollars; and in the year 1794, 153,856 dollars. The bank of New Hampshire was established in 1792, with a capital of 60,000 dollars; by an act of assembly the stock-holders can increase it to 200,000 dollars specie, and 100,000 dollars in any other estate. The only college in the state is at Hanover, called Dartmouth college, which is amply endowed with lands, and is in a flourishing situation. The principal academics are those of Exeter, New Ipswich, Atkinson, and Amherst.

A brief, and we must add (as will be seen by comparison with this) very unsatisfactory account of New Hampshire is given by our author under article HAMPSHIRE. It contains, however, a list of all the principal towns and settlements in the state, which see. Also for many particular details relative to its history, see Index to additional matter respecting Massachusetts; likewise NEW ENGLAND and UNITED STATES.]

[NEW HAMPSTEAD, a township in Orange county, New York, bounded *e.* by Clarkstown, and *s.* by the state of New Jersey. It was taken from Haverstraw, and incorporated in 1791. By the state census 1796, there were 245 of its inhabitants qualified electors.]

[HEW HAMPTON, a post-town of New Hampshire; situated in Strafford county, on the *w.* side of lake Winnipiseogee, nine miles *s. e.* of Plymouth, and nine *n. w.* of Meredith. The township was incorporated in 1777, and contains 652 inhabitants.]

[NEW HANOVER, a maritime county of Wilmington district, N. Carolina, extending from Cape Fear river *n. e.* along the Atlantic ocean. It contains 6831 inhabitants, including 3738 slaves. Chief town, Wilmington.]

[NEW HANOVER, a township in Burlington county, New Jersey, containing about 20,000 acres of improved land, and a large quantity that is barren and uncultivated. The compact part of the township is called New-mills, where are about 50 houses, 27 miles from Philadelphia, and 13 from Burlington.]

[NEW HANOVER, a township in Morgan county, Pennsylvania.]

[NEW HARTFORD, a small post-town in Litchfild county, Connecticut, 13 miles *n. e.* of Litchfield, 19 *w.* by *n.* of Hartford.]

[NEW HAVEN County, Connecticut, extends along the sound between Middlesex county on the *e.* and Fairfield county on the *w.*; about 30 miles long from *n.* to *s.* and 28 from *e.* to *w.* It is divided into 14 townships. It contained in 1756, 17,955 free persons, and 226 slaves; in 1774, 25,896 free persons, and 925 slaves; and in 1790, 30,397 free persons, and 433 slaves.]

[NEW HAVEN City, the seat of justice in the above county, and the semi-metropolis of the state. This city lies round the head of a bay which makes up about four miles *n.* from Long Island sound. It covers part of a large plain which is circumscribed on three sides by high hills or mountains. Two small rivers bound the city *e.* and *w.* It was originally laid out in squares of 60 rods; many of these squares have been divided by cross streets. Four streets run *n. w.* and *s. e.* and are crossed by others at right angles. Near the centre of the city is the public square, on and around which are the public buildings, which are a state-house, two college edifices, and a chapel, three churches for Congregationalists, and one for Episcopalians; all which are handsome and commodious buildings. The college edifices, chapel, state-house, and one of the churches are of brick. The public square is encircled with rows of trees, which render it both convenient and delightful. Its beauty, however, is greatly diminished by the burial-ground, and several of the public buildings, which occupy a considerable part of it. Many of the streets are ornamented with rows of trees on each side, which give the city a rural appearance. The prospect from the steeples is greatly variegated and extremely beautiful. There are between 3 and 400 neat dwelling-houses in the city, principally of wood. The streets are sandy but clean. Within the limits of the city are 4000 souls. About one in 70 die annually. Indeed as to pleasantness of situation and salubrity of air, New Haven is hardly exceeded by any city in America. It carries on a considerable trade with New York and the W. India islands. The exports for one year, ending September 30, 1794, amounted to the value of 171,868 dollars. Manufactures of card-teeth, linen, buttons, cotton, and paper, are carried on here. Yale college, which is established in this city, was founded in 1700, and remained at Killingworth until 1707, then at Saybrook until 1716, when it was removed and fixed at New Haven. It has its name from its principal benefactor Governor Yale. There are at present six college domiciles, two of which, each 100 feet long and 40 wide, are inhabited by the students, containing 32 chambers each, sufficient for lodging 120 students; a chapel 40 by 50 feet, with a steeple 130 feet high; a dining-hall 60 by 40 feet; a house for the president, and another for the professor of divinity. In the chapel is lodged the public library, consisting of about 3000 volumes, and the philosophical apparatus, as complete as most others in the United States, and contains the machines necessary for exhibiting experiments in the whole course of experimental philosophy and astronomy. The museum, to which additions are constantly making, contains many natural curiosities. From the year 1700 to 1793, there had been educated and graduated at this university about 2303. The number of students is generally 150. The harbour, though inferior to New London, has good anchorage, with three fathom and four feet water at common tides, and 2½ fathom at low water. This place and Hartford are the seats of the legislature alternately. It is 36 miles *s. w.* by *s.* of Hartford, 36 from New London, 62 from New York, 105 from Boston, and 131 *n. e.* of Philadelphia. Lat. 41° 16′ *n.* Long. 72° 53′ *w.*]

[NEW HAVEN, a township in Addison county, Vermont, on Otter creek or river, containing 723 inhabitants.]

[NEW HEBRIDES, a cluster of islands in the Pacific ocean, so called by Capt. Cook in 1794—the same as the archipelago of the Great Cylades of Bougainville, or the Terra Austral of Quiros, which see.]

[NEW HOLDERNESS, a township in Grafton county, New Hampshire; situated on the *e.* side of Pemigewasset river, about three miles *e.* by *s.* of Plymouth. It was incorporated in 1761, and contains 329 inhabitants.]

[NEW HOLLAND, a town of Pennsylvania, Lancaster county, in the midst of a fertile country. It contains a German church and about 70 houses. It is 15 miles *e. n. e.* of Lancaster, and 41 *w. n. w.* of Philadelphia.]

[NEW HUNTINGTON, a mountainous township in Chittenden county, Vermont, on the *s. w.* side of Onion river, containing 136 inhabitants.]

[NEW INVERNESS, in Georgia, is situated near Darien on Alatamaha river. It was built by the Scotch highlanders, 160 of whom landed here in 1735.]

[NEW IPSWICH, a township in Hillsborough county, New Hampshire, on the *w.* side of Souhegan river, near the *s.* line of the state. It was incorporated in 1762, and contains 1241 inhabitants. There is an academy, founded in 1789, having a fund of about 1000*l.* and has generally about 40 or 50 students. It is about 24 miles *s. e.* of Keene, and 52 *w. s. w.* of Portsmouth.]

NEW JERSEY, a province of N. America, formerly belonging to England, and now one of those composing the United States. It was founded in 1682, and ceded to Lord George Catcret, and some other English gentlemen, who gave it the name of New Jersey, from the estates which the family of this name possess in an island so called.

The continual disputes which lasted for many years between the settlers and proprietors, brought this province to a miserable state. It was divided into two parts, with the titles of E. and W. Jersey; and in the reign of Queen Anne they were united. It is bounded *n.* by New York, *e.* and *s. e.* by the Atlantic, *w.* and *s. w.* by the river and bay of Delaware; between 38° 56′ and 41° 22′ *n.* lat. and between 75° 44′ and 75° 40′ long. from the meridian of London, and is 143 miles long and 62 wide.

Before the formation of the United States and the establishment of the independence, it was a royal government with a council of assembly nominated by the king, the province also nominating deputies to represent the people. For some time the authority of the governor of New York extended also over New Jersey.

The climate is, for the most part, more temperate here than in the former, or even than in New England, from its more *s.* situation. It produces all kinds of vegetable productions, cattle, swine, and skins, and exports wheat, barley, flour, oxen, fish, some butter, flax seed, beer, barrelled herrings, and harness, to the W. Indies, receiving in exchange sugar, rum, and other effects; and to England it sends skins, hides, tobacco, fish, pitch, oil, and whale-bone, and other productions; taking in exchange crockery ware and clothes. As its towns are inland, its articles of commerce are also of an inland quality. In one spot there were 150 to 200 families, which, although for the most part Dutch, lived subject to the English government in great peace and tranquillity.

There are in this province two iron mines, one in the river Passaick, the other in the upper part of the Raritan.

E. Jersey, which is the largest and most populous part, extends from *s.* to *n.* nearly 100 miles, the length of the coast of the river Hudson, from the bay of Little Egg to the part of the aforesaid river, which is in lat. 41°, and is divided into that of S. and that of W. Jersey by a line of division which passes from Egg bay to the river Cheswick and *e.* arm of the Raritan. The width of this part is very irregular, it being, in some parts, more contracted than at others. It is, however, looked upon as the best part of the two Jerseys, and is divided into the following counties:

Monmouth,	Essex,
Middlesex,	Bergen.

W. Jersey is not so much cultivated or so populous as the former, but the convenience, offered by its large lakes for commerce, gives it rather a favourable distinction. In this part, six counties were judiciously erected by Dr. Cox; but his successors pulled down the system, and now there is only one, called Cape May, which is a piece of land or *e.* point at the entrance of the bays of Delaware and of Egg, which separates the two Jerseys; and here there are several scattered houses, the principal of them being Cox's-hall. The fall of the river Passaick deserves particular description. It is a part where the waters become confined in a channel of 40 yards across, and where, with an immense rapidity, they rush to fall down a precipice of 70 perpendicular feet. The greater part of the inhabitants of this province are fishermen, employed in catching whales, which abound in the bay of Delaware; and this bay, with the river of the same name, have on their shores all the part of New Jersey running from *s.* to *e.* and to *s. w.* as also all the plantations, which, from being united, are called cities. The river Mauricius, between that of Coanzi and Cape May, is the largest in the whole country; and this last river, although small, is very deep and navigable for small vessels. Ten or 12 miles up the same is a city of its name, with about 80 families. In this division are the counties of Burlington, Gloucester, Salem, Cumberland, Cape May, Hunterdon, Morris, and Sussex. In this province there is no established religion, but it has 22 churches, 57 meeting-houses for Scotch and English presbyterians, 22 for Dutch, 39 Quakers meetings, 220 meetings of Anabaptists, seven of Lutherans, one of Moravians, one of Separatists; and its population is composed of 13,000 souls of all sexes and ages, including Negro slaves.

[In giving what we conceive a fuller and correcter view of this state, we shall not be afraid of entering into some trifling repetitions.

New Jersey contains about 8320 square miles, equal to 5,324,800 acres. It is divided into 13 counties, viz. Cape May, Cumberland, Salem, Gloucester, Burlington, Hunterdon, and Sussex; these seven lie from *s.* to *n.* on Delaware river; Cape May and Gloucester extend across to the sea; Bergen, Essex, Middlesex, and Monmouth, lie from *n.* to *s.* on the *e.* side of the state; Somerset and Morris are inland counties.

The number of inhabitants in 1796 was 184,139, of whom 11,423 were slaves; and by the census of 1810, the total population amounted to 245,562 souls. The most remarkable bay is Arthur kull or Newark bay, formed by the union of Passaick and Hackinsac rivers. The rivers in this state, though not large, are numerous. A traveller in passing the common road from New York to Philadelphia, crosses three considerable rivers, viz. the Hackinsac and Passaick, between Bergen and Newark, and the Rariton by Brunswick. Passaick is a very crooked river. It is navigable about 10 miles, and is 230 yards wide at the ferry. The cataract, or great falls, in this river, is one of the greatest natural curiosities in the state. The river is about 40 yards wide, and moves in a slow gentle current, until coming within a short distance of a deep cleft in a rock which crosses the channel, it descends and falls above 70 feet perpendicularly, in one entire sheet. One end of the cleft, which was evidently made by some violent convulsion in nature, is closed: at the other the water rushes out with incredible swiftness, forming an acute angle with its former direction, and is received into a large bason, whence it takes a winding course through the rocks, and spreads into a broad smooth stream. The cleft is from four to 12 feet broad. The falling of the water occasions a cloud of vapour to arise, which, by floating amidst the sun-beams, presents rainbows to the view, which adds beauty to the tremendous scene. The new manufacturing town of Patterson is erected upon the great falls in this river. Rariton river is formed by two considerable streams, called the *n.* and *s.* branches; one of which has its source in Morris, the other in Hunterdon county. It passes by Brunswick and Amboy, and, mingling with the waters of the Arthur Kull sound, helps to] form the fine harbour of Amboy. Bridges have lately been erected over the Passaick, Hackinsac, and Rariton rivers, on the post-road between New York and Philadelphia. These bridges will greatly facilitate the intercourse between these two great cities.

The counties of Sussex, Morris, and the *n.* part of Bergen, are mountainous. As much as five-eighths of most of the *s.* counties, or one-fourth of the whole state, is almost entirely a sandy barren, unfit in many parts for cultivation. All the varieties of soil, from the worst to the best kind, may be found here. The good land in the *s.* counties lies principally on the banks of rivers and creeks. The barrens produce little else but shrub oaks and yellow pines. These sandy lands yield an immense quantity of bog iron ore, which is worked up to great advantage in the iron works in these counties. In the hilly and mountainous parts which are not too rocky for cultivation, the soil is of a stronger kind, and covered in its natural state with stately oaks, hickories, chesnuts, &c. and when cultivated produces wheat, rye, Indian corn, buek-wheat, oats, barley, flax, and fruits of all kinds common to the climate. The land in this hilly country is good for grazing, and farmers feed great numbers of cattle for New York and Philadelphia markets. The orchards in many parts of the state equal any in the United States, and their cider is said, and not without reason, to be the best in the world. The markets of New York and Philadelphia receive a very considerable proportion of their supplies from the contiguous parts of New Jersey. These supplies consist of vegetables of many kinds, apples, pears, peaches, plums, strawberries, cherries, and other fruits; cider in large quantities, butter, cheese, beef, pork, mutton, and the lesser meats. The trade is carried on almost solely with and from those two great commercial cities, New York on one side and Philadelphia on the other; though it wants not good ports of its own.

Manufactures here have hitherto been inconsiderable, not sufficient to supply its own consumption, if we except the articles of iron, nails, and leather. A spirit of industry and improvement, particularly in manufactures, has, however, of late greatly increased. The iron manufacture is, of all others, the greatest source of wealth to the state. Iron works are erected in Gloucester, Burlingtou, Sussex, Morris, and other counties. The mountains in the county of Morris give rise to a number of streams, necessary and convenient for these works, and at the same time furnish a copious supply of wood and ore of a superior quality. In this county alone are no less than seven rich iron mines, from which might be taken ore sufficient to supply the United States; and to work it into iron there are two furnaces, two rolling and slitting mills, and about thirty forges, containing from two to four fires each. These works produce annually about 540 tons of bar-iron, 800 tons of pigs, besides large quantities of hollow ware, sheet iron, and nail-rods. In the whole state it is supposed there is yearly made about 1200 tons of bar-iron, 1200 ditto of pigs, 80 ditto of nail-rods, exelusive of hollow ware, and various other castings, of which vast quantities are made.

The inhabitants are a collection of Low Dutch, Germans, English, Scotch, Irish, and New Englanders, and their descendants. National attachment and mutual convenience have generally induced these several kinds of people to settle together in a body, and in this way their peculiar national manners, customs, and character, are still preserved, especially among the poorer class of people, who have little intercourse with any but those of their own nation. The people of New Jersey are generally industrious, frugal, and hospitable. All the religious denominations live together in peace and harmony; and are allowed by the constitution of the state to worship Almighty God agreeably to the dictates of their own consciences. The college at Princetown, called Nassau hall, has been under the care of a succession of presidents, eminent for piety and learning; and has furnished a number of civilians, divines, and physieiaus of the first rank in America. It has considerable funds, is under excellent regulations, and has generally from 80 to 100 students, principally from the *s.* states. There are academies at Freehold, Trenton, Mackinsac, Orangedale, Elizabethtown, Burlington, and Newark; and grammar-schools at Springfield, Morristown, Bordentown, and Amboy.

There are a number of towns in this state nearly of equal size and importance, and none that has more than 300 houses compactly built. Trenton is one of the largest, and the capital of the state. The other principal towns are Brunswick, Burlington, Amboy, Bordentown, Princetown, Elizabethtown, Newark, and Morristown.

This state was the seat of war for several years, during the bloody contest between Great Britain and America. Her losses both of men and property, in proportion to the population and wealth of the state, was greater than of any other of the Thirteen States. When General Washington was retreating through the Jerseys, almost forsaken by all others, her militia were at all times obedient to

[his orders; and, for a considerable length of time, composed the strength of his army. There is hardly a town in the state that lay in the progress of the British army that was not rendered signal by some enterprise or exploit.

Governors of New Jersey from the surrender of the Government by the Proprietors in 1702.

* Edward Viscount Cornbury, 1702 to 1708, removed and succeeded by
* John Lord Lovelace, 1708 to 1709, died and the government devolved to

Lieut. Gov. Richard Ingoldsby, 1709 to 1710, when came in

* Brigadier Robert Hunter, 1710 to 1720, who resigned in favour of
* William Burnet, 1720 to 1727, removed and succeeded by
* John Montgomery, 1728 to 1731, died and was succeeded by
* William Crosby, 1731 to 1736, died and the government devolved to

John Anderson, president of the council, 1736, by whose death about two weeks after, the government devolved to

John Hamilton, president of the council, 1736 to 1738.

Those marked * were governors in chief, and down to this time were governors of New York and New Jersey, but from 1738 forward, New Jersey has had a separate governor.

* Lewis Morris, 1738 to 1746, died and the government devolved to

John Hamilton, president, 1746, by whose death it devolved to

John Reading, president, 1746 to 1747.

* Jonathan Belcher, 1747 to 1757, died and the government again devolved to

John Reading, president, 1757 to 1758.

Thomas Pownall, then governor of Massachusetts, being lieutenant-governor, arrived on the death of Governor Belcher, but continued in the province a few days only.

* Francis Bernard, 1758 to 1760, removed to Boston and succeeded by
* Thomas Boone, 1760 to 1761, removed to S. Carolina and succeeded by
* Josiah Hardy, 1761 to 1763, removed and succeeded by
* William Franklin, 1763 to 1776, removed and succeeded by
* William Livingston, 1776 to 1790, died and succeeded by
* William Patterson, 1791.]

[New Jersey Company's Grant of Lands, lies on the *e.* side of Mississippi river, *s.* of the Illinois, and *n. w.* of the Army lands, which form the tract shaped by the confluence of Ohio with Mississippi.]

[New Kent, a county of Virginia, bounded on the *s.* side by Pamouky and York rivers. It is about 33 miles long and 12 broad, and contains 6239 inhabitants, including 3700 slaves. New Kent court-house is 28 miles from Richmond, and 25 from Williamsburg.]

[New Lebanon, a post-town in Dutchess county, New York, celebrated for its medicinal springs. The compact part of this town is pleasantly situated partly in an extensive valley and partly on the declivity of the surrounding hills. The spring is on the *s.* side, and near the bottom of a gentle hill, but a few rods *w.* of the Massachusetts *w.* line; and is surrounded with several good houses, which afford convenient accommodations for the valetudinarians who visit these waters. Concerning the medicinal virtues of this spring, Dr. Waterhouse, professor of the theory and practice of physio at Harvard university, and who visited it in the summer of 1791, observes, "I confess myself to be at a loss to determine the contents of these waters by chemical analysis, or any of the ordinary tests. I suspect their impregnation is from some cause weakened. Excepting from their warmth, which is about that of new milk, I never should have suspected them to come under the head of medicinal waters. They are used for the various purposes of cookery, and for common drink by the neighbours; and I never could discover any other effects from drinking them than what we might expect from rain or river water of that temperature. There was no visible change produced in this water by the addition of an alkali, or by a solution of alum; nor was any effervescence raised by the oil of vitriol; neither did it change the colours of gold, silver, or copper; nor did it redden beef or mutton boiled in it; nor did it extract a black tincture from galls; neither did it curdle milk, the whites of eggs, or soap. The quality of the waters of the pool at Lebanon is, therefore, very different from those of Saratoga. These are warm and warmish, those very cold, smart, and exhilarating. Frogs are found in the pool of Lebanon, and plants grow and flourish in and around it; but plants will not grow within the vapour of those of Saratoga, and as for small animals, they soon expire in it. Hence we conclude that that *spiritus mineralis* which some call aerial acid, or fixed air, abounds in the one but not in the other. Yet the Lebanon pool is famous for having wrought many cures, especially in rheumatisms, stiff joints, scabby cruptions, and even in visceral obstructions and indigestions; all of which is very probable. If a person who has brought on a train of chronic complaints by intemperance in eating and drinking, should swallow four or five quarts of rain or river water in a day, he would not feel so keen an appetite for animal food, or thirst for spirituous liquors. Hence such a course of water-drinking will open obstructions, rinse out impurities, render perspiration free, and thus remove that unnatural load from the animal machine, which causes and keeps up its disorders. Possibly, however, there may be something so subtle in these waters as to elude the scrutinizing hand of the chemists, since they all allow that the analysis of mineral waters is one among the most difficult things in the chemical art." A society of Shakers inhabit the *s.* part of the town in view of the main stage-road which passes through this town. Their manufactures of various kinds are considerable, and very neat and excellent. It is about 23 miles *e.* by *s.* of Albany, 112 *n.* by *e.* of New York, and six *w.* of Pittsfield.]

[NEW LONDON, a maritime county of Connecticut, comprehending the *s. e.* corner of it, bordering *e.* on Rhode Island, and *s.* on Long Island sound, about 30 miles from *e.* to *w.* and 24 from *n.* to *s.* It was settled soon after the first settlements were formed on Connecticut river; and is divided into 11 townships, of which New London and Norwich are the chief. It contained in 1756, 22,844 inhabitants, of whom 829 were slaves; in 1790, 33,200, of whom 586 were slaves.]

[NEW LONDON, a city, port of entry, and post-town, in the above county, and one of the most considerable commercial towns in the state. It stands on the *w.* side of the river Thames, about three miles from its entrance into the sound, and is defended by fort Trumbull and fort Griswold, the one on the New London, the other on the Groton side of the Thames. A considerable part of the town was burnt by Benedict Arnold in 1781. It has since been rebuilt. Here are two places of public worship, one for Episcopalians, and one for Congregationalists, about 300 dwelling houses, and 4600 inhabitants. The harbour is large, safe, and commodious, and has five fathoms water; high water at full and change, 54 minutes after eight. On the *w.* side of the entrance is a light-house, on a point of land which projects considerably into the sound. The exports for a year ending September 30, 1794, amounted to 557,453 dollars. In that year 1000 mules were shipped for the W. Indies. It is 12 miles *s.* of Norwich, 38 *s. e.* by *s.* of Hartford, 36 *e.* of New Haven, and 162 *n. e.* by *e.* of Philadelphia. Lat. 41° 19′ *n.* Long. 72° 10′ *w.* The township of New London was laid out in lots in 1648, but had a few English inhabitants two years before. It was called by the Indians Nameag or Towawog, and from being the seat of the Pequot tribe, was called Pequot. It was the seat of Sassacus, the grand monarch of Long island, and part of Connecticut and Narraganset.]

[NEW LONDON, a small township in Hillsborough county, New Hampshire, incorporated in 1779, and contains 311 inhabitants. It lies at the head of Blackwater river, and about three miles from the *n. e.* side of Sunapee lake.]

[NEW LONDON, a post-town of Virginia, and the chief town of Bedford county. It stands upon rising ground, and contains about 130 houses, a court-house and gaol. There were here in the late war several workshops for repairing fire-arms. It is 87 miles *w.* by *s.* of Richmond, and 87 *w.* of Petersburgh.]

[NEW MADRID, in the *n.* part of Louisiana, is a settlement on the *w.* bank of the Mississippi, commenced some years ago, and conducted by Colonel Morgan of New Jersey, under the patronage of the Spanish king. The spot on which the city was proposed to be built is situated in lat. 36° 30′ *n.* and 45 miles below the mouth of Ohio river. The limits of the new city of Madrid were to extend four miles *s.* and two *w.* from the river; so as to cross a beautiful, living, deep lake, of the purest spring-water, 100 yards wide, and several miles in length, emptying itself, by a constant and rapid narrow stream, through the centre of the city. The banks of this lake, called St. Annis, are high, beautiful, and pleasant; the water deep, clear, and sweet, and well stored with fish; the bottom a clear sand, free from woods, shrubs, or other vegetables. On each side of this delightful lake, streets were to be laid out, 100 feet wide, and a road to be continued round it, of the same breadth; and the streets were directed to be preserved for ever, for the health and pleasure of the citizens. A street 120 feet wide, on the bank of the Mississippi, was laid out; and the trees were directed to be preserved for the same purpose. Twelve acres, in a central part of the city, were to be preserved in like manner, to be ornamented, regulated, and improved by the magistracy of the city for public walks; and 40 half-acre lots for other public uses; and one lot of 12 acres for the king's use. We do not hear that this scheme is prosecuting, and conclude it is given up. The country in the vicinity of this intended city is represented as excellent, and, in many parts, beyond description. The natural growth consists of mulberry, locust, sassafras, walnut, hickory, oak, ash, dog-wood, &c. with one or more grape-vines running up almost every tree; and the grapes yield, from experiments, good red wine in plenty, and with little labour. In some of the low grounds grow large cypress trees. The climate is said to be favourable to health, and to the culture of fruits of various kinds, particularly for garden vegetables. The prairies or meadows are fertile in grass, flowering plants, strawberries, and when cultivated produce good crops of wheat, barley, Indian corn, flax, hemp, and tobacco, and are easily tilled. Iron and lead mines and salt springs, it is asserted, are found in such plenty as to afford an abundant supply of these necessary articles. The banks of the Mississippi, for many leagues in extent, commencing about 20 miles above the mouth of the Ohio, are a continued chain of lime-stone. A fine tract of high, rich, level land, *s. w.*, *w.*, and *n. w.* of New Madrid, about 25 miles wide, extends quite to the river St. Francis.]

[NEW MARLBOROUGH, a township in Ulster county, New York. See MARLBOROUGH.]

[NEW MARLBOROUGH, Berkshire county, Massachusetts. It is 23 miles *s.* of Lenox.]

[NEW MARLBOROUGH, a town in King George's county, Virginia, on the *w.* side of Patowmac river, 10 miles *e.* of Falmouth.]

[NEW MEADOWS River, in the district of Maine, a water of Casco bay, navigable for vessels of a considerable burden a small distance. See CASCO Bay.]

[NEW MEXICO. See MEXICO, NEW.]

[NEW MILFORD, a post-town of Connecticut, Litchfield county, on the *e.* side of Housatonick river, about 16 miles *n.* of Danbury, 13 *s. w.* of Litchfield, and 45 *w. s. w.* of Hartford.]

[NEW NORTH WALES. See WALES, and NEW BRITAIN.]

[NEW ORLEANS, the metropolis of Louisiana, was regularly laid out by the French in the year 1720, on the *e.* side of the river Mississippi, in lat. 30° *n.* and long. 90° 12′ *w.*; 18 miles from Detour des Anglois, or English Turn, and 117 from the Bella island, and 78 from the mouths of the Mississippi. All the streets are perfectly straight, but too narrow, and cross each other at right angles. There were, in 1788, 1100 houses in this town, generally built with timber frames, raised about eight feet from the ground, with large galleries round them, and the cellars under the floors level with the ground: any subterraneous buildings would be constantly full of water. Most of the houses have gardens. In March 1788, this town, by a fire, was reduced in five hours to 200 houses. It has since been rebuilt, and at present contains, according to Mr. Ashe, near 15,000 inhabitants. They are a mixture from all nations, but chiefly France and Spain. Those from the other American states constitute, according to Mr. Ashe, by far the worst part of the population.

The side of the town next the river is open, and is secured from the inundations of the river by a raised bank, generally called the Levee, which extends from the English Turn to the upper settlements of the Germans, a distance of more than 50 miles, with a good road all the way. There is reason to believe that in a short time New Orleans may become a great and opulent city, if we consider the advantages of its situation, but a few leagues from the sea, on a noble river, in a most fertile country, under a most delightful and wholesome climate, within two weeks sail of Mexico, and still nearer the French, Spanish, and British W. India islands, with a moral certainty of its becoming a general receptacle for the produce of that extensive and valuable country on the Mississippi, Ohio, and its other branches; all which are much more than sufficient to ensure the future wealth, power, and prosperity of this city. The vessels which sail up the Mississippi haul close alongside the bank next to New Orleans, to which they make fast, and take in or discharge their cargoes with the same ease as at a wharf. Its commerce, since its acquisition with the rest of Louisiana by the United States, has very considerably increased; nor, indeed, are the whole of the surrounding districts in a less flourishing state of population than the capital itself, since by the census of 1810, the inhabitants of that portion of country, comprised under the title of the Territorial Government, amounted to 76,556 souls.

A letter from New Orleans, dated August 21, 1812, gives the following account of a serious storm, with which this city has lately been visited.

"On Wednesday night last, about 10 o'clock, a gale commenced, occasionally accompanied with rain and hail, and which continued with a most dreadful violence for upwards of four hours. As we have never witnessed any thing to equal it, neither do we believe the imagination can picture to itself a scene more truly awful and distressing than that which its consequences present. The market-honse, a large and solid building, entirely demolished; its brick columns, of two feet diameter, swept down as though their weighty construction presented no obstacle whatever to the overwhelming element. The roof carried off from the church of the convent, the fence surrounding which, as also the trees in the garden, many whereof are remarkably large, levelled to the ground. The tin covering of the theatres, nailed on in such a manner, as would certainly have resisted any ordinary force, twisted and torn off as though it were mere paper. A great part of the brick wall surrounding the garrison beat down. It would be impossible to particularise all the damage that has been done; we believe, however, we may assert, that there is not a building in the city or fauxbourgs, but what has been more or less injured.

" But the scene presented to us on visiting the shore, who shall attempt to describe? The level almost entirely destroyed; the beach covered with fragments of vessels, merchandise, trunks, &c. and here and there the eye falling upon a mangled corpse. All the shipping below town high and dry in the woods. All the river craft, barges, market-boats, &c. entirely crushed to atoms. As far as we have heard from the country, the ravages have been terrible; the planters dwellings, sugar-houses, &c. demolished; and we have reason to fear that nearly the whole crop of sugar will be lost."]

[NEW PALTZ, a township in Ulster county, New York; bounded *e.* by Hudson river, *s.* by Marlborough and Shawangunk. It contains 2309 inhabitants, including 302 slaves. The compact part of it is situated on the *e.* side of Wall kill, and contains about 250 houses and a Dutch church. It is 10 miles from Shawangunk, nine *s.* of Kingston, 13 *s. w.* of Rhinebeck, and 67 *n.* of New York.]

[NEW PROVIDENCE Island. See PROVIDENCE.]

[NEW RIVER, a river of Tennessee, which rises on the *n.* side of the Alleghany mountains, and running a *n. e.* course enters Virginia, and is called KANHAWAY; which see.]

[NEW ROCHELLE, a township in W. Chester county, New York, on Long Island sound. It contained 692 inhabitants, of whom 89 were slaves, in 1790. In 1796, there were 100 of the inhabitants qualified electors. It is six miles *s. w.* of Rye, and 20 *n. e.* of New York city.]

[NEW SALEM, or PEQUOTTINK, a Moravian settlement, formed in 1786, on the *e.* side of Huron river, which runs *n.* into lake Erie.]

[NEW SALEM, a township in Hampshire county, Massachusetts; bounded *e.* by the *w.* line of Worcester county. It was incorporated in 1753, and contains 1543 inhabitants. It is 56 miles *w.* by *n.* of Boston.]

[NEW SALEM, a township in Rockingham county, New Hampshire, adjoining Pelham and Haverhill.]

[NEW SAVANNAH, a village in Burke county, Georgia, on the *s. w.* bank of the Savannah, 85 miles *s. s. e.* of Augusta.]

[NEW SHOREHAM. See BLOCK Island.]

[NEW SMYRNA Entrance, or MOSKITO Inlet, on the coast of Florida, is about 11 leagues *n. n. w.* one quarter *w.* from cape Canaverel.]

[NEW SOUTH WALES. See WALES, and NEW BRITAIN.]

[NEW SPAIN. See MEXICO.]

[NEW STOCKBRIDGE. See STOCKBRIDGE, NEW.]

[NEW SWEDELAND was the name of the territory between Virginia and New York, when in possession of the Swedes; and was afterwards possessed, or rather claimed, by the Dutch. The chief town was called Gottenburgh.]

[NEW THAMES River. See THAMES.]

[NEW UTRECHT, a small maritime town of New York, situated in King's county, Long Island, opposite the Narrows, and seven miles *s.* of New York city. The whole township contains 562 inhabitants, of whom 76 are qualified electors, and 206 slaves.]

[NEW WINDSOR, a township of Ulster county, New York, pleasantly situated on the *w.* bank of Hudson river, just above the high lands, three miles *s.* of Newburgh, and six *n.* of W. point. It contains 1819 inhabitants, of whom 261 are qualified electors, and 117 slaves. A valuable set of works in this town for manufacturing scythes was destroyed by fire. In 1795, the legislature granted the unfortunate proprietor, Mr. Boyd, 1500*l.* to enable him to re-establish them. The compact part of the town contains about 40 houses and a Presbyterian church: 48 miles *n.* of New York. The summer residence of Governor Clinton was formerly at a rural seat, on the margin of the river, at this place.]

[NEW WRENTHAM, district of Maine, a township six miles *e.* of Penobscot river, adjoining Orrington, and 15 miles from Buckston.]

[NEW YEAR'S Harbour, on the *n.* coast of Staten Land island, at the *s.* extremity of S. America, affords wood and good water; was discovered January 1, 1775; hence its name. Lat. 54° 49′ *s.* Long. 64° 11′ *w.*]

[NEW YEAR'S Islands, near the above harbour, within which is anchorage at *n.* half *w.* from the harbour, at the distance of two leagues from it.]

[NEW YORK, one of the United States of America, is situated between lat. 40° 33′ and 45° *n.* and between long. 73° 10′ and 80° *w.*: is about

[311 miles in length, and 265 in breadth; bounded *s. e.* by the Atlantic ocean, e. by Connecticut, Massachusetts, and Vermont, *n.* by Upper Canada, *s. w.* and *w.* by Pennsylvania, New Jersey, and lake Erie. It is subdivided into 21 counties, as follows, viz. New York, Richmond, Suffolk, West Chester, Queen's, King's, Orange, Ulster, Dutchess, Columbia, Rensselaer, Washington, Clinton, Saratoga, Albany, Montgomery, Herkemer, Onondago, Otsego, Ontario, and Tioga. In 1790, this state contained 340,120 inhabitants, of whom 21,324 were slaves. Since that period the counties of Rensselaer, Saratoga, Herkemer, Onondago, Otsego, and Tioga have been taken from the other counties. In 1796, according to the state census, there were 195 townships, and 64,017 qualified electors. Electors in this state are divided into the following classes:

Freeholders to the value of 1000*l.* . . .	36,338
Do. to the value of 20*l.* and under 100*l.*	4,838
Do. who rent tenements of 40*l.* per annum	22,598
Other freeholders	243
	64,017

By the census of 1810, its population amounted to 959,220 souls.

It is difficult to ascertain accurately the proportion the number of electors bears to the whole number of inhabitants in this state. In the county of Herkemer the electors to the whole number of inhabitants was, in 1795, nearly as one to six, but this proportion will not hold through the state. In 1790, the number of inhabitants in the state was, as already mentioned, 340,120, of whom 41,785 were electors. In 1795, the number of electors was 64,017, which, if the proportion between the electors and the whole number of inhabitants be the same, gives, as the whole number of inhabitants in 1795, 530,177, an increase, in five years, of 190,057.

The chief rivers are Hudson, Mohawk, and their branches. The rivers Delaware and Susquehannah rise in this state. The principal lakes are Otsego, Oneida, George, Seneca, Cayuga, Salt, and Chautaughque. The principal bay is that of York, which spreads to the *s.* before the city of New York. The legislature of New York, stimulated by the enterprising and active Pennsylvanians, who are competitors for the trade of the *w.* country, have lately granted very liberal sums, towards improving those roads that traverse the most settled parts of the country, and opening such as lead into the *w.* and *n.* parts of the stale, uniting as far as possible the establishments on Hudson's river, and the most populous parts of the interior country, by the nearest practicable distances. By late establishments of post-roads a safe and direct conveyance is opened between the most interior *w.* parts of this state, and the several states in the union: and when the obstructions between Hudson's river and lake Ontario are removed, there will not be a great deal to do to continue the water communication by the lakes and through Illinois river to the Mississippi.

New York, to speak generally, is intersected by ridges of mountains extending in a *n. e.* and *s. w.* direction. Beyond the Alleghany mountains, however, the country is level, of a fine rich soil, covered in its natural state with maple, beech, birch, cherry, black walnut, locust, hickory, and some mulberry trees. On the banks of lake Erie are a few chesnut and oak ridges. Hemlock swamps are interspersed thinly through the country. All the creeks that empty into lake Erie have falls, which afford many excellent mill-seats. The lands between the Seneca and Cayuga lakes are represented as uncommonly excellent, being most agreeably diversified with gentle risings, and timbered with lofty trees, with little underwood. The legislature have granted a million and a half acres of land, as a gratuity to the officers and soldiers of the line of this state. This tract forms the military townships of the county of Onondago. See MILITARY Townships, and ONONDAGO.

East of the Alleghany mountains, which commence with the Kaat's kill, on the *w.* side of Hudson's river, the country is broken into hills with rich intervening valleys. The hills are clothed thick with timber, and when cleared afford fine pasture; the valleys, when cultivated, produce wheat, hemp, flax, pease, grass, oats, Indian corn, &c. Of the commodities produced from culture, wheat is the principal. Indian corn and pease are likewise raised for exportation; and rye, oats, barley, &c. for home consumption.

The best lands in the state, along Mohawk river and *n.* of it and *w.* of the Alleghany mountains, but a few years ago were mostly in a state of nature, but have been of late rapidly settling. In the *n.* and unsettled parts of the state are plenty of moose, deer, bears, some beavers, martins, and most other of the inhabitants of the forest, except wolves.

The Ballstown, Saratoga, and New Lebanon medicinal springs are much celebrated; these are noticed under their respective heads. The salt made from the Salt springs here is equal in goodness to that imported from Turk's island. The weight of a bushel of the salt is 156 lb. A spring is reported to have been discovered in the Susquehannah country, impregnated with nitre, from] [which salt-petre is made in the same manner that common salt is made from the Onondago springs. Large quantities of iron ore are found here. A silver mine has been worked at Phillipsburg, which produced virgin silver. Lead is found in Herkemer county, and sulphur in Montgomery. Spar, zinc or spelter, a semi-metal, magnez, used in glazings, pyrities of a golden hue, various kinds of copper ore, and lead and coal mines, are found in this state; also petrified wood, plaster of Paris, ising-glass in sheets, tales, and crystals of various kinds and colours, flint, asbestos, and several other fossils. A small black stone has also been found, which vitrifies with a small heat, and it is said makes excellent glass.

The chief manufactures are iron, glass, paper, pot and pearl ashes, earthen ware, maple sugar and molasses, and the citizens in general manufacture their own clothing. This state, having a short and easy access to the ocean, commands the trade of a great proportion of the best settled and best cultivated parts of the United States. Their exports to the W. Indies are, biscuit, peas, Indian corn, apples, onions, boards, staves, horses, sheep, butter, cheese, pickled oysters, beef, and pork. But wheat is the staple commodity of the state, of which no less than 677,700 bushels were exported so long ago as the year 1775, besides 2555 tons of bread, and 2828 tons of flour. The increase since has been in proportion to the increase of the population. In wheat and flour more than a million bushels are now annually exported. W. India goods are received in return for the above articles. Besides the articles already enumerated, are exported flax-seed, cotton, wool, sarsaparilla, coffee, indigo, rice, pig-iron, bar-iron, pot-ash, pearl-ash, furs, deer-skin, logwood, fustic, mahogany, bees-wax, oil, Madeira wine, rum, tar, pitch, turpentine, whale-fins, fish, sugars, molasses, salt, tobacco, lard, &c.; but most of these articles are imported for re-exportation. The exports to foreign parts, for the year ending September 30, 1791, 1792, &c. consisting principally of the articles above enumerated, amounted as follows: in 1791, to 2,505,465 dollars 10 cents;—1792, 2,535,790 dollars 25 cents;—1793, 2,932,370 dollars:—1794, 5,442,183 dollars 10 cents;—1795, 10,304,580 dollars 78 cents. This state owned in 1792, 46,626 tons of shipping, besides which she finds employment for about 40,000 tons of foreign vessels.

There are in this state two handsomely endowed and flourishing colleges, viz. Columbia, formerly King's college, in the city of New York, and Union college, at Schenectady. See New York City, and Schenectady. Besides these, there are dispersed in different parts of the state, 14 incorporated academies, containing in the whole as many as 6 or 700 students. These, with the establishment of schools, one at least in every district of four square miles, for the common branches of education, must have the most beneficial effects on the state of society. The sums granted by the legislature of this state for the encouragement of literature since the year 1790, have been very liberal, and is evincive of the wisest policy. In March 1790, the legislature granted to the regents of the university, who have by law the superintendance and management of the literature of the state, several large and valuable tracts of land, on the waters of lakes George and Champlain, and also Governor's island in the harbour of New York, with intent that the rents and income thereof should be by them applied to the advancement of literature. At the same time they granted them 1000*l.* currency, for the same general purpose. In April 1792, they ordered to be paid to the regents, 1500*l.* for enlarging the library, 200*l.* for a chemical apparatus, 1200*l.* for erecting a wall to support the college grounds, and 5000*l.* for erecting a hall and an additional wing to the college; also 1500*l.* annually for five years to be discretionally distributed among the academies of the state; also 750*l.* for five years, to be applied to the payment of the salaries of additional professors. In their sessions since 1795, the sums they have granted for the support of the colleges, academies, and of common schools throughout the state, have been very liberal.

The religious sects or denominations in the state are, English Presbyterians, Dutch Reformed, Baptists, Episcopalians, Friends or Quakers, German Lutherans, Moravians, Methodists, Roman Catholics, Shakers, a few followers of Jemima Wilkinson at Geneva, and some Jews in the city of New York.

The treasury of this state is one of the richest in the union. The treasurer of the state reported to the legislature in January 1796, that the funds amounted to 2,119,068 dollars 33 cents, which yields an annuity of 234,218 dollars. Besides the above immense sum, there was at that period in the treasury 134,207*l.* 19*s.* 10¼*d.* currency. The ability of the state, therefore, is abundantly competent to aid public institutions of every kind, to make roads, erect bridges, open canals, and push every kind of improvement to the most desirable length. The body of the Six Nations of Indians inhabit the w. part of this state. See Six Nations.]

[The English language is generally spoken throughout the state, but is not a little corrupted by the Dutch dialect, which is still spoken in some counties, particularly in King's, Ulster, Albany, and that part of Orange which lies s. of the mountains. But as Dutch schools are almost, if not wholly discontinued, that language, in a few generations, will probably cease to be used at all. And the increase of English schools has already had a perceptible effect in the improvement of the English language.

Besides the Dutch and English, there are in this state many emigrants from Scotland, Ireland, Germany, and some few from France. Many Germans are settled on the Mohawk, and some Scots people on the Hudson, in the county of Washington. The principal part of the two former settled in the city of New York; and retain the manners, the religion, and some of them the language of their respective countries. The French emigrants settled principally at New Rochelle, and on Staten island, and their descendants, several of them, now fill some of the highest offices in the United States. The w. parts of the states are settled and settling principally from New England. There are three incorporated cities in this state, New York, Albany, and Hudson.]

[New York County, in the above state, comprehending the island of New York or Mahattan, on which the metropolis stands, and the following small islands: Great Barn, Little Barn, Manning's, Nutten, Bedlow's, Bucking, and Oyster islands. It contained, in 1790, 33,131 inhabitants, including 2369 slaves. In 1796, the number of inhabitants amounted to about 70,000, of whom 7272 were qualified electors.]

[New York City is situated on the s. w. point of York island, at the confluence of Hudson and E. rivers, and is the metropolis of the state of its name, and the second in rank in the union. The length of the city on E. river is upwards of two miles, and rapidly increasing, but falls short of that distance on the banks of the Hudson. Its breadth on an average is about a mile; and its circumference four or five miles. The plan of the city is not perfectly regular, but is laid out with reference to the situation of the ground. The ground which was unoccupied before the peace of 1783, was laid out in parallel streets of convenient width, which has had a good effect upon the parts of the city lately built. The principal streets run nearly parallel with the rivers. These are intersected, though not at right angles, by streets running from river to river.

The government of the city (which was incorporated in 1696) is now in the hands of a mayor, alderman, and common-council. The city is divided into seven wards, in each of which there is chosen annually by the people an alderman and an assistant, who, together with the recorder, are appointed annually by the council of appointment. The mayor's court, which is held from time to time by adjournment, is in high reputation as a court of law. A court of session is likewise held for the trial of criminal causes. The situation of the city is both healthy and pleasant. Surrounded on all sides by water, it is refreshed with cool breezes in summer, and the air in winter is more temperate than in other places under the same parallel.

A want of good water is a great inconvenience to the citizens, there being few wells in the city. Most of the people are supplied every day with fresh water, conveyed to their doors in casks, from a pump near the head of Queen street, which receives it from a spring almost a mile from the centre of the city. This well is about 20 feet deep and four feet diameter. The average quantity drawn daily from this remarkable well, is 110 hogsheads of 130 gallons each. In some hot summer days 216 hogsheads have been drawn from it; and what is very singular, there is never more or less than about three feet water in the well. The water is sold commonly at three-pence a hogshead at the pump. Several proposals have been made by individuals to supply the citizens by pipes, but none have yet been accepted.

New York has rapidly improved within the last 20 years, and land, which then sold in that city for 50 dollars, is now worth 1500; but it is a place of too much importance, in a political point of view, to be treated with a general description. Much has been written concerning it by late travellers, and we shall divide the remarks we have collected from their works, (especially from that of Mr. Lambert, to whom we have been indebted for much of the information we have given concerning the United States) under the following heads, viz.

Chap. I.

The military.—The harbour.—The Broadway.—Bowery road.—Shops.—Hotels.—Public buildings.—The park.—The theatre.—Vauxhall.—Ranelagh.—Wharfs.—Places of worship.—Public buildings.—King's or Columbia college.—State prison.—Courts of law.—Board of health.—Quarantine station.—Chamber of commerce.—Inspectors of lumber, &c.—Commerce of New York.—Increase of commerce.—Market places.—Abundance of provisions.—Articles

[*brought to market.—Fly market.—Bare market.—Price of commodities at New York.—Charitable institutions.—The ladies' society for the relief of poor widows with small children.—The Cincinnati.—Medical society.—Protestant Episcopal society.—Columbia college.—Newspapers.—Literary fair.*

CHAP. II.

Number of deaths at New York.—Mode of living.—The yellow fever.—Population of New York.—Deaths.—Church-yards.—Funerals.—Society of New York.—Elegant women.—Personal attractions.—Education.—Thirst after knowledge.—Arts and sciences.—Literature.—Taste for reading.—Salmagundi.—The Echo.—Barlow's Columbiad.—Smoking.—Style of living at New York.—Marriages.—Christmas-day.—Recommendations of the clergy.—New-year's day.—Political parties.—Duels.—Lat. and Long.

CHAP. I.

IT is well known that the 25th of November is the anniversary of the evacuation of New York by the British troops at the peace of 1783. The militia, or rather the volunteer corps, are accustomed on this day to be assembled from different parts of the city, on the grand battery by the water-side, so called from a fort having been formerly built on the spot, though at present it is nothing more than a lawn for the recreation of the inhabitants, and for the purpose of military parade. The troops do not amount to 600, and are gaudily dressed, in a variety of uniforms, every ward in the city having a different one: some of them with helmets appear better suited to the theatre than the field. The general of the militia and his staff are dressed in the national uniform of blue, with buff facings. They also wear large gold epaulets and feathers, which altogether has a very showy appearance. The gun-boats, which are stationed off the battery, fire several salutes in honour of the day, and the troops parade through the streets leading to the water-side. They then go through the forms practised on taking possession of the city, manœuvring and firing fens-de-joye, &c. as occurred on the evacuation of New York. One of the corps consists wholly of Irishmen, dressed in light green jackets, white pantaloons, and helmets.

York island (or, as it is sometimes called, Manhattan) is separated from the continental part of the state of New York by the Haerlem river. Its length is about 16 miles, and its breadth varies from a quarter to a mile and a half. The bay is about nine miles long and three broad, without reckoning the branches of the rivers on each side of the town. From the ocean at Sandy hook to the city, is not more than 28 miles. The water is deep enough to float the largest vessels. Ships of 90 guns have anchored opposite the city. There they lie land-locked, and well secured from winds and storms; and fleets of the greatest number have ample space for mooring. During the revolutionary war, New York was the great rendezvous for the British fleet; from the time of its surrender in 1776 to the peace of 1783, our ships of war passed all seasons of the year here in security.

It has been often observed that the cold of winter has less effect upon the water of New York harbour, than in several places further to the *s.* When Philadelphia, Baltimore, and Alexandria are choked up by ice in severe winters, as in that of 1804, New York suffers scarcely any inconvenience from it. This is owing partly to the saltness of the sound and the bay; while the Delaware, Patapsco, and Patowmac, at the respective cities above mentioned are fresh, and consequently more easy to freeze. The water at New York differs but little in saltness from the neighbouring Atlantic. The openness of the port is also to be ascribed in part to the greater ebb and flow of the tide. Another reason of the greater fitness of New York for winter navigation is the rapidity of the currents. The strength of these in ordinary tides, and more especially when they are agitated by storms, is capable of rending the solidity of the ice, and reducing it to fragments. And although the whole harbour was covered by a bridge of very compact ice in 1780, to the serious alarm of the British garrison, the like has never occurred since. The number of vessels that entered from foreign ports only into this port in 1795 amounted to 941. The islands in the vicinity of New York are Long island, Staten island, Governor's, Bedlow's and Ellis's islands. The first is of very considerable extent, being 120 miles in length, and about eight miles in breadth. It is a fertile and well cultivated piece of land; inhabited chiefly by the descendants of the old Dutch settlers.

The Broadway and Bowery road are the two finest avenues in the city, and nearly of the same width as Oxford street in London. The first commences from the grand battery situate at the extreme point of the town, and divides it into two unequal parts. It is upwards of two miles in length, though the pavement does not extend above a mile and a quarter; the remainder of the road consists of straggling houses which are the commencement of new streets, already planned out. The Bowery road commences from Chatham] [street which branches off from the Broadway to the right, by the side of the park. After proceeding about a mile and a half it joins the Broadway, and terminates the plan which is intended to be carried into effect for the enlargement of the city. Much of the intermediate spaces between these large streets, and from thence to the Hudson and East rivers, is yet unbuilt upon, or consists only of unfinished streets and detached buildings.

The houses in the Broadway are lofty and well built. They are constructed in the English style, and differ but little from those of London at the *w.* end of the town; except, that they are universally built of red brick. In the vicinity of the battery, and for some distance up the Broadway, they are nearly all private houses, and occupied by the principal merchants and gentry of New York; after which, the Broadway is lined with large commodious shops of every description, well stocked with European and India goods; and exhibiting as splendid and varied a show in their windows, as can be met with in London. There are several extensive book-stores, print-shops, musie-shops, jewellers, and silversmiths; hatters, linen-drapers, milliners, pastry cooks, coachmakers, hotels, and coffee-houses. The street is well paved, and the foot-paths are chiefly bricked. In Robinson street, the pavement before one of the houses, and the steps of the door, are composed entirely of marble.

The city hotel is the most extensive building of that description in New York; and nearly resembles in size and style of architecture the London tavern in Bishopgate street. The ground-floor of the hotel at New York is, however, converted into shops, which have a very handsome appearance in the Broadway. Mechanic hall is another large hotel at the corner of Robinson street, in the Broadway. It was erected by the society of mechanics and tradesmen, who associated themselves for charitable purposes, under an act of the legislature in 1792. There are three churches in the Broadway; one of them, called Grace church, is a plain brick building, recently erected: the other two are St. Paul's and Trinity; both handsome structures, built with an intermixture of white and brown stone. The adjoining church-yards, which occupy a large space of ground, railed in from the street, and crowded with tomb-stones, are far from being agreeable spectacles in such a populous city. At the commencement of the Broadway, near the battery, stands the old government-house, now converted into offices for the customs. Before it is a small lawn railed in, and in the centre is a stone pedestal, upon which formerly stood a leaden statue of George III. In the revolutionary war it was pulled down by the populace, and made into bullets.

The city hall, where the courts of justice are held, is situated in Wall street, leading from the coffee-house slip by the water side into the Broadway. It is an old heavy building, and very inadequate to the present population and wealth of New York. A court-house on a larger scale, and more worthy of the improved state of the city, is now building at the end of the park, between the Broadway and Chatham street, in a style of magnificence, unequalled in many of the larger cities of Europe. The exterior consists wholly of fine marble, ornamented in a very neat and elegant style of architecture, and the whole is to be surmounted by a beautiful dome, which, when finished, will form a noble ornament to that part of the town, in which are also situated the theatre, mechanic hall, and some of the best private houses in New York. The park, though not remarkable for its size, is, however, of service, by displaying the surrounding buildings to a better advantage; and is also a relief to the confined appearance of streets in general. It consists of about four acres planted with elms, planes, willows, and catalpas; and the surrounding foot-walk is encompassed by rows of poplars: the whole is inclosed by a wooden paling. Neither the park nor the battery are very much resorted to by the fashionables of New York, as they have become too common. The genteel lounge is in the Broadway, from eleven to three o'clock, during which time, it is as much crowded as the Bond street of London: and the carriages, though not so numerous, are driven to and fro with as much velocity. The foot-paths are planted with poplars, and afford an agreeable shade from the sun in summer. About three years ago the inhabitants were alarmed by a large species of caterpillar, which bred in great numbers on the poplars, and were supposed to be venomous; various experiments were tried, and cats and dogs were made to swallow them; but it proved to be a false alarm, though the city for some time was thrown into the greatest consternation.

The theatre is on the *s. e.* side of the park, and is a large commodious building. The outside is in an unfinished state, but the interior is handsomely decorated, and fitted up in as good style as the London theatres, upon a scale suitable to the population of the city. It contains a large coffee room, and good sized lobbies; and is reckoned to hold about 1200 persons. The scenes are well painted and numerous; and the machinery,] [dresses, and decorations, are elegant and appropriate to the performances, which consist of all the new pieces that come out on the London boards, and several of Shakspeare's best plays. The only fault is, that they are too much curtailed, by which they often lose their effect; and the performances are sometimes over by half past 10, though they do not begin at an earlier hour than in London. The drama had been a favourite in New York before the revolution. During the time the city was in our possession, threatrical entertainments were very fashionable; and the characters were mostly supported by officers of the army. After the termination of the war, the play-house fell into the hands of Messrs. Hallam and Henry, who for a number of years exerted themselves with much satisfaction to please the public. After the death of Mr. Henry, the surviving manager formed a partnership with a favourite and popular performer, under the firm of Hallam and Hodgkinson. Their efforts were soon after aided by the addition of Mr. W. Dunlap. After some time Hallam and Hodgkinson withdrew from the concern, and Mr. Dunlap commenced sole manager. In this capacity he continued till 1804. During his management of the theatrical concerns, he brought forward many pieces of his own composition, as well as several translations from the German. He is now publishing his dramatic works in 10 volumes. Mr. Cooper succeeded him in the direction of the theatre, and in his hands it at present remains. The theatre has been built about 10 years, and of course embraces every modern improvement.

New York has its Vauxhall and Ranelagh; but they are poor imitations of those near London. They are, however, pleasant places of recreation for the inhabitants. The Vauxhall garden is situated in the Bowery road about two miles from the city hall. It is a neat plantation, with gravel walks adorned with shrubs, trees, busts, and statues. In the centre is a large equestrian statue of General Washington. Light musical pieces, interludes, &c. are performed in a small theatre situate in one corner of the gardens: the audience sit in what are called the pit and boxes, in the open air. The orchestra is built among the trees, and a large apparatus is construced for the display of fire-works. The theatrical corps of New York is chiefly engaged at Vauxhall during summer. The Ranelagh is a large hotel and garden, generally known by the name of Mount Pitt, situated by the water side, and commanding some extensive and beautiful views of the city and its environs.

A great portion of the city, between the Broadway and the E. river, is very irregularly built; being the oldest part of the town, and of course less capable of those improvements which distinguish the more recent buildings. Nevertheless, it is the chief seat of business, and contains several spacious streets crowded with shops, stores, and warehouses of every description. The water side is lined with shipping which lie along the wharfs, or in the small docks called slips, of which there are upwards of 12 towards the E. river, besides numerous piers. The wharfs are large and commodious, and the warehouses, which are nearly all new buildings, are lofty and substantial. The merchants, ship-brokers, &c. have their offices in front on the ground floor of these warehouses. These ranges of buildings and wharfs extend from the grand battery, on both sides the town, up the Hudson and E. rivers, and encompass the houses with shipping, whose forest of masts gives a stranger a lively idea of the immense trade which this city carries on with every part of the globe. New York appears to him the Tyre of the new world.

New York contains 33 places of worship, viz. nine Episcopal churches, three Dutch churches, one French church, one Calvinist, one German Lutheran, one English Lutheran, three Baptist meetings, three Methodist meetings, one Moravian, six Presbyterian, one Independent, two Quakers' and one Jews' synagogue.

Besides the public buildings which we have mentioned, there are numerous banks, insurance companies, commercial and charitable institutions, literary establishments, &c. The new state prison is an establishment worthy of imitation in England. By the law of New York, treason, murder, and the procuring, aiding, and abetting any kind of murder, are the only crimes punishable by death. The mode of execution is the same as in England. All other offences are punished by imprisonment for a certain period in the state prison. This building is situated at Greenwich, about two miles from the city hall, on the shore of the Hudson river. The space inclosed by the wall is about four acres, and the prison is governed by seven inspectors appointed by the state council. They meet once a month, or oftener, together with the justices of the supreme court, the mayor and recorder of the city, the attorney-general, and district attorney. The inspectors make rules for the government of the convicts, and other persons belonging to the prison; and appoint two of their own body to be visiting inspectors monthly. The board of inspectors have charge of the prison, and appoint a keeper or deputy, and as many assistants as they find to be ne-]

[cessary. The salaries of the keepers are paid out of the treasury of the state. The inspectors, or rather the agents of the prison, are empowered to purchase clothing, bedding, provisions, tools, implements, and raw or other materials for the employment of the convicts, and keep accounts of the same: also to open an account with each convict, charging him with his expences, and crediting him with his labour: and if there should be any balance due to the convict at the time of his discharge, to give him a part or the whole of it; but if the whole should not be given to him, to convey the residue to the credit of the state. If a convict on entering the prison is unacquainted with any trade, he has the choice of learning one most agreeable to him. It is said, that a certain man who became a shoe-maker in that prison, came out, at the end of his time, with several hundred dollars in pocket. Hence the country is benefited; and individuals, instead of being made worse in prison, are rendered useful members of society.

The expence of conveying and keeping the convicts is always paid by the state. They are dressed in uniforms of coarse cloth, according to their classes and conduct, and kept at some kind of work. For profane cursing, swearing, indecent behaviour, idleness, negligence, disobedience of regulations, or perverse conduct, the principal keeper may punish the convicts by confinement in the solitary cells, and by a diet of bread and water, during such term as any two of the inspectors advise. For the greater security, there is a detachment of firemen allotted to the prison, also an armed guard consisting of a captain, a serjeant, two corporals, a drummer, a fifer, and twenty privates.

The laws are administered by the following courts of justice.

I. The court for the trial of impeachments, and the correction of errors. Since the removal of the seat of government to Albany, this court is now held in that place. It is the court of dernier resort, and consists of the president of the senate, for the time being, and the senators, chancellor, and judges of the supreme court, or the major part of them.

II. The court of chancery. This court, consisting of the chancellor, is held twice a year at least in New York, and twice in the city of Albany, and at such other times as the chancellor may think proper. Appeals lie from the decisions of the chancellor to the court for the correction of errors.

III. The supreme court. This court consists of a chief justice, and four puisne judges, and there are four stated and regular terms. The court appoints circuit courts to be held in the vacation in the several counties, before one of the judges, for the trial of all causes before a jury. Questions of law which arise on the facts, are argued before the whole court. Writs of error may be brought on the judgments of the supreme court, to the court for the correction of errors.

IV. The court of exchequer. The junior justice in the supreme court, or, in his absence, any other of the pnisne judges, is, *ex officio*, judge of the court of exchequer. This court is held during the terms of the supreme court, and at the same places. It hears and determines all causes and matters relating to forfeitures for recognizances or otherwise, fines, issues, amercements, and debts due to the people of the state.

V. The courts of oyer and terminer, and general gaol delivery. These courts are held pursuant to an act of the legislature, without a special commission, by one or more of the justices of the supreme court; together with the mayor, recorder, and aldermen of the city, or any three of them, of whom a justice of the supreme court must always be one. They have the power to hear and determine all treasons, felonies, and other crimes and misdemeanors, and to deliver the gaols of all prisouers confined therein.

VI. The court of common pleas, commonly called the mayor's court. This is held before the mayor, aldermen, and recorder, or before the mayor and recorder only. This court hears and determines all actions, real, personal, or mixed, arising within the city of New York, or within the jurisdiction of the court. Where the sum demanded is above 250 dollars, the cause may be removed at any time before the trial, into the supreme court. A writ of error lies from all judgments of this court to the supreme court.

VII. The court of general sessions of the peace. This court is also held by the mayor, recorder, and aldermen, of whom the mayor or recorder must always be one. Courts of special sessions of the peace may also be held at any time the common council may direct, and may continue as long as the court may think proper for the dispatch of business. These courts have the power to hear and determine all felonies and offences committed in the city of New York. There is also a court of special sessions for the trial of petty offences; which consists of the mayor, recorder, and aldermen.

VIII. The court of probates. Since the removal of the seat of government to Albany the] [judge of this court is required to reside in that city. He has all the powers of jurisdiction relative to testamentary matters, which were formerly exercised by the governor of the colony, as judge of the prerogative court, except as to the appointment of surrogates.

IX. Court of surrogate. Surrogates are appointed for each county by the council of appointment, one of which resides and holds his court in the city of New York. They have the sole and exclusive power to take proof of the last wills and testaments of persons deceased, who at the time of their death were inhabitants of the city, in whatever place the death may have happened; to issue probates, and grant letters of administration of the goods, chatties, and credits of persons dying intestate, or with the wills annexed. Appeals from the orders and decrees of the surrogate lie to the court of probates.

X. District court of the United States. This court, consisting of a single judge, has four regular sessions in a year, and special sessions are held as often as the judge thinks necessary. It has exclusive original jurisdiction of civil causes, of admiralty and maritime jurisdiction, including all seizures under the laws of impost, navigation, or trade of the United States, on the high seas, and in the navigable waters, as well as seizures on land within other waters, and all penalties and forfeitures arising under the laws of the United States. It has also jurisdiction, exclusive of the state courts, of all crimes and offences cognizable under the authority of the United States, committed within the district, or upon the high seas, where no other punishment than whipping, not exceeding 30 stripes, a fine not exceeding 100 dollars, or a term of imprisonment not exceeding six months, is to be inflicted. It also has concurrent jurisdiction with the courts of the state, where an alien sues for a tort only, in violation of the laws of nations, or treaties of the United States; and where the United States sue, and the matter in dispute does not exceed 100 dollars. It has a jurisdiction over the state courts of all suits against consuls and vice-consuls.

XI. The circuit court of the United States for the district of New York, in the second circuit, is held in the city on the 1st of April and the 1st of September in each year. It consists of one of the judges of the supreme court of the United States, and the judge of the district court. It has original cognizance of all civil suits, where the matter in dispute exceeds 500 dollars, and the United States are plaintiffs, or an alien is the party; or the suit is between citizens of different states. It has exclusive cognizance of all crimes and offences cognizable under the authority of the United States, except where it is otherwise provided by law; and a concurrent jurisdiction with the district court of the crimes cognizable therein.

Of late years a board of health has been established at New York, under an act of the legislature, and a variety of regulations are enjoined, for the purpose of preventing the introduction of malignant fevers. A station is also assigned on Staten island, where vessels perform quarantine: the buildings which constitute the hospital are separated from each other, and are capable of accommodating upwards of 300 sick. The situation is extremely pleasant, and well adapted to the purpose.

There are five banks and nine insurance companies: one of the latter is a branch of the Phœnix company of London. There is a chamber of commerce in New York, which has for its object the promotion and regulation of mercantile concerns; and is also a charitable institution for the support of the widows and children of its members. The origin of this institution is of a singular nature; and proves that non-intercourse acts in America are not of recent origin. The following is an account of it.

On the 5th of April 1768, 20 merchants met in the city of New York, and formed themselves into a voluntary association, which they called "The New York chamber of commerce." On the 2d of May 1769, they received a message of thanks from the house of assembly to the merchants of the city and colony, for their patriotic conduct in declining the importation of goods from Great Britain at that juncture. The words on this occasion were the following: "I have it in charge from the general assembly, to give the merchants of this city and colony the thanks of the house, for their repeated, disinterested, public-spirited, and patriotic conduct, in declining the importation or receiving of goods from Great Britain, until such acts of parliament as the general assembly had declared unconstitutional, and subversive of the rights and liberties of the people of this colony, should be repealed." On the 13th of March 1770, during the administration of Dr. Colden, as lieutenant-governor of the province, a charter was granted to the society, by the name of "The corporation of the chamber of commerce in the city of New York in America." They are enabled to hold property not exceeding a clear yearly value of 3000*l.* sterling per annum. The objects are to enable them the better to carry into execution, encourage, and promote, by just and] [lawful ways and means, such measures as tend to promote and extend just and lawful commerce; and to provide for such members as may be hereafter reduced to poverty, their widows and children.

The merchants, in their address to the governor, for his condescension in allowing the charter, observed, among other things, that they are thereby enabled to execute many plans of trade, which, as individuals, they could not before accomplish; and promised themselves many and great advantages to the colony from their incorporation. The chamber, by its charter, is authorised to make regulations for the government of its officers and members, and for regulating all its other affairs, with penalties for the violation of them. They are also empowered to appoint a committee of five members, at each monthly meeting, to adjust and determine all mercantile disputes which may be referred to them: and the secretary is directed to cause the names of this monthly committee to be published in one of the public newspapers, for the information of those who may wish to submit any disputes to their decision. No person can be admitted as members, but merchants and insurance-brokers.

The committees must report to the chamber, at the next stated meeting after their time of service is ended, the several objects of dispute which have been referred to their decision, with the names of the parties, together with the arguments and principles upon which their adjudications have been founded, in order that they may be recorded by the secretary. If the members of the chamber refuse to submit all disputed matters of accounts between each other, to the final arbitration and determination either of a monthly committee, or such members as may be chosen by the parties, they may be punished by expulsion.

Bills of exchange drawn upon any of the W. India islands, Newfoundland, or other foreign possessions in America, and returned protested for non-payment, are liable to 10 per cent. damages, on demand, at the current exchange, when the bill with the protest is presented either to the drawer or indorser thereof. Bills of exchange drawn on any part of Europe, and returned protested, are liable to 20 per cent. damages. The chamber has also published regulations for estimating the tonnage of bulky articles, for correcting mistakes in freight, and for fixing inland and foreign commissions. By an act of the state legislature, passed in the year 1784, all the privileges granted in the charter were fully confirmed and perpetuated.

Inspectors are appointed by the state council to examine lumber, staves, and heading, pot and pearl ashes, sole leather, flour and meal, beef and pork, previous to exportation. Persons shipping the above articles without having them inspected, are liable to heavy penalties.

New York is esteemed the most eligible situation for commerce in the United States. It almost necessarily commands the trade of one half New Jersey, most of that of Connecticut, part of that of Massachusetts, and almost the whole of Vermont, besides the whole fertile interior country, which is penetrated by one of the largest rivers in America. This city imports most of the goods consumed between a line of 30 miles *e.* of Connecticut river, and 20 miles *w.* of the Hudson, and between the ocean and the confines of Canada, a considerable portion of which is the best peopled of any part of the United States; and the whole territory contains upwards of 1,000,000 people, or one-fifth of the inhabitants of the union. Besides, some of the other states are partially supplied with goods from New York. But in the staple commodity, flour, Pennsylvania and Maryland have exceeded it, the superfine flour of those states commanding a higher price than that of New York; not that the quality of the grain is worse, but because greater attention is paid in those states to the inspection and manufacture of that article. In the manufacture likewise of iron, paper, cabinet works, &c. Pennsylvania exceeds not only New York, but all her sister states. In times of peace, however, New York will command more commercial business than any town in the United States. In time of war it will be insecure without a marine force; but a small number of ships will be able to defend it from the most formidable attacks by sea.

The commerce of New York, before the late embargo, was in a high state of prosperity and progressive improvement. The merchants traded with almost every part of the world, and though at times they suffered some privations and checks from the belligerent powers of Europe, yet their trade increased, and riches continued to pour in upon them. They grumbled, but nevertheless pursued their prosperous career, and seldom failed in realizing handsome fortunes. What a mortifying stroke, then, was the embargo! a measure which obliged them to commit a sort of commercial suicide in order to revenge themselves of a few lawless acts, which might have been easily avoided if the merchants had speculated with more prudence. The amount of tonnage belonging to the port of New York in 1806 was 183,671 tons; and the number of vessels in the] harbour on the 25th of December 1807, when the embargo took place, was 537. The moneys collected in New York for the national treasury, on the imports and tonnage, have for several years amounted to one-fourth of the public revenue. In 1806, the sum collected was 6,500,000 dollars, which, after deducting the drawbacks, left a nett revenue of 4,500,000 dollars; which was paid into the treasury of the United States, as the proceeds of one year. In the year 1808, the whole of this immense sum had vanished! In order to shew how little the Americans have suffered upon the aggregate from Berlin decrees and orders of council; from French menaces, and British actions; it is only necessary to state, that in 1803 the duties collected at New York scarcely amounted to 4,000,000 of dollars; and that at the period of laying on the embargo, at the close of the year 1807, they amounted to nearly 7,000,000 dollars. After this it is hardly fair to complain of the violation of neutral rights!

Every day, except Sunday, is a market-day in New York. Meat is cut up and sold by the joint or in pieces, by the licensed butchers only, their agents, or servants. Each of these must sell at his own stall, and conclude his sales by one o'clock in the afternoon, between the 1st of May and the 1st of November, and at two, between the 1st of November and the 1st of May. Butchers are licensed by the mayor, who is clerk of the market. He receives for every quarter of beef sold in the market, six cents; for every hog, shoat, or pig above 14lbs. weight, six cents; and for each calf, sheep, or lamb, four cents; to be paid by the butchers and other persons selling the same. To prevent engrossing, and to favour housekeepers, it is declared unlawful for persons to purchase articles to sell again, in any market or other part of the city, before noon of each day, except flour and meal, which must not be bought to be sold again until four in the afternoon; hucksters in the market are restricted to the sale of vegetables with the exception of fruits. The sale of unwholesome and stale articles of provisions, of blown and stuffed meat, and of measly pork, is expressly forbidden. Butter must be sold by the pound, and not by the roll or tub. Persons who are not licensed butchers, selling butchers' meat on commission, pay treble fees to the clerk of the market.

The markets are abundantly supplied with every thing in its season, which the land and water affords. In an enumeration made a few years ago by several gentlemen of experience, it appeared that the number of different species of wild quadrupeds brought to market in the course of the year, in whole or in part, alive or dead, was eight; amphibious creatures, five; shell fish, 14; birds, 51; and of fishes proper, 62. Their names are as follow. Quadrupeds: bear, deer, racoon, ground-hog, opossum, squirrel, rabbit, hare. Amphibious: green-turtle, hawksbill, loggerhead, snapper, terrebin. Shell fish: oyster, lobster, prawn, crab, sea crab, cray fish, shrimp, clam, sea clam, soft clam, scollop, grey mussel, black mussel, perriwinkle. Birds: wild goose, brant, black duck, grey duck, canvas back, wood duck, wigeon, teal, broad-bill duck, dipper, sheldrake, old-wife, coote, hell-diver, whistling-diver, redhead, loon, cormorant, pilestart, sheerwater, curlew, merlin, willet, woodcock, English snipe, grey snipe, yellow-legged snipe, robin snipe, dovertie, small-sand snipe, green plover, grey plover, kildare, wild turkey, heath hen, partridge, quail, meadow hen, wild pigeon, turtle dove, lark, robin, large grey snow bird, small blue snow bird, blue jay, yellow tail, clape blackbird, woodpecker, blue crane, white crane. Fishes: salmon, codfish, blackfish, streaked bass, sea bass, sheepshead, mackarel, Spanish mackarel, horse mackarel, trout, pike, sunfish, lucker, chub, roach, shiner, white perch, yellow perch, black perch, sturgeon, haddock, pollock, hake, shad, herring, sardine, sprat, manhaden, weakfish, smelt, mullet, bonetto, kingfish, silverfish, porgey, skipjack, angel fish, grunt's tusk, red drum, black drum, sheepshead drum, dogfish, killifish, bergall, tommycod, red gurnard, grey gurnard, spearings, garfish, frost fish, blow fish, toad fish, hallibut, flounder, sole, plaice, shait, stingray, common eel, conger eel, lamprey.

The principal market in New York is called the Fly market. A name which might, perhaps, lead a stranger to expect a market swarming with flies. This, however, is not the real meaning of the term. This part of the city, *s.e.* of Pearl street, was originally a salt-meadow, with a creek running through it, from where Maiden lane now is, to the bay or East river; forming such a disposition of land and water, as was called by the Dutch *Vlaie*, a valley or wet piece of ground: when a market was first held there it was called the Vlaie market, from which has originated the name of Fly market.

On the *w.* side of the city in Greenwich street, and between it and the Hudson river, is the market of the second importance. This is known by as odd and whimsical a name as the former. It arose in the following manner: During the time the city was in the hands of the British troops in [the revolutionary war, a considerable portion of the buildings in that neighbourhood was burnt down. Soon after the peace a market was established there, and in the progress of improvement it happened that the market-house was finished long before the streets were rebuilt, or the generality of inhabitants re-established. As there were for a considerable time but few housekeepers or purchasers, so there was but a small number of sellers of produce to frequent this public place; which led the citizens to distinguish it by the name of Bare market, or the market at which there was little or nothing brought for sale; and the name is continued to this day, though it is now situated in the heart of the town, and the supplies are steady and abundant. Besides these two large markets, there are four others, somewhat smaller, but always well stocked with provisions of every description.

The price of several commodities before the embargo was as follows, in sterling money: beef 6½*d.* per lb.; mutton 5*d.*; veal 7*d.*; butter 10*d.*; bread, the loaf of 2¼ lb. 7*d.*; cheese 7*d.*; turkies 7*s.* each; chickens 20*d.* per couple; oysters 7*d.* per dozen; flour 27*s.* per barrel of 196lbs.; brandy 4*s.* 6*d.* per gallon; coffee 1*s.* 6*d.* per lb.; green tea 5*s.*; best hyson 10*s.*; coals 70*s.* per chaldron; wood 20*s.* per cord; a coat 7*l.* 10*s.*; waistcoat and pantaloons 4*l.* 10*s.*; hat 54*s.*; pair of boots 54*s.*; washing 3*s.* 6*d.* per dozen pieces. Price of lodging at genteel boarding houses, from 1*l.* 11*s.* 6*d.* to 3*l.* 3*s.* per week. After the embargo took place, the price of provisions fell to nearly half the above sums, and European commodities rose in proportion. The manufactures of America are yet in an infant state; but in New York there are several excellent cabinet-makers, coach-makers, &c. who not only supply the country with household furniture and carriages, but also export very largely to the W. Indies, and to foreign possesssons on the continent of America. Their workmanship would be considered elegant and modern in London, and they have the advantage of procuring mahogany and other wood much cheaper than we.

Game laws are not wholly unknown in America. There is an act in force for the preservation of heath hens and other game, which was passed in the year 1791. This statute makes it penal to kill any heath hen, within Queen's or Suffolk counties, or any partridge, quail, or woodcock, within Queen's, King's, and New York counties, in the following manner. Heath hen, partridge, and quail are protected by the law from the 1st of April to the 5th of October, and woodcock from the 20th of February to the 1st of July; they who violate the law are liable to a penalty of two dollars and a half for every bird. There is also a society established called the Brush Club, for the purpose of detecting poachers, and interlopers upon private property. Laws are also passed for the protection of deer; persons violating them are subject to penalties of seven dollars and a half; 25 dollars if the deer are killed within 30 rods of any road or highway.

There are 31 benevolent institutions in New York. The names of them are as follows: Tammany society, freeschool, provident society, mutual benefit society, benevolent society, Albion benevolent society, ladies' society for the relief of poor widows with small children, fire department, New York manufacturing society, society of mechanics and tradesmen, the dispensary, lying-in hospital, sailors' snug harbour, marine society, manumission society, kine-pock institution, city hospital, alms house, house carpenters' society, Bellevue hospital, marine hospital at Staten island, humane society, masonic society, containing 13 lodges, German society, society of unitas fratrum, first Protestant Episcopal charity school, St. George's society, St. Patrick's society, St. Andrew's society, the New England society, the Cincinnati. Most of these institutions are mere benefit societies, resembling those which are so numerous in England. The ladies' society for the relief of poor widows with small children merits, however, particular notice, since it is an institution most honourable to the character of the amiable women of that city; and is worthy of imitation in Great Britain.

This association, of which gentlemen cannot be members, though they may be contributors, was commenced in November 1797, and organised the 29th December following. At their first stated meeting in April 1798, it was reported that 98 widows, with 223 children had been brought through the severity of winter, with a degree of comfort, who without this interposition would probably have gone to the alms house, or have perished. Relief is given in necessaries, but never in money, without a vote of the directresses at their board. It is not granted in any case until after the applicants shall be visited at their dwellings by one of the managers, and particular inquiry made into their character and circumstances. Immorality excludes from the patronage of the society; neither is relief given to any applicant who refuses to put out at service or to trades, such of her children as are fit, and to place the younger ones, of proper age, at a charity school; unless] [in very particular cases, of which the board judges.

The managers are required to exert themselves to create and maintain habits of industry among their applicants, by furnishing them, as far as possible, with suitable employment. White and checked linen has been extensively distributed among the poor widows who could not find employment elsewhere, to be made into shirts, on hire, and afterwards sold by the society at first cost. The ladies were incorporated by an act of the legislature on April 2, 1802, and are allowed to hold an estate of 50,000 dollars, applicable only to the relief of poor widows with small children. Their affairs are managed by a board of direction, composed of a first and second directress, a secretary, treasurer, and not less than six, nor more than twelve managers, two-thirds of whom make a quorum. Husbands of married women who are members or officers of this corporation, are not liable for any loss occasioned by the neglect or misfeasance of their wives, nor for any subscription or engagement of their wives, except in the case of their having received from their wives money or property belonging to the corporation.

The New York manufacturing society was originally established for the purpose of furnishing employment for the honest and industrious poor; and for several years, spinning, weaving, and some other branches of business, were carried on at their manufactory in Vesey street. But the experiment did not answer the expectations of the stockholders, and the society discontinued their operations; so that it may now be considered as dissolved.

The marine society is established for the purpose of improving maritime knowledge, and to assist indigent and distressed masters of vessels, their wives, and orphans. They may hold property not exceeding the yearly value of 3000*l.* sterling.

The manumission society has for its object the mitigation of the evils of Negro slavery, to assist free blacks unlawfully kept in slavery, to prevent kidnapping, and to better the condition of Negroes, by teaching them reading, writing, and accounts. They have a free school for black children, whose number is about 100.

The humane society is established for a different purpose to that of London, being devoted to the relief of distressed debtors confined in the city prison, and for supplying soup to the distressed poor throughout the city, either gratuitously, or for the small consideration of three halfpence a quart.

The society of unitas fratrum, or united brethren, has for its object the propagation of the gospel among the heathen, and is composed chiefly of Moravians.

The society of the Cincinnati was established at the close of the revolutionary war. Many of the officers who had meritoriously served their country, on laying down their commissions, returned to their original calling, or some other department of civil life. A respectable number of these, struck with the resemblance of their situation to that of the great Roman dictator Cincinnatus, associated themselves into a body of military friends, which they denominated the society of the Cincinnati. This corps of heroic gentlemen still preserves its original organisation, and holds meetings from time to time, to commemorate public events, perform deeds of beneficience, and to hold converse on the defence of the country.

There are also two other societies not noticed in the preceding enumeration; these are the medical society, and the Protestant Episcopal society for promoting religion and learning in the state of New York.

The first is a corporate body, and was established in 1806, by virtue of a law to incorporate medical societies for the purpose of regulating the practice of physio and surgery in the state. By this statute it is declared lawful for these physicians and surgeons (not less than five), who were then authorised by law to practise in their several professions, to assemble in their respective counties, and to incorporate themselves by choosing a president, vice-president, secretary, and treasurer; and depositing in the clerk's office a copy of all their proceedings within the 20 days immediately succeeding the first Tuesday of July, or their other time of meeting. Each county society may hold an estate, real or personal, to the amount of 1000 dollars. A county society, thus organised, is empowered to examine all students, who shall present themselves for that purpose, and to grant them diplomas, which allow the possessor to practise physic and surgery all over the state. Such a society may also appoint a board of censors, consisting of not less than three, nor more than five, whose duty it is to examine students, and report their opinion thereon, in writing, to the president. After the 1st of September 1806, all persons practising physic and surgery without having undergone an examination, and received a diploma, are debarred from collecting any debts incurred by such practice, in any court of law.

The Protestant Episcopal society for promoting]

[religion and learning in the state of New York, is established for the following objects:—The members are to be in amity with the Protestant Episcopal church; to adopt measures for insuring a sufficient number and succession of pions and learned ministers of the gospel, attached to the doctrines and discipline of the Protestant Episcopal church; to afford assistance to such young men as are of good character and competent abilities, but in circumstances which do not admit of prosecuting the study of divinity without aid; to encourage those who may distinguish themselves by extraordinary attainments; to receive all donations for pious purposes, and to superintend the application of them; to provide funds for establishing a theological library, for the establishment of schools, and for providing one or more fellowships in Columbia college. In a word, to pursue a system of measures whereby the situation of the clergy may be rendered respectable, the church obtain a permanent support, and learning and piety be generally diffused throughout the state.

Columbia college was incorporated in the year 1754. The institution was then called King's college, and was intended for the instruction and education of youth in the learned languages, and liberal arts and sciences. And for their further encouragement the college was authorised to confer such degrees upon the students and other persons, as are usually granted in the English universities. Under these powers there have been two faculties established in the college, viz. the faculty of the arts, and a faculty of medicine.

The former consists of a president, who is also a professor of moral philosophy; of a professor of classical literature, who also gives lectures on Grecian and Roman antiquities; of a professor of mathematics, natural philosophy, and astronomy, who likewise teaches geography and chronology; and of a professor of logic, rhetoric, belles lettres, &c.

The faculty of physic is composed of a professor of anatomy and surgery; of midwifery and clinical medicine; of botany and materia medica; of the theory and practice of physio, and of chemistry. The annual commencement is the first Wednesday in August. Lectures are regularly delivered on all these literary, scientific, and professional subjects; and the professors labour with zeal and ability in their several departments. There are some rare books and valuable apparatus belonging to their institution. Since the revolution the seminary has been so far altered, as was necessary to adapt it to the new state of affairs; it is now called Columbia college. The trustees have the power of filling up all vacancies in their body, occasioned by death, removal, or resignation. The income of the college is about 1500*l.* but is expected to increase with the renewal of some of their expiring leases of land. To this college Mr. Joseph Murray, an eminent counsellor at law, left his large library, and almost the whole of his fortune, amounting to 10,000*l.*

There are upwards of twenty newspapers published in New York, nearly half of which are daily papers; besides several weekly and monthly magazines or essays. The high price of paper, labour, and taxes in Great Britain, has been very favourable to authorship, and the publication of books in America. Foreign publications are also charged with a duty of 13 per cent.; and foreign rags are exempted from all impost. These advantages have facilitated the manufacture of paper, and the printing of books in the United States; both of which are now carried on to a very large extent. The new works that appear in America, or rather original productions, are very few; but every English work of celebrity is immediately reprinted in the States, and vended for a fourth of the original price. The booksellers and printers of New York are numerous, and in general men of property. Some of them have published very splendid editions of the bible, and it was not a little gratifying to the American patriot to be told, that the paper, printing, engraving, and binding, were all of American manufacture. For several years past, a literary fair has been held alternately at New York and at Philadelphia. This annual meeting of booksellers has tended greatly to facilitate intercourse with each other, to circulate books throughout the United States, and to encourage and support the arts of printing and paper-making.

A public library is established at New York, which consists of about 10,000 volumes, many of them rare and valuable books. The building which contains them is situated in Nassau street, and the trustees are incorporated by an act of the legislature. There are also three or four public reading-rooms, and circulating libraries, which are supported by some of the principal booksellers, from the annual subscriptions of the inhabitants. There is a museum of natural curiosities in New York, but it contains nothing worthy of particular notice.

CHAP. II.

IT does not appear that the malignant or yellow fever made very great ravages among the inhabitants in 1805, the last time of its appearance in]

[New York; for the deaths very little exceeded the preceding and subsequent years.

In 1804 the deaths were 2064
1805 - - - - - 2352
1806 - - - - - 2252

Of the above number, 51 were suicides; and according to the statement of Dr. Mitchill, upwards of one-third of the deaths are occasioned by consumption and debility. To the influence of moisture and the sudden changes of the weather, has been attributed the prevalence of nervous disorders and debility, among a great number of the inhabitants of the United States. Much may, no doubt, be ascribed to those causes; but it is thought the mode of living has a more immediate effect upon the human frame than even the climate of the country. The higher and middling classes of the Americans who reside chiefly in the great towns, or their neighbourhood, live, generally speaking, in a more luxurious manner than the same description of people in England. Not that their tables are more sumptuously furnished on particular occasions, than ours; but that their ordinary meals consist of a greater variety of articles, many of which, from too frequent use, may perhaps become pernicious to the constitution. The great consumption of green tea, which we reckon the most unwholsome, in consequence (as it is said) of its being dried upon copper, is most likely very injurious to the constitution. The Americans use scarcely any other than this tea, while in England, the souchong, and other black teas, are most in request. The constant use of segars by the young men, even from an early age, may also tend to impair the constitution, and create a stimulus beyond that which nature requires, or is capable of supporting. Their dread of the yellow fever has induced a more frequent use of tobacco of late years; but it is now grown into a habit that will not be readily parted with. The other classes of the community who reside in the interior, and back parts of the country, are often obliged to live upon salt provisions the greatest part of the year, and sometimes on very scanty fare; besides which, they generally dwell in miserable log huts, incapable of defending them effectually from the severity of the weather. Those who have the means of living better are great eaters of animal food, which is introduced at every meal, together with a variety of hot cakes, and a profusion of butter: all which may more or less tend to the introduction of bilious disorders, and perhaps lay the foundation of those diseases which prove fatal in hot climates. The effects of a luxurious or meagre diet are equally injurious to the constitution, and together with the sudden and violent changes of the climate, may create a series of nervous complaints, consumption, and debility, which in the states bordering on the Atlantic, carry off at least one third of the inhabitants in the prune of life.

The malignant or yellow fever generally commences in the confined parts of the town, near the water side, in the month of August or September. It is commonly supposed to have been introduced by the French refugees from St. Domingo during the French revolution; though some are of opinion that it originated in the States; and many physicians were puzzling their brains about its origin, at a time when they ought to have been devising means to stop its ravages. As soon as this dreadful scourge makes its appearance in New York, the inhabitants shut up their shops, and fly from their houses into the country. Those who cannot go far, on account of business, remove to Greenwich, a small village situate on the border of the Hudson river, about two or three miles from town. Here the merchants and others have their offices, and carry on their concerns with little danger from the fever, which does not seem to be contagious beyond a certain distance. The banks and other public offices also remove their business to this place: and markets are regularly established for the supply of the inhabitants. Very few are left in the confined parts of the town except the poorer classes and the Negroes. The latter not being affected by the fever, are of great service at that dreadful crisis; and are the only persons who can be found to administer the hazardous duties of attending upon the sick, and burying the dead. Upwards of 26,000 people removed from the interior parts of the city, and from the streets near the water side, in 1805. Since then, the town has happily been free from that dreadful scourge; and from the salutary regulations which have since been adopted, it is to be hoped, that it will never make its appearance again. The finest cities in America were no doubt preserved from depopulation, during the prevalence of the fever, by the timely retreat of the inhabitants into the country. It were to be wished that the same practice was permitted in Spain, and other parts of the continent, which are sometimes visited by pestilential fevers, instead of surrounding the towns by a cordon of troops, and cutting off all communication between the unfortunate inhabitants and the country.

The following census of the population of New York was taken in 1807, and laid before the mayor, aldermen, and commonalty of the city in 1808.]

Census of the Electors and total Population of the City of New York.

Wards.	Free Persons.	Slaves.	Total Inhabitants.	Electors possessed of Freeholds of the Value of 100*l.* and upwards.	Ditto possessed of Freeholds of 200*l.* and under 100*l.*	Ditto not possessed of Freeholds, but who rent Tenements of the yearly Value of 40*s.*	Ditto who were Freemen on the 14th October, 1775.	Total Electors.
First	7,584	370	7,954	374	—	707	5	1,086
Second	7,424	127	7,551	355	—	687	—	1,042
Third	7,303	406	7,709	337	1	779	1	1,118
Fourth	9,089	147	9,236	351	—	976	4	1,331
Fifth	12,603	136	12,739	462	4	1,429	6	1,901
Sixth	9,749	112	9,861	258	—	1,163	6	1,427
Seventh	19,363	124	19,487	413	5	2,718	4	3,140
Eighth	5,959	108	6,067	302	6	715	—	1,023
Ninth	2,680	246	2,926	158	4	174	3	339
Total	81,754	1776	83,530	3,010	20	9,348	29	12,407

Of the preceding number of inhabitants 42,881 are females, and 40,649 are males: making a total of 83,530. In 1805 the population of New York was 75,770, thus in the course of one year and ten months there has been an increase of inhabitants to the amount of 7760; and within the same period, the number of slaves has decreased 272. The following table exhibits the population of this city at different periods from its earliest settlement.

In the year 1697 there were 4,302 inhabitants.
1756 . . . 15,000
1771 . . . 21,863
1786 . . . 23,614
1791 . . . 33,131
1801 . . . 60,489
1805 . . . 75,770
1807 . . . 83,530

Hence it appears that the population of New York has, in a period of 20 years from 1786 to 1805, more than tripled itself; and should the population continue to increase at the rate of five per cent. per annum, it will, in 1855, amount to 705,650, a population nearly equal to that of Paris. At this day it is equal to the whole number of inhabitants in the state of New York fifty years ago.

There are about 4000 Negroes and people of colour in New York, 1700 of whom are slaves. These people are mostly of the Methodist persuasion, and have a chapel or two of their own with preachers of their colour; though some attend other places of worship according to their inclination. All religious sects in the United States are upon an equal footing, no one has any established prerogative above another; but in any place, on particular occasions, where precedence is given to one over another, the Episcopal church, or that sect which is most numerous, generally takes the lead.

If any estimate can be formed of the salubrity of the climate, and the healthiness of the inhabitants of a town, by the number of deaths, London must be reckoned to have the advantage of New York in those respects. The amount of deaths in the former city is about a fiftieth part of its population, while in New York it is at least one thirtieth; the number of deaths ranging between 2500 and 3000 per annum. We are, however, more inclined to attribute this great mortality to improper diet and mode of living, than to the insalubrity of the climate. The church-yards and vaults are also situate in the heart of the town, and crowded with the dead. If they are not prejudicial to the health of the people, they are, at least, very unsightly exhibitions. One would think there was a scarcity of land in America, by seeing such large pieces of ground in one of the finest streets of New York occupied by the dead. But even if no noxious effluvia were to arise (and we rather suspect there must in the months of July, August, and September), still the continual view of such a crowd of white and brown tomb-stones and monuments as is exhibited in the Broadway, must, at the sickly season of the year, tend very] [much to depress the spirits, which should rather be cheered and enlivened; for at that period much is effected by the force of imagination. There is a large burying ground a short distance out of town; but the cemeteries in the city are still used at certain periods of the year.

They bury their dead within 24 hours; a custom probably induced by the heat of the climate during the summer months; but we see no reason why it should be extended to the winter months, which are cold enough to allow of the dead being kept for three or four days, if nothing else prevents it.

Funerals at New York, as well as in almost every other part of the United States, are attended by a numerous assemblage of the friends and acquaintances of the deceased, who are invited, by advertisements in the newspapers, to attend their departed friend to the grave; it is common to see upwards of 500 people attending on such occasions, and the larger the number the more the deceased is supposed to be respected and valued. We cannot help thinking, however, that these numerons meetings savour somewhat of ostentation, though certainly there is no parade of hearses, nodding plumes, and mourning coaches. The people attend, for the most part, in their ordinary dress, except those who are nearly related, or particularly intimate with the deceased. The clergyman, physician, and chief mourners, wear white scarfs, which it is also the custom to wear on the following Sunday. The deceased is interred with or without prayers, according to the faith he professed.

The society of New York consists of three distinct classes. The first is composed of the constituted authorities and government officers; divines, lawyers, and physicians of eminence; the principal merchants, and people of independent property. The second comprises the small merchants, retail dealers, clerks, subordinate officers of the government, and members of the three professions. The third consists of the inferior orders of the people. The first of these associate together in a style of elegance and splendour little inferior to Europeans. Their houses are furnished with every thing that is useful, agreeable, or ornamental; and many of them are fitted up in the tasteful magnificence of modern style. The dress of the gentlemen is plain, elegant, and fashionable; and corresponds in every respect with the English costume. The ladies in general seem more partial to the light, various, and dashing drapery of the Parisian belles, than to the elegant and becoming attire of our London beauties, who improve upon the French fashions. But there are many who prefer the English costume, or at least a medium between that and the French.

The young ladies of New York are in general handsome, and almost universally fine, genteel figures. Fair complexions, regular features, and fine forms, seem to be the prevailing characteristics of the American fair sex. They do not, however, enjoy their beauty for so long a period as English women, neither do they possess the blooming countenance and rosy tinge of health so predominant among English women. Their climate is, however, not so favourable to beauty as that of England, in consequence of the excessive heat, and violent changes of the weather peculiar to America.

Most travellers who have visited America have charged the ladies of the United States, universally, with having bad teeth. This accusation is certainly very erroneous, when applied to the whole of the fair sex, and to them alone. That the inhabitants of the state are often subject to a premature loss of teeth, is allowed by themselves, and the cause has even been discussed in the papers read before the American philosophical society; but it does not particularly attach to the females, who are, in truth, much more exempt from that misfortune than the men.

Much has also been said of the deficiency of the polite and liberal accomplishments among both sexes in the United States. Whatever truth there may have formerly been in this statement, we do not think there is any foundation for it at present, at least in New York, where there appears to be a great thirst after knowledge. The riches that have flowed into that city, for the last 20 years, have brought with them a taste for the refinements of polished society; and though the inhabitants cannot yet boast of having reached the standard of European perfection, they are not wanting in the solid and rational parts of education; nor in many of those accomplishments which ornament and embellish private life. It has become the fashion in New York to attend lectures on moral philosophy, chemistry, mineralogy, botany, mechanics, &c. and the ladies in particular have made considerable progress in those studies. Many young men who were so enveloped in business as to negleet or disdain the pursuit of such liberal and polite acquirements, have been often laughed from the counting-house to the lecture-room by their more accomplished female companions. The desire for instruction and information, indeed, is not confined to the youthful part of the community; many married ladies and their families may be] [seen at philosophical and chemical lectures, and the spirit of inquiry is becoming more general among the gentlemen. The majority of the merchants, however, still continue more partial to the rule of three, than a dissertation upon oxygen or metaphysics. Most of them have acquired large fortunes by their regular and plodding habits of business, and loath to part with any portion of it, at their time of life, in the purchase of knowledge, or the encouragement of the arts and sciences. Some, it must be allowed, are exceptions; and others, if they will not partake of instruction themselves, are not sparing of their money in imparting it to their children. The immense property which has been introduced into the country by commerce, has hardly had time to circulate and diffuse itself through the community. It is at present too much in the hands of a few individuals, to enable men to devote the whole of their lives to the study of the arts and sciences. Farmers, merchants, physicians, lawyers, and divines, are all that America can produce for many years to come: and if authors, artists, or philosophers, make their appearance at any time, they must, as they have hitherto done, spring from one of the above professions.

Colleges and schools are multiplying very rapidly all over the United States; but education is in many places still defective, in consequence of the want of proper encouragement and better teachers. A grammar-school has recently been instituted at New York, for the instruction of youth, upon a similar plan to the great public schools in England. This seminary, says an American writer, is founded on the principle of training the students to become sound and accurate classical scholars, according to the old plan of acquiring the elements of ancient learning by grammar; discarding the learning by rote. The success of this institution will compel the colleges to adopt a less superficial and defective plan of instruction; and it will follow that when once liberal and sound education is permanently introduced, literature will revive; the trading spirit will be checked or modified; literary rewards and honours will flow rapidly, and the public will eventually become the promoters of genius and learning, by creating an extensive demand for books.

A taste for reading has of late diffused itself throughout the country, particularly in the great towns; and several young ladies have displayed their abilities in writing. Some of their novels and fugitive pieces of poetry and prose are written with taste and judgment. Two or three at New York have particularly distinguished themselves. It seems, indeed, that the fair sex of America have within these few years been desirous of imitating the example of the English and French ladies, who have contributed so much to extend the pleasures of rational conversation and intellectual enjoyment. They have cast away the frivolous and gossiping tittle tattle, which before occupied so much of their attention; and assumed the more dignified and instructive discourse upon arts, sciences, literature, and moral philosophy.

Many of the young men, too, whose minds have not been wholly absorbed by pounds, shillings, and pence, have shewn that they possess literary qualifications and talents, that would, if their time and fortune permitted, rank them among some of the distinguished authors of Europe. The most prominent of their late productions is the Salmagundi, published in monthly essays at New York. This little work has been deservedly a great favourite with the public, and bids fair to be handed down with honour to posterity. It possesses more of the broad humour of Rabelais and Swift, than the elegant morality of Addison and Steele, and therefore less likely to become a classical work; but as a correct picture of the people of New York, and other parts of the country, though somewhat heightened by caricature, and as a humorons representation of their manners, habits, and customs, it will always be read with interest by a native of the United States.

A publication called the Echo is a smart production of detached poetry, commenced for the purpose of satirizing the vices and follies of the political intriguers of the day, who broached their revolutionary dogmas through the medium of the public prints. Several other publications of merit have originated in America, and are well known in England. Mr. Barlow's Columbiad has lately made its appearance in a very splendid form. It is an enlargement of his vision of Columbus.

Dancing is an amusement that the New York ladies are passionately fond of, and they are said to excel those of every other city in the union. Many of the young ladies are well accomplished in music and drawing, and practise them with considerable success; but they do not excel in those acquirements as they do in dancing. Among the young men those accomplishments are but little cultivated. Billiards and smoking seem to be their favourite amusement. A segar is in their mouth from morning to night, when in the house, and not unfrequently when walking the street. A box full is constantly carried in the coat pocket, and handed occasionally to a friend, with a degree of interesting familiarity and nonchalance. Billiards] [are played with two red balls. This is called the American game, and differs in no other respect from the mode of playing in England. New York contains several excellent tables.

The style of living in New York is fashionable and splendid, many of the principal merchants and people of property have elegant equipages, and those who have none of their own, may be accommodated with handsome carriages and horses at the livery stables; for there are no coach stands. The winter is passed in a round of entertainments and amusements; at the theatre, public assemblies, philosophical and experimental lectures, concerts, balls, tea and card parties, cariole excursions out of town, &c. The American cariole, or sleigh, is much larger than that of Canada, and will hold several people. It is fixed upon high runners, and drawn by two horses in the curricle style. Parties to dinner and dances are frequently made in the winter season when the snow is on the ground. They proceed in carioles a few miles out of town to some hotel or tavern, where the entertainment is kept up to a late hour, and the parties return home by torch light.

Marriages are conducted in the most splendid style, and form an important part of the winter's entertainments. For some years it was the fashion to keep them only among a select circle of friends; but of late the opulent parents of the new-married lady have thrown open their doors, and invited the town to partake of their felicity. The young couple, attended by their nearest connections and friends, are married at home in a magnificent style, and if the parties are Episcopalians, the bishop of New York is always procured, if possible; as his presence gives a greater zest to the nuptials. For three days after the marriage ceremony, the new-married couple see company in great state, and every genteel person who can procure an introduction may pay his respects to the bride and bridegroom. It is a sort of levee; and the visitors, after their introduction, partake of a cup of coffee or other refreshment, and walk away. Sometimes the night concludes with a concert and ball, or cards, among those friends and acquaintance who are invited to remain.

Several young ladies in New York have fortunes of 100 or 150,000 dollars; and often bestow their hand upon a favourite youth, who has every thing to recommend him but money. Unhappy marriages are by no means frequent, and parents are not apt to force the inclinations of their children from avaricious motives. Summer affords the inhabitants the diversions of hunting, shooting, fishing, and horse-racing; excursions upon the water to the island in the bay, and to Sandy hook, and a variety of beautiful tours within 20 miles of the city. Among the most distinguished are those of New Utrecht, Rockaway, Islip, the Passaick falls, and Kingsbridge. A place called Ballston, within 200 miles of New York, in the interior of the state, contains some mineral springs; and of late years has become a fashionable place of resort for invalids. Like most places of that kind in England, it is visited by the gentry, who go there more for amusement and fashion than to drink the waters. Ballston possesses but few natural attractions, except its mineral springs.

The inhabitants of New York are not remarkable for early rising, and little business seems to be done before nine or ten o'clock. Most of the merchants and people in business dine about two o'clock, others, who are less engaged, about three; but four o'clock is usually the fashionable hour for dining. The gentlemen are partial to the bottle, but not to excess; and at private dinner parties they seldom sit more than two hours drinking wine.

In consequence of there being no established form of worship, the clergy are accustomed only to *recommend* to the people the religious observance of certain festivals. The following is one of their resolutions for Christmas day, 1807.

" In common council, December 21, 1807. The following communication having been received from the reverend clergy of this city:

" A number of the clergy, of different denominations, of this city, at a meeting held on Wednesday the 16th inst. having taken into consideration the merciful dispensations of Divine Providence towards this city, during the last season, and also the present aspect of public affairs:

" Resolved, That it is proper to take public and solemn notice of the divine goodness, and as a people, to implore the continued protection, and those temporal and spiritual blessings, which are so essential to our welfare.

" Resolved, That it be recommended to the several congregations under our pastoral care, to set apart Friday the 25th instant, as a day of solemn thanksgiving and prayer; and that abstaining from all kinds of servile labour and recreations on that day, they come together to acknowledge the mercy of God, in again exempting us from the scourge of pestilence, to praise him for the multiplied favours of his gracious providence, to beseech him to preserve us in peace, and to continue and extend our national prosperity; and above all, to pray for the sanctifying influences of the]

[Holy Spirit on our churches, and that we may be favoured with all spiritual and heavenly blessings in Christ Jesus.

"Signed by order of the meeting,

"John Rodgers, Chairman."

"Resolved, That the board unite in the recommendation of the reverend clergy of this city, upon the above occasion, and accordingly recommend, that Friday the 25th day of December be observed and set apart as a day of public and special thansgiving and prayer to Almighty God, for his benevolent dispensations of mercy to this city: and we accordingly recommend to our fellow-citizens, that they carefully abstain from all recreations and secular employments on that day.

"By the common council,

"John Pintard, Clerk.

"*New York, Dec.* 22, 1807."

The shops are accordingly shut, the people attend at public worship, and the day is religiously and strictly observed. It is not, however, to be understood, that roast beef and plum-pudding, turkey and chine, mince pies, &c. smoke on the American tables as they do in England on that festival; though, perhaps, those Americans who yet retain a spice of the English character about them, may continue the old practice of their ancestors.

New year's day is the most important of the whole year. All the complimentary visits, fun, and merriment of the season seem to be reserved for this day; though much is now worn away by the innovations of fashion. Many of the shops are shut up; and the Presbyterians and a few other religious dissenters, attend public worship. The mayor of the city, and others of the constituted authorities, advertise, two or three days before, that they will reciprocate the compliments of the season, with the inhabitants at their house on new year's day.

The bakers on this day distribute to their customers small cakes made in a variety of shapes and figures; and the newspaper editors greet their readers with a poetical retrospect of the events of the old year: it accords with their political principles, and is generally a severe party philippic. New York, like the other large cities of the union, is a prey to the violent spirit of the two parties, who are known under the titles of federalists and democrats. The newspapers are almost equally divided between the two, to whose views they are of course subservient, and have the effect of keeping up a continual warfare, in which they belabour each other, their rulers, and the English and French nations, without mercy. "Every day," as Mustapha Rubadub observes in Salmagundi, (the work to which we have before alluded) "have these slang-whangers made furious attacks on each other, and upon their respective adherents, discharging their heavy artillery, consisting of large sheets, loaded with scoundrel! villain! liar! rascal! numskull! nincompoop! dunder-head! wiseacre! blockhead! jackass! and I do swear by my beard, though I know thou wilt scarcely credit me, that in some of these skirmishes the grand bashaw himself has been wofully pelted! yea, most ignominiously pelted! and yet have these talking desperadoes escaped without the batinado!"

The drinking of toasts at public dinners is a very common method of venting party spleen in America, and of drinking destruction to their enemies. The newspapers publish long lists of these toasts the next day, as so many poofs of patriotism and virtue; and take a pride in shewing how brilliantly their partisans can blackguard public characters in their cups.]

[New York, an Indian town of the Creek nation; situated on Tallapoose river, in Georgia; and so named by Col. Ray, a New York British loyalist.]

[New York Island, on which the city of that name stands, is about 15 miles long, and does not extend two in any part in breadth. It is joined to the mainland by a bridge called King's bridge, 15 miles *n.* of New York city.]

[NEWARK, a township in Essex county, in Vermont.]

[Newark Bay, in New Jersey, is formed by the confluence of Passaick and Hackensack rivers from the *n.* and is separated from that part of North river opposite to New York city, by Bergen neck on the *e.* which neck, also, with Staten island on the *s.* of it, form a narrow channel from the bay to North river *e.* Newark bay also communicates with Rariton bay, at the month of Rariton river, by a channel in a *s.* by *w.* direction along the *w.* side of Staten island. The water passage from New York to Elizabeth Town point, 15 miles, is through this bay.]

[Newark, a post-town of New Jersey, and capital of Essex county, is pleasantly situated at a small distance *w.* of Passaick river, near its mouth in Newark bay, and nine miles *w.* of New York city. It is a handsome and flourishing town, celebrated for the excellence of its cider, and is the seat of the largest shoe manufacture in the state: the average number made daily throughout the] year, is estimated at about 200 pairs. The town is of much the same size as Elizabeth town, and is six miles *n. e.* of it. There is a Presbyterian church of stone, the largest and most elegant building of the kind in the state. Besides these is an Episcopal church, a court-house, and gaol. The academy which was established here in June 1792, promises to be a useful institution. In Newark and in Orange, which joins it on the *n. w.* there are nine tanneries, and valuable quarries of stone for building. The quarries in Newark would rent, it is said, for 1000*l.* a year, and the number of workmen limited. This town was originally settled by emigrants from Branford, Connecticut, as long ago as 1662.]

[NEWARK, a village in Newcastle county, Delaware; situated between Christiana and White Clay creeks, nine miles *w.* of New-Castle, and 10 *s. w.* of Wilmington.]

[NEWARK, a town lately laid out by the British in Upper Canada, on the river which connects lake Erie and Ontario, directly opposite Niagara town and fort.]

[NEWBERN, one of the *e.* maritime districts of N. Carolina; bounded *e.* and *s. e.* by the Atlantic, *s. w.* by Wilmington, *w.* by Fayette, *n. w.* by Hillsborough, *n.* by Halifax, and *n. e.* by Edenton district. It comprehends the counties of Carteret, Jones, Craven, Beaufort, Hyde, Pitt, Wayne, Glasgow, Lenoir, and Johnston; and contains 55,540 inhabitants, including 15,900 slaves.]

[NEWBERN, the capital of the above district, is a post-town and port of entry; situated in Craven county, on a flat, sandy point of land, formed by the confluence of the rivers Neus on the *n.* and Treat on the *s.* Opposite to the town, the Neus is about a mile and a half, and the Trent three quarters of a mile wide. Newborn is the largest town in the state, contains about 400 houses, all built of wood except the palace, the church, the gaol, and two dwelling-houses, which are of brick. The palace was erected by the province before the revolution, and was formerly the residence of the governors. It is large and elegant, two stories high, with two wings for offices, a little advanced in front towards the town; these wings are connected with the principal building by a circular arcade. It is much out of repair; and the only use to which this once handsome and well furnished building is now applied, is for schools. One of the halls is used for a school, and another for a dancing room. The arms of the king of Great Britain still appear in a pediment in front of the building. The Episcopalian church is a small brick building with a bell. It is the only house for public worship in the place. The court-house is raised on brick arches, so as to render the lower part a convenient market-place; but the principal marketing is done with the people in their canoes and boats at the river side. In September 1791, near one third of this town was consumed by fire. It carries on a considerable trade to the W. Indies and the different states, in tar, pitch, turpentine, lumber, corn, &c. The exports in 1794 amounted to 69,615 dollars. It is 77 miles *n. e.* from Raleigh, 54 *s.* by *w.* of Edenton, 78 *n. e.* by *n.* of Wilmington, 120 *s.* of Petersburgh in Virginia, and 305 *s. s. w.* of Philadelphia. Lat. 35° 17′ 30″ *n.* Long. 77° 18′ *w.*]

[NEWBURGH, a township in Ulster county, New York; bounded *e.* by Hudson's river, and *s.* by New Windsor, and contains 2365 inhabitants; of whom 373 are electors, and 57 slaves. The compact part of the town is neatly built, and pleasantly situated on the *w.* bank of the Hudson, 50 miles *n.* of New York, opposite Fish Kill landing, five miles from Fish Kill, 19 from Goshen, and 13 *s.* from Poughkeepsie. It consists of between 50 and 60 houses and a Presbyterian church, situated on a gentle ascent from the river. The country *n.* is well cultivated, and affords a rich prospect. Vessels of considerable burden may load and unload at the wharfs, and a number of vessels are built annually at this busy and thriving place.]

[NEWBURY, a county of Ninety-six district, S. Carolina, which contains 9342 inhabitants, of whom 1144 are slaves. Newbury court-house is 37 miles from Columbia, and 23 from Laurens court-house.]

[NEWBURY, a township in York county, Pennsylvania.]

[NEWBURY, the capital of Orange county, Vermont, pleasantly situated on the *w.* side of Connecticut river, opposite to Haverhill, in Grafton county, New Hampshire, and from which it is five miles distant. It contains about 50 houses, a gaol, a court-house, and a handsome church for Congregationalists with a steeple, which was the first erected in Vermont. The court-house stands on an eminence, and commands a pleasing prospect of what is called the Great Oxbow of Connecticut river, where are the rich intervale lands called the Little Coos. Here a remarkable spring was discovered, about 20 years since, which dries up once in two or three years. It has a strong smell of sulphur, and throws up continually a peculiar kind of white sand; and a thick yellow scum rises upon the water when settled. This is the more noticeable as the water of the ponds and rivers in

Vermont is remarkably clear and transparent. It is 87 miles *n. e.* of Bennington, and 287 *n. e.* by *n.* of Philadelphia. Lat. 44° 5′ *n.* Long. 72° 2′ *w.* Number of inhabitants 873.]

[NEWBURY, a township in Essex county, Massachusetts, incorporated in 1635; situated on the *s.* bank of Merrimack river, and contains 3972 inhabitants. It formerly included Newbury port, and with Merrimack river encircles it. It is divided into five parishes, besides a society of Friends or Quakers. Dummer academy, in this township, is in a flourishing state; it was founded by Lieutenant-governor Dummer in 1756, opened in 1763, and incorporated in 1782. The inhabitants are principally employed in husbandry. The land, particularly in that part of the town which lies on Merrimack river, and is here called Newbury Newton, is of a superior quality, under the best cultivation, and is said by travellers to be little inferior to the most improved parts of Great Britain. Some of the high lands afford a very extensive and variegated view of the surrounding country, the rivers, the bay, and the sea-coast, from cape Ann to York in the district of Maine. Some few vessels are here owned, and employed in the fishery, part of which are fitted out from Parker river. It rises in Rowley, and after a course of a few miles, passes into the sound which separates Plumb island from the mainland. It is navigable about two miles from its month. A woollen manufactory has been established on an extensive scale in Byfield parish, and promises to succeed. This township is connected with Salisbury by Essex Merrimack bridge, about two miles above Newbury port, built in 1792. At the place where the bridge is erected, an island divides the river into two branches: an arch of 160 feet diameter, 40 feet above the level of high water, connects this island with the main on the opposite side. The whole length of the bridge is 1030 feet; its breadth 34; its contents upwards of 6000 tons of timber. The two large arches were executed from a model invented by Mr. Timothy Palmer, an ingenious house-wright in Newbury port. The whole is executed in a style far exceeding any thing of the kind hitherto essayed in this country, and appears to unite elegance, strength, and firmness. The day before the bridge was opened for the inspection of the public, a ship of 350 tons passed under the great arch. There is a commodious house of entertainment at the bridge, which is the resort of parties of pleasure, both in summer and winter.]

[NEWBURY Port, a port of entry and post-town in Essex county, Massachusetts; pleasantly situated on the *s.* side of Merrimack river, about three miles from the sea. In a commercial view it is next in rank to Salem; but it suffered considerably of late, by a fire which broke out on the evening of the 31st of May 1811, and which consumed 200 houses, stores, &c. the loss being stated at 2,000,000 of dollars. It contains 4837 inhabitants, although it is, perhaps, the smallest township in the state, its contents not exceeding 640 acres. It was taken from Newbury, and incorporated in 1764. The churches, six in number, are ornamented with steeples; the other public buildings are the court-house, gaol, a bank, and four public school-houses. To the honour of this town, there are in it 10 public schools, and three printing offices. Many of the dwelling houses are elegant. Before the war there were many ships built here; but some years alter the revolution the business was on the decline: it now begins to revive. The Boston and Hancock continental frigates were built here, and many privateers during the war. The harbour is safe and capacious, but difficult to enter. See MERRIMACK River. The marine society of this town, and other gentlemen in it, have humanely erected several small houses on the shore of Plumb island, furnished with fuel and other conveniencies, for the relief of shipwrecked mariners. Large quantities of rum are distilled in Newbury port, there is also a brewery; and a considerable trade is carried on with the W. Indies and the *s.* states. Some vessels are employed in the freighting business, and a few in the fishery. In November 1790, there were owned in this port, six ships, 45 brigantines, 39 schooners, and 28 sloops; making in all 11,870 tons. The exports for a year, ending September 30, 1794, amounted to 363,380 dollars. A machine for cutting nails has been lately invented by Mr. Jacob Perkins of this town, a gentleman of great mechanical genius, which will turn out, if necessary, 200,000 nails in a day. Newbury port is 32 miles *n. n. e.* of Boston, 16 *s.* by *w.* of Portsmouth, nine *n.* of Ipswich, and 264 *n. e.* of Philadelphia. The harbour has 10 fathoms water: high water at full and change 15 minutes after 11 o'clock. The light-house on Plumb island lies in lat. 42° 47′ *n.* and long. 70° 47′ *w.*]

[NEWENHAM, Cape, is the *n.* point of Bristol bay, on the *n. w.* coast of N. America. All along the coast the flood tide sets strongly to the *n. w.* and it is high water about noon on full and change days. Lat. 58° 42′ *n.* Long. 162° 24′ *w.*]

NEWFOUNDLAND, a large island of the N. sea, in N. America; discovered by John Gabot in 1494, who took possession of it for the English, and to these it at present belongs. It is of a triangular figure, and is 930 miles in circumference. On the *n.* it is separated from the land and country of Labrador or New Britain by the straits of Belle-isle; surrounded on the *w.* by the gulf of S. Lawrence, and *s.* and *e.* by the Atlantic ocean. The most *s.* part of the island is cape Race, which is in lat. 46° 45′, the most *w.* is cape Anguille in 47° 54′, and its most *n.* point is in lat. 51° 40′. This island is full of mountains covered with firs; so that it is only passable in such parts as where the inhabitants have cut paths through the middle of its woods. The trees seldom exceed 18 or 20 feet in height, excepting those which grow in the valleys, being sheltered from the winds; and here they will rise to 40 feet.

The cold is excessive in the winter, and the frosts, which are very severe, begin about November, when after a short time all the ports and bays become frozen. With these the whole of the island is surrounded, and they are very large and well sheltered by the mountains, so that vessels may lie in them in perfect security. Some are a league and an half or two leagues deep, and nearly half a league wide; and into them flow several rivers and streams of sweet water which descend from the mountains; many are so close together as to be separated merely by a point of land; there are very few that are two leagues apart from each other, and thus is the whole coast a continued line of ports, although in the very principal only are there any settlements or towns, and this too where the natural advantages of the country have induced the inhabitants to form establishments.

The population, with respect to the extent of coast, is very small. The cod-fishery is here the only occupation; and there are large store-houses where they preserve their tackle and accoutrements against the season when the fisheries commence, and which they use for laying up their merchandize, which they export, either on their account, or by foreign vessels, taking in exchange or payment such goods as these vessels, which are very numerous, may bring. In every settlement there is a battery, for its defence in time of war, the coast being much frequented by pirates.

This country was first peopled by a race of Indian savages, who retired to the continent, sometimes however visiting their old abodes. They lived by hunting, there being foxes, bears, and other quadrupeds here, the same as in Canada, but these animals, being in great request on account of their skins, are not so numerous as they were. In spite of the severity of the climate the inhabitants are not without flocks, but the difficulty of preserving them through the winter is great. In the gardens nothing is produced but a few pot-herbs, all other necessary fruits being brought from the other colonies of Europe. Although cod-fish is caught along the whole coast of Newfoundland, all parts are not equally abundant in these fish: they lie mostly in sandy bottoms, are found less in sea-weedy places, and never in rocky parts; the best depth for them is a little above 30 fathoms. As soon as a vessel anchors here, the crew form cabins on the shore, which soon have the appearance of a small village; and at the water's edge they build a kind of wharf, where arrive the innumerable fishing-boats; the above habitations being allowed to remain for the next season, the lawful property of the first comer. With regard to the above fishery, the necessaries being provided, the boats divide themselves into companies, each having their respective crews with the different services entrusted to each man, some being employed in the actual taking of the fish, others in cutting it open, others in salting it, and others in heaping it up. The fishermen leave the coast at day-break, and do not return before the evening, unless, indeed, their boats should be filled. They catch the fish with an hook, and every boat goes well provided with these and lines to guard against losses. As soon as they arrive at the shore, it is heaped up in piles, and turned and salted and cleansed for some succeeding days, after which, when quite dry, it is done up in small packets with the skins outward; though it still continues to be turned and salted till the time of embarkation. As the boats are continually out, the fatigue of the fishermen is very great, and they frequently go with little rest for nights and days.

The great bank of Newfoundland is a large heap of sand 580 miles long, and 233 wide, the depth of the water varying from 15 to 60 fathoms, and the bottom strewed with shells, and abounding with small fish serving as food for the cod, the numbers of which here are incredible; though some idea may be formed when it is known that 500 vessels were laden annually with it for some time past, and that, although the present consumption be much greater, their abundance is not found to be the least diminished; and it is indeed doubted whether this fishery is not a mine of greater wealth than those either of Mexico or Peru.

[In illustration of what our author has advanced, we have to observe, that in 1785, Great Britain and the United States, at the lowest computation, used to employ 3000 sail of small craft in this fishery; on board of which, and on shore to cure and pack the fish, were upwards of 100,000 hands; so that this fishery is not only a very valuable branch of trade to the merchant, but a source of livelihood to many thousands of poor people, and a most excellent nursery to the royal navy. This fishery is computed to increase the national stock 300,000*l.* a year in gold and silver, remitted for the cod sold in the North, in Spain, Portugal, Italy, and the Levant. Not only plenty of cod, but several other species of fish, are caught in almost equal abundance along the shores of Nova Scotia, New England, and the isle of Cape Breton; and very profitable fisheries are carried on upon all their coasts.

This island, after various disputes about the property, was entirely ceded to England by the treaty of Utrecht, in 1713; but the French were left at liberty to dry their nets on the *n.* shores of the island; and by the treaty of 1763, they were permitted to fish in the gulf of St. Lawrence, but with this limitation, that they should not approach within three leagues of any of the coasts belonging to England. The small islands of St. Pierre and Miquelon, situated to the *s.* of Newfoundland, were also ceded to the French, who stipulated to erect no fortifications on these islands, nor to keep more than 50 soldiers to enforce the police. By the last treaty of peace, the French are to enjoy the fisheries on the *n.* and on the *w.* coasts of the island; and the inhabitants of the United States are allowed the same privileges in fishing, as before their independence. The chief towns in Newfoundland are, Placentia, Bonavista, and St. John's; but not above 1000 families remain here in winter. A small squadron of men of war are sent out every spring to protect the fisheries and inhabitants, the admiral of which, for the time being, is governor of the island; besides whom, there is a lieutenant-governor, who resides at Placentia.

In June of this year, 1812, the British had on the Halifax, Newfoundland, and W. India stations, three sail of the line, 21 frigates, 19 sloops of war, and 18 smaller vessels, making a total of 61 armed vessels.

Vessels, it has been stated, lie in the bays and harbours of this island in perfect security, being well sheltered, except at the entrance, by the mountains. Some of these bays, (the whole circuit of the island being full of them) it should appear, are a league or two leagues in length, and near half a league in breadth; and it is a subject of curious inquiry for the philosopher to determine the causes of their contiguity and depth, and the consequent narrowness of the slips by which they are separated. The towns and villages are only on the larger and more commodious bays. The number of fowls called penguins, are certain marks for the bank of Newfoundland, and are never found off it; these are sometimes seen in flocks, but more usually in pairs. The French used to employ in this fishery 264 ships, tonnage 27,439; and 9403 men. Total value 270,000*l.* sterling.

The spaces of ground called ships rooms in Newfoundland, were by an act passed last year, 1811, exempted from the clause hitherto attached to them by the first comers, and are now let out as private property, for building dwelling houses and store houses, and for other uses necessary to the trade and fishery. It was also lawful under the same act for the governor to institute surrogate courts in the adjacent islands.]

[NEWICHWAWANICK. See PISCATAQUA.]

[NEWINGTON, a township; formerly part of Portsmouth and Dover, in Rockingham county, New Hampshire, five miles distant from the former. It contains 542 inhabitants.]

[NEWLIN, a township in Chester county, Pennsylvania.]

[NEWMANSTOWN, Pennsylvania; situate in Dauphin county, on the *e.* side of Mill creek. It contains about 30 houses, and is 14 miles *e.* by *n.* of Harrisburg, and 72 *n. w.* by *w.* of Philadelphia.]

[NEWMARKET, a township in Rockingham county, New Hampshire, *n.* of Exeter, of which it was formerly a part, and 10 miles *w.* of Portsmouth. It was incorporated in 1727, and contains 1137 inhabitants. Fossil shells have been found near Lamprey river in this town, at the depth of 17 feet; and in such a situation as that the bed of the river could never have been there. The shells were of oysters, muscles, and clams intermixed.]

[NEWMARKET, a village in Frederick county, Maryland, on the high road to Frederickstown, from which it lies nearly 13 miles *w. s. w.* and about 30 miles *n. w.* of the Federal city.]

[NEWMARKET, a village in Dorchester county, Maryland, three miles *n. e.* of Indian town, on Choptank river, nine *n. e.* of Cambridge, and as far *n. w.* of Vienna.]

[NEWMARKET, a town in Virginia, Amherst county, on the *n.* side of James river, at the mouth of Tye river. It is a small place, contains a tobacco warehouse; is 68 miles above Richmond.]

[NEWNHAM Cape. See NEWENHAM.]

[NEWPORT, a township in Cheshire county, New Hampshire, *e.* of Claremont. It was incorporated in 1761, and contains 780 inhabitants.]

[NEWPORT, a township of Nova Scotia in Hants county, on the river Avon. The road from Halifax runs part of the way between this township and

Windsor; and has settlements on it at certain distances.]

[NEWPORT, a maritime county of the state of Rhode Island, comprehending Rhode island, Canonicut, Block, Prudence, and several other small islands. It is divided into seven townships, and contains 14,300 inhabitants, including 366 slaves.]

[NEWPORT, the chief town of this county, and the semi-metropolis of the state of Rhode Island, stands on the *s. w.* end of Rhode island, about five miles from the sea. Its harbour (which is one of the finest in the world) spreads *w.* before the town. The entrance is easy and safe, and a large fleet may anchor in it and ride in perfect security. It is probable this may, in some future period, become one of the man-of-war ports of the American empire. The town lies *n.* and *s.* upon a gradual ascent as you proceed *e.* from the water, and exhibits a beautiful view from the harbour, and from the neighbouring hills which lie *w.* upon the main. West of the town is Goat island, on which is fort Washington. It has been lately repaired and a citadel erected in it. The fort has been ceded to the United States. Between Goat island and Rhode island is the harbour. Newport contains about 1000 houses, built chiefly of wood. It has 10 houses for public worship, four for Baptists, two for Congregationalists, one for Episcopalians, one for Quakers, one for Moravians, and one for Jews. The other public buildings are, a state-house, and an edifice for the public library. The situation, form, and architecture of the state-house, give it a pleasing appearance. It stands sufficiently elevated, and a long wharf and paved parade lead up to it from the harbour. Front of Water street is a mile in length. Here is a flourishing academy, under the direction of a rector and tutors, who teach the learned languages, English grammar, geography, &c. A marine society was established here in 1572, for the relief of distressed widows and orphans, and such of their society as may need relief. This city, far famed for the beauty of its situation and the salubrity of its climate, is no less remarkable for the great variety and excellent quality of fresh fish which the market furnishes at all seasons of the year. No less than 60 different kinds have been produced in this market. The excellent accommodations and regulations of the numerous packets which belong to this port, and which ply thence to Providence and New York, are worthy of notice. They are said, by European travellers, to be superior to any thing of the kind in Europe. This town, although greatly injured by the late war, and its consequences, has a considerable trade. A cotton and duck manufactory have been lately established. The exports for a year, ending September 30, 1794, amounted to 311,200 dollars. It was first settled by Mr. William Coddington, afterwards governor, and the father of Rhode island, with 17 others, in 1639. It is 23 miles *s.* by *e.* of Providence, 10 *s.* of Bristol, 54 *s. w.* by *s.* of Boston, 75 *e. n. e.* of New Haven, and 201 *n. e.* by *e.* of Philadelphia. Lat. 41° 25′ *n.* Long. from Greenwich 71° 14′ 30″.]

[NEWPORT, a small post-town in Newcastle county, Delaware; situated on the *n.* side of Christiana creek, three miles *w.* of Wilmington. It contains about 200 inhabitants, and carries on a considerable trade with Philadelphia in flour. It is six miles *n. e.* by *n.* of Christiana bridge, and 28 *s. w.* of Philadelphia.]

[NEWPORT, a township in Luzerne county, Pennsylvania.]

[NEWPORT, a small post-town in Charles county, Maryland, 11 miles *s. e.* of port Tobacco, 50 *s.* by *w.* of Baltimore.]

[NEWPORT. See ISLE OF WIGHT County, Virginia.]

[NEWPORT, a very thriving settlement in Liberty county, Georgia; situated on a navigable creek, 34 miles *s.* of Savannah, and seven or eight *s. w.* from Sunbury. This place, commonly known by the name of Newport Bridge, is the rival of Sunbury, and commands the principal part of the trade of the whole country. A post-office is kept here.]

[NEWTON, a pleasant township in Middlesex county, Massachusetts; situated on Charles river, and is nine miles *w.* of Boston. It was incorporated in 1791, and contains 1360 inhabitants.]

[NEWTON, a small town in Chester county, Pennsylvania, 22 miles from Philadelphia.]

[NEWTON, a township in Rockingham county, New Hampshire, on Powow river, adjoining Amesbury in Massachusetts, 10 or 12 miles *s.* of Exeter, and 20 from Portsmouth. It was incorporated in 1749, and contains 530 inhabitants.]

[NEWTOWN, a post-town in Fairfield county, Connecticut, nine miles *e. n. e.* of Danbury, 20 *w. n. w.* of New Haven, 34 *s. w.* of Hartford, and 59 *n. e.* of New York. The town stands pleasantly on an elevated spot, and was settled in 1708.]

[NEWTOWN, on Staten island, New York, is three miles *n. e.* of Old town, as far *e.* of Richmond, and nine *s. w.* of New York.]

[NEWTOWN, a township in Queen's county, New York, includes all the islands in the sound opposite the same. It is about eight miles *e.* of

New York, and contains 2111 inhabitants, including 533 slaves.]

[NEWTOWN, a township in W. Chester county, New York; of whose inhabitants 276 are electors.]

[NEWTOWN, a township in Tioga county, New York, lies between the *s.* end of Seneca lake and Tioga river; having Chemung township *e.* from which it was taken, and incorporated in 1792. In 1796, 169 of its inhabitants were electors.]

[NEWTOWN, a township in Gloucester county, New Jersey.]

[NEWTOWN, the seat of justice in Sussex county, New Jersey, is about 10 miles *e.* of Sandyston.]

[NEWTOWN, the capital of Bucks county, Pennsylvania. It contains a Presbyterian church, a stone gaol, a court-house, an academy, and about 50 houses. It was settled in 1725, and is eight miles *w.* of Trenton in New Jersey, and 19 *n. e.* by *n.* of Philadelphia. There are two other townships of this name, the one in Delaware county, the other in that of Cumberland.]

[NEWTOWN, a small town of Virginia, situated in Frederick county, between the *n.* and *s.* branches of Shenandoah river; seven miles *s.* of Winchester, and 104 *n. n. w.* of Richmond.]

NEXAPA, a jurisdiction and *alcaldia mayor* of Nueva España, in the province and bishopric of Oaxaca, one of the best peopled, largest, and most lucrative. It comprehends also the district of Los Mistepeques, where there is a lieutenant and *alcalde mayor.* It enjoys different temperatures from being in the *sierra,* though some parts of it consist of a *llano* or plain land, watered by several rivers, which render it extremely fertile in cochineal, indigo, and sugar-cane, of which it has a great commerce, and which causes it to be one of the most considerable *alcaldias* in the kingdom. The capital is the settlement of San Pedro de Quiechapi, and the other settlements of the jurisdiction are,

S. Baltasar,
S. Francisco,
S. Pedro Lespi,
S. Domingo,
S. Thomas Quiri,
Santiago Lachivea,
S. Juan Xanaguecho,
Sta. Catalina,
Quiquitane,
S. Pedro,
S. Juan,
S. Lorenzo,
Néxapa,
S. Bartolomé Yautepec,
S. Juan de la Xarcia,
Sta. Ana,
S. Juan de Lachixila,
S. Juan Beca,
S. Pedro Acatlan,
Santiago Tuctla,
Sta. Cruz,
S. Pedro,
Sta. Maria Lagicojani,
S. Agustin Mistepec,
S. Joseph Lachiguiri,
S. Andres Mixtepec,
Santiago,
S. Juan Tepalcaltepec,
S. Pablo Topiltepec,
Sta. Maria,
S. Pedro Martir,
Sta. Cruz Huilotepec,
Chiltepec,
Sta. Lucia,
Santiago Tecolotepec,
Santiago, 2,
Sta. Maria Coatlan,
Iscuintepec,
S. Lucas Cocatlan,
Sta. Margarita,
S. Miguel Quezaltepec,
S. Juan Mazatlan,
Sta. Maria Nizagui,
Sta. Maria Totolapa,
Zoquitlan,
Candelaria,
Zuchiltepec,
S. Matias,
S. Lorenzo,
Acatepec,
S. Lucas Hiscotepec,
S. Juan, 2,
Santo Tomas,
S. Domingo,
S. Andres,
Sta. Maria Quieguelani,
Santiago, 3,
S. Juan, 3,
S. Juan Xicula,
S. Pedro Ocotepec,
Sta. Maria Cacalotepec,
Sta. Maria Acatlazinto,
Santiago Malacatepec,
Chimaltepec,
Santiago Xilotepec,
S. Sebastian.

NEXAPA, SANTIAGO DE, a town of this jurisdiction, situate in a flourishing plain, through which passes the royal road from Mexico to the kingdom of Guatemala; inhabited by 27 families of Indians, and 15 of Negroes and Mulattoes. It contains a convent of the religious order of S. Domingo, and its population was formerly numerous, but it suffered much by the epidemic distemper, called there *matlazuaga,* in 1736. Five leagues *w.* of its capital.

NEXAPAM, SAN ANTONIO DE, a settlement of the province and kingdom of Guatemala; situate in the valley of this name. It is large and inhabited by 1730 Indians, who speak the Xachiquel idiom; but amongst the above are counted the Indians of the two settlements annexed to its curacy, called San Bernabé Acatenango and San Pedro Yepocapa. The curacy of this settlement belonged to the religious order of San Francisco, before it was put under the clergy by decree of his Majesty, with the exception, however, of the *reduccions* made by the missionaries.

NEXAPAM, another settlement, in Nueva España, close to which runs a river rising from a volcano and passing near the settlement. It is said of this river by the *Fr.* Juan de Torquemada, that it only runs between the hours of seven and eight in the day, and loses itself near the mountain of San Juan.

NEXPA, a settlement of the head settlement of the district of Xoxutla, and *alcaldía mayor* of Cuernavaca, in Nueva España.

NEXQUIPAYAC, SAN CHRISTOVAL DE, a settlement of the head settlement of the district and *alcaldía mayor* of Tezcoco in Nueva España; situate in a plain fertile in wheat and other seeds, which the natives cultivate. It is reduced to 58 families of Indians and six of Spaniards. Two leagues *n.* of its capital.

NEXTIPAC, a settlement of the province of Mexico in the time of the Mexican Indians, who established themselves in it, having fled from those of Mexiltcatzinco. It stood about half a league from the capital of Mexico, on which account it was abandoned.

NEXTLALPAN, a settlement of the *alcaldía mayor* of Tula in Nueva España; annexed to the curacy of this settlement, from whence it lies one league to the *n.* containing 143 Indian families.

[NEYBE. See NEIVA.]

NIAGARA, a large and abundant river of the country of the Iroquees Indians in Canada and N. America. It is, properly speaking, the great river of St. Lawrence, which runs from the lake Erie and enters the lake Ontario by a large channel of 20 miles in length, and at 19 miles in this course, it forms the celebrated falls of its name, the largest known in the world, where the water rushes down a precipice of 140 feet. The French geographer De l'Isle, by the relation of the Baron de la Hontán and the Father Hennepin, makes it 600 feet, and Mr. Bowen above 700, but this exaggeration arises from the difficulty of measuring it, as it is not possible to approach very near to, or to regard it otherwise than by a profile view: what we have given respecting it is the best received and believed by the most intelligent of those who have seen it. The river, at this cascade, is nearly half a league wide, and just before it comes to it, the stream is so rapid that animals attempting to cross it are sometimes hurried away by its impetuosity and precipitated down the abyss to certain destruction.

At the top of the cascade and in the centre of the river, is an island which divides the falling water into two large sheets, and when it has reached the bottom, it dashes up with a white foam like snow, and is in constant agitation just as if it were boiling. The vapours ascending from it have the appearance of a thick smoke, but when the sun shines rainbows are formed of the most beautiful colours.

[Niagara river receives Chippeway or Welland river from the *w.* and Tonewanto creek from the *e.* and embosoms Great and Navy islands. Fort Slusher stands on the *e.* side of this river near Navy island The falls, in this river, are opposite fort Slusher, about seven or eight miles *s.* of lake Ontario, and form the greatest curiosity which this, or indeed any other country, affords. In order to have a tolerable idea of this stupendous fall of water, it will be necessary to conceive that part of the country in which lake Erie is situated, to be elevated above that which contains lake Ontario about 300 feet; the slope which separates the upper and lower country is generally very steep, and in many places almost perpendicular; it is formed by horizontal strata of stone, great part of which is lime-stone. The slope may be traced by the *n.* side of lake Ontario, near the bay of Torento, round the *w.* end of the lake; thence the direction is generally *e.* Between lake Ontario and lake Erie it crosses the strait of Niagara and the Gennessee river; after which it becomes lost in the country towards Seneca lake. It is to this slope the country is indebted both for the cataract of Niagara and the great falls of Gennessee. The cataract of Niagara, some have supposed, was formerly at the *n.* side of the slope near the landing; and that from the great length of time, and the quantity of water, and distance which it falls, the solid stone is worn away for about seven miles up towards lake Erie; but for this latter opinion, observes General Lincoln, who visited and examined these falls in 1794, "on a careful examination of the banks of the river, there appears to be no good foundation."

There is a chasm down which the water rushes with a most astonishing noise and velocity, after it makes the great pitch. Here the fancy is constantly engaged in the contemplation of the most romantic and awful prospect imaginable; when the eye catches the falls, the contemplation is instantly arrested, and the beholder admires in silence. The river is about 742 yards wide at the falls. The perpendicular pitch of this vast body of water produces a sound that is frequently heard at the distance of 20 miles, and in a clear day, and fair wind, 40 and even 50 miles. A perceptible tremulous motion in the earth is felt for several rods round. Just below the Great pitch, the water and foam may be seen puffed up in large spherical figures; they burst at the top, and project a column of the spray to a prodigious height, and then subside, and are succeeded by others which burst in like manner. This appearance is most remarkable about half way between the island that divides the falls and the *w.* side of the strait, where the largest column of water descends. The descent into the chasm of this stupendous cataract is very difficult, on account of the great height of the banks; but when once a person has descended, he may go up to the foot of the falls, and take shelter behind the descending column of water, between that and the precipice, where there is a space sufficient to contain a number of people in perfect safety, and where conversation may be held without interruption from the noise, which is less here than at a considerable distance. On Christmas 1795, a severe shock of an earthquake was felt here, and by which a large piece of the rock that forms the famous cataract was broken off.

Whatever else is curious in this stupendous fall has been accurately stated by Alçedo.]

NIAGARA, a fort built by the French in 1687, near the former river, under the direction of Mr. Denonville, governor of New France, in spite of the opposition made by the English and the governor of New York, Colonel Dongan. Mr. de Troye, with a detachment, was nominated governor, but the greater part of his men dying from the badness of the climate, it was abandoned and ruined. In 1721 it was rebuilt by Mr. de Joncayre, who also met with some opposition from the English, who at last took it, being headed by William Johnson, in 1759. [It was delivered up to the United States, according to the treaty of 1794, by the British in 1796.

Niagara is now a post-town as well as fort, and is situate on the *e.* side of Niagara river, at its entrance into lake Ontario, and opposite to Newark in Canada. Niagara fort is a most important post, and secures a greater number of communications through a large country, than probably any other pass in interior America. . It is about nine miles below the cataract, 63 *n. w.* of Williamsburgh on Gennessee river, 266 *n. w.* of Philadelphia, and 305 *w.* by *n.* of Boston. Lat. 43° 16′ *n.* Long. 79° 4′ *w.* Although it is a degree *n.* of Boston, yet the season is quite as mild here as at that town, and vegetation quite as early and forward. It is thought that the climate meliorates in the same latitude as one proceeds from the Atlantic *w.*]

NIBEQUETEN, a river of the kingdom of Chile. It is abundant, and rises in the *cordillera*, and enters the Biobio, gives its name to a tribe of valorous Indians, amongst the Araucanos, who dwell on its shores.

NICAGUA, a small river of the island of S. Domingo, which rises in the *e.* head of the island, and enters the sea in the great bay of Samaná.

NICARAGUA, a province and government of the kingdom of Guatemala in N. America; bounded *n.* by the province of Guatemala, *s.* by that of Costa Rica, *e.* by the N. sea, and *w.* by the S. sea. It is 50 leagues long from *e.* to *w.* and nearly as many wide from *n.* to *s.* of an hot temperature, and the most woody part of Nueva España, although not without many *llanuras.*

This province has very few rivers, and is subject to tempests in the winter; extremely fertile, and abounding in all the productions that can be mentioned, except wheat, so that it provides itself with flour from the provinces of Peru. It has large breeds of neat cattle, swine, and goats, but particularly of mules and horses, carrying on a great trade in these with the kingdom of Tierra Firme, supplying the drovers of Costa Rica, who are employed in carrying goods from Panamá to Portobelo; but this trade was much greater when the galleons used to arrive. To the above it adds the considerable branches of commerce of cotton, honey, *pita*, wax, maize, *agi*, and French beans. It produces also indigo, sugar, cochineal, and *cacao*, quantities of fish and fine salt, the whole being sold here at a very reasonable price. In the woods are found excellent sorts of timber, namely, Brazil wood, and some *zeibas*, so large that 10 men with their arms extended cannot encompass them. In this province there is likewise found amber, turpentine, pitch, *naptha*, and various balsams and medicinal drugs much esteemed in Europe.

It abounds in deer and animals of the chase of all kinds, as well as in birds; but it is not without snakes, vipers, scorpions, bats, lizards, mice, and mosquitoes of various sorts, which render in some degree disagreeable a country which some Spaniards of consequence have called the paradise of Mexico, and others, with greater justness, the paradise of Mahomet.

This province was conquered by Gil Gonzalez Davila and Francisco Fernandez de Cordoba. It has a large lake called Del Desaguadero, seen into the waters of the great lake Nicaragua empty themselves; and at three leagues from this lake is a very lofty volcano, continually vomiting smoke and stones; also at the distance of four leagues is another lake which is small and round, and may rather be denominated a well, as from the surface of its water to the top of its bank there is no less a distance than 2000 yards, and although the descent is nearly perpendicular, yet will the Indians go down to fetch water, climbing by certain holes which they have made in the rock, and ascend with the pitcher on their heads with a velocity truly surprising. This is one of the provinces of the greatest number of inhabitants, and the natives are ingenious and diligent in the pursuit of the arts, in which they excel the other Americans, especially in their silversmiths and musicians, to which employments they have a natural turn. The capital is the city of Nicaragua, in lat. 11° 16′ *n.* and long. 85° 4′ *w.*

Bishops who have presided in Nicaragua.

1. Don Diego Alvarez Osorio, native of America, although we know not of what settlement; he was chanter of the church of Panamá when he was elected first bishop of this diocese, in 1531.

2. Don *Fr.* Antonio de Valdivieso, of the order

of S. Domingo, native of Villa-hermosa in the archbishopric of Burgos; presented to the bishopric of Nicaragua; he died of some blows he received from Juan Bermejo, one of the partisans of the two rebel brothers the Contreras, who had robbed the treasury, whilst manifesting his zeal in the king's cause, in 1549.

3. Don *Fr.* Gomez Fernandez de Cordoba, of the order of San Gerónimo, native of Cordoba, of whom we have treated amongst the bishops of Guatemala, to which place he was promoted in 1574.

4. Don Fernando de Menavias, of the same order as the former, a preacher of great repute he died here.

5. Don *Fr.* Antonio de Zayas, of the order of San Francisco, native of Eeija; presented to the bishopric of Nicaragua in 1574; entered to take possession in 1577, according to Gil Gonzalez Davila; although the *Fr.* Antonio Daza says that at this time the bishop of that place was *Fr.* Gerónimo Villa Carrillo, of the order of San Francisco.

6. Don *Fr.* Domingo de Ulloa, of the order of S. Domingo, of the house of the Marquises of Mota, collegiate in the college of San Gregorio de Valladolid and its rector, prior of various convents of his order, vicar-general of the province of Castilla, and presented by his Majesty Philip II. to the bishopric of Nicaragua in 1584, and promoted to the church of Popayán, 1591.

7. Don *Fr.* Gerónimo de Escobar, native of Toledo, of the order of San Agustin, a celebrated preacher; elected bishop of Nicaragua in 1592, but, after having embarked to go to its church, he was forced to put into the port of Cadiz, where he died.

8. Don *Fr.* Antonio Diaz de Salcedo, of the order of San Francisco; promoted from the bishopric of Cuba to this in 1597.

9. Don *Fr.* Gregorio Montalvo, of the order of S. Domingo, native of Coca in the bishopric of Segovia, prior of his convent at Placencia; elected bishop, and afterwards promoted to Yucatán.

10. Don Pedro de Villareal, native of Andujar, visitor of the archbishopric of Granada, and presented to the bishopric of Nicaragua, where he died in 1619.

11. Don *Fr.* Benito de Valtodano, of the order of San Benito, collegiate of the college of San Vicente of Salamanca, abbot of San Claudio, visitor of its order; elected bishop in 1620; he died in 1627.

12. Don *Fr.* Agustin de Hinojosa, of the order of San Francisco, native of the court of Madrid, guardian of his convent at Sevilla, jubilee lecturer, *definidor*, and preacher of great fame; elected bishop of Nicaragua in 1630; he died in Villanueva de la Serena before that he passed to his destination in the following year of 1631.

13. Don Juan de Baraona Zapata, also native of Madrid, where he studied in the imperial college arts and philosophy, and in the university of Salamanca canons and laws, graduating as licentiate and doctor; chaplain of the royal chapel of Alcalá, an honest, pious, and charitable man, and a great observer of silence, presented to the bishopric in 1631; he died before he departed for it, in 1632.

14. Don *Fr.* Hernando Nuñez Sagredo, of the order of La Santisima Trinidad, native of Rodilla in the archbishopric of Burgos, lecturer in his convent of Toledo at Alcalá and Valladolid, *calificador* of the inquisition of Cuenca and of the supreme council of the same, minister of his order in the convents of Santa Maria del Campo, Segovia, Cuenca, and Burgos, provincial and vicar-general of the province of Castilla, presented to the bishopric of Nicaragua in 1633; he died in 1639.

15. Don *Fr.* Alonso Breceño, of the order of San Francisco, native of Santiago of Chile, a lecturer and philosopher of great talents, twice jubilist and grand theologist, guardian of the college of Lima, *definidor* of the province, commissary and visitor of those of Charcas and Chile, vicar-general in Xauxa and Caxamarca, guardian of this convent, *definidor* of the province of Lima: he assisted at the general chapter in Rome, was nominated *calificador* of the holy office, presented to the bishopric of Nicaragua in 1644, of which he took possession in 1646; promoted to the bishopric of Charcas in 1659.

16. Don Andres de las Navas Quevedo, of the order of Nuestra Señora de la Merced, native of the city of Baza: after having received different prelacies in his order, he was presented to the bishopric of Nicaragua in 1667, and promoted to that of Guatemala in 1682.

17. Don *Fr.* Diego Morcillo Rubio, of Auñon, of the order of La Santisima Trinidad Calzada, native of Villa Rolledo in La Mancha, a man of great virtues and powers of government; elected bishop of Nicaragua, from whence he was promoted to that of La Paz; afterwards archbishop of Charcas and Lima, where he by a special authority from his Majesty was endowed with the viceroyship of that kingdom, and where he governed with address.

18. Don *Fr.* Benito Garrat, premostratensian canon of the order of San Noberto; nominated bishop of Nicaragua in 1708.

19. Don *Fr.* Andres Quiles Galindo, of the order of San Francisco, native of Zelaya in the bishopric of Mechoacán: he studied in the university of Mexico, and in the Colegio Maximo of San Pedro and San Pablo, Latin, rhetoric, philosophy, and theology; after adopting the religion, he maintained the professorships for 15 years, was consultor and *calificador* of the holy office; destined to be *pro-ministro* provincial to Europe, when he was elected bishop of Nicaragua, in 1718.

20. Don *Fr.* Dionisio de Villavisencio, of the order of San Agustin, in 1725.

21. Don Domingo Antonio Zeratain, chanter of the church of La Puebla de los Angeles in 1736.

22. Don Isidro Marin Bullon y Figueroa, of the order of Alcántara, rector of its college at Salamanca, of the lap and cloister of that university, honorary chaplain to his Majesty in 1743; he died in 1749.

23. Don Pedro Agustin Morel of Santa Cruz, dean of the holy church of Santiago of Cuba; elected bishop of Nicaragua in 1749, and promoted to that of Santiago in 1753.

24. Don Joseph Florez de Rivera, elected bishop of this church of Nicaragua in 1753; he died in 1757.

25. Don. *Fr.* Mateo de Navia y Bolaños, of the order of San Agustin, native of Lima, master in his religion; immediately upon his coming to Europe he was presented to the bishopric of Nicaragua in 1757; he died in the city of Granada in 1762, whilst on the visitation.

26. Don Juan de Vilches y Cabrera, dean of the same holy church of Nicaragua, elected bishop in 1763; he died in 1774.

27. Don Esteban Lorenzo de Tristan, native of Jaen in Andalucia, nominated chanter of the holy church of Guadix, and before he took possession, elected to that of Nicaragua in 1775; afterwards promoted to the church of Durango in 1783.

28. Don Juan Felix de Villegas, native of Cobreces in the bishopric of Santander, elected bishop of Nicaragua in 1784, being then inquisitor of Cartagena.

Nicaragua, a lake of fresh water of the above province, being in extent 120 miles long, and 41 wide, navigable by the largest vessels, as it is of an immense depth. On its coasts are many estates of large cattle, and in each a small port for the canoes and vessels which run in to lade with the productions of the country, and which are employed for the expediting traffic.

In this lake are several isles, and it enters the sea by the *e.* through a channel called the river San Juan, or Del Desaguadero, of 64 miles direct distance long. In this channel sail flat-bottomed vessels of the size of bilanders; also very large canoes laden with tallow and other effects, which they carry to Portovelo, 236 miles from the port of S. Juan de Nicaragua; and in the time when the galleons arrived, they carried, under a licence, clothes and other articles for the supply of the province, though not without great risk from the attacks of the Zambos and Mosquitos Indians, who used to be making continual depredations on this lake, and who still infest it, as also on the coast of Honduras, near to which they live dispersed on the numerous islands. The above vessels make this voyage under a necessity of discharging their burthen in these shallow parts, called the *raudales*, since there is not depth of water for them otherwise. On one of these *raudales* is situate the castle of Nuestra Señora de la Concepcion, upon a mountain of living rock, and although this castle be not of any considerable size, it serves to guard the pass of the river against an enemy: it is furnished with 36 cannon, and has a very well constructed mound, from whence, although the enemy should take the fortress, such an attack might be continued against him as to make him abandon his purpose. At the water's edge is a platform with six cannons, and on the land side it is fortified by a ditch and estacade which reaches as far as the river. It is ordinarily defended with 100 men, besides 16 artillery-men, a constable, 40 musketeers, a governor, chaplain, lieutenant, and 20 militia-men, for the management of the *champanes* or barks, two of which are posted every night on guard above and below the fort upon the river; also with 18 slaves, men and women, to do the cooking, &c. of the garrison, a supply of maize, meat, vegetables, fowl, and other things being sent from the city of Granada, 60 leagues distance, and a six months supply being always reserved.

The temperature here is very sickly, the rain falling continually; and thus it is usual every two years, or earlier if necessary, for the governor to demand at the capital of Guatemala a fresh supply of 50 men, to restore the loss occasioned by the fatality of the place; and the governor of the province has strict injunctions to send whatever number may be required.

This castle is called the *antemural*, or great wall and barrier, of the kingdoms of Nueva España and of Peru; for should an enemy make his way up this river, as was twice effected, namely, by the pirates Lolonois and John Morgan, they might go on to occupy Nueva España, and, having established themselves in the port of Realejo, which is 30 leagues from the city of Granada, to make themselves masters of the S. sea; where also, by the facilities offered by abundance of fine timber, and of every other requisite, save that of iron bolts and nails, they might soon construct a noble fleet. The castle of La Concepcion has, for these reasons, been an object of great jealousy with the Spanish government.

NICASIA, a settlement of the province and *corregimiento* of Lampa in Peru, distinct from the following.

NICASIO, a settlement of the same province and kingdom as the former.

NICE, BERNARDO DE, a settlement of the province and *corregimiento* of Caxamarca in Peru; annexed to the curacy of Chalique.

[NICHOLA, or NICHOLA TOWN Gut, on the *n. e.* coast of the island of St. Christopher's.]

[NICHOLAS, Cape ST. the *n. w.* extremity of the island of St. Domingo, in the W. Indies. It is four miles *w.* of the town of its name, but more commonly called the Mole, 40 miles *e.* of cape Mayzi, at the *e.* end of the island of Cuba, and 94 miles *n. e.* by *n.* of cape Dame Marie, and, with this last cape, forms the entrance into the large bay called the Bight of Leogane.

In the beginning of July last, a severe shock of an earthquake was experienced at cape Nicholas Mole, which threw down eight houses. Two lives were lost. See The MOLE.]

[NICHOLAS, Port ST. on the coast of Peru in S. America, lies *n.* of port St. John, about a league to leeward of the river Masca, and six leagues *s. s. e.* of port Cavallo. It is safer than St. John's harbour, but affords neither wood nor water.]

[NICHOLAS. See NICOLAS.]

NICHOLSON, a fort of the English, in the province and colony of New York; situate on the shore of the river Hudson, near the confines of the country of the Iroquees Indians.

[NICKAJACK, an Indian town on the *s. e.* side of Tenessee river, at the point of a large bend, about 33 miles *n. e.* of the Creeks crossing place. Half way between these lies the Crow town, on the same side of the river.]

NICKER, a small island of the N. sea, inhabited by the English, and one of those called the Virgin isles. It lies between the Anegada and the Virgen Gorda, on which it is dependent.

NICLETON, a small river of the island of San Christoval, one of the Antilles. It runs *e.* and enters the sea on the coast, running from *n. w.* to *s. e.* in the district of the parish of Cinq Combles.

NICODEL, a small river of Canada in N. America. It runs *n. w.* and enters the S. Lawrence, opposite the Three Rivers.

NICOLAO, BAXO, a shoal or isle of the N. sea, near the coast of this rhumb, of the island of Cuba, between cape Blanco and that of La Cruz.

NICOLAS, S. a settlement of the head settlement of the district of Tantima, and *alcaldía mayor* of Tampico, in Nueva España, of a warm and moist temperature; situate amongst uncultivated woods. It contains 83 families of Indians, who cultivate much cotton, of which they make several kinds of woven stuffs. It is three leagues from its head settlement, and 15 *s.* of the capital.

NICOLAS S. another settlement, in the head settlement of the district and *alcaldía mayor* of Tepeaca of the same kingdom. It contains 27 families of Indians, and lies a little more than two leagues from its capital.

NICOLAS, S. another, of the head settlement and *alcaldía mayor* of Marinalco in the same kingdom; situate at a league and an half's distance from the foot of a very lofty mountain.

NICOLAS, S. another, of the head settlement of the district and *alcaldía mayor* of Guejozingo in the same kingdom. It contains 54 families of Indians, and lies *s.* of its capital.

NICOLAS, S. another, of the head settlement of Armadillo, and *alcaldía mayor* of Sau Luis de Potosi, in the same kingdom. It contains 32 families of Indians, whose trade and employment is reduced to the dressing of hides and making of harness and riding equipage. Three leagues from its head settlement.

NICOLAS, S. another, of the same *alcaldía mayor* and kingdom as the former. Six leagues to the *e.* of Santa Maria del Rio.

NICOLAS, S. another of the same, which is the *real* of silver mines of the province of Ostimuri, formerly a large and rich town, but at present reduced to great poverty. Seven leagues *e. n. e.* of the *real* de Rio Chico.

NICOLAS, S. another, of the missions which were held by the Jesuits in the province of Tepeguana and kingdom of Nueva Vizcaya, on the shore of the river Las Nasas.

NICOLAS, S. another, of the province and government of Darien, and kingdom of Tierra Firme; situate on the coast, on the shore of the Rio Grande de Tuira, near the gulf of San Miguel.

NICOLAS, S. another, called De la Barranquilla, in the province and government of Cartagena, and Nuevo Reyno de Granada; situate in the extremity or point of the island in which that city stands, and at the entrance or mouth of the Rio Grande de la Magdalena.

NICOLAS, S. another, of the same province and kingdom as the former, in the district of Zinú;

situate on the shore of the river of its name, and near its mouth or entrance into the sea.

Nicolas, S. another, of the province and government of Venezuela in the same kingdom; situate in an extensive *llanura*, which extends from the coast on the shoré of the river Aroa, and is almost to the *n.* of the town of San Felipe.

Nicolas, S. another, of the province and government of Antioquia in the same kingdom; situate on the Rio Grande de la Magdalena.

Nicolas, S. another, of the province and government of Moxos in the kingdom of Quito; a *reduccion* of Indians of this nation, made by the Jesuits, to the *s.* of the mountains of Oro, and on the shore and at the source of the river Bauras or Guazumuri.

Nicolas, S. a town of the province and government of Buenos Ayres; situate on a small river, about 130 miles *n. w.* of Buenos Ayres, in lat. 33° 19′. Long. 60° 25′ 4″.

Nicolas, S. a settlement of Indians, also of the province and government of Buenos Ayres; situate on a small branch of the river Piratiny, on the *s.* side of the Uruguay. Lat. 28° 12′ *s.* Long. 55° 19′ 53″ *w.*

Nicolas, S. another, called Mole de S. Nicolas, a parish of the French, in the part they possess in the island S. Domingo; situate at the *w.* extremity of the island, by the cape of its name.

Nicolas, S. another, of the province and *corregimiento* of Cuenca in the kingdom of Quito; annexed to the curacy of the settlement of Delee.

Nicolas, S. another, of the Nuevo Reyno de Leon in N. America, near the town of Cadereita.

Nicolas, S. a large river of Nueva España, called thus from an estate of this name on its shores. It rises 10 leagues *n.* of the settlement of Mascota, in the *alcaldía mayor* of Ostotipac, and runs into the S. sea, through the valley of Vanderas, at the cape of Corrientes; its mouth being of the settlement of Ostotipac, 20 leagues to the *w.*

Nicolas, S. another, a small river of Canada in N. America. It runs *w.* between those of Marquet and Sable, and enters the lake Michigán.

Nicolas, S. a bay on the *n.* coast of the strait of Magellan, between cape Galand and the bay of Pico, according to the voyage and description of Nodales.

Nicolas, S. another port, on the *n.* coast of the river St. Lawrence in Canada, between the port S. Pancras and Trinité bay.

[Nicolas, S. See S. Nicholas.]

NICOPERAS, Asperezas, some rough and impassable mountains of the province and *captainship* of Rey in Brazil.

NICOYA, a province and *alcaldía mayor* of the kingdom of Guatemala in N. America: bounded *e.* by the province of Costarica, *n.* by the lake of Nicaragua, *w.* and *s.* by the Pacific ocean. It is of limited extent, and is looked upon as a district of the province of Nicaragua, the governor of it being nominated by the *alcaldía mayor*. The population is contained in only three settlements; which are Cantrén, Orotina, and Chorote, besides the capital, which is the town of the same name, situate on the shore of the river Capanso, near its entrance into the S. sea.

This province produces much maize, honey, pulse, and herbs, with which, by means of the sea, it carries on a great trade with Tierra Firme. It has a very good port and dock, where many fine vessels have been built. Here is also gathered much cotton, of which various stuffs are made, being dyed with the juice of the *caracol* caught in the bay of Las Salinas, and which cannot be washed out, and is much esteemed in all parts; cotton thread, which is likewise made here, is dyed in the same manner. In the above port are found pearls of a very fine quality. In lat. 9° 46′ *n.* Long. 84° 55′ 30″ *w.*

[NICTAU, a river of Nova Scotia, which waters the township of Annapolis; on its banks are quantities of bog and mountain ore. A bloomery has been erected in the town.]

[NICUESA, Gulf of, is on the *e.* coast of the country of Honduras, on the Spanish main, having cape Gracias a Dios for its *n.* limit, and cape Blanco on the *s.*; Catharine, or Providence, is due *e.* from it.]

NICULLIPAI, a small river of the kingdom of Chile. It runs *s.* very near the coast, and enters the Valdivia near its entrance into the sea.

NIEBE Bay. See Neiva.

NIERUIN, a settlement of the province and government of Santa Marta in the Nuevo Reyno de Granada; situate *n. e.* of the valley of Tenerife.

NIEUA, Nuestra Senora de, a small city of the province and government of Mainas in the kingdom of Quito; founded by Captain Juan de Salinas, in 1541, on the shore of the river of its name, to the *s. w.* of the Marañon. It is destroyed, and nothing but its ruins remain.

Nieua. The aforesaid river rises in the centre of the mountains of the province, and runs nearly due *n.* till it enters the Marañon or Amazonas between the narrow pass of Guaracayo and the Pongo of Manseriche.

[NIEVA Island lies *s. w.* of Mistake bay, and on the *n. e.* side of Hudson's straits.]

[NIEVA TERRA, near the *e.* end of Hudson's straits, in N. America, in lat. 62° 4′ *n.* and long. 67° 7′ *w.* and has high water on the spring-tide days at 50 minutes past nine o'clock.]

NIEVE, BAHIA DE MUCHA, or Bay of Much Snow, on the coast of the strait of Magellan, and at the third narrow pass called the Passage.

NIEVES, NUESTRA SENORA DE LAS, a settlement of the province and government of Mainas and kingdom of Quito.

NIEVES, SANTA MARIA DE LAS, another settlement, of the head settlement of the district and *alcaldía mayor* of Guejozingo in Nueva España. It contains 60 families of Indians, and is a very short league *w.* of its capital.

NIEVIS. See NEVIS.

NIGANDARI, a settlement of the province and *corregimiento* of Caxamarquilla in Peru.

NIGANICHE, a small island of the N. sea, near the *e.* coast of Cape Breton, between the port of Achepe and cape Fume.

NIGANICHE, a large and convenient bay of the same coast.

[NIGUA, a river on the *s.* side of the island of St. Domingo. Its mouth is seven leagues *e.* of the Nisao. The rivers Nigua and Jayna are not very far apart. But as they advance from their springs they recede from each other, the former running *w.* from the latter. Between them lies an extensive and fertile plain. The quantity of pure gold that was dug from its cavities, its sugar, cocoa, indigo, and other plantations, paid duties of a greater amount than those now paid by all the Spanish part of the island put together. All these rivers might be easily rendered navigable. The parish and small town of Nigua contain about 2500 persons; partly free people of colour.]

NIGUAS, a settlement of the province and government of Esmeraldas in the kingdom of Quito. It is small, situate in a wood of an hot and moist climate; surrounded by some small rivers, in which are caught excellent skates, which are carried to be sold at Quito. It produces many and delicate plantains, is annexed to the curacy of Mindo. In lat. 3° 8′ *n.*

NIGUAS, another settlement, in the same province and kingdom; situate to the *w.* 12 leagues from the capital, on the *n.* shore of the river Coca; annexed to the curacy of Yambe. Its territory is full of woods; and it produces abundance of wild wax, *zarzaparilla*, plantains, and some tobacco and cotton. In lat. 41′ 52″ *n.*

NIGUATA, a port of the coast of the province and government of Venezuela in the Nuevo Reyno de Granada; between those of Guaira and Caraeoli. On its shore is a small settlement and a fort for the defence and security of merchant vessels.

NIGUE, a point on the coast of the kingdom of Chile, between the mouths of the rivers Tolten and Queuli.

NIJAQUE, a settlement of the province and *corregimiento* of Chachapoyas in Peru; annexed to the curacy of Soritor.

NIKESA, a river of the colony and government of Surinam, in the part of Guayana possessed by the Dutch. It runs *n.* making many windings, and enters the sea very close to the river Corentin.

NILHAUE, a large, fertile, and beautiful valley of the district of Chanco in the kingdom of Chile, between the river of its name and that of Martaquino.

NILHAUE. The aforesaid river runs to *n. n. w.* and enters the sea near the *quebrada* of Lora.

NILCOS, a port of the N. sea, on the coast of the gulf of Urabá; of the province of Darien and kingdom of Tierra Firme; the only port in that part capable of receiving large vessels. It lies towards the *e.* near San Sebastian de Buena Vista, in lat. 6° 50′ *n.*

NIMAIMA, a settlement of the *corregimiento* of Panches in the Nuevo Reyno de Granada; of an hot temperature, and abounding in sugar canes, plantains, *yucas*, and some tobacco. It is poor; its population of Indians is scanty, and the Spaniards are very few. Sixteen leagues *w.* of Santa Fé.

NINACACA, a settlement of the province and *corregimiento* of Tarma in Peru.

NINDASOS, a barbarous nation of Indians of the province of Guanuco in Peru. It is divided into various tribes, who wander about through the woods without fixed abode. They are bounded *n.* by the Guatahuagas and *e.* by the Panataguas.

[NINETY-SIX, a district of the upper country of S. Carolina, *w.* of Orangeburg district, and comprehends the counties of Edgefield, Abbeville, Laurens, and Newbury. It contains 33,674 white inhabitants, sends 12 representatives and four senators to the state legislature, three of the former and one of the latter for each county, and one member to congress. It produces considerable quantities of tobacco for exportation. Chief town, Cambridge, or as it was formerly called, Ninety-six, which is 48 miles *w.* by *n.* of Columbia, 127 *n. w.* of Charleston, and 49 *n.* of Augusta in Georgia. In May 1781, this town was closely besieged by General Greene, and bravely defended by the British, commanded by Colonel Cruger.]

NINHUE, a settlement of Indians of the kingdom of Peru; situate at the source of the river Biobio.

NIO, a settlement of the missions which were held by the Jesuits in the province and government of Cinaloa in N. America.

NIOUE, a settlement of Indians of the province of Sagadahock in N. America; situate on the shore of the river Penobscot.

NIOURE, Bay of, on the *e.* coast of the lake Ontario, of the province and country of the Iroquees Indians.

NIPE, a settlement of the French, in their part of the island of St. Domingo; on the *n.* coast, at the *w.* head, and on the shore of the river of its name.

NIPE. This river runs *n.* and enters the sea opposite the island of Goanava.

[NIPEGON, a large river which empties into lake Superior from the *n.* It leads to a tribe of the Chippewas, who inhabit near a lake of the same name. Not far from the Nipegon is a small river, that, just before it enters the lake, has a perpendicular fall, from the top of a mountain of 600 feet. It is very narrow, and appears like a white garter suspended in the air.]

NIPES, a bay on the *n.* coast of the island of Cuba, between port Altabonita and the river Platanos; with a settlement between the points of Mulas and Maisi.

NIPISIGUIT, a river of Nova Scotia or Acadia, which rises from lake Nipisigouche, runs *e.* for many leagues, and enters the sea in the bay of Chaleurs.

[NIPISIGUIT, a small village of New Brunswick, on the *s.* side of Chaleur bay, inhabited by Roman Catholics; above 12 leagues *w.* of Caraquit island; between which and point Masanette, are the capes of Poiquchaw. At this village a number of coasting traders touch during the summer, where they purchase of the inhabitants codfish and salmon, as also feathers, peltry, and some furs.]

[NIPISSINS, Indians inhabiting near the head waters of the Ottowas river. Warriors 300.]

NIPISSING, a small lake of the province and country of the Iroquees Indians in N. America; formed by the river François, and running out by a large arm into the Utawas.

NIQUE, a river of the province and government of Darien, and kingdom of Tierra Firme, which rises in the centre of the same province and enters the river Cupá.

NIRUA DEL COLLADO, a town of the province and government of Venezuela, in the Nuevo Reyno de Granada; founded in 1553 from the fugitives of the city of Las Palmas, which was abandoned on account of the invasion of the infidel Indians. This town was rebuilt in the neighbourhood of the mines called Villa Rica, after which its situation was thrice removed; but such was the distress it experienced from the repeated attacks of the Indians, and so great were the difficulties of procuring Negroes to work the mines, that there remained of this unfortunate settlement nothing but the name, when it was at last founded by Francisco Faxardo in 1560, on the spot where it now stands, two leagues from the port of Guaira.

[The environs of this city (says Depons) are fertile, but the air is unwholesome, and the inhabitants are subject to agues, which always end fatally. There are not more than four or five white families. All the offices in the *cabildo* are held by the *Sambos*. The lieutenant "de justicia mayor," appointed by the governor, is the only person who can be a white. The city appears completely in decay. The population is about 3200 souls, chiefly *Sambos*, who are the offspring of the Indians and Negroes. They are robust, strong, and healthy, but lazy, addicted to drunkenness, theft, and every species of vice. Nirua is in lat. 10° and long. 71° 10′ from Paris. It is 48 leagues from Caracas.]

[NISAO, a river which rises in the centre of the island of St. Domingo, and falls into sea on the *s.* side, and on the *w.* side of the point of its name; seven leagues *w.* of Nigua river.]

[NISQUEUNIA, a settlement in the state of New York, above the city of Albany. This is the principal seat of the society called Shakers. A few of this sect came from England in 1774; and a few others are scattered in different parts of the country.]

NISUCO, or NISEICO, as some call it, a river of the province and government of Yucatán, which runs into the sea close to the island of Cozumél.

NITAIIAURITS, a settlement of Indians of S. Carolina; situate on the shore of the river Albama or Cousas.

NITO, a settlement of the province and government of Honduras, the spot where the fair or market of the whole province used to be celebrated, and consequently very rich. It was conquered by by Gil Gonzalez Davila, who pulled down the greater part of it and built it up anew. It stood upon the sea-coast, and formed a pleasing retreat to Hernan Cortes, after all his perils in his journey from Mexico, undertaken to chastise the rebellious

Christóval de Olid, who, before he arrived, had suffered death at the hands of Francisco de las Casas.

[NITTANY Mountain, in Pennsylvania, is between the Juniatta and the *w.* branch of Susquehannah river.]

[NIVERNOIS, a large bay at the *e.* end of lake Ontario.]

[NIXONTON, a post-town of N. Carolina, and capital of Pasquotank county, lies on the *n.* water of Albemarle sound, and contains a court-house, gaol, and a few dwelling-houses. It is 12 miles *e.* of Edenton.]

NIZAQUI, a settlement of the *alcaldía mayor* of Nexapa in the province and bishopric of Oaxaca, and kingdom of Nueva España; situate on the middle of a lofty plain. Of a cold temperature, and inhabited by 62 families of Indians, devoted solely to the commerce of cochineal. Twelve leagues *e.* of the capital.

NIZAO, a settlement of the island of Cuba; situate on the *n.* coast, between the Caragaya and the Jagua Grande.

Nizao, a river of the island of S. Domingo, which rises in the mountains of the centre of the same, runs *s.* and enters the sea at the point of its name, between the point of Palenque and the river Bani.

Nizao. The aforesaid point is on the *s.* coast of the same island, between the points Salina and Palenque.

NOADAN, a river of the province and government of Vera Cruz in Nueva España, which runs *w.* and enters the sea between the settlements of Almerí and Zempoala.

NOANAMA, San Joseph de, a settlement of the province and government of Chocó in the Nuevo Reyno de Granada; situate on the shore of the river S. Juan.

NOASI, a settlement of the province and government of Tucumán in Peru, of the jurisdiction of Santiago del Estero; situate on the shore of the river Choromoros.

NOBANI, a settlement of the head settlement of the district of Teotalzinco and *alcaldía mayor* of Villalta in Nueva España; of an hot temperature. It contains 17 families of Indians, and is 18 leagues *n.* of its capital.

[NOBLEBOROUGH, a township in Lincoln county, district of Maine, incorporated in 1788, and contains 516 inhabitants. It is 10 miles *s. e.* of Newcastle.]

[Nobleborough, a township in the *n. e.* part of Herkemer county, New York; situated on the *n. w.* side of Canada creek.]

NOCAIMA, a settlement of the jurisdiction of the town of Honda in the Nuevo Reyno de Granada; of an hot temperature, abounding in cotton, sugar-canes, maize, *yucas*, plantains, &c. annexed to the curacy of the settlement of La Vega.

NOCATABURI, a settlement of the province of Taraumara and kingdom of Nueva Vizcaya; situate at the source of the river Hiaqui.

NOCHIHA, a settlement of the province of Itza in the kingdom of Guatemala.

NOCHITLAN, a settlement of the *alcaldía mayor* of Tixtlan in Nueva España; of a hot temperature, very fertile and pleasant, and abounding in fruit and sugar-canes. It contains 233 families of Indians.

NOCHIZTLAN, a jurisdiction and *alcaldía mayor* of Nueva España, in the province and bishopric of Oaxaca. It is very fertile in cochineal and cotton, and a place of great traffic, as lying in the direct and high road from Mexico to Oaxaca. Its jurisdiction consists of the following settlements,

Nochiztlan, the capital, which is of the same name, contains in it a convent of the religious order of S. Domingo, 30 families of Spaniards, *Mustees*, and Mulattoes, and 134 of Mistecos Indians, engaged in the cultivation and commerce of grain, and in the manufacture of woven cotton stuffs. It is 155 miles *e.* with an inclination to the *s.* of Mexico, in lat. 17° 14′. Long. 97° 36′. The settlements are,

Santa Cruz Mitlatongo,	Guautla,
Xaltepec,	Texultepéc,
Santiago Mitlatongo,	Tiltepec,
Tilantongo,	San Juan Tamazula,
Santiago Yucunduche,	Cachuapa.

Nochiztlan, another settlement of the head settlement of the district and *alcaldía mayor* of Cuquio in the same kingdom. Its population is very large, and it is three leagues *n.* one-quarter to *n. e.* of its head settlement.

[NOCKAMIXON, a township in Buck's county, Pennsylvania.]

NOCUPETAJO, a settlement of the *alcaldía mayor* of Cinagua in Nueva España. It conatins 21 families of Indians, who trade in large cattle and maize, which they grow. Its population consisted formerly of more than 4000 families, and it was fertilized by a river which passed through it; but it is said that the inhabitants having ill treated and beaten their curate after having stripped him, received the vengeance of heaven by the river drying up; so that their fields became parched and barren, and a noxious heat arose, which caused an epidemical distemper, which soon swept off this numerous people. Thirty-seven leagues *s. e.* of its capital.

[NODDLE'S Island, a small, pleasant, and fertile island in Boston harbour, Massachusetts. It is about two miles *e. n. e.* of the town, on the Chelsea shore. It is occupied as a farm, and yields large quantities of excellent hay.]

[NODDWAY, a river or rather a long bay which communicates with James bay, at the *s. e.* extremity of Rupert's river.]

NOEL, a settlement of Nova Scotia or Acadia in N. America; situate on the shore of the Basin des Mines, in the interior part of the bay of Fundy.

NOGALES, San Juan Baptista de, a settlement of the head settlement of the district of Maltrata, and *alcaldía mayor* of Orizava, in Nueva España. It contains 124 families of Indians and 50 of *Mustees*, Mulattoes, and Negroes, including those of its wards, which are at about a league and a half's distance; the greater part of the inhabitants employing themselves as drovers. One league and a half from its head settlement.

Nogales, another settlement, of the province and government of Tucumán in Peru, of the jurisdiction of the capital; on the shore of the river Choromoros.

NOGUERA, a settlement of the missions which were held by the Jesuits in the province and government of Cinaloa.

NOGUNCHE, a settlement of the Indians of the province and *corregimiento* of Itata in the kingdom of Chile; situate on the coast, near the mouth of the river Itata.

NOHUKUN, or Rio Grande, a great river of the province and government of Yucatán, which runs *e.* and enters the sea in the gulf of Honduras, passing through the city of Salamanca.

[NOIR, Cape, on the *s. w.* coast of the island of Tierra del Fuego, at the entrance of the straits of Magellan. Lat. 54° 30′ *s.* Long. 73° 13′ *w.*]

[Noir, Cape, or Black Cape, on the *n.* side of Chaleur bay, is about seven leagues *w. n. w.* of Bonaventure.]

NOIRE, a river of the province and government of Neiva in the Nuevo Reyno de Granada, which runs *w.* between those of Cobo and Otáz, and enters the Grande de la Magdalena.

Noire, another, a small river of S. Carolina, in the county of Craven. It runs *s. e.* and unites itself with the Blackmingo to enter the Pedi.

Noire, another, also a small river of the same province, which runs *e.* and enters the Congari.

Noire, another, a small river of the province and government of Louisiana. It runs *s. e.* between those of Ailes and Quiovecovet, and enters the Mississippi.

Noire, another, a small river of Canada, which runs *s. w.* and enters the lake Michigan at the end of the *e.* coast.

Noire, another, a small river of the province and country of the Iroquees Indians in New France. It runs *n.* and enters the lake Ontario.

Noire, a cape or point of land on the *e.* coast of the island of Newfoundland, close to S. Francis.

[NOIX, Isle au, or Nut Isle, a small isle of 50 acres, near the *n.* end of lake Champlain, and within the province of Lower Canada. Here the British have a garrison containing 100 men. It is about five miles *n. n. e.* of the mouth of La Cole river, 20 *n.* of isle La Motte, and 12 or 15 *s.* of St. John's.]

[NOLACHUCKY, a river in the *e.* part of the state of Tennessee, which runs *w. s. w.* into French Broad river, about 26 miles from Holstein river. Near the banks of this river Greenville college is established.]

[NOLIN Creek, a branch of Green river in Kentucky. The land here is of an inferior quality.]

[NOMAN'S Land Island lies a little *s. w.* of Martha's Vineyard, and is about three miles long and two broad. It belongs to Duke's county, Massachusetts. Lat. 41° 14′ *n.* Long. 70° 45′ 30″ *w.*]

NOMBRE de Dios, a town of the province and bishopric of Guadalaxara in N. America. It is populous and rich from the abundance of the silver mines in its district. It has a very good parish church, besides a convent of the order of San Francisco. It has this name, because, when Pedro de Espinareda came to preach the gospel to these Indians, by order of St. Francis, he said, "Let us begin this work in the name of God;" and from this time this title was always given to the settlement, the which, from its concourse of inhabitants, was raised into a city. It is situate a little *n.* of the tropic of Cancer. One hundred and seventy miles *n.* of the city of Guadalaxara, in long. 103° 7′. Lat. 24°.

[Nombre de Dios is (according to Humboldt) in the intendancy of Durango, on the road from the famous mines of Sombrerete to Durango; and he states its population at 6800 souls.]

Nombre de Dios, another city, formerly in the province and kingdom of Tierra Firme, with a good port in the N. sea, discovered by Admiral

Christóval Colon at the same time as was that of Portobelo, and founded by Diego de Albitez in 1510. It is of bad temperature, moist, and rainy; for which reason, and also because the port of Portobelo was preferable, the city was removed to this last-mentioned place, by order of Philip II. in 1585, by Don Iñigo de la Mota, when the former city became reduced to a miserable village, its port being frequented by foreign vessels, which carried on a contraband trade. The English pirate Francis Drake sacked the city in 1598. The Admiral Don Francisco Cornejo had off the coast a combat with two Dutch frigates, in 1724, and the Count de Clavijo, who was commander of the vessels for guarding the coast, had also two other engagements in the following years of 1725 and 1726. The English admiral Hosier blockaded in this port, for a whole year, some galleons under the command of the General Don Blas de Leso, in 1538. It is five leagues from Portobelo.

Nombre de Dios, a settlement of the missions which are held by the religious order of San Francisco, in the province of Taraumara, and kingdom of Nueva Vizcaya; situate 12 leagues *w. n. w.* of the town and *real* of mines of San Felipe de Chiguagua.

Nombre de Dios, another city, founded in the strait of Magellan by Pedro Sarmiento, in 1582; but it had only existed three years when all its inhabitants perished of hunger, except Fernando Gomez, who was taken up by Thomas Cavendish, who passed that strait in 1587. Since that time the port has been called De Hambre or Famine.

Nombre de Dios, a river of the province and *corregimiento* of Arequipa in Peru; called also Tambapalla, since it traverses the valley of this name. It runs *w.* and enters the sea opposite the island of Chile.

NOMSCOT, a small river of the province of New Hampshire in N. America; one of New England. It rises from a lake, runs *s.* forming in its course various other lakes, and enters the Amariscoggin.

[NONESUCH, a river of Cumberland county, district of Maine. It passes to the sea through the town of Scarborough; and receives its name from its extraordinary freshets.]

[Nonesuch, a harbour at the *e.* end of the island of Antigua. The road is foul and full of rocks; and it has not more than six or eight feet water, except in one place, which is very difficult.]

NONET, a port of the *s.* coast of the island of S. Domingo, and *w.* head, in the part of the French, between points Cascajo and Abacú.

NONO, a settlement of the province and government of Esmeraldas in the kingdom of Quito.

NONURA, a small island of the S. sea, near the coast of the province and *corregimiento* of Piura in Perú, to the *n.* of that of Lobos. It is barren and uninhabited, in lat. 5° 48′ *s.*

Nonura, a point of land on the same coast and province.

NOODLE, a small island of the N. sea, in Boston bay.

[NOOHEVA, one of the Ingraham or Marquesas islands, said to be the parent of them all; situate about 10 leagues *s. w.* of Ooahoona. Capt. Roberts named it Adams; it is the same which Ingraham called Federal island. The lat. of the body of the island is 8° 48′ *s.* and nearly in the same meridian with Wooapo, between 139° 53′ and 140° 4′ *w.* long. from Greenwich. All accounts of the natives concurred, says Captain Roberts, in representing it as populous and fruitful, and to have a large bay with good anchorage.]

[NOORT Point, on the coast of Chile, is the *n.* point of the bay or port of Coquimbo, the other is called point Tortugas.]

[NOOTKA or King George's Sound, on the *n. w.* coast of N. America, is very extensive. That part of it where the ships under Capt. Cook anchored, lies in lat. 49° 36′ *n.* and long. 126° 42′ *w.* from Greenwich. Capt. Cook judged the sound to occupy a degree and a half in latitude, and two of longitude, exclusive of its arms and branches unexplored. The whole seuml is surrounded by high land, in many places broken and rugged, and in general covered with wood to the very top. The natives were very numerous, and were in possession of iron and beads; which probably were conveyed to them across the continent from Hudson's bay. They are rather below the middle size, and besmear their bodies with red paint, but their faces are bedaubed with various colours.

Notwithstanding the accurate information which we owe to the English and French navigators, it would still be interesting to publish the observations of M. Moziño on the manners of the Indians of Nootka. These observations embrace a great number of curious subjects, viz. the union of the civil and ecclesiastical power in the person of the princes or tays; the struggle between Quautz and Matlox, the good and bad principle by which the world is governed; the origin of the human species at an epocha when stags were without horns, birds without wings, and dogs without tails; the

Eve of the Nootkians, who lived solitary in a flowery grove of Yucuatl, when the god Qunutz visited her in a fine copper canoe; the education of the first man, who, as he grew up, past from one small shell to a greater; the genealogy of the nobility of Nootka, who descend from the oldest son of the man brought up in a shell, while the rest of the people (who even in the other world have a separate paradise called Pinpulu) dare not trace their origin farther back than to younger branches; the calendar of the Nootkians, in which the year begins with the summer solstice, and is divided into 14 months of 20 days, and a great number of intercalated days added to the end of several months, &c. &c.

The strait De Fuego encompasses the large cluster of islands among which this sound is situated. See FUCA, PINTARD, WASHINGTON Islands, and NORTH-WEST Coast; also Index to new matter respecting MEXICO, Chap. XI.

All pretensions to this sound were abandoned by Spain in favour of the court of London, by a treaty signed at the Escurial on the 28th October 1790; and it was formally taken possession of by Lieutenant Pearce of the British navy, in 1795, in the name of his Britannic Majesty.]

NOPALLAN, a province of Nueva España in the time of the gentilism of the Indians; conquered by Moctheutzuma in the 12th year of his reign, and six before the entrance of the Spaniards.

NOPALUCA, a settlement and head settlement of the district of the *alcaldia mayor* of Tepeaca in Nueva España; situate on the top of a lofty and extensive plain. Is is of a cold and dry temperature, scanty of water, having no other than such as is preserved in two cisterns made for this purpose. It contains 10 families of Spaniards, 63 of *Mustees* and Mulattoes, and 176 of Mexican Indians. In its district are 17 estates, in the tillage of which the inhabitants are employed, as also in making saltpetre from a lake close to the settlement. It happened here in the year 1740, that the earth experienced a trembling shock for the space of three months unremittingly. Six leagues to the *n.* one-fourth to the *n. e.* of its capital.

NOPSA, a settlement of the province and *corregimiento* of Tunja in the Nuevo Reyno de Granada. It is of a cold temperature, abounding in wheat and other fruits of a cold climate. It is very dangerous to be out in the evening air on account of the vapours which exhale from the lakes with which the settlement is surrounded. In the district is an estate of the religious order of San Agustin, where there is a chapel, in which is venerated an image of Nuestra Señora de Belen. It contains more than 100 white and as many black inhabitants; also as many Indians. Seven leagues *n.* of Tunja.

NOQUETS, a river of New France or Canada in N. America. It runs *e.* and enters the bay of Puants of the lake Michigan.

NOQUETS, a bay on the *w.* coast of the lake Michigan, in the same province.

NOQUETS, a barbarous nation of Indians of Canada in N. America, dwelling by the gulf or bay of its name. It once was on the shore of the lake Superior, but established themselves in the former place when they had fled from a war in which they were almost all exterminated. It consists now of nothing but some dispersed families.

NORD, or NORTH, Islands of the, some islands of the *w.* coast of Cape Breton. They are two, and situate opposite to the *e.* point of St. John's island.

NORD, a cape or point of land, the *e.* extremity of the island S. John in Nova Scotia or Acadia.

NORD, another, on the *e.* coast of Hudson's bay.

NORD, another, the extremity of Cape Breton, which looks upon Newfoundland.

NORD, a small river of Virginia, which runs *n. e.* in the county of Albemarle.

[NORD, RIO DEL, or RIO BRAVO. See NORTH River, in the gulf of Mexico.]

NORD, another, of the same province, called the *n.* branch or arm.

NORDESTE, or NORTH-EAST, a point or extremity of the island of Jamaica, which looks upon S. Domingo, between Long bay and Cold bay.

NORFIELD, a city of the province of Massachusetts, one of those of New England, on the shore of the river Connecticut.

[NORFOLK, a populous maritime county of Massachusetts, lately taken from the *s.* part of Suffolk county, and lies to the *s.* around the town and harbour of Boston. It contains 20 townships, of which Dedham is the seat of justice. Number of inhabitants 24,280.]

[NORFOLK, a populous county of Virginia, bounded *n.* by James's river, which divides it from Warwick. It contains 14,524 inhabitants, including 5345 slaves.]

[NORFOLK, a port of entry and post-town and seat of justice in the above county, on the *e.* side of Elizabeth river, immediately below the confluence of the *e.* branch. It is the most considerable commercial town in Virginia. The channel of the river is from 350 to 400 yards wide, and at common flood tide has 18 feet water up to the town. The harbour is safe and commodious, and large enough to contain 300 ships. It was burnt on the 1st of January 1776, by the Liverpool man of war, by order of the British governor Lord Dunmore, and the loss amounted to 300,000*l.* sterling. It now contains about 500 dwelling-houses, a court-house, goal, an Episcopal and Methodist church, a theatre, and an academy. In 1790, it contained 2959 inhabitants, including 1294 slaves. The town is governed by a mayor and several aldermen. It carries on a brisk trade to the W. Indies, Europe, and the different states, and constitutes, with Portsmouth, which stands on the opposite side of the river, a port of entry. The exports for one year, ending Sept. 30th, 1794, amounted to 1,660,752 dollars. A canal of 16 miles in length is now cutting from the *n.* branch of Albemarle sound in N. Carolina, to the waters of the *s.* branch of Elizabeth river. It will communicate with Elizabeth river nine miles from Norfolk. Merchant vessels of the largest size may go within a mile from the mouth of the canal; and here, the water being fresh, the worm, which does such damage to vessels in Norfolk and Portsmouth, will not affect them. It is 74 miles *e. s. e.* of Richmond, 29 from Williamsburgh, 13 *e.* of Suffolk, and 195 *s.* by *w.* of Philadelphia. Lat. 36° 55′ *n.* Long. 76° 23′ *w.*]

[Norfolk, a township in Litchfield county, Connecticut, 15 miles *n.* of Litchfield, on the Massachusetts line.]

NORI, a large, fertile, and beautiful valley of the government of Antioquia in the Nuevo Reyno de Granada, between the rivers Cauca and Tonusco.

NORIA, a settlement of the province and government of Tucumán in Peru; situate on the shore of the river Dulce.

[NORMAN Cape, on the *w.* coast of Newfoundland island, is on the gulf of St. Lawrence, and the *w.* entrance of the narrow bay of Manco, 20 leagues from cape Ferrol. Lat. 51° 39′ *n.* Long. 55° 58′ *w.* High water at full and change days at nine o'clock.]

NORONA, an island of the N. sea, opposite the Brazil coast, discovered in 1517, by Fernando Noroña, a Portuguese, who gave it this name. It is two leagues long, and has two very good ports, one to the *n.* defended by three forts, and another to the *n. w.* defended by two, the one of which is in a lofty and inaccessible spot. The Portuguese abandoned this island as useless and barren, and it was taken possession of by the French company of the W. Indies, but was recovered by the former, who fortified it. It produces nothing, and the food is brought from Pernambuco. The coasts are full of rocks and shoals which render its access difficult. The principal settlement is San Pablo, distant a mile and a half from the sea, being the residence of the Portuguese governor, before that it was ceded by these to the Spaniards, its present possessors. The fort of Los Remedios is the best. The island is 70 leagues *e.* of the coast, in lat. 38° 31′.

NOROSI, a settlement of the province and government of Santa Marta in the Nuevo Reyno de Granada; situate on the shore of the cape of La Loba, where the river Grande de la Magdalena communicates with the Colorado.

[NORRIDGEWALK, or Norridgewock, a post-town in Lincoln county, on Kennebeck river, Maine, incorporated in 1788, and contains 376 inhabitants. It is 12 miles *w.* of Canaan, and 160 *n.* by *e.* of Boston. The Indian town of this name stood about 40 miles above fort Halifax, where Kennebeck river, as you ascend it, after taking a *s. w.* course, turns to the *n.* and forms a point where the town stood. It was destroyed by a party under Colonal Harman, in 1724.]

[NORRITON, the principal town in Montgomery county, Pennsylvania, is about 15 miles *n. w.* of Philadelphia, on the *n.* bank of the Schuylkill, having about 20 houses, a court-house and gaol, and a handsome edifice of stone for the preservation of records, and an observatory.

This town was the residence of that celebrated philosopher and philantrophist, Dr. David Rittenhouse. In his observatory, near his mansion-house, he was interred, agreeably to his request, June 1796. His tomb-stone contains nothing but his name and the simple record of the days and years of his birth and death. "Here," says the elegant writer of his eulogy, Dr. Rush, "shall the philosophers of future ages resort to do homage to his tomb, and children yet unborn shall point to the dome which covers it, and exultingly say, 'There lies our Rittenhouse.'"]

NORTE, a large and abundant river of Nuevo Mexico, the last boundary of the known lands or countries of N. America, being also called the river Colorado. It runs towards the *s.* and enters the sea at the *n.* end of the gulf of California or Mar Roxo de Cortes, in lat. 32° 35′ *n.*

[For further account of this river, see Kingdom of Mexico, Nuevo, (new matter).]

Norte, a bay of the island of S. Domingo, in the *e.* rhumb and at the cape of Samaná, between the river Limones and port Gozier.

Norte, a settlement and garrison of the province of Taraumara and kingdom of Nueva Vizcaya in N. America, where there is a captain and sufficient number of troops to restrain the incursions of the infidel Indians.

Norte, a cape or point of land on the coast of the province and country of Las Amazonas, one of those which form the mouth or entrance of the river Marañon or Amazon, and that which looks to the *n.* from whence it is thus called. Lat. 1° 49′ 30″ *n.* Long. 49° 48′ *w.*

Norte, another, of the island Margarita; and it is one of those extremities which form the bay here.

Norte, a port of the island in which is the above point.

NORTH, a small river of the province of Massachusetts, which runs *e.* and enters the sea close to cape Cod.

North, another, also a small river in the district of Carteret in S. Carolina. It runs *s.* and enters the sea.

[North America comprehends all that part of the continent of America which lies *n.* of the isthmus of Darien, extending *n.* and *s.* from about the 9° of *n.* lat. to the *n.* pole, and *e.* and *w.* from the Atlantic to the Pacific ocean, between the 52° and 168° of *w.* long. from Greenwich. Beyond the 70° *n.* lat. few discoveries have been made. North America was discovered in 1495, in the reign of Henry VII. by John Cabot, a Venetian, and was then thickly inhabited by Indians. In July 1779, Captain Cook proceeded as far as lat. 71°, when he came to a solid body of ice from continent to continent. The vast tract of country, bounded *w.* by the Pacific ocean, *s.* and *e.* by California, New Mexico and Louisiana, the United States, Canada and the Atlantic ocean, and extending as far *n.* as the country is habitable, (a few scattered British, French, and some other European settlements excepted), is inhabited wholly by various nations and tribes of Indians. The Indians also possess large tracts of country within the Spanish, American, and British dominions. Those parts of N. America, not inhabited by Indians, belong (if we include Greenland) to Denmark, Great Britain, the American states, and Spain. Spain claims E. and part of W. Florida, and all *w.* of the Mississippi, and *s.* of the *n.* boundaries of Louisiana, New Mexico, and California. Great Britain claims all the country inhabited by Europeans, lying *n.* and *e.* of the United States, except Greenland, which belongs to Denmark. The remaining part is the territory of the Sixteen United States. The particular governments, provinces, and states of N. America, are exhibited in the table at the beginning of this work, vol. i.

On casting one's eyes upon the map of N. America, it is immediately perceived that the English still possess vast possessions on the continent, the most important parts of which are Canada and Nova Scotia. Masters of Canada, they command the navigation of the river St. Lawrence, from Montreal to Quebec. This river is navigable for large ships, which transport either to England or to the English Antilles, articles of the first necessity. The navigation of these rivers is protected by the maritime station of Halifax, considered as the capital of Nova Scotia. This is not the only advantage which the English derive from the possession of Canada; it affords them the means of eluding the non-intercourse laws passed by the American congress. From Montreal to fort St. Jean the distance by land is very inconsiderable, therefore this town has become an entrepot for English goods, which by lake Champlain are easily introduced into the states of Vermont and New York.

We proceed to insert *An alphabetical list of the mountains of N. America, a corresponding list of those of Spanish America being given under article* Mountains; and conclude the article, of which we treat, with *A summary account of the first discovery and settlement of N. America, arranged in chronological order. N. B. The discoveries respecting Spanish N. America will be found under article* Mexico, (*new matter, Chap.* XI.)

Alphabetical list of the mountains of N. America, a corresponding list of those of Spanish America being given under article Mountains.

Acha, sierra de	Blacklog mountains
Agamont hill	Bonabeag hills
Allegany or Apallachian mountains	Bostan, volcano de
	Bradeard mountains
Almagre, sierra de	Broad mountains
Amilpas, volcano de la	Brushy mountains
Ararat mountains	Burros, cerro de las
Bald mountains	Camaron, alta del
Baker mountains	Carcay, sierra de
Barigan, sierra de	Carieboef mountains
Battle hills	Carmilo, sierra del
Bearstooth hill	Chanate, sierra del
Beaver hills	Chesnut mountains
Blue mountains	Chigni, sierra de]

[Clinches mountains
Clara, cerro de
Cola del Aguila, sierra de la
Colima, volcano de
Coronel, cerro del
Cosinas, sierra de las
Cumberland mountains
Diablo, sierra de
Elias, St. mountain
Encomiendo, sierra de la
Enfado, sierra de
Evits mountains
Fairweather mountains
Flattop mountains
Florida, sierra de la
Gauley mountains
Genela, cerro de la
Grandfather's mountains
Grun mountains
Grullas, sierra de las
Guacaros, sierra de las
Guadalupe, sierra de
Guatemala, volcano de
Hart mountain
Hood mountain
Horn mountain
Iron mountains, Great
Jack's mountains
Jackson's mountains
Jere mountains
Jurillo or Juruyo, volcano de
Iztacibath, sierra de
Katskill mountain
King mountain
Laurel mountains
Long mountains
Lucerno, sierra del
Mahony mountains
Maiz, cerro del
Maraval, volcano de
Martinez, cerro de
Miguel, St. volcano de
Mimbres, sierra de los
Mixes, cerro de los
Montague mountains
Nevada, sierra
North mountains
Nunic mountains
Obscura, sierra de
Occonee mountains
Organos, sierra de los
Orizabo pico
Orosi, volcano de
Palma, sierra de
Papagayo, volcano de
Penobscot hills
Perpetua, cerro de
Peter's mountans
Piernas de Dona Maria, sierra de las
Pilares, sierra de
Plata, sierra de la
Popocateptl de la cordillera Inahuac
Powel's mountains
Rainier mountains
Rattlesnake mountains
Rocky mountains
Sacatuoluca, volcano de
Sacramento, sierra de
Saddle mountains
San Saba, montanas de
Sapanza, cerro de
Savage mountains
Scuttock hills
Sel Geme, montanas de
Sentualtepec pico
Sewel mountains
Shade mountains
Shavungunk mountains
Sideling mountains
Sincoque, cerro de
Slate mountains
Soconusco, volcan de
Sonsonate, volcano de
Sopotitlan, volcano de
South mountains
Stone mountains
Suchetepec, volcano
Tacon mountains
Tancitaro, pic de
Tecapa, volcano de,
Tenonco, volcano de
Timpingos, sierra de
Tlica, volcano de
Tuscarora mountains
Tussey mountains
Valle, cerro del
Varu, volcano de
Verde, sierra
Vergines, volcans de las
Viego, volcans del
Volcano mountain
Wambacho, volcano de
White mountains
White Oak mountains
Wills mountains
Yellow mountains

A summary account of the first discovery and settlement of N. America, arranged in chronological order. N. B. The discoveries respecting Spanish America will be found under article MEXICO, (*new matter*).

North America was discovered in the reign of Henry VII. a period when the arts and sciences had made very considerable progress in Europe. Many of the first adventurers were men of genius and learning, and were careful to preserve authentic records of such of their proceedings as would be interesting to posterity. These records afford ample documents for American historians. Perhaps no people on the globe can trace the history of their origin and progress with so much precision as the inhabitants of N. America; particularly that part of them who inhabit the territory of the United States.

The fame which Columbus had acquired by his first discoveries on this *w.* continent, spread through Europe, and inspired many with the spirit of enterprise. As early as 1495, four years only after the first discovery of America, John Cabot, a Venetian, obtained a grant or commission from Henry VII. to discover unknown lands and annex them to the crown. (See Hazard's Historical Collection, vol. i. p. 9, where this grant is recited at large. It is dated A. D. 1495.)

In the spring of 1496, he sailed from England with two ships, carrying with him his three sons. In this voyage, which was intended for China, he fell in with the *n.* side of Terra Labrador, and coasted *n.* as far as the 67° of latitude.

1497. The next year he made a second voyage to America with his son Sebastian, who afterwards proceeded in the discoveries which his father had begun. On the 24th of June he discovered Bonavista, on the *n. e.* side of Newfoundland. Before his return he traversed the coast from Davis's straits to cape Florida.

1502. Sebastian Cabot was this year at Newfoundland; and on his return, carried three of the natives of that island to King Henry VII.

1513. In the spring of 1513, John Ponce sailed from Porto Rico *n.*. and discovered the continent in lat. 30° 8′ *n.* He landed in April, a season when the country around was covered with verdure, and in full bloom. This circumstance induced him to call the country Florida, which, for many years, was the common name for N. and S. America.

1516. In 1516, Sir Sebastian Cabot and Sir Thomas Pert, explored the coast as far as Brazil in S. America.

This vast extent of country, the coast of which]

[was thus explored, remained unclaimed and unsettled by any European power (except by the Spaniards in S. America) for almost a century from the time of its discovery.

1524. It was not till the year 1524 that France attempted discoveries on the American coast. Stimulated by his enterprising neighbours, Francis I. who possessed a great and active mind, sent John Verazano, a Florentine, to America, for the purpose of making discoveries. He traversed the coast from lat. 28° to 50° *n.* In a second voyage, some time after, he was lost.

1525. The next year Stephen Gomez, the first Spaniard who came upon the American coast for discovery, sailed from Groyn in Spain, to Cuba and Florida, thence *n.* to cape Razo, in lat. 46° *n.* in search of a *n.* passage to the E. Indies.

1534. In the spring of 1534, by the direction of Francis I. a fleet was fitted out at St. Malo's in France, with design to make discoveries in America. The command of this fleet was given to James Cartier. He arrived at Newfoundland in May of this year; thence he sailed *n.*; and on the day of the festival of St. Lawrence, he found himself in about lat. 48° 30′ *n.* in the midst of a broad gulf, which he named St. Lawrence. He gave the same name to the river which empties into it. In this voyage he sailed as far *n.* as lat. 51°, expecting in vain to find a passage to China. (In Hazard's Historical Collections, vol. i. p. 19, is a commission from Francis I. to James Cartier or Quartier, for making an establishment in Canada, dated October 17, 1540. Probably this commission was given him in consequence of his former discoveries.)

1535. The next year he sailed up the river St. Lawrence 300 leagues, to the great and swift fall. He called the country New France; built a fort, in which he spent the winter, and returned in the following spring to France.

1539. On the 12th of May 1539, Ferdinand de Soto, with 900 men, besides seamen, sailed from Cuba, having for his object the conquest of Florida. On the 30th of May he arrived at Spirito Santo, from whence he travelled *n.* to the Chickasaw country, in about lat. 35° or 36°. He died and was buried on the bank of Mississippi river, May 1542, aged 42 years. Alverdo succeeded him.

1542. In 1542, Francis la Roche, Lord Robewell, was sent to Canada by the French king, with three ships and 200 men, women, and children. They wintered here in a fort which they had built, and returned in the spring. About the year 1550, a large number of adventurers sailed for Canada, but were never after heard of. In 1598, the king of France commissioned the Marquis de la Roche to conquer Canada, and other countries not possessed by any Christian prince. We do not learn, however, that La Roche ever attempted to execute his commission, or that any further attempts were made to settle Canada during this century.

January 6, 1548-49. This year King Henry VII. granted a pension for life to Sebastian Cabot, in consideration of the important services he had rendered to the kingdom by his discoveries in America. (See Hazard's Hist. Coll. vol. i. p. 23. Hackluyt calls this, "the large pension granted by King Edward VI. to Sebastian Cabot, constituting him grand pilot of England.") Very respectable descendants of the Cabot family now live in the commonwealth of Massachusetts.

1562. The admiral of France, Chatillon, early in this year, sent out a fleet under the command of John Ribalt. He arrived at cape Francis on the coast of Florida, near which, on the 1st of May, he discovered and entered a river, which he called May river. It is more than probable that this river is the same which we now call St. Mary's, which forms a part of the *s.* boundary of the United States. As he coasted *n.* he discovered eight other rivers, one of which he called Port Royal, and sailed up it several leagues. On one of the rivers he built a fort, and called it Charles, in which he left a colony under the direction of Captain Albert. The severity of Albert's measures excited a mutiny, in which, to the ruin of the colony, he was slain. Two years after, Chatillon sent Rene Laudonier with three ships to Florida. In June he arrived at the river May, on which he built a fort, and, in honour to his king, Charles IX. he called it Carolina.

In August, this year, Captain Ribalt arrived at Florida the second time, with a fleet of seven vessels, to recruit the colony, which, two years before, he had left under the direction of the unfortunate Captain Albert.

The September following, Pedro Melandes, with six Spanish ships, pursued Ribalt up the river on which he had settled, and overpowered him in numbers, cruelly massacred him and his whole company. Melandes, having in this way taken possession of the country, built three forts, and left them garrisoned with 1200 soldiers. Laudonier and his colony on May river, receiving information of the fate of Ribalt, took the alarm and escaped to France.

1567. A fleet of three ships was this year sent from France to Florida, under the command of Dominique de Gourges. The object of this ex-]

[pedition was to dispossess the Spaniards of that part of Florida which they had cruelly and unjustifiably seized three years before. He arrived on the coast of Florida, April 1568, and soon after made a successful attack upon the forts. The recent cruelty of Melandes and his company excited revenge in the breast of Gourges, and roused the unjustifiable principle of retaliation. He took the forts, put most of the Spaniards to the sword, and having burned and demolished all their fortresses, returned to France. During the 50 years next after this event, the French enterprised no settlements in America.

1576. Captain Frobisher was sent this year to find out a *n.w.* passage to the E. Indies. The first land which he made on the coast was a cape, which, in honour to the queen, he called Queen Elizabeth's Foreland. In coasting *n.* he discovered the straits which bear his name. He prosecented his search for a passage into the W. ocean, till he was prevented by the ice, and then returned to England. (Hazard's Hist. Coll. vol. i. p. 23.)

June 11th, 1578. In 1578, Sir Humphrey Gilbert obtained a patent from Queen Elizabeth, for lands not yet possessed by any Christian prince, provided he would take possession within six years. With this encouragement he sailed for America, and on the first of August 1583, anchored in Conception bay. Afterwards he discovered and took possession of St. John's harbour, and the country *s.* of it. In pursuing his discoveries he lost one of his ships on the shoals of Sablon, and on his return home, a storm overtook him, in which he was unfortunately lost, and the intended settlement was prevented.

1584. This year two patents were granted by Queen Elizabeth, one to Adrian Gilbert, (Feb. 6) the other to Sir Walter Raleigh, (March 25) for lands not possessed by any Christian prince. (Hazard's Hist. Coll. vol. i. p. 28 and 33.) By the direction of Sir Walter, two ships were fitted and sent out under the command of Philip Amidas and Arthur Barlow, with 107 passengers. In June 1585 they arrived on the coast, and anchored in a harbour seven leagues *w.* of the Roanoke. This colony returned to England in June 1586. On the 13th of July, they, in a formal manner, took possession of the country, and, in honour of their virgin queen, Elizabeth, they called it Virginia. Till this time the country was known by the general name of Florida. After this Virginia became the common name for all N. America.

1586. This year, Sir Walter Raleigh sent Sir Richard Greenville to America, with seven ships. He arrived at Wococon harbour in June. Having stationed a colony of more than 100 people at Roanoke, under the direction of Captain Ralph Lane, he coasted *n. e.* as far as Chesapeak bay, and returned to England.

The colony under Captain Lane endured extreme hardships, and must have perished, had not Sir Francis Drake fortunately returned to Virginia, and carried them to England, after having made several conquests for the queen in the W. Indies and other places.

A fortnight after, Sir Richard Greenville arrived with new recruits; and although he did not find the colony which he had before left, and knew not but they had perished, he had the rashness to leave 50 men at the same place.

1587. The year following, Sir Walter sent another company to Virginia, under Governor White, with a charter and 12 assistants. In July he arrived at Roanoke. Not one of the second company remained. He determined, however, to risk a third colony. Accordingly he left 115 people at the old settlement, and returned to England.

This year (Aug. 13) Manteo was baptized in Virginia. He was the first native Indian who received that ordinance in that part of America. He, with Towaye, another Indian, had visited England, and returned home to Virginia with the colony. On the 18th of August, Mrs. Dare was delivered of a daughter, whom she called Virginia. She was born at Roanoke, and was the first English child that was born in N. America.

1590. In the year 1590, Governor White came over to Virginia with supplies and recruits for his colony; but, to his great grief, not a man was to be found. They had all miserably famished with hunger, or were massacred by the Indians.

1602. In the spring of this year, Bartholomew Gosnold, with 32 persons, made a voyage to N. Virginia, and discovered and gave names to cape Cod, Martha's Vineyard, and Elizabeth islands, and to Dover cliff. Elizabeth island was the place which they fixed for their first settlement. But the courage of those who were to have tarried, failing, they all went on board and returned to England. All the attempts to settle this continent which were made by the Dutch, French, and English, from its discovery to the present time, a period of 110 years, proved ineffectual. The Spaniards only, of all the European nations, had been successful. There is no account of there having been one European family, at this time, in all the vast extent of coast from Florida to Greenland.

1603. Martin Pring and William Brown were this year sent by Sir Walter Raleigh, with two [small vessels, to make discoveries in N. Virginia. They came upon the coast, which was broken with a multitude of islands, in lat. 43° 30′ *n.* They coasted *s.* to Cape Cod bay; thence round the cape into a commodious harbour, in lat. 41° 25′, where they went ashore and tarried seven weeks, during which time they loaded one of their vessels with sassafras, and returned to England.

Bartholomew Gilbert, in a voyage to S. Virginia, in search of the third colony which had been left there by Governor White in 1587, having touched at several of the W. India islands, landed near Chesapeak bay, where, in a skirmish with the Indians, he and four of his men were unfortunately slain. The rest, without any further search for the colony, returned to England.

France, being at this time in a state of tranquillity, in consequence of the edict of Nantz in favour of the Protestants, passed by Henry IV. (April 1598), and of the peace with Philip, king of Spain and Portugal, was induced to pursne her discoveries in America. Accordingly the king signed a patent (see Hist. Coll. vol. i. p. 45) in favour of De Mons, (Nov. 8, 1603) of all the country from lat. 40° to 46° *n.* under the name of Acadia. The next year De Mons ranged the coast from St. Lawrence to cape Sable, and round to cape Cod.

1605. In May 1605, George's island and Pentecost harbour were discovered by Captain George Weymouth. In May he entered a large river in lat. 43° 20′, (variation 11° 15′ *w.*), which Mr. Prince, in his Chronology, supposes must have been Sagadahock; but from the latitude, it was more probably the Piscataqua. Captain Weymouth carried with him to England five of the natives.

1606. April 10th this year, James I. by patent, (see Hist. Coll. vol. i. p. 50) divided Virginia into two colonies. The *s.* included all lands between lat. 34° and 41° *n.* This was styled the first colony, under the name of S. Virginia, and was granted to the London company. The *n.* called the second colony, and known by the general name of N. Virginia, included all lands between lat. 38° and 45° *n.* and was granted to the Plymouth company. Each of these colonies had a council of 13 men to govern them. To prevent disputes about territory, the colony which should last place themselves was prohibited to plant within 100 miles of the other. There appears to be an inconsistency in these grants, as the lands lying between the 38° and 41° are covered by both patents.

Both the London and Plymouth companies enterprised settlements within the limits of their respective grants. With what success will now be mentioned.

Mr. Piercy, brother to the earl of Northumberland, in the service of the London company, went over with a colony to Virginia, and discovered Powhatan, now James river. In the mean time the Plymouth company sent Captain Henry Challons, in a vessel of 55 tons, to plant a colony in N. Virginia; but in his voyage he was taken by a Spanish fleet and carried to Spain.

1607. The London company this spring sent Captain Christopher Newport, with three vessels, to S. Virginia. On the 26th of April he entered Chesapeak bay, and landed, and soon after gave to the most *s.* point the name of Cape Henry, which it still retains. Having elected Mr. Edward Wingfield president for the year, they next day landed all their men, and began a settlement on James river, at a place which they called James Town. This is the first town that was settled by the English in N. America. The June following, Captain Newport sailed for England, leaving with the president 104 persons. In August died Captain Bartholemew Gosnold, the first projector of this settlement, and one of the council. The following winter James town was burnt.

During this time, the Plymouth company fitted out two ships under the command of Admiral Rawley Gilbert. They sailed for N. Virginia on the 31st of May, with 100 planters, and Captain George Popham for their president. They arrived in August, and settled about nine or 10 leagues to the *s.* of the mouth of Sagadahock river. A great part of the colony, however, disheartened by the severity of the winter, returned to England in December, leaving their president, Captain Popham, with only 45 men.

It was in the fall of this year that the famous Mr. Robinson, with part of his congregation, who afterwards settled at Plymouth in New England, removed from the *n.* of England to Holland, to avoid the cruelties of persecution, and for the sake of enjoying "purity of worship and liberty of conscience."

This year a small company of merchants at Dieppe and St. Malo's founded Quebec, or rather the colony which they sent built a few huts there, which did not take the form of a town until the reign of Lewis XIV.

1608. Sagadahock colony suffered incredible hardships after the departure of their friends in December. In the depth of winter, which was extremely cold, their store-house caught fire and was consumed, with most of their provisions and lodgings. Their misfortunes were increased soon] [after by the death of their president. Rawley Gilbert was appointed to succeed him.

Lord Chief Justice Popham made every exertion to keep this colony alive, by repeatedly sending them supplies. But the circumstance of his death, which happened this year, together with that of president Gilbert's being called to England to settle his affairs, broke up the colony, and they all returned with him to England.

The unfavourable reports which these first unfortunate adventurers propagated respecting the country, prevented any further attempts to settle N. Virginia for several years after.

1609. The London company, last year, sent Captain Nelson, with two ships and 120 persons, to James town; and this year, Captain John Smith, afterwards president, arrived on the coast of S. Virginia, and by sailing up a number of the rivers, discovered the interior country. In September, Captain Newport arrived with 70 persons, which increased the colony to 200 souls.

Mr. Robinson and his congregation, who had settled at Amsterdam, removed this year to Leyden; where they remained more than 11 years, till a part of them came over to New England.

The council for S. Virginia having resigned their old commission, (the second charter of Virginia bears date May 23, 1609. Hist. Coll. vol. i. p. 58) requested and obtained a new one; in consequence of which they appointed Sir Thomas West, Lord de la War, general of the colony; Sir Thomas Gates, his lieutenant; Sir George Somers, admiral; Sir Thomas Dale, high-marshal; Sir Ferdinand Wainman, general of the horse; and Captain Newport, vice-admiral.

June 8. In June, Sir Thomas Gates, Admiral Newport, and Sir George Somers, with seven ships, a ketch and a pinnace, having 500 souls on board, men, women, and children, sailed from Falmouth for S. Virginia. In crossing the Bahama gulf, on the 24th of July, the fleet was overtaken by a violent storm, and separated. Four days after, Sir George Somers ran his vessel ashore on one of the Bermuda islands, which, from this circumstance, have been called the Somer Islands. The people on board, 150 in number, all got safe on shore; and there remained until the following May. The remainder of the fleet arrived at Virginia in August. The colony was now increased to 500 men. Captain Smith, then president, a little before the arrival of the fleet, had been very badly burnt by means of some powder which had accidentally caught fire. This unfortunate circumstance, together with the opposition he met with from those who had lately arrived, induced him to leave the colony and return to England; which he accordingly did the last of September. Francis West, his successor in office, soon followed him, and George Piercy was elected president.

1610. The year following, the S. Virginia or London company sealed a patent to Lord de la War, constituting him governor and captain-general of S. Virginia. He soon after embarked for America with Captain Argal and 150 men in three ships.

The unfortunate people, who, the year before, had been shipwrecked on the Bermuda islands, had employed themselves during the winter and spring, under the direction of Sir Thomas Gates, Sir George Somers, and Admiral Newport, in building a sloop to transport themselves to the continent. They embarked for Virginia on the 10th of May, with about 150 persons on board; leaving two of their men behind, who chose to stay; and landed at James town on the 23d of the same month. Finding the colony, which at the time of Captain Smith's departure consisted of 500 souls, now reduced to 60, and those few in a distressed and wretched situation, they with one voice resolved to return to England; and for this purpose, on the 7th of June, the whole colony repaired on board their vessels, broke up the settlement, and sailed down the river on their way to their native country.

Fortunately, Lord de la War, who had embarked for James town the March before, met them the day after they sailed, and persuaded them to return with him to James town, where they arrived and landed the 10th of June. The government of the colony of right devolved upon Lord de la War. From this time we may date the effectual settlement of Virginia. Its history from this period will be given in its proper place.

As early as the year 1607 and 1608, Henry Hudson, an Englishman, under a commission from King James, in the employ of certain merchants, made several voyages for the discovery of a *n. w.* passage to the E. Indies. In 1609, upon some misunderstanding, he engaged in the Dutch service, in the prosecution of the same design, and on his return ranged along the sea-coast of what has since been called New England, (which three years before was granted by King James to his English subjects, the Plymouth company) and entered Hudson's river, giving it his own name. He ascended this river in his boat as far as what has since been called Aurania or Albany. In 1613, the Dutch W. India company sent some persons to this river, to trade with the Indians; and as early as 1623, the Dutch had a trading] [house on Connecticut river. In consequence of these discoveries and settlements, the Dutch claimed all the country extending from cape Cod to cape Henlopen along the sea-coast, and as far back into the country as any of the rivers within those limits extend. But their claim has been disputed. This extensive country the Dutch called New Netherlands, and in 1614 the states-general granted a patent to sundry merchants for an exclusive trade on Hudson's river, who the same year (1614) built a fort on the *w.* side near Albany. From this time we may date the settlement of New York, the history of which will be annexed to a description of the State.

Conception Bay, on the island of Newfoundland, was settled in the year 1610, by about 40 planters under Governor John Guy, to whom King James had given a patent of incorporation.

Chaplain, a Frenchman, had begun a settlement at Quebec 1608. St. Croix, Mount Mansel, and port Royal were settled about the same time. These settlements remained undisturbed till 1613, when the Virginians, hearing that the French had settled within their limits, sent Captain Argal to dislodge them. For this purpose he sailed to Sagadahock, took their forts at Mount Mansel, St. Croix, and port Royal, with their vessels, ordnance, cattle and provisions, and carried them to James town in Virginia. Quebec was left in possession of the French.

1614. This year Captain John Smith, with two ships and 45 men and boys, made a voyage to N. Virginia, to make experiments upon a gold and copper mine. His orders were, to fish and trade with the natives, if he should fail in his expectations with regard to the mine. To facilitate this business, he took with him Tantum, an Indian, perhaps one that Captain Weymouth carried to England in 1605. In April he reached the island Monahigan, in lat. 43° 30′. Here Captain Smith was directed to stay and keep possession with ten men, for the purpose of making a trial of the whaling business, but being disappointed in this, he built seven boats, in which 37 men made a very successful fishing voyage. In the mean time the captain himself with eight men only, in a small boat, coasted from Penobscot to Sagadahock, Acocisco, Passataquack, Tragabizanda, now called cape Ann, thence to Acomac, where he skirmished with some Indians; thence to cape Cod, where he set his Indian, Tantum, ashore, and left him, and returned to Monahigan. In this voyage he found two French ships in the bay of Massachusetts, who had come there six weeks before, and during that time had been trading very advantageously with the Indians. It was conjectured that there were, at this time, 3000 Indians upon the Massachusetts islands.

In July, Captain Smith embarked for England in one of the vessels, leaving the other under the command of Captain Thomas Hunt, to equip for a voyage to Spain. After Captain Smith's departure, Hunt perfidiously allured 20 Indians (one of whom was Squanto, afterwards so serviceable to the English) to come on board his ship at Patuxit, and seven more at Nausit, and carried them to the island of Malaga, where he sold them for 20*l.* each, to be slaves for life. This conduct, which fixes an indelible stigma upon the character of Hunt, excited in the breasts of the Indians such an inveterate hatred of the English, as that, for many years after, all commercial intercourse with them was rendered exceedingly dangerous.

Captain Smith arrived at London the last of August, where he drew a map of the country, and called it New England. From this time N. Virginia assumed the name of New England, and the name Virginia was confined to the southern colony.

Between the years 1614 and 1620, several attempts were made by the Plymouth company to settle New England, but by various means they were all rendered ineffectual. During this time, however, an advantageous trade was carried on with the natives.

1617. In the year 1617, Mr. Robinson and his congregation, influenced by several weighty reasons, meditated a removal to America. Various difficulties intervened to prevent the success of their designs until the year 1620, when a part of Mr. Robinson's congregation came over and settled at Plymouth. At this time commenced the settlement in New England.

The particulars relating to the first emigration to this *n.* part of America, the progress of its settlement, &c. will be given in the history of New England, to which the reader is referred.

In order to preserve the chronological order in which the several colonies, now grown into independent states, were first settled, it will be necessary that we should just mention, that the next year (1621) after the settlement of Plymouth, Captain John Mason obtained of the Plymouth council a grant of a part of the present state of New Hampshire. Two years after (1623), under the authority of this grant, a small colony fixed down near the mouth of Piscataqua river. From this period we may date the settlement of New Hampshire.

1627. In 1627, a colony of Swedes and Finns came over and landed at cape Henlopen; and af-]

[terwards purchased of the Indians the land from cape Henlopen to the falls of Delaware, on both sides the river, which they called New Swedeland Stream. On this river they built several forts, and made settlements.

1628. On the 19th of March 1628, the council for New England sold to Sir Henry Roswell, and five others, a large tract of land lying round Massachusetts bay. The June following, Captain John Endicot, with his wife and company, came over and settled at Naumkeag, now called Salem. "Among others who arrived at Naumkeag, were Ralph Sprague, with his brethren, Richard and William, who, with three or four more, by Governor Endicot's consent, undertook a journey through the woods above 12 miles westward, till they came to a neck of land called Mishawum, between Mystic and Charles rivers, full of Indians, named Aberginians. Their old sachem being dead, his eldest son, called by the English John Sagamore, was chief; a man of gentle and good disposition, by whose free consent they settled here; where they found but one English house thatched and pallisadoed, possessed by Thomas Walford, a smith." (Prince's Chronicle, p. 174.)

"June 1629, Mr. Thomas Graves removed from Salem to Mishawum, and with the governor's consent called it Charlestown. He laid the town out in two-acre lots, and built the great house, which afterwards became the house of public worship. Mr. Bright, minister." (Ibid. p. 188.)

Naumkeag then was the first English settlement which was made in Massachusetts bay. Plymouth, indeed, which is now included in the commonwealth of Massachusetts, was settled eight years before, but at this time it was a separate colony, under a distinct government; and continued so until the second charter of Massachusetts was granted by William and Mary in 1691; by which Plymouth, the province of Main and Sagadahock, were annexed to Massachusetts.

June 13, 1633. In the reign of Charles I. Lord Baltimore, a Roman Catholic, applied for, and obtained a grant of, a tract of land upon Chesapeak bay, about 140 miles long and 130 broad. Soon after this, in consequence of the rigour of the laws of England against the Roman Catholics, Lord Baltimore, with a number of his persecuted brethren, came over and settled it, and in honour of Queen Henrietta Maria, they called it Maryland.

The first grant of Connecticut was made by Robert Earl of Warwick, president of the council of Plymouth, to Lord Say and Seal, to Lord Brook and others, in the year 1631. (Hazard's Hist. Coll. p. 318.) In consequence of several smaller grants made after by the patentees to particular persons, Mr. Fenwick, in 1635, made a settlement at the mouth of Connecticut river, and called it Saybrook. About the same time (1636) a number of people from Massachusetts bay came and began settlements at Hartford, Wethersfield, and Windsor, on Connecticut river. Thus commenced the English settlement of Connecticut.

Rhode island was first settled in consequence of religious persecution. Mr. Roger Williams, who was among those who came early over to Massachusetts, not agreeing with some of his brethren in sentiment, was very unjustifiably banished the colony, and went with 12 others, his adherents, and settled at Providence in 1635. From this beginning arose the colony, now state of Rhode Island.

1664. On the 20th of March 1664, Charles II. granted to the Duke of York what is now called New Jersey, then a part of a large tract of country, by the name of New Netherlands. Some parts of New Jersey were settled by the Dutch as early as about 1615.

1669. In the year 1662, Charles II. granted to Edward Earl of Clarendon, and seven others, almost the whole territory of the three *s.* states, N. and S. Carolina and Georgia. Two years after he granted a second charter, enlarging their boundaries. The proprietors, by virtue of authority vested in them by their charter, engaged Mr. Locke to frame a system of laws for the government of their intended colony. Notwithstanding these preparations, no effectual settlement was made until the year 1669, (though one was attempted in 1667) when Governor Sayle came over with a colony and fixed on a neck of land between Ashley and Cooper rivers. Thus commenced the settlement of Carolina, which then included the whole territory between lat. 29° and 36° 30′ *n.* together with the Bahama islands, lying between lat. 22° and 27° *n.*

1681. The royal charter for Pennsylvania was granted to William Penn on the 4th of March 1681. The first colony came over the next year and settled under the proprietor, William Penn, who acted as governor from October 1682 to August 1684. The first assembly in the province of Pennsylvania was held at Chester, on the 4th of December 1682. Thus William Penn, a Quaker, justly celebrated as a great and good man, had the honour of laying the foundation of the present populous and very flourishing state of Pennsylvania.

The proprietary government in Carolina was attended with so many inconveniencies, and occa-]

[sioned such violent dissensions among the settlers, that the parliament of Great Britain was induced to take the province under their immediate care. The proprietors (except Lord Granville) accepted of 22,500*l*. sterling from the crown for the property and jurisdiction. This agreement was ratified by act of parliament in 1729. A clause in this act reserved to Lord Granville his eighth share of the property and arrears of quit-rents, which continued legally vested in his family till the revolution in 1676. Lord Granville's share made a part of the present state of N. Carolina. About the year 1729, the extensive territory belonging to the proprietors was divided into N. and S. Carolina. They remained separate royal governments until they became independent states.

For the relief of poor indigent people of Great Britain and Ireland, and for the security of Carolina, a project was formed for planting a colony between the rivers Savannah and Alatamaha. Accordingly, application being made to King George II. he issued letters patent, bearing date June 9th, 1732, for legally carrying into execution the benevolent plan. In honour of the king, who greatly encouraged the plan, they called the new province Georgia. Twenty-one trustees were appointed to conduct the affairs relating to the settlement of the province. The November following, 115 persons, one of whom was General Oglethorp, embarked for Georgia, where they arrived; and landed at Yamacraw. In exploring the country, they found an elevated pleasant spot of ground on the bank of a navigable river, upon which they marked out a town, and from the Indian name of the river which passed by it, called it Savannah. From this period we may date the settlement of Georgia.

The country now called Kentucky was well known to the Indian traders many years before its settlement. They gave a description of it to Lewis Evans, who published his first map of it as early as the year 1752. James Macbride, with some others, explored this country in 1754. Colonel Daniel Boon visited it in 1769.

1773. Four years after, Colonel Boon and his family, with five other families, who were joined by 40 men from Powle's valley, began the settlement of Kentucky, which is now one of the most growing colonies, perhaps, in the world, and was erected into an independent state, by act of congress, December 6th, 1790, and received into the union, June 1st, 1792. The settlement of Kentucky was made in violation of the treaty, in 1768, at fort Stanwix, which expressly stipulates, that this tract of country should be reserved for the *w.* nations to hunt upon, until they and the crown of England should otherwise agree. This has been one great cause of the enmity of those Indian nations to the Virginians.

The tract of country called Vermont, before the late war, was claimed both by New York and New Hampshire. When hostilities commenced between Great Britain and her colonies, the inhabitants considering themselves as in a state of nature, as to civil government, and not within any legal jurisdiction, associated and formed for themselves a constitution of government. Under this constitution, they have ever since continued to exercise all the powers of an independent state. Vermont was not admitted into union with the other states till March 4th, 1791; yet we may venture to date her political existence as a separate government from the year 1777, because, since that time, Vermont has to all intents and purposes been a sovereign and independent state. The first settlement in this state was made at Bennington as early as about 1764.

The extensive tract of country lying *n. w.* of the Ohio river, within the limits of the United States, was erected into a separate temporary government, by an ordinance of congress passed the 13th of July 1787.

Thus we have given a summary view of the first discoveries and progressive settlement of N America in a chronological order.

The following recapitulation will comprehend the whole in one view.

Names of places.	*When settled.*	*By whom.*
Quebec, ...	1608	By the French.
Virginia, ...	June 10, 1610	By Lord de la War.
Newfoundland, ...	June 1610	By Governor John Guy.
New York, } New Jersey, }	about 1614	By the Dutch.
Plymouth, ...	1620	By part of Mr. Robinson's congregation.
New Hampshire, ...	1623	By a small English colony near the mouth of Piscataqua river.]

[Names of places.	When settled.	By whom.
Delaware, Pennsylvania,	1627	By the Swedes and Finns
Massachusetts bay,	1628	By Captain John Endicot and company.
Maryland,	1633	By Lord Baltimore, with a colony of Roman Catholics.
Connecticut,	1635	By Mr. Fenwick, at Saybrook, near the mouth of Connecticut river.
Rhode island,	1635	By Mr. Roger Williams and his persecuted brethren.
New Jersey,	1664	Granted to the Duke of York by Charles II. and made a distinct government, and settled some time before this by the English.
South Carolina,	1669	By Goveroor Sayle.
Pennsylvania,	1682	By William Penn, with a colony of Quakers.
North Carolina,	about 1728	Erected into a separate government, settled before by the English.
Georgia,	1732	By General Oglethorp.
Kentucky,	1773	By Colonel Daniel Boon.
Vermont,	about 1764	By emigrants from Connecticut and other parts of New England.
Territory *n. w.* of Ohio river,	1787	By the Ohio and other companies

The above dates are from the periods when the first permanent settlements were made.

[NORTH CAROLINA, one of the United States, is bounded *n.* by Virginia, *e.* by the Atlantic ocean, *s.* by S. Carolina, and *w.* by the state of Tennessee. It lies between lat. 33° 50′ and 36° 30′ *n.* and between long. 76° and 83° 34′ *w.* being about 386 miles in length, and 160 in breadth, containing about 31,000 square miles. The districts of this state are classed in three divisions, viz. the *e.* districts, Edenton, Newbern, aud Wilmington; the middle districts, Fayetteville, Hillsborough, and Halifax; and the *w.* districts, Morgan and Salisbury. The *e.* districts are on the sea-coast, extending from the Virginia line *s.* to S. Carolina. The five others cover the whole state *w.* of the maritime districts; and the greater part of them extend across tho state from *n.* to *s.* These districts are subdivided into 58 counties, which contained, in 1790, 393,751 inhabitants, of whom 100,571 were slaves; and by the census of 1810, the total population of the state amounted ty 563,526 souls. The chief rivers of N. Carolina are Chowan and its branches, Roanoke, Tar, Neus, and Cape Fear or Clarendon. Most of these and the smaller rivers have bars at their mouths; and the coast furnishes no good harbours except Cape Fear. There are two remarkable swamps in this state, the one in Curritück county, the other on the line between this state and Virginia. See CURRITUCK County and DISMAL. The most remarkable sounds are Albemarle, Pamlico, and Core sounds; the capes, Lookout, Hatteras, and Fear, which are described under their respective names. Newbern is the largest town in the state; the other towns of note are Edenton, Wilmington, Halifax, Hillsborough, Salisbury, and Fayetteville; each of which have been, in their turns, the seat of the general assembly. Raleigh, situated near the centre of the state, has lately been established as the metropolis. N. Carolina, in its whole width, for 60 miles from the sea, is a dead level.

A great proportion of this tract lies in forest, and is barren. On the banks of some of the rivers, particularly of the Roanoke, the land is fertile and good. Interspersed through the other parts are glades of rich swamp, and ridges of oak-land, of a black, fertile soil. Sixty or 80 miles from the sea the country rises into hills and mountains, as in S. Carolina and Georgia. Wheat, rye, barley, oats, and flax, grow well in the back hilly country; Indian corn and pulse of all kinds, in all parts. Cotton and hemp are also considerably cultivated here, and might be raised in much greater plenty. The cotton is planted yearly; the stalk dies with the frost. The labour of one man will produce 1000 pounds in the seeds, or 250 fit for manufacturing. A great proportion of the produce of the back country, consisting of tobacco, wheat, Indian corn, &c. is carried to market in S. Carolina and Virginia. The *s.* in-]

[terior counties carry their produce to Charlestown, and the *n.* to Petersburgh in Virginia. The exports from the lower parts of the state are tar, pitch, turpentine, rosin, Indian corn, boards, scantling, staves, shingles, furs, tobacco, pork, lard, tallow, bees-wax, myrtle-wax, and a few other articles, amounting in the year ending September 30, 1791, to 524,548 dollars. Their trade is chiefly with the W. Indies and the *n.* states.

In the flat country near the sea-coast the inhabitants, during the summer and autumn, are subject to intermitting fevers, which often prove fatal, as bilious or nervous symptoms prevail. The *w.* hilly parts of the state are as healthy as any part of America. This country is fertile, full of springs and rivulets of pure water. Autumn is very pleasant, both in regard to the temperature and serenity of the weather, and the richness and variety of the vegetable productions which the season affords. The winters are so mild in some years that autumn may be said to continue till spring. Wheat harvest is in the beginning of June, and that of Indian corn early in September.

The large natural growth of the plains, in the low country, is almost universally pitch-pine, which is a tall handsome tree, far superior to the pitch-pine of the *n.* states. This tree may be called the staple commodity of N. Carolina. It affords pitch, tar, turpentine, and various kinds of lumber, which, together, constitute at least one-half of the exports of this state. No country produces finer white and red oak for staves. The swamps abound with cypress and bay trees. The latter is an evergreen, and is food for the cattle in winter. The misletoe is common in the back country. This is a shrub which differs in kind, perhaps, from all others. It never grows out of the earth, but on the tops of trees. The roots (if they may be so called) run under the bark of the tree, and incorporate with the wood. It is an evergreen resembling the garden box-wood.

The late war, by which N. Carolina was greatly injured, put a stop to several iron-works. There are four or five furnaces in the state that are in blast, and a proportionable number of forges.

The *w.* parts of this state, which have been settled within the last 50 years, are chiefly inhabited by Presbyterians from Pennsylvania, the descendants of people from the *n.* of Ireland, and are exceedingly attached to the doctrines, discipline, and usages of the church of Scotland. They are a regular industrious people. The Moravians have several flourishing settlements in the upper part of this state. The Friends or Quakers have a settlement in New-garden, in Guilford county, and several congregations at Pequimins and Pasquotank. The Methodists and Baptists are numerous and increasing. The general assembly of N. Carolina, in December 1789, passed a law incorporating 40 gentlemen, five from each district, as trustees of the universty of N. Carolina. The state has given handsome donations for the endowment of this seminary. The general assembly, in December 1791, loaned 5000*l.* to the trustees, to enable them to proceed immediately with their buildings. There is a very good academy at Warenton, another at Williamsborourgh in Granville, and three or four others in the state, of considerable note.

N. Carolina has had a rapid growth. In the year 1710, it contained but about 1200 fencible men. In 1794, the number was estimated at about 50,000. It is now, in point of numbers, the fourth state in the union. By the constitution of this state, which was ratified in December 1796, all legislative authority is vested in two distinct branches, both dependent on the people, viz. a senate and house of commons, which, when convened for business, are styled the general assembly. The senate is composed of representatives, one from each county, chosen annually by ballot. The house of commons consists of representatives chosen in the same way, two for each county, and one for each of the towns of Edenton, Newbern, Wilmington, Salisbury, Hillsborough, Halifax, and Fayetteville.

We have in vain endeavoured to seek for more early and authentic information concerning the first settlement of this state than that given by Alcedo, under the article Carolina. It should appear by that authority, that the country was discovered as far back as 1512, by Ponce de Leon, a Spaniard. It is certain, however, that the history of N. Carolina is less known than that of any other of the states. From the best recent accounts that history affords, the first permanent settlement in N. Carolina was made about the year 1710, by a number of Palatines from Germany, who had been reduced to circumstances of great indigence by a calamitous war. The infant colony remained under the general government of S. Carolina, till about the year 1729, when seven of the proprietors, for a valuable consideration, vested their property and jurisdiction in the crown; and the colony was erected into a separate province, by the name of N. Carolina, and its present limits established by an order of George II.]

[NORTH CASTLE, a township of New York, in W. Chester county, *n.* of Mount Pleasant, and the White plains on the borders of Connecticut.

In 1790, it contained 2478 inhabitants. In 1796, there were 173 of the inhabitants qualified electors. It is 10 miles from White plains, and 20 from Ridgefield in Connecticut.]

[NORTH-EAST, a small river which empties in at the head of Chesapeak bay, about five miles below Charlestown; only noticeable for the quantity of herrings caught in it.]

[NORTH-EAST Town, a township in Dutchess county, New York, about 90 miles *n.* of New York city, between Rhynbec and Connecticut *w.* line. In 1790, it contained 3401 inhabitants. In 1796, there were in it 391 qualified electors.]

[NORTH EDISTO Inlet, on the coast of S. Carolina, is 11 miles from Stono Inlet, and three *e. n. e.* from S. Edisto.]

[NORTH HAMPTON, a township of New Hampshire, in Rockingham county, which contains 657 inhabitants, taken from Hampton and incorporated in 1742.]

[NORTH HAVEN, a township of Connecticut, situated in New Haven county, on the *e.* side of E. river, eight miles *n.* by *e.* of New Haven, and 32 *s.* by *w.* of Hartford. It was settled in 1660 by 35 men, principally from Saybrook. This town is the birth-place of that learned, pious, and excellent man, Dr. Ezra Stiles, late president of Yale college.]

[NORTH HEMPSTEAD, a township in Queen's county, Long island, New York, bounded *e.* by Oyster bay, *n.* by the sound, and *s.* by S. Hempstead. In 1790, it contained 2696 inhabitants, of whom 507 were slaves. In 1796, 232 of the inhabitants were qualified electors. The soil is but indifferent.]

[NORTH HUNTINGTON, a township in Westmoreland county, Pennsylvania.]

[NORTH Island, on the coast of S. Carolina, lies on the *n.* side of Winyah harbour.]

[NORTH KINGSTOWN, a town in Washington county, Rhode Island, which carries on a considerable trade in the fisheries, besides some to the W. Indies. Its harbour is called Wickford, on the *w.* side of Narraganset bay, opposite the *n.* end of Connecticut island. It is about 10 miles *n. w.* of Newport, and 16 *s.* of Providence. The township contains 2907 inhabitants.]

[NORTH Mountain, one of the ridges of the Alleghany mountains, which extends through Virginia and Pennsylvania. There is a curious syphon fountain in Virginia, near the intersection of Lord Fairfax's boundary with the N. mountain, not far from Brock's gap, on the stream of which is a grist-mill, which grinds two bushels of grain at every flood of the spring.]

[NORTH Reef, off the island of St. Domingo, in the W. Indies, lies in lat. 20° 33′ *n.* and long. 69° 12′ *w.*]

[NORTH River, in New York. See HUDSON's River.]

[NORTH River, in Massachusetts, for its size, is remarkable for its depth of water, being in some places not more than 40 or 50 feet wide, yet vessels of 300 tons are built at Pembroke, and descend to Massachusetts bay, 18 miles distant, as the river runs. It rises in Indian Head pond in Pembroke, and runs a serpentine course between Scituate and Marshfield. The river is navigable for boats to the first fall, five miles from its source. Thence to the nearest waters which run into Taunton river, is only three miles. A canal to connect the waters of these two rivers, which communicate with Narraganset and Massachusetts bays, would be of great utility, as it would save a long and dangerous navigation round cape Cod.]

[NORTH River, a very considerable river of New Mexico in N. America, which rises in the *n.* part of it, and directs its course to the *s. e.* and empties into the gulf of Mexico, at the *w.* end, in and about lat. 26° 12′ *n.*]

[NORTH River, a branch of Fluvanna river in Virginia. See COW AND CALF Pasture.]

[NORTH SALEM, a township in W. Chester county, New York, bounded *s.* by Salem, *e.* by Connecticut, *n.* by Dutchess county, and *w.* by the middle of Croton river. In 1790, it contained 1058 inhabitants, including 58 slaves. In 1796, 162 of the inhabitants were qualified electors.]

[NORTH Sea is a name that has been given by geographers to various parts of the oceans, where they happen to wash the *n.* parts of the American continent or islands. Thus, the gulf of Mexico, and the Atlantic ocean further to the *e.* from their waters washing the *n.* coast of Mexico or New Spain in N. America, and Tierra Firme in S. America, have been distinguished by this name. It has also been applied to the *s.* part of the gulf of Mexico, in particular by the Spaniards, on their crossing the isthmus of Darien from the *n.* to the *s.* coast, in opposition to the Pacific ocean, to which they gave the name of the S. Sea. The Atlantic ocean also on the *e.* coast of N. America has been sometimes called the N. Sea; which appellation has also been given to the Frozen ocean, from its bounding N. America on the *n.* See article MAR.]

[NORTH SOUND Point is the projecting point of land on the *n. e.* side of the island of Antigua in the W. Indies, and is about *s. s. e.* from Long island.]

[North-West Coast of America. The country on the *n. w.* part of the continent of America, lying on the Pacific ocean, is thus denominated. According to accounts given by voyagers to this coast, the vast country lying upon it, with very little deviation, has the appearance of one continued forest, being covered with pines of different species, and these intermixed with alder, birch, witch-hazel, &c. besides various kinds of brushwood; and the valleys and low grounds afford wild currants, gooseberries, raspberries, and various flowering shrubs. On the coast are many islands, spacious bays, commodious harbours, and mouths of navigable rivers: among the former are Washington or Queen Charlotte's islands, extending from lat. 51° 42′ to 54° 18′ *n.*—*w.* long. from Greenwich, 129° 54′ to 133° 18′. Here are Nootka sound, Admiralty bay, and port Mulgrave, Prince William's sound, Cook's river, the peninsula of Alaska, and the islands surrounding it, Bristol bay, and Norton sound; which last lie *s. e.* of Behring's straits.

The coast is inhabited by numerous but small tribes of Indians; each tribe appearing to be independent, and governed by its own chief. They differ from each other in their language and customs, and are frequently at war. It is impossible to ascertain with any degree of certainty the number of inhabitants; but they have been computed at 10,000, from Nootka sound to Cook's river, an extent of about 1000 miles.

The natives are for the most part short in stature; their faces, men and women, are in general flat and round, with high cheek bones and flat noses, and their teeth white and regular. Their complexions are lighter than the *s.* Indians, and some of their women have rosy cheeks. Both sexes are fond of ornamenting themselves with beads and trinkets, and they generally paint their hands and faces. They have a custom of making a longitudinal slit in the under lip, between the mouth and chin, some of them as large as the mouth, in which they wear a piece of bone, wood, or ivory, fitted with holes in it, from which they suspend beads as low as the chin. There appears to be a greater uniformity in the dress of the different tribes than in their ornaments. The aperture or second month, above the chin, seems confined to the men of Cook's river and Prince William's sound; whilst the wooden ornament in the under lip is worn by the women only, in that part of the coast from port Mulgrave to Queen Charlotte's islands. The inhabitants wholly subsist by fishing and hunting. Their clothing is made of the skins of animals and birds. They live in a very dirty manner, and are a complete picture of filth and indolence. The chief object of civilized nations in navigating this coast hitherto, has been to traffic with the natives for furs; which they give in exchange for pieces of iron, nails, beads, pen-knives, and other trifling trinkets. These furs are carried to China, and disposed of to a great profit. The skins obtained are those of the sea-otter, racoon, pine-martin, laud beaver, earless mammot, &c. The other articles which might be procured are *ginseng*, copper, oil, spars, &c. with great quantities of salmon. From 1785 to Feb. 1788, there had arrived at China from this coast nine vessels of different nations. Six of these had furs, which sold for 96,842 dollars; the cargo of two French ships was 54,837 dollars; and 17,000 skins imported by the Spaniards remained unsold. What furs the Russians procure is not known, as they never carry them to Canton.

An inland sea has been lately discovered in this country. Mr. Etches, who fitted out ships from England, has lately discovered, that all the *w.* coast of America from lat. 48° to 57° *n.* is not a continued tract of land, but a chain of islands which had never been explored, and that those concealed the entrance to a vast inland sea, like the Baltic or Mediterranean in Europe, and which seems likewise to be full of islands. Amongst these Mr. Etches's ship, the Princess Royal, penetrated several hundred leagues in a *n. e.* direction, till they came within 200 leagues of Hudson's bay; but as the intention of the voyage was merely commercial, they had not time fully to explore the archipelago just mentioned, nor did they arrive at the termination of this new Mediterranean sea. The islands, of which upwards of 50 were visited, were inhabited by tribes of Indians, who appeared very friendly, and well disposed to carry on a commerce. In consequence of an expedition undertaken in 1787, Captain J. Kendrick, of the ship Columbia, while prosecuting an advantageous voyage with the natives for furs, purchased of them, it is said, for the owners, a tract of delightful country, comprehending 4° of lat. or 240 miles square. The deeds are said to be in China, and registered in the office of the American consul: the agents in London are authorised to treat with any gentleman or association for the purchase of a tract of land no where exceeded for fertility and climate, and which may perhaps, by a prudent management of some wise constitution, become of the utmost importance. See Index to new matter respecting Mexico, Chap. XI.]

[North-West River, a branch of Cape Fear or Clarendon river, in N. Carolina. It is formed by the junction of Haw and Deep rivers; and it is 300 yards wide at Ashwood, 80 or 90 miles above the capes, even when the stream is low, and within its banks. See CAPE FEAR River. On the *w.* side of this river, about 40 miles above Ashwood, in the banks of a creek, five or six feet below the sandy surface, are to be seen projecting out many feet in length, trunks of trees entirely petrified.]

[NORTH-WEST Territory. See TERRITORY.]

[NORTH YARMOUTH, a post-town of the district of Maine in Cumberland county, on a small river which falls into Casco bay. It is 11 miles *w.* by *s.* of Brunswick, 10 *n.* by *e.* of Portland, and 96 *n. n. e.* of Boston. The township is extensive, was incorporated in 1713, and contains 1978 inhabitants. Cassen's river divides it from Freeport on the *n. e.*]

NORTHAM, a city of the province and colony of New England in N. America.

[NORTHAMPTON, a large uneven county of Pennsylvania; situated in the *n. e.* corner of the state, on Delaware river, which separates it from the state of New Jersey and New York. It is divided into 27 townships, and contains 24,250 inhabitants.]

[NORTHAMPTON, a township in Buck's county, Pennsylvania.]

[NORTHAMPTON, a town in Northampton county, Pennsylvania, on the *s. w.* bank of Lehigh river, five or six miles *s. w.* of Bethlehem.]

[NORTHAMPTON, a county of Halifax district, N. Carolina, bounded *n.* by the state of Virginia, containing 9981 inhabitants, including 4409 slaves.]

[NORTHAMPTON, a maritime county of Virginia; situated on the point of the peninsula which forms the *e.* side of the entrance into Chesapeak bay. It has the ocean *e.* and Accomack county on the *n.* Its *s.* extremity is cape Charles, in lat. 37° 13′ *n.* and long. 75° 57′ *w.* off which is the small island called Smith's Island. This county contains 6889 inhabitants, including 3244 slaves. The lands are low and sandy.]

[NORTHAMPTON Court-house, in the above county, where a post-office is kept, is 30 miles *s.* by *w.* of Onancock court-house, 33 *n. e.* of Norfolk, and 164 *s.* of Philadelphia.]

[NORTHAMPTON, a respectable post-town and capital of Hampshire county, Massachusetts; situated within a bend of Connecticut river, on its *w.* side, 55 miles *n.* of Hartford in Connecticut, and 76 *w.* of Boston. It contains a spacious Congregational church, a court-house, gaol, and about 250 dwelling-houses, many of which are genteel buildings. Its meadows are extensive and fertile; and it carries on a considerable inland trade. This township was incorporated in 1685, and contains 1628 inhabitants.]

[NORTHAMPTON, a township in Burlington county, New Jersey, which contains about 56,000 acres, half of which is under improvement, the other half is mostly pine barren. The chief place of the township is called Mount Holly. It contains about 150 houses, an Episcopal church, a Friends' meeting-house, and a market-house. It is 16 miles from Trenton, and 17 from Philadelphia. See MOUNT HOLLY.]

[NORTHBOROUGH, a township in Worcester county, Massachusetts, formerly the *n.* part of Westborough. It was incorporated in 1760, and contains 619 inhabitants. It is 10 miles *e.* of Worcester county, Massachusetts, taken from Uxbridge, which bounds it on the *s.* It was incorporated in 1772, and contains 569 inhabitants. Blackstone river runs through this town. It is 12 miles *s.* by *e.* of Worcester, and 35 *s. w.* of Boston.]

NORTHERN Archipelago consists of several groups of islands, which are situated between the *e.* coast of Kamtschatka, and the *w.* coast of the continent of America.

Mr. Coxe observes, that "the first project for making discoveries in that tempestuous sea which lies between Kamtschatka and America, was conceived and planned by Peter I." Voyages with that view were accordingly undertaken at the expence of the crown; but when it was discovered that the islands in that sea abounded with valuable firs, private merchants immediately engaged with ardour in similar expeditions; and within a period of 10 years, more important discoveries were made by these individuals, at their own private cost, than had hitherto been effected by all the efforts of the crown. The investigation of useful knowledge has also been greatly encouraged by the late empress of Russia; and the most distant parts of her vast dominions, and other countries and islands, have been explored, at her expence, by persons of abilities and learning, in consequence of which several discoveries have been made.

Some of the islands of this archipelago are only inhabited occasionally, and for some months in the year, and others are very thinly peopled; but others have a great number of inhabitants, who constantly reside in them. The inhabitants of these islands are, in general, of a short stature, with strong and robust limbs, but free and supple. They have lank, black hair, and little beard, flattish faces, and fair skins. They are for the most part well made, and of strong constitutions, suitable to the boisterous climate of their isles.

The Fox islands, one of the groups, are so called from the great number of black, grey, and red foxes, with which they abound. The dress of the inhabitants consists of a cap and a fur coat, which reaches down to the knee. Some of them wear common caps of a parti-coloured bird skin, upon which they leave part of the wings and tail. On the fore part of their hunting and fishing caps, they place a small board like a screen, adorned with the jaw bones of sea bears, and ornamented with glass beads, which they receive in barter from the Russians. At their festivals and dancing parties they use a much more shewy sort of caps. They feed upon the flesh of all sorts of sea animals, and generally eat it raw; but if at any time they choose to dress their victuals, they make use of a hollow stone; having placed the fish or flesh therein, they cover it with another, and close the interstices with lime or clay; they then lay it horizontally upon two stones, and light a fire under it. The provision intended for keeping is dried without salt in the open air. Their weapons consist of bows, arrows, and darts, and for defence they use wooden shields.

The most perfect equality reigns among these islanders. They have neither chiefs nor superiors, neither laws nor punishments. They live together in families, and societies of several families united, which form what they call a race, who, in case of an attack or defence, mutually help and support each other. The inhabitants of the same island always pretend to be of the same race; and every person looks upon his island as a possession, the property of which is common to all the individuals of the same society. Feasts are very common among them, and more particularly when the inhabitants of one island are visited by those of the others. The men of the village meet their guests beating drums, and preceded by the women, who dance. At the conclusion of the dance, the hosts serve up their best provisions, and invite their guests to partake of the feast. They feed their children when very young with the coarsest flesh, and for the most part raw. If an infant cries, the mother immediately carries it to the sea-side, and, whether it be summer or winter, holds it naked in the water until it is quiet. This custom is so far from doing the children any harm, that it hardens them against the cold, and they accordingly go barefooted through the winter without the least inconvenience. They seldom heat their dwellings; but, when they are desirous of warming themselves, they light a bundle of hay, and stand over it; or else they set fire to train-oil, which they pour into a hollow stone. They have a good share of plain natural sense, but rather slow of understanding. They seem cold and indifferent in most of their actions; but let an injury or even suspicion only rouse them from this phlegmatic state, and they become inflexible and furious, taking the most violent revenge, without any regard to the consequences. The least affliction prompts them to suicide; the apprehension of even an uncertain evil often leads them to despair; and they put an end to their days with great apparent insensibility.]

[NORTHFIELD, a township in Orange county, Vermont; between 20 and 30 miles *w* of Newbury, in the *w.* part of the county.]

[NORTHFIELD, a thriving township in the *n.* part of Hampshire county, Massachusetts; situate on the *e.* side of Connecticut river, 23 miles *n.* of Northampton, 69 *n. w.* by *w.* of Boston. It contains 868 inhabitants. The town was incorporated in 1763, and some years after desolated by the Indians. The inhabitants returned again in 1685, but it was soon after destroyed a second time. In 1713 it was again rebuilt, and one-third of the township was taken off, and incorporated by the name of Hinsdale. Fort Drummer was in the vicinity of this town.]

[NORTHFIELD, a small town in Rockingham county, New Hampshire, taken from Canterbury, on the *e.* side of Merrimack river, and incorporated in 1780. It contained 606 inhabitants.]

[NORTHFIELD, a township in Richmond county, Staten island, New York; containing 1021 inhabitants, including 133 qualified electors, and 133 slaves.]

[NORTHLINED Lake, in N. America, is about 160 miles *s.* of the head of Chesterfield inlet; is full of islands, and about 80 miles long, and 25 broad.]

[NORTHPORT, a township in Hancock county, district of Maine, taken from the *n.* part of Duck-trap plantation, and incorporated in 1796.]

[NORTHUMBERLAND, a town in Grafton county, New Hampshire; situate on the *e.* side of Connecticut river, at the mouth of the Upper Amonoosuck. It was incorporated in 1779, and contains 117 inhabitants.]

[NORTHUMBERLAND, a county of Pennsylvania; bounded *n.* by Lycoming, *s.* and *w.* by Dauphin and Mifflin counties. It is divided into 16 townships, and in 1790 contained 17,161 inhabitants. The county of Lycoming has, since the census of 1796, been taken from it, but the county is supposed to contain nearly as many inhabitants as before; a great number of people having emigrated to this part of the state. Chief town, Sunbury.]

[NORTHUMBERLAND, a flourishing post-town in the above county; situate on the point of land formed by the junction of the *e.* and *w.* branches of the Susquehannah. It is laid out regularly, and contains about 120 houses, a Presbyterian church, and an academy. It is two miles *n.* by *w.* of Sunbury, and 96 *n. w.* by *w.* of Philadelphia.]

[NORTHUMBERLAND, a county of Virginia; bounded *e.* by Chesapeak bay, and *w.* by Richmond. It contains 9163 inhabitants, including 4460 slaves. The court-house, where a post-town is kept, is eight miles from Kinsale, 17 from Lancaster court-house, and 53 from Fredericksburgh.]

[NORTHUMBERLAND, a county of Pennsylvania. There is iron ore in this county; also a salt-spring.]

[NORTHWOOD, an interior and elevated township in Rockingham county, New Hampshire, in which, and on its borders, are a number of small ponds, whose waters feed Piscataqua and Suncook rivers. It was incorporated in 1773; contains 744 inhabitants, and is about 59 miles *n. w.* of Portsmouth. Crystals and crystalline spars are found here.]

[NORTON, a township in Essex county, Vermont; situate on the Canada line, having Canaan *e.* and Holland on the *w.*]

[NORTON, a township of Massachusetts; situate in Bristol county, and 33 miles *s.* of Boston. It was incorporated in 1711, and contains 1428 inhabitants. The annual amount of the nail manufacture here is not less than 300 tons. There is also a manufacture of ochre, which is found here, similar to that at Taunton.]

[NORTON, a settlement on the *n. e.* coast of Cape Breton island.]

[NORTON'S Sound, on the *n. w.* coast of N. America, extends from cape Darby on the *n. n. w.* to cape Denbigh or cape Stephen's on the *s.* or *s. e.* Lat. 64° 50′ *n.*]

NORVEL, a settlement of the island Barbadoes, in the district of the parish of S. Luke.

NORVES, a canal in the straits of Magellan, at the *n.* entrance of the third narrow pass, called the Passage.

[NORWALK, a pleasant post-town in Fairfield county, Connecticut; situated on the *n.* side of Long Island sound. It contains a Congregational and Episcopal church, which are neat edifices, and between 40 and 50 compact houses. It is 14 miles *w.* by *s.* of Fairfield, 27 *s. w.* by *w.* of New Haven, 40 *n. e.* of New York, and 108 from Philadelphia. Lat. 41° 8′ *n.* Long. 73° 25′ *w.* The township is situated in a fertile wheat country, and was settled in 1651. Here are iron works and a number of mills. It has a small trade to New York and the W. Indies.]

[NORWAY, a township of New York, in Herkemer county, incorporated in 1792. By the state census of 1796, it contained 2164 inhabitants, of whom 353 were electors.]

[NORWAY, a new township in Cumberland county, district of Maine, incorporated 1797.]

[NORWICH, a considerable township in Windsor county, Vermont, on the *w.* side of Connecticut river, opposite to Dartmouth college. It contains 1158 inhabitants.]

[NORWICH, a township in Hampshire county, Massachusetts, 24 miles *s. w.* of Northampton, and 86 *w.* of Boston. It was incorporated in 1773, and contains 742 inhabitants.]

[NORWICH, a city and post-town of Connecticut, and of the second rank in New London county; situated at the head of navigation on Thames river, 10 miles *n.* of New London, and 34 *s. e.* of Hartford. This commercial city has a rich and extensive back country, and avails itself of its happy situation on a navigable river, which affords a great number of convenient seats for mills, and water machines of all kinds. The inhabitants manufacture paper of all kinds, stockings, clocks and watches, chaises, buttons, stone and earthen ware, oil, chocolate, wire, bells, anchors, and all kinds of forge-work. The city contains about 450 dwelling houses, a court-house, and two churches for Congregationalists, and one for Episcopalians, and about 3000 inhabitants. The city is in three detached, compact divisions, viz. Chelsea, at the landing, the Town, and Bean hill; in the latter division is an academy, and in the Town is an endowed school. The courts of law are held alternately at New London and Norwich. This town was settled in 1660, by 35 men, principally from Saybrook. It is 170 miles *n. e.* of Philadelphia. Lat. 41° 29′ *n.* Long. 72° 3′ *w.*]

[NORWICH, a township in Tioga county, New York, taken from the towns of Jerico and Union, and incorporated in 1793. It is settled principally by people from Connecticut; is bounded *s.* by Oxford, and lies 50 miles *w.* of Cherry valley. By the state census of 1796, 129 of its inhabitants were electors.]

NOSACO, a settlement of the province and *captainship* of Pernambuco in Brazil; situate on the point of land to the *n.* forming the mouth of the river Grande de S. Francisco.

NOTAYE, a river of the province of Guayana, in the part possessed by the French.

[NOTCH, The, a pass in the *w.* part of the White mountains in New Hampshire; the narrowest part of which is but 22 feet wide, between two perpendicular rocks. It is 25 miles from the Upper Coos. From the height above it a brook descends, and meanders through a meadow, formerly a beaver pond. It is surrounded by rocks, which, on one side, are perpendicular, and on the others, rise in an angle of 45 degrees, a strikingly picturesque scene. This defile was known to the Indians, who formerly led their captives through it to Canada; but it had been forgotten or neglected till the year 1771, when two hunters passed through it. There is a road this way now to the Upper Coos.]

[NOTCH, Cape, is the *w.* point of Goodluck bay, in the straits of Magellan. Lat. 53° 33′ *s.*]

NOTOWAY, a river of the province and colony of Virginia in N. America. It runs nearly due *e.* and enters the Blackwater.

[NOTTAWAY, a small river of Virginia, which runs *e.* by *s.* and receives Blackwater on the line of N. Carolina; thence pursuing a *s.* by *w.* course of about 10 miles, it joins the Meherrin; the confluent stream then assumes the name of Chowan river, and empties into Albemarle sound.]

[NOTTAWAY, a county of Virginia, bounded *n.* and *n. w.* by Amelia, from which it was taken in the year 1788. See AMELIA.]

[NOTTINGHAM, a township in Rockingham county, New Hampshire, 12 miles *n.* of Exeter, and 18 *n. w.* of Portsmouth. It was incorporated in 1722, and contains 1068 inhabitants.]

[NOTTINGHAM, WEST, a township in Hillsborough county, New Hampshire; situated on the *e.* side of Merrimack river, 40 miles distant from Portsmouth, was incorporated in 1746, and contains 1064 inhabitants. It has Massachusetts line for its *s.* boundary, which divides it from Dracut, and is about 35 miles *n. n. w.* of Boston.]

[NOTTINGHAM, a township in Chester county, Pennsylvania.]

[NOTTINGHAM, the most *n.* town of Burlington county, New Jersey; situate on the *e.* bank of Delaware river, between Bordentown and Trenton.]

[NOTTINGHAM, a town in Prince George's county, Maryland; situate on Patuxent river, nearly 11 miles *n. e.* of Piscataway, and 16 *s. e.* of the Federal city.]

NOTUCO, a volcano of the kingdom of Chile, near the river Solivinokon.

NOU, a settlement of the missions which were held by the Portuguese Carmelites, in the territory and country of Las Amazonas; situate on the banks of the river Negro.

NOUA, a small island of the N. sea, near the coast of Brazil, opposite the great island of Marajo, or Joanes, near that of Maguary.

NOUILLO, a settlement of the province and government of Santa Marta in the Nuevo Reyno de Granada; situate on the shore of the river Guazar.

NOUITA, SAN GERONIMO DE, a city and capital of the province and government of Choco in the Nuevo Reyno de Granada. The greater part of the houses are built of wood, and roofed with straw, without regularity or symmetry. Nearly the whole of the population is of Negroes, Mulattoes, *Mustees*, and Zambos; for although there be certain families of distinction, they dwell rather in the neighbourhood of the gold mines, to superintend the slaves who are working them. It is of an hot, moist, and unhealthy climate; situate in a *llanura* surrounded on all sides with woods. It produces nothing, and is provided with food, clothes, and other necessaries from the province of Popayán, at a very dear rate. See article CHOCO.

NOUVELLE, a large and abundant river of New France or Canada in N. America. It runs many leagues *s. w.* through the country of the Huron Indians, between lakes Huron and Erie, and enters the S. Clare.

[NOUVELLE, LA, commonly called East Nouvelle, lies on the *n.* side of Chaleur bay. It is a small river, about four leagues from port Daniel.]

[NOUVILLE, LA GRANDE, or WEST NOUVILLE, on the *n.* side of Chaleur bay, is above one league from Carleton, where is also a custom-house, and a respectable mercantile house.]

[NOVA SCOTIA, formerly called New Scotland, a British province of N. America; separated on the *n. e.* from Cape Breton island by the gut of Canso; on the *n.* it has a part of the gulf of St. Lawrence, and the straits of Northumberland, which divide it from the island of St. John's; on the *w.* it has New Brunswick and the bay of Fundy; on the *s.* and *s. e.* the Alantic ocean. Its length is about 235 miles from cape Sable on the *s. w.* to cape Canso on the *n. e.* Its extreme breadth is 88 miles; but between the head of Halifax-harbour and the town of Windsor, at the head of the *s. e.* arm of the basin of Minas it is only about 22 miles broad. It contains 8,789,000 [acres; of which three millions have been granted, and two millions settled and under improvement.

Nova Scotia is accommodated with many spacious harbours, bays, and coves of shelter, equal to any in the world. The chief of these are Canso, Halifax, on Chebucto bay, Chedabucto, Frederick, George, Torbay, Charlotte, King's, Barrington, Townsend, St. Mary's, Annapolis Royal, the basin of Minas, the bay of Fundy; and a vast number of capes, lakes, and rivers, which are described under their respective names.

The most remarkable mountains are the highland of Aspotageon, and the Ardeis mountain. The *s.* shores present to the eye of a stranger rather an unfavourable appearance, being in general broken and stony; but the innumerable islands along its coasts, coves, and harbours, though generally composed of rocky substances, appear designed by nature for the drying of fish, being covered with materials for fish-flakes and stages; and there is land sufficient for pastures and gardens, to serve the purposes of fishermen. As you advance into the back country, it wears a more promising appearance; and at Cornwallis, Windsor, Horton, Annapolis, Cumberland, Cobequid, Pictou, and along the *n.* shores of the province, there are extensive, well improved farms. The gradual improvement in husbandry, which has been encouraged by the laudable and successful experiments of the argricultural society, lately established here, afford some good ground to expect that Nova Scotia may become a flourishing colony. The lands in general, on the sea-coast, the county of Lunenburgh excepted, and a few hills of good land, are rocky, and interspersed with swamps and barrens. The growth in general is a mixture of spruce, hemlock, pine, fir, beech, and some rock-maple, which furnish an inexhaustible supply for ship-building and other purposes.

The coast abounds with fish of various kinds, as cod, salmon, mackerel, herring, alewives, trout, &c. and being near to the banks of Newfoundland, Quero, and Sable bunks, fisheries, under proper management and regulations, might be carried on with certainty of success. There are mines of coal at Cumberland, and on the E. river, which falls into Pictou harbour. There is plenty of bog and mountain ore in Annapolis township, on the borders of Nietan river, and a bloomery is erected there. Copper has been found at cape D'Or, on the *n.* side of the basin of Minas. The forts in this province are fort Edward, Cumberland, and Cornwallis. Nova Scotia is divided into eight counties, viz. Hants, Halifax, King's, Annapolis, Cumberland, Sunbury, Queen's, and Lunenburg. These are subdivided into above 40 townships. The whole population of Nova Scotia, New Brunswick, and the islands adjoining, is estimated at about 50,000. The amount of imports from Great Britain to this country, at an average of three years, before the new settlements, was about 26,500*l.* The articles exported in exchange are timber and the produce of the fishery, which at a large average amounts to 38,000*l.* Nova Scotia was confirmed to Great Britain in 1760. Halifax is the metropolis. See New Brunswick, Canada, &c.

In concluding this article we think it necessary to insert the following memorials and authentic documents transmitted to the British government in 1804, as throwing a steady light on the views and resources of this colony, and particularly on the nature of its interests as contrasted with those of the United States. We shall give them therefore entire, not merely as objects of vague and general information, but of abstruse inquiry.

'*To the Right Hon. Lord Hobart, one of his Majesty's principal secretaries of state, &c. &c.*

'*The petition of the merchants and other inhabitants of Halifax, in the province of Nova Scotia,*

'Humbly sheweth,

'That the trade of this province arises principally from the fish caught on its coasts, great quantities of which are exported annually by your petitioners to the W. India islands. That in the pursuit of this commerce, your petitioners are rivalled by the citizens of the American states, to whom the ports of those islands are ever open, and who are exempt from duties and other expences to which your petitioners are liable. Your petitioners have heard, that in the existing negociation, relative to the twelfth article of the treaty with America, the Americans aim at a further extension of their trade with the British W. India islands, which, if obtained, would utterly ruin the already declining fisheries of the British colonies, whence the nation has long derived much wealth, and its navy a supply of hardy seamen.

'That the coasts of this province, as well as the gulph of St. Lawrence, and the islands of Newfoundland and Cape Breton, abound with fish of the most valuable sorts; so that with encouragement these colonies would satisfy, to its utmost extent, the demand of the W. India islands for dry and pickled fish.

'Your petitioners, therefore, most humbly pray, that your Lordship, and his Majesty's other ministers, would take the premises, and the annexed memorial, into consideration, and would protect the trade and fisheries of his Majesty's subjects] [in these colonies against the views of the Americans, by granting to the British colonists the exclusive privilege of supplying their fellow subjects in the W. Indies with the article of fish caught on the coasts of N. America.

(Signed) William Sabatier,
William Smith,
George Grassic,
James Fraser,
William Lyon,
Committee appointed by the merchants and other inhabitants of Halifax, Nova Scotia.
'*Halifax, Nova Scotia, March 23d*, 1804.'

'*Memorial and statement of the case referred to in the annexed petition.*

'As every British province and island in these *n.* climates is individually able to furnish the W. India islands with some essential article of consumption, which in whole, or in part, is deficient in others, the petitioners, in the following statement, have extended their observations beyond the limits of the single province in which they reside.

'The W. India islands require to be supplied with the undermentioned articles, viz.

'From the fisheries.—Dried cod fish, barrel or pickled fish, viz. salmon, herring (of various species), and mackerel and oil.

'Forest.—Lumber, viz. squared timber, scantling, planks and boards, shingles, clapboards, hoops, and oak staves.

'Agriculture.—Biscuits and flour, Indian corn and meal, pork, beef, butter, cheese, potatoes, and onions; live stock, viz. horses, oxen, hogs, sheep, and poultry.

'Mines.—Coals.

'Of these articles, the following are produced by the several colonies. New Brunswick produces, in the greatest abundance, lumber of every kind, except oak staves; it yields already many of the smaller articles which serve to complete a cargo, and its shores abound with various fish fit for pickling. Nova Scotia produces lumber of all sorts, except oak staves, but in a lesser degree than New Brunswick; horses, oxen, sheep, and all the other productions of agriculture, except wheat and Indian corn; the *e.* and *n.* parts of the province abound in coal, and its whole coast yields inexhaustible quantities of cod fish, and others fit for pickling.

'Cape Breton and Prince Edward islands; the former yields coal in abundance, its fisheries are considerable; but without dealing directly with the W. Indies, they serve to increase the exports of Nova Scotia. Both these islands supply Newfoundland with cattle, and with due encouragement would rival some of the more opulent colonies in articles of agriculture; their fisheries also may be greatly extended, as the whole circuit of these islands abounds in fish.

'Canada can supply any quantity of oak staves, as well as flour and Indian corn, for six months in the year. Newfoundland yields little lumber, but its trade in dried cod-fish has hitherto, in a great measure, supplied all Europe and the W. Indies, and it is capable of still greater extension.

'The petitioners have therefore no hesitation in affirming, that these mother colonies are able to supply the W. Indies with dried fish, and every species of pickled fish, for their consumption; and that at no very distant period they could also supply all the other articles herein before enumerated, except, perhaps, flour, Indian meal and corn, and oak staves.

'Having stated the foregoing facts, the petitioners beg leave to request the attention of his Majesty's ministers to the peculiar circumstances of this province, the permanent establishment of which took place about 54 years ago; for previous to the settlement of Halifax, there were few inhabitants in it, and but little trade. The mother country, sensible of the favourable situation of this colony for fisheries, that its harbours are seldom more than a few miles from each other, and that its extensive sea-coast teems every season with shoals of fish of the most useful sorts, made every effort to establish them. The fisheries, however, until the close of the American war, languished from one cause only—the want of inhabitants. The influx of inhabitants at that time, and since, has promoted industry and domestic comfort, and a race of people born on the soil have become attached to it. The clearing of the lands, and other causes, have improved the climate; and by a late survey of the interior of the province, it is discovered that the lands are not only better than had been imagined, but superior to the greater part of the rest of N. America.

'The present situation of this province with regard to its trade, resembles that of New England at the close of the 17th century; and unless checked at this crisis, it has the most reasonable expectation of a more rapid increase than the latter ever experienced.

'Encouraged by the prospect before them, and conscious of the abuses that have crept into the fisheries, the petitioners are looking forward to the aid of the provincial legislature, and to other means, for correcting those abuses and for esta-]

[blishing and improving the fisheries, that great source of wealth to the parent state, the colonial husbandman, and merchant; but they perceive, with regret, that their efforts will prove ineffectual, unles the citizens of the United States, according to the ancient policy of Great Britain towards foreigners, are wholly or partially excluded from the islands, or a permanent equivalent is granted to the colonists.

'The American legislature has rejected the 12th article of the late treaty: the citizens of the United States would have been excluded from the W. Indies, if the governors of those islands had not, under the plea of necessity, by proclamation, admitted them. In this trade the Americans possess the following advantages over the colonies.

'First,—In the islands of Barbadoes, Antigua, Saint Kitt's, and Jamaica, a stranger's duty of two and a half, or more, per cent. is imposed on imports, and in the island of Saint Vincent, British subjects exclusively are subject to a duty of three per cent. which must be paid in specie, and to procure which a forced sale is frequently made of part of the cargo to great disadvantage. From this duty the Americans, being invited by proclamation, are exempt.

'Second,—During the late and present war, the citizens of the United States, being neutrals, have not been burthened with the heavy charge of insurance against the enemy, which to the colonists has increased the premium 10 per cent. to the smaller islands, and 12½ per cent. to Jamaica.

'Third,—The *n.* states have granted a bounty of near 20*s.* per ton on vessels in their fisheries.

'From those circumstances, so unable are the petitioners to contend with the Americans in the W. India markets, that they derive greater advantage by selling their fish at an inferior price in the United States; whence the Americans re-export them to the W. India islands under the abovementioned advantages, so as to make a profit even on their outward voyage.

'It is well known, and in an ample report made to congress in the years 1790 and 1791, by the now president of the United States, then their secretary of state, it was set forth, that the fisheries of New England were on the verge of ruin, and he recommended, what was afterwards adopted, the grant of a bounty to counterbalance the disadvantages the trade then laboured under. At that period the fisheries of Nova Scotia made a rapid increase; the whale fishery alone from the port of Halifax consisted of 28 sail of ships and brigs from 60 to 200 tons burthen: but the succeeding war and other unfavourable circumstances soon destroyed this important branch of the fishery. By the aid of bounties from the state legislature, the American fisheries recovered their former vigour, and are now carried on with great spirit, increasing their trade with the W. Indies to an incredible extent; considerable numbers of our best fishermen have emigrated from Newfoundland and this province to the United States, within a few months, and more are daily following them: thus it appears evident, that a wise policy, steadily pursued, will preserve a sinking trade, and that this province is not wanting in exertion, when favourable opportunities for it are offered.

'Should the Americans obtain by treaty an indulgence of their trade in fish with the W. Indies, it will prove the ruin of that of the British *n.* colonies, and draw away from them their most industrious inhabitants. The islands will then depend on foreign states for supplies of all the articles before enumerated; and if at any time hereafter differences should take place between Great Britain and the American States, from what quarter, it may be asked, are the islands to obtain their supplies; the ruined trade and fisheries of those colonies may prove, too late, the fatal policy of throwing into the hands of foreigners a trade, which, with a little encouragement, might have been almost, if not entirely, confined to British subjects.

'From these considerations the justice and policy of giving encouragement to the *n.* colonies are evident. Should the stranger's duty, imposed in the islands, be taken off: should a bounty equal to that granted by the state legislature be allowed, and the present war succeeded by a peace, then may the W. India islands receive from these colonies supplies of all kinds of dried or pickled fish, on terms as advantageous as they are now furnished with them from a foreign state. It is obvious that the Americans and the W. India planters have a mutual interest in the free trade to the islands, but the planters have no right to expect supplies from a neutral nation in time of war, merely because it affords them at a cheaper rate than the British colonies; they should bear the inconveniences of war as well as their fellow-subjects, who have been driven into these *n.* regions by their zealous loyalty in support of the happy constitution under which they now live. The supplies required by the islands cannot greatly increase; while the *n.* colonies, from their great extent and growing population, will every year be more and more able to furnish those supplies. The islands are, in a measure, limited in their extent: but the *n.* colonies are almost unbounded.

The inhabitants of those colonies have acquired]

[their present condition, which, at best, is mediocrity, by a continued exertion of industry and frugality, under a climate and a soil which yield their blessings to persevering exertion alone. The W. India planters have ever been in a different situation, and can afford to wait a reasonable time for the accomplishment of those expectations which are justly entertained by the colonists; in the interim they ought to give a fair equivalent for the articles of which they stand in need, and not expect, at an inferior price, commodities whose value the imperious circumstances of the times have tended to enhance. The *n.* colonists have struggled with all the difficulties incident to a young country, and they are now arrived at a period, when, if duly encouraged, they may be enabled to reap the fruits of their honest labour; but restricted in their trade to the Mediterranean by an ancient regulation, which obliges them to land their cargoes in some English European port, before they can proceed on homeward-bound voyages, and burthened also in the manner here stated in the W. India trade, the petitioners cannot contend with the Americans, but look forward with the most distressful prospects to means of procuring a future subsistence, unless his Majesty, in his goodness, shall be pleased to afford them protection and relief. They, therefore, anxiously hope, that the observations contained in this memorial may not appear unworthy of the attention of his Majesty's ministers, but that whatever temporary indulgences may be granted to the American citizens, the British colonists, agreeably to their former solicitations on that subject, may be permitted to return to America, without entering at any port in Great Britain.'

'My Lord, *Halifax, Jan.* 30, 1805.

'We the committee of the merchants and inhabitants of Halifax, Nova Scotia, who presented to Lord Hobart, your Lordship's predecessor, a petition, praying that the British colonists might have the exclusive right of supplying his Majesty's W. India islands with fish, have lately seen, in print, a letter written to your Lordship by G. W. Jordan, Esq. colonial agent for Barbadoes, containing observations on our petition, and the memorial annexed to it; we think it our duty briefly to answer those observations, and to enforce the object of our petition.

'Mr. Jordan's first remark is founded on a misconception or perversion of the allegation of the petitioners: we assert in our memorial, "that in the islands of Barbadoes, Antigua, Saint Kitt's, and Jamaica, a stranger's duty of two and a half per cent. is imposed on imports, and that in the island of Saint Vincent, British subjects exclusively are subject to a duty of three per cent.:" no charge is therefore made that the duty is not general in the island of Barbadoes; the charge is clearly confined to the single island of Saint Vincent.

'We are not alarmed, my Lord, at the reference made by Mr. Jordan to papers which were not intended for his inspection, but for private information only; since those papers contain no other facts than such as can be proved. The practice in the W. India islands of keeping the ports always open to the Americans, amounts, in our apprehension, to the grant of a free trade; and that goods of foreign manufacture are by these means introduced into the islands no one who is at all acquainted with the character and practices of the American traders can doubt. We lament that even in these colonies, into whose ports no American vessels are admitted, except fishing vessels, which by treaty are allowed to resort to our coasts, such quantities of foreign goods do find admittance, that it is to be feared more than half the E. India goods consumed in this province is supplied from the neighbouring states of America.

'We do not, as Mr. Jordan is pleased to assert, claim a right of selling our own commodities at our own prices in time of war; but we contend that, when the article of fish is furnished from the *n.* colonies in abundance, although increased in price by the war expences, the W. India colonists ought not, on that account, to require or permit the introduction of it from foreign states, and in foreign bottoms; especially as the fish is generally paid for in the produce of the islands, of which the planters take care to raise the price in proportion. That these *n.* colonies can supply the islands with their whole consumption of fish, and at reasonable prices, can be easily proved, and that they are, therefore, entitled to do so, exclusively, Mr. Jordan himself admits.

'The right of the W. India colonists to obtain from the American States all articles of the first necessity, which they cannot adequately obtain from the dominions of Great Britain, is not disputed by us; but we assert that the article of fish can be adequately obtained from the British colonies. That the allowing supplies to be imported in American bottoms has been destructive to the British carrying trade, has been lately demonstrated by a very able writer on the subject; and that the indulgences granted to the Americans have injured the fisheries, and greatly reduced the tonnage and seamen employed in these colonies, we can assert from our own sad experience. An in-]

[spection into the imports and exports of the island of Jamaica for one year, as laid before their house of assembly, and published in the Jamaica almanack for the last year, will shew how large a portion of the W. India carrying trade is engrossed by the Americans.

'If, my Lord, we have stated in our memorial, that it is, now, more advantageous for the merchants of this colony to dispose of their fish in the United States, than to send it to the W. India islands;—we have made it a subject of complaint; and at the same time have set forth the reasons why the Americans rival us in that trade. Were our commerce with the islands placed on a fair foundation, the same British ships would convey our fish thither, which now carry it to the American markets. But burthened as that trade is with insurance against the enemy, and confined as it is, and ought to be, to a fair dealing in legitimate merchandise, we contend in those ports with the Americans at every disadvantage.

'Had Mr. Jordan fairly observed on our petition and memorial, he would not have asserted that the positive affirmation in the former, "that these provinces can supply the W. Indies with fish," was shaken by a subsequent observation, "that, under certain circumstances, the trade and fisheries of these colonies would be ruined, which, with encouragement, might be almost, if not entirely, confined to British subjects." The observation refers expressly to the trade in all the articles enumerated in the memorial; the affirmation is confined to the single article of fish. One reading of the paragraph referred to will entirely refute Mr. Jordan's remark.

'Having already, my Lord, observed, that the increase of the price of fish, occasioned by war, is no just ground for the introduction of that article, from foreign ports, and in foreign vessels, we shall not follow Mr. Jordan in the curious inference he undertakes to draw from our admission, that, in war time, the Americans can undersell us in fish. So little are we disposed to require an extravagant price for our fish, that we most readily would accede to Mr. Jordan's proposal, of fixing the maximum price of cod-fish at eight dollars, in time of war; and, indeed, could we obtain even three-fourths of that price, generally, during the war, the fisheries would soon flourish again, and the islands be at all times amply supplied with fish.

'On the two facts with which Mr. Jordan closes his observations, we shall only remark, that the former is conceded by us as to the flour and grain imported into Nova Scotia from the United States; and it is perfectly consistent with our memorial, in which we confess that this province is deficient in the articles of wheat and corn. The other fact we must dispute; and although we are not provided with documents to ascertain the tonnage employed between the British N. American provinces and the W. India islands, for the particular year 1791, yet we are furnished with returns of the tonnage employed in the trade to and from the W. India islands for the year 1792, and entered at the custom-house in Halifax, being for one only of the two districts into which this province is divided, and which we beg leave to insert, as follows:

1792.	*Outwards.*		*Inwards.*
Spring quarter, .	886	. .	719
Midsummer ditto,	1436½	. .	3605
Michaelmas ditto,	2397	. .	585
Christmas ditto,	1770	- -	1862¼
Tons	6489½		6571¼

'It is therefore incredible, that, in the year 1791, only 4837 tons were employed in the trade between all the British *n.* provinces and the W. India islands, when, in the subsequent year, it appears by an authentic return, that in one district, of one province, upwards of 6000 tons were actually engaged in that commerce.

'Here, my Lord, we conclude our observations on Mr. Jordan's letter; nor shall we presume to intrude on your Lordship's patience further than to state one fact, which must demonstrate the efficiency of the British colonies, or at least of British shipping, to supply the demands of the W. India markets. From the year 1785 to the year 1794, American ships were excluded from the W. India islands, yet they were, during that period, so well provided with articles of the first necessity, that ships from these colonies were frequently unable to find a sale for their cargoes in our own islands, and were obliged to resort to foreign islands for a market. By returns collected from the merchants of this province, engaged in the W. India trade, we find that the prices obtained by them for cod-fish, from the year 1785 to the year 1792 inclusive, never exceeded five dollars per quintal, and sometimes fell short of half that sum. In the year 1793, we meet with a single instance of cod-fish selling for six dollars; but the common price, even in that first year of the war, was not more than three dollars and a half per quintal. The cheapness, therefore, of this article, clearly proves the abundance of it in the W. India islands, and consequently, that the allowing] the Americans to import fish in American ships was not a measure of necessity.

We have the honour to be, with the greatest respect, your Lordship's most obedient and most humble servants,

(Signed), William Sabatier,
William Smith,
George Grassic,
James Fraser, and
William Lyon.

The Right Hon. Lord Camden, &c. &c. &c.

[NOVOYA, a parish of the province and government of Buenos Ayres; situate on a river of the same name, about 40 miles *s. e.* of Sta. Fé, in lat. 32° 17′ 43″ *s.* Long. 60° 4′ 34″.]

[NOXAN, or Noxonton, or Nox Town, a town of Newcastle county, Delaware, 21 miles *n.* of Dover, and nine *s.* by *s. w.* of St. George's town.]

NOXTEPEC, a settlement and head settlement of the district of the *alcaldía mayor* of Tasco in Nueva España. It contains 65 families of Indians, and in its boundaries are various cultivated estates and breeding farms, in which dwell 60 other families of Spaniards, *Mustees*, and Mulattoes. Five leagues *n. n. w.* of its capital.

NUBE, or Nuble, a large and abundant river of the kingdom of Chile, flowing down from the *cordillera* of the Andes. It runs *w.* washing the environs of the ancient city of San Bartolomé de Chillan, and united afterwards with that of Itala, runs to empty itself into the Pacific ocean, in lat. 36° *s.*

[NUBLADA, an island in the Pacific ocean, with three small ones *n.* of it and near to it, *w.* by *s.* of cape Corientes, on the coast of Mexico, and *e.* of Roca Portida. Lat. 16° 40′ *n.* Long. 122° 30′ *w.*]

NUCARAY, or Nucuray, a rapid river of the province and country of Las Amazonas. It runs from *n.* to *s.* near the source of the river Amazonas. In the woods on its borders dwell the Maynas, Zimarrones, and other barbarians of the nation of the Umuranas. It enters the Marañon by its *n.* shore, in lat. 4° 18′ *s.*

[NUCHVUNK, a place in New Britain, the resort of walrusses in winter; with the teeth of these animals the Indians head their darts. Lat. 60° *n.*]

NUE, a small river of N. Carolina, which runs *e.* and enters the Hughes.

NUECES, an abundant river of the province and government of Texas in N. America. After running many leagues, it enters the sea in the gulf of Mexico.

NUESTRA Senora, Los Ceros de, a bay of the coast of Peru, in the province and *corregimiento* of Atacama.

[Nuestra Senora de la Paz, an episcopal see and town of Peru; situate on a large plain about 20 miles to the *e.* of the *cordillera* of Acama, 33 miles *s. e.* of Laguna Titicaca, 86 *n. w.* of Cochabamba, and 233 *w.* of the city of Sta. Cruz de la Sierra. Lat. 17° 30′ *s.* Long. 68° 26′ *w.*]

[Nuestra Senora de la Vittoria, a town of Mexico. Lat. 18° *n.* Long. 92° 35′ *w.*]

NUEVA Andalucia. See Andalucia.

Nueva Galicia. See Galicia.

Nueva Vizcaya. See Vizcaya.

NUEVAS, a port of the island of Cuba.

NUEVILLA, a settlement of the island of Cuba; situate on the *n.* coast.

[NUEVO Baxo, a bank called by the British the New Bear, being about 150 miles *s.* of the *w.* end of the island of Jamaica, in lat. 15° 49′ *n.* It has a key, two cables length long and one and one-third broad; stretching *e.* by *n.* and *w.* by *s.* The British find this a good station in a Spanish war, as most ships come this way from the Spanish main, going to the Havannah.]

[Nuevo Santander, capital of the province of the same name, does not admit the entry of vessels drawing more than from eight to ten *palmas* of water. The village of Sotto la Marina, to the *e.* of Santander, might become of great consequence to the trade of this coast, could the port be remedied. At present the province of Santander is so desert, that fertile districts of 10 or 12 square leagues were sold there in 1802 for 10 or 12 francs.]

Nuevo Pueblo, a settlement in the province and *captainship* of Rey in Brazil; situate on the coast of the Rueou de Turotetama.

Nuevo Pueblo, another settlement, in the province and government of Veragua and kingdom of Tierra Firme.

Nuevo Pueblo, a river of the *n.* coast of the straits of Magellan, which runs *e.* and enters the bay of Abrigada.

Nuevo Pueblo, another, of the province and government of Tucumán in Peru. It runs *n.* and enters the Parapiti.

Nuevo Pueblo, another, of the island of Jamaica, which enters the sea between ports New and Dry.

Nuevo Pueblo, a rocky shoal near the coast of the province and government of Cartagena in the Nuevo Reyno de Granada, near that city and *n.* of Salmedina.

Nuevo Pueblo, another, near the coast of the province and government of Yucatán.

Nuevo Pueblo, another, of the N. sea, near that of La Vivora, to the *s*.

Nuevo Reyno de Granada. See Granada.

Nuevo Reyno de Leon. See Leon.

Nuevo Mexico. See Mexico.

NUGALAPA, San Juan de, a settlement of the head settlement of the district of the province and *alcaldía mayor* of Suchitepeques in the kingdom of Guatemala. It contains 80 families of Indians, who speak the Satubil idiom, and is annexed to the curacy of its head settlement.

NUISI, a river of the province and country of Las Amazonas, in the part possessed by the Portuguese. It rises between the Caquitá and Urubecchi, runs *e.* for many leagues, and enters the Negro.

NUITO, Santa Maria de, a settlement of the head settlement of the district of Pinotepa, and *alcaldía mayor* of Xicayan, in Nueva España. It is of a mild temperature, contains 18 families of Indians, who cultivate cochineal, tobacco, and seeds, and is six leagues *n.* of its head settlement.

NULPE, or Nulpi, a river of the province and government of Esmeraldas in the kingdom of Quito. It rises in the mountains of the province of Pasto, to the *n.* of Mayasquier, runs from *s. w.* to *n. w.* and from the settlement of Caiquier begins to be navigable for canoes or small barks, and enters the Mira, in lat. 1° 21′ *n.*

NUMARAN, Santiago de, a settlement of the head settlement of the district of Puruandiro, and *alcaldía mayor* of Valladolid, in the province and bishopric of Mechoacán; situate on a *llanura* on the shore of the river Patiquato. It is annexed to the curacy of Penxamo in the jurisdiction of the town of Leon, contains 10 families of Spaniards and 90 of Indians, and is 28 leagues from Pasquaro.

NUMBALLE, or Namballe, a river of the province and government of Jaen de Bracamoros in the kingdom of Quito. It receives the waters of the river Canche, and runs from *w.* to *e.* to enter by the *w.* part into the Chinchipe, opposite the settlement of Pamaca, in lat. 5° 16′ *s.* On its *s.* shore is a small settlement to which it gives its name, and which is inhabited solely by Indians.

NUNOA, a settlement of the province and *corregimiento* of Santiago in the kingdom of Chile, in the district of which are four chapels of ease.

NUNTIALI, a settlement of Indians of N. Carolina, in the territory of the Moyens Cherokees.

NUNUALCO, Santiago de, a settlement and head settlement of the district of the province and *alcaldía mayor* of San Vicente de Austria, in the kingdom of Guatemala. It contains 1700 Indians, counting those of the small settlements annexed to its curacy, all speaking the Mexican language.

NUNURA. See Nonura.

NURE, or Nuri, a settlement of the province of Ostimuri in N. America; situate on the shore and at the source of the river Hiaqui.

NURES, a nation of Indians, reduced to the Catholic faith, in the province of Cinaloa of N. America. It dwells near the nation of the Rebomes, whose example it followed, when converted, seeking of their own accord that the Jesuits should come amongst them. They were formed into a settlement, which now contains only 200 inhabitants, though it was formerly much larger.

NURST, a settlement of the island of Barbadoes, in the district of the parish of Todos Santos.

O

[OACHATE Harbour, near the *s.* point of Ulietea, one of the Society islands, in the S. Pacific ocean, *n. w.* of Otaheite. Lat. 16° 55′ *s.* Long. 151° 24′ *w.*]

[OAHAHA, a river of Louisiana, which empties into the Mississippi from the *n. w.* in lat. 38° 48′ *n.* and seven miles *n.* of Riviere au Beuf.]

[OAHOONA, one of the Ingraham isles, which is said to be the northernmost of all this cluster. It lies about 10 leagues *n. e.* of Nooheeva. To this island Captain Roberts gave the name of Massachusetts. Captain Ingraham had before called it Washington.]

OAITIPIHA or Aitepcha Bay, situate near

the *n. e.* end of the lesser peninsula of the island of Otaheite, has good anchorage in 12 fathoms. Lat. 17° 46′ *s.* Long. 149° 14′ *w.*]

[OAK Bay, or the DEVIL's Head, in the bay of Fundy, is nine leagues *s. s. e.* of Moose island. It is very high land, and may be seen at 10 or 12 leagues distance.]

[OAK Island, a long narrow island on the coast of N. Carolina, which with Smith's island forms the *s. w.* channel of Cape Fear river. See BALD Head, and Cape FEAR.]

[OAKFUSKEE. See TALLAPOOSE River.]

[OAKFUSKIES, an Indian tribe in the *w.* part of Georgia. The warrior Mico, called the White Lieutenant, has the sole influence over 1000 gun-men.]

[OAKHAM, a township in Worcester county, Massachusetts, 15 miles *n. w.* of Worcester, and 45 *w.* of Boston. It was incorporated in 1762, and contains 772 inhabitants.]

[OAKMULGEE River is the *s.* great branch of the beautiful Alatamaha in Georgia. At the Oakmulgee fields it is about 300 or 400 yards wide. These rich and fertile fields are on the *e.* side of the river, above the confluence of the Oconee with this river; these two branches are here about 40 miles apart. Here are wonderful remains of the power and grandeur of the ancients of this part of America, consisting of the ruins of a capital town and settlement, vast artificial hills, terraces, &c. See ALATAMAHA River.]

OAPAN, a settlement and head settlement of the district of the *alcaldía mayor* of Tixtlan in Nueva España; situate in a mild temperature, and inhabited by 212 families.

OAS, SANTA ROSA DE, a settlement of the missions which are held by the regulars of the Jesuits in the province and government of Mainas, of the kingdom of Quito; founded in 1665, on the shore of a river which enters the Napo.

OAS, a nation of Indians of the province and country of Las Amazonas, one of the missions which were held by the Jesuits; discovered and reduced to the faith by the Father Ramon de Santa Cruz, who founded a settlement on the shore of the river which empties itself into the Napo, in 1666.

[OATARA, a small woody island on the *s. e.* of Ulietea island, in the S. Pacific ocean; between three and four miles from which, to the *n. w.* are two other small islands in the same direction as the reef, of which they are a part.]

OAXACA, a province and *alcaldía mayor* of Nueva España; situate on the narrowest part of the continent, bounded *n.* and *s.* by the N. and S. seas, *e.* by the province and bishopric of Chiapa, *n. e.* by the province of Tabasco, *s. e.* by that of Soconusco, in an oblique line from this port in the S. sea to the former port in the N. sea, and in the opposite rhumbs, by the province of Tlaxcala and bishopric of La Puebla de los Angeles, its line of division from one sea to the other forming a figure of S. Its length from *e.* to *w.* by the coast of the S. sea is 96 leagues, namely, from the river Taquelamama to the port Soconusco, and 50 leagues by the *n.* coast, from the river Alvarado to that of Goazacoalco; its width is also about 50 leagues, that is to say, from the isthmus of the aforesaid river Alvarado to the port Aguatulco in the S. sea.

It is of a mild climate and fertile territory, and abounding particularly in mulberry-trees, which are finer here than in any other province of America. The greater part of it is mountainous, with the exceptiou of the valley of its name, although it has nevertheless large breeds of cattle. It produces sugar, cotton, wheat, *cacao*, plantains, and other vegetable productions, and has rich mines of gold, silver, and lead; gold being also found in the sands of its rivers. It produces likewise a quantity of cochineal, cinnamon, and crystal; and were its inhabitants industrious, it would be one of the richest provinces in America; but they lead an idle life, neither do the Indians in any degree cultivate the soil, owing to the ease with which they pick up the gold, which, however, is generally the employment of the women.

At the commencement of the conquest of the kingdom of Nueva España it was very populous, but its inhabitants have much fallen off. A great part of this province belongs to the estates of the house of Hernan Cortés, being granted to him by the Emperor Charles V. with the the title of Marques del Valle de Oaxaca. In the mountain of Cocola, dividing this province from Tlaxcala, are found mines of gold, silver, crystal, and vitriol, as also various kinds of precious stones.

This province was conquered by Juan Nuñez de Mercado, who was page to Hernan Cortes in 1521, and it was peopled in 1528 by Juan Sedeño and Hernando de Badajos. Its bishopric comprehends 21 *alcaldías mayores* and jurisdictions, which are,

Antequera,	Atlatlauca,
Quatro Villas	Miahuatlan,
Chichicapa,	Nexapa,
Guejolotitlan,	Xicayan,
Ixtepexi,	Teutitlan,
Tepozcoluca,	Nochiztlan,
Cuicatlan,	Yanguitlan,
Ixquintepec,	Teosaqualco,

2

Tecoquilco, Teutila,
Huameluca, Villalta.
Tehuantepec,

The above province was organised in 1535. The capital is the city of ANTEQUERA, to which is also given the name of the province, [both of which see.]

[The name of the province of Oaxaca, which other geographers less correctly call Guaxaca, is derived from a Mexican name of the city and valley of Huaxyacac, one of the principal places of the Zapotec country, which was almost as considerable as Teotzapotlan, their capital. The intendancy of Oaxaca is one of the most delightful countries in this part of the globe. The beauty and salubrity of the climate, the fertility of the soil, and the richness and variety of its productions, all minister to the prosperity of the inhabitants; and this province has accordingly from the remotest periods been the centre of an advanced civilization.

To give its description according to Humboldt, it is bounded on the *n.* by the intendancy of Vera Cruz, on the *e.* by the kingdom of Guatemala, on the *w.* by the province of Puebla, and on the *s.* for a length of coast of 11 leagues, by the Great ocean. Its extent exceeds that of Bohemia and Moravia together; and its absolute population is nine times less; consequently its relative population is equal to that of European Russia.

The mountainous soil of the intendancy of Oaxaca forms a singular contrast with that of the provinces of Puebla, Mexico, and Valladolid. In place of the strata of basaltes, amygdaloid, and porphyry with grünstein base, which cover the ground of Anahuac from the 18° to the 22° of lat. we find only granite and gneiss in the mountains of Mixteca and Zapoteca. The chain of mountains of trap formation only recommences to the *s. e.* on the *w.* coast of the kingdom of Guatemala. We know the height of none of these granitical summits of the intendancy of Oaxaca. The inhabitants of this fine country consider the Cerro de Senpualtepec, near Villalta, from which both seas are visible, as one of the most elevated of these summits. However, this extent of horizon would only indicate a height of 2350 metres. The visual horizon of a mountain of 2350 metres (7709 feet) of elevation has a diameter of 3° 20′. The question has been discussed, if the two seas could be visible from the summit of the Nevado de Tolca? The visual horizon of this has 2° 21′ or 58 leagues of radius, supposing only an ordinary refraction. The two coasts of Mexico nearest to the Nevado, those of Coyuca and Tuspan, are at a distance of 54 and 64 leagues from it. It is said that the same spectacle of the two seas may be enjoyed at La Ginetta, on the limits of the bishoprics of Oaxaca and Chiapa, at 12 leagues distance from the port of Tehuantepec, on the great road from Guatemala to Mexico.

The vegetation is beautiful and vigorous throughout the whole province of Oaxaca, and especially half way down the declivity in the temperate region, in which the rains are very copious from the mouth of May to the month of October. At the village of Santa Maria del Tule, three leagues *e.* from the capital, between Santa Lucia and Tlacochiguaya, there is an enormous trunk of cupressus disticha (*sabino*) of 36 metres, or 118 feet, in circumference. This ancient tree is consequently larger than the cypress of Atlixco, the dragonnier of the Canary islands, and all the boababs (Adansoniæ) of Africa. But on examining it narrowly, M. Anza observes, that what excites the admiration of travellers is not a single individual, and that three united trunks form the famous sabino of Santa Maria del Tule.

The intendancy of Oaxaca comprehends two mountainous countries, which from the remotest times went under the names of Mixteca and Tzapoteca. These denominations, which remain to this day, indicate a great diversity of origin among the natives. The old Mextecapan is now divided into Upper and Lower Mixteca (Mixteca Alta y Baxa). The *e.* limit of the former, which adjoins the intendancy of Puebla, runs in a direction from Ticomabacca, by Quaxiniquilapa, towards the S. sea. It passes between Colotopeque and Tamasulapa. The Indians of Mixteca are an active, intelligent, and industrious people.

If the province of Oaxaca contains no monuments of ancient Aztec architecture equally astonishing from their dimensions as the houses of the gods (*teocallis*) of Cholula, Papantla, and Teotihuacan, it contains the ruins of edifices more remarkable for their symmetry and the elegance of their ornaments. The walls of the palace of Mitla are decorated with *grecques*, and labyrinths in mosaic of small porphyry stones. We perceive in them the same design which we admire in the vases falsely called Tuscan, or in the frieze of the old temple of Deus Redicolus, near the grotto of the nymph Egeria at Rome. Humboldt caused part of these American ruins to be engraved, which were very carefully drawn by Colonel Don Pedro de la Laguna, and by an able architect, Don Luis Martin. If we are justly struck with the great] [analogy between the ornaments of the palace of Mitla and those employed by the Greeks and Romans, we are not on that account to give ourselves lightly up to historical hypotheses, on the possibility of the existence of ancient communications between the two continents. We must not forget, that under almost every zone, mankind take a pleasure in a rythmical repetition of the same forms which constitute the principal character of all that we call *grecques*, meanders, labyrinths, and arabesques.

The village of Mitla was formerly called Miguitian, a word which means in the Mexican language, a place of sadness. The Tzapotec Indians call it Leoba, which signifies tomb. In fact, the palace of Mitla, the antiquity of which is unknown, was, according to the tradition of the natives, as is also manifest from the distribution of its parts, a palace constructed over the tombs of the kings. It was an edifice to which the sovereign retired for some time on the death of a son, a wife, or a mother. Comparing the magnitude of these tombs with the smallness of the houses which served for abodes to the living, we feel inclined to say with Diodorus Siculus, (lib. i. c. 51) that there are nations who erect sumptuous monuments for the dead, because, looking on this life as short and passing, they think it unworthy the trouble of constructing them for the living.

The palace, or rather the tombs of Mitla, form three edifices symmetrically placed in an extremely romantic situation. The principal edifice is in best preservation, and is nearly 40 metres, or 131 feet, in length. A stair formed in a pit leads to a subterraneous apartment of 27 metres in length and eight in breadth, viz. 88 feet by 26. This gloomy apartment is covered with the same grecques which ornament the exterior walls of the edifice.

But what distinguishes the ruins of Mitla from all the other remains of Mexican architecture, is six porphyry columns, which are placed in the midst of a vast hall and support the cieling. These columns, almost the only ones found in the new continent, bear strong marks of the infancy of the art. They have neither base nor capitals. A simple contraction of the upper part is only to be remarked. Their total height is five metres, or 16.4 feet; but their shaft is of one piece of amphibolous porphyry. Broken down fragments, for ages heaped together, conceal more than a third of the height of these columns. On uncovering them, M. Martin found their height equal to six diametres, or 12 modules. Hence the symmetry would be still lighter than that of the Tuscan order, if the inferior diameter of the columns of Mitla were not in the proportion of 3 : 2 to their upper diameter.

The distribution of the apartments in the interior of this singular edifice, bears a striking analogy to what has been remarked in the monuments of Upper Egypt, drawn by M. Denon and the *savans* who compose the institute of Cairo. M. de Laguna found in the ruins of Mitla curious paintings representing warlike trophies and sacrifices.

The intendancy of Oaxaca has alone preserved the cultivation of cochineal (*coccus cacti*), a branch of industry which it formerly shared with the provinces of Puebla and New Galicia. M. Humboldt asserts, that all the *vaynilla* consumed in Europe comes from this province and Vera Cruz. — Oaxaca, he adds, furnishes annually 32,000 *arrobas* of cochineal, which, at 75 dollars, are worth 2,400,000 dollars.

The family of Hernan Cortes (as Alçedo has observed) bears the title of Marquis of the Valley of Oaxaca. The property is composed of the four Villas del Marquesado, and 49 villages, which contain 17,700 inhabitants.

The population of the intendancy of Oaxaca amounted, in 1803, to 534,800 souls; the extent of surface in square leagues is 4447, giving 120 inhabitants to the square league.

The most remarkable places of this province are, Oaxaca, or Antequera, (which see), Tehuantepec, San Antonio de los Cues.

The mines of this intendancy worked with the greatest care are, Villalta, Zolago, Yxtepexi, and Totomostla.]

Bishops who have presided in Oaxaca.

1. Don Juan Lopez de Zarate, canon of the holy church of Oviedo, licentiate in theology, and of great knowledge in both kinds of law: he left the reputation of having been a good preacher, made the visitation of the bishoprics, and regulated the parishes, bringing as curates some monks of the province of Mexico, there not being sufficient clergy; in the which undertaking he suffered great hardships, not only from an anxiety of mind, but the heat of the climate and difficulties of the roads: he passed over to Mexico to assist at the first council, where he died in 1554.

2. Don *Fr.* Bernardo de Alburquerque, of the order of S. Domingo, native of the town of this name: he studied in the university of Alcalá, where he made equal progress in science as in virtue: he sought the habit of lay-brother in the

convent of Salamanca, but his eloquence having been overheard in a certain dispute, he was commanded by the prior to begin his noviciate as a priest: he was one of the first religious men that passed over to America, and here he learnt the Zapotecan tongue: he obtained various prelacies, until he became provincial, and the Emperor Charles V. presented him to the bishopric of this diocese in 1555; but he regularly observed the rules of his order, was extremely charitable, and gave his episcopal palace to be converted into a monastery for nuns, which was called De Santa Catalina de Sena; he died, as it is supposed, a saint, in 1579.

3. Don *Fr.* Bartolomé de Ledesma, a monk of the same order as the former, native of Niera in the bishopric of Salamanca: he passed over to Peru as confessor to the viceroy Don Martin Enrriquez, and from thence to Mexico; was first professor of theology in his university, presented to the bishopric of Oaxaca in 1581: he founded a college dedicated to the apostle S. Bartolomé: he was charitable, and carried his disinterestedness to a great pitch, so much so that it was with the most repeated persuasions that the Indians could prevail upon him on a certain occasion to accept of a small pot of balsam; and although he did accept it, he made them, in return, a present of a cup for their altar: he assisted at the third Mexican council, and it was in his time that occurred the prodigy of the cross of Guatulco; he died in 1604.

4. Don *Fr.* Baltasar de Covarrubias, of the order of San Agustin, native of Mexico, where he distinguished himself for his virtue and literature; presented by Philip III. to the mitre of Nuestra Señora de la Asuncion of the river La Plata, in 1601; promoted from thence to the mitre of Cazeres in the Philippine isles, and afterwards to this in 1605, from whence he passed to that of Mechoacán in 1608.

5. Don Juan de Cervantes, native of Mexico: he began his studies in that university, and followed them up in the university of Salamanca, where he graduated as doctor in theology, was treasurer of the church of La Puebla, canonical lecturer and archdeacon of the church of his native place, professor of writing in its university, governor of the archbishopric, judge in ordinary, *calificador* of the holy office, and elected bishop of this church in 1608; he preached to his parishioners with great effect, and was very charitable, built a chapel in the church, where he might place the miraculous image of Guatulco, sending a bit of it, about the size of a hand, richly adorned, to the pontiff Paul V.; he died on the eve of the exaltation of the cross in 1614.

6. Don *Fr.* Juan Bartolomé de Bohorques, of the order of S. Domingo, native of Mexico: he was lecturer of philosophy and theology, master in his religion, rector of the college of S. Luis de la Puebla, prior of the convent of Izucar and provincial: he graduated as doctor in the university of Mexico, passed over to Spain, where he was elected bishop of Venezuela, and promoted to this church in 1617. In his time occurred the miraculous appearance of the Virgin de la Soledad at Oaxaca: he died in 1633.

7. Don Leonel de Cervantes, also native of Mexico: he passed over to Spain to study in the university of Salamanca, where he obtained the degree of doctor in sacred canons, was *maestre-escuela* and archdeacon of the holy church of Santa Fé, provisor and vicar-general of the archbishoprics of Don Bartolomé Lobo and Don Fernando Arias de Ugarte; presented to the bishopric of Santa Marta in 1620, where he remained five years, and from thence promoted to Guadalaxara in 1615, and afterwards to this; but before he could enter it he died at Mexico in 1637.

8. Don Bartolomé de Benavente y Benavides, native of Madrid: he studied in the university of Siguenza, and graduated as licentiate in canons and doctor in theology; was made canon of the church of Lima, as also *maestre-escuela*, archdeacon, commissary of crusade, visitor-general of its archbishopric, and professor of that university; presented to this bishopric, where he entered in 1639; he visited the whole diocese, and died in 1652.

9. Don *Fr.* Diego de Evia y Valdes, of the order of San Benito, native of the principality of Asturias; presented to the bishopric of the church of Durango in Nueva Vizcaya, and promoted to this church of Antequera, of which he took possession in 1654; he died two years after, in 1656.

10. Don Alonso de las Cuevas Davalos, native of Mexico, a man distinguished by his birth and virtues, and who by his literature had attained to the first prizes in the universities: he was magisterial canon, treasurer, and archdeacon of La Puebla, and the same in the metropolitan church of Mexico, which he left to become prelate of this church: he was charitable in the extreme, and considered a model for bishops, promoted to the archbishopric of Mexico in 1664; he died before he could take possession.

11. Don *Fr.* Tomas de Monterroso, of the order of S. Domingo, master in the same order, a man who had obtained great fame in the professorships

and in the pulpits at Madrid, and the king having here heard him preach a sermon, immediately appointed him to this bishopric, of which he took possession in 1665; he founded the seminary college, and died in 1678.

12. Don Nicholas del Puerto, native of the town of Santa Catalina de las Minas, in the archbishopric of Mexico, collegiate of the college of San Ildefonso of this city, doctor of sacred canons in its university, and advocate of the royal audience, where he was admitted, through his eloquence, canonical doctor of the metropolitan church, commissary-general of crusade in that kingdom. The steps which he took at the time of the expiration of the bull, not only brought down upon him the approbation of the court, but induced the king to present him to this bishopric, of which he took possession in 1679: he founded the professorship of grammar and philosophy, and two of theology, in the seminary college, and left to the same his grand library; he died in 1681.

13. Don Isidro Sariñana, native of the city of Mexico, doctor in sacred theology, professor of sacred writings in its university. To the study of letters he added the exercise of the virtues, as well in the parish of Santa Cruz as of Sagrario: he was then canonical lecturer, chanter, and archdeacon of that church, *calificador* of the holy office, and synodical examiner of the archbishopric, and bishop of Oaxaca, where he acquired great esteem for his fine discernment: he was an eloquent preacher, prudent, humble, affable, benign, pacific, charitable, and zealous for the public weal; he died in 1696.

14. Don *Fr.* Angel Maldonado, of the order of San Bernardo, native of Oeaña, doctor and professor of theology in the university of Alcalá, master in his religion: he was presented to the bishopric of Honduras, and promoted to this church in 1702: he affected a love of great poverty, rebuilt the cathedral, which entirely ruined him, refused the promotions to the bishoprics of Mechoacán and Orihuela in Spain, to which he was promoted, frequently visited his diocese, and died to the universal regret in 1728.

15. Don *Fr.* Francisco de Santiago y Calderon, of the order of La Mercced, native of Torralva in the bishopric of Cuenca, lecturer of arts in the convent of Huete, and of sacred theology in the colleges of Salamanca and Alcalá, rector and *comendador* of the college of Madrid, provincial of Castilla, visitor of the convents of Galicia and Asturias, theologist of La Nunciatura, decreed by the council of the Indies to be bishop of Guatemala in 1728, and afterwards promoted to this bishopric, of which he took possession in 1730: he consecrated its cathedral, added to it two fine towers and a clock, and, having paid the visitation, given great alms, and nicely fulfilled the duties of his station, he died in 1736.

16. Don Tomas Montaño, native of Mexico, collegiate of the royal college of San Ildefonso, prebendary and synodical examiner of the bishopric of Mechoacán, *medio-racionero*, canon, treasurer, chanter, archdeacon, and dean, of the metropolitan of his native place, in the university of which he was professor of theology, and thrice rector; elected bishop of this church of Oaxaca in 1737; he made his entry the following year, and governed only three years, in which short period he completely gained the affection of his flock by his kind treatment and disinterestedness: he shewed his zeal both in the pulpit as in conferences, which he was accustomed to hold with the clergy one day in every week: he endowed a fellowship in the college of San Ildefonso for some poor nobleman, and also established a mass of grace for the last day in the year in the college of the abolished order of Jesuits; he died in 1742.

17. Don Diego Felipe Gomez de Angulo, native of Burgos, advocate of the royal audiences of Guatemala and Mexico: he obtained through his merits a curacy in the first of these two cities, where he was also provisor, afterwards dean of the holy cathedral of Puebla, and for a long time governor of the bishopric. Being presented to the bishopric of Antequera in 1745, he began his mission by redressing grievances and succouring the needy: he also etablished an holy jubilee: his affability, kindness, and general virtue, caused him to be esteemed by all, and he died in 1752.

18. Don Buenaventura Blanco y Helguero, native of Valladolid, collegiate in the *mayor collegio* of San Ildefonso de Alcalá, canonical doctor of the holy church of Calahorra, visitor, provisor, and vicar-general of that bishopric, and elected to this of Antequera in 1754: he was a most charitable prelate, and left behind him evident marks of his zealousness and ability, and died full of virtues in 1764.

19. Don Miguel Anselmo Alvarez de Abreu y Valdes, native of Teneriffe, one of the Canaries, doctor in sacred canons in the university of Sevilla, secretary of the chamber and government of the most illustrious Señor Don Domingo Guerra, bishop of Segovia and confessor of the Queen Dowager, *racionero*, confessor, and prior of the holy cathedral church of Canarias, judge of the reverend apostolical chamber, and of the holy tribunal of crusade, judge, examiner, and visitor of

the aforesaid bishopric, auxiliary to that of the Puebla de los Angeles; elected to that of Comayagua, and to this of Antequera in 1765; he died in 1774.

20. Don Joseph Gregorio de Ortigosa, native of Viguera in the diocesse of Calahorra; nominated bishop of Oaxaca in 1775.

[Oaxaca, or Guaxaca, a principal town of the intendancy of this name, the ancient Huaxyacac, called Antequera at the beginning of the conquest. Thiery de Menonville only assigns 6000 inhabitants to it; but by the enumeration in 1792 it was found to contain 24,000. See Antequera.]

OBACATIARAS, a barbarous nation of Indians of Brazil, who live in some islands at the entrance of the mouth of the river San Francisco. They are at present far from numerous; their arms are bows and arrows, and they live by fishing and the chase.

[OBED'S River, in Tennessee, runs *s. w.* into Cumberland river, 290 miles from its month, by the course of the stream. Thus far Cumberland river is navigable for large vessels.]

OBEITA, or Ubeyta, a country, anciently very populous and rich, in the old province of Tensa in the Nuevo Reyno de Granada, in the *e.* part, and below the dominion of the king of Tunja. This country was entered by Ximinez de Quesada in 1537, who made a great booty. It is at present much reduced, consisting of nothing but a poor settlement of Indians.

OBERABA, a lake of the province and government of Paraguay, near the shore of this river.

[OBION, a navigable river of Tennessee, which runs *s. w.* into the Mississippi, 14 miles *s.* of Reelsfoot rivers. It is 70 yards broad, 17 miles from its mouth.]

OBIRABASI, a river of the province and country of Las Amazonas. It rises in the territory of the Anamaris Indians, runs *n.* and enters the Madera.

OBISPO, a settlement of the province and *corregimiento* of Copiapo in the kingdom of Chile; situate near the coast, between the ports of Huasco and Totoral.

Obispo, another settlement, in the province and government of Maracaibo and Nuevo Reyno de Granada; situate on the shore of the river S. Domingo, *n. n. e.* of the city of Barinas Nueva.

Obispo, a shoal of the sound of Campeche and government of Yucatán, near the coast.

Obispo, a *farallon* or isle on the coast of the province and *corregimiento* of Quillota in the kingdom of Chile, between the point of Caramilla and the port of Castello Viejo.

[OBITEREA, an island 100 leagues *s.* of the Society islands. Lat. 22° 40′ *s.* Long. 150° 50′ *w.* It contains no good anchorage, and the inhabitants are averse to the intrusion of strangers.]

OBRAGILLO, a settlement of the province and *corregimiento* of Canta in Peru, annexed to the curacy of its capitol.

OBRANDIUE, a settlement of the province and *captainship* of Rio Grande in Brazil; situate on the coast, between the shoal of Las Salinas and the point of Piedras.

OBSCURO, a river of the province and government of Veraguay, and kingdom of Tierra Firme. It rises in the mountains close to the *n.* coast, runs *w.* and enters the sea not far from the mouths of the Toro.

OCABAMBA, a large, fertile, and beautiful valley of the province and *corregimiento* of Calca and Lares in Peru, of the district of the second; bounded by the infidel Chuncos Indians. It has no formal town or settlement in it, but is full of sugar and cocoa estates, inhabited by very many families. It is irrigated by the river Pilcomayo, which abounds in fine fish, such as shad, dories, and pejerreyes, or king-fish. In spiritual concerns it is under the bishop of Cuzco, who nominates two priests with the titles of *beneficiados*, who administer the sacraments in the chapels of the said estates. At some distance are some *ranchos*, or temporary habitations, of the Chunchos Indians, who were beginning to be reduced to the faith by the religious order of San Francisco; but just as there was every ground for hoping that they would do well, they burst out into an insurrection, and relapsed into their pristine gentilism, in 1744, putting to death many of the Neofitos, and causing others to fly.

OCABAYA, a settlement of the province and *corregimiento* of Sicacica in Peru, annexed to the curacy of Pasca.

OCABITA, a province, large and rich in the time of the Indian gentilism, of the Nuevo Reyno de Granada, to the *e.* of the city of Santa Fé. It is named from a cacique or chief who was then its governor, and was taken possession of in 1537 by Juan de San Martin, but the natives rose up in 1541, fortifying themselves on the top of a great rock, where they held out for a long time against the attacks of different Spanish captains, until that, persuaded by the eloquence of Alonso Marin, they capitulated and made a peace, which they never after broke. This nation is, at the present day, nearly extinct, and nothing remains of it but a miserable village bearing its name.

OCACOCK, a small island of the N. sea, near

the coast of the province of S. Carolina, between the islands of Hatteros and cape Core. This island forms the *e.* side of the entrance into Pamlico sound, which entrance or inlet bears the same name.

[OCAMARE, See OCUMARE.]

OCASTA, a city of the province and government of Santa Marta in the Nuevo Reyno de Granada; situate in the *llanura* of Hacari, from whence it is also called Santa Ana de Hacari; founded by Francis Hernandez in 1572, and translated to the spot where it now stands in 1576, in the province of the Carates Indians. It is small, but cheerful and beautiful; its territory is very fertile and pleasant, and the wheat and sugar are of excellent quality. It has a very good parish church, a convent of San Francisco, and another of San Agustin. In it dwell different families of rank and distinction, and the women are in general of nice appearance. It lies about 60 miles *s. s. e.* of Laguna Zapatosa, to the *e.* of the Grande de la Magdalena, on the *n.* shore of the river Oro, and having a good embarking place in the river Lebrija, near where this communicates with the Magdalena. This city is 218 miles *n.* with a slight inclination to the *e.* Lat. 8° 2′ *n.* Long. 73° 20′ *w.*

OCASTA, a settlement of the province and *corregimiento* of Lucanas in Peru, annexed to the curacy of Laramate.

OCASTA, a river of the province of Mexico and kingdom of Nueva España. It runs into the sea at the bay of Mexico, between the river Alvarado and the *sierras* of San Martin.

OCATLAN, a settlement of the head settlement of the district and *alcaldía mayor* of Barca in the kingdom of Nueva Galicia, close to its capital on the *w.*

OCAY, a small river of the province and government of Buenos Ayres in Peru. It runs *w.* and enters the Uruguay, between the rivers Guating and De Aguas.

[OCCOA, or OCOA, a bay on the *s.* side of the island of St. Domingo, into which fall the small rivers Sipicepy and Ocoa. It lies *e.* of Neybe or Julienne bay, and is bounded *s. e.* by point Salinas, and *w.* by the *e.* point at the mouth of Bya river. Spanish ships of war anchor in this bay. Point Salinas is 22 leagues *w.* of the city of St. Domingo.]

[OCCOA, a bay near the *e.* end of the island of Cuba, in the windward passage, about 20 miles *e.* of Guantanamo bay.]

[OCCOCHAPPO, or BEAVER Creek, in the Georgia *w.* territory, empties through the *s. w.* bank of Tennessee river, just below the muscle shoals. There is a portage of only about 50 miles from this creek to the navigable waters of Mobile river. The mouth of this creek is in the centre of a piece of ground, the diameter of which is five miles, ceded by the *s.* Indians to the United States for the establishment of trading posts.]

[OCCONEACHEY Islands, two long narrow islands at the head of Roanoke river in Virginia, just below where the Staunton and Dan unite and form that river.]

OCCOQUAN, a small river of the province of Virginia in N. America. It runs *e.* and enters the Patowmack.

OCHANACHE, JESUS DE, a settlement of the conversion and missions held there by the religious order of San Francisco, in the province of Caxamarquilla or Pataz, and kingdom of Peru; situate on the shore of the river Gibitas, near its entrance into the Maranon.

[OCHILLA Island, more properly ORCHILLA, which see.]

OCHOA, a settlement of the province and *corregimiento* of Quillota, and kingdom of Chile; situate on the shore of the river Quillota.

OCHOA, a river of the province and government of Honduras, which runs *n.* and enters the sea between those of Cangrejo and Pico de Gata.

OCKA, a settlement of Indians of S. Carolina; situate on the shore of the river Albama, where the English have an establishment and fort for its defence.

OCLAYAS, a small river of the province and government of Tucumán in Peru, which runs *w.* and enters the Bermejo. On its shores the fathers and missionaries Gaspar Ossario and Antonio Ripario, of the Jesuits, were murdered by the infidel Indians.

OCMULGI, a river of the province of Georgia in N. America. It runs in a very abundant stream to *s. e.* and enters the Alatahama, close to the settlement of Fourche.

OCOA, a small river of S. Domingo, which runs into the sea seven leagues from Nizao, and nine from the town of Azua.

OCOBAMBA, a settlement of the province and *corregimiento* of Andahuailas in Peru, annexed to the curacy of Ongoy.

OCOBAMBA, another settlement, of the province and *corregimiento* of Castro-Virreyna, annexed to the curacy of Cordoba.

OCOCO, a settlement of the same province and kingdom as the former, and also annexed to that curacy.

OCOI, a small river of the province and government of Paraguay in Peru, which runs *w.* and enters the Parama between the Piracaby and the Cay.

OCOLCO, a settlement of the province of Mexico in Nueva España. In the time of the Indians it was the place where the workmen used to be, who undertook to adorn with feathers and precious metals the palace of the emperor; it was near the city of Tezcuco, but now no longer exists.

OCONA, a settlement of the province and *corregimiento* of Cumana in Peru, situate near the sea, having a creek in which much fish is caught.

OCONA, a river of the province and *corregimiento* of Chumbivilcas in the same kingdom; it rises in the *cordillera* here, runs to that of Condesuyos de Arequipa, where, swollen with the waters it receives from various other streams, it changes its name, and enters, with a large body, into the sea at the former port.

OCONAHUA, a settlement of the head settlement of the district and *alcaldía mayor* of Izatlan in Nueva España.

OCONGATE, a settlement of the province and *corregimiento* of Quispicanchi in Peru, annexed to the curacy of the settlement of Cerca in the province of Paucartambo.

OCONI, a settlement of Indians in the province of Georgia, and N. America; situate on the shore and at the source of the river of its name, where the English have an establishment for their commerce with the Indians, and a fort for their defence.

OCONI. The aforesaid river runs *s. e.* and enters the Alatahama or Georgia. On its shores were formerly many small settlements of Indians; but in the war waged against them by the English in 1715, they were forced to retire to the river Chatahotchi.

[The Oconi is the *n.* main branch of Alatamaha river, Georgia. It is in many places 250 yards wide. Its banks abound with oak, ash, mulberry, hickory, black walnut, elm, sassafras, &c.]

OCONORI, a settlement of the missions which were held by the Jesuits in the province and government of Cinaloa.

OCOPA, SANTA MARIA DE, a college of the missionaries of the order of San Francisco, in the province and *corregimiento* of Jauja in Peru; situate in a desert spot at the foot of some lofty and dry mountains; its temperature is dry, and although the hoar frosts are frequent in June and July, yet is it not so cold here as in Europe; but at this time the strong winds are very disagreeable, and in November, December, January, and February, there are constantly great tempests with much rain. This college was founded by the *Fr.* Francisco de San Joseph, in 1724, for the purpose of establishing missions for the many infidel Indians who have been reduced since 1709. Here, having obtained a grant of the site, he built a college capable of containing 40 monks; also a church, entirely of stone, with 11 altars, and adorned with rich ornaments and precious jewels, all of which were procured by the aims of the devout, and the great zeal of the missionaries.

This college enjoys the title of Relicario del Peru; it has by it two houses of entertainment for the numerous people who flock to it, they having sometimes, in holy week, amounted to upwards of 1000. From this college the missionaries issue forth amongst the mountains to reduce to the faith the infidels, and great has been the fruit of their labours, even to the present day. Convinced of the use of this institution, the pious King Charles III. afforded to it much assistance; and in 1578, it was erected by his Holiness Clement XIII. into the college *De Propaganda Fide*.

It is about 116 miles to the *e.* of Lima, 33 *s. e.* of the city of Tarma, and five to the *e.* of the river Jauja; bounded *n.* by the province of Tarma, *e.* by the *serranías* of the settlements of Comas and Andamarca, and by the mountains of the infidels, *w.* by the province of Yauyos, and *s.* by that of Huanta, in the bishopric of Guamanga. Lat. 12° 2′ *s.* Long. 75° 2′ *w.*

OCOPI, SANTA ROSA DE, a settlement of the province and government of La Guayana or Nueva Andalucia, in the Nuevo Reyno de Granada; one of the missions held there by the order of San Francisco; founded in 1723 by the *Fr.* Joseph de Vega, who assembled together a certain number of Chaymas Indians, to whom were added, in 1732, by the *Fr.* Matias Garcia, 20 families of the Guaraunos, the whole, at the present day, amounting to 650 persons. The situation is beautiful, the territory level, at a small distance from the river of its name. It abounds in palms of *moriche*, and in the fruits of the country, as also in cattle, having excellent pastures. The natives are very devout and fond of their religion.

OCORO, a settlement of the province and *corregimiento* of Guanta in Peru; annexed to the curacy of Colcabamba.

OCOSCONI, a settlement of the missions of, and a *reduccion* of Indians made by, the Jesuits in the province and government of Cinaloa.

OCOTELULCO, a settlement and head settlement of the district of the province of Tlaxcala, in the time of the republic of the Indians, and one of the settlements which assisted Hernan Cortes in the chastisement of the city of Cholula.

OCOTENANGO, a settlement of the province and *alcaldía mayor* of Zedales in the kingdom of Guatemala.

OCOTEPEC, Santo Tomas de, a settlement of the head settlement of the district and *alcaldía mayor* of Tepozcolula, in the province and bishopric of Oaxaca and kingdom of Nueva España; of a mild temperature. It contains 216 families of Indians, including those of the wards of its district. Its precincts are extremely fertile in seeds and vegetable productions, and especially in cochineal, in which consists its principal commerce. Ten leagues *w.* of its capital.

Ocotepec, San Salvador de, a settlement in the head settlement of the district and *alcaldía mayor* of Cuernavaca in the same kingdom.

Ocotepec, San Pedro de, another settlement of the head settlement of Xiculá, and *alcaldía mayor* of Nexapa, in the same kingdom; of a cold temperature. It contains 150 families of Indians, and is five leagues *n. e.* of its head settlement.

Ocotepec, another, of the *alcaldía mayor* of Villalta in the same kingdom. It is of a cold temperature, contains 41 families of Indians, and is nine leagues *e.* of its capital.

Ocotepec, another, of the province and *alcaldía mayor* of Los Zoques in the kingdom of Guatemala.

OCOTEQUILA, a settlement of the head settlement of the district of Acantepec, and *alcaldía mayor* of Tlapa, in Nueva España. It contains 25 families of Indians, and is two leagues to the *w.* of Clasivilungo.

OCOTIC, a settlement of the head settlement of the district and *alcaldía mayor* of Cuquio in Nueva España. Three leagues *w.* of its head settlement.

OCOTITLAN, a settlement of the head settlement of the district and *alcaldía mayor* of Tlapa in Nueva España. It contains 48 families of Indians, and is 12 leagues *w. n. w.* of its capital.

OCOTLAN, S. Domingo de, a settlement and head settlement of the district of the *alcaldía mayor* of Antequera, in the province and bishopric of Oaxaca and kingdom of Nueva España; of a mild temperature. It contains 1130 families of Indians, including those of two wards in its district. These Indians are particularly devoted to the culture of cochineal and the land, having most abundant harvests. Six leagues *s.* of its capital.

Ocotlan, another settlement, in the head settlement of the district and *alcaldía mayor* of Zapopan in the same kingdom. It is inhabited by some *Mustees*, Mulattoes, and Indians, who are given to agriculture.

Ocotlan, another, with the dedicatory title of San Francisco, in the head settlement of the district of Coronango, and *alcaldía mayor* of Cholula, in the same kingdom. It contains 87 families of Indians, and is a league and a half *n.* of its capital.

OCOUICA, Santa Clara de, a settlement of the head settlement of the district of S. Andres de Cholula, and *alcaldía mayor* of this name, in Nueva España. It contains 76 families of Indians, and is a league and a half *s.* of its capital.

OCOZINGO, a town and capital of the province and *alcaldía mayor* of Los Zedales in the kingdom of Guatemala; situate in a beautiful valley, which is watered by several streams, and thus rendered fertile in maize, honey, cattle, birds, some *cacao*, and *achote;* and it would be much more productive were the natives at all industrious. It serves as a boundary or frontier against the infidel Indians. Some of the inhabitants have sown wheat and sugar-canes, and they have yielded beyond all expectation.

OCRAMANE, a point of land on the coast of the river S. Lawrence, *n. e.* of the city of Tadousac.

[OCRECOCK Inlet, on the coast of N. Carolina, leads into Pamlico sound, and out of it into Albemarle sound, through which all vessels must pass that are bound to Edenton, Washington, Bath, or Newbern. It lies in lat. 34° 55′ *n.* A bar of hard sand crosses the inlet, on which is 14 feet water at low tide. The land on the *n.* is called Ocrecock, that on the *s.* Portsmouth. Six miles within the bar there is a hard sand shoal, which crosses the channel called the Swash. On each side of the channel are dangerous shoals, sometimes dry. Few mariners, however well acquainted with the inlet, choose to go in without a pilot; as the bar often shifts during their absence on a voyage. It is about 7¼ leagues *s. w.* half *w.* of cape Hatteras.]

OCROS, a settlement of the province and *corregimiento* of Vilcas-Huaman in Peru; annexed to the curacy of Vizchongo.

Ocros, another, in the province and *corregimiento* of Caxatambo in the same kingdom.

OCSABAMBA, a river of the province and *corregimiento* of Chumbivilcas in Peru. It rises in the *cordillera* of the Andes, *s.* of the settlement of Cocha, and enters the Apurimac.

OCTAGROS, a settlement of Canada in N. America, in the country of the Renards or Foxes, on the shore of the river Maskoutens.

OCTIBEA, a settlement of the Indians of S. Corolina; situate at the source of the river Soulahue.

OCTLATLAN, an ancient province of the kingdom of Guatemala, in the time of the Indians, but now confounded with other provinces.

OCTOHATCHI, a small river of the province of Georgia in N. America. It runs *s. e.* and enters the Ocmulgi between the Echecouna and the Togosa.

OCTORARA, a small river of the province and colony of Maryland. It runs *s. w.* and enters the Susquehanna.

OCTOYAS, a river of the province and government of Tucumán in Peru, of the district of the city of Jujui. It runs *e.* and enters the Bermejo.

OCTUPA, a settlement of the province of Tlaxcala in Nueva España, near the city of La Puebla de los Angeles.

OCUAPA, San Francisco de, a settlement and head settlement of the district of the *alcaldía mayor* of Acayuca in Nueva España. It is the head settlement of the district of the Ahualulcos Indians, contains four families of Spaniards, 20 of Mulattoes, and as many of Indians. Forty-three leagues *s.* of its capital.

OCUBIRI, a settlement of the province and *corregimiento* of Lampa in Peru.

OCUILA, with the surname of Santiago de, a head settlement of the district of the *alcaldía mayor* of Marinalco in Nueva España. It is situate on the shore of a mountain, at the top of which is a convent of the religious order of S. Agustin, and six settlements subject to its jurisdiction. Its population thus collectively consists of 424 families of Indians, and six of Spaniards and *Mustees*. Its temperature is hot and moist, and its commerce consists in wheat, maize, French beans, and other seeds; also in very many fruits peculiar to the country; likewise in coal and wood, which are carried for sale to Mexico. Three leagues *n.* of its capital.

OCUITECOS, a nation of Indians of the kingdom of Nueva España, in the jurisdiction at present under the name of Tasco, conquered and united to the empire of Mexico by the Emperor Itzcohuatl.

OCUITUCO, a very considerable settlement of the district of the *alcaldía mayor* of Coautla in Nueva España. It contains 60 families of Spaniards, 12 of *Mustees*, and 110 of Mexican Indians, with a good convent of the religious order of San Agustin, the first founded in that kingdom, and as such, one of the first houses of that order. This district was formerly a separate *alcaldía mayor*, and was afterwards united to that with which it is at present incorporated. The settlement is five leagues *e. n. e.* of the capital.

OCUMARE, or Ocamare, Sabana de, a settlement of the province and government of Venezuela in the Nuevo Reyno de Granada; situate on the coast, about 28 miles *s. s. e.* of the city of Caracas, and about 22 *n. n. w.* of the settlement of Alta Gracia, between the mountains of Alta Gracia and the river Tuy.

[This port lies five leagues *e.* of Puerto Cabello. It affords good anchorage, and is sheltered from the *n.* wind. On the *e.* point of the bay is a fort of eight 12 and 8-pounders. Also a village of the same name, which lies a league from the bay. This village is watered by a river named Ocumare, which falls into the bay at the foot of the fort.]

Ocumare, Morno de, a mountain of the same province and government, on an island close upon the coast, between point Barburata and port Choroni.

OCUMUCHO, a settlement of the head settlemaut of the district and *alcaldía mayor* of Periban in Nueva España; of a cold and moist temperature, and containing 190 families of Indians. Three leagues *e.* of the settlement of Patamba.

OCURI, a settlement and *asiento* of silver mines of the province and *corregimiento* of Chayanta or Charcas in Peru; annexed to the curacy of Pitantora.

OCUYOACAQUE, San Martin de, a principal and head settlement of the district of the *alcaldía mayor* of Metepec in Nueva España. It contains 334 families of Indians, and is the head of the curacy, to which various others are annexed. Two leagues *e.* of the capital.

ODUCHAPA, a river of the province and *corregimiento* of Loxa in the kingdom of Quito. It runs *w.* on the confines of the province of Cuenca, and enters the river Leon, in lat. 3° 26′ *s.*

Oduchapa, a small settlement of this province; situate on the *n.* shore of the former river.

[OENEMACK, the *s.* point of Bistol bay, on the *n. w.* coast of N. America. Lat. 51° 30′ *n.* Long. 160° 30′ *w.*]

OESTE Point, a *w.* extremity of the island of Tortuga, near the island S. Domingo.

Oeste, another, in the island called Caicopequeño, of the islands denominated Caicos.

OEUFS, or Eggs, an island in the river S. Lawrence, in the province of New France or Canada in N. America. It is small, and lies at the mouth or entrance of the river Trinidad.

OFOGOULAS, a settlement of Indians of the

nation thus called, in the province and government of Louisiana in N. America, on the shore of the Mississippi.

OGECHI, a settlement of Indians of S. Carolina; situate at the source of the river of this name, where the English have a fort and establishment for their commerce.

[OGECHEE, a river of Georgia, 18 miles *s.* of Savannah river, and whose courses are nearly parallel with each other. It empties into the sea opposite the *n.* end of Ossabaw island, 18 miles *s.* of Savannah. Louisville, Lexington, and Georgetown, are on the upper part of this river.]

OGERON, PRISION DE L', a settlement of the French, on the *n.* coast of this part of the island S. Domingo, between the river Tuerto and port Marge.

[OGLETHORPE, a new county on the *n.* side of Alatamaha river, *w.* of Liberty county. Fort Telfair is in the *s. e.* corner of this county, on the Alatamaha.]

[OHAMANENO, a small but good harbour, on the *w.* side of Ulietea, one of the Society islands, in the S. Pacific ocean. Lat. 16° 45′ *s.* Long. 151° 38′ *w.* The variation of the compass in 1777 was 6° 19′ *e.*]

[OHAMENE Harbour, a fine bay on the *e.* side of Otaba, one of the Society islands. It passes in by a channel between the two small islands Toahouta and Whennuaia. Within the reef it forms a good harbour, from 25 to 16 fathoms water, and clear ground.]

[OHERURUA, a large bay on the *s. w.* part of the island of Otaba, one of the Society islands, and the next harbour to the *n.* from Apotopoto bay. There is anchorage from 20 to 25 fathoms, and has the advantage of fresh water. The breach in the reef which opens a passage into this harbour is a quarter of a mile broad, in lat. 16° 38′ *s.* Long. 151° 30′ *w.*]

[OHETEROA, one of the Society islands, which is about 12 miles long and six broad, inhabited by a people of very large stature, who are rather browner than those of the neighbouring islands. It has no good harbour nor anchorage. Lat. 22° 27′ *s.* Long. 150° 47′.]

[OHETUNA, a harbour on the *s. e.* side of Ulietea, one of the Society islands.]

[OHEVAHOA, an island in the S. Pacific ocean. Lat. 9° 41′ *s.* Long. 139° 2′ *w.*]

[OHIO, a most beautiful river, separates the N. W. territory from Kentucky on the *s.* and Virginia on the *s. e.* Its current gentle, waters clear, and bosom smooth and unbroken by rocks and rapids, a single instance only excepted. It is one quarter of a mile wide at fort Pitt, 500 yards at the mouth of the Great Kanhaway, 1200 yards at Louisville, and at the rapids half a mile, but its general breadth does not exceed 600 yards. In some places its width is not 400, and in one place particularly, far below the rapids, it is less than 300. Its breadth, in no one place, exceeds 1200 yards; and at its junction with the Mississippi, neither river is more than 900 yards wide. Its length, as measured according to its meanders by Captain Hutchins, is as follows:—From fort Pitt to

Log's town	18½
Big Beaver creek . .	10¾
Little Beaver creek .	13½
Yellow creek . . .	11¾
Two creeks	21¾
Long reach	53¾
End of Long reach .	16½
Muskingum	26½
Little Kanhaway . .	12¼
Hockhocking . . .	16
Great Lanhaway . .	82½
Guiandot	43¾
Sandy creek	14½
Sioto, or Scioto . . .	48¾
Little Miami . . .	126¼
Licking creek . . .	8
Great Miami . . .	26¾
Big Bones	32½
Kentucky	44¼
Rapids	77¼
Low country . . .	155¾
Buffalo river . . .	64½
Wabash	97¼
Big cave	42¾
Shawanee river . . .	52½
Cherokee river . . .	13
Massac	11
Mississippi	46
	1188

In common winter and spring floods, it affords 30 or 40 feet water to Louisville, 25 or 30 feet to La Tarte's rapids, 40 above the mouth of the Great Kanhaway, and a sufficiency at all times for light batteaux and canoes to fort Pitt. The rapids are in lat. 38° 6′.

The inundations of this river begin about the last of March, and subside in July, although they frequently happen in other months; so that boats which carry 300 barrels of flour from the Monongahela, or Youhiogany, above Pittsburgh, have seldom long to wait for water. During these floods, a first-rate man of war may be carried from Louis-]

ville to New Orleans, if the sudden turns of the river and the strength of its current will admit a safe steerage. It is the opinion of some well informed gentlemen, that a vessel properly built for the sea, to draw 12 feet water, when loaded, and carrying from 12 to 1600 barrels of flour, may be more easily, cheaply, and safely navigated from Pittsburgh to the sea, than those now in use; and that this matter only requires one man of capacity and enterprise to ascertain it. A vessel intended to be rigged as a brigantine, snow, or ship, should be double-decked, take her masts on deck, and be rowed to the Ibberville, below which are no islands, or to New Orleans, with 20 men, so as to afford reliefs of 10 and 10 in the night. Such a vessel, without the use of oars, it is supposed, would float to New Orleans from Pittsburgh in 20 days. If this be so, what agreeable prospects are presented to the citizens in the *w.* country! The rapids at Louisville descend about 10 feet in the distance of a mile and a half.

The bed of the river is a solid rock, and is divided by an island into two branches, the *s.* of which is about 200 yards wide, but impassable in dry seasons. The bed of the *n.* branch is worn into channels by the constant course of the water, and attrition of the pebble-stones carried on with that, so as to be passable for batteaux through the greater part of the year. Yet it is thought that the *s.* arm may be most easily opened for constant navigation. The rise of the waters in these rapids does not exceed 20 or 25 feet. There is a fort situated at the head of the falls. The ground on the *s.* side rises very gradually. At fort Pitt the river Ohio loses its name, branching into the Monongahela and Alleghany.]

[OHIO Rapids lie in lat. 38° 6′ *n.* 705 miles below Pittsburg to the *s. w.* including the windings, but only 290 in a direct line, and 482 miles including the windings, and 180 in a direct line, from the confluence of the Ohio with the Mississippi. They are occasioned by a ledge of rocks that stretch across the bed of the river Ohio. The situation of the rapids is truly delightful. The river is full a mile wide, and the fall of the water, which is a constant cascade, appears as if nature had designed it to shew how inimitable and stupendous are her works. The town of Louisville commands a grand view of the rapids.]

[OHIO, the *n.* westernmost county of the state of Virginia, bounded *e.* by Washington county in Pennsylvania, and *n. w.* by the river Ohio, which divides it from the N. W. territory. It contains 5212 inhabitants, including 281 slaves. Chief town, Liberty.]

[OHIO COMPANY'S Purchase, in the N. W. territory, is a tract of excellent land, situated on the *n.* bank of the Ohio, *e.* of Colonel Syme's purchase.

At Cincinnati there is an office for the sale of lands, and in 1806 no less than 17,000 contracts, at the rate of two dollars per acre, were entered there, bearing the names of persons from all quarters of Europe as well as America.

In this tract there were about 2500 inhabitants in 1792; but it proved not more pre-eminent in fertility, than in industry and morals. It was admitted, as a state, into the union in 1803. Mr. Ashe does not mention the amount of its population, but we find Dr. Holmes states it to have been in that year upwards of 76,000; and it appears by the census of 1810, that its inhabitants amounted to 230,760 souls.]

[OHIO, Territory N. W. of the. See TERRITORY.]

[OHIO, Territory S. of the. See TENNESSEE.]

[OHIOPE, a small *n.* tributary stream of Alatamaha river in Oglethorpe county, Georgia.]

[OHIOPIOMINGO, a tract of land so called in the state of Kentucky, situated in Nelson county, on Ohio river, and *s. w.* of Salt river.]

[OHIOPYLE Falls, in Youghiogany river, are about 20 feet perpendicular height, where the river is 80 yards wide. They are 30 or 40 miles from the mouth of this river, where it mingles its waters with the Monongahela.]

[OHITAHOO, an island in the S. Pacific ocean. Lat. 9° 55′ *s.* Long. 139° 6′ *w.*]

OIBA, SAN MIGUEL DE, a settlement of the jurisdiction of the town of San Gil, in the province and *corregimiento* of Tunja, and Nuevo Reyno de Granada. It is of an hot temperature, but healthy, though badly situate, as lying in an hollow. It was entirely of Indians, but an intercourse with the whites had produced a considerable race of *Mustees*, and the few Indians that remained afterwards attached themselves to the settlement of Chitaraque. The inhabitants of this should amount, at the present day, to 900. They are a proud and haughty set, and put one another to death with great barbarity. Its territory produces much sugar-cane, maize, *yucas*, plantains, and other fruits. It is 19 miles *s. s. w* of San Gil, somewhat less from Socorro, and about half as much from Tirano, Charala, and Zimacota, its district being bounded by the four last mentioned settlements.

OICATA, a settlement of the same province and kingdom as the former. It is of a very cold temperature and subject to much wind, but abounding in wheat, maize, barley, *papas*, and other productions of a cold climate. Its natives breed cattle, and make a tolerable number of woven manufactures of their wool. It contains 50 housekeepers and 140 Indians, and is one league *n.* of the capital, close to the settlements of Chibata, Combita, and Tnta.

[OIL Creek, in Alleghany county, Pennsylvania, issues from a spring, on the top of which floats an oil, similar to that called Barbadoes tar, and empties into Alleghany river. It is found in such quantities, that a man may gather several gallons in a day. The troops sent to guard the *w.* posts halted at this spring and collected some of the oil, and bathed their joints with it. This gave them great relief from the rheumatic complaints, with which they were afflicted. The waters, of which the troops drank freely, operated as a gentle cathartic.]

OINARE, a large and rapid river of the province and government of Venezuela in the Nuevo Reyno de Granada. It runs from *s.* to *n.* and runs into the sea, opposite the island Orchilla, in lat. 10° 5′ *n.*

OINGSTOWN, or OISTINTOWN, a city of the island of Barbadoes, one of the Smaller Antilles, in the district and parish of Christ-church. It took its name from a Mr. Oistin, a person of large landed estate there, and one of its first settlers.

This town is small and consists only of a large street, at the extremity of which there is celebrated at the end of every week a considerable fair or market. In the *e.* part is the church, which is a very good one, and the town stands on a large bay, which facilitates its commerce, and although the room for anchorage is small, owing to the number of rocky shoals, there is 18 fathom of water to the *n.* and *s.* of the city, at two or three cables length from shore.

The town is defended by forts well furnished with artillery, and along the whole length of its coast an intrenchment is thrown up to prevent an enemy's landing. The vessels lie safe in the above port from the *n. w.*, *n. e.*, and *s. e.* winds, but are much exposed to the *w.* and *s.* Half a league to *w.* of the city is fort Maxwel, and at a league's distance, on point Nedhans, is Charles fort.

OISEAUX, or BIRDS, Isles of, in the gulf of S. Lawrence in N. America, one of those called Magdalena, to the *e.* of the isle of Brion.

[OISTINS Bay is near the *s.* extremity of the island of Barbadoes, in the W. Indies. It is formed to the *s. e.* by Kendal's point. The bay is well defended by forts. The town of Oistins stands on this bay.]

OJATATAS, or OJATAES, a nation of Indians of the province and government of Tucumán in Peru, discovered by the Colonel Don Antonio Tixera, in 1710, at an entrance he made to reduce the province of Chaco, when this nation manifested such a liking for the Spaniards, that they almost all, with one consent, became reduced to the Catholic faith. They used to dwell near the river San Francisco, and were afterwards removed to under the government of Buenos Ayres, to hinder them from returning to their idolatry, as has been the case with other nations of their district. These have thus been always constant to the faith.

OJIBAR, a settlement of the district of Babahoyo in the province and government of Guayaquil, and kingdom of Quito. It is 28 leagues from the capital, and in it ordinarily reside during the winter the curate and lieutenant of the district.

OJITOS, TRES, a settlement of the province and government of Sonora in N. America; situate on the shore of the gulf of California or Mar Roxo de Cortés.

OJO-CALIENTE, a settlement of the head settlement of the district and *alcaldía mayor* of Sierra de Pinos in Nueva España. Twelve leagues *w. s. w.* of its capital.

OJO-CALIENTE, another settlement, of the province of Taraumara, and kingdom of Nneva Vizcaya.

OJO-CALIENTE, another, of the missions which were held by the religious order of San Francisco, in the kingdom of Nuevo Mexico.

OKELOUSA, a settlement of Indians of S. Carolina; situate at the head of Buffalo river.

OLA, a settlement of the jurisdiction and *alcaldía mayor* of Natá in the province and kingdom of Tierra Firme; situate two leagues *n. e.* of the above town.

OLAGA, a river of the province and government of Maracaibo in the Nuevo Reyno de Granada, which rises in a *llanura* between the lakes Zapatosa and Maracaibo, runs *e.* and enters the Atole.

OLANCHO, SAN JORGE DE, a town of the province and government of Honduras. It is very poor and scantily peopled. Forty leagues from Comayagua.

OLANDA. See HOLLAND.

OLAYA, SANTA, a settlement of the province and *corregimiento* of Guarochiri in Peru.

[OLD Cape FRANÇOIS forms the *n.* point of Ecossoise or Cosbeck bay, on the *n. e.* part of the island of St. Domingo. All the French ships coming from Europe or the Windward islands, and bound to the *n.* or *w.* part of St. Domingo island, are obliged to come in sight of the cape Samana, (near 27 leagues *s. e.* by *e.* of this cape), or at least of Old cape François, on account of the dangers of shoals to the *e.* It is about five leagues *e.* of cape De la Roche.]

[OLD FORT Bay is situated at the *s.* end of the island of St. Lucia in the W. Indies, having St. Mary's island and bay to the *e.*]

[OLD FOUT Islands, in Esquimaux bay, on the coast of Labrador in N. America. Lat. 51° 24′ *n.* Long. 57° 48′ *w.*]

[OLD Harbour, on the *s.* coast of the island of Jamaica in the W. Indies, is to the *w.* of port Royal. There are a number of shoats and islands in the entrance to it. Under some of them there is safe riding, in from six to eight fathoms.]

[OLD MAN's Creek, in New Jersey, empties into Delaware river, about four miles below Penn's neck, and separates the counties of Salem and Gloucester.]

[OLD MEN's Port lies *n.* of Lima river in Peru, eight or nine miles *n.* of Cadavayllo river.]

[OLD ROAN, a town and harbour in the island of Antigua in the W. Indies.]

[OLD ROAD Bay, on the *s. w.* coast of the island of St. Christopher's in the W. Indies, between Church gut *w.* and Bloody point *e.* There is from five to 15 fathoms near the shore, and the least towards the fort.]

[OLD ROAD Town, on this bay, lies between E. and Black rivers, and is a port of entry.]

[OLD Town, or FRANK's OLD Town, on Juniatti river. See FRANKSTOWN.]

[OLD Town, in the state of New York, is situated on Staten island, three miles *s. w.* of New town, and 12 *s. w.* of New York city.]

[OLD Town, a small post-town of Maryland, situated in Alleghany county, in lat. 39° 36′, on the *n.* bank of Patowmack river, and *w.* side of Saw Mill run, 10 miles *s. e.* of Cumberland, 89 *w.* by *n.* of Baltimore, and 161 from Philadelphia.]

[OLD Town, in N. Carolina, near Brunswick.]

[OLD Town, a small town of Georgia, lying on the Ogeechee river, 85 miles *n. w.* by *w.* of Savannah.]

[OLEOUT, a small creek, which empties into the *e.* branch of Susquehannah, five miles *n. e.* of the mouth of Unadilla river.]

OLINALA, a settlement and head settlement of the district of the *alcaldia mayor* of Tlapa in Nueva España. It contains 10 families of Spaniards, 200 of *Mustees* and Mulattoes, and 162 of of Tlapanocos Indians.

OLINDA, a city, the capital of the province and *captainship* of Pernambuco in Brazil; situate on a lofty spot near the sea, surrounded with forts and wells or drains, which render the territory fertile, pleasant, and delightful. It stands upon four hills, the declivity of which forms an amphitheatre extremely pleasing to the sight.

It has very sumptuous buildings, and such is the grand church which belonged to the Jesuits, and which cost 120,000*l.* French, being one of the finest buildings in all America, and having every thing to render it perfect. Here are the following convents, namely, of the barefooted Carmelites, of San Benito of the Recogimiento de Nuestra Señora de la Concepcion, for ladies; also the parish of San Pedro Martir, the churches of Nuestra Señora del Rosario, De Guadalupe, De San Sebastian, De San Juan, which is a regular confraternity, and De Nuestra Señora de Monserrat; also a magnificent hospital and a sumptuous palace of the governor of the province, not to mention many superior and elegant private houses. It is garrisoned by two regiments of regular infantry, and has 3000 housekeepers.

This town was extremely opulent at the time that it was taken and destroyed by the Dutch in 1630, and the ruins which now remain speak what must have been its ancient splendour.

On one side of the city runs the large river Berberibe, of a rich and crystalline water, forming close to it a port called Baradero, where the sea runs in for upwards of half a league; and to take advantage of the river water, there is a large bridge constructed, over which runs an aqueduct. Here, also, on the bridge, there is a beautiful house of refreshment, where the populace frequently go in large societies to amuse themselves; and at the end of the bridge to the *s.* is a convent of the Carmelites in a retired and pretty spot. At a league's distance from the port of Baradero, along a strait isthmus, is the town of San Antonio de Arecife, which is the port of this capital, the intermediate space between the one and the other town being defended with many castles and batteries. The city is 76 miles *s.* of the city of Paraiba, in lat. 8° 12′ 30″ *s.* and long. 35° 5′ *w.*

OLINTEPEC, a settlement of the head settlement of the district of Tutepec, and *alcaldia mayor* of Xicayan, in the province and bishopric of Oaxaca, and kingdom of Nueva España. It contains 14 families of Indians, who live by the cultivation of cochineal, cotton, seeds, and tobacco. Fourteen leagues *n.* of the capital.

OLINTEPEC, another settlement, with the dedicatory title of San Juan, in the province and *alcal-*

día mayor of Gueguetenango in the kingdom of Guatemala; annexed to the curacy of Totonicapa, of Indians of the Quiché nation. It is of the doctrinal establishments of the order of San Francisco.

OLINTLA, a settlement of the head settlement of the district of Teutalpan, and *alcaldía mayor* of Zacatlan, in the same kingdom as the former.

OLITA, a settlement of the head settlement of the district and *alcaldía mayor* of Acaponeta, in the same kingdom as the former; situate on the shore of the river Cañas, nine leagues *s. w.* of its capital.

OLIVAR, a settlement of the province and *corregimiento* of Truxillo in Peru, founded in the valley of Virú, near a chasm or *quebrada*, six leagues from the settlement of S. Pedro.

OLIVARES, an island of the S. sea, near the coast lying between the river La Plata and the straits of Magellan, within the port Deseado. It is large, of an oval figure, and very near the coast. It was named by the naval captain Don Joaquin de Olivares, who discovered it in 1746, whilst reconnoitring that coast by the king's order, accompanied by Father Joseph Cardiel of the Jesuits.

OLLACACHI, a settlement of the *corregimiento* of the district of Las Cinco Leguas, of the city of Quito

OLLACHEA, a settlement of the province and *corregimiento* of Carabaya in Peru, in the vicinity of which is a mountain called Ucuntaya, where, in 1709, was discovered a silver mine, so rich as at first to yield 4700 marks each *caxon*. These riches were the leaders to great animosities, tumults, and deaths; and whilst the archbishop and viceroy Don *Fr.* Diego Morcillo was endeavouring to suppress the same, the top of the mine fell in one night, and was so completely covered up that there was no one of sufficient fortune to undertake working it anew.

OLLANTAI, a settlement of the province and *corregimiento* of Calca and Lares in Peru.

OLLEROS, a settlement of the province and *corregimiento* of Guailas, in the same kingdom as the former; annexed to the curacy of Reguay.

Olleros, another settlement, with the dedicatory title of San Miguel, in the province and *corregimiento* of Chachapoyas in the same kingdom.

Olleros, another, of the province and *corregimiento* of Guarochiri in the same kingdom.

OLLUCOS, a river of the province and government of Popayán in the Nuevo Reyno de Granada, to the *s.* It rises from the lake Guanacas in the *n.* part, and, after collecting in its course the waters of several others, enters the Magdalena by the *w.* shore. Its waters are always cold, dark-coloured, and very unwholesome.

OLMECAS, an ancient nation of infidel Indians, who passed over from the province of Mexico to establish themselves in that of Guatemala, conquering this country and driving its natives from out it, according to the tradition there, and also the account by the *Fr.* Juan de Torquemada, in his book called the Indian Monarchy.

OLMOS, a settlement of the province and *corregimiento* of Piura in Peru.

OLTO, a settlement of the province and *corregimiento* of Luya and Chillaos in Peru.

OLTOCUITLA, San Juan de, a settlement and head settlement of the district of the province and *alcaldía mayor* of San Salvador in the kingdom of Guatemala. Its population is composed of 1500 Mexican Indians, including those of two other settlements of its district.

OLUTLA, San Juan de, a town of the head settlement of the district and *alcaldía mayor* of Acayauca in Nueva España, of an hot and moist temperature; its commerce being wheat, maize, and French beans, its productions, and more particularly thread of *pita*, which is sold to the other jurisdictions. It contains 97 families of Indians, and is one league *s. e.* of its capital or head settlement.

OMACHA, a settlement of the province and *corregimiento* of Chilques and Masques in Peru.

OMAGUA, San Joaquin de la Grande, a settlement of the missions which were held by the Jesuits in the province and government of Mainas, and kingdom of Quito. It contains more than 600 Indians; situate to the *n.* of the river Marañon or Amazonas, in lat. 4° 9′ 17″ *s.*

OMAGUACAS, a nation of Indians of the province and government of Paraguay, who having been reduced to the Catholic faith by the Jesuits, returned to their apostacy, putting to death the missionaries and Spaniards, and twice destroying the city of Xuxuy in the province of Tucumán, and so continued their hostilities for a period of 30 years, until that a commission to subjugate them was given to Don Francisco Arganaez, who reduced them to request anew some missionaries; and there were, accordingly, appointed Father Gaspar de Monrroy, and the brother Juan de Toledo, who in a very short space of time converted 600, and finally the whole of them, with the exception of Cacique Piltipicon, who, although baptized in his youth, was, as a man, possessed of the most abominable and vicious dispositions. He never failed to have a severe enmity against the Spaniards, and was constantly endeavouring to put to death the missionaries and the priests, and burn the churches, until that the Father Monrroy, zealous in the cause of religion, and armed only with his confidence, marched to find him, and having met with him, addressed him in the following words; "The interest I have taken in your happiness has made me offer myself naked before you to an almost certain death: you cannot gain much honour in killing me, a disarmed man. If, persuaded by my solicitations, you spare me, the fruit of our conversation will be yours, but if I die, an immortal crown in heaven awaits me." It pleased God to touch the heart of the barbarian, he laid down his arms, silently followed his adviser, and became a steady and devout Catholic. In 1696 all the new converts near this nation were conveyed to Tucumán, where they remained steady to the faith. The ex-jesuit Coleti confounds these Omaguacas with the Omaguas of the Marañon.

OMAGUAS, a barbarous nation of Indians, the most numerous and extensive of any known in America, with different names according to the places it inhabits. There are some who live in the province of Venezuela, between the rivers Napo, Curaray, Negro, and Putumayo, and who lived formerly in the islands of the river Marañon. They are warlike, strong, faithful, courteous, and of docile disposition; are at continual war with the other nations, and have always been the conquerors. They were reduced to the Catholic religion in 1686, by Father Samuel Fritz, a German, of the order of the Jesuits, an extraordinary missionary and great mathematician. He found amongst these Indians some degree of civilization: they covered their nakedness, and lived in society, (a circumstance which materially assisted their conversion), and were convinced by the light of nature of the propriety and truth of the doctrine preached to them, and of the evils which must be consequent upon their existing state of life. Amongst their customs, the most singular was that of compressing the head of the new-born infant between two boards, so as to bring it to the shape of a mitre: this being a mark which distinguished them from all other nations. The settlement of San Joaquin is the principal, and contains more than 600 inhabitants.

OMAGUASIETE, a barbarous nation of Indians, of the province and country of Las Amazonas in S. America. They are of the Omagnas, and in their language the name means true Omaguas.

OMARA, a small river of the province and country of Las Amazonas, in the Portuguese possessions. It rises between the rivers Yurbay and Utay, runs *n. n. w.* and enters the right arm of the second, a little before it runs into the Marañon or Amazon.

OMANA, another river of this name, of the province and *captainship* of Seara and kingdom of Brazil, distinct from the former. It runs *n.* and unites itself nearly at its mouth, where it enters the sea with the Hiperugh.

OMAS, a settlement of the province and *corregimiento* of Yauyos in Peru.

OMASUYOS, a province and *corregimiento* of the kingdom of Peru; bounded *n.* and *n. w.* by the province of Asangaro, *n. e.* and *e.* by that of Larecaja, the heights and *cordilleras* called De Acama running between, *s.* by the city of La Paz and province of Pacajas, and extending *s. w.* by the lake Umamarca, a large arm of the principal lake of Titacaca, since they communicate by the strait of Tiquina; bounded also *w.* by the Great lake, there being in this part a long strip of land. It is 40 leagues long trom *n.* to *s.* and from 16 to 20 wide; its temperature is cold, although the settlement on either side of the Great lake, which runs the whole length of this province, is somewhat mild. Here breed all kinds of cattle and some wild animals, horses, sheep, swine, *huanacos*, *vicuñas*, foxes, *vizcachas*, and of the wools of the sheep the Indians make baizes, cloths, &c. and from the wool of the native sheep, or *huanacos*, various woven stuffs for clothing.

This province has different rivers, which flow down from the *cordillera*, and enter the Great lake, and which swell amazingly in the rainy seasons. The productions are *papas*, *ocas*, bark, *cañahua*, and barley, and in some parts more temperate are cultivated cabbages and flowers. Near Tanahucas are seen the vestiges of a large castle and fortress of the times of the gentilism. It is of stone, and looks upon the Great lake. The lake abounds in fish, and water-fowl which breed in the rushes which abound on the shores, islands, and bays. Between the settlement of Huaico and that of Cambuco is a ruined chapel, which was dedicated to San Bartolomé, and where, according to ancient tradition, this holy apostle appeared to preach to the gentiles; and on the grand altar of the chapel of the last of the above-mentioned settlements is a large cross of very strong wood, splinters of which are sought by the faithful as relics and preservatives against many dangers: this cross, it is thought, was left by the said apostle to one of his disciples.

Between the settlements of Huarina and Pucarani is a field, called La Batalla, a name which it has retained ever since the conflicts there of Gonzalo Pizzarro with the king's army, which was commanded by Diego Centeno, in 1547. In the settlement of Copacavana, situate on the long strip of land on the other side of the lake, is venerated a most miraculous image of Nuestra Señora de la Candelaria, placed there in 1583. It has a sumptuous and magnificent temple of beautiful architecture, and richly adorned, the same being the sanctuary of the greatest worship and devotion in Peru.

In the settlement of Huarina is a crucifix, with the dedicatory title of Señor de las Batallas, which was sent by the Emperor Charles V. and the miraculous influence of which attracts the devotion of many people of all those provinces.

In this province are gold mines, but they are not worked; and in former times there were some of silver which yielded abundantly. To this province belongs the island of Titacaca, the most celebrated and notable in the lake. This is four leagues from Copacavana, 3½ leagues long, and having in it 13 clefts or ravines, in which spring the most delicious waters, also some well cultivated estates, where, from the more mild state of the temperature, are produced some seeds, flowers, and fruits, as well as some cattle, wild rabbits, and doves. By what can be seen in this island, which was anciently held sacred, the account of the grandeur and sumptuousness of the temple of the sun, and the palace of the Inca, also of the fabulous origin of the founder of the empire, likewise the assertion that on the arrival of the Spaniards these buildings were destroyed, and that with their riches they were thrown into the lake; such account and assertion, we repeat, are not warranted, there not appearing the least vestige whatever to support them. This, however, is by no means the case with regard to Cuzco, Tiahuanaco, and other parts; the ruins of such ancient and splendid edifices still remaining there, clearly proving that they once existed. This lake may be navigated by vessels of any size, however large; and some years since a corregidor built a bark, in which he crossed over to visit the neighbouring provinces.

The inhabitants of this province should amount to 45,000, divided into eight settlements, the capital of the curacy, and with 10 settlements annexed. The *repartimiento* made by the *corregidor* was 96,605 dollars, and it used to pay an *alcabala* or centage of goods for sale of 772 dollars annually. The capital is the settlement of Achacache; and the other settlements are,

Huascho,	Ancoraimes,
Carabuco,	Santa Catalina,
Beleu,	Icacachi,
Tiquina,	Pucarani,
Huarina,	Lasca,
Santiago,	Copacavana,
Llocllo,	Ancomaya,
Tajara,	Nuestra Señora de las
Santa Lucia,	Peñas.

OMATE, a very lofty mountain of the province of Moquehua in Peru, on the top of which is a volcano, which exploded in 1600, inundating with its lava near the whole of the province, and leaving barren many pieces of territory which were before cultivated: indeed the devastation reached as far as Arequipa, a distance of 20 leagues; nor was the mischief there inconsiderable, since it lost many of its settlements, which were completely destroyed.

Omate, a settlement in the province aforesaid; annexed to the curacy of Puquina.

[OMEE, a corrupt name for the Miami of the Lake; which see. The Miami towns on its banks are called the Omee towns, or Au-Mi, by the French Americans, as a contraction of Au Miami.]

[Omee Town, one of the Miami towns; situate on a pleasant point formed by the junction of the rivers Miami and St. Joseph. This town stood on the bank of the latter, opposite the month of St. Mary's river, and was destroyed in General Harmar's expedition, in 1790.]

OMEREQUE, a settlement of the province and government of Mizque in Peru.

OMETEPEC, a small island, situate on the lake of Nicaragua.

OMETEPETL, an island of the lake of Maracaibo, the largest in this lake. Its name in the language of the country signifies two *sierras*, from two lofty mountains found in it; and making the same it is necessary to pass two leagues by water. Here are abundance of cotton, *centli*, *agi*, French beans, pumpkins, and many other fruits of a warm climate; also small cattle and little monkeys with white heads.

OMILTECAS, an ancient nation of Indians of Nueva España, not now existing, as being confounded with other nations. It was conquered and united to the Mexican empire by the Emperor Axayacatl.

OMOA, San Fernando de, a settlement of the province of Yucatán, in the government and by the gulf of Honduras; situate on the shore of a large convenient bay, with good soundings, and with a river of excellent fresh water, which is constantly taken in by vessels which come here to load with indigo, and other productions of the province. The climate is so sickly that it is, generally speaking, reputed the worst in America. It is of a hot and moist temperature, and is much infested with mosquitoes and other insects. Notwithstanding these disadvantages, the importance of this bay, as lying between the gulf of Honduras and that of Triunfo de la Cruz, induced the government to build on it a castle. This work was undertaken by Lieutenant-general Don Joseph Vazquez Priego in 1752, but he died in a few days after his arrival, and his example was followed by most of those who had attended him; but in 1775, during the presidency of Don Martin de Mayorga, it was at last finished by the engineer Don Joseph Firminor, and for its garrison a battalion of eight companies of infantry was formed in Europe in 1777, four companies of which were to be on duty at the castle, whilst the other was, alternately, at rest in the city of San Pedro Lula. The commander of these troops was the Lieutenant-colonel Don Joseph de Estacheria; but he, fearful of his health, so arranged matters that during the war the castle was defended by some dragoons instead of his own men.

OMOBAMBA, a settlement of the province and *corregimiento* of Larecaja in Peru; annexed to the curacy of Charazani.

OMONESES, Rio de los, a river of the province and *captainship* of Seara in Brazil. It enters the sea between the *sierra* Salada and the port of Mello.

[OMPOMPANOOSUCK, a short, furious river of Vermont, which empties into the Connecticut at Norwich, opposite to Dartmouth college. Its course is *s. e.*; its breadth not more than 40 or 50 yards.]

ONA, a settlement of the province and *corregimiento* of Loxa in the kingdom of Quito; situate on the great road from Loxa to Popayán. Thirty-six miles *n.* by *e.* of the city of Loxa, and 34 *s.* by *w.* of Cuenca.

ONABAS, a settlement of the province of Ostimuri in Nueva España, on the shore and at the source of the river Hiaqui, between the settlements of Tonichi and Nure.

ONAKANNA, a river of the county of Hudson in N. America. It runs *n.* and enters the sea in the bay of its name.

ONAPA, a settlement of the province of Ostímuri. Ten leagues *n.* of the river Chico.

[ONATIAYO, or Oneatoyo, an island in the S. Pacific ocean. Lat. 9° 58′ *s.* Long. 138° 51′ *w.*]

ONAUAS, a barbarous nation of Indians of the province and government of Cinaloa in N. America; they dwell in the *sierras* of the Nebomas Altos.

ONCARI, a river of the province and country of Las Amazonas, in the territory or part occupied by the Portuguese. It becomes an arm of the Madera, and then flows into one of the main large pools or takes of this province.

[ONDA. See Vincent de la Pazes.]

ONDORES, a large settlement of the province and *corregimiento* of Tarma in Peru; situate on the shore of the lake Chinchaycocha, between this and the river Pari; near it is another small settlement annexed to its curacy, called Ullumayo, where there is a fort and garrison to restrain the Chunchos Indians, who border close upon this part. This settlement is 19 miles *n. w.* of the city of Tarma.

[ONEEHOW, one of the Sandwich islands in the N. Pacific ocean, called also Neeheeheow, about five or six leagues to the *w.* of Atooi. There is anchorage all along the coast of the island. It produces plenty of yams, and a sweet root called *tee.* Lat. 21° 50′ *n.* Long. 160° 15′ *w.*]

ONEIDA Lake is about 16 miles *w.* of old fort Stanwix, now called Rome, state of New York, and is 20 miles long, and narrow. It is connected with lake Ontario on the *w.* by Oswego river, and with fort Stanwix by Wood creek.]

ONEIDAS, a barbarous nation of Indians of Canada in N. America. They are worthy of note for the answer which they made to the deputies of the United States, when these solicited their alliance to sustain the war against the English; and the sum of it was to say, that a war between brothers was a thing entirely new and strange to them, as by their traditions no such thing was ever related; that parties in such a situation should pray to heaven to remove from them the clouds of darkness which shadowed their eyes, a reasoning, forsooth, not unworthy of more enlightened nations.

[Their principal village, Kahnonwolohale, is about 20 miles *s. w.* of Whitestown. These Indians, for a number of years past, have been under the pastoral care of the Rev. Mr. Kirkland, who, with the Rev. Mr. Sarjeant, have been chiefly supported in their mission by the society established in Scotland for promoting Christian knowledge.

This nation receives an annuity from the state of New York of 3552 dollars for lands purchased of them in 1795, and an annuity of about 628 dollars from the United States. With these annuities, (which operate as a discouragement to industry) together with the corn, beans and potatoes raised by the squaws, and the fish and game caught by the men, afford them a barely tolerable subsistence. They are a proud nation, and affect to despise their neighbours, the Stockbridge and Brotherton Indians, for their attention to agriculture; but they already begin to feel their dependence on them, and are under a necessity of purchasing provisions of them. The nation is divided into three tribes or clans, by the names of the Wolf, the Bear, and the Turtle. They have their name from their pagan deity, which some few of the nation still worship, and which is nothing more than a misshapen, rude, cylindrical stone, of about 120 pounds weight, in their language called Oneida, which signifies the upright stone. Formerly this stone was placed in the crotch of a tree, and then the nation supposed themselves invincible. These Indians are all of mixed blood; there has not been a pure Oneida for several years past.]

[ONEMACK Point is the *s. w.* point of the continent of N. America, on the *n. w.* coast, and the *s.* limit of Bristol bay. It is 82 leagues *s. s. w.* of cape Newenham, or the *n.* point of that extensive bay; and in lat. 54° 30′ *n.*]

ONEOCHQUAGE, a settlement of the province of Pennsylvania, situate on the shore of the *e.* arm of the river Susquehanna.

ONGOI, a settlement of the province and *corregimiento* of Andahuallas in Peru.

ONGOL, a pleasant and fertile *llanura* of the kingdom of Chile, on which is situate the city of Los Confines, called also Ongol or Angol: it is washed on the *n.* by the rivers Claro and Puchangi, and on the *s.* by the Vergara. It is of a very benign and delightful climate, but little cultivated for want of inhabitants.

ONGOLMO, an extensive, fertile, and beautiful valley of the kingdom of Chile, in the territory of the infidel Araucanos Indians, celebrated for the assembly made in it by Caupolican and the other caciques, to treat on points of war against the Spaniards.

ONGONOSAQUI, a settlement of the province and *corregimiento* of Caxamarquilla in Peru.

ONGOS, a settlement of the province and *corregimiento* of Castro Virreyna, in the same kingdom as the former, annexed to the curacy of Vinac in the province of Yauyos.

ONGUISIA, a small river of the province and government of Mainas in the kingdom of Quito. It runs *w.* and enters the Putumayo.

[O-NIMAMOU, a habour on the *s. e.* coast of Ulietea, one of the Society islands, in the S. Pacific ocean. It is *n. e.* of Ohetuna harbour, on the same coast.]

ONINGO, a settlement of Indians of the province and country of the Iroquees, in N. America, and in the district of the nation of the Mesagues.

[ONION, Cape, on the *s. w.* side of Newfoundland island, is about four leagues *w.* of Quirpon island, or the *n.* point of that extensive island.]

[ONION River, in the state of Vermont, formerly called French River, and by the Indians Winoski, rises in Cabot, about nine miles to the *w.* of Connecticut River, and is navigable for small vessels five miles from its mouth, in lake Champlain, between the towns of Burlington and Colchester, and for boats between its several falls. It is one of the finest streams in Vermont, and runs through a most fertile country, the produce of which for several miles on each side of the river is brought down to the lake at Burlington. It is from 20 to 30 rods wide 40 miles from its mouth, and its descent in that distance is 172 feet, which is about four feet to the mile. Between Burlington and Colchester this river has worn through a solid rock of lime-stone, which in some time of remote antiquity must have formed at this place a prodigious cataract. The chasm is between 70 and 80 feet in depth at low water, and in one place 70 feet from rock to rock, where a wooden bridge is thrown across. At Bolton there is a chasm of the same kind, but somewhat wider, and the rock is at least 130 feet in height. From one side several rocks have fallen across the river, in such a manner as to form a natural bridge at low water, but in a situation to be an object of curiosity only. It was along this river that the Indians formerly travelled from Canada, when they made their attacks on the frontier settlements on Connecticut river.]

ONNEYOUTHS, a canton of Indians, or tribe of the nation of the Iroquees, in Canada. The French were at war with them for a long time, and these, being tired out with their great losses, sent some deputies to the French to make terms of peace, and to request that they would send amongst them some missionaries to instruct them in religion; and to this end were destined the Fathers Brugas, Francisi and Garnier, in 1668. Their conversion lasted but a short time, for they soon returned to their idolatry, and again waged war with the French, and were sadly worsted by the Marquis de Vaudrevil in 1696, and were, in short, obliged to renew the peace.

ONNONTAGUES, a canton of Iroquees Indians, or tribe of this nation, in Canada; who became voluntarily reduced to the faith, requesting missionaries in 1655. To them were sent by the French the Fathers Chaumont and Dablon, Jesuits; and in the following year an establishment was formed to trade with them for skins, which was furnished by 50 men under Monsieur Dupies, an officer of the garrison of Quebec, and appointed by the governor of this place.

ONOCUTURE, a settlement of the *llanos* of Cazanare and Meta in the Nuevo Reyno de Granada; composed of Indians of the Aguas nation, but who abandoned it in 1666, removing their abode to a spot between the rivers Meta, Atanari, and Casiriacuti.

ONOHUALCO, an ancient province of Nueva España in the time of the Indian gentilism, anciently peopled by Quetzalcohuatl, chief of the Tultecas. It was very extensive, and comprehended the provinces at present called Campeche, Yucatán, and Tabasco.

[ONONDAGO Castle, on the Onondago Reservation lands in the state of New York, is 25 miles *w. s. w.* of Oneida castle.]

[ONONDAGO or SALT Lake, in the state of New York, is about five miles long and a mile broad, and sends its waters to Seneca river. The waters of the salt springs here are capable of producing immense quantities of salt. One person near the lake boiled down at the rate of 50 bushels a week, in the year 1792, which he sold for 5*s.* a bushel; but any quantity may be made, and at a less price. These springs are in the state reservation, and are a great benefit to the country, every part of which is so united by lakes and rivers as to render the supply of this bulky and necessary article very easy.]

[ONONDAGO, or OSWEGO, a river of New York, which rises in the Oneida lake, and runs *w.* into lake Ontario at Oswego. It is boatable from its mouth to the head of the lake, 74 miles, except a fall which occasions a portage of 20 yards, thence batteaux go up to Wood creek, almost to fort Stanwix, or Rome, whence there is a portage of a mile to Mohawk river. Toward the head of this river salmon are caught in great numbers.]

[ONONDAGO, a county of New York state; consisting of military lands divided into 11 townships, viz. Homer, Pompey, Manlius, Lysander, Marcellus, Ulysses, Milton, Scipio, Aurelius, Ovid, and Romulus. Some of these comprehend other towns, as will be noticed under their respective names. The country is bounded *w.* by Ontario county, and *n.* by lake Ontario, the Onondago river, and Oneida lake. The county courts are held in the village of Aurora, in the township of Scipio. This county is admirably situated for inland navigation, being intersected by the two navigable rivers Seneca and Oswego, having besides five lakes and a number of creeks. For an account of the reserved lands, see MILITARY Townships. There were 1323 of the inhabitants qualified to be electors in 1796, as appears by the state census.]

[ONONDAGO, formerly the chief town of the Six Nations; situated in a very pleasant and fruitful country, and consisted of five small towns or villages, about 30 miles *s. w.* of Whitestown.]

[ONONDAGOES, a tribe of Indians who live near Onondago lake. About 20 years since they could furnish 260 warriors. In 1779, a regiment of men was sent from Albany, by Gen. I. Clinton, who surprised the town of this tribe, took 33 prisoners, killed 12 or 14, and returned without the loss of a man. A part of the Indians were then ravaging the American frontiers. This nation, which now consists of 450 souls, receives annually from the state of New York 2000 dollars; and from the United States about 450 dollars.]

[ONSLOW, a maritime county of Wilmington district, N. Carolina, *w.* of cape Lookout. It contains 5387 inhabitants, including 1748 slaves. Chief town Swansborough.]

[ONSLOW, a township of Nova Scotia, Halifax county, at the head of the basin of Minas, 35 miles *n. e.* of Windsor, and 46 *n.* by *w.* of Halifax. It was settled by emigrants from New England.]

[ONTARIO, one of that grand chain of lakes which divide the United States from Upper Canada. It is situated between lat. 43° 9′ and 44° 10′ *n.* and long. 76° 20′ and 80° *w.* Its form is nearly elliptical; its greatest length is from *s. w.* to *n. e.* and its circumference about 600 miles. The division line between the state of New York and Canada, on the *n.* passes through this lake and leaves within the United States 2,390,000 acres of the water of lake Ontario, according to the calculation of Mr. Hutchins. It abounds with fish of an excellent flavour, among which are the Oswego bass, weighing three or four pounds. Its banks in many places are steep, and the *s.* shore is covered principally with beech-trees, and the lands appear good. It communicates with lake Erie by the river Niagara. It receives the waters of Genessee river from the *s.* and of Onondago or Oswego at fort Oswego, from the *s. e.* by which it communicates through Oneida lake, and Wood creek, with the Mohawk river. On the *n. e.* the lake discharges itself into the river Cataraqui, (which at Montreal takes the name of St. Lawrence) into the Atlantic ocean. It is asserted that these lakes fill once in seven years! but the fact is doubted. The islands are all at the *e.* end, the chief of which are Wolf, Amherst, Gage, and Howe islands.]

[Ontario, a large, fertile county of New York, comprehending the Genessee country, and bounded *n.* by the lake of its name. It is well watered by Genessee river, its tributaries, and a number of small lakes. Here are eight townships, viz. Genessce, Erwine, Jerusalem, Williamsburg, Toulon, Seneca, Bloomfield, and Canadaqua, or Kanandaigua, which is the last chief town, situated at the *n. w.* corner of Canandarqua lake, 15 miles *w.* of Geneva, and 23 *n. e.* of Williamsburg. This county was taken from Montgomery in 1789, and in 1790 contained 1075 inhabitants, including 11 slaves. Such has been the emigration to this country, that there were, in 1796, 1258 of the inhabitants who were qualified to be electors.]

ONTUEROS. See Ciudad Real.

ONZAGA, a settlement of the province and *corregimiento* of Tunja in the Nuevo Reyno de Granada. Its district enjoys various temperatures, and it accordingly produces all the different fruits peculiar to its climates, such as maize, plantains, sugar-canes, *arracachas*, &c. It is situate in the high road leading to San Gil, Socorro, Jiron, and Ocaña, is very reduced and poor, containing 50 housekeepers and 30 Indians, and is 24 leagues from its capital.

[ONZAN, a cape or point on the *n.* coast of Brazil, opposite to cape St. Lawrence, forming together the points of Laguariba river; the latter cape being on the *w.* side of the river. The river is 10 leagues *s. e.* by *e.* of Bahia Baxa.]

ONZOLES, a large and abundant river of the province and government of Esmeraldas in the kingdom of Quito. It runs from *s.* to *n.* and enters the San Miguel where this runs into the S. sea, in lat. 58° *n.*

[OONALASHKA, one of the islands of the *n.* archipelago, on the *n. w.* coast of America, the natives of which have the appearance of being a very peaceable people, being much polished by the Russians, who also keep them in subjection. There is a channel between this and the land to the *n.* about a mile broad, in which are soundings from 40 to 27 fathoms. Lat. 53° 40′ *n.* Long. 167° 20′ *w.*]

OPAHUACHO, a settlement of the province and *corregimiento* of Parinacochas in Peru, annexed to the curacy of Pacca.

OPAN, a settlement of the province and *corregimiento* of Cuenca in the kingdom of Quito; annexed to the curacy of Ozogues. In its district is a large estate, called Chuquipata.

OPANO. See Upano.

OPECKON, a river of the province and colony of Virginia in N. Carolina. It runs *n. n. e.* and enters the Patowmack.

OPICO, San Juan de, a settlement and head settlement of the district of the *alcaldía mayor* of Santa Ana in the kingdom of Guatemala. It has three settlements annexed to its curacy.

OPITAMA, a settlement of the province and government of Antioquia in the Nuevo Reyno de Granada, near the source of the river Cauca.

OPLOCA, a settlement of the province and *corregimiento* of Chichas and Tarija in Peru; of the district of the former, and annexed to the curacy of Tupisa.

OPOPEO, a settlement of the *alcaldía mayor* of Valladolid in the province and bishopric of Mechoacán, and kingdom of Nneva España; annexed to the curacy of Cobre. It contains 34 Indian families, and is half a league distant from Cobre.

OPPON, some very lofty mountains which run from the province and government of Popayán to that of Santa Marta in the Nuevo Reyno de Granada, being a branch of the grand *cordillera* of the Andes. They are inhabited by some barbarous nations of the Panches Indians, and the first who passed over them was Gonzalo Ximenez de Quesada, in 1536. They are rugged and full of thick woods; and from them descend to the *w.* many streams, which form the source of the grand river Magdalena. Some confound these mountains of Oppon with those of Bogotá, which divide the Nuevo Reyno de Granada from the *llanos* or plains of the Orinoco, and run in a direction from *s. w.* to *n. e.*

Oppon, a river of the same province and kingdom as are the above mountains. It rises in the valley of Alferez, runs *n.* and, forming a lake, follows a short course to enter the Magdalena.

Oppon, another, a small river, in the province and government of Mainas of the same kingdom, in the *sierras* of Perija. It runs *e.* and enters the great lake of Maracaibo by the *w.* side.

[OPPS, a village in Northampton county, Pennsylvania, six miles *s. e.* of Bethlehem, and about seven *n.* by *e.* of Quaker's town.]

OPOSTURA, a settlement of the province of Ostimuri in Nueva España; situate 40 leagues to the *n.* of the *real* of mines of the Rio Chico.

OPTOQUE. See OTOQUE.

OPUTU, a settlement of the province and government of Ostimuri in Nueva España. Forty-two leagues *n. n. e.* of the *real* of mines of the Rio Chico.

[OR, Cape D', in Nova Scotia, is situate on the *n.* side of the basin of Minas. Some small pieces of copper have been found here.]

[ORA CABECA Bay, on the *n.* side of the island of Jamaica in the W. Indies, has a strong fort on the *e.* side, and Salt gut *w.*; at both these places is good anchorage for large vessels.]

ORACHICHE, a settlement of the province and government of Venezuela, and kingdom of Granada, *e.* of the city of Barquisimeto.

ORADADA, a point of the coast in the province and government of Cartagena, and Nuevo Reyno de Granada; one of those which form the bay of Zipato.

[ORANAI, or RANAI, one of the Sandwich islands in the N. Pacific ocean, nine miles from Mowee and Morotoi. The *s.* point is in lat. 20° 51′ *n.* and long. 157° 1′ *w.*]

ORANAYA, a fall of the river Madera, in the province and country of Las Amazonas. It lies between the rivers Erena and Guiaparana on the *s.* and the Chamari and Mamoroni on the *n.*

[ORANG'S Key, one of the Bahama islands, in the W. Indies. Lat. 24° 36′ *n.* Long. 79° 12′ *w.*]

ORANGE, a cape or point of land of the coast of the N. sea, in the province and government of Guayana, in the part which was possessed by the French; it forms the *e.* side of the river Oyapok or Vincent Pincon's bay. Lat. 4° 18′ 30″ *n.* Long. 51° 13′ *w.*

ORANGE, another point of the *s.* coast of the straits of Magellan, close to the point Anegada, which is one of those forming the great bay of Lomas.

ORANGE, a small port on the coast of the island of Newfoundland, in the Blanche bay.

ORANGE, a shoal or isle of the N. sea, near the coast of S. Domingo, in the part possessed by the French between the point Pasqual and the island of Jaquin.

ORANGE, a settlement of the Dutch, in their possessions in Guayana, and in the colony of Surinam; situate on the coast between the rivers Surinam and Marowine.

[ORANGE, a bay on the *n. e.* coast of the island of Jamaica, *e. n. e.* of the high mountain, a little within land, under which is Crawford's town. Also a bay at the *n. w.* end of the same island, between Green island *n.* and N. Negril harbour *s.* or *s. w.*]

[ORANGE Key, or Cay, a small island of Orange bay, at the *n. w.* end of the island of Jamaica.]

[ORANGE, a county of Vermont, which in 1790 contained 10,529 inhabitants. Since that time several other counties have been erected out of it. It is bounded *w.* by part of Addison and Chittenden counties, and *e.* by Connecticut river. It now contains 20 townships. The county town, Newbury, and the townships *s.* of it, viz. Bradford, Fairlee, and Thetford, front Connecticut river. It is high land, and sends numerous streams in opposite directions, both to Connecticut river and to lake Champlain.]

[ORANGE, a township on the *n.* line of the above county, in the *n. e.* corner of which is Knox's mountain.]

[ORANGE, formerly Cardigan, a township in Grafton county, New Hampshire, which gives rise to an *e.* branch of Mascomy river. It was incorporated in 1796, contains 131 inhabitants, and is 20 miles *e.* of Dartmouth college.]

[ORANGE, a township of Massachusetts; situated on the *e.* line of Hampshire county, on Miller's river, 94 miles *n. w.* by *w.* of Boston. It was incorporated in 1783, and contains 784 inhabitants.]

[ORANGE, a mountainous and hilly county of New York, which contains all that part of the state bounded *s.* by the state of New Jersey, *w.* by the state of Pennsylvania, *e.* by the middle of Hudson's river, and *n.* by an *e.* and *w.* line from the middle of Murderer's creek. It is divided into eight townships, of which Goshen is the chief, and contains 18,492 inhabitants, of whom 2098 are electors, and 966 slaves. In this county are raised large quantities of excellent butter, which is collected at Newburgh and New Windsor, and thence transported to New York. On the *n.* side of the mountains in this county is a very valuable tract called the Drowned Lands, containing about 40 or 50,000 acres. The waters which descend from the surrounding hills, being but slowly discharged by the river issuing from it, cover these vast meadows every winter, and render them extremely fertile; but they expose the inhabitants of the vicinity to intermittents. Walkill river, which passes through this tract and empties into Hudson's river, is, in the spring, stored with very large eels in great plenty. The bottom of this river is a broken rock; and it is supposed that for 2000*l.* the channel might be deepened so as to drain off the waters, and thereby redeem from the floods a large tract of rich land for grass, hemp, and Indian corn.]

[ORANGE, called also ORANGEDALE, a town in Essex county, New Jersey, containing about 80 houses, a Presbyterian church, and a flourishing academy, and lies *n. w.* of Newark, adjoining.]

[ORANGE, a county of Hillsborough district, N. Carolina; bounded *n.* by Caswell county, and *s.* by Chatham. The rivers Haw and Enoe in this county have rich lands on their borders. It contains 12,216 inhabitants, of whom 2060 are slaves. Chief town, Hillsborough.]

[ORANGE, a county of S. Carolina, in Orangeburg district.]

[ORANGE, a county of Virginia, bounded *n.* by Culpepper, and *s.* by Albemarle. It contains 9921 inhabitants, including 4421 slaves. The court-house is situated 17 miles from Culpepper court-house, and 30 from Charlotteville.]

[ORANGEBURG, a district of S. Carolina, bounded *s. w.* by Savanna river; *e.* by the river Santee, and *n. e.* by the Congaree, which divide it from Camden district; *s.* by Beaufort, and *s. e.* by Charleston district. It contains 18,513 inhabitants; of whom 5931 are slaves. Sends to the state legislature 10 representatives and three senators; and with the district of Beaufort, one member to congress. It is divided into four counties, viz. Lewisburg, Orange, Lexington, and Winton.]

[ORANGEBURG, a post town of S. Carolina, and capital of the above district, is on the *e.* side of the *n.* branch of Edisto river. It has a court-house, gaol, and about 50 houses; distant 60 miles *n. n. w.* of Charlestown, and 33 *s.* of Columbia.]

[ORANGETOWN, or GREENLAND, a plantation in Cumberland county, Maine, *n. w.* of Waterford. One branch of Songo river rises in the *n.* part of this plantation, within about three miles of Amariscoggin river, where there is a pond two miles long, called Songo Pond, from thence the stream runs *s.* It is very difficult to effect roads through this mountainous country; some of the mountains affording precipices 200 feet perpendicular. The sides of the mountains and valleys are fertile, produce good crops, and in some instances afford wild onions, which resemble those that are cultivated. Winter rye, which is the chief produce, has amounted to 20 bushels an acre. The country in the neighbourhood formerly abounded with variety of game, viz. moose, deer, bears, beaver, racoon, sable, &c.; but since it has been inhabited game has become scarce; deer are extirpated from the vicinity; some moose remain among the mountains, and a few beaver, that are too sagacious to be taken by the most crafty hunter. Since the deer have been destroyed, the wolves have wholly left this part of the country.]

[ORANGETOWN, in Orange county, New York, is situated on the *w.* side of the Tappan sea, opposite Philipsburgh, and about 27 miles *n.* of New York city. The township is bounded *e.* by Hudson's river, and *s.* by the state of New Jersey. It contains 1175 inhabitants, of whom 162 are electors, and 203 are slaves.]

[ORANGETOWN, in Washington county, Maine, is 19 miles distant from Machias.]

ORANO, a settlement of the province and government of Santa Marta in the Nuevo Reyno de Granada, of the division and district of the Rio del Hacha; situate on the coast.

ORAPE, a river of the province and government of Venezuela, in the same kingdom as the former settlement. It rises *s.* of the city of Nirúa, and shortly after unites itself with the Coronel to enter the Tinaco.

ORAPU, a river of the province and government of Guayana, in the French possessions.

ORATORIO, a settlement of the province and government of Tucumáu in Peru; situate on the shore of the river Dulce.

ORAUIA, a settlement of the missions which were held by the Jesuits, in the province and government of Mainas, and kingdom of Quito; situate on the *s.* shore of the river Napo, about 100 miles before this river enters Marañon or Amazon, in the country and territory of the Abijiras Indians.

ORCHILLA, or URCHILLA, a small island of the N. sea, situate near the coast of the province and government of Venezuela in the Nuevo Reyno de Granada, opposite the mouth of the river Unare. It is of a low territory and semicircular figure, desert and uncultivated, the soil being barren, although it has some good pastures as well on the *e.* as the *w.* side, where there are two hills forming points or promontories, and having a few goats browsing on their sides.

In the part looking *s. w.* the shore is very bold, running down as though it were a wall, so much so that ships may come up almost close to the land. In the *n. w.* part it is nothing but a barren wild, not only destitute of trees but of every kind of plant and herbage. It is very scant of water, and the only animals it produces are goats and lizards. Around this island are several small isles or rocks, the largest of which are Tortuga and Roxa.

[The navigation about these islands is very difficult and dangerous. As they stretch from *e.* to *w.* and consequently lie all very nearly in the same latitude, the mariner, without local knowledge, when once entangled among them, cannot, from his solar observations alone, determine exactly one from another. The currents likewise in this sea are not only very variable, but violent also. In the course of a few dark hours a vessel is swept down upon a danger which journals and observations had concurred in placing still far distant; and, besides this, vessels have been known to have drifted upon Orchilla in a dead calm, notwithstanding all the efforts of the crew in their boats to keep her off. It is therefore thought that it would be far preferable for vessels bound to Curaçoa to keep well to the *n.* until near Buenayre; or otherwise at once to penetrate and keep to leeward of the whole chain, even should they see the mainland of America. After making cape Codera, or the high mountains above La Guayra, the course is simple, and the prevailing winds favourable, for Curaçoa. The same observations will apply to vessels bound to La Guayra: they should endeavour, as soon as possible, to get a sight of the continent of America, to the *e.* of the island of Margarita, the passage between this island and the main being full of danger, without an experienced pilot on board.] Orchilla is in long. 66° 9′ *w.* Lat. 11° 52′ *n.*

ORCHOCOCHA, a lake of the province and *corregimiento* of Yauyos in Peru, from whence rises the river Pisco.

ORCO-PAMPA, a settlement of the province and *corregimiento* of Condesuyos in Peru; annexed to the curacy of Andahua.

ORCOS, a *llanura* of the province and *corregimiento* of Cuzco in Peru, from whence it is distant six leagues. Towards the *s.* it has a lake celebrated on account of a tradition, that into it was thrown the wonderful gold chain, called in the Inca tongue *curi-huale*, which was made by command of the Emperor Huayana-Capac to grace the festivities of his first-born, and who was from this circumstance called Huascar-Inca, whereas his name would otherwise have been Cusi-Huallpa. Many Spaniards have endeavoured to make canals whereby to empty this lake and to find the above treasure, but always without effect, on account of its great depth, this being more than 25 fathoms. From this lake the river Pisco rises.

[ORDADO Rock, near the coast of Peru, is four miles *s.* by *e.* of port Calloa. Near it are some smaller ones, and round them from nine to 16 fathoms water.]

ORDONEZ, a settlement of the province and government of Cartagena in the Nuevo Reyno de Granada; situate on the sea-coast in Tolú bay. It belongs to the district and jurisdiction of this town.

[OREAHOU, or OREEHOU, a small elevated island, close to the *n.* side of Oneeheow, one of the Sandwich islands; with which it is connected by a reef of coral rocks. It contains about 4000 inhabitants. Lat. 22° 2′ *n.* Long. 160° 8′ *w.*]

[OREGAN River. See RIVER OF THE WEST.]

OREGUATUS, a barbarous nation of Indians of the province and country of Las Amazonas, who dwell on the shore of the river Madera.

OREJONES, a barbarous nation of Indians of the province and government of Paraguay in Peru, called also Yaraces, this being the name of a large island inhabited by them in the lake of Los Xarayes, the said lake being more than nine leagues from *n.* to *s.* These Indians are called Orejones from their having ears extremely large, and drawn down by weighty ornaments suspended to them. They are very numerous, dwell in the islands about here, and also on the continent to the *n. w.*

ORELLANA. See MARANON.

ORELLUDOS, a river of the province and country of Las Amazonas, which rises in the territory of the Indians of this name, runs *s. s. e.* and turning to the *s.* enters at the end of its course into the Yupura or Caquetá.

ORELLUDOS, a river of the island of Guadalupe, one of the Antilles, which rises in the mountains to the *e.* runs to this rhumb, and enters the sea between the rivers Grand Carbet and Grand Bananiers.

ORFELEINS, Bank of, in the gulf of St. Lawrence. See ORPHANS,

[ORFORD, a township in Grafton county, New Hampshire; situated on the *e.* bank of Connecticut river, about 11 miles *n.* of Hanover, and opposite to Fairlee in Vermont. It was incorporated in 1761, and contains 540 inhabitants. The soap-rock, which has the property of fuller's earth in cleansing cloth, is found here; also alum ore, free-stone fit for building, and a grey-stone, in great demand for mill-stones, reckoned equal in quality to the imported burr stones.]

[ORFORD Cape, in the *n.* westernmost point of the large island to the *w.* of Falkland's sound in the Falkland's islands, in the S. Atlantic ocean, and *s. e.* of cape Percival.]

[ORFORD, a cape on the *n. w.* coast of N. America; situate between cape Mendocino and cape Flattery, in lat. 42° 54′ *n.* and long. 124° 31′ *w.*]

ORGANOS, some rocky shoals of the N. sea, on the *n.* coast of the island of Cuba, close to the shoals of S. Isabel.

ORGAOS, Sierra dos, some mountains of the province and *captainship* of Rio Janeiro in Brazil, which run from *n. n. e.* to *s. s. w.* following the course of the river Paraiba del Sur, near the coast and cape Trio.

ORI, a river of the province and government of Guayana or Nueva Andalucía, which rises in the country of the Armacotos Indians, runs *n. e.* and enters the Paraguay.

Ori, another, a small river in the same province and government, which rises in the territory of the Caribes Indians, at a great distance from the former river, and enters the Arivi.

ORIA, a river of the province and government of Veragua in the kingdom of Tierra Firme. It rises in the mountains to the *s.* and running to this rhumb, enters the sea in a bay a little before point Mala.

ORIBANTES, or Sierra, a river of the province and government of Maracaibo in the Nuevo Reyno de Granada. It rises at the foot of the *sierra* Nevada, to the *s.* of the city of Merida, runs in a serpentine course for many leagues, and changing its name to Apure, enters much swollen into the Orinoco.

[**ORICARO**, Sebastian de, more properly called Ocumare; which see.]

ORIGINAL, a small river of Canada in N. America, which runs *n.* and enters the lake Superior, between the point Carbet and the river Tonnagane.

ORIGUECA, a large and rich settlement of the province and government of Santa Marta in the Nuevo Reyno de Granada; situate in an extensive *llanura.* It was peopled with Taironas Indians, but their numbers have been so much diminished that it is now a very mean place.

ORINO, an ancient settlement of the same province and kingdom as the former, in a pleasant and fertile spot; once a large town of Guajiros Indians, but now consisting of a few straggling houses of some wretched inhabitants.

ORINOCA, a settlement of the province and *corregimiento* of Carangas in Peru, and of the archbishopric of Charcas; annexed to the curacy of Andamarca.

ORINOCO, a large, navigable, and most abundant river of the Nuevo Reyno de Granada, and S. America, one of the four largest rivers on the continent. It rises in the *sierras* Nevadas to the *n.* of the lake Parime, in the province of Guayana, according to the discovery made by order of the court by Admiral Don Joseph de Iturriaga, and by the informations received from the Caribes Indians, proving erroneous the origin given to it by the Father Joseph Gumilla, the Jesuit, in his book entitled "Orinoco Illustrado," as also the origin given it by the ex-jesuit Coleti, namely, in the province of Mocoa, in lat. 1° 21′ *n.* [The fact is, that according to the more recent and best accounts, it should appear to rise in the *sierra* Ibermoqueso, from a small lake called Ipava, which is, agreeably with the account of our author, in the province of Guayana.]

This river runs more than 600 leagues, receiving in its extended course an exceeding number of other rivers, which swell it to an amazing size, and it proceeds to empty itself into the sea opposite the island of Trinidad, by seven different months, forming various isles, namely, the Orotomecas or Palomas, so called from a barbarous nation of Indians of this name inhabiting them.

The Orinoco bears the name of Iscaute until it passes through the country of the Tames Indians; where it receives by the *w.* side the rivers Papamene and Plasencia, and acquires then the name of that district, which it changes at passing through the settlement of San Juan de Yeima into that of Guayare, and then to that of Barragan, just below where it is entered by the abundant stream of the Meta, and before it is joined by the Cazanare, of equal size. It receives on the *n.* side the rivers Pau, Guaricu, Apuré, Cabiari, Sinaruco, Guabiaris, Irricha, San Carlos, and others; and by the *s.* those of Benituari, Amariguaca, Cuchivero, Caura, Aroi, Caroni, Aquiri, Piedras, Vermejo or Colorado, and others of less note; and being rendered thus formidable with all the above, it at last becomes the Orinoco.

Its shores and islands are inhabited by many barbarous nations of Indians, some of whom have been reduced to the Catholic faith by the Jesuits, who had founded some flourishing missions, until the year 1767; when, through their expulsion from the Spanish dominions, these Indians passed to the charge of the Capuchin fathers.

The Orinoco is navigable for more than 200 leagues for vessels of any size, and for canoes and small craft from its mouth as far as Tunja or San Juan de los Llanos. It abounds exceedingly in all kinds of fish; and on its shores, which are within the ecclesiastical government of the bishop of Puerto Rico, are forests covered with a great variety of trees and woods, and inhabited by strange animals and rare birds; the plants, fruits, and insects being the same as those on the shores of the Marañon. This last mentioned river communicates with the Orinoco by the river Negro, although this was a problem much disputed until acknowledged by the discovery made by the Father Ramuel Roman, the Jesuit, in 1743.

The principal mouth of the Orinoco was discovered by Admiral Christopher Columbus in 1498, and Diego de Ordaz was the first who entered it, he having sailed up it in 1531. The sounding between fort San Francisco de la Guayana and the channel of Limon is 65 fathoms, measured in 1734 by the engineer Don Pablo Dias Faxardo, and at the narrowest part it is more than 80 fathoms deep; in addition to which, in the months of August and September, the river is accustomed to rise 20 fathoms at the time of its swelling or overflow, which lasts for five months; and the natives have observed that it rises a yard higher every 25 years.

The flux and reflux of the sea is clearly distinguishable in this river for 160 leagues. In the part where it is narrowest stands a formidable rock in the middle of the water, of 40 yards high, and upon its top is a great tree, the head of which alone is never covered by the waters, and is very ufeful to mariners as a mark to guard against the rock. Such is the rapidity and force with which the waters of this river rush into the sea, that they remain pure and unconnected with the waters of the ocean for more than 20 leagues distance. Its principal mouth, called De Navios, is in lat. 8° 9′ *n.*

[The Orinoco is remarkable for its rising and falling once a year only; for it gradually rises during the space of five months, and then remains one month stationary, alter which it falls for five months, and in that state continues for one month also. These alternate changes are regular, and even invariable. Perhaps the rising of the waters of the river may depend on the rains which constantly fall in the mountains of the Andes every year about the month of April; and though the height of the flood depends much upon the breadth or extent of the bed of the river, yet in one part where it is narrowest, it rises (as Alçedo has correctly observed) to the astonishing height of 120 feet. The mouth of the river is *s.* by *e.* of the gulf of Paria, in lat. 8° 50′ *n.* and long. 60° *w.* and opposite to the island of Trinidad. It is large and navigable, and has many good towns on its banks, that are chiefly inhabited by the Spanish, and is joined also on the *e.* side by the lake Casipa. There are two other islands at its mouth, the entrance to which is also somewhat dangerous, as there is frequently a dreadful conflict between the tide of the ocean and the current of the river, that must, for the reasons assigned, sometimes run very rapidly. It is true that the river, including its windings, takes a course of about 1380 miles. It may be considered as having many mouths, which are formed by the islands that lie before its opening towards the ocean; yet there are only two that are considered as of any use for the purposes of navigation. These are the channels of Sabarima and Corobana, otherwise called Caribbiana. The latter lies in a *s.* by *w.* direction, and is also divided into two distinct channels that afterwards meet again at the island of Trinidad in the mouth of the grand river. But pilots pretend to say, that the mouth of this great river begins from the river Amugora, reaching from thence to the river Sabarima, and from thence about to the river Caribbiana; and some accounts state its mouths to be upwards of 40 in number, as if it were a collection of many rivers, all uniting at the mouth of the great river, and assisting to convey the main stream of that river into the ocean.

The *w.* passage or channel of the river Orinoco, called by the Spaniards the Gulf of Paria, lies between cape Salinas on the main, and the *n. w.* point of the island of Trinidad. It contains several islands, which divide the stream of the river into several branches, particularly the Boco Grande, or Great month, which is the easternmost, being about gun-shot wide, but having no soundings, with 300 fathoms, and the Boco Pequeño, or Little month, which is the westernmost, being almost as wide as the other, and having ground at from 50 to 60 fathoms. At New cape Araya, on the *n.* side of the mouth of this river, are salt-pits, which yield the finest salt in the world. In some maps the head-waters are called Inirchia.

A more diffuse and particular account of this mighty river will be found under the following heads, which we have translated and selected from the work of Depons, and other writers, viz.

The seven principal mouths of the Orinoco.---The navigation of the Orinoco up to St. Thomas.---Enchanting variety of its banks.---Importance of this river.---Further account of its waters, and its annual swell.---Its tides, and peculiar animals inhabiting it.---Table of latitudes and longitudes of these parts.

It is presumed that the course of this river, for the first 100 leagues, is *n. e.* and *s.* In this part it leaves the imaginary lake of Parima 60 leagues from its left bank. The rivers which flow into the Orinoco give it, before it has run these 100 leagues from its source, as rapid a current and as great a body of water as any of the most considerable rivers. From the Esmeraldas to San Fernando de Atabapa, its course is from *e.* to *n. e.* Between these places is the canal of Casiquiari, which forms the communication between it and the Amazonas, by the river Negro.]

[At about 100 miles from the sea, the Orinoco, like the Nile, forms a sort of fan, scattered with a number of little islands, which divide it into several branches and channels, and oblige it to discharge itself through this labyrinth into the sea by an infinite number of mouths, lying *n. e.* and *s. w.* and extending more than 170 miles. These islands increase so on the coast that the mouths of the Orinoco are very numerous, but very few of them are navigable. It is computed that these openings amount to near 50, and only seven of them admit the entrance of vessels, and these must not be of a large burden. An idea of the prudence and skill requisite for the navigation of these mouths may be formed by what daily happens amongst the Guayanos Indians, who, although born on the islands, and from subsisting solely on fish, are so accustomed to the intricacies of the different channels, yet frequently lose themselves, and are obliged to allow the current to carry them out to sea, and then to re-enter, not without the most minute observations and endeavours to ascertain the proper passage. It even requires a considerable skill to find the current; for the numerous channels have such different directions that in the greater part of them no current at all is perceptible, and in the others the eddies or the winds give the currents a direction up the river instead of down. The compass is frequently of no use, and when a person is once lost, he is often obliged to wander several days among the Guayanos islands, conceiving he is ascending the river when he is descending, or that he is descending when he is ascending; and at length, he probably finds himself at the very point from which he set out.

The first of the mouths which are navigable is 25 miles *s. e.* of the entrance of the Guarapiche river, in the province of Cumaná. It is one of those which empty their water in the gulf of Paria. It is called the Great Manamo, in contradistinction to the Little Manamo, which runs in the same channel with it, nearly to the sea, and is navigable for shallops.

The second month is 20 miles *n. e.* of the first, and is called the Pedernales. It runs from the *e.* of the island of Guarisipa, and falls into the sea three leagues *s. w.* of Soldiers island, which is situated at the *s.* entry of the gulf of Paria. It is only navigable for canoes, or at the most for shallops.

The third mouth is called Capuro; it is an arm of the channel of Pedernales, from which it branches off at 30 miles from the sea. Its mouth is in the southernmost part of the gulf of Paria, 31 miles *s. e.* of that of the channel of Pedernales. The navigation is hardly fit for any vessels but canoes and shallops.

Macareo is the name of the fourth mouth; it enters the sea, six leagues *s.* of Capuro, and is the channel of communication between Guayana and Trinidad, and every thing concurs to give it this advantage exclusively. It is navigable for moderate-sized vessels, its channel is exceedingly straight and clear, and it falls into the sea opposite the point and river Erin in Trinidad.

The fifth mouth is very little frequented, on account of the difficulty of the navigation and the ferocity of the Indians inhabiting its banks. They are called Mariusas, and have given their name to this fifth passage of the Orinco. This mouth is 35 miles *e. s. e.* of the fourth.

Between Mariusas and the sixth mouth are several outlets to the sea, which are navigable by the tide or by the floods.

Twenty-five miles more to the *s. e.* is what is called the Great Mouth of the Orinoco; it bears the name of Mouth of Vessels, because it is the only one which admits of ships of 200 or 300 tons burden. Its extent is six leagues, but it is far from being every where of an equal depth.

Navigation of the Orinoco up to St. Thomas.— The grand mouth of the Orinoco is formed by cape Barima to *s. s. e.* which is in 8° 54′ lat. *n.* and the island of Cangrejos, lying *w. n. w.* of the cape. They are 25 miles from each other, but the breadth of the navigable part of the passage is not quite three. The depth of water on the bar, which lies a little farther out to sea than the cape, is, at ebb, 17 feet.

Immediately on passing the bar, the depth, on the side of the island, is four or six fathoms, whilst on the side of the cape, it is not more than 1½. The flats extend from Cangrejos seven leagues into the sea, but from cape Barima they do not extend more than two leagues.

Nearly one league from Barima is a river of the same name, which discharges itself into the Orinoco. The entrance is by a narrow channel 1½ fathom deep. On the same shores, *s.* of the Orinoco, and two leagues higher up than this river, is the mouth of the Amaruco, which crosses a great part of the most *e.* territory of Guayana, occupied by the Capuchins of Catalonia. Shallops can sail 10 or 15 leagues up. It is *s.* of the island and cape of Cangrejos, which forms, as has been before observed, the *n.* coast of the mouth of Navios or Vessels.

Three leagues above Cangrejos is the island of Arenas, which is small and of a sandy soil. It is from 12 to 15 feet under water in spring tides. In] [the *s.* part of it is a channel, which is often altered by the sand, of which the bottom is composed. Before ascending half a league there are two points, called by the Spaniards Gordas. That on the *n.* side has a flat which runs out a little, but not enough to obstruct the navigation.

Proceeding along the *s.* shore of the Orinoco, eight leagues above Barima is the river Araturo, the source of which bounds the savannas of the missionaries. Its mouth is very narrow, but it is navigable for 10 leagues. It communicates by different arms with the river Amacuro to the *e.* and with the Aguirre to the *w.* There is much wood on its banks, and some small islands, bearing its name, opposite its mouth. On the *n.* side is the channel called Cocuma. It discharges itself into the sea.

Eleven leagues above Barima is the island of Pagayos, in the middle of the Orinoco, but nearest to its right bank. Its soil is white mud, it is covered with mangles, and at flood tide it is 11 feet under water. It was formerly much larger than it is at present, and is observed to diminish sensibly. Immediately above the island of Pagayos, is that of Juncos. It is the most *e.* of the Itamaca islands, which occupy a space of 18 leagues in the Orinoco. They divide the river into two branches; the *s.* branch being called Itamaca, and the *n.* Zacoopana. Both of these are navigable, but the *s.* branch, although the least, has by far the most water.

We will describe the Itamaca branch to the *w.* point of the chain of islands, and afterwards give a description of that of Zacoopana.

The *e.* entrance of the Itamaca branch, which is 900 fathoms wide, is formed by the island of Juncos and cape Barima Zanica, which juts out from the right bank of the Orincoo. A creek, called Carapo, runs from the cape in-shore, and afterwards joins the river Arature.

A little higher up is the mouth of the river Aguirre. Its source is in the tract of the missionaries of the Catalanian Capuchins. Its mouth is very broad, and the depth, at 10 or 12 leagues from the Orinoco, is three fathoms. It was once much more navigable than it is at present, but very trifling repairs would be sufficient to restore it to its former state. As this river does not pass through any cultivated country, it is only frequented by those who resort to its banks for wood. The trees on each side are so high, that the sail cannot be used, and vessels consequently avail themselves of the title.

Two leagues from the mouth of this river, in the midst of the Orinoco, is the little island of Venado, and on the *s.* bank of the Orinoco, eight leagues above the Aguirre, is the creek of Caruzina. It proceeds from the Orinoco, runs by the back of the mountains, and thence takes its course *s. e.* thus forming of the bank of the Orinoco an island, on which the Guayanos Indians have built a hamlet, subject to the Indian Gemericabe. This creek or branch has plenty of water at its entrance, but the point of the rising grounds of Itamaca causes it to be hardly navigable for half a league. This creek spreads into an infinite number of branches, and therefore it might be of great use to agriculture, the neighbouring land lying too high for inundation. The Spaniards have recently entertained the project of driving away the Guayanos Indians, of building villages, and of erecting batteries for the defence of the Orinoco.

The river Itamaca, on the *n.* side of the Orinoco, is next to be described. Its mouth is narrow, but deep, having from 16 to 18 feet water. There is a bank in the Orinoco running across the mouth of the Itamaca, with the exception of a very narrow passage, which requires, especially at low water, great precaution in the navigation. This river, six miles from its mouth, divides into two branches, the first of which goes to the *w.* and runs through the valleys formed by the mountains, the other runs to the savanna, near the mission of Polomar. The river is navigable up to where it thus branches off, for small craft and boats. The *w.* point of the Itamaca islands is 2¼ leagues from the river.

We will here again descend the Orinoco to where the Itamaca and Zacoopana branches unite, for the purpose of describing the latter, and afterwards reascend in making the tour of the island of Juncos, leaving it to our left.

From the *e.* point of Juncos runs out a flat to the *n.* making a very narrow but deep passage for vessels, which should keep close to the *n.* coast. Within the *e.* point of the island of Juncos is that of Pericos, which has very lately disappeared. It formed two channels, that to the *s.* was almost choked by the sand, that to the *n.* was narrow, and afforded but a difficult passage for vessels. This island was small and sandy, it was seen at ebb tide, and in the swellings of the Orinoco. Its disappearance was not occasioned by any earthquake or extraordinary inundation.

Four leagues above the point where was once the island of Pericos, is the isle of Hogs, which we leave to the right, because it inclines to the *n.* The navigable channel continues to the *s.*; it has, however, between it and the shore a narrow creek na-]

[vigable for small vessels. A league to the *w.* of the isle of Hogs is the channel Laurent, on the *n.* side of the Orinoco. From its mouth proceeds a shallow which crosses half of the Zacoopana channel. The Laurent, at its mouth, has the appearance of a large river, but at a very little distance to the *n.* it forms many ramifications, all of which are so shallow, that only by one can small vessels find egress to the sea. At the entrance of the Laurent channel there is a small island of the same name, from which proceeds a flat which extends to the mouth of Mateo, which crosses the Itamaca branch. Musquito island, situated near the *s.* shore, has from its *e.* and *w.* points flats extending more than a league. In the middle of the river is the channel, half of a league broad.

From the mouth of the Abacayo channel runs a shallow extending to the island of Palomas. On the *n.* coast are two channels which fall into the sea. Another flat runs from the island of Palomas, and reaches to the westernmost point of the Itamaca islands.

At the mouth of the channel of the island of Zacoopana commences a flat, running two leagues to the *w.* and often filling half of the river. Between this flat and another which proceeds from the island of Palomas is the passage for vessels. Here the Orinoco, or rather that part of it which discharges itself into the sea by the mouth of Vessels, forms only one channel, eight leagues *w.* In this space is seen the mouth of a lake, on the *s.* shore, at a little distance from the river. It extends to the foot of the mountain of Piacoa. From the middle of the Orinoco to the *s.* are seen the mountains of Meri.

We now come to the chain of little islands which divide the channel of Piacoa from the river. They extend 12 leagues from *s.* to *w.* On the *n.* bank is the mouth of the Little Paragoan, from which runs a flat extending to the Great Paragoan. The two channels called Paragoan unite before falling into the sea.

Above the Great Paragoan is detached the arm known under the name of Mouth of Pedernales, and which the Orinoco throws towards the coast of Trinidad. It forms the divers channels from the Orinoco to that island, and proceeds from the Orinoco at a league from the *e.* point of Yaya. There is here a flat which crosses half the river.

A league and a half up the river are the Red bogs. This is the first place where, the Orinoco re-appearing to the *n.* is seen Tierra Firme, and land entirely secured from the water. Opposite is a shallow, which runs along the *s.* coast, nearly half a league from *e.* to *w.* The passage for vessels is here along the two banks, but the *n.* bank is the better of the two. In the middle of these bogs there is a very narrow channel called Guaritica, by which shallops can pass in the flood tides, or during the swelling of the river, to a lake which is close to it.

A league higher up, on the *n.* bank, is the mouth of the Guarapo channel. During summer it has but very little water, but nevertheless for several years vessels carried on a contraband trade in mules, oxen, and the productions of Cumana and Venezuela, giving in exchange dry goods.

This channel, excepting at its mouth, is very deep, and admits of the navigation of large vessels, but on account of the high mountain by which it runs, they are obliged to use the oar or to be towed. Two leagues above Guarapo, is the island of Araya; it is of a moderate size, and is close to the *n.* coast.

Towards the *s.* coast are seen the cascades of Piacoa, they are formed by three or four ridges which extend from the middle of the channel to the *s.* coast, but there is sufficient water on the *n.* coast for large vessels. On this coast was formerly the mission of Piacoa and the Catalanian Capuchins. Here is excellent pasture, very fertile land, good water, regular winds, and a good situation for agriculture.

After having reveiwed the three islands of Arciba, the next is that of Iguana, it is more than half a league from the *n.* bank. The river continues navigable on the *s.* side. In summer, on the *n.* side, are banks of sand which have very little water, but in winter there are no obstructions. From the *w.* point of the island of Iguana, the small mountain of Naparenia is only one league. It indeed appears to be nothing more than a high rock.

All this coast as far as the isles of Iguana and Araya is full of sand-banks.

The Simon's channel, lying on the *s.* coast, has at its mouth the ruins of a small fort. From hence is seen the island of St. Vicente, having a flat on the *e.* part, which crosses the channel unto a little below the fortress, but which at full tide is of no inconvenience. This is the spot where once stood the ancient capital of Guayana before it was transferred to Angostura. The distance described is therefore 50 leagues, and it is consequently 40 leagues hence to St. Thomas. The Spaniards, when they transferred the capital 40 leagues higher up, thought proper to leave the forts they destined for the defence of Guayana, on the site of the old town. They are now seen at the foot of a] [small hill, one is called St. Francis and the other El Padastro. By the side of these are two small lakes, named El Zeibo and Baratello. Half a league lower than St. Francis is the little rivulet of Usupamo, having a lake near its mouth.

Nearly half a league above the old town, in the centre of the river, is the large rock of Morocoto, it is rather nearer the *s.* bank than the *n.* and is visible in the summer, but under water during the winter. Not far from this rock is the island of Mares, and on the *s.* side is the rock of the same name, and another called Hache. The channel *n.* of this island is preferable to that on the other side. Three leagues higher on the *s.* side, is point Aramaya, which is merely a jutting rock. Opposite this point are the three little islands of San Miguel: they are all of stone, with a little sand in summer. When the river is swelling they are nearly under water. On the right bank, opposite the village of San Miguel, are two islands called Chacarandy, from the wood with which they are covered; they are divided by only a narrow channel. The island of Faxardo is in the middle of the river, opposite the mouth of the river Caroni. It is 3000 fathoms long and 1387 broad. The *w.* side is subject to inundations. On the right bank, and a league above this island, is the island of Torno. It is separated from the mainland by a small channel; and on the *w.* point there are rocks, and a flat running out to five leagues.

Point Cardinal is on the *s.* side of the island, three leagues above Faxardo. Nearly a quarter of a league from this point is a chain of rocks stretching to opposite Gurampo. During winter but one of these is visible, but in summer three are discernible opposite Gurampo. There is a port formed by point Cardinal, called Patacon. Gurampo is a number of rocks lying five leagues above the island of Faxardo, on the *n.* coast. These rocks form a port bearing the same name. A shallow runs from this port nearly *n.* and *s.* with E. point Cardinal, and having on the *w.* extremity three rocks, under water in winter. The island of Taquache lies half a league from Gurampo, on the left bank. It is 1½ league from *e.* to *w.*

On the opposite side of the river is the island of Zeiba, four leagues long and more than one league broad. The channel separating it from the mainland has very little water, excepting in the winter. Between the mainland to the *n.* and the island of Taguache, there is a channel navigable at all seasons.

The river Cucazana on the *e.* point has a flat, running a little to the *w.* and occupying half of the river. At the mouth is the island of the same name, which nearly joins that of Taguache. It has also a flat on the *w.* point which is in many places visible during summer.

The Mamo channel has at its mouth a flat reaching nearly to the middle of the river, and seven leagues below the capital is another, lying *n.* and *s.* with the island of Mamo, and having from the mouth of January to April only eight feet water. Vessels are obliged to be lightened in order to pass, which is the case with another channel which forms the island of Mano.

After this bar is passed, are numerous rocks on the coast and in the middle of the river. The Currucay points are but jutting rocks, and lie three leagues above port St. Anne. Nearly opposite these points, in the middle of the river, is a large rock named La Pierre du Rosaire. Between this and the coast are several others. To the *n.* of the Pierre du Rosaire is a channel very narrow on account of the rocks lying under water, and stretching to nearly the coast. Vessels run great risks in summer, and in winter the current is so violent that if the wind dies away, they are in danger of being wrecked against the Pierre du Rosaire. A league above this is a point of rocks on the *n.* shore, and some distance from this are three ridges near each other, and bearing *s.* of the *e.* point of the island of Panapana.

The island of Panapana is a league above point Des Lapins, separated from the *s.* shore by a channel moderately wide, but very shallow in summer. At the *e.* and *w.* points there are flats with very little water on them. That of the *w.* point ascends more than a league, and inclines always to the *s.* Between this island, which is 1¼ league long, and the *n.* coast, is the principal channel of the Orinoco. It is rather narrow and of little depth, excepting when the river experiences its swellings.

Two leagues higher up is the narrowest part of the Orinoco, called by the Spaniards Angosturita. Two rocks *n.* and *s.* form this strait. A little higher up, and nearly in the centre, is a large rock called *Lavadero*, that is, Washing-place. It is visible only in summer. Between this and the *s.* coast there is a little island of stones, opposite which the river Maruanta discharges itself. Point Tinco to the *n.* and point Nicasio to the *s.* are also formed of rocks.

St. Thomas, the capital of Spanish Guayana, is the next place. It is situate at the foot of a small hill on the right bank of the river. There is a fort for its protection on the opposite side. This place is called Port Raphael, and is the passage of communication between Guayana and the province of]

[Venezuela and Cumaná. Between port St. Raphael and the city, is an island called Del Medio, from being in the centre of the river. It is a rock under water in winter, but the *n.* side is dry during summer. The principal channel is between this island and the city. It has at ebb tide 200 feet of water, and about 50 more at flood.

In summing up this description, it is to be observed, that from the junction of the river Apure with the Orinoco to St. Thomas's, they reckon 80 leagues. In all this space no other important river falls into the Orinoco on the *s.* save the Caura and Caucapusia. It is however true, that from its source it receives almost all the rivers by its left shore, and from the Apure it receives others which ensure it from thence to Guayana all the commerce of the *s.* plains. The navigation of all the upper part of the Orinoco is very far from being as easy and safe as the size of the river would make one imagine. Scattered with islands which obstruct the channel, and which throw its bed sometimes to the right bank, and sometimes to the left; filled with rocks of all sizes and heights, of which some are consequently even with the water, and others of a depth more or less alarming according to the season; subject to terrible squalls; the Orinoco cannot be navigated but by good pilots, and with vessels of a certain construction and size; though, be it observed, that this description relates here, peculiarly to the navigation from Guayana to the Orinoco, and from the mouth of the Meta to the capital.

Enchanting variety of its banks.—The naturalist must be enraptured with the navigation of this river. Its banks are frequently bordered by forests of majestic trees, which are the resort of birds of the most beautiful plumage and exquisite melody. Various species of monkeys contribute by their cries, their leaps, and gambols, to the embellishment of the enchanting scenery. The savages inhabiting the woods, content in sharing the possession with the wild beasts, are fed by the same fruits as the birds and quadrupeds, living in perfect harmony with them, neither inspiring fear nor feeling apprehension. In some parts, the eye, no longer confined in its view by the foliage of the forest, roves over enchanting plains, which burst upon the sight in luxuriant verdure, covered with excellent pasture, and extending 20 or 30 leagues.

Importance of this river.—Volume and rapidity of its water, and its annual swell.—The Orinoco, excepting the Amazonas, is the largest river in the world. Mr. de Humboldt observes, that the mouth of the Amazonas is much more extended than that of the Orinoco, but the latter river is of equal consideration with respect to the volume of water which it has in the interior of the continent, for at 200 leagues from the sea, it has a bed of from 2500 to 3000 fathoms, without the interruption of a single isle. Its breadth before St. Thomas is 3850 fathoms, and its depth, at the same place, according to the measurement made by order of the king in 1734, in the month of March, the season when its waters are at the lowest, was 65 fathoms.

This river, like the Nile and others, has an annual swell. This commences regularly in April and ends in August. All the mouth of September it remains with the vast body of water it has acquired the five preceding months, and presents a spectacle astonishingly grand. With this encrease of water it enlarges, as it were, its natural limits, making encroachments of from 20 to 30 leagues on the land. The rise of the river is, opposite to St. Thomas, 30 fathoms, but it is greater in proportion to the proximity to the sea; it is perceptible at 350 leagues from its mouth, and never varies more than one fathom. It is pretended in the country, that there is every 25 years a periodical extraordinary rise of an additional fathom. The beginning of October the water begins to fall, leaving imperceptibly the plains, exposing in its bed a multitude of rocks and islands. By the end of February it is at its lowest ebb, continuing so till the commencement of April. During this interval, the tortoises deposit themselves on the places recently exposed, but which are still very humid; it is then that the action of the sun soon develops in the egg the principles of fecundity. The Indians resort from all parts with their families, in order to lay in a stock of food, drying the tortoises and extracting an oil from their eggs, which they either make use of for themselves or sell.

The water of the Orinoco is potable, and even some medicinal virtues are attributed to it.

Its tides, and peculiar animals inhabiting it.—Though the tide is very strong at the mouth of the river, it is so broken and obstructed by the numerous channels through which it passes, that before the town of St. Thomas it is scarcely perceptible, or rather there is no tide at all so high up, excepting in summer, or when the wind blows from the sea. The Orinoco abounds in fish of various descriptions, but these, although they bear the same name as the fishes of Europe, are found not to correspond precisely with them in their nature or quality. The amphibious animals are also curious and worthy of notice. For an account, however, of the most peculiar both of the one and the other of the inhabitants of these waters, see the articles]

[CARRIBE, CAYMAN, IGUANA, CHIQUIRE, LAPA, WATER-DOG, DORMOUSE, MANATI, &c. in the vocabulary of provincial terms at the end of this work.

Table of the latitudes and longitudes.—For the table of latitudes and longitudes of the most important places in these parts, see the end of the general preface.]

[ORINOCO, LITTLE. See MOCOMOCO.]

ORIO, a river of the province and government of Panamá, in the kingdom of Tierra Firme in S. America. It divides the jurisdiction of this province from that of Veraguas, and is the boundary of the isthmus. It runs from *n. w.* to *s. e.* and to the *w.* of the Punta Mala enters the Pacific sea, in lat. 7° 25′ *n.*

ORISKUNI, a small river of the province and country of the Iroquees Indians in N. America, which runs *w.* then turns its course to *n.* and enters the Mohawks.

ORISTAN, a city of the island of Jamaica, founded by the Spaniards in 1510 on the *s.* coast, but which has not existed since that the island became in the possession of the English.

ORITO, or LONITO-YACU, a river of the province and country of Las Amazonas, which runs *s.* through the woods to the *n.* of this river. Near its source dwell many barbarous nations of Mainas, Zimarrones, and Umuranas Indians, and it enters the Marañon in lat. 4° 5′ 10″ *s.*

ORITUCO, a river of the province and government of Venezuela, and Nuevo Reyno de Granada. It rises in the mountains of the city of Altagracia, runs, forming a curve, to the *w.* and enters the Guarico.

ORIZABA, a jurisdiction and *alcaldía mayor* of Nueva España, belonging to the bishopric of La Puebla de los Angeles; bounded *e.* by the town of Cordoba, *n.w.* by Vera Cruz, Antigua, *s.w.* by the province of Thehuacan, and *n. w.* by the mountains of Tepeaca. Its extent is a little more than seven leagues from *e.* to *w.* and five in width from *n.* to *s.* It is of an hot and moist temperature, very fertile in tobacco, which is its principal article of commerce, and the greater part of its inhabitants are drovers, employed by the traffic of the neighbouring provinces. The population consists of the following settlements:

S. Miguel Thomatlan,	San Martin Atlahuilco,
Naranjal,	San Andres Nexapa,
San Juan de Atlaca,	S. Francisco Necoxtla,
Temilolacan,	S. Juan Acolzingo,
San Pedro Thequilan,	S. Pedro Maltrata,
Tenango,	Huiluapan,
Santa Maria Aquila,	San Juan Bautista Nogales,
Ixtazoquitlan,	Santa Maria Ixhuatlan.
S. Juan del Rio,	
Santiago Tilapán,	

The capital is the settlement of the same name; situate in a spacious plain of a league long from *e.* to *w.* and half a league wide from *n.* to *s.* It is of an hot and moist temperature, and one of the best settlements in the whole kingdom for its opulence, pleasantness, abundance of provisions, and disposition of its houses. These are built so as to form straight streets, the principal of which is that called La Real, and which is upwards of a quarter of a league long.

This town is fertilized by the abundant rivers with which it is surrounded. One of these rises in the *sierra* from a volcano, and, running for some distance, incorporates itself with the Tuzpango, which runs by the *s.* side, and from these two are thrown out many arms, which run in different directions.

The parish church is a costly and modern building, and a fine piece of architecture. It has a chapel of ease, which is a magnificent temple of Nuestra Señora de Guadalupe; two convents of monks, namely, of barefooted Carmelites and San Juan de Dios, the latter of which is very useful in this town, because, as standing at the usual entrance of European travellers from Vera Cruz, who arrive sick, it affords its relief and hospitality readily and kindly offered by the monks, its inhabitants. They have, indeed, two curious infirmaries for this charitable purpose, the one appropriated to the laity, the other to the clergy.

The population is composed of 510 families of Spaniards, 300 of *Mustees*, 220 of Mulattoes, and 800 of Mexican Indians, who gain their livelihood by cultivating maize, French beans, large vetches, *ajonjoli*, and fruits, and maintaining thereby a commerce. The trade of the Spaniards consists in clothes, and native and European merchandise.

This country produces much leaf-tobacco, some years as much as 2000 *cargas*, or loads, which are carried to Mexico, La Puebla, and other cities and settlements of the kingdom, leaving a revenue to this of 100,000 dollars annually. In this trade many Mulattoes and *Mustees* are employed; some, however, in mechanical works, and others as drovers. The above are formed into four companies of militia of 100 men each, and there are two companies of Spaniards, one of infantry, the other of horse, and all these are obliged to march to Vera Cruz upon necessity, and at 50 hours notice.

This town, which is very large, has three wards

or hermitages, and in the parish is venerated a miraculous image of the child Jesus lost, which was brought here from Genoa about the middle of the 16th century, and concerning which the following account is related by Don Joseph Villaseñor in his "Teatro Americano," namely, "that a certain viceroy passing through the town, taking advantage of his authority, and induced by the veneration in which he held the image, to carry it away with him, in spite of the tears and entreaties of the inhabitants, whilst he was departing, was taken suddenly extremely ill; and that a certain chaplain who had been picked out by the people, seizing the opportunity, went to the viceroy and informed him, that heaven had heard the prayers of the disconsolate people, and that it would not suffer any one with impunity to take away that image which had been such a consolation to them in their necessities and afflictions: it was further said, that the viceroy immediately delivered back the child, and that he as quickly convalesced." This image is held to the present day in the greatest reverence.

[Orizaba, according to Humboldt, of the intendancy of Vera Cruz, lies a little to the *n.* of the Rio Blanco, which discharges itself into the Laguna d'Alvarado. It has been long disputed if the new road from Mexico to Vera Cruz should go by Xalapa or Orizaba. Both these towns having a great interest in the direction of this road, have employed all the means of rivalry to gain over the constituted authorities to their respective sides. The result was, that the viceroys alternately embraced the cause of both parties, and during this state of uncertainty no road was constructed. Within these few years, however, a fine causeway was commenced from the fortress of Perote to Xalapa, and from Xalapa to L'Encero.

Orizaba is 120 miles *e.* by *s.* of Mexico, in lat. 18° 48′ *n.* Long. 97° 7′ *w.*]

ORIZABA, another settlement, the head settlement of the district of the *alcaldía mayor* of Iximiquilpan in the same kingdom, in the district of which are many approximate wards; amongst all of which are 945 families of Othomies Indians, and 80 of Spaniards, *Mustees*, and Mulattoes, whose spiritual necessities are attended to by only two priests, by far too short a number to fulfil the duties required; so that many are obliged to go without their assistance. Nearly all these settlements are of a mild temperature, and fertilized with the waters from the river of the *sierra* of Mextitlan, by which also are irrigated many gardens, orchards, and cultivated fields. The inhabitants make charcoal, fishing-tackle, and rigging; they also procure honey from the *magueyes*, which they cultivate. This head settlement of the district has in its division six estates, namely, Juan Dó, Domingo, Azuchitlen, La Florida, Vetza, and San Pablo, in the which they gather great harvests of seed and grain, owing to the fertility procured by the aforesaid river.

ORLEANS, NEW, a city of the province and government of Louisiana in N. America; situate between the *e.* shore of the river Mississippi and the Fish. Thirty-three miles from the sea. See NEW ORLEANS.

[ORLEANS, the middle of the three *n.* counties of Vermont. A part of lake Memphremagog projects into the *n.* part of it from Canada. It contains 23 townships. It is very high land, and sends its waters in almost every direction of the compass. Clyde, Barton, and Black rivers empty into lake Memphremagog; the waters of many branches of Missiscoui, La Moelle, and Onion rivers, rising here, fall into lake Champlain; those of Mulhegan and Pasumpsick empty into Connecticut river.]

[ORLEANS, a township in the county of Barnstable, Massachusetts; taken from the *s.* part of Eastham, and incorporated 1797.]

[ORLEANS, Isle of, is situated in the river St. Lawrence, a small distance below Quebec, and is remarkable for the richness of its soil. It lies in the middle of the river; the channel is upon the *s.* side of the island, the *n.* side not having depth of water at full tide, even for shallops. The *s. w.* end of the island is called Point Orleans. The coast is rocky for a mile and a half within the *s.* channel, where there is a careening place for merchant ships. Round point Levi, and along the *s. e.* side of the river, the shore is rocky, but the middle of the bason is entirely free.]

ORLEANS, a French fort of the province and government of Louisiana, on the shore of the river Missouri, opposite the settlement of this name.

ORNE, a settlement of the province and government of Venezuela in the Nuevo Reyno de Granada.

ORO, SAN JUAN DEL, a town of the province and *corregimiento* of Carabaya in Peru; founded by the fugitive Spaniards of the parties of Pizarro and Almagro, and who, after penetrating through woods and chasms, established themselves here, allured by the richness of the country. They all became opulent, and having obtained a special privilege from the viceroy Don Antonio de Mendoza, some of them passed over to España, received honours and rewards of the emperor, and at last growing haughty and intoxicated with their good fortune, began to have parties and dissensions amongst each other, so that from being a very flourishing settlement, with a population of upwards of 3000, this has become so reduced as to contain not more than six Spanish families. It is just at the source of the river Inambari.

Oro, a settlement and *real* of silver mines, of the province of Tepeguana, and kingdom of Nueva Vizcaya, on the bank of the stream of Parral.

Oro, a town of the province and country of Las Amazonas, in the territory of Mato Groso; situate at the source of the river Maloques. To the *n.* are some rich gold mines, from which it takes its name.

Oro, a river of the province and government of Santa Marta in the Nuevo Reyno de Granada. It rises *w.* of the city of Salazar de las Palmas, and enters the Lebrija.

Oro, another, of the province of Pamplona in the same kingdom, which empties itself into the sea; and is thus called from gold being found on its shores.

Oro, another, of the province and government of Neiva in the same kingdom. It runs *s.s.w.* and enters the Magdalena, between those of Olaz and Neiva.

Oro, another, of the province and government of Darien, and kingdom of Tierra Firme, which runs into the sea between the island of La Laguna and the river Francisa.

Oro, another, with the dedicatory title of Fino, in the territory and country of the Guayazas Indians in Brazil. It is small, runs *n.n.e.* and enters the head of the Tocantines.

Oro, an island of the N. sea, on the coast of the province and government of Darien in the kingdom of Tierra Firme. It is opposite the point which forms the bay and port of Calidonia to the *w.*

Oro, some mountains of the province and government of Moxos in the kingdom of Quito, which run from *w.* to *e.* from the river Baures to that of Serre, to the *n.* of the settlement of the missions of San Nicolas.

Oro, another *sierra* or *cordillera* of mountains, of the province and government of Buenos Ayres in Peru. They run *s.s.e.* near the coast of the river La Plata.

OROATA, a small lake of the province and country of Las Amazonas; formed from a small river which enters the Madera by the *w.* side.

OROCOMA, an ancient and extensive province to the *s.* of the province of Venezuela and Nuevo Reyno de Granada, between the river San Pedro to the *e.* the mountains of Tucuyo to the *n.* those of Bogotá to the *w.* and the *llanos* of Cazanare to the *s.*; bounded *n.* by the nation of the Cuicas Indians, and *e.* by a tribe of the nation of the Panches. It is nearly depopulated, as it is subject to continual inundations. The climate is hot, moist, and unhealthy, but it abounds in excellent pastures.

OROCOPICHE, a small river of the province and government of Guayana or Nueva Andalucia. It rises *s.* of the city of Nueva Guayana, runs *n.* and enters the Orinoco opposite that city.

[ORODADA Pena, on the coast of Peru, is two leagues due *n.* of Lobos de Payta, and two *s.* by *w.* of Payta.]

OROKUPIANAS, a nation of barbarian Indians of the province and country of Las Amazonas, who inhabit, with various other nations, the *sierras* and mountains on the shore of the river Basururu, the which empties itself into the Marañon or Amazon by the *n.* coast, 32 leagues from the mouth of the Cuchiguara.

[OROMCOTO, a river of New Brunswick, which empties into St. John's river. By this passage the Indians have a communication with Passamaquoddy bay.]

ORONAS, Sierras de, some mountains of the province and government of Darien, and kingdom of Tierra Firme, near the *s.* coast. They run between the rivers Chepo and Francisca.

[ORONDOCKS, an Indian tribe who live near Trois Rivieres, and could furnish 100 warriors about 20 years ago.]

OROPESA, a town of the province and *corregimiento* of Cochabamba in Peru; founded in a beautiful, fertile, and extensive valley of the name of Cochabamba, and by which name the settlement is also known, by order of the viceroy Don Francisco de Toledo in 1575, on the ruins of another town which had been founded in 1565 by Pedro de Cardenas, and which bore his title. This town is watered by a small river, called Sahacá, which fertilizes the neighbouring gardens and orchards, and then enters the Cachimayu. It has, besides the parish church, two convents of San Francisco, one of the Observers, and the other of the Recoletans, a convent of San Agustin, another of La Merced, an hospital of San Juan de Dios, two monasteries of nuns, the one of Santa Clara, the other of the barefooted Carmelites. Its population is composed of 17,000 souls in communion, amongst whom are many rich and noble families, descended from the ancient conquerors of Peru, and from some illustrious houses in Spain. In some foreign geographical charts this town is wrongly called Oropalza. Eight miles *n.* of Cochabamba, and 89 *n. n. w.* of Chuquisaca or La Plata, in lat. 18° 11′ *s.* Long. 67° 18′ *w.*

OROPESA, a settlement of the same kingdom as the former, in the province and *corregimiento* of Quispicanchi, distant half a league from the lake called La Mohina, which is more than a league long, and an half wide, and in which there is a quantity of *totora* and reed-mace, some fish and aquatic fowl. One end of it extends to the foot of the mountain called Rumicolca, where there are to be seen the ruins of the palace of the Emperor Huasca-Inca; and there is a tradition, that in the centre of this mountain were secreted the immense treasures of the 11 monarchs of Peru, when the Spaniards entered; and this report has induced many to spend large sums in attempting to discover the fortunate spot, but to no purpose, nothing having been found but caves and openings which they call *chinganas*, and different channels for carrying off the water. This settlement has, besides the parish church, two others well adorned, with the titles of Nuestra Señora de la Estrella and La Virgen de la Hermita; situate 10 miles *e.* of Cuzco, in lat. 13° 42′ *s.* Long. 71° 6′ *w.*

OROPESA, another settlement, of the province and *corregimiento* of Aimaraes in the same kingdom; situate on the shore of the river Pachachaca.

OROPESA, a river of the same province and kingdom as the former settlement. It rises in the province of Cotabambas, to the *e.* of the settlement of Pituhuanca, runs inclining to *n. w.* and enters the Pachachaca.

OROPI, a large lake of the province and country of Las Amazonas, in the territory of the Guaranacaos Indians. It is formed from a waste water of a river which runs *w.* and then enters the Madera.

OROPOTO, a settlement and *asiento* of rich gold mines in the province and *corregimiento* of Asangaro, and kingdom of Peru.

OROPUCHE, a settlement of the province of Barcelona and government of Cumaná, on the shore of a river which enters the Huere, to the *s.* of the town of Aragua, and *n.* of the town of Pao, about an equal distance from each.

OROPUCHE, a river of the island of Trinidad, which runs *e.* and enters the sea.

OROQUARAS, a barbarous nation of Indians, but little known, of the province and country of Las Amazonas. They dwell in the woods to the *s.* of the river Marañon, 45 leagues below the mouth of the river Cayari.

OROTINA, a settlement of the province and government of Nicoya, and kingdom of Guatemala. It is one of the principal there, and well peopled with Indians, who are of a good disposition and very friendly to the Spaniards. It is distant from the capital seven leagues by sea and 20 by land.

OROYA, a settlement of the province and *corregimiento* of Tarma in Peru; annexed to the curacy of the capital.

[ORPHAN'S Bank, a fishing bank of the *s. e.* point of Chaleur's bay, on the *n. e.* coast of New Brunswick, in N. America. On it is from 75 to 30 fathoms water.]

[ORPHAN'S Island, a settlement belonging to Hancock county, district of Maine, having 104 inhabitants.]

[ORRINGTON, a plantation in Hancock county, district of Maine, having 477 inhabitants. It lies on the *e.* side of Penobscot river, 16 miles above Buckstown, and 180 *n. n. e.* of Boston.]

ORTEGA, SAN JOAQUIN DE, a settlement of the province of Tucumán in Peru, of the district of the country of Gran Chaco; a *reduccion* of the Morampas Indians made by the missionaries of the Jesuits, and now under the charge of the religious order of San Francisco.

ORTEGA, another settlement, of the province and government of Popayán in the Nuevo Reyno de Granada.

ORTEZ, a settlement of the province and government of Venezuela in the kingdom of Nnevo Granada; situate on the shore of a river which enters the Guarico to the *s.* of the town of San Sebastian.

ORTEZ, a small river of the province and government of Buenos Ayres in Peru, which runs *n.* and enters the river La Plata.

ORTEZ, a shoal or sand-bank, at the entrance of the mouth of the river La Plata.

ORTOCUNA, a settlement of the province and *corregimiento* of Xauja in Peru.

[ORUA, ORUBO, or ARUBA, the most *w.* of the Caribbee islands in the W. Indies, called by the Spaniards Las Islás de Sotovento. It is on the coast of the Spanish main. Lat. 12° 31′ *n.* Long. 70° 7′ *w.*]

ORUBA. [See ORUA.]

ORUBILLA, another, a small island of the N. sea, to the *w.* of the former.

ORUILIERES, a river of the province of Guayana, in the French possessions. It enters the Oyapoco.

ORUNA, San Joseph de, a city and capital of the island and government of Trinidad; founded on a mountain in a strong and advantageous situation by Gonzalo Ximenez de Quesada in 1591, at two leagues from the sea. It belongs to the bishoprick of Puertorico; and in its parish church is seen the sepulchre of Nicolas de Labrit, a French bishop, killed by the Caribes Indians, by the Caño de Aquire, not far from the coast, whilst instructing them in the faith. In this city there is a convent of the religious order of San Francisco.

ORUOILLA, a settlement of the province and *corregimiento* of Lampa in Peru.

ORURO, a province and *corregimiento* of Peru; bounded *n.* by the province of Sicasica, *e.* by that of Cochabamba, *s.* and *s. e.* by that of Paria, and *w.* and *n. w.* by that of Pacajes. It is of a cold and dry temperature, and very subject to tempests. Its productions are *papas*, *quinua*, and some barley. It has breeds of large and native cattle, and much gunpowder is made here from the abundance of saltpetre, although not so much as in former times, when its gold and silver mines were in a flourishing state. At present these are in great decay, and the greater part are filled with water, which, on account of the want of declivity in the territory, it is impossible to drain, and on this account the population is daily diminishing. The whole of the province does not count more than 8000 souls. It is 18 leagues from *e.* to *w.* and 20 from *n.* to *s.* Its *corregidor* had a *repartimiento* of 35,527 dollars; and it used to pay an *alcabala* of 284 dollars annually.

The capital is the town of the same name; founded in a beautiful. valley, of nine miles long, the greater part being swampy and abounding in saltpetre, with the name of San Felipe de Asturia. In 1590, were re-opened some of the rich mines which were begun to be worked by the Indians in the time of their Incas; and amongst the best of these was the mine called Pie de Gallo, which is, however, at the present day abandoned, the mine of Popo, and a few others, only being being worked, although these alone yield yearly 600 bars of silver of about 200 marks each bar; and in this consists the commerce of the place.

It has five convents, namely of San Francisco, Santo Domingo, San Agustin, La Merced, San Juan de Dios, and a college which belonged to the Jesuits; also four parishes for its numerous population, with the titles of San Felipe, San Miguel de la Rancheria, San Ildefonso de Paria and Sepulturas. Eighty-five miles *s.* with a slight inclination to the *e.* of La Paz, and 70 *n. n. w.* of Potosi, in lat. 18° 48′ *s.* Long. 68° *w.*

[ORWELL, a township of Vermont, the *n.* westernmost in Rutland county, and situated on the *e.* side of lake Champlain. It contains 778 inhabitants. Mount Independence stands in this township opposite Ticonderoga, in the state of New York. Near mount Independence is a chalybeate spring.]

OSABAW, a small island of the N. sea, near the coast of the province of Georgia. It forms with the island of Wasa a strait of its name.

[OSAGE, Grand, a nation of Indians of N. America, who claim the country within the following limits, viz. commencing at the mouth of a *s.* branch of the Osage river, called Neangua, and with the same to its source; thence *s.* to intersect the Arkansas about 100 miles below the three forks of that river; thence up its principal branch to the confluence of a large *n.* branch, lying a cousiderable distance *w.* of the Great Saline, and with that stream nearly to its source; thence *n.* towards the Kansas river, embracing the waters of the upper portion of the Osage river, and thence obliquely approaching the same to the beginning. The climate of the country they inhabit is delightful, and the soil fertile in the extreme. The face of the country is generally level and well watered; the *e.* part of the country is covered with a variety of excellent timber; the *w.* and middle country consists of high prairies. Their territory embraces within its limits four salines, which are, in point of magnitude and excellence, unequalled by any known in N. America; there are also many others of less note. The principal part of the Great Osage nation have always resided at their villages, on the Osage river, since they have been known to the inhabitants of Louisiana. About five years since, nearly one-half of this nation, headed by their chief the Big-track, emigrated to three forks of the Arkansas, near which, and on its *n.* side, they established a village, where they now reside. The Little Osage nation formerly resided on the *s. w.* side of the Missouri, near the mouth of Grand river; but being reduced by continual warfare with their neighbours, were compelled to seek the protection of the Grent Osage, near whom they now reside. There is no doubt but their trade will increase: they could furnish a much larger quantity of beaver than they do. Two villages on the Osage river might be prevailed on to remove to the Arkansas and the Kansas higher up the Missouri; and thus leave a sufficient scope of country for the Shawnees, Dillewars, Miames, and Kickapoos. The Osages cultivate corn, beans, &c.]

Osage, a river which rises in the territory of

the aforesaid Indians, runs *n. e.* and enters the Missouri.

OSATAMA, a small settlement of the *corregimiento* of Pasca in the Nuevo Reyno de Granada; annexed to the curacy of Fusagusagá.

OSBORN, a settlement of the island of Barbadoes, in the district and parish of S. Thomas.

OSSEY. See PAXAROS.

OSIACURI, a settlement of the province and government of Cartagena in the Nuevo Reyno de Granada; situate in the vicinity of the road which leads from that capital to the river Grande de la Magdalena, between the settlement of Piojon and Malambo. Thirty-five miles *n. e.* of Cartagena, on the *w.* side of R. Magdalena.

OSIPEE, a small river of the province of Continent, one of the four of New England in N. America. It rises from a small lake, runs *e.* and enters the Saco.

[OSNABURG, a small island in the S. Pacific ocean, having the appearance of the roof of a house. It is about four leagues in circuit; is high land; full of cocoa trees; has no anchoring place, and scarcely affords landing for a boat. It was discovered by Captain Wallis, and is called Maitea by the natives. Lat. 17° 40′ *s.* Long. 148° 6′ *w.*]

[OSNABURG, another island in the same sea, discovered by Captain Carteret. Lat. 22° 4′ *s.* Long. 148° 36′ *w.*]

[OSNABURG House, a settlement of the Hudson's bay company, in N. America; situated at the *n. e.* corner of lake St. Joseph, 122 miles *w.* by *s.* of Gloucester house. Lat. 51° 4′ *n.* Long. 90° 15′ *w.*]

OSNO, SAN MIGUEL DE, a settlement of the province and *corregimiento* of Guanta in Peru; annexed to the curacy of Tambos.

OSNO, another settlement, with the dedicatory title of San Salvador, to distinguish it from the former, in the same province and kingdom, and also annexed to that curacy.

OSO, RIO DEL, a river in the province of Nuevo Mexico and N. America.

OSORNO, a city of the kingdom of Chile, founded by D. Andres Hurtado de Mendoza, marquis of Cañete, in 1558, on the shore of the river Bueno, 24 miles from the S. sea, 212 *s.* of the city of La Concepcion, and 34 from the garrison of Valdivia. Its territory was barren in vegetable productions, but abundant in gold mines, the which made it a rich and beautiful town, inhabited by many illustrious families. It had two convents, one of San Francisco, the other of S. Domingo, and a monastery of the nuns of Santa Clara. The Charaucabis and Arucanos Indians, who made an insurrection here in 1599, destroyed and burnt the town, putting to death the Spaniards, and taking away the women to marry with them. After this lamentable fall it has never since been rebuilt, and nothing but its ruins remain. It stood in lat. 40° 20′ *s.*

OSORNO, a mountain or volcano of the same kingdom, to the *e.* of the former city, in the *cordillera* of the Andes of that kingdom. Sixty-seven miles *e.* by *s.* of the city of its name, in lat. 40° 36′ *s.*

OSORNO, a canal between the continent of the same kingdom and the *n.* point of the isle of Chiloe, at the entrance to the *ancud* or archipelago of Chiloe, the which Mr. Martiniere calls the lake of Anand, in the article Osorno.

OSPA, a settlement of the province of Florida in N. America.

OSPINO, a settlement of the province and government of Venezuela in the Nuevo Reyno de Granada; founded a few years since.

[OSSABAW Sound and Island, on the coast of the state of Georgia. The sound opens between Wassaw island on the *n.* and Ossabaw island on the *s.* and leads into the river Ogeechee.]

[OSSIPEE, or OSAPY, a township, mountain, and pond, in New Hampshire, in Stafford county, near the *e.* line of the state. The town was incorporated in 1785, and has 139 inhabitants. The lake lies *n. e.* of Winnipiseogee lake, between which and Ossipee lake is Ossipee mountain, described in the account of New Hampshire. Its waters run *e.* and joined by South river, form Great Ossipee river, which empties into Saco river, near the division line between York and Cumberland counties, in Maine, between Limerick and Gorham.]

[OSSNOBIAN, or ASSENEBOYNE Indians, a tribe found about the source of Ossnobian or Asseneboyne river, far *w.* of lake Superior. They are said by the Moravian missionaries to live wholly on animal food, or at least to confine themselves to the spontaneous productions of nature; giving those who dig the ground the appellation of slaves. Bread is unknown to them. A traveller who lived some months in their country offered to some a few remnants of bread, which they chewed and spit out again, calling it rotten wood. These Indians, as well as those numerous nations who inhabit the country from lake Superior, towards the Shining mountains, are great admirers of the best hunting horses, in which the country abounds. The horses prepared by them for hunters have large holes cut above their natural nostrils, which they say makes them longer winded than others not thus prepared.

. The Ossnobians have no permanent place of abode, but live wholly in tents, made of buffalo and other hides, with which they travel from one place to another, like the Arabs; and as soon as the food for their horses is expended, they remove and pitch their tents in another fertile spot; and so on continually, scarcely ever returning to the same spots again.]

OSTIA, a settlement of the province and government of Cartagena in the Nuevo Reyno de Granada; situate in one of the islands which are formed by the arms of the river Cauca.

[OSTICO, a small lake in Onondago county, New York, partly in the *s. e.* corner of Marcellus, and *n. w.* corner of the township of Tully. It sends its waters from the *n.* end, which is eight miles *s. w.* of Onondago castle, by a stream 16 miles long, to Salt lake.]

OSTIMURI, a province of Mexico, in the government of Sonora, with the title of San Ildefonso. It begins on the other side of the river Mayo, seven leagues from the *real* of Los Alamos, so as that from thence to the river Chico it is 40 leagues from *n.* to *w.* bounded by the river Grande de Hiaquis. It is very fertile in maize, pease, French beans, and pulse, in the summer time; for in the winter the river rises to such a height as to inundate the greater part of the territory, not only destroying the crops, but even the settlements. It has many mines of gold and silver, which metals are of base alloy, and are but little coveted.

This province is peopled by different nations of Indians, who were reduced to the faith by the Jesuits, who founded the following settlements:

Rio Chico, the capital,	S. Marcial,
Ostimuri, formerly the capital,	S. Joseph,
	Nacori,
Bethlem,	Cumpas,
Ruan,	Thesico,
Potan,	Tonichi,
Bocon,	Onapa,
Cocarin,	Aribethechi,
Todos Santos,	Bacanora,
Nacozari,	Saguaripa,
S. Xavier,	Las Juntas,
Opostura,	Tacupero,
Oputú,	S. Marcos,
Comoripa,	S. Miguel,
Zuaque,	Tecoripa,
Yecora,	Matape,
S. Nicolas,	Guazabas.
Onabas,	

[OSTINES, or CHARLESTOWN, a considerable town in the island of Barbadoes.]

OSTIONES, a port of the S. sea, in the province and government of Chocó, and Nuevo Reyno de Granada, between the port of Buenaventura and the point Arena.

OSTITAN, SAN PEDRO DE, a small settlement or ward of the head settlement of the district of Moloacan, and *alcaldia mayor* of Acayuca, in Nueva España, close to the settlement of Huamanguillo.

OSTOGERON, a settlement of Indians of the province and colony of Pennsylvania, N. America; situate on the shore of the *e.* arm of the river Susquehannah.

OSTOTIPAC, or TEPIC, a province and *alcaldia mayor* of Nueva Galicia, and bishopric of Guadalaxara, in N. America. It is of limited extent and hot temperature, but abounding in cattle, and producing large crops of cotton, maize and coco, and plenty of salt, these being the articles of its commerce. The capital is the settlement of the same name, inhabited by a large popolation of Indians, *Mustees*, and Mulattoes, who live in the estates, and the *ranchos* of its district; it is 100 miles to the *w.* one quarter to the *n. w.* of Guadalaxara, in 104° 45′ long. 21° 37′ *n.* lat. The other settlements of this jurisdiction are reduced to the following:

S. Sebastian, Mascota, Talpa.

OSTOTIPAC, another settlement, of the jurisdiction and *alcaldia mayor* of Otumba in Nueva España; inhabited by 144 families of Indians, and being one league *s. e.* of its capital.

OSTOTIPAC, another, with the dedicatory title of Santa Maria, in the head settlement of the district and *alcaldia mayor* of Tepeaca in the same kingdom as the former. It contains only 13 families of Indians, and is a league and an half from its capital.

OSTOTIPAQUILLO, a jurisdiction and *alcaldia mayor* of the kingdom of Nueva Galicia, and bishopric of Guadalaxara, in N. America, and the most reduced of any there, being equally without productions and inhabitants. It is of a hot temperature, and yields nothing but some sugar cane, of which honey, the only branch of its commerce, is made. Its population is reduced to the two settlements of Cacalutla and San Francisco, besides the capital, which consists of 60 families of Indians. It is 25 leagues *n. w.* of the city of Guadalaxara.

OSTOTITLAN, a settlement of the head settlement of the district and *alcaldia mayor* of Toluca in Nueva España; it contains 58 families of Indians, and is a little to the *w.* of its capital.

OSTOZINCO, a settlement of the head settlement of the district of Acantepec, and *alcaldia mayor* of Tlapa, in Nueva España. It contains 50 families of Indians, and is three leagues and a half from its head settlement.

OSTRAS, Rio de las, a river of the province and *captainship* of Rio Janeiro in Brazil, which runs *s. s. e.* and enters the sea opposite the isle of Ancora.

OSTUA, a settlement of the head settlement and district of San Pedro de Metapas, and *alcaldia mayor* of Santa Ana, in the kingdom of Guatemala, annexed to the curacy of that head settlement.

OSTUMA, a settlement of the head settlement of the district of Santiago Nunualco, in the province and *alcaldia mayor* of San Vicente de Austria, and kingdom of Guatemala; annexed to the curacy of its head settlement.

OSTUMCALCO, San Juan de, a principal and head settlement of the district of the *alcaldia mayor* and province of Quezaltenango in the kingdom of Guatemala. It contains in its district 5200 Indians of the Quihe, Kazchiquel, and Zotohil nations, and was one of the doctrinal establishments of the religious order of San Francisco.

[**OSWEGATCHIE** River and Lake, in Herkemer county, New York. The river empties into the river St. Lawrence, or Catariqui. Oswegatchie lake is about 19 miles long from *s. w.* to *n. e.* and two broad, and sends its waters *n. e.* into the river of its name. It is about 10 miles *s. e.* of the Thousand lakes, near the entrance into lake Ontario. There is a fort of the same name, situated on the Cataraqui river, 62 miles *n. e.* of Kingston on lake Ontario.]

[**OSWEGATCHIES**, an Indian tribe residing at Swagatchey, on the river St. Lawrence, in Canada. They could furnish about 100 warriors 20 years since.]

[**OSWEGO**, a navigable river of New York, which conveys the waters of Oneida and a number of small lakes, into lake Ontario. It is more commonly coiled Onondago; which see.]

Oswego, a fort of the English, built in the territory and country of the Iroquees Indians, on the *s.* shore of the lake Ontario, and at the entrance or mouth of the river Onondago, or Oswego, where the former carry on a great commerce with the Indians in hides, giving in exchange all kinds of iron ware, brandy, and other articles; which traffic begins in the mouth of May, and lasts till the end of July. [This fort was taken by the British from the French in 1756, and confirmed to them by the peace of 1763. It was delivered up to the United States, July 14, 1796. It is about 110 miles *e.* by *n.* of Niagara, in lat. 43° 23′ *n.* Long. 76° 41′ *w.*]

[**OTABALO.** See Otavalo.]

[**OTAHA**, one of the Society islands in the S. Pacific ocean, whose *n.* end is in lat. 16° 26′ *s.* and long. 151° 30′ *w.* It has two good harbours. See Ohamene and Oherurua.]

OTAHITI, or Otaheiti, an island of the S. sea, which gives name to various others, discovered by the English captain, Samuel Wallis, in 1767, commander of the ship Dolphin, in the reign of George III. king of England, and for whom he took possession of it.

The viceroy of Peru, Don Manuel Arval, sent the pilot, Don Joseph Amich, to reconnoitre these islands in 1772; but he being prevented from the badness of the weather to effect his object, the king determined to send some missionaries amongst these barbarians, to reduce them to the faith; and accordingly, in 1774, there went out to this end the Fathers Geronimo, Clot, and Narciso Gonzalez, of the order of San Francisco, taking with them some Indians who had been catechized and baptized at Lima; and these being established in the aforesaid island of Otahiti, which is the principal, remained there till 1775, when the Captain Don Cayetano arrived in the Aguila frigate with provisions for them; and then the missionaries finding that no troops were sent for their protection, determined to retire, after merely making a few observations respecting the islands, as to their productions and the manners of the natives; and it is from their accounts that we shall relate the following particulars.

These islands, it seems, are well peopled with infidels, and in each of them is a cacique or lord, whom they style *eriri;* also in the great islands are many subordinate to one superior, who is called *eririattu.* The temperature of them all is hot and moist, so that they are well covered with trees, and shady; they produce many *cocos*, plantains, *ñames*, and another fruit which serves as bread. They have swine and turkeys, aud grow much sugar-cane.

The natives are corpulent and well made, of a brown mulatto colour, with long and crisp hair, which they anoint with oil of *cocos.* They go constantly naked, and wear only a swathe round the waist, passing one end between the thighs; the chiefs wear a small *poncho* or matted cloak of very fine palm, and some very delicate woven stuffs which the women make from the barks of trees, and of which specimens are to be seen in the royal cabinet of natural history at this court. These Indians are pacific, cheerful, jovial, and docile, manifest great genius in the building of their houses and canoes, and in the manufacture of their woven stuffs. They, however, use the barbarous custom of sacrificing to their false idols. Whilst the aforesaid missionaries were amongst them, the criri fell sick, and to implore with greater success for his speedy recovery, they sacrificed to their deity four of their unfortunate prisoners. They form amongst themselves societies, wherein they mutually engage to stand by and assist each other in any difficulties; but it is indispensable that those admitted should be without male children, and this ordinance has been the cause of parents continually putting their infant sons to death.

[Otahiti consists of two peninsulas, which are connected by a low neck of land, about two miles over; the circumference of both peninsulas is somewhat more than 90 miles. The whole island is surrounded by a reef of coral rocks, within which the shore forms several excellent bays and harbours, where there is room and depth of water for any number of the largest ships. The face of the country, except that part of it which borders upon the sea, is very uneven; it rises in ridges that run up into the middle of the island, and there forms mountains, that may be seen at the distance of 60 miles. Between these ridges and the sea is a border of low land, extending along all the coast, except in a few places, where the ridges rise directly from the sea. This border is of different breadths, but no where more than a mile and a half. There are several rivers much larger than could be expected from the extent of the island; among the rocks through which these precipitate their waters from the mountains, not the least appearance of minerals is to be found. The stones shew evident tokens of having been burnt. Traces of fire are also manifest in the very clay upon the hills. It may therefore not unreasonably be supposed, that this and the neighbouring islands are either shattered remains of a continent, which were left behind when the rest was sunk by the explosion of a subterraneous fire, or have been torn from rocks under the bed of the sea, by the same cause, and thrown up in heaps to an height which the waters never reach. The low lands between the foot of the ridges and the sea, and some of the interjacent valleys, are the only parts of the island that are inhabited. Here indeed it is populous. The houses do not form villages or towns, but are ranged along the whole border, at the distance of about 50 yards from each other.

When the island was first discovered, hogs, dogs, and poultry were the only tame animals; ducks, pigeons, paroquets, with a few other birds and rats, the only wild animals. The breed of hogs has been greatly improved by some of a larger kind, that were left by the Spaniards in 1774. Goats were first introduced by Captain Cook in 1773; to these the Spaniards have added some, and they are now in such plenty, that every chief of any note has them. Cats were left by Captain Cook, and European dogs of several sorts by the Spaniards. In 1777, the stock of new animals received the important addition of a turkey cock and hen; a peacock and hen; a gander and three geese; a drake and four ducks; a horse and mare; a bull and three cows. A bull and a ram had been also left by the Spaniards. Beasts of prey or noxious reptiles, there are none.

The vegetable productions are bread-fruit, cocoa-nuts, bannanas of 13 sorts, and all excellent; plantains; a fruit resembling an apple; sweet potatoes, yams, and cocoas.

The people exceed the middle size of Europeans in stature. In their dispositions, notwithstanding the charge of infanticide, before alleged against them; they are brave, open, and generous, without either suspicion or treachery. Except a few traces of natural cunning, and some traits of dissimulation, equally artless and inoffensive, they possess the most perfect simplicity of character. Their actions are guided by the immediate impulse of the reigning passion. Their passions are the genuine effusions of the heart, which they have never been taught to disguise or repress, and are therefore depictured by the strongest expressions of countenance and gesture. Their feelings are lively, but in no case permanent; they are affected by all the changes of the passing hour, and reflect the colour of the time, however frequently it may vary. Their vivacity is never disturbed by anxiety or care, insomuch that when brought to the brink of the grave by disease, or when preparing to go to battle, their faces are unclouded by melancholy or serious reflection. Their language is soft and melodious; it abounds with vowels, and is easily pronounced. It is rich in beautiful and figurative expressions, and admits of that inverted arrangement of words, which distinguishes the ancient from most modern languages. It is so copious, that for the bread-fruit alone they have above twenty names. Add to this, that besides the common dialect, they often expostulate in a kind of stanza or recitative, which is answered in the same manner.

The two peninsulas formerly made but one kingdom. They are now divided into two, under the]

[names of Opureanou or Otaheitenooe, and Tirabou; although Otoo, the sovereign of the former, still possesses a nominal superiority over the latter, and is styled king of the whole island. To him also the island of Eimeo is subject. These kingdoms are subdivided into districts, each with its respective chief. The number of inhabitants in 1774 was estimated by Captain Cook at 204,000. Wars are frequent between the two kingdoms, and perhaps between separate districts of each. The inhabitants of Eimeo are often excited by some powerful chief to assert their independence. The power and strength of this and the neighbouring islands lie entirely in their navies; and all their decisive battles are fought on the water. Otahiti alone is supposed to be able to send out 1720 war canoes, and 68,000 able men. The chief of each district superintends the equipping of the fleet in that district; but they must all pass in review before the king, so that he knows the state of the whole before they assemble to go on service. Otahiti lies in about 17° 40′ of *s.* lat. and 149° 25′ of *w.* long.]

The archipelago consists of 23 islands, the names of which are the following:

S. Simon,	Opijá,
S. Quintin,	Tajaá,
Todos Santos,	Oyataa,
Matutarua,	Oaginé,
S. Cristoval,	Tupá,
Otahiti,	Obayó,
Morea,	Guayopé,
Genúa,	Aynayú,
Tapuamanú,	Atiú,
Mavavá,	Tatupá,
Tirá,	Quemaura.
Paraporrá,	

[OTAKOOTAI, or Okatootaia, a small island in the S. Pacific ocean, four leagues from Wateeoo, and about three miles in circuit. Lat. 19° 50′ *s.* Long. 158° 23′ *w.*]

OTALLUC, a river of the province and *corregimiento* of Ambato in the kingdom of Quito. It rises in the mountains of Avitahua, runs from *n.* to *s.* and enters the Pastaza by the *n.* side, in lat. 1° 30′ *s.*

OTANAUIS, a barbarous nation of Indians, of the province and country of Las Amazonas, who dwell with many other nations on the shores of the river Napo.

OTAO, a settlement of the province and *corregimiento* of Guarochiri in Peru; annexed to the curacy of Casta.

OTAOS, a settlement of the missions which were held by the Jesuits in the province of Topia, and kingdom of Nueva Vizcaya.

OTATAI, a small river of the province and *captainship* of Marañan in Brazil. It rises near the coast, runs *n.* between the Grande de Paraguay and the Camindey, and enters the sea in the low coast.

OTATITLAN, San Andres de, a settlement of the head settlement of the district of Tlacotalpan, and *alcaldía mayor* of Cozamaloapan, in Nueva España, at the distance of a league from the river Grande de Alvarado, in the middle of a lofty mountain. It contains 19 families of Indians and a beautiful temple, in which is venerated a miraculous image of Christ crucified, with the same title as has the settlement; and of which the following account is extant, namely, that more than 170 years ago an Indian, who had cut a piece of cedar, was desirous of making a cross, and was looking out for a person who might undertake the work, together with the image of our Lord; that there arrived at his house two handsome youths, who professed themselves sculptors, and offered to do what he required; that he put them into an apartment where the wood was, paid them for their hire, and left with them some food; when, returning the following day to see what they had done, he found to his surprise the youths flown, the money and the food untouched, and the image most beautifully and perfectly finished. This settlement is 13 leagues *e.* of its head settlement.

OTAVALO, a province and *corregimiento* of the kingdom of Quito; bounded *e.* by the mountain of Cayambe, *n. e.* by that of the town of Ibarra, *n. w.* by Esmeraldas, and *s.* by the district of the *corregimiento* of the city of Quito. It is 12 leagues long from *n. w.* to *s. e.* and running in width from *s.* to *n.*

It is watered by the river Batan, which rises, as well as other different streams that fertilize it, in the mountains, and becoming united with the rest, forms the river Blanco. Although, as we have made appear, its territory is not of much extent, it is covered with cultivated estates and manufactories where they make linens peculiar to the country, or, as they are sometimes called, *tucuyos*, carpets, quilts, and other articles, all of white cotton, the which are much esteemed throughout the kingdom.

The mode of sowing the wheat and barley in this province is very singular, for instead of scattering the seed they make small holes and pour in a certain quantity from their hands; a practice proved by experience amongst them to be very advantageous, and to yield from 100 to 150 grains for one. In the estates are enclosures for breeding horses, as also dairies, where they make a large portion of cheese, to carry for sale to the other jurisdictions. The pasture in these farms being very fine, and abounding in excellent cattle, and although it is not in want of sheep, yet are these less common than other animals.

In the territory of this province are two lakes, one of which is called S. Pablo, from a settlement of that name on its shore, the same being a league long, and half a league wide, and abounding in geese, herons, *gallaretas* or ducks, and covered with the reed called *totora;* moreover receiving its waters from the mountain Mojanda, and having issue out from it one of the arms which form the river Blanco. The other lake, which is at a small distance off, is of the same size, and is called Cuicocha, from being upon the mountain of this name on a small table just before the extreme top of the said mountain. In the middle of this latter lake are two islands, in which breed many *cuyes*, or white rabbits, and deer, who swim from the island, and when pursued by the hunters, regain it for serenity in the same manner. In this lake are found some small fish no bigger than prawns and without scales: the Spaniards call them *prenadillas*, and esteem them so much that they are carried as a rarity to Quito for sale.

The settlement of Cayambe, situate in the middle of an extensive *llano*, or plain, is backed by some lofty mountains of those *cordilleras* called Cayamburo, which vie with Chimboraso, and are taller than any lying between that spot and Quito, from which place they are discernible. These mountains being constantly covered with snow, make the temperature of the valley cold and unpleasant, assisted in no small degree by the strong winds which continually blow here.

The inhabitants of this province are divided into the seven following settlements:

Cayambe,	San Pablo,
Tabacundo,	Tocache,
Atontaqui,	Urcuqui.
Cotacache,	

The capital, which is the town and *asiento* of the same name, is in a fine situation, of a cold temperature, and abounding in cattle, with which it supplies the other settlements. The natives are rather inclined to the manufacture of cotton stuffs, in which they have a great trade, than to the cultivation of the land. It has two parishes, and a good convent of the monks of San Francisco. Its population amounts to about 18 or 20,000 souls. [It is in lat. 13′ 3″ *n.* and long. 78° 5′ *w.* It is 30 miles *n. e.* of Quito, and 167 *s. s. w.* of Popayán, on the royal road between those places.]

OTAZ, Jesus Nazareno de, a settlement of the province and government of Neiva in the Nuevo Reyno de Granada, on the shore of a small river called Caño de Otáz. It is of the same temperature, and produces the same fruits as the other settlements of this jurisdiction, but in less quantity, from the want of people, its inhabitants amounting to only 40 Indians.

The aforesaid river runs to *s. s. w.* and enters the Grande de la Magdalena, between those of Norte and Oro.

[OTCHHER, a bay on the *n.* coast of S. America, to the *w.* of the river or creek called Urano, and *e.* of cape Caldero.]

OTEAPA, a settlement of the head settlement of the district of Tenanzitlan, and *alcaldía mayor* of Acayuca, in Nueva España, containing 69 families of Indians. It is eight leagues to the *e.* one quarter to *s. e.* of its head settlement.

[OTEAVANOOA, a large and spacious harbour and bay, on the *s. w.* coast of the island of Bolabola, one of the Society islands. Lat. 16° 18′ *s.* Long. 151° 43′ *w.*]

OTEQUET, a settlement of the province and *corregimiento* of Chancay in Peru; annexed to the curacy of Ignari.

OTER, a small river of the province and colony of Virginia, which runs *s. e.* and enters the Staunton.

Oten, a small island of the province of Georgia, one of those called the Georgian; situate near that of Scabrouks.

OTERREZUCA, a settlement of the jurisdiction of the Villa de Honda in the Nuevo Reyno de Granada.

OTHOVES, a barbarous nation of Indians, of the province and government of Louisiana in N. America, who dwell near the shores of the river Missouri. They are not numerous.

[OTISFIELD, a plantation in Cumberland county, district of Maine, *e.* of Bridgetown in York county. A stream from Songo pond passes through the *w.* part of this town, on its way to Sebago. It is very free of ragged hills and mountains. The greatest part of it affords a growth of beech, maple, ash, bass, and birch, and is good land. It contains 197 inhabitants.]

OTOCA, a settlement of the province and *corregimiento* of Lucanas in Peru.

OTOCTATA, a settlement of Indians, of the province and government of Louisiana in N. America, on the shore of the river Panis.

[OTOGAMIES, an Indian nation in the N. W. territory, who inhabit between the lake of the Woods and Mississippi river. Its warriors amount to 300.]

OTOLUA, a settlement of the province and government of Popayán in the Nuevo Reyno de Granada, on the shore of a river at a small distance from the city of Buga to the *n. w.* and which river divides the settlement from this city.

OTOMACOS, a nation of Indians, reduced, for the greater part, to the faith, and dwelling in the vicinity of the Orinoco and Nuevo Reyno de Granada. They are of such extravagant and rare customs in their natural state of gentilism, that they deserve particular note. At the first dawn of light they all start from their sleep, and begin distressing cries and shouts for their dead, the which last till day-light; and then they begin to dance and sing and amuse themselves till twelve at night, thus allowing themselves no more than three hours sleep. During the morning they go and throw themselves into the river, and then assemble at the doors of their captains, and there pick out those who are to go either fishing, or hunting on the mountains the wild-boars, or to employ themselves in tilling the ground; after these are selected, the rest are allowed to go and amuse themselves.

It is truly curious to see them play at tennis; their dexterity is wonderful, and some of them will throw themselves along the ground to meet the ball, and will repel it with their shoulder. This game is well ordered: they have regular umpires, and there is much betting on the two sides, or parties, which regularly amount to 12 each.

The women, in the mean time, occupy themselves in making very fine crockery-ware, and in weaving curious garments and nets, sacks, &c. of hemp, or *pita*, which they procure from the Mauriche, as also in making pavilions or tents to defend them whilst asleep from the swarms of mosquitoes with which they are infested. About midday they give over their labour; and also amuse themselves in playing at ball, and with no less dexterity than their husbands.

These Indians have, in their play, a way of cutting and lacerating themselves with iron spikes to such a degree, that in order to stop themselves from bleeding to death they are obliged to plunge into the cold river, and will there fill up the wounds with sand; a practice which, instead of being fatal to them, makes them robust and hardy. They are fond of eating earth; and this food is, no doubt, prevented from being fatal to them through the quantity of oil or grease of the alligator, with which they mix it up.

This is the only nation of Indians of this kingdom who permit polygamy. The young men are forced to marry old widows, and the old men, on losing their wives, may marry young women; since they assert that it is madness to put two foolish unexperienced people together.

These Indians were, formerly, very numerous, and at constant war with the Caribes, until that these, aided by the Dutch, brought fire-arms into the field, when the Otomacos were so discomfited, as to betake themselves to unknown and distant parts.

They make their bread of certain roots, which they permit to putrefy in water, and then mix it up with the earth and grease, as we have before observed. This nation is bounded *n. e.* by the nation of the Paos, and *n. w.* by that of the Irauros. They began to be reduced to the Catholic faith by the Jesuits in 1732.

OTOMIES, a nation of civilised Indians of Nueva España; thus called as being descendants of Otomiel, sixth son of Iztac Micuatl, a noble of the Seven Caves, and of one of the most numerous nations of that kingdom.

They became united to the republic of Tiaxcala, when they fled from the war made against them by the emperor of Mexico, who wished to subject them to his dominion; and when the same declared war against this republic, the greatest confidence was put in these Indians, and they were placed at the very frontiers of Mexico: also in reward for their services they were endowed with great honours, and the first families of Tlaxcala gave them their daughters to wife; nor have the Mexicans been ever able to shake their allegiance to this republic.

At the entrance of the Spaniards, they were induced to serve in the conquest of that empire, and after this they changed their name from Otomies to Chichimecas.

OTONCAPULCO, a small settlement of the province of Mexico and kingdom of Nueva España, where Cortés, after the fatigues of battle, rested himself the night that he retired from Mexico, and where he afterwards constructed a temple, with the title of Nuestra Señora. See REMEDIOS.

OTONTEPEC, SAN JUAN DE, a settlement of the head settlement of the district of Tantoyuca, and *alcaldía mayor* of Tampico, in Nueva España. It contains 69 families of Indians, and is 18 leagues *w.* of its head settlement.

OTOPARI, a large river of the kingdom of Peru, which rises between the Inambari and Cuchivara, near the province of Paucartambo. It runs with different names for an infinite number of leagues through unknown countries of infidel Indians, making one large curve, until that, directing its course to *n.* in the territory or country of

Las Amazonas, it enters the Marañon or Amazon by two arms.

OTOPUN, a settlement of Indians, of the missions which are held by the religious order of S. Domingo, in the territory and district of the city of San Christoval, in the Nuevo Reyno de Granada; situate on the shore of the river Apure. It is of an hot temperature, abounding in *cacao*, maize, *yucas*, and other fruits of a warm climate; but of so small a population as to contain no more than 50 Indians.

OTOQUE, a small island of the S. sea, in the gulf of Panamá, near the point of Chame, on the coast of Tierra Firme. It is very delightful, and well cultivated with vegetable productions, and of an hot though healthy temperature. In lat. 8° 37′ *n.* Long. 79° 25′ *w.*

OTOQUILLO, a small island of the same province and kingdom as the former, and situate near to it.

OTOTITLAN, a settlement of the missions which were held by the Jesuits, in the province of Topia and kingdom of Nueva Vizcaya, on the shore of the river Tabala.

[OTSEGO, a county of New York, on the *s.* side of Mohawk river, opposite the German flats. The head waters of Susquehannah, and the Cookquago branch of Delaware, intersect this county. Here are also the lakes Otsego and Caniaderago, which send their waters, in an united stream, to the Susquehannah. It contains nine townships, viz. Kortright, Harpersfield, Franklin, Cherry Valley, Dorlach, Richfield, Otsego, Burlington, and Unadilla. It contained, a few years ago, about 1000 inhabitants; but such has been the rapid settlement of this county, that in January 1796 it contained 3237 inhabitants qualified to be electors. In 1791, when this county was but thinly settled, as many as 300 chests of maple sugar were manufactured here, 400lbs. each. The courts are held at Cooperstown in the township of Otsego.]

[OTSEGO, a township and lake in the county above described. The township was taken from Unadilla, and incorporated in 1796. On the *e.* the township encloses lake Otsego, which separates it from Cherry Valley. Lake Otsego is about nine miles long, and little more than a mile wide. The lands on its banks are very good, and the cultivation of it easy. In 1790, it contained 1702 inhabitants, including eight slaves. By the state census of 1796, there were 490 of its inhabitants electors.]

[OTTAWAS, an Indian nation in the N. W. territory, who inhabit the *e.* side of lake Michigan, 21 miles from Michilimackinack. Their hunting grounds lie between lakes Michigan and Huron. They could furnish 200 warriors 20 years ago. A tribe of these also lived near St. Joseph's, and had 150 warriors. Another tribe lived with the Chippewas, on Saguinam bay, who together could raise 200 warriors. Two of these tribes, lately hostile, signed the treaty of peace with the United States, at Greenville, August 3d, 1795. In consequence of lands ceded by them to the United States, government has agreed to pay them in goods, 1000 dollars a year, for ever.]

[OTTAWAS, a large river of Canada, which empties into the St. Lawrence at the lake of the Two Mountains, 11 miles from Montreal. The communication of the city of Montreal with the high lands, by this river, if not impracticable, is at least very expensive and precarious, by reason of its rapids and falls.]

[OTTER Bay, on the *s.* coast of the island of Newfoundland, is between Bear bay and Swift bay, and near cape Raye, the *s. w.* point of the island.]

[OTTER Creek, called by the French Riviere à Lotris, a river of Vermont, which rises in Bromley, and pursuing by its course a *n.* direction about 90 miles, empties into lake Champlain at Ferrisburgh; and in its course receives about 15 small tributary streams. In it are large falls at Rutland, Pittsford, Middlebury, and Vergennes. Between the falls the water is deep and navigable for the largest boats. Vessels of any burden may go up to the falls at Vergennes, five miles from its mouth. The head of this river is not more than 30 feet from Batten kill, which runs in a contrary direction, and falls into Hudson's river. Its mouth is three miles *n.* of Bason harbour.]

[OTTER Creek, a small stream which empties into Kentucky river, in the state of that name, and *e.* of Boonsborough.]

[OTTER's Head, a small peninsula, projecting from the *n. e.* shore of lake Superior, and *n. w.* of Michipicoton island.]

[OTTOES, Indians of N. America. They have no idea of an exclusive possession of any country, nor do they assign themselves any limits. It would appear that they would not object to the introduction of any well-disposed Indians; they treat the traders with respect and hospitality, generally. In their occupations of hunting and cultivation, they are the same with the Kauzas and Osage. They hunt on the Saline and Nimmehaw rivers, and in the plains *w.* of them. The country in which they hunt lies well; it is extremely fertile and well watered; that part of it which borders on the Nimmehaw and Missouri possesses a good portion of timber: population rather increasing. They have always resided near the place where their village is situated, and are the descendants of the Missouris.]

OTUMBA, a province and *alcaldía mayor* of Nueva España, and one of the smallest and poorest, although formerly one of the richest, from the abundance of cochineal which was found here. Its jurisdiction consists of three settlements, namely, Goatlanzinco, Axapusco, Ostotipac, and some wards or small settlements annexed to them.

The capital is of the same name, formerly a large and good town, as may be seen by the remains and ruins of its walls. It consists now of only 10 or 12 houses of Spanish families, and 406 of Indians, employed in cultivating maize, barley, and other seeds; as also in the breeding of pigs. It has a convent of the religious order of San Francisco, governed by the curate until that it was resigned to some nuns. The water was brought to the town by an aqueduct, made at great expence and with much art, a work which proved that this jurisdiction was formerly capable of going to a great expence. In the middle of the chief square is a pyramidical stone, remarkable for its height, as being of only one piece.

This town is the place where the viceroys resign the staff and the command to the successor, and on this occasion it is thronged with all the prelates and chiefs who come to compliment their new master; and indeed, this is the only time that it can be said to be inhabited, as it is in itself one of the most barren and unpeopled towns of the kingdom. Twenty-six miles *n. e.* of Mexico, in long. 98° 44′ *w.* Lat. 19° 40′ 30″ *n.*

OTUMBA, a beautiful and extensive valley of the former province, celebrated for the victory which was gained by Hernan Cortes against the whole power of the Mexican empire, when this was obliged to retreat from that court, and re-establish itself in Tlaxcala. It is at the foot of some very lofty mountains which give it its name.

OTUSCO, a settlement of the province and *corregimiento* of Caxamarca la Grande in Peru.

OTUSTLA, a settlement of the province and *alcaldía mayor* of Chiapa in the kingdom of Guatemala, of the district of that city.

OTUTO, a settlement of the province and *corregimiento* of Guamachuco in Peru, at the source of the river Bamba.

OTZANDERKET, a small lake of Canada, between the salt lakes and the river Femmes-Blanches.

OTZOLOTEPEC, S. BARTOLOME DE, a settlement and head settlement of the district of the *alcaldía mayor* of Metepec in Nneva España. It contains 380 families of Indians, and is the head of the curacy, to which as many other Indians are annexed. Three leagues *s. e.* of its capital.

OTZOLOTEPEC, another settlement, with the dedicatory title of Santa Maria, which is the head settlement of the district of the *alcaldía mayor* of Miahuatlan, in the same kingdom; of a cold and moist temperature from being in the *sierra*. It contains 970 families of Indians, including those of its wards, who cultivate maize and other seeds, although its principal trade is in cochineal. Eighteen leagues *s. e.* of its capital.

OTZOLOTEPEC, another, of the head settlement of the district of Puxmecatan, and *alcaldía mayor* of Villalta, in the same kingdom; containing 10 families of Indians. Twenty-six leagues from its capital.

OTZOLOTEPEC, another, with the dedicatory title of San Juan, the head settlement of the district of the *alcaldía mayor* of Miahuatan; situate in the *sierra*. Thirty leagues from the capital.

OTZOLOTEPEC, another, with the dedicatory title of Santa Maria, the head settlement of the district of the *alcaldía mayor* of Antequera, in the province and bishopric of Oaxaca in Nueva España. It contains 30 families of Indians, who live by cultivating cochineal, wheat, and other seeds, for commerce. Thirty leagues *e. s. e.* of its capital.

OTZULUMA, SANTA MARIA DE, a settlement and head settlement of the district and *alcaldía mayor* of Tampico in Nueva España; of a dry and hot temperature. It contains a convent of the order of San Francisco, and its population is composed of 214 families of Guastecos Indians, and 62 of Mulattoes, who sow maize, and breed some large cattle, of which they make dried meat, their only article of trade, save that of the same cattle alive, which they take annually to sell in the neighbouring jurisdictions, and from the product of which they procure themselves such articles as they may require, and which their territory does not afford.

This part of the country is much infested with poisonous insects, no place more so. Here are three cultivated estates and grazing farms for large cattle; and formerly it used to buy quantities of salt brought in the vessels from Campeche, with which to cure their fish and meats; but this system of trade has gone to decay from the very increased price of the salt, and since the coast here affords none. Fourteen leagues *n. w.* of its capital, and 80 from Mexico.

[OUABASH. See WABASH River.]

OUACHAS, a lake of the province and government of Louisiana in N. America, on the coast and island formed by the rivers Mississippi and Chetimachas. It empties itself into the sea by two months near the bay of Ascension.

OUACHETAS, a river of the same province and government as the former lake. It rises from another lake, between the river Negro and Missisippi, runs *s.* and turning *w.* with many windings, enters the former river.

OUADEBA, a river of the same province and government as the former settlement. It rises from three lakes to the *s.* of lake Superior, runs *w.* and turning its course to *s. w.* enters the Mississippi.

OUADEUAMENISSOUTE, or River of ST. PETCR, in the same province and government as the former. It rises from cape Tinton, runs *e.* and enters the Verde or Green river.

OUADOUGEOUNATON, a settlement of Indians of the province and government of Louisiana, in the territory of the Sioux of the West.

OUAGARON, a river of the province and country of the Iroquees Indians in Canada, which runs *s.* and enters the Catarakuy.

OUAINCO, a settlement of Indians of the province and government of Louisiana in N. America; situate on the shore of Rouge river.

[OUAIS'S Bay and River are about two leagues round the *n.* point of the island of Cape Breton, in the gulf of St. Lawrence, and *s. s. w.* of the island of Limbach.]

[OUANAMINTHE, a French parish and village on the *n.* side of the island of St. Domingo, about a league and a half *w.* of Daxabon, in the Spanish part, from which it is separated by the river Massacre, six leagues from the mouth of the river, and five *s. e.* of fort Dauphin.]

OUANARI, a river of the province of Guayana or Nueva Andalucia, in the French possessions. It enters the sea between the Aprobague and the Oyapoco.

OUANARI, some mountains of this province, which run as far as the sea-coast.

OUANDO, a settlement of the province and government of Popayán in the Nuevo Reyno de Granada.

OUAOUACHE, a river of Canada in N. America, called also S. Gerome and Handsome river. It is the Ohio, and was discovered by the Fathers Marquete and San Joliet, Frenchmen, of the abolished order of Jesuits, in 1673, they having been the first who navigated it.

OUAPITOUÇAN, an island of the gulf of St. Lawrence, near the coast of the country and land of Labrador.

[OUAQUAPHENOGAW, or EKANFANOKA, is a lake or rather marsh, between Flint and Oakmulgee rivers in Georgia, and is nearly 300 miles in circumference. In wet seasons it appears like an inland sea, and has several large islands of rich laml; one of which the present generation of Creek Indians represent as the most blissful spot on earth. They say it is inhabited by a peculiar race of Indians, whose women are incomparably beautiful. They tell that this terrestrial paradise has been seen by some enterprising hunters, when in pursuit of their game, who being lost in inextricable swamps and bogs, and on the point of perishing, were unexpectedly relieved by a company of beautiful women, whom they call Daughters of the Sun, who kindly gave them such provisions as they had with them, consisting of fruit and corn cakes, and then enjoined them to fly for safety to their own country, because their husbands were fierce men and cruel to strangers. They further say, that these hunters had a view of their settlements, situated on the elevated banks of an island, in a beautiful lake; but in all their endeavours to approach it, they were involved in perpetual labyrinths, and, like enchanted land, still as they imagined they had just gained it, it seemed to fly before them; and having quitted the delusive pursuit, they with much difficulty effected a retreat. They tell another story concerning this sequestered country, which seems not improbable, which is, that the inhabitants are the posterity of a fugitive remnant of the ancient Yamases, who escaped massacre after a bloody and decisive battle between them and the Creeks, (who it is certain conquered and nearly exterminated that once powerful people), and here found an asylum remote and secure from the fury of their proud conquerors. The rivers St. Mary and Sitilla, which fall into the Atlantic, and the beautiful Little St. Juan, which empties into the bay of Appalachi at St. Mark's, are said, by Bartram, to flow from this lake.]

OUAQUEZUPI, a river of the province and *captainship* of Marañan in Brazil, which flows down from the mountains of the *w.* and runs *e.* until it enters the Miari. Near its source dwell some barbarian Tocantines Indians, and in its vicinity are cultivated sugar-canes, of which sugar is made here.

OUARABICHE, a rapid river of the province of Nueva Andalucia, which runs from *e.* to *w.* passing through the country inhabited by the Saimagoes Indians, and after a course of 25 leagues becomes divided into two arms, which branch into 20 or more, entering the Atlantic sea, opposite the cape or fort Gallo, of the island of Trinidad, in lat. 9° 49′ *n.*

OUARIPANA, a river of the province and country of Las Amazonas, and part possessed by the Portuguese. It rises in the territory of the Moruas Indians, runs *e.* and turning its course to *s. s. e.* enters the Marañon, between those of Irupura and Putumayo, very near the settlement of San Christoval.

OUAS, a settlement of the province and *corregimiento* of Guamalies in Peru; annexed to the curacy of Puchas.

OUASA, a river of the province and government of Guayana, in the part possessed by the French. It enters the Couripi.

[OUASIOTO Mountains are situated *n. w.* of the Laurel mountains in Virginia. They are 50 or 60 miles wide at the gap, and 450 in length *n. e.* and *s. w.* They abound in coal, lime, and freestone. Their summits are generally covered with good soil, and a variety of timber, and the intervale lands are well watered.]

OUATESAOU, a small river of the country or land of Labrador, which runs *s.* between the Salmon and Misina rivers, and enters the sea in the gulf of S. Lawrence.

OUATIROU, a settlement of the island of Jamaica; situate in the bay of Kozo, with a good fort. The French, under Mr. Ducase, took it after great resistance, from the English, in 1694.

OUEJAS, Rio de las, a river in the province and government of Buenos Ayres. It is an arm of the river Salado, which issues from the lake of Christal, runs *s. s. e.* and enters the Paraná.

Ouejas, another river, in the province and government of Popayán, and Nuevo Reyno de Granada, which rises near the city of Cali, and enters the Grande de la Magdalena, although Mr. Bellin asserts that it enters the Cauca.

OUELLE, a small river of Canada in N. America, which rises from a small lake, runs *w.* and enters the S. Lawrence.

[It has its source in mountains to the *s.* and falls into the aforesaid river, near 100 miles below Quebec. For several miles before it joins the St. Lawrence, it runs through a level and very fertile country; and the tide flows up for a considerable way, so as to make it navigable for small vessels. This district is well cultivated, and very populous. The neighbouring parishes of Kamouraska and St. Ann's are also populous, and well cultivated.

The configuration of this part of the country is very curious. In the middle of rich plains you see a number of small hills covered with wood; they rise like so many rocks in the ocean. On approaching and examining them narrowly, you find that they are literally bare rocks, of primitive granite; full of fissures, in which pine trees have taken root, and grown to a considerable size, so as to cover the rocks. It is probable the great river at some former period covered this part of the country, when these hills were so many islands; and that the rich soil which now surrounds them, is a deposition from its waters. The probability of this conjecture is strengthened by the circumstance, that the islands of Kamouraska, still insulated only at high water, resemble in every respect the rocky hills surrounded by the fertile fields.]

[OUEPAS, a town on the coast of Costa Rica, on the N. Pacific ocean, and *s.* of Carthago.]

OUETACARES, a barbarous nation of Indians of the kingdom of Brazil, who wander about the woods and mountains. They are ferocious, cruel, and of terrible aspect, entirely naked, both men and women, cannibals, and have a language entirely different from any other nation of that kingdom, with all of whom they are at continual war. Their hatred to the Portuguese is inveterate, notwithstanding they have some commercial dealings with them; but on these occasions both parties meet armed, and all that can be said in favour of these Indians is, that they are faithful to their engagements.

[OUIATANON, a small stockaded fort in the N. W. territory, on the *w.* side of the Wabash river, in lat. 40° 20′ *n.* and long. 86° 28′ *w.* and said to be about 120 miles *s. w.* of fort St. Joseph. This was formerly a French post. Thus far the Wabash is navigable 412 miles from its mouth, including its windings, for batteaux drawing three feet water. A silver mine has been discovered here. The neighbouring Indians are the Kickapoos, Musquitons, Pyankishaws, and a principal part of the Ouiatanons. The whole of these tribes could furnish, about 20 years ago, 1000 warriors. The fertility of soil and diversity of timber in this country are the same as in the vicinity of post St. Vincent.]

OUIGNES, Bay of, in the island of S. Christopher, one of the Antilles, on the *n. w.* coast, and in the part possessed by the French before that the island was ceded to the English. It lies between the cape Enragé and the bay of Papillons.

[OUINEASKE or Shelburne Bay, on the *e.* side of lake Champlain, sets up *s. e.* through the town of Burlington, in Vermont, into the *n.* part of Shelburne.]

[OUISCONSING, a navigable river of the

N. W. territory, which empties into the Mississippi, in lat. 41° 56′ and long. 89° 45′, where are villages of the Sack and Fox tribes of Indians. This river has a communication with Fox river, which, passing through Winnebago lake, enters Puan bay in lake Michigan. Between the two rivers there is a portage of only two miles. On this river and its branches reside the Indians of its name. Warriors 300.]

OUITCHAGENE, a small lake of New France or Canada, in the country of the Petit Mustassins Indians.

OULAMANITIE, a small river of the same province as the former, which runs *w.* between those of Bucies and Margurite, and enters the lake Michigan.

[OULIONT, a village of the state of New York, on the post-road from Hudson to the Painted post. It is 27 miles *w.* of Delaware, and 37 *n. e.* of Union, on Susquehannah river, and lies on the *n.* side of a creek of its name which empties into Unadilla river.]

OUMACHIS, a small river, also of the same province as the former, in the country and territory of the Algenovins Indians. It runs *s. e.* and enters the lake S. Peter, formed by the river S. Lawrence.

OUMAMIS, a nation of Indians of Canada in N. America.

OUMANIOUETS, a small lake of the same province, formed from the waste-waters of the rivers Beauharnois and Miskovaskane.

OUMAS, a barbarous nation of Indians of New France or Canada, who dwell two leagues from the part where the Mississippi divides itself. Its natives are very well disposed to the French.

OURAMANI, a river of the province of Virginia in N. America. It runs *w.* and enters the Illinois. On its shores copper mines have been discovered.

OURANGABENA, a lake of Canada, on the confines of Nova Scotia; formed from the river S. Francis, and entering the S. George.

OURO, Corigo de, a village or settlement of the Portuguese, in the kingdom of Brazil; situate on the shore of the river Manuel Alz, not far from the Tocantines.

Ouro, a river in the same district as the former settlement. It is small, rises near that of Manuel Alz, runs nearly due *w.* and enters the Tocantines close to the settlement of its name.

OURS, or Bears, Cape of the, on the *s.* coast of the island of S. John, of Nova Scotia.

Ours, some mountains or *cordilleras* of the province and government of Louisiana, which run from *w.* to *e.* from the shore of the Mississippi; so called from abounding in bears.

Ours, a river of the province of N. Carolina. It is small, runs in a serpentine course *s. w.* and enters the Pelisipi.

Ours, another, a small river of Canada, which runs *n. e.* then turns *s. e.* and enters the lake Eric.

[Ours, a bay, with the additional title of Blancs, on the *s.* coast of Newfoundland, towards the *w.* extremity. See Bear Cove.]

Ours, a port of the province of Nova Scotia or Acadia in N. America.

OUTAGAMIS, a barbarous nation of Indians of Canada in N. America. They were of ferocious customs, and interrupted the commerce between the French and the other nations. They are also called Fox Indians, and were begun to be won over to the English by the Iroquees. They made war against the French in 1712, besieging a fort which these had in the part called the Strait, the commander of which was Mr. du Buisson, who manifested such resistance against their attack that they were at last obli ed to sue for peace, through the numbers they had lost. At the present day they are almost extinct.

Outagamis, a river of Canada in N. America. It runs through the country of the Indians of its name, expanding itself as wide as a lake, and enters another river called Kitchigamini.

OUTARDES, a river of New Britain or country of Labrador in N. America. It is large, rises from a small lake, and enters the river S. Lawrence.

OUTAUES, a nation of Indians, of the same country as the former, almost entirely destroyed by the Iroquees; and the few remaining wander about on the shores of the Mississippi. They were reduced to the Catholic religion by the Fathers Drevilletes and Garreau, Frenchmen, of the so- society of the Jesuits, in 1656.

[OUTER Buoy, in Hudson's bay, lies in lat. 51° 38′ *n.* and five miles *e.* of N. bluff.]

[Outer Island, on the coast of Labrador, is in the cluster called St. Augustin's Square; *s. w.* of Sandy island, and *e.* of Inner island.]

[OUTIMACS, a tribe of Indians in the N. W. territory, residing between lakes Michigan and St. Clair. Warriors 200.]

OUYAPE, a settlement of Indians of the province and government of Louisiana, on the shore of the river Mississippi; 550 leagues from the mouth or entrance of which the French have a fort for their defence.

[OUYATOISKA Bay and River, on the coast of Esquimaux, or *n.* shore of the gulf S. Lawrence, is to the *w.* of Natachquoin river.]

[OVEN'S-MOUTH Bay, in the district of Maine, lies on the *s.* side of Booth-bay township, in Lincoln county, 12 miles from the shire town.]

[OVID, a township of New York, in Onondago county. It was incorporated in 1794; is separated from Milton on the *e.* by Cayuga lake, and comprehends all the lands in the county on the *w.* side of Seneca lake. The centre of the township is 20 miles *s.* of the *w.* side of the ferry on Cayuga lake. In 1796, there were 107 of its inhabitants qualified to be electors.]

[OWASCO, a lake, partly in the towns of Aurelius and Scipio, in Onondago county, New York. It is about 10 miles long and one broad, and communicates with Seneca river on the *n.* by a stream which runs through the town of Brutus. The high road from Kaat's kill *w.* passes towards Cayuga ferry, near the *n.* end of the lake.]

[OWEGO, a post-town in Tioga county, New York, on the *e.* branch of the Susquehannah, 14 miles *w.* of Union, 30 *n. e.* of Athens, at Tioga point, and 144 from Philadelphia. In 1796, 170 of its inhabitants were electors.]

[OWEGO Creek, in Tioga county, serves as the *e.* boundary of the township of its name. It has several small branches which unite and empty through the *n.* bank of the *w.* branch of Susquehannah river, about 18½ miles *w.* of the mouth of Chenengo river.]

OWENDOES, a settlement of Indians of Canada in N. America, on the shore of the river Bever. Here the English had an establishment, one of the first formed by them on the Ohio.

[OWHARREE, a harbour on the *n.* part of the *w.* coast of Houaheine, one of the Society islands, 25 leagues *n. w.* by *w.* of Otaheite island. Lat. 16° 44′ *s.* Long. 151° 5′ *w.*]

[OWHYHEE, the largest of the Sandwich islands, is about 300 miles in circumference; between 18° 40′ and 20° 20′ *n.* lat. and between 154° 50′ and 156° 10′ *w.* long. from Greenwich. The extensive mountain, named Mouna Roa, on the *s. e.* part of the island, is 16,020 feet high. It consists of three peaks which are perpetually covered with snow, (though within the tropics), that are visible 40 leagues out at sea. At the *s.* end of the island is a village called Kaoo-A-poona, on the *s. e.* side; Aheedoo is on the *n. e.* part of the island, Amakooa on the *n.* end, Tirooa on the *n. w.* side, where is the bay of Toyabyah, and on the *w.* side, *n. w.* of Kaoo, is the bay of Kara-kakooa. It has the same productions as the Society and Friendly islands, and about 150,000 inhabitants, who are naturally mild, friendly, and hospitable to strangers. The sea abounds with a great variety of excellent fish. The celebrated navigator Capt. James Cook lost his life here, by an unfortunate and momentary jealousy of the natives.]

[OWL'S HEAD, a head-land on the *w.* side of Penobscot bay, in the district of Maine. It has a good harbour on the larboard hand as you go to the *e.* The harbour makes with a deep cove; has four fathoms water, and muddy bottom. It is open to the *e.* to *n.* and *e. n. e.* winds; but in all other winds you are safe. The tide of flood sets to the *e.* and the tide of ebb *s. w.* through the Muscle ridges.]

[OX, a river of Louisiana. See RED RIVER.]

[OXBOW, GREAT, a bend of the river Connecticut, about the middle of the township of NEWBURY, in Vermont, which see. It contains 450 acres of the finest meadow-land in New England.]

[OXFORD, a township in Worcester county, Massachusetts. It contains 1000 inhabitants; is 12 miles *s.* of Worcester, and 44 *s. w.* of Boston.]

[OXFORD, a village in Bristol county, Massachusetts. See NEW BEDFORD.]

[OXFORD, a parish in the *n.* part of Derby in Connecticut, containing 140 families. Seventeen miles *n. w.* of Newhaven.]

[OXFORD, a post-town of New York, in Tioga county, 22 miles *n. e.* of Union, and 16 *w.* of Butternuts. This township lies between Jericho and Union, and is bounded *n.* on Norwich, and *w.* by the tract called the Chenengo triangle. It was incorporated in 1793. Here is an incorporated academy.]

[OXFORD, a township of New Jersey; situated in Sussex county, on the *e.* bank of Delaware river, 13 miles *n. e.* of Easton in Pennsylvania. It contains 1905 inhabitants, including 65 slaves.]

[OXFORD, a township of Pennsylvania; situated in Philadelphia county. There is one of the same name in Chester county.]

[OXFORD, a port of entry, on the *e.* shore of Chesapeak bay, in Talbot county. Its exports in 1794 amounted to 6956 dollars. It is 12 miles *s.* of Easton, and about 47 *s. e.* of Baltimore.]

[OXFORD, a small post-town of N. Carolina, 30 miles from Hillsborough.]

OXIBA, a river of the province and government of Guayaquil in the kingdom of Quito. It also takes the names of Caluma and Caracol from the settlements through which it passes.

OXITLAN, San Lucas de, a settlement and head settlement of the district and *alcaldia mayor* of Teutitla in Nueva España; of a hot temperature. It contains 60 families of Indians, who trade in cotton and *vainilla*, and is 10 leagues *s.* of its capital.

OYA, a settlement of the kingdom of Nueva Vizcaya in N. America; situate near the garrison of Conchos.

OYAC, a river of the province and government of Guayana, in the French possessions.

OYACACHI, a small and poor settlement of Indians of the kingdom of Quito; situate on the *w.* shore of the river Suno; annexed to the curacy of the settlement of Quinche. It is celebrated for a wonderful image of our Lady, the devotion of which began from the fame of its miracles in 1591. This image was removed to Quinche in 1640. The climate of Oyacachi is cold and moist; situate amongst woods almost uninhabitable; and the few Indians residing in it gain their livelihood by sawing planks, which they carry to sell at Cayamba and the other immediate settlements. In lat. 10′ 7″ *s.*

OYACATLA, a settlement of Indians of the province of Misteca in Nueva España, in the time of the Indian gentilism, but now not existing.

OYADAIBUISC, Santiago de, a settlement of the province and government of Sonora in N. America; situate in the country of the Cocomaricopas Indians, on the shore of the grand river Gila, between the town of San Felipe and the settlement of S. Simon de Tucsani.

OYAMBARO, a settlement of the kingdom of Quito, on the *llano* or plain of Yaruqui.

OYAPAPU, or Orapapu, a small river of the province and government of Guayana, which runs *e.* and enters the Arny.

OYAPO, a river of the same province and government as the former, in the part occupied by the French.

OYAPOC, a river of the same province as the former, in the district of the French, who have built a fort on its shore, called S. Luis. Some geographers wrongly confound this river with that of Vicente Pinzon. It is one of the largest rivers in that territory, and enters the sea in a kind of bay of four leagues wide, and into which other rivers also flow. The point which forms the bay on the *e.* side is called Cape Orange, which is distinguished at a great distance off. This river is two leagues wide at its mouth, and there is anchorage of four fathoms. At the distance of one league from its entrance is an island, called De Biches, which is covered at high tides, and to go up the river you pass to the *w.* of it, the other side not being navigable on account of the sand banks there. Six leagues up the river is a bay or port, where very near the shore there is six fathoms of water, and where is the fort of which we have spoken, built in 1726. The territory in its vicinity is very fertile and well cultivated. The Dutch established themselves there in 1676, and the French were some time before they drove them out. Three leagues from the island are other small isles, and from thence the river becomes gradually narrower and shallower, until it is scarcely more than seven or eight feet deep.

Oyapoc, a settlement of the French, in the same province, and on the shore of the former river.

OYES, or Geese, River of the, in the county of Lunenburgh, of the province of Virginia, to the *s. e.* It is small and enters the Staunton.

Oyes, an island in the gulf of S. Lawrence, at the entrance, close to the point of Raye, of the *s.* coast of Newfoundland.

OYOLO, a settlement of the province and *corregimiento* of Parinacochas in Peru.

OYON, a settlement of the province and *corregimiento* of Caxatambo, in the same kingdom as the former; annexed to the curacy of Churin.

[O-YONG-WONGEYK, on lake Ontario, at Johnson's landing-place, about four miles *e.* of fort Niagara.]

[OYSTER Bay, a township of New York; situated in Queen's county, Long island, extending from the sound *s.* to the Atlantic ocean, and includes Lloyd's neck, or Queen's village, and Hog's island. It contains 4097 inhabitants, of whom 611 are electors, and 381 slaves.]

[Oyster Bay, a harbour for small vessels in the *s. w.* limits of the town of Barnstable, in Barnstable county, Massachusets; which see. It affords excellent oysters; hence its name.]

[Oyster Beds, in Delaware bay, lie opposite Nantuxet bay.]

[Oyster Point, on the coast of S. Carolina, where the water does not ebb till an hour and a half after it begins to ebb at the bar of Ashley river, near Charlestown. It is best to go in an hour and an half before high water.]

[Oyster Pond, a part of the waters of the Atlantic ocean, which set up *w.* into Long island, in the state of New York, between the *n.* easternmost point of the island called Oyster Pond Point, and Gardner's island. Off the point are two small isles, one of which is called Plumb Island.]

[Oyster River, a *w.* branch of Piscataqua

river in New Hampshire; which see. Durham stands on its *s.* side, near its junction with the main stream at Helton's point.]

[OZAMA, one of the largest rivers of the island of St. Domingo in the W. Indies, and on which the city of St. Domingo is situated. It is navigable nine or 10 leagues from *s.* to *n.* One may judge of the enormous volume of water which the continent stream of Isabella and Ozama sends to the sea, by the red colour it gives it in the time of the floods, and which is perceivable as far as the eye can distinguish. There is a rock at the mouth which prevents the entrance of vessels drawing more than 18 or 20 feet of water. The river for a league is 24 feet deep; and its banks are 20 feet perpendicular; but *n.* of the city this height is reduced to four feet. This real natural bason has a bottom of mud or soft sand, with a number of careening places. It seldom overflows its banks, except in very extraordinary inundations. The road before the mouth of the Ozama is very indifferent, and lies exposed from *w. s. w.* to *e.* It is impossible to anchor in it in the time of the *s.* winds; and the *n.* winds drive the vessels from their moorings out into the sea, which here runs extremely high. See DOMINGO City. The mouth of the river is in lat. 18° 18′ *n.* and long. 72° 38′ *w.* from Paris.]

OZAMA, a large and abundant river of the island S. Domingo, which rises in the mountains, runs *w.* and enters the sea, having at its mouth the capital of the island.

OZCOTICA, a settlement of Indians of the kingdom of Nueva España, in the time of the gentilism of the Indians, but no longer existing. It was one of those destined to maintain the provisions of the Casa Real.

OZELOTLAN, a settlement of the head settlement of the district of Chinameca, and *alcaldía mayor* of the province of San Miguel, in the kingdom of Guatemala; annexed to the curacy of that head settlement.

OZICALA, SAN JUAN DE, a very considerable head settlement of the district and *alcaldía mayor* of San Miguel in the kingdom of Guatemala. Its district consists of 1450 Indians of the nations Ulúa and Popoluca, the which are divided into nine settlements.

OZIER, a port of the coast of the river Mississippi in the province and government of Louisiana; discovered by Hernando de Soto, conqueror of Florida, in 1541.

OZIERS, Isles of, situate near the coast of the river S. Lawrence in Canada. They are many, and all small, at the month of the river Des Outardes.

OZOCOTLAN, a settlement of the province and kingdom of Guatemala.

OZOGOCHE, a river of the province and *corregimiento* of Alausi in the kingdom of Quito. It rises from the lake Mactallan, runs *n.* and uniting itself with the Guamote, in lat. 1° 54′ *s.* forms the Achambo, receiving first the waters of the lake Colay-cocha. Its waters then proceed to fertilize the province of Riobamba.

OZOMATLAN, a settlement of the *alcaldía mayor* of Tixtlan in Nueva España; situate on the other side of the river of Las Balzas. It contains 48 families of Indians, and is one league from the settlement of Hostotipan.

OZTLOTLAUCHAN, a settlement of the kingdom of Tezcuco in Nueva España, in the time of the gentilism of the Indians.

OZTOLOAPAN, SAN MARTIN DE, a settlement of the head settlement of the district of San Francisco del Valle, and *alcaldía mayor* of Zultepec, in Nueva España. It was formerly very numerous, since it counted 800 families of Indians, but was almost depopulated by an epidemical disorder. It is of great extent, has a good parish church, and is 12 leagues *w.* of the *real* of the Mines.

OZTOMATACAS, a barbarous nation of Indians of Nueva España, made war against and conquered by Motezuma, the last emperor of that kingdom.

OZTOPALCO, a settlement of the province and *alcaldía mayor* of Tezcuco in Nueva España; situate near the capital.

OZTOTIPAC, a settlement of the province and *alcaldía mayor* of Tezcuco in Nueva España. It was the capital of a noble in the time of the gentilism. See OSTOTIPAC.

OZUANAS, a barbarous nation of Indians, but little known, dwelling in the vicinities of the river Yotan and living by the chase, their arms being bows and arrows.

OZUMAZINTLA, a settlement of the province and *alcaldía mayor* of Los Zoques in the kingdom of Guatemala.

OZUMBA, a settlement of the *alcaldía mayor* of Chalco in Nueva España. It contains 278 families of Indians, some Spaniards, and a convent of the religious order of San Francisco. Four leagues from its capital.

END OF THE THIRD VOLUME.

HARDING and WRIGHT, Printers, St. John's Square.